Evolution and Biogeography

The Natural History of the Crustacea Series

Functional Morphology and Diversity (Volume 1)
Edited by Les Watling and Martin Thiel

Lifestyles and Feeding Biology (Volume 2)
Edited by Martin Thiel and Les Watling

Nervous Systems and Control of Behavior (Volume 3)
Edited by Charles Derby and Martin Thiel

Physiology (Volume 4)
Edited by Ernest S. Chang and Martin Thiel

Life Histories (Volume 5)
Edited by Gary Wellborn and Martin Thiel

Reproductive Biology (Volume 6)
Edited by Rickey Cothran and Martin Thiel

Developmental Biology and Larval Ecology (Volume 7)
Edited by Klaus Anger, Steffen Harzsch, and Martin Thiel

Evolution and Biogeography (Volume 8)
Edited by Gary C. B. Poore and Martin Thiel

Evolution and Biogeography

The Natural History of the Crustacea
Volume 8

EDITED BY GARY C. B. POORE AND MARTIN THIEL

OXFORD
UNIVERSITY PRESS

OXFORD
UNIVERSITY PRESS

Oxford University Press is a department of the University of Oxford. It furthers the University's objective of excellence in research, scholarship, and education by publishing worldwide. Oxford is a registered trade mark of Oxford University Press in the UK and certain other countries.

Published in the United States of America by Oxford University Press
198 Madison Avenue, New York, NY 10016, United States of America.

CIP data is on file at the Library of Congress
ISBN 978-0-19-063784-2

9 8 7 6 5 4 3 2 1

Printed by Integrated Books International, United States of America

PREFACE

This is the eighth volume of a 10-volume series on *The Natural History of the Crustacea*. Our volume examines *Evolution and Biogeography*, and it follows *Volume 1: Functional Morphology and Diversity; Volume 2: Lifestyles and Feeding Biology; Volume 3: Nervous Systems and Control of Behavior; Volume 4: Physiology; Volume 5: Life Histories; Volume 6: Reproductive Biology*; and *Volume 7: Developmental Biology and Larval Ecology*.

The first part of this volume is entirely dedicated to the explanation of the origins and successful establishment of the Crustacea in the oceans. In their chapter, Stefan Richter and Christian Wirkner introduce the ur-crustacean, the ancestor of all crustaceans, while Thomas Hegna and his colleagues present evidence of the first crustaceans, or pancrustaceans as they are more precisely called, in the fossil record. Shane Ahyong explains the evolution and radiation of Crustacea from their Cambrian ancestors. Heather Bracken-Grissom and Joanna Wolfe discuss conflicted phylogenies clouding the pancrustacean conundrum, while Björn von Reumont and Gregory Edgecombe explore the common origins of crustaceans and insects.

In the second part of the volume, the biogeography of the Crustacea is explored in order to infer how they conquered different biomes globally while adapting to a wide range of aquatic and terrestrial conditions. Claudia Halsband and team undertake a review of the biogeography of the oceans worldwide. Torben Riehl and colleagues relate how crustaceans conquered the ocean depths since the Mid-Paleozoic. Benny Chan and collaborators describe the biogeography and host usage of coral-associated crustaceans, focusing on barnacles, copepods, and gall crabs. Joanna Legeżyńska and her colleagues examine the colonization of polar regions by crustaceans, while Joseph Staton and collaborators explore crustacean communities in coastal and estuarine environments. Kristina von Rintelen, together with her colleagues, explains the biodiversity of crustaceans in standing waters, especially in ancient lakes, while Shane Ahyong and Chao Huang provide an overview of the colonization, adaptation, radiation, and diversity of crustaceans in other freshwater habitats. Renée Bishop and colleagues dive into subterranean and anchialine waters to describe crustacean biodiversity and conservation in these less visited environments. In the chapter on terrestrial crustaceans, Spyros Sfenthourakis and Stefano Taiti join their knowledge on slaters and pill-bugs with that of Jim Lowry and Alan Myers on leaf-hopping amphipods.

A final section examines more general patterns and processes and looks to the future. Marcelo Rivadeneira and Gary Poore generate an updated synthesis of latitudinal gradients of diversity of marine crustaceans. Anna McCallum and Torben Riehl reach into the abyss to reveal bathymetric patterns of crustaceans from the shallows into the deep ocean. In an overview of phylogeography,

Christine Ewers-Saucedo and John Wares assess the impact of reproductive strategy, population connectivity, and vicariance on distribution. And finally, Karolina Bącela-Spychalska and colleagues review anthropogenic habitat changes and species introductions and dare to explore how human pressure will impact on crustacean populations in future.

Collectively, these 18 chapters provide a thorough exposition of present knowledge across the major themes in evolution and biogeography of crustaceans. They do this by summarizing what is known and providing novel analyses of patterns. We expect this volume will be valuable to scholars and students of the evolution and biogeography of crustaceans, and hope its syntheses and thoughtful ideas spur new avenues of research on these topics within the Crustacea and beyond.

ACKNOWLEDGMENTS

We thank our contributors for graciously sharing their time, energy, knowledge, and insights in order to make this volume possible—it has been both a pleasure and honor to work with each of them. Our editorial assistants, Annie Mejaes, Mika Tan, Tim Kiessling, Miles Abadilla, and Miguel Angel Penna-Díaz, were impeccably skilled and organized and always kept us moving forward. We thank our external reviewers for their valuable and generous feedback. Finally, we express our appreciation to our publisher, Oxford University Press, for its commitment to this project.

Editing of this book was generously supported by Universidad Católica del Norte, Chile.

CONTRIBUTORS

EDITORS
Gary C. B. Poore
Museums Victoria
Melbourne, Australia

Martin Thiel
Facultad Ciencias del Mar, Universidad Católica del Norte
Coquimbo, Chile

AUTHORS
Shane T. Ahyong
Australian Museum
Sydney, Australia
School of Biological, Earth and Environmental Sciences, University of New South Wales
Kensington, Australia

Karolina Bącela-Spychalska
Department of Invertebrate Zoology and Hydrobiology, Faculty of Biology and Environmental Protection, University of Łódź
Łódź, Poland

Karen M. Barkel
Department of Biological Sciences, Ferris State University
Big Rapids, MI, United States of America

Renée E. Bishop
Penn State Worthington Scranton, The Pennsylvania State University
Dunmore, PA, United States of America

Stephen A. Borgianini
Department of Natural Sciences, University of South Carolina Beaufort
Bluffton, SC, United States of America

Heather Bracken-Grissom
Department of Biological Sciences, Florida International University-Biscayne Bay Campus
North Miami, FL, United States of America

Simone Nunes Brandão
Centro de Biociências, Universidade Federal do Rio
Grande do Norte, Lagoa
Nova, RN, Brazil

Brandão Nunes s Simone
Programa de Pós-Graduação em Sistemática e Evolução—Centro de Biociências, Programa de Pós-Graduação em Geodinâmica e Geofísica
Universidade Federal do Rio Grande do Norte, Laboratorio de Geologia e Geofisica Marinha e Monitoramento Ambiental-GGEMMA, Departamento de Geologia
Brazil

Angelika Brandt
Department Marine Zoology, Section Crustacea, Senckenberg Research Institute and Natural History Museum
Frankfurt, Germany
Institute for Ecology, Evolution and Diversity, Goethe-University of Frankfurt FB 15
Frankfurt am Main, Germany

Brian A. Canada
Department of Computer Science, University of South Carolina Beaufort
Bluffton, SC, United States of America

Benny K. K. Chan
Biodiversity Research Center, Academia Sinica
Taipei, Taiwan

Yu-Rong Cheng
Department of Fisheries Production and Management, National Kaohsiung University of Science and Technology
Kaohsiung City, Taiwan

Claude De Broyer
Royal Belgian Institute of Natural Sciences Operational Directorate Taxonomy and Phylogeny
Brussels, Belgium

Patricio De los Ríos
Laboratorio de Ecología Aplicada y Biodiversidad, Escuela de Ciencias Ambientales, Facultad de Recursos Naturales, Universidad Católica de Temuco
Temuco, Chile

Gregory D. Edgecombe
Department of Earth Sciences, The Natural History Museum
London, United Kingdom

Christine Ewers-Saucedo
Zoological Museum, of the Christian-Albrechts University
Kiel, Germany

Will P. Goodall-Copestake
Centre for Ecology and Hydrology
Edinburgh, United Kingdom

Michał Grabowski
Department of Invertebrate Zoology and Hydrobiology, Faculty of Biology and Environmental Protection, University of Łódź,
Łódź Poland

Claudia Halsband
Akvaplan-niva, Fram Centre for Climate and the Environment
Tromsø, Norway

Thomas A. Hegna
Department of Geology, Western Illinois University
Macomb, IL, United States of America

Chao Huang
School of Biological, Earth and Environmental Sciences, University of New South Wales
Kensington, Australia

William Humphreys
Western Australian Museum
Welshpool, Australia

Damià Jaume
Department of Biodiversity and Conservation, Mediterranean Institute for Advanced Studies (IMEDEA)
Mallorca, Spain

Ksenia Kosobokova
Shirshov Institute of Oceanology, Russian Academy of Sciences
Moscow, Russia

Joanna Legeżyńska
Department of Marine Ecology, Institute of Oceanology Polish Academy of Sciences
Sopot, Poland

James K. Lowry
Australian Museum Research Institute
Sydney, Australia

Javier Luque
Department of Biological Sciences
University of Alberta
Edmonton, Canada
Smithsonian Tropical Research Institute
Panamá, República de Panamá
Department of Geology and Geophysics, Yale University
New Haven, CT, United States of America

Enrique Macpherson
Centro de Estudios Avanzados de Blanes (CEAB-CSIC)
Girona, Spain

Anna McCallum
Marine Sciences, Museums Victoria
Melbourne, Australia

Alan A. Myers
School of Biological, Earth and Environmental Sciences, University College Cork
Cork, Ireland

Gary C. B. Poore
Marine Sciences, Museums Victoria
Melbourne, Australia

Björn M. von Reumont
University of Leipzig, Institute of Biology, AG Molecular Evolution and Animal Systematics
Leipzig, Germany
Department of Life Sciences, The Natural History Museum
London, United Kingdom

Stefan Richter
Institut für Biowissenschaften, Allgemeine und Spezielle Zoologie, Universität Rostock
Rostock, Germany

Torben Riehl
Department Marine Zoology, Section Crustacea, Senckenberg Research Institute and Natural History Museum
Frankfurt, Germany

Marcelo M. Rivadeneira
Laboratorio de Paleobiología, Centro de Estudios Avanzados en Zonas Áridas (CEAZA)
Coquimbo, Chile
Facultad Ciencias del Mar, Universidad Católica del Norte
Coquimbo, Chile

Spyros Sfenthourakis
Department of Biological Sciences, University of Cyprus
Nicosia, Cyprus

Joseph L. Staton
School of Science and Mathematics, University of South Carolina Beaufort
Bluffton, SC, United States of America

Stefano Taiti
Consiglio Nazionale delle Ricerche, Istituto di Ricerca sugli Ecosistemi Terrestri
Florence, Italy

Kristina von Rintelen
Museum für Naturkunde, Leibniz Institute for Evolution and Biodiversity Science
Berlin, Germany

Thomas von Rintelen
Museum für Naturkunde, Leibniz Institute for Evolution and Biodiversity Science
Berlin, Germany

Peter Ward
British Antarctic Survey
Cambridge, United Kingdom

John P. Wares
Department of Genetics, Franklin College of Arts and Sciences, University of Georgia
Athens, GA, United States of America

Jan Marcin Węsławski
Department of Marine Ecology, Institute of Oceanology, Polish Academy of Sciences
Sopot, Poland

Christian S. Wirkner
Institut für Biowissenschaften, Allgemeine und Spezielle Zoologie, Universität Rostock
Rostock, Germany

Joanna M. Wolfe
Department of Earth, Atmospheric and Planetary Sciences, Massachusetts Institute of Technology
Cambridge, MA, United States of America

Kingsley J. H. Wong
Biodiversity Research Center, Academia Sinica
Taipei, Taiwan

CONTENTS

1

WHAT THE UR-CRUSTACEAN LOOKED LIKE

Stefan Richter and Christian S. Wirkner

Abstract

Crustaceans are a paraphyletic assemblage within arthropods. Hexapoda (insects) are nested within crustaceans, with the Remipedia the most likely sister group to Hexapoda. Together, crustaceans and hexapods comprise the monophyletic Tetraconata (also called Pancrustacea). Herein, we "reconstruct" the last common ancestor of crown group Tetraconata, calling it the ur-crustacean. We base our reconstruction on knowledge of extant crustaceans. We tentatively suggest that the ur-crustacean displayed certain characters: The ur-crustacean was a free-living marine species with a distinct head and equipped with two pairs of sensory limbs (antennule and antenna), mandibles, and two more pairs of mouthparts (maxillule and maxilla). We suggest that no further segments were fused to the head and that no maxilliped was present. The ur-crustacean may or may not have possessed a carapace. Its brain was complex, with an extended olfactory system, possibly a central complex, and a lateral protocerebrum containing at least two optical neuropils. The protocerebrum was connected to a nauplius eye as well as to compound eyes. The ur-crustacean might have had a uniformly segmented trunk posterior to its five-segmented head or (less probably) may have possessed two tagmata, a limb-bearing thorax and a limb-less abdomen. It had a heart that might have extended right through the trunk independently of tagmatization. Its thoracopodal appendages were true arthropodal (consisting of podomeres) with a protopod (probably subdivided into coxa and basis), an exopod, and an endopod. Larval development started with a nauplius larva (probably an orthonauplius).

INTRODUCTION

Crustaceans exhibit a remarkably high level of morphological disparity, which extends to major aspects of their general body organization (Schram 2013). Their tagmatization ranges from taxa with an almost uniform trunk (as in Remipedia) to taxa whose trunk is clearly separated into a limb-bearing

Evolution and Biogeography. Edited by Gary C. B. Poore and Martin Thiel.

thorax and a limb-less abdomen (e.g., Copepoda, Cephalocarida, Mystacocarida) and taxa with a functional differentiation between two limb-bearing trunk tagmata (as in Malacostraca). In some taxa the head is separated from the thorax (e.g., Cephalocarida, Mystacocarida, Copepoda), whereas in other taxa the head and parts of the thorax are fused to form a cephalothorax (notably in Decapoda, Euphausiacea, and Peracarida). Limb structure varies as well: in some groups limbs are phyllopodous and in some stenopodous, some limbs possess chelae, some limbs are transformed into mouthparts, others into filter baskets, and others into reproductive structures. Some taxa possess a carapace, which can be bivalved (as in ostracods and "clam shrimps") or univalved (as in many malacostracans) or transformed into a brood pouch (some Cladocera), while in other taxa the carapace is absent entirely (e.g., Copepoda, Anaspidacea, Amphipoda, Isopoda) (Olesen 2013). Lifestyle varies too, from pelagic in oceans and fresh water (e.g., Copepoda, Cladocera) to benthic (e.g., Decapoda Reptantia). Some groups are sessile (e.g., Cirripedia Thoracica), some parasitic (e.g., Rhizocephala, many copepods). Without knowing anything about crustacean phylogeny, it would be difficult to imagine that such groups as barnacles, water fleas, and ostracods were closely related to decapods and isopods. And indeed, crustaceans do not form a monophylum, i.e., they cannot be traced back to one exclusive common ancestor that is not also the ancestor of other arthropods. The latest phylogenomic studies show that some taxa are more closely related to the hexapods (insects) than to other crustaceans, with Remipedia likely being the sister group to Hexapoda (Schwentner et al. 2017). It might be time that we accepted crustacean paraphyly and banished the proper noun Crustacea from classifications (see also Schram 2013). But the colloquial use of the term "crustaceans" (plural and no capital), which might still be of value to the community of carcinologists (or crustaceanologists), should not be ruled out (see Richter et al. 2009). For the purpose of the present chapter, we use the latest and most comprehensive phylogenetic analysis by Schwentner et al. (2017) (Fig. 1.1).

Crustaceans and hexapods together form the monophyletic taxon Tetraconata (also called Pancrustacea) (Zrzavy and Stys 1997, Dohle 2001). The exact relationships to other arthropods and of its constituents are currently under debate (see Chapters 3, 4, and 5 in this volume). Tetraconata are part of Arthropoda (in some classifications Euarthropoda), which also include spiders, scorpions and allies (Chelicerata), and millipedes and centipedes (Myriapoda). Arthropod monophyly is well supported by morphology, in particular that of the limbs (arthropodia), which are composed of sclerotized podomeres separated by arthrodial membranes, and that of the sclerotized exoskeleton—both character complexes are absent in the closest relatives to arthropods, Onychophora and Tardigrada. Giribet and Edgecombe (2013) listed a significant number of additional characters that support arthropod monophyly. Within arthropods, the sister group of Tetraconata is Myriapoda. Together, the two form the Mandibulata. Although the monophyly of Mandibulata is not beyond doubt, recent phylogenomic analyses make it look increasingly likely, and there is also strong morphological support, including the name-giving character, the mandible group (Edgecombe 2010, von Reumont et al. 2012, Schwentner et al. 2017).

RECONSTRUCTING THE UR-CRUSTACEAN

The aim of this chapter is to explore what the ur-crustacean might have looked like. On the basis of the foregoing it should be clear that the last common ancestor of all crustaceans was the last common ancestor of Tetraconata (Fig. 1.1.) and that it looked like a "crustacean" of some kind and not like an insect. We intend to reconstruct the ur-crustacean by looking at a selection of characters applicable to extant crustaceans, but we also include in our considerations the sister group to Tetraconata, Myriapoda. We use the term "character" as Hennig (1966) introduced it: a "series of states" (transformation series) in which "character states" represent the manifestations present in extant crustacean taxa. For each "transformation series" (character) we need to decide which character state was the original one, i.e., which was present in the last common ancestor

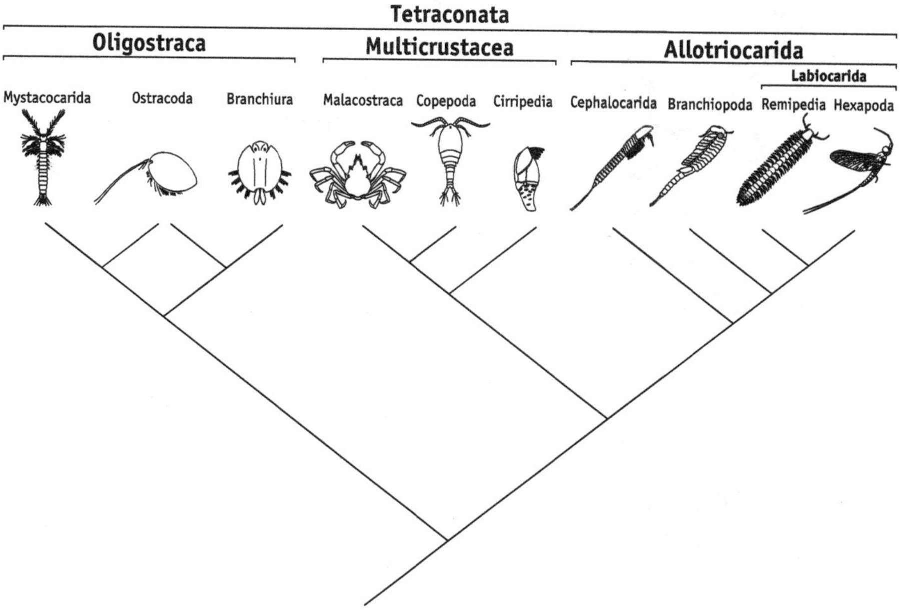

Fig. 1.1.
Crustaceans and hexapods build the monophyletic Tetraconata. Crustaceans are paraphyletic. Remipedia might be the sister group of Hexapoda. The high-ranked crustacean taxa are Oligostraca, Multicrustacea, and Allotriocarida. The ur-crustacean as defined herein is the last common ancestor of all extant crustaceans and hexapods. Phylogenetic relationships based on Schwentner et al. (2017).

of Tetraconata, or ur-crustacean as we are calling it here. It is possible, of course, that none of the character states present in extant crustaceans represent the original condition. Our approach refers to the crown group of Tetraconata (see Fig. 1.1). In fact, the reason we prefer the term Tetraconata to Pancrustacea is that the prefix "pan" is sometimes used in systematics to refer to a more inclusive taxon, i.e., the crown group and all fossil stem lineage representatives (see Lauterbach 1989, Meier and Richter 1992).

The term Crustacea (sometimes with the addition "s.l.") has itself been used to refer to a more inclusive taxon, one containing several fossil taxa, mostly from the Orsten fauna (e.g., Maas et al. 2003, Stein et al. 2005, Waloszek et al. 2007) and possibly myriapods (Lauterbach 1988, Edgecombe 2017). Focusing on the crown group of Tetraconata has the advantage of allowing us also to refer to characters not detectable in fossil taxa, such as most parts of the internal anatomy, or gene expression data.

To obtain a complete understanding of the origin of crustaceans, it would be necessary to know about selection pressures and the paleo-environment of the ur-crustacean. However, the conditions in which the ur-crustacean evolved remain unresolved; all we know for certain is that the ur-crustacean was marine and probably lived in the Ediacaran (Schwentner et al. 2017). To say more would be pure speculation.

Our approach to reconstructing the ur-crustacean is different from (in that it clearly refers to a phylogenetic hypothesis) but in some respects akin to Hessler and Newman's (1975) comparison of various crustaceans, particularly those they considered "primitive": cephalocarids, leptostracan malacostracans, and notostracan branchiopods. Notably, Hessler and Newman compared character by character and did not include the entire taxon. Both approaches are reminiscent of Goethe's ideas on the *Urpflanze* and the *Urthier*, which—as well as constituting the first use of the prefix ur—proceed from empirical evidence to conclusions about the "type" (Fig. 1.2).

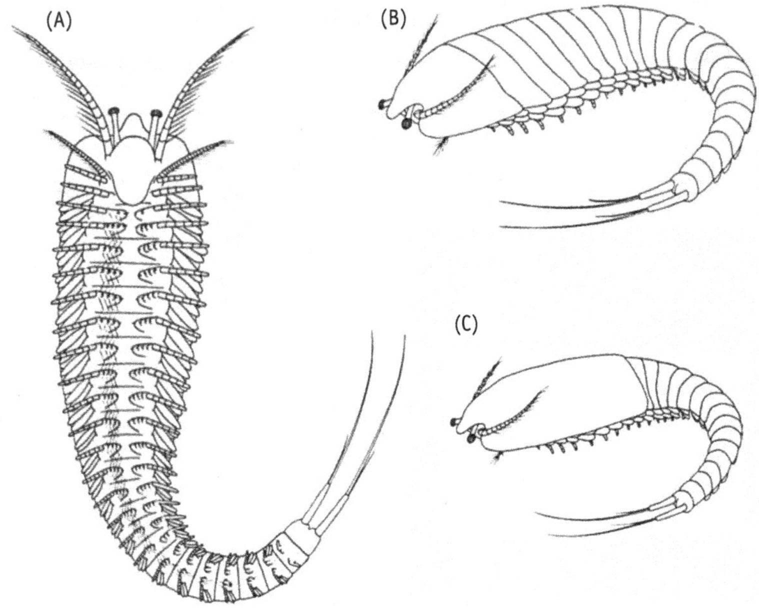

Fig. 1.2.
The ur-crustacean as "constructed" by Hessler and Newman (1975). The two authors disagreed about the presence or absence of a carapace (B + C).

When Hessler (1992: 315) later wrote that cephalocarids "best personify what the ur-crustacean must have looked like," he was clearly aware that some cephalocarid features are highly derived. For Schram (1983: 23), the description of the first remipede by Yager (1981) reopened "the debate about the origin and early evolution of the phylum." Schram (p. 28) felt that the paddle-like biramous trunk appendages in particular, and feeding localized at the mouthparts "illustrates the basic antiquity to the remipede structural plan." In the following sections, we discuss all major external morphological and some anatomical features of the central nervous and the hemolymph vascular systems.

The Crustacean Head

To permit the formation of a "head" from five (or six) independent segments, condensation and integration took place in various character complexes. Examples include the cerebralization of the central nervous system, the integration of sensory function and food uptake in the anterior appendages, and the covering of anterior segments by a continuous dorsal shield. The borders between these complexes, however, do not necessarily correspond to each other (Richter et al. 2013).

In Cephalocarida, a particularly interesting case, a dorsal head shield covers the segments bearing the antennules, antennae, mandibles, maxillules, and maxillae and clearly defines the border between the head and the trunk, although the maxillae and the first trunk limbs closely resemble each other (Hessler 1964). In the central nervous system, a well-demarcated brain (consisting of a proto-, deuto-, and tritocerebrum) can be distinguished from a subesophageal ganglion, and the latter from the first thoracic ganglion (Stegner and Richter 2011).

Possession of two pairs of sensory appendages, antennules and antennae (appendages of the deuto- and tritocerebral segments), is the most obvious diagnostic character for crustaceans (Schram 2013). Based on their position within crustaceans, the antennae in hexapods must have been reduced. Interestingly, the corresponding appendage (the tritocerebral appendage) has also been lost in myriapods, which used to be taken as a strong argument for the monophyly of Atelocerata (Tracheata), i.e., Myriapoda + Hexapoda, which is now contradicted by all analyses based on molecular data.

A separate head is present in Cephalocarida, Branchiopoda, and Mystacocarida and within Malacostraca in Leptostraca and Bathynellacea (for discussion of the latter two, see Richter and Scholtz 2001). In other taxa, the anterior tagma, known as the cephalothorax, additionally incorporates one or more thoracic segments (Casanova 1991). A cephalothorax incorporating one thoracic segment only is present in Branchiura, Copepoda, the malacostracan taxa Anaspidacea, Lophogastrida, Amphipoda and Isopoda, and in Remipedia. In all these taxa except Branchiura, the first thoracopod is transformed into a maxilliped (Gruner 1993). Interestingly, a maxilliped is also present in mystacocarids, a group in which the head is separated from the thorax. We were able to show, however, that the Hox gene regime in mystacocarids differs from that in the other taxa with a maxilliped (Fritsch and Richter 2017). In the light of the conditions in Branchiura (cephalothorax but no maxilliped) and Mystacocarida (no cephalothorax but a maxilliped), both representatives of the Oligostraca, we tentatively suggest that in the ur-crustacean neither a cephalothorax nor a maxilliped was present.

One of the characters related to the head is the carapace, which according to the traditional definition by Calman (1909) is a cuticular fold originating from the posterior dorsal region of the head. The carapace has been recently reviewed by Olesen (2013), so we restrict this paragraph to a few comments only. In general, the carapace consists of a thin duplication of the epidermis. The two layers are usually attached to each other, and both secrete a cuticle. The external side of the carapace is sclerotisized or often calcified in order to perform its protective role. In many cases, the internal layer, i.e., the one facing the body, is equipped with a delicate cuticle only and has been suggested as playing a role in respiration (Wirkner and Richter 2013). A carapace is present in phyllopod branchiopods, ostracods, branchiurans, cirripedes, and most malacostracans. The carapace is bivalved in "clam shrimps," ostracods, and cirripedes and is univalved in branchiurans and malacostracans. In adult Malacostraca, the carapace (when present) is fused with the tergites of a variable number of segments commensurate with the number of segments included in the cephalothorax (Casanova 1987, 1991). A carapace is absent in Mystacocarida, Copepoda, anostracan branchiopods, Cephalocarida, and Remipedia. Based on the taxonomic distribution of the carapace, and on its variety, it is unclear whether or not a carapace was present in the ur-crustacean. The carapace must either have evolved several (three or more) times independently or have been lost several (three or more) times.

In crustaceans, as well as in hexapods and myriapods, the second post-antennulary limb is modified to form a mouthpart, the mandible (Snodgrass 1938) (Fig 1.3). In crustaceans, the mandibular gnathal edge is formed by the coxa. This is obvious in taxa such as certain copepods and mystacocarids (e.g., Gruner 1993, Olesen 2001), where both parts of the protopod (coxa and basis) are present along with the endopod and the exopod. However, in cases where no mandibular palp (podomeres distal of the *corpus mandibularis*) is present, a debate exists on what forms the gnathal edge.

It has been suggested that whole-limb mandibles (where the gnathal edge is formed by the distal tip of the limb), as thought to be the case in myriapods and hexapods, are fundamentally different from the gnathobasic mandible in crustaceans (e.g., Manton 1964, 1973, Fryer 1996, Kraus and Kraus 1994). This idea was even used as one of the main arguments in support of arthropod polyphyly (Manton 1964, 1973). Today, there is no doubt that mandibles and their gnathal edges are homologous, that hexapods and myriapods do not possess a whole-limb mandible, and that the "tip" of the (gnathobasic) mandible, i.e., the palp, is simply reduced. This is supported by the temporary and very restricted presence or total absence of *Distal-less* (*Dll*) expression in the *corpus mandibularis* in embryonic stages of certain malacostracans, hexapods, myriapods, (Popadic et al. 1996, Scholtz et al. 1998) and the branchiopod *Cyclestheriahislopi*, despite (as is the case for all branchiopods) no palp being present in the adults (Richter 2002).

Manton (1964, 1977) has suggested that mandibles like those found in anostracans and other branchiopods that grind with a "sweeping and rolling" action and whose entire gnathal edge is formed by

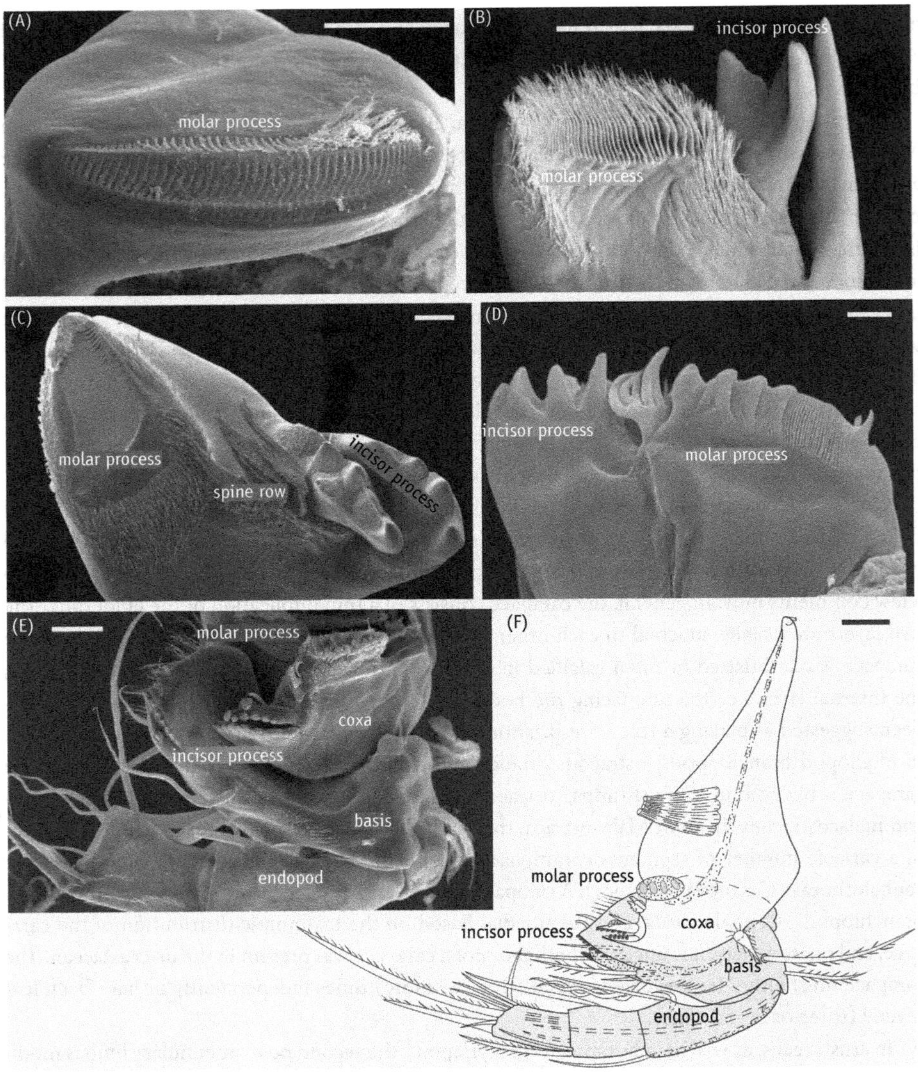

Fig. 1.3.
Mandibles in different arthropod species. (A) *Cyclestheria hislopi* (Cladoceromorpha, Branchiopoda), right mandible, scale 50 μm. Edgecombe et al. (2003). (B) *Speleonectes* cf. *tulumenis* (Remipedia), right mandible, distal tip, anterior view, scale 100 μm. Richter et al. 2002. (C) *Boreomysis nobilis* (Mysida, Malacostraca), left mandible, medial view, scale 100 μm. Richter et al. (2002). (D) *Hanseniella* sp. (Symphyla, Myriapoda), left mandible, distal tip of gnathal lobe, anterior view, scale 10 μm. Richter et al. 2002. (E) *Spelaeoecia capax* (Halocypridina, Copepoda), left mandible, adult male, medial view, scale 1 mm. (F) *Spelaeoecia capax* (Halocypridina, Copepoda), left mandible, adult male, medial view, scale 1 mm. Richter and Kornicker (2006).

the *pars molaris* represent the original crustacean condition (Fig. 1.3A). But, like the absence of the palp in adults, it is much more likely that this represents an autapomorphy of Branchiopoda (Richter 2004).

The gnathal edges of other crustacean mandibles (and those of some hexapod and myriapod mandibles) commonly consist of three parts: an incisor process (*pars incisiva*), a molar process (*pars molaris*), and in some cases a spine row and a *lacinia mobilis* in between (Fig. 1.3B–E). Structural details of the *pars incisiva* and *pars molaris* in various mandibulate taxa provide additional corroboration for the basic assumption of homology based on positional correspondence (i.e., location on the distal and proximal sections of the gnathal edge, respectively). The *pars incisiva* in certain myriapods, hexapods, and crustaceans is a blade-like process with a row of teeth. The *pars molaris* typically has a flattened surface, which is fringed by a row of spines. These details of structure, combined with the distribution of the *pars incisiva* and *pars molaris* across the major lineages of mandibulates, suggest that these elements are homologous and were thus present in the common ancestor of the Mandibulata (see Edgecombe et al. 2003 for more details) as well as in the ur-crustacean. The *lacinia mobilis* has also been suggested to be homologous between various mandibulate taxa (Fanenbruck and Wägele 2000, Bitsch and Bitsch 2004). Contrarily, Richter et al. (2002) argued that the *lacinia mobilis* present in certain Malacostraca (mostly in Peracarida) and similar structures in other mandibulate taxa (the *prostheca* of Hexapoda, the internal tooth of Diplopoda, the *lacinia mobilis* of Symphyla and Remipedia) are not homologous and do not belong to the ground pattern of Mandibulata and thus were not present in the ur-crustacean.

To get a better idea of what kind of mandible might have been present in the ur-crustacean, we need to take a closer look at representatives of the Oligostraca. Representatives of the Mystacocarida (see e.g., Olesen 2001) and Ostracoda (e.g., Richter and Kornicker 2006) possess mandibles with well-developed endopods and exopods and a coxa whose gnathal edge exhibits various kinds of dentition. Particularly interesting are the mandibles of the Halocypridina (Mydocopida, Ostracoda), where a row of terminal cusps is present on the distal edge of the basis (Fig. 1.3E–F), implying (on the basis of structure and position, which is comparable to that of the *pars incisiva*) that there might be a second edge for manipulating food (Richter and Kornicker 2006). The idea that this might represent the condition in the ur-crustacean is interesting but difficult to prove. The *lacinia mobilis* on the coxal gnathal edge in Halocypridina is probably convergent with the one in Peracarida, despite the remarkable similarities between them (Richter and Kornicker 2006).

Brain and Eyes

Like all other arthropods, crustaceans possess a tripartite syncerebrum (the brain) consisting of a protocerebrum, deutocerebrum, and tritocerebrum. While the segmental origin of the neuromeres and their sensory input is clear in the case of the deutocerebrum/1st antenna and tritocerebrum/2nd antenna, the origin of the protocerebrum remains enigmatic. The protocerebrum may consist of one neuromere or two (traditionally called the archicerebrum and the prosocerebrum), and the appendage belonging to the protocerebral region/segment could be either the labrum or the frontal filaments. Recent reviews dealt with the history and current state of this discussion (Richter et al. 2013, Scholtz 2016).

The most prominent components of the protocerebrum are the central complex and the optic lobe. The central complex, an intricate cluster of neuropils and tracts that occurs in the brain of various hexapods and malacostracans, has been one of the most studied topics in arthropod neuroanatomy over the past few decades (Harzsch 2006, 2007, Homberg 2008, Strausfeld 2012, Pfeiffer and Homberg 2014, Sandeman et al. 2014, Boyan et al. 2015).

A central complex has also been documented in some non-malacostracan taxa, namely anostracan Branchiopoda (Harzsch and Glötzner 2002), Remipedia (Fanenbruck et al. 2004, Fanenbruck and Harzsch 2005, Stemme et al. 2012), and Copepoda (Andrew et al. 2012). Several other crustacean taxa, e.g., Mystacocarida (Brenneis and Richter 2010) and Cephalocarida (Stegner and Richter 2011), have been reported to lack the neuropils of a central complex completely or to exhibit only certain components of a central complex; this is the case, for example, in phyllopod Branchiopoda (Fritsch and Richter 2010, 2012, 2016). However, for most non-malacostracan taxa including Ostracoda, Copepoda (with the exception of the study by Andrew et al. 2012), and Thecostraca (see review by Stegner et al. 2014), solid recent evidence is unavailable. The presence of a central complex in the ur-crustacean, therefore, remains uncertain.

Crustaceans (and other arthropods) with compound eyes possess an optic lobe with a specific number of optic neuropils. In general, there are four optic neuropils in malacostracans and hexapods and only two in the remaining crustacean taxa (Strausfeld 2012, Loesel et al. 2013). While there is little doubt about the homology of the two distal optic neuropils (the lamina and medulla) in malacostracans and hexapods, the homology between them and the two neuropils present, e.g., in branchiopods, traditionally also called the lamina and medulla, has been questioned (Strausfeld 2005, 2009, 2012). Strausfeld (2005, 2009, 2012) suggested a scenario based on the idea that the tetraconate ancestor (the ur-crustacean) possessed only two optic neuropils connected by uncrossed fibers, concluding that the "lamina" described in branchiopods is homologous to the lamina plus medulla in malacostracans (for more details, see Strausfeld 2012, Loesel et al. 2013). It is important to remember that all the conclusions reached by Strausfeld and others for non-malacostracans were based primarily on the conditions in branchiopods. The only studies into the neuroarchitectural structure of the two optic neuropils and their connecting fibers in non-branchiopods and non-malacostracan crustaceans date from decades ago. Though there is no doubt that myodocopid ostracods (Hartmann 1967), branchiurans (Madsen 1964), and thecostracan cypris larvae (Harrison and Sandeman 1999) do possess two optic neuropils, the structural details of these neuropils are almost completely unknown. It does not even appear to be certain whether the connection between the two neuropils is chiasmic or straight. In our view, a better structural understanding of the optical neuropils in other crustaceans would be highly desirable. Nevertheless, the presence of an optic lobe with two neuropils most plausibly represents the condition in the ur-crustacean.

The name Tetraconata refers to the number of partitions in the crystalline cone, which in most crustacean and almost all hexapod taxa is four (Paulus 1979). The myriapod taxon Scutigeromorpha also possesses four cone cells, but these are divided, meaning that the cone consists of about eight partitions (Müller et al. 2003).

This notwithstanding, the general assumption that crustacean ommatidia have exactly two corneagenous cells, four crystalline cone cells, and eight retinula cells might be too simple a view. Not all crustacean taxa exhibit this pattern, and in light of the paraphyly of crustaceans, the question of the exact pattern in the ur-crustacean remains open. In addition, certain key taxa such as Cephalocarida, Mystacocarida, Copepoda, and Remipedia do not possess compound eyes at all. Other taxa possess different numbers of ommatidial components, e.g., five crystalline cone cells in the Onychocaudata (Richter et al. 2007, Olesen and Richter 2013) or two cone cells in many Malacostraca (Xenommacarida *sensu* Richter 1999). Nevertheless, the presence of tetrapartite cones most probably represents the condition in the ur-crustacean. Of potential importance is the fact that the cone cell nuclei lie proximally to the basement membrane in Branchiura (Hallberg 1982) and myodocopid Ostracoda (Andersson 1979) (Fig. 1.4A, B), a condition similar to that in Scutigeromorpha (Myriapoda) (Müller et al. 2003) and one that might represent the state for the ur-crustacean.

A point of interest pertaining particularly to compound eyes is that, as first put forward by Paulus (1979) and now considered textbook knowledge, only hexapods possess pigments in their corneagenous cells (which is why they are called primary pigment cells), whereas in crustaceans these cornea-forming cells are pigment-free. However, both a cirripede and a branchiuran species are known in which a pair of pigmented cells are found in a position corresponding to that of corneagenous cells. Furthermore, these cells continue as distal pigment cells (Hallberg 1982, Hallberg and Elofsson 1983) (Fig. 1.4A, C). This could indicate that "primary pigment cells" might not have evolved in hexapods but were already present in the ur-crustacean. In certain ostracods, however, both distal pigment cells and corneagenous cells are present (Fig. 1.4B). Within the various crustacean taxa, the ability of the corneagenous cells to produce pigment would then have been lost, perhaps several times. It must be kept in mind, however, that these conclusions are based on data from very few species, studied using limited techniques.

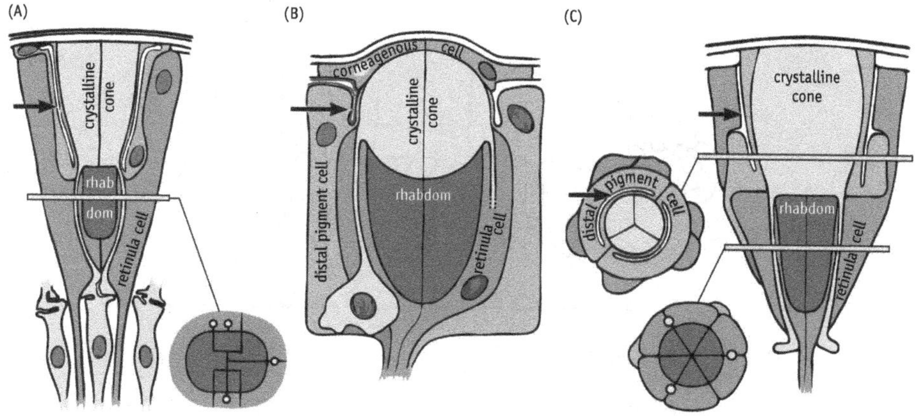

Fig. 1.4.
Ommatidia in three different maxillopodan species. (A) *Argulus foliaceus* (Branchiura). Overview with transverse section through the rhabdom. Retinula cell between two cone cell processes. (B) *Cypridina norvegica* (Ostracoda). Overview. (C) *Balanus crenatus* (Cirripedia). Overview with two transverse sections through the crystalline cone and the distal pigment cells and through the rhabdom. Three cone cell processes are present. Arrows (A, B, C) indicate extracellular space between the distal pigment cells in all three species. After Richter (2002). See color version of this figure in centerfold.

Nauplius Eye

The nauplius eye is the second type of light-sensory organ in crustaceans. During ontogeny, the nauplius eye develops before the compound eyes appear, and while it is the only light-sensory organ in nauplius larvae (hence the name), it often persists until the adult stage. The combination of three or four median eye cups into a single organ is unique in crustaceans (Lauterbach 1983; see also Richter 2002). However, a nauplius eye (actually a median eye of any kind) is completely absent in Mystacocarida, Cephalocarida, and Remipedia. Most of our knowledge of the microscopic anatomy and ultrastructure of nauplius eyes comes from Elofsson (1965, 1966), though a number of TEM studies have since been carried out. Elofsson (2006) stressed that nauplius eyes evolved independently, pointing out the differences between the everse nauplius eyes of Malacostraca and the inverse eyes of other crustaceans. Reimann and Richter (2007) demonstrated the presence of a tapetum layer between the absorbing pigment layer and the rhabdomeres in branchiopod crustaceans. In the remaining crustacean taxa (e.g., Branchiura—Elofsson 1966; Copepoda—Fahrenbach 1964, Lacalli 2009; Cirripedia-Thoracica—Kauri 1962, Takenaka et al. 1993; Ostracoda-Podocopa—Andersson and Nilsson 1981; Myodocopa—Andersson 1979) a tapetum layer made up of reflective crystals and consisting of two tapetal cells per nauplius eye cup has been described, making it very different from the tapetum found in Branchiopoda. The distribution of the non-branchiopod tapetum-bearing nauplius eye might suggest that this was the kind that existed in the ur-crustacean.

Tagmatization

A striking character complex in crustaceans and arthropods in general is tagmatization (Fusco and Minelli 2013). Between the various crustacean taxa, however, tagmatization varies considerably. The pleon (malacostracans) and abdomen (non-malacostracans) are clearly different tagmata, not only because of the presence versus absence of limbs (Fig. 1.5). Within Malacostraca, trunk tagmatization is almost uniform. Malacostracans possess a trunk divided into an eight-segmented

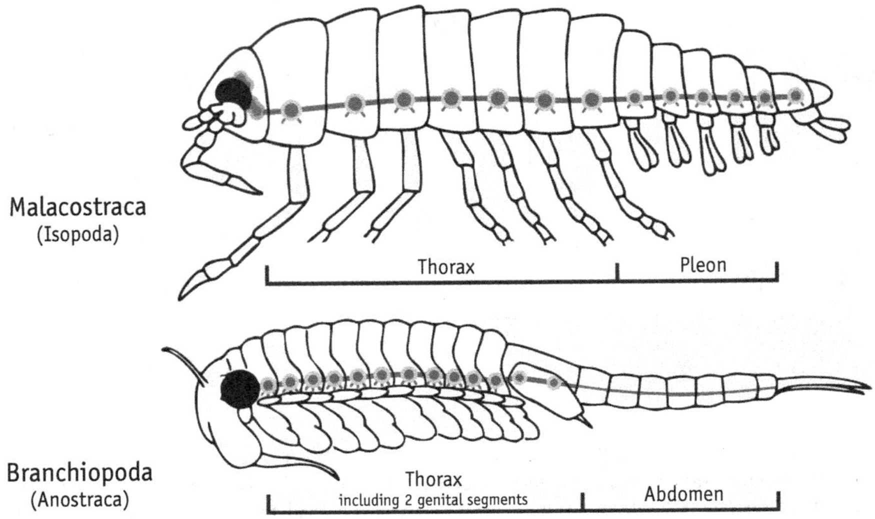

Fig. 1.5.
Tagmatization in malacostracan and branchiopod crustaceans. See color version of this figure in centerfold.

thorax and a six-segmented pleon, and both tagmata are limb-bearing (see below for Leptostraca). In many non-malacostracan crustaceans, a limb-bearing thorax and a limb-less abdomen exist. Both the number of thoracic segments and the number of abdominal segments vary. In some cases, the abdomen does not display any kind of segmentation (e.g., in Branchiura, Cirripedia, Ostracoda). Unlike the hexapod abdomen, the crustacean abdomen displays neither anlagen nor derivatives of limbs (Gruner and Scholtz 2004). Largely on the basis of differences in Hox-gene expression patterns, Schram and Koenemann (2004) argued that the non-malacostracan abdomen, the hexapod abdomen, and the malacostracan pleon are very different units, defining the crustacean (i.e., non-malacostracan) abdomen as the body region that lies posterior to the expression of Hox-gene *abdominal-B* and shows no Hox-gene expression in itself.

A more inclusive definition, this time encompassing reference to the nervous system, was suggested by Deutsch and Mouchel-Vielh (2003; based on Deutsch 2001), who additionally characterized the crustacean abdomen by its absence of segmental ganglia and commissures. Taxa in which ganglia are present in the thorax but not in the (segmented) limb-less abdomen include Branchiopoda (Harzsch and Glötzner 2002, Fritsch and Richter 2010, 2012, Frase and Richter 2016,) and Copepoda (Park 1966; see Boxshall 1992). In Branchiura, the "abdomen" is unsegmented (Gruner 1993).

In the tagmatized trunk (composed of an eight-segmented thorax and six-segmented pleon) of Eumalacostraca, all the segments bear limbs and are equipped with a pair of ganglia, which give rise to the nerves in the appendages (Hanström 1928, Chaudonneret 1978). Ganglia may be fused in the midline or anteroposteriorly (e.g., forming a sub-esophageal ganglion). Leptostraca differ from their sister group Eumalacostraca in featuring an additional limb-less seventh segment that apparently lacks a ganglion. However, Manton (1934) found that the sixth and seventh pleonal ganglia in Leptostraca fuse developmentally to form the sixth pleonal segment. Lauterbach (1975) suggested that the seventh limb-less segment represents the original abdomen.

Any strong correspondence between segmental appendages and the presence of ganglia and commissures in the ventral nerve cord (Deutsch and Mouchel-Vielh 2003) is contradicted by the finding of ganglia and commissures in the last two abdominal segments of Mystacocarida (Brenneis and Richter 2010) and Cephalocarida (Stegner et al. 2014). In Cephalocarida, the abdominal ganglia differ significantly from the thoracic ganglia in shape, size, and commissural pattern (Stegner et al. 2014), while abdominal ganglia 1 to 3 in Mystacocarida correspond broadly to thoracic ganglia 2 to 5 (Brenneis and Richter 2010).

A limb-less (and clearly segmented) abdomen appears in representatives of Oligostraca (i.e., Mystacocarida), Multicrustacea (i.e., Copepoda, Ascothoracida), and Allotriocarida (i.e., Cephalocarida, Branchiopoda), which might indicate that an abdomen was already present in the ur-crustacean. However, in the homonomous trunk of Remipedia and Myriapoda as the sister group to Tetraconata, all segments bear limbs and are equipped with a pair of ganglia, which give rise to appendage nerves (Hanström 1928, Harzsch et al. 2005). The very different kinds of posterior segmented tagmata present in crustaceans cause us to interpret them as having evolved independently.

Heart and Arterial System

The hemolymph vascular system in crustaceans displays a huge degree of disparity (Wirkner et al. 2013, Wirkner and Richter 2013), but the arterial system is only highly complex in Malacostraca (Fig. 1.6) (Wirkner and Richter 2010). Most other taxa possess a simple system which, in addition to the heart, might include an anterior and a posterior aorta and cardiac arteries. A heart and the entire vascular system is absent in mystacocarids, podocopid ostracods, and most copepods. A heart that runs almost the entire length of the trunk is present in Remipedia, certain Malacostraca (e.g., Stomatopoda), Cephalocarida and Anostraca, among others, which implies that heart extension is not related to tagmatization. Remipedes, for example, possess a homonomous trunk, malacostracans possess a bi-partite thorax (i.e., thorax and pleon), and cephalocarids and anostracans possess a thorax and an abdomen. Based on the recent work of Göpel and Wirkner (2018), which summarizes the distribution of various character states related to the vascular system, we suggest that the ur-crustacean had a tubular heart extending the length of the trunk and that it was equipped with segmentally arranged pairs of ostia and some cardiac arteries.

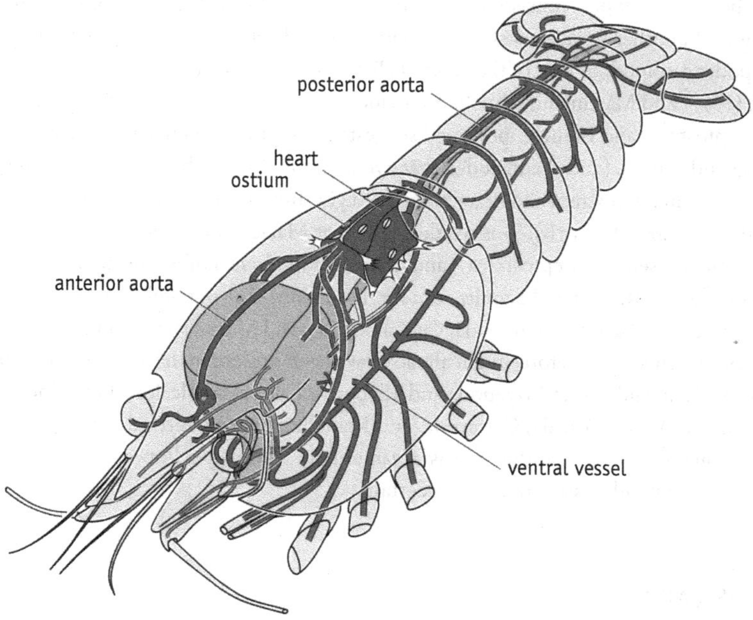

Fig. 1.6.
Schematic diagram of the complex HVS in the marbled crayfish *Procambarus fallax f. virginalis* (Malacostraca). Only the proximal parts of the arteries of the right body side are shown. From Scholz et al. (2018). See a color version of this figure in the centerfold.

Limbs

Most crustaceans possess "typical," i.e., stenopodous or narrow, arthropod limbs. Crustacean (post-antennulary) limbs are mostly biramous and consist of a protopod, an endopod made up of cylindrical podomeres, and an exopod. A biramous limb with an undivided protopod was probably already present in the arthropod ancestor (Boxshall 2004). Waloszek et al. (2007) suggested that the phylogenetic precursor of the coxa was a separately movable setated endite (the "proximal endite") nested within the ample joint membrane medially below the basipod (or undivided protopod) (see also Haug et al. 2013). Boxshall (2004, 2013) hypothesized that the protopod (Waloszek's basipod) was subdivided by the appearance of a transverse articulation into the proximal coxa and the distal basis, presenting support for this scenario from gene expression data (see Williams 2013). A divided protopod in (some) post-maxillary limbs is present in the oligostracan taxon Mystacocarida (Olesen 2001) and in certain ostracods (Hartmann 1966) and branchiurans (Gruner 1993). This might indicate that a divided protopod was present in the ur-crustacean.

Phyllopodous (or folicaeous) limbs are quite different from stenopodous limbs (Fig. 1.7). In phyllopodous limbs, the proximal endite is simply the proximal-most endite of the series of endites on the median margin of the protopod (Fig. 1.7D). It may be capable of independent motion from the rest of the limb (Boxshall 2013). Hessler and Newman (1975) derived their ur-crustacean limb from the phyllopodous limbs in branchiopods, cephalocarids, and leptostracans (Fig. 1.7A). When Schram (1986) and Schram and Hof (1998) proposed a taxon comprising Branchiopoda, Cephalocarida, and Leptostraca on the basis of the shared character "phyllopodous limbs," Olesen (2009) noted that a character such as "trunk limbs being phyllopodous" is too general to be relevant to crustacean phylogeny.

However, Olesen (2009) did in fact argue in favor of the homology of branchiopod (i.e., Anostraca + Phyllopoda) phyllopodous limbs (contrary to Preuss 1951, 1957), implying that the stenopodous limbs in the raptorial water-fleas evolved secondarily from phyllopodous limbs (Olesen et al. 2001). Olesen (2009) also emphasized the similarity of the limbs of the Cambrian *Rehbachiella*, which were probably only partly phyllopodous (see Walossek 1993). Cephalocarida possess limbs whose protopod bears an endopod consisting of five podomeres and a partly foliaceous exopod. The latter is the origin of another foliaceous structure with marginal setae, which has been interpreted as a pseudepipod, an epipod, or part of the exopod (Fig. 1.7C) (see Richter 2002 for discussion). While Walossek (1993) and Martin and Christiansen (1995) emphasized the differences in limb structure between Leptostraca and Branchiopoda and suggested that phyllopodous limbs evolved independently, Pabst and Scholtz (2009) concluded that overall similarities in the general developmental pattern of the limbs in Leptostraca and Branchiopoda might support the homology of the phyllopodous limbs in the two taxa (Fig. 1.7B). This would imply that Malacostraca originally had phyllopodous limbs (as still present in Leptostraca) and that stenopodous limbs are an autapomorphy of Eumalacostraca (see Richter and Scholtz 2001 for discussion). The possibility of homology between the phyllopodous limbs in Branchiopoda, Cephalocarida, and Malacostraca still exists.

Phyllopodous limbs, therefore, might already have been present in the ancestor of Multicrustaca + Allotriocarida, meaning that Copepoda and Thecostraca, for example, lost them. The limbs in the oligostracan taxa Mystacocarida, Ostracoda, and Branchiura are all derived in their own way. This might additionally indicate that the ur-crustacean did not possess phyllopodous limbs, though the exact kind of limb it did possess remains uncertain.

NAUPLIUS LARVA

The nauplius larva constitutes the hatching stage in most crustacean taxa (Martin et al. 2014). In some groups (e.g., copepods, ostracods, certain branchiopods, ascothoracids, cirripedes, remipedes; for

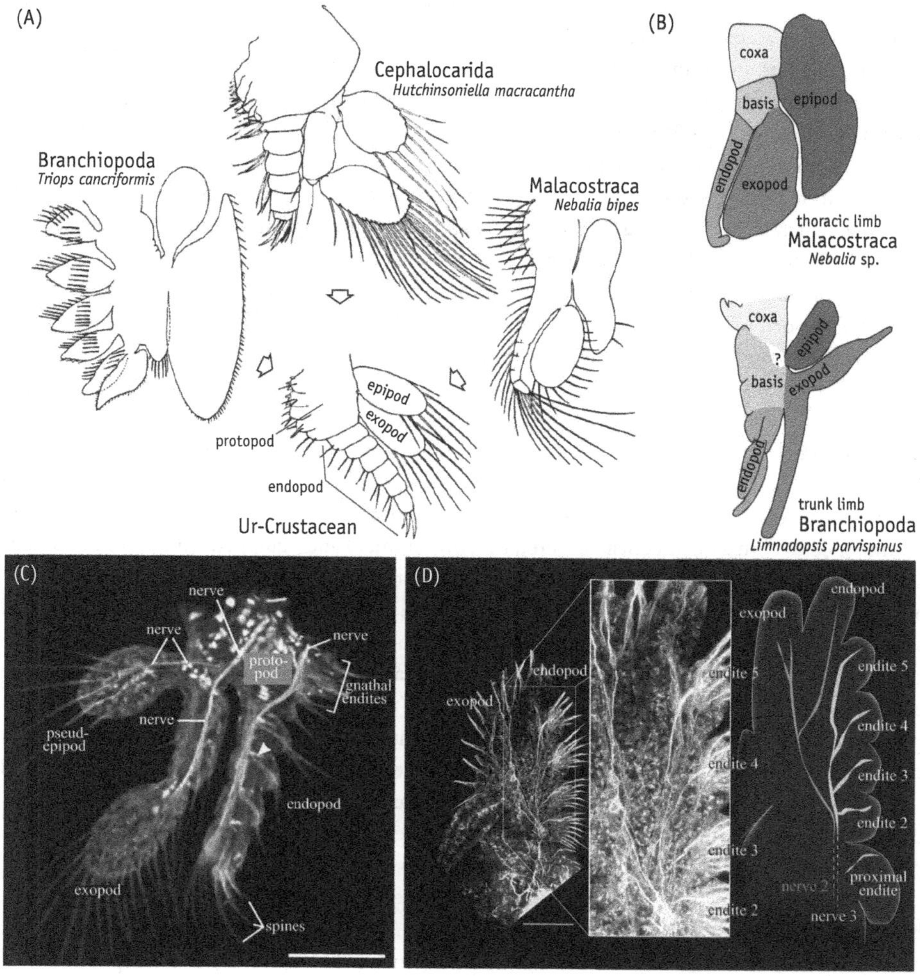

Fig. 1.7.
Phyllopodous limbs in crustaceans. (A) Derivation of the ur-crustacean trunk limb morphology from that of a notostracan branchiopod, a cephalocarid, and a leptostracan malacostracan. After Hessler and Newman (1975). (B) Schematic representations of phyllopodous limbs. In *Nebalia* the epipod is connected to coxa and basis. After Pabst and Scholtz (2009). (C) Innervation of right trunk limb in *Hutchinsoniella macracantha.* Anterior aspect. Arrowhead indicates the insertion site of a podomere muscle. After Stegner et al. (2014). (D) Innervation of trunk limbs in *Leptestheria dahalacensis* (Branchiopoda; 3rd trunk limb of early juvenile). After Kaji et al. (2014).

dendrobranchiates and euphausiaceans, see below), pelagic life starts as an orthonauplius with only three pairs of limbs (antennule, antenna, mandible). In others it starts as a metanauplius, which additionally displays the anlagen of more posterior appendages. In both cases, further development is anameric, with additional segments and limb anlagen appearing in the succeeding stages, which are separated from each other by larval moults (e.g., Stegner and Richter 2015).

The (always lecithotrophic) malacostracan nauplius, only present in dendrobranchiate decapods and in Euphausiacea, has been suggested to be a secondary hatching stage (Scholtz 2000), which implies that the original hatching stage in Malacostraca was probably some kind of zoea larva (Jirikowski et al. 2013). In other malacostracan taxa, a so-called "egg nauplius" appears during embryonic development in what appears to be a recapitulation of the free-swimming nauplius, though the concept of recapitulation traditionally only involves the epidermal part of the embryo (and not, for example, the musculature). Jirikowski et al. (2015) emphasized the importance of the last embryonic stage before hatching for our understanding of the evolution of nervous system development (see Richter et al. 2016 for more details). In those taxa that do hatch as nauplius larva, this embryonic stage represents an "egg nauplius" comparable to that in malacostracans. For obvious reasons, the sensory and motor nervous system must be functional at the moment of hatching, which implies that the "naupliar nervous system" develops in the embryonic phase (e.g., Frase and Richter 2013, Stegner and Richter 2015, Richter et al. 2016).

Based on its taxonomic distribution, an orthonauplius was likely present in the ur-crustacean. Whether this orthonauplius was planktotrophic or lecithotrophic is difficult to decide, because in most taxa both modes occur (Martin et al. 2014).

CONCLUSION

The ur-crustacean (i.e., the last common ancestor of Tetraconata/Pancrustacea) was a free-living marine species with a distinct head covered by a dorsal shield and equipped with two pairs of sensory limbs (antennule and antenna), mandibles, and two more pairs of mouthparts (maxillule and maxilla). It is likely that no further segments were fused to the head and that no maxilliped was present. The ur-crustacean may or may not have possessed a carapace. Its mandible consisted of a subdivided protopod with a coxal gnathal edge bearing *pars incisiva* and *pars molaris*, and an endopod and exopod originating from the distal part of the protopod, the basis. Its brain was complex, with an extended olfactory system, possibly a central complex, and a lateral protocerebrum containing at least two optical neuropils. The protocerebrum was connected to a nauplius eye as well as to compound eyes. The ommatidia possessed tetrapartite crystalline cones, possibly with cone cell nuclei proximally of the basement membrane, and pigmented corneagenous cells were present in the ommatida.

The ur-crustacean might have had a uniformly segmented trunk posterior to its five-segmented head or (less probably) may have possessed two tagmata, a limb-bearing thorax and a limb-less abdomen. It did have a heart, which might have extended through the trunk independently of tagmatization. Its thoracopodal appendages were true arthropodal (consisting of podomeres) with a protopod (probably subdivided into coxa and basis), an exopod, and an endopod. Larval development started with a nauplius larva (probably an orthonauplius).

ACKNOWLEDGMENTS

We are grateful to Martin Thiel and Gary Poore for inviting us to write this chapter and for their patience in waiting for it to actually be finished. We are grateful to all our former students for providing much of the data cited in this chapter. We thank Lucy Cathrow for improving the English.

REFERENCES

Andersson, A., and D.-E. Nilsson. 1981. Fine structure and optical properties of an ostracode (Crustacea) nauplius eye. Protoplasma 107:361–374.

Andersson, A. 1979. Cerebral sensory organs in ostracodes (Crustacea). Ph.D. dissertation, University of Lund, Lund, Sweden.

Andrew, D.R., S.M. Brown, and N. Strausfeld. 2012. The minute brain of the copepod *Tigriopusca lifornicus* supports a complex ancestral ground pattern of the tetraconate cerebral nervous systems. The Journal of Comparative Neurology 520:3446–3470.

Bitsch, C., and J. Bitsch. 2004. Phylogenetic relationships of basal hexapods among the mandibulate arthropods: a cladistic analysis based on comparative morphological characters. Zoologica Scripta 33:511–550.

Boxshall, G.A. 1992. Copepoda. Pages 347–384 *in* F.W. Harrison and A.G. Humes, editors. Microscopic Anatomy of Invertebrates, Vol. 9: Crustacea. Wiley-Liss, New York.

Boxshall, G.A. 2004. The evolution of arthropod limbs. Biological Reviews 79:253–300.

Boxshall, G.A. 2013. Arthropod limbs and their development. Pages 241–268 *in* A. Minelli, G. Boxshall, and G. Fusco, editors. Arthropod Biology and Evolution—Molecules, Development, Morphology. Springer, Heidelberg, Germany.

Boyan, G., L. Williams, and Y. Liu. 2015. Conserved patterns of axogenesis in the panarthropod brain. Arthropod Structure & Development 44:101–112.

Brenneis, G., and S. Richter. 2010. Architecture of the nervous system in Mystacocarida (Arthropoda, Crustacea)—an immunohistochemical study and 3D reconstruction. Journal of Morphology 271:169–189.

Calman, W.T. 1909. Crustacea. Pages 1–346 *in* E.R. Lancester, editor. A Treatise on Zoology. A&C Black, London.

Casanova, B. 1987. Relations entre les tergites thoraciques et la carapace dans la serie des Mysidacés, Euphausiacés et Décapodes Dendrobranchiata (Crustacés). Comptes Rendus de l'Académie des Sciences—Series III—Sciences de la Vie 305:655–660.

Casanova, B. 1991. Origine protocéphalique antennaire de la carapace chez les Leptostracés, Mysidacés et Eucarides (Crustacés). Comptes Rendus de l'Académie des Sciences—Series III—Sciences de la Vie 312:461–468.

Chaudonneret, J. 1978. Phylogenese du systeme nerveux annelido-arthropodien. Bulletin de la Societé Zoologique de France 103:69–95.

Deutsch, J. S., and E. Mouchel-Vielh. 2003. Hox genes and the crustacean body plan. BioEssays 25:878–887.

Deutsch, J. 2001. Are Hexapoda members of the Crustacea? Evidence from comparative developmental genetics. Pages 41–49 *in* T. Deuve, editor. Origin of the Hexapoda. Annales de la Société entomologique de France (NS), 35.

Dohle W. 2001. Are the insects terrestrial crustaceans? A discussion of some new facts and arguments and the proposal of the proper name "Tetraconata" for the monophyletic unit Crustacea + Hexapoda. Pages 85–103 *in* T. Deuve, editor. Origin of the Hexapoda. Annales de la Société entomologique de France (NS), 35.

Edgecombe, G.D. 2010. Arthropod phylogeny: an overview from the perspectives of morphology, molecular data and the fossil record. Arthropod Structure & Development 39:74–87.

Edgecombe, G.D. 2017. Inferring arthropod phylogeny: fossils and their interaction with other data sources. Integrative and Comparative Biology 57:467–476.

Edgecombe, G.D., S. Richter, and G.D.F. Wilson. 2003. The mandibular gnathal edges: homologous structures across Mandibulata? African Invertebrates: A Journal of Biodiversity Research 44:115–135.

Elofsson, R. 1965. The nauplius eye and frontal organs in Malacostraca (Crustacea). Sarsia 19:1–54.

Elofsson, R. 1966. The nauplius eye and frontal organs of the non-Malacostraca (Crustacea). Sarsia 25:1–128.

Elofsson, R. 2006. The frontal eyes of crustaceans. Arthropod Structure & Development 35:275–291.

Fahrenbach, W.H. 1964. The fine structure of a nauplius eye. Zeitschrift für Zellforschung und Mikroskopische Anatomie 62:182–197.

Fanenbruck, M., and S. Harzsch. 2005. A brain atlas of *Godzilliognomus frondosus* Yager, 1989 (Remipedia, Godzilliidae) and comparison with the brain of *Speleonectes tulumensis* Yager, 1987 (Remipedia, Speleonectidae): implications for arthropod relationships. Arthropod Structure & Development 34:343–378.

Fanenbruck, M., and J.W. Wägele. 2000. New light on the ancestry of Tracheata: cephalic anatomy of *Godzilliognomus frondosus* (Godzilliidae, Remipedia). Zoology 103:101.

Fanenbruck, M., S. Harzsch, and W. Wägele. 2004. The brain of the Remipedia (Crustacea) and an alternative hypothesis on their phylogenetic relationships. Proceedings of the National Academy of Sciences USA 101:3868–3873.

Frase, T., and S. Richter. 2013. The fate of the onychophoran antenna. Development Genes and Evolution 223:247–251.

Frase, T, and S. Richter. 2016. Nervous system development in the fairy shrimp *Branchinella* sp. (Crustacea: Branchiopoda: Anostraca)—insights into development and evolution of the branchiopod brain and its sensory organs. Journal of Morphology 277:1423–1446.

Fritsch, M., and S. Richter. 2010. The formation of the nervous system during larval development in *Triops cancriformis* (Bosc) (Crustacea, Branchiopoda): An immunohistochemical survey. Journal of Morphology 271:1457–1481.

Fritsch, M. S., and S. Richter. 2012. Nervous system development in Spinicaudata and Cyclestherida (Crustacea, Branchiopoda)—comparing two different modes of indirect development by using an event pairing approach. Journal of Morphology 273:672–695.

Fritsch, M., and S. Richter. 2016. Branchiopoda. Pages 514–521 *in* A. Schmidt-Rhaesa, S. Harzsch, and G. Purschke, editors. Structure and Evolution of the Invertebrate Nervous Systems. Oxford University Press, New York.

Fritsch, M., and S. Richter. 2017. Unexpected UBX expression in the maxilliped of the mystacocarid crustacean *Derocheilocharis remanei*—evidence for a different way of making a maxilliped? Development Genes and Evolution 227:289–296.

Fryer, G. 1996. Reflections on arthropod evolution. Biological Journal of the Linnean Society 58:1–55.

Fusco, G., and A. Minelli. 2013. Arthropod segmentation and tagmosis. Pages 197–222 *in* A. Minelli, G. Boxshall, and G. Fusco, editors. Arthropod Biology and Evolution—Molecules, Development, Morphology. Springer, Heidelberg, Germany.

Giribet G., and G.D. Edgecombe. 2013. The Arthropoda: a phylogenetic framework. Pages 17–40 *in* A. Minelli, G. Boxshall, and G. Fusco, editors. Arthropod Biology and Evolution—Molecules, Development, Morphology. Springer, Heidelberg, Germany.

Göpel, T., and C.S. Wirkner. 2018. Morphological description, character conceptualization and the reconstruction of ancestral states exemplified by the evolution of arthropod hearts. PLoS ONE 13:e0201702.

Gruner, H.E., and G. Scholtz. 2004. Segmentation, tagmata, and appendages. Pages 13–57 *in* J. Forest, J.C. von Vaupel-Klein, and F.R. Schram, editors. Treatise on Zoology—Anatomy, Taxonomy, Biology. The Crustacea Revised and Updated from the Traité de Zoologie. Koninklijke Brill, Leiden, The Netherlands.

Gruner, H.E. 1993. Arthropoda. Lehrbuch der Speziellen Zoologie. Gustav Fischer Verlag, Jena, Germany.

Hallberg, E., and R. Elofsson. 1983. The larval compound eye of barnacles. Journal of Crustacean Biology 3:17–24.

Hallberg, E. 1982. The fine structure of the compound eye of *Argulus foliaceus* (Crustacea: Branchiura). Zoologischer Anzeiger 208:227–236.

Hanström, B. 1928. Vergleichende Anatomie des Nervensystems der wirbellosen Tiere unter Berücksichtigung seiner Funktion. Springer Verlag, Berlin.

Harrison, P.J.H., and D.C. Sandeman. 1999. Morphology of the nervous system of the barnacle cypris larva (*Balanus amphitrite* Darwin) revealed by light and electron microscopy. Biological Bulletin 197:144–158.

Hartmann, G. 1966. Ostracoda. Pages 1–216 *in* H. Bronns, editor. Klassen und Ordnungen des Tierreichs, Arthropoda, Crustacea, Ostracoda. Geest und Portig, Leipzig, Germany.

Hartmann, G. 1967. Ostracoda. Pages 217–408 *in* H. Bronns, editor. Klassen und Ordnungen des Tierreichs, Arthropoda, Crustacea, Ostracoda. Geest und Portig, Leipzig, Germany.

Harzsch, S., and J. Glötzner. 2002. An immunohistochemical study on structure and development of the nervous system in the brine shrimp *Artemia salina* Linnaeus, 1758 (Branchiopoda, Anostraca) with remarks on the evolution of the arthropod brain. Arthropod Structure & Development 30:251–270.

Harzsch, S. 2006. Neurophylogeny: architecture of the nervous system and a fresh view on arthropod phylogeny. Integrative and Comparative Biology 6:162–194.

Harzsch, S. 2007. Architecture of the nervous system as a character for phylogenetic reconstructions: examples from the Arthropoda. Species, Phylogeny & Evolution 1:33–57.

Harzsch, S., M. Wildt, B. Battelle, and D. Waloszek. 2005. Immunohistochemical localization of neurotransmitters in the nervous system of larval *Limulus polyphemus* Linnaeus, 1758 (Chelicerata,

Xiphosura): evidence for a conserved architecture of the protocerebrum in Euarthropoda. Arthropod Structure & Development 34:327–342.

Haug, J. T., C. Castellani, C. Haug, D. Waloszek, and A. Maas. 2013. A *Marrella*-like arthropod from the Cambrian of Australia: a new link between "Orsten"-type and Burgess Shale assemblages. Acta Palaeontologica Polonica 58:629–639.

Hennig, W. 1966. Phylogenetic Systematics. University of Illinois Press, Chicago.

Hessler, R.R., and W.A. Newman. 1975. A trilobitomorph origin for the Crustacea. Fossils and Strata 4:437–459.

Hessler, R.R. 1964. The Cephalocarida: comparative skeleto-musculature. Memoirs of the Connecticut Academy of Arts & Sciences 16:1–97.

Hessler, R.R. 1992. Reflections on the phylogenetic position of the Cephalocarida. Acta Zoologica 73:315–316.

Homberg, U. 2008. Evolution of the central complex in the arthropod brain with respect to the visual system. Arthropod Structure & Development 37:347–362.

Jirikowski, G., S. Richter, and C. Wolff. 2013. Myogenesis of Malacostraca—the "egg-nauplius" concept revisited. Frontiers in Zoology 10:76.

Jirikowski, G.J., C. Wolff, and S. Richter. 2015. Evolution of eumalacostracan development—new insights into loss and reacquisition of larval stages revealed by heterochrony analysis. EvoDevo 6:4.

Kaji, T., M. Fritsch, M. Schwentner, J. Olesen, and S. Richter. 2014. Male claspers in clam shrimps (Crustacea, Branchiopoda) in the light of evolution: a case study on homology versus analogy. Journal of Experimental Biology Part B: Molecular and Developmental Evolution 322:269–280.

Kauri, T. 1962. On the frontal filaments and nauplius eye in *Balanus*. Crustaceana 4:131–142.

Kraus, O., and M. Kraus. 1994. Phylogenetic system of the Tracheata (Mandibulata): on "Myriapoda"–Insecta interrelationships, phylogenetic age and primary ecological niches. Verhandlungen des Naturwissenschaftlichen Vereins in Hamburg (NF) 34:5–31.

Lacalli, T.C. 2009. Serial EM analysis of a copepod larval nervous system: naupliar eye, optic circuitry, and prospects for full CNS reconstruction. Arthropod Structure & Development 38:361–375.

Lauterbach, K.E. 1975. Über die Herkunft der Malacostraca (Crustacea). Zoologischer Anzeiger 194:165–179.

Lauterbach, K.E. 1983. Zum Problem der Monophylie der Crustacea. Verhandlungen des Naturwissenschaftlichen Vereins Hamburg 26:293–320.

Lauterbach, K.E. 1988. Zur Position angeblicher Crustacea aus dem Ober-Kabrium im phylogenetischen System der Mandibulata (Arthropoda). Verhandlungen des Naturwissenschaftlichen Vereins Hamburg 30:409–467.

Lauterbach, K.E. 1989. The pan-monophylum—a practical help in phylogenetic systematics. Zoologischer Anzeiger 223:139–156.

Loesel, R., H. Wolf, and M. Kenning. 2013. Architectural principles and evolution of the arthropod central nervous system. Pages 299–342 *in* A. Minelli, G. Boxshall, and G. Fusco, editors. Arthropod Biology and Evolution—Molecules, Development, Morphology. Springer, Heidelberg, Germany.

Maas, A., D. Waloszek, and K.J. Müller. 2003. Morphology, ontogeny and phylogeny of the Phosphatocopina (Crustacea) from the Upper Cambrian "Orsten" of Sweden. Fossils and Strata 49:1–238.

Madsen, N. 1964. The anatomy of *Argulus foliaceus* Linné—with notes on *Argulus coregoni* Thorell and *Argulus africanus* Thiele. Kungliga Fysiografiska Sälskabets Handlingar (NF) 74:1–32.

Manton, S.M. 1934. V. On the embryology of the crustacean *Nebalia bipes*. Philosophical Transaction of the Royal Society of London B 223:163–238.

Manton, S.M. 1964. Mandibular mechanisms and evolution of arthropods. Philosophical Transaction of the Royal Society of London B 247:1–183.

Manton, S.M. 1973. Arthropod phylogeny—a modern synthesis. Journal of Zoology 171:111–130.

Manton, S.M. 1977. The Arthropoda: Habits, Functional Morphology and Evolution. Clarendon Press, Oxford.

Martin, J.W., and J.C. Christiansen. 1995. A morphological comparison of the phyllopodous thoracic limbs of a leptostracan (*Nebalia* sp.) and a spinicaudate conchostracan (*Leptestheria* sp.), with comments on the use of Phyllopoda as a taxonomic category. Canadian Journal of Zoology 73:2283–2291.

Martin, J.W., J. Olesen, and J.T. Hoeg. 2014. Atlas of Crustacean Larvae. Johns Hopkins University Press, Baltimore.

Meier, R., and S. Richter. 1992. Suggestions for a more precise usage of proper names of taxa. Ambiguities related to the stem lineage concept. Zeitschrift für Zoologische Systematik und Evolutionsforschung 30: 81–88.

Müller, C., J. Rosenberg, S. Richter, and V.B. Meyer-Rochow. 2003. The compound eye of *Scutigera coleoptrata* (Linnaeus, 1758) (Chilopoda: Notostigmophora): an ultrastructural re-investigation that adds support to the Mandibulata-concept. Zoomorphology 122:191–209.

Olesen, J., and S. Richter. 2013. Onychocaudata (Branchiopoda: Diplostraca), a new high-level taxon in branchiopod systematics. Journal of Crustacean Biology 33:62–65.

Olesen, J. 2001. External morphology and larval development of *Derocheilocaris remanei* Delamare-Deboutteville and Chappuis, 1951 (Crustacea, Mystacocarida), with a comparison of crustacean segmentation and tagmosis patterns. Biologiske Skrifter 53, Det Kongelige Danske Videnskabernes Selskab, C.A. Reitzels Forlag.

Olesen, J. 2009. Phylogeny of Branchiopoda (Crustacea)—character evolution and contribution of uniquely preserved fossils. Arthropod Systematics & Phylogeny 67:3–39.

Olesen, J. 2013. The crustacean carapace: Morphology, function, development, and phylogenetic history. Pages 103–139 *in* L. Watling and M. Thiel, editors. The Natural History of the Crustacea, Vol. 1: Functional Morphology and Diversity. Oxford University Press, New York.

Olesen, J., S. Richter, and G. Scholtz. 2001. The evolutionary transformation of phyllopodous to stenopodous limbs in the Branchiopoda (Crustacea)—Is there a common mechanism for early limb development in arthropods? International Journal of Developmental Biology 45:869–876.

Pabst, T., and G. Scholtz. 2009. The development of phyllopodous limbs in Leptostraca and Branchiopoda. Journal of Crustacean Biology 29:1–12.

Park, T.S. 1966. The biology of a calanoid copepod: *Epilabidocera amphitrites* McMurrich. Cellule 66:129–251.

Paulus, H.F. 1979. Eye structure and the monophyly of the Arthropoda. Pages 299–383 *in* A.P. Gupta, editor. Arthropod Phylogeny. Van Nostrand Reinhold, New York.

Pfeiffer, K., and U. Homberg. 2014. Organization and functional roles of the central complex in the insect brain. Annual Review of Entomology 59:165–84.

Popadic, A., D. Rusch, M., Peterson, B.T., Rogers, and T.C. Kaufman. 1996. Origin of the arthropod mandible. Nature 380:395

Preuss, G. 1951. Die Verwandtschaft der Anostraca und Phyllopoda. Zoologischer Anzeiger 147:50–63.

Preuss, G. 1957. Die Muskulatur der Gliedmaßen von Phyllopoden und Anostraken. Mitteilungen aus dem Zoologischen Museum Berlin 33:221–256.

Reimann, A., and S. Richter. 2007. The nauplius eye complex in "conchostracans" (Crustacea, Branchiopoda: Laevicaudata, Spinicaudata, Cyclestherida) and its phylogenetic implications. Arthropod Structure & Development 36:408–419.

Richter, S., and Kornicker L.S. 2006. The mandibles of a halocyprid ostracod (Halocypridae; Halocypridina; Ostracoda)—a new record of mandibular gnathal edges with a "lacinia-mobilis." Journal of Crustacean Biology 26:113–118.

Richter, S.,and G. Scholtz. 2001. Phylogenetic analysis of the Malacostraca (Crustacea). Journal of Zoological Systematics and Evolutionary Research 39:113–136.

Richter, S. 1999. The structure of the ommatidia of the Malacostraca—a phylogenetic approach. Verhandlungen des naturwissenschaftlichen Vereins Hamburg (NF) 38:161–204.

Richter, S. 2002. The Tetraconata concept: hexapod–crustacean relationships and the phylogeny of Crustacea. Organisms, Diversity and Evolution 2:217–237.

Richter, S. 2004. A comparison of the mandibular gnathal edges in branchiopod crustaceans—implications for the phylogenetic position of the Laevicaudata (Crustacea, Branchiopoda). Zoomorphology 123:31–44.

Richter, S., G.D. Edgecombe, and G.D.F. Wilson. 2002. The lacinia mobilis and similar structures—a valuable character in arthropod phylogenetics? Zoologischer Anzeiger 241:339–361.

Richter, S., J. Olesen, and W.C. Wheeler. 2007. Branchiopod phylogeny—a combined analysis using morphology and six molecular loci. Cladistics 23:301–336.

Richter, S., O.S. Møller, and C.S. Wirkner. 2009. Advances in crustacean phylogenetics. Arthropod Systematics & Phylogeny 67:275–286.

Richter, S., M. Stein, T. Frase, and N. U. Szucsich. 2013. The arthropod head. Pages 223–240 *in* A. Minelli, G. Boxshall, and G. Fusco, editors. Arthropod Biology and Evolution—Molecules, Development, Morphology. Springer, Heidelberg, Germany.

Richter, S., T. Stach, and A. Wanninger. 2016. Perspective—nervous system development in bilaterian larvae: testing the concept of "primary larvae." Pages 313–324 *in* A. Schmidt-Rhaesa, S. Harzsch, and G. Purschke, editors. Structure and Evolution of the Invertebrate Nervous Systems. Oxford University Press, New York.

Sandeman, D., M. Kenning, and S. Harzsch. 2014. Adaptive trends in malacostracan brain form and function related to behavior. Pages 11–48 *in* C. Derby and M. Thiel, editors. The Natural History of the Crustacea. Vol. 3: Crustacean Nervous System and Control of Behaviour. Oxford University Press, New York.

Scholtz, G. 2000. Evolution of the nauplius stage in malacostracan crustaceans. Journal of Zoological Systematics and Evolutionary Research 38:175–187.

Scholtz, G. 2016. Heads and brains in arthropods: 40 years after the "endless dispute." Pages 402–410 *in* A. Schmidt-Rhaesa, S. Harzsch, and G. Purschke, editors. Structure and Evolution of Invertebrate Nervous Systems. Cambridge University Press, Cambridge, UK.

Scholtz, G., B. Mittmann, and M. Gerberding. 1998. The pattern of distal-less expression in the mouthparts of crustaceans, myriapods and insects: new evidence for a gnathobasic mandible and the common origin of Mandibulata. International Journal of Developmental Biology 42:801–810.

Scholz, S., S. Richter, and C.S. Wirkner. 2018: Constant morphological patterns in the hemolymph vascular system of crayfish (Crustacea, Decapoda). Arthropod Structure & Development 47:248–267.

Schram, F.R., and C.H.J. Hof. 1998. Fossils and the interrelationships of major crustacean groups. Pages 233–302 *in* G.D. Edgecombe, editor. Arthropod Fossils and Phylogeny. Columbia University Press, New York.

Schram, F.R., and S. Koenemann. 2004. Developmental genetics and arthropod evolution: On body regions of Crustacea. Pages 75–92 *in* G. Scholtz, editor. Crustacean Issues 15, Evolutionary Developmental Biology of Crustacea. A.A. Balkema, Lisse, The Netherlands.

Schram, F.R. 1983. Remipedia and crustacean phylogeny. Pages 23–28 *in* F.R. Schram, editor. Crustacean Issues 1, Crustacean Phylogeny. A.A. Balkema, Rotterdam, The Netherlands.

Schram, F.R. 1986. Crustacea. Oxford University Press, New York.

Schram, F.R. 2013. Comments on crustacean biodiversity and disparity of body plans. Pages 1–33 *in* L. Watling and M. Thiel, editors. The Natural History of the Crustacea Volume 1: Functional Morphology and Diversity. Oxford University Press, New York.

Schwentner, M., D.J. Combosch, J.P. Nelson, and G. Giribet. 2017. A phylogenomic solution to the origin of insects by resolving crustacean-hexapod relationships. Current Biology 27:1818–1824.

Snodgrass, R.E. 1938. Evolution of the Annelida, Onychophora and Arthropoda. Smithsonian Miscellaneous Collections 97:1–159.

Stegner, M.E.J., and S. Richter. 2011. Morphology of the brain in *Hutchinsoniella macracantha* (Cephalocarida, Crustacea). Arthropod Structure & Development 40:221–243.

Stegner, M.E.J., G. Brenneis, and S. Richter. 2014. The ventral nerve cord in Cephalocarida (Crustacea): New insights into the ground pattern of Tetraconata. Journal of Morphology 275:269–294.

Stegner, M.E.J. and S. Richter. 2015. Development of the nervous system in Cephalocarida (Crustacea): early neuronal differentiation and successive patterning. Zoomorphology 134:183–209.

Stegner, M.E.J., M. Fritsch, and S. Richter. 2014. The central complex in Crustacea. Pages 361–384 *in* J.W. Wägele and T. Bartholomäus, editors. Deep Metazoan Phylogeny: The Backbone of the Tree of Life. De Gruyter, Berlin.

Stein, M., D. Waloszek, and D. Maas. 2005. *Oelandacaris oelandica* and the stem lineage of Crustecea. Pages 55–72 in S. Koenemann and R.A. Jenner, editors. Crustacean Issues 16, Crustacea and Arthropod Relationships. Taylor & Francis, Boca Raton, FL.

Stemme, T., T. M. Iliffe, G. Bicker, S. Harzsch, and S. Koenemann. 2012. Serotonin immunoreactive interneurons in the brain of the Remipedia: new insights into the phylogenetic affinities of an enigmatic crustacean taxon. BMC Evolutionary Biology 12:168.

Strausfeld, N.J. 2005. The evolution of crustacean and insect optic lobes and the origins of chiasmata. Arthropod Structure & Development 34:235–256.

Strausfeld, N.J. 2009. Brain organization and the origin of insects: an assessment. Proceedings of the Royal Society B 276:1929–1937.

Strausfeld, N.J. 2012. Arthropod Brains—Evolution, Functional Elegance, and Historical Significance. The Belknap Press of Harvard University Press, Cambridge, MA.

Takenaka, M., A. Suzuki, T. Yamamoto, M. Yamamoto, and M. Yoshida. 1993. Remodeling of the nauplius eye into the adult ocelli during metamorphosis of the barnacle, *Balanus amphitrite hawaiiensis.* Development, Growth & Differentiation 35:245–255.

von Reumont B.M., R.A. Jenner, M.A. Wills, E. Dell'Ampio, G. Pass, I. Ebersberger, B. Meyer, S. Koenemann, T.M. Iliffe, A. Stamatakis, O. Niehuis, K. Meusemann, and B. Misof. 2012. Pancrustacean phylogeny in the light of new phylogenomic data: support for Remipedia as the possible sister group of Hexapoda. Molecular Biology and Evolution 29:103–1045.

Walossek, D. 1993. The Upper Cambrian *Rehbachiella* and the phylogeny of Branchiopoda and Crustacea. Lethaia 26:318–318.

Waloszek D., A. Maas, J. Chen, and M. Stein. 2007. Evolution of cephalic feeding structures and the phylogeny of Arthropoda. Palaeogeography Palaeoclimatology Palaeoecology 254:273–287.

Williams, T.A. 2013. Mechanisms of limb patterning in Crustacea. Pages 74–102 *in* M. Thiel and L. Watling, editors. The Natural History of Crustacea, Vol. 1: Functional Morphology and Diversity. Oxford University Press, New York.

Wirkner, C.S., M. Tögel, and G. Pass. 2013. The arthropod circulatory system. Pages 343–391 *in* A. Minelli, G. Boxshall, and G. Fusco, editors. Arthropod Biology and Evolution—Molecules, Development, Morphology. Springer, Heidelberg, Germany.

Wirkner, C.S., and S. Richter. 2010. Evolutionary morphology of the circulatory system in Peracarida (Malacostraca; Crustacea). Cladistics 26:143–167.

Wirkner, C.S., and S. Richter. 2013. Circulatory system and respiration. Pages 376–412 *in* M. Thiel and L. Watling, editors. The Natural History of Crustacea, Vol. 1: Functional Morphology and Diversity. Oxford University Press, New York.

Yager, J. 1981. Remipedia, a new class of Crustacea from a marine cave in the Bahamas. Journal of Crustacean Biology 1:328–333.

Zrzavý, P., and J. Štys. 1997. The basic body plan of arthropods: insights from evolutionary morphology and developmental biology. Journal of Evolutionary Biology 10:353–367.

2

THE FOSSIL RECORD OF THE PANCRUSTACEA

Thomas A. Hegna, Javier Luque, and Joanna M. Wolfe

Abstract

Fossils are critically important for evolutionary studies as they provide the link between geological ages and the phylogeny of life. The Pancrustacea are an incredibly diverse clade, representing over 800,000 described extant species, encompassing a variety of familiar and unfamiliar forms, such as ostracods, tongue worms, crabs, lobsters, shrimps, copepods, barnacles, branchiopods, remipedes, and insects. Having colonized nearly every environment on Earth, from hydrothermal vents to terrestrial habitats, they have a diverse fossil record dating back to the Cambrian (540–485 Ma). The quality of the fossil record of each clade is variable and related to their lifestyle (e.g., free-living versus parasitic, benthic versus pelagic) and the degree of mineralization of their cuticle. We review the systematics, morphology, preservation, and paleoecology of pancrustacean fossils; each major clade is discussed in turn, and, where possible, fossil systematics are compared with more recent data from molecular phylogenetics. We show that the three epic clades of the Pancrustacea—Allotriocarida, Multicrustacea, and Oligostraca—all have Cambrian roots, but the diversification of those clades did not take place until the Middle and Late Paleozoic. We also address the potential affinities of three "problematic" clades: euthycarcinoids, thylacocephalans, and cyclids. We conclude by assessing the future of pancrustacean paleobiology, discussing new morphological imaging techniques and further integration with growing molecular phylogenetic data.

INTRODUCTION

It has been over 35 years since the last review of the crustacean fossil record (Schram 1982), which predated the recognition of Crustacea as a paraphyletic group with respect to Hexapoda (now united in the taxon Pancrustacea). Although some groups have been reviewed in the intervening years (Feldmann 2003, Van Damme and Kotov 2016), there has been no wholesale review of the

Evolution and Biogeography. Edited by Gary C. B. Poore and Martin Thiel.

fossil record of Pancrustacea. Many significant advances relating to the pancrustacean fossil record, and revolution in molecular phylogenetics and phylogenomics, perhaps most profound in its effects, have provided new avenues to test paleontological hypotheses and new questions for paleontology. The discovery of new fossils has advanced our understanding of pancrustacean evolution, in particular at new sites of exceptional preservation like the Herefordshire Lagerstätte (Siveter 2008a), which have extended the first occurrence of several lineages backward in time. New ways of looking for fossils (Harvey et al. 2012, Harvey and Pedder 2013) have also contributed novel insights into crustacean evolution. What follows is a discussion of pancrustacean origins and a taxonomic review (outlined in Table 2.1) of the pancrustacean fossil record (summarized in Figure 2.1). The organization is drawn from Oakley et al. (2013).

IDENTIFYING CAMBRIAN PANCRUSTACEAN ORIGINS

Molecular clocks clearly indicate that the earliest pancrustacean divergences occurred in the Cambrian (540–485 Ma) (Oakley et al. 2013, Lozano-Fernández et al. 2016; see Chapter 5), thus inferring a Cambrian or Ediacaran (635–540 Ma) stem lineage. Major impediments to the resolution of early pancrustacean evolution come from a lack of universal autapomorphies of "Crustacea" as well as historical conflict between molecular and morphological phylogenies in inferring the internal relationships of pancrustaceans (reviewed by Schram 2013; see Chapter 4). Proposed characters defining crown "crustaceans" include a five-segmented head bearing two pairs of antennae, gnathobasic mandibles, and two pairs of differentiated maxillae; early development involving a nauplius larva; the presence of a proximal endite on post-antennular limbs; multi-annulated exopods on post-antennular limbs; and post-antennular limbs with no more than five podomeres on the endopod (see Chapter 5). None of these characters is universal. The second antenna and nauplius are missing from hexapods. Gnathobasic mandibles are also present in myriapods. The proximal endite and multi-annulated post-annenular exopods are observed in larvae only (Boxshall 2004). Several Cambrian fossils that possess these larval characters lack differentiated cephalic limbs, casting doubt on their membership in crown Pancrustacea. It is clear that none of the characters above can alone distinguish a crown-group pancrustacean fossil; phylogenetic investigations are required to test specific combinations of expressed characters.

Cambrian Pancrustaceans

The oldest likely fossils (~514 Ma) of crown-group pancrustaceans are two three-dimensionally phosphatized larval fossils from the Early Cambrian "Orsten type" deposits of South China, *Yicaris dianensis* (Fig. 2.2A; Zhang et al. 2007) and possibly *Wujicaris muelleri* (Zhang et al. 2010, Wolfe and Hegna 2014, Wolfe et al. 2016). *Yicaris dianensis* in particular bears epipodites on its trunk limbs, a convincing character found only in crown pancrustaceans. Younger fossils of similar preservation from the Orsten type locality in the Late Cambrian (~497 Ma) of Sweden have already acquired characters asserting their membership within crown-group Allotriocarida (*Rehbachiella kinnekullensis;* Müller 1983) and crown-group Ichthyostraca (*Boeckelericambria pelturae*; Walossek and Müller 1994), as previously analyzed (Wolfe and Hegna 2014, Wolfe et al. 2016). The exact phylogenetic positions of two other Orsten fossils, *Bredocaris admirabilis* and *Skara anulata* (Müller 1983), are less certain (Wolfe and Hegna 2014) but almost certainly within the pancrustacean crown group.

Additionally, disarticulated small carbonaceous fossils have been discovered from several Early to Middle Cambrian localities, representing mandibular molar surfaces (Fig. 2.2B) and other feeding limbs (e.g., Harvey et al. 2012). Many of the fragments are exceptionally well preserved and

Table 2.1. Placement of All Taxa Mentioned in Chapter 2

- **Non-Pancrustacea** (*Cambropachycope clarksoni, Martinssonia elongata, Musacaris gerdgeyeri, Oelandocaris oelandica, Priscansermarinus barnetti*)
    - **Bradoriida**
    - **Phosphatocopina**
- **Pancrustacea** (*Bredocaris admirabilis, Skara anulata, Wujicaris muelleri, Yicaris dianensis*)
  - **Hymenostraca** (*Branchiocaris pretiosa, Canadaspis perfecta, Ercaicunia multinodosa, Nereocaris exilis, Tokummia katalepsis, Waptia fieldensis*)
    - **Oligostraca**
      - **Ostracoda**
        - **Myodocopa** (*Luprisca incuba*)
          - **Halocyprida**
      - **Myodocopida** (*Colymbosathon ecplecticos, Nymphatelina gravida*)
          - **Leperditicopida** (*Leperditia baltica*)
      - **Podocopa**
          - **Palaeocopida** (*Homeokiesowia epicopa, Manawa staceyi*)
          - **Platycopida**
          - **Podocopida** (**Cypridoidea**, *Mutilus speyeri*)
      - **Ichthyostraca**
        - **Branchiura**
        - **Pentastomida** (*Boeckelericambria pelturae, Invavita piratica*)
    - **Multicrustacea**
      - **Malacostraca**
        - **Phyllocarida** (*Cascolus ravitis, Cinerocaris magnifica*)
          - **Archaeostraca** (*Arenosicaris inflata, Ceratiocaris cf. macroura, Echinocaris socialis, Nahecaris stuertzi*)
          - **Hoplostraca** (**Sairocariidae**)
          - **Leptostraca** (*Rhabdouraea bentzi*)
        - **Peracarida**
          - **Amphipoda**
          - **Bochusacea**
          - **Cumacea**
          - **Isopoda**
          - **Lophogastrida**
          - **Mictacea**
          - **Mysida**
          - **?Pygocephalomorpha** (*Hoplita ginsburgi, Tealliocaris walloniensis*)
          - **Spelaeogriphacea**
          - **Stygiomysida**
          - **Tanaidacea** (*Anthracocaris scotica*)
          - **Thermosbaenacea**
        - **Hoplocarida** (*Eopterum devonicum*)
          - **Aeschronectida**
          - **Stomatopoda**

(*continued*)

Table 2.1. ***(Continued)***

        - **Archaeostomatopoda**
        - **Palaeostomatopoda**
        - **Unipeltata**
          - **Pseudosculdidae**
          - **Sculdidae** (*?Sculda pennata/spinosa*)
          - **Verunipeltata**
    - **Syncarida**
      - **Anaspidacea**
      - **Bathynellacea**
      - **Palaeocaridacea**
    - **Eucarida**
      - **Euphausiacea** (*Anthracophausia, Archangeliphausia*)
        - **Bentheuphausia** (*Bentheuphausia amblyops*)
        - **Euphausiidae**
      - **Decapoda** (**Angustidontidae**, *Imocaris colombiensis, Imocaris tuberculata, Palaeopalaemon newberryi, Protoclytiopsis antiqua*)
        - **Dendrobranchiata**
          - **Penaeoidea** (*Aciculopoda mapesi*, **Penaeidae**)
        - **Pleocyemata**
          - **Meiura** (*Eocarcinus praecursor, Platykotta akaina*)
            - **Anomura**
            - **Brachyura**
              - **Dromiacea**
              - **Dakoticancroidea**
          - **Achelata**
            - **Palinura (Palinuridae)**
          - **Astacidea**
          - **Caridea** (*Amphionides reynaudii*)
          - **Gebiidea (Upogebiidae)**
          - **Glypheidea (Erymoidea, Glypheidae)**
          - **Polychelida**
          - **Stenopodidea** (*Devonostenopus pennsylvaniensis*)
  - **Copepoda (Ectinosomatidae)**
  - **Thecostraca**
    - **Acrothoracica**
    - **Ascothoracida**
    - **Rhizocephala**
    - **Thoracica** (*Cyprilepas holmi, Emersonius cybosyrinx, Illilepas damrowi, Rhamphoverritor reduncus*)
- **Allotriocarida**
  - **Branchiopoda** (*Rehbachiella kinnekullensis*)
    - **Anostraca** (*Lepidocaris rhyniensis*)

Table 2.1. ***(Continued)***

		Calmanostraca (*Jeholops hongi*)	
			Notostraca
			Kazacharthra (*Panacanthocaris ketmenica*)
		Laevicaudata	
		Onychocaudata	
			Spinicaudata
			Cyclestherida
			Cladocera
	Cephalocarida		
	Hexapoda		
	Remipedia (*Cryptocaris hootchi, Tanazios dokeron, Tesnusocaris goldichi*)		
Incertae sedis			
	Cyclida (*Halicyne ornata, Maastrichtiocaris rostrata, Schramine montanaensis*)		
	Euthycarcinoidea (*Euthycarcinus kessleri*)		
	Thylacocephala (*Austriocaris, Dollocaris ingens*)		
Invalid Taxonomic Names			
	Amphionidacea (*constituent taxa now within Caridea*)		

Note: Taxa placed in parentheses with higher taxonomic groups indicate that they are either stem-group representatives of that group or they have an ambiguous placement within the group. Bold names are higher taxonomic groups.

bear extraordinary similarities to appendages of extant pancrustaceans, especially branchiopods (Harvey et al. 2012), copepods (Harvey et al. 2012, Harvey and Pedder 2013), and ostracods (Harvey et al. 2012). However, these fossils are highly fragmentary, thus the persistence of highly adaptive feeding-related characters over geological time, and their membership within crown-group pancrustaceans remains debatable (Wolfe et al. 2016).

Until recently, a convincing stem group (extinct group most closely related to the common ancestor of all living species) has not been apparent for Pancrustacea. The Silurian (~440–420 Ma) species *Tanazios dokeron* (Siveter et al. 2007a; Fig. 2.3D) was initially suggested to belong in the stem group, but this has been refuted by anatomical evidence and phylogenetic analysis (Boxshall 2007, Legg et al. 2013; see below). A stem group, however, has recently been reconstructed on the basis of new three-dimensional fossil material and phylogenetic analyses (Zhai et al. 2019). In a redescription of the Early Cambrian bivalved species *Ercaicunia multinodosa* (Fig. 2.2C), which was not previously known to preserve any limbs, the second antenna and organization of the head limbs was revealed, as was the structure of the trunk limbs (Luo et al. 1999; Zhai et al. 2019). Phylogenetic analysis has revealed a clade of bivalved fossil euarthropods comprising a new pancrustacean stem group, including *E. multinodosa* (Zhai et al. 2019), and other notable Cambrian fossils from the Chengjiang and Burgess Shale faunas, including *Branchiocaris pretiosa, Canadaspis perfecta, Nereocaris exilis, Tokummia katalepsis*, and *Waptia fieldensis*. All these fossils share a bivalved carapace covering only the anterior part of the body and a many-segmented limbless abdomen. While the organization of the head is challenging to interpret for flattened Burgess Shale fossils (obscured by the bivalved carapace), the morphology of *E. multinodosa* implies that all members of the stem group may have borne at least four pairs of differentiated head limbs.

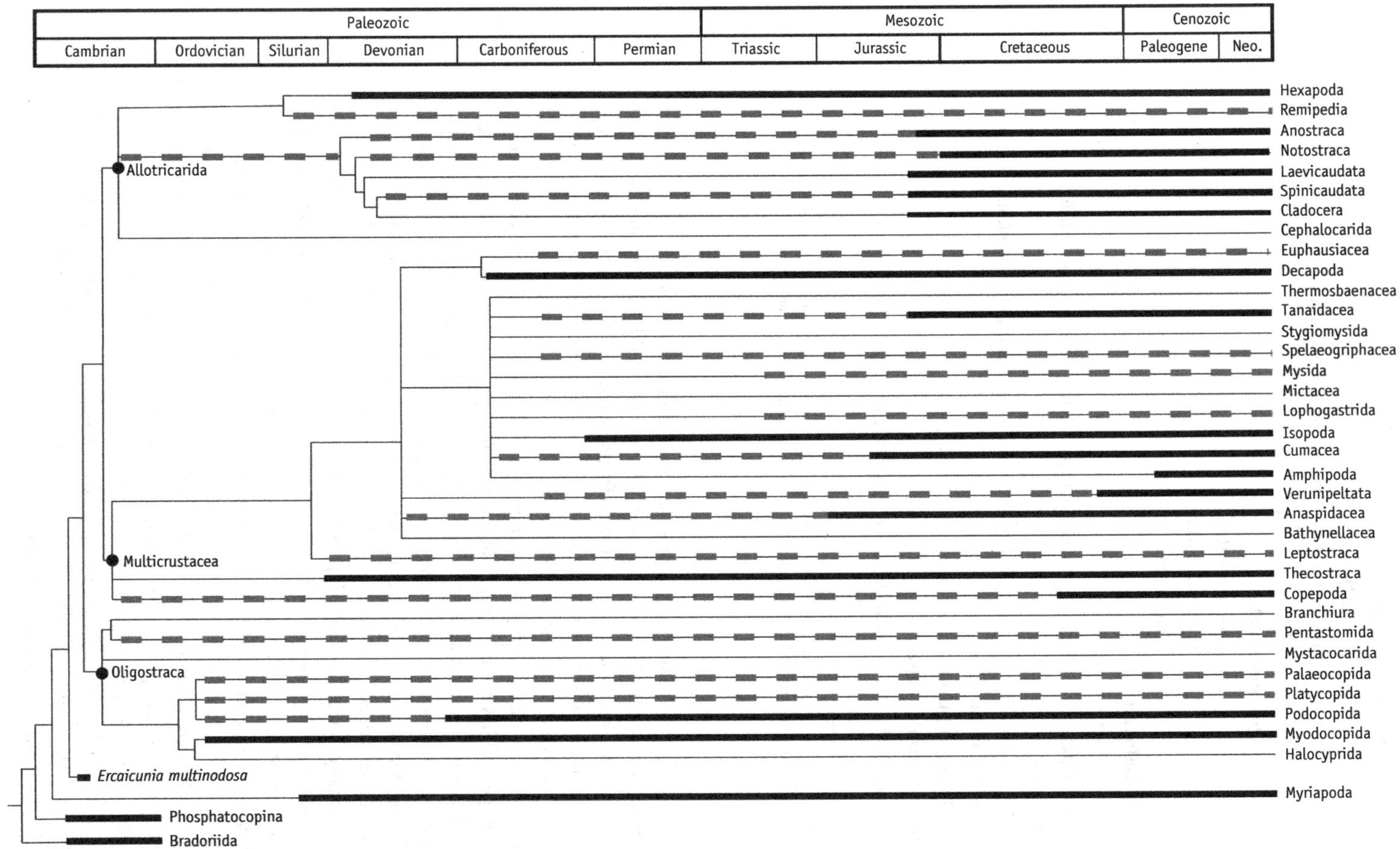

Fig. 2.1.
A time-scaled phylogeny of pancrustaceans showing stem-group lineages (thick dotted lines) and crown-group lineages with known fossils (thick bold lines) or unknown fossils (thin lines). Scale bar is ~20 million years.

Cambrian Fossils That Are Not Pancrustaceans

Fossils most frequently invoked as members of the pancrustacean stem group represent a collection of larval taxa from Cambrian Orsten-type faunas (e.g., Maas et al. 2006, Haug et al. 2013). The most common are members of the fossil clade Phosphatocopina, which were previously considered ostracods (Hou et al. 2010) or the immediate sister group of crown pancrustaceans (Maas et al. 2003). The phosphatocopine maxillule and maxilla, however, are undifferentiated from the trunk limbs, unlike in other crown pancrustaceans (Wolfe and Hegna 2014). Phylogenetic analysis has placed phosphatocopines within (Hou et al. 2010, Oakley et al. 2013) and outside (Schram and Koenemann 2004) of the pancrustacean crown, as well as outside the mandibulate crown (Legg et al. 2013). Given their volatility in phylogenetic placement and caveats associated with larval development (Wolfe and Hegna 2014), we tend to believe that phosphatocopines are immediate members of the mandibulate stem (Legg et al. 2013).

Other larval fossils from Orsten-type faunas, including *Martinssonia elongata* (Müller and Walossek 1986), *Cambropachycope clarksoni* (Walossek and Müller 1990), *Musacaris gerdgeyeri* (Haug et al. 2010b), and *Oelandocaris oelandica* (Müller 1983), likely form a paraphyletic sister grade to phosphatocopines plus mandibulates. This topology implies that the proximal endite of trunk limbs (e.g., Waloszek et al. 2007) is symplesiomorphic for mandibulates and lost in myriapods.

Bradoriids, abundant bivalved fossil arthropods, were attributed to ostracods based on carapace shape prior to discovery of preserved appendages (Hou et al. 2010). The head appendages revealed mandibles, maxillules and maxillae, that were undifferentiated from trunk limbs, effectively rejecting bradoriids from the mandibulate crown group (Hou et al. 2010, Oakley et al. 2013). This clade has an uncertain placement in phylogenetic analyses and may indeed represent the deepest split between the mandibulate and chelicerate stem groups (Briggs et al. 2016).

OLIGOSTRACA

Ostracoda

Extant Ostracoda comprise at least 8,000 marine and 2,000 freshwater species described, half of an estimated ~20,000 species total, occupying nearly every aquatic or semi-aquatic environment worldwide (Martens et al. 2008). Typical ostracods have a bivalved, hinged carapace completely enclosing a body with reduced segmentation and locomotory antennules and antennae. These are divided into two clades: Podocopa (subdivided into Podocopida (Fig. 2.2D), Platycopida, and Palaeocopida (Fig. 2.2E), with the latter known in the Recent only from the single species *Manawa staceyi;* Swanson 1989) and Myodocopa (subdivided into Myodocopida (Fig. 2.2F) and Halocyprida). Contemporary morphological (Legg et al. 2013) and total evidence (Oakley et al. 2013) phylogenies suggest monophyly of crown Ostracoda. An additional ~30,000 to 55,000 fossil species are estimated (Horne et al. 2002). The copious fossil record of ostracod carapaces makes them an important proxy for studies of regional biostratigraphic zonation, paleoenvironments, and climate change; such research employs assemblages, populations, and individual carapaces (e.g., Boomer et al. 2003, Börner et al. 2013).

Ostracods bodies are enveloped by a bivalved carapace, often of calcium carbonate (calcite). The morphology of the hinge joining the valves and scars on the inside of the valves (representing attachment points for adductor muscles that hold the valves closed) are often used as important fossil characters. It must be cautioned, however, that the shape of the carapace alone can be misleading for phylogenetic placement of ostracods due to convergent morphologies (Siveter 2008b, Tinn and Oakley 2008, Siveter et al. 2013). While playing a part in the systematics of extant ostracods, limb morphology,

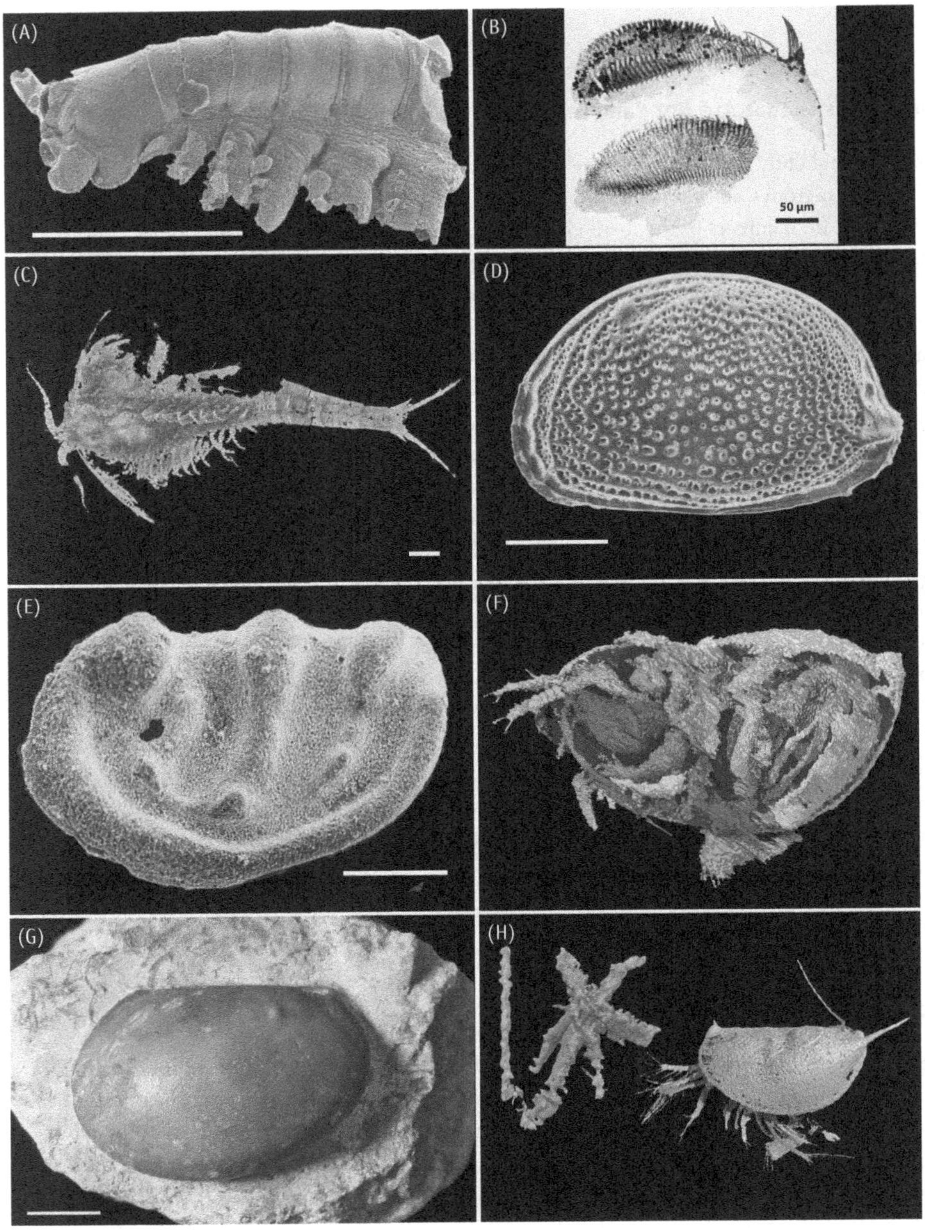

Fig. 2.2.
Cambrian fossils and Oligostracans. (A) *Yicaris dianensis* (Key Laboratory for Palaeobiology, Yunnan University (YKLP) 10840, holotype) from the Lower Cambrian of Yunnan, China. Scale bar is 300 µm. (B) Branchiopod-type mandibles (Geological Survey of Canada (GSC) 135369, 135370) from the Middle and Late Cambrian of Saskatchewan, Canada. Scale bar is 300 µm. (C) A virtual reconstruction of *Ercaicunia multinodosa* (YKLP 16201) from the Lower Cambrian of Yunnan, China. Scale bar is 1 mm. (D) Left valve of a female specimen of the podocopid ostracod *Mutilus speyeri* (British Museum of Natural History IO 4977) from Pliocene-Pleistocene aged sediments off the south coast of Turkey. Scale bar is 250 µm. (E) Left valve of a female specimen of the palaeocope ostracod *Homeokiesowia epicopa* (British Museum of Natural History OS 6695) from the Ordovician of Wales. Scale bar is 500 µm. (F) A virtual reconstruction of the fossil myodocope ostracod *Colymbosathon ecplecticos* from the Silurian of Herefordshire, England. (G) The myodocope-like leperditicopid *Leperditia baltica* (Yale Peabody Museum Invertebrate Paleontology (YPMIP) 7300) from the Late Silurian of Gotland, Sweden. Scale bar is 5 mm. (H) A virtual reconstruction of the fossil pentastomid *Invavita piratica*, shown isolated (left) and on its host *Nymphatelina gravida* (Ostracoda, Myodocopa) (right), both from the Middle Silurian (Wenlock) of Herefordshire, England. A, from Zhang et al. (2007), with permission from Springer Nature; B, from Harvey et al. (2012), with permission from PNAS; D, from Doruk (1973); E, from Siveter (1982); F, from Siveter et al. (2003); G, courtesy of the Yale Peabody Museum; H, from Siveter et al. (2015). Figures 2.2D–F, H courtesy of D. J. Siveter. See color version of this figure in centerfold.

rarely preserved in fossils, is far more important for modern forms. Indeed, discoveries of exceptionally preserved specimens (bearing appendages) have entirely excluded the Cambrian phosphatocopines and bradoriids from ostracods. Exceptional preservation of limbs permits phylogenetic analysis confirming the placement of several taxa in crown group orders and families (Oakley et al. 2013). No fossil Platycopida, stem group Palaeocopida, or Halocyprida are known to preserve soft parts.

Among the Podocopida (Fig. 2.2D), exceptionally preserved fossils are known from the Early Devonian (~420–390 Ma) of Ukraine (Olempska et al. 2012), the Early Cretaceous (~145–100 Ma) Romualdo Formation of the Santana Group (preserving >800 specimens; Smith 2000), Paleogene (Eocene, ~56–34 Ma) Baltic amber (Keyser and Friedrich 2016), Neogene (Miocene, ~23–5 Ma) Chiapas amber (Serrano-Sánchez et al. 2015), a Neogene (Miocene) cave in Australia (Matzke-Karasz et al. 2014), and Neogene (Miocene) paleolakes of Antarctica (Williams et al. 2008) and California (Wilkinson et al. 2010). Giant filamentous sperm resembling those of Recent Cypridoidea were preserved in the Miocene (Matzke-Karasz et al. 2014). All known fossil and extant podocopids are benthic and lack compound eyes. Although the Miocene (~23–5 Ma) samples represent the earliest limb-bearing freshwater podocopids, carapace-only records ascribed to several lineages (Cypridoidea, Cytheroidea, and Darwinuloidea) suggest fresh water was invaded 9 to 12 times, with the earliest examples in the Pennsylvanian (~320–300 Ma) and possibly in the Devonian (~420–360 Ma) (Martens et al. 2008).

Among the Myodocopida (Fig. 2.2F), fewer specimens with limbs have been preserved, with examples known from the Late Ordovician Beecher's Bed (~450 Ma) (Siveter et al. 2014; Fig. 2.3A), Early Silurian Herefordshire fauna (~425 Ma) (Siveter et al. 2003, 2007b, 2010, 2013), Early Triassic of Spitzbergen (~250 Ma) (Weitschat 1983), and several Jurassic-Cretaceous (~200–66 Ma) localities of Russia and Benelux (Kornicker et al. 2006). Brooded embryos are known from Beecher's Bed and Herefordshire, thus parental care in myodocopids is at least as old as the Ordovician (Siveter et al. 2007b, 2014). A male copulatory appendage (Fig. 2.2F) was also preserved in the Silurian (Siveter et al. 2003), although most Herefordshire ostracods are described from single specimens, rendering studies of sexual dimorphism impossible. All crown-group myodocopids, fossil and extant, are marine. The earliest myodocopids were likely nektobenthic, as many are today (supported by Ordovician specimens from dysoxic benthic sediments; Siveter et al. 2014). Invasion of the pelagic realm occurred in the Silurian at the latest, a conclusion supported by the preservation in Herefordshire specimens of gills, differentiated swimming appendages, and lateral compound eyes (Perrier et al. 2015).

The fossil Leperditicopida (Ordovician-Devonian, 485–360 Ma) were a bivalved clade compared to myodocopids by Vannier et al. (2001), on the basis of adductor muscle scars, carapace ultrastructure, and carapace ridges interpreted as circulatory in function (Fig. 2.2G). Recently, leperditicopid soft parts were described from the early Silurian (~440 Ma) of Waukesha, Wisconsin (Jones 2016). Appendages, especially the flagelliform seventh appendage, as well as brooded embryos, strongly support a relationship to Myodocopida (Jones 2016). Placement within crown group Myodocopa is significant, as some of the earliest leperditicopids were found in brackish facies, in contrast to the exclusively marine habitats of all extant myodocopes.

Another fossil clade previously allied to ostracods are the Ordovician-Mississippian (485–320 Ma) Eridostracina, which are known for multilayered carapaces, as they did not molt during growth (Olempska 2012, Jones and Olempska 2013). There are several characters suggesting a relationship to ostracods (particularly Palaeocopida), including the adductor muscle scars and structure of the valve hinge (Olempska 2012). However, the only extant taxa with molt retention are within the branchiopod clade Onychocaudata. In the absence of soft-part preservation, it is uncertain whether eridostracines should be considered ostracods, or if the clade itself represents a polyphyletic assemblage of arthropods that convergently evolved molt retention.

Ichthyostraca

Pentastomida

Modern-day adult pentastomids are respiratory parasites of mammals, "reptiles," and birds, with fishes as an intermediate host for some taxa (see Riley 1986). They have an anterior sucking mouth with two hook pairs and a vermiform body (Riley 1986). Fossil pentastomids are exclusively marine, whereas modern forms are primarily terrestrial. Fossil pentastomids are only found associated with invertebrates, whereas very few modern pentastomids utilize invertebrates (insects) primarily as intermediate hosts (Riley 1986, Christoffersen and De Assis 2013). Modern pentastomids are not known to infect ostracods. Fossil forms with unknown hosts could not have had the lifestyle of a modern pentastomid because of the absence of terrestrial vertebrates. Host switching must have occurred. Or, as observed by Siveter et al. (2015), pentastomids may have followed their vertebrate hosts onto land during the Middle Paleozoic (~540–250 Ma). The mere fact of their preservation suggests that fossil pentastomids were either external or free-living for at least part of their life cycle, as internal parasites have a minuscule preservation potential.

Four pentastomid genera are preserved via "Orsten"-style phosphatization from the Upper Cambrian of Sweden (Walossek and Müller 1994) and the Lower Ordovician of Newfoundland, Canada (Walossek et al. 1994). The youngest fossil pentastomid, *Invavita piratica* (Siveter et al. 2015) (Silurian, Herefordshire Lagerstätte, 425 Ma), is the only one preserved with its host, on the outside of ostracod valves (Fig. 2.2H). No evidence of pentastomids exists between the Devonian (~420–360 Ma) and modern era, giving them a fossil record gap of over 420 million years. Molecular and spermatological evidence supports a Branchiura-Pentastomida sister relationship (Riley 1986, Oakley et al. 2013).

Branchiura

Modern branchiurans are external fish parasites recognized by a flattened, bilobed dorsal shield with paired compound eyes and distinctively modified maxillules as either suckers or hooks (Schram 1986). Despite the deep origins of their immediate sister group, branchiurans have no recognized fossil record, although other parasitic arthropod groups have been observed in association with fishes in Mesozoic (250–66 Ma) Plattenkalks and nodule faunas (Cressey and Boxshall 1989, Nagler et al. 2016). Diligent examination of exceptionally preserved fish may yet turn up remains of branchiurans. It has been speculated that branchiurans are related to an enigmatic group of fossil arthropods called the cyclids (discussed below), but this fossil relationship is untested with respect to the molecular hypothesis of Ichthyostraca.

MULTICRUSTACEA

Malacostraca

Phyllocarida

Phyllocarids, the total group (crown + stem) of extant leptostracans, are marine, scavenging bivalved arthropods in a world where bivalved carapaces are notoriously misleading as to their true taxonomic affinities. The bivalved carapace often hides the details of the ventral morphology, leaving remains that are depauperate in synapomorphies. Underneath their carapace, they have stalked eyes, a mobile rostrum, and phyllopodous thoracopods. Historically, the Phyllocarida has been divided into four extinct groups and one extant group.

The Hoplostraca are a small distinctive group of Carboniferous (~360–300 Ma) phyllocarids distinguished by their short, triangular-to-axe–shaped carapace (Schram 1986). Based on their only known family, Sairocariidae, they are thought to be too specialized to have given rise to the hoplocaridans (i.e., mantis shrimp; Schram 1973, 1982). The hoplostracans have not been studied recently, and they have been absent from recent phylogenetic analyses of phyllocarid arthropods (e.g., Collette and Hagadorn 2010b). They may be an autapomorphic group of archaeostracan phyllocarids.

The Archaeostraca (Fig. 2.4A) are by far the most speciose of the fossil phyllocarids. Traditionally, the archaeostracans and leptostracans (see below) have been regarded as sister groups (Wolfe et al. 2016). Archaeostracans have been the subject of several phylogenetic analyses (Rode and Lieberman 2002, Collette and Hagadorn 2010b, Bergmann and Rust 2014), but they all assume archaeostracan monophyly and use only modern leptostracans as outgroups. Archaeostracan monophyly, however, has never been phylogenetically tested, and there is precedent for questioning it (Rolfe 1981, Oakley et al. 2013, Jones et al. 2015, Wolfe et al. 2016). Archaeostracans may prove to be paraphyletic because malacostracans and/or eumalacostracans may have originated from an archaeostracan-like ancestor (Jones et al. 2015). For example, the phylogenetic analysis of Oakley et al. (2013) recovered the archaeostracan *Nahecaris stuertzi* variably as either a stem-group leptostracan or a stem-group malacostracan. Jones et al. (2015) suggested on the basis of the possession of an antennal scale that *Ceratiocaris* cf. *macroura* was a stem eumalacostracan. These results taken together suggest possible para- or polyphyly of archaeostracans.

The oldest archaeostracans are from the upper Cambrian (*Arenosicaris inflata* from Wisconsin; Collette and Hagadorn 2010a). The youngest known archaeostracans are Permian (~300–250 Ma) (Kobayashi 1937, Feldmann et al. 2004). Both Feldmann et al. (2004) and Collette and Hagadorn (2010b) mention the presence of Late Triassic (~250–200 Ma) archaeostracans, but in fact, the supposed Late Triassic archaeostracan is a thylacocephalan, *Austriocaris*. Archaeostracans are absent from the record after the end-Permian mass extinction.

The Leptostraca essentially have no fossil record. The only fossil assigned to the leptostracan lineage is *Rhabdouraea bentzi* (Malzahn 1958; see Schram and Malzahn 1984), but it is fragmentary and has neither been adequately differentiated from the archaeostracan lineage nor put into context with the leptostracan stem lineage. The leptostracan total group may have diverged from the malacostracan lineage during or before the Silurian (~440–420 Ma); exceptionally preserved members of the leptostracan stem lineage, such as *Cinerocaris magnifica* (Briggs et al. 2003) (Fig. 2.3B) and *Cascolus ravitis* (Siveter et al. 2017), have been reported from the Herefordshire Lagerstätte.

Peracarida

Peracarida contains a diverse set of taxa, amphipods, bochusaceans, cumaceans, isopods, lophogastrids, mictaceans, mysids, spelaeogriphaceans, stygiomysids, tanaidaceans, and thermosbaenaceans. All have three tagmata, a carapace not covering one or more of the posterior thoracomeres, a mandible in adults bearing a "*lacinia mobilis*," and a female brood pouch or marsupium formed by oostegites. There is broad agreement from both morphological and combined molecular/morphological analyses that most of these taxa form a monophyletic group (Wolfe et al. 2016). However, molecular analyses question the place of Mysida within the peracarid pantheon (Spears et al. 2005, Meland and Willassen 2007, Jenner et al. 2009). These molecular analyses find Mysida to be a sister group to a configuration of Decapoda + Stomatopoda. There is an additional extinct group, the pygocephalomorphs (Fig. 2.4B), which have been regarded as peracarids due to their possession of oostegites (flaps emerging from the coxae of thoracic

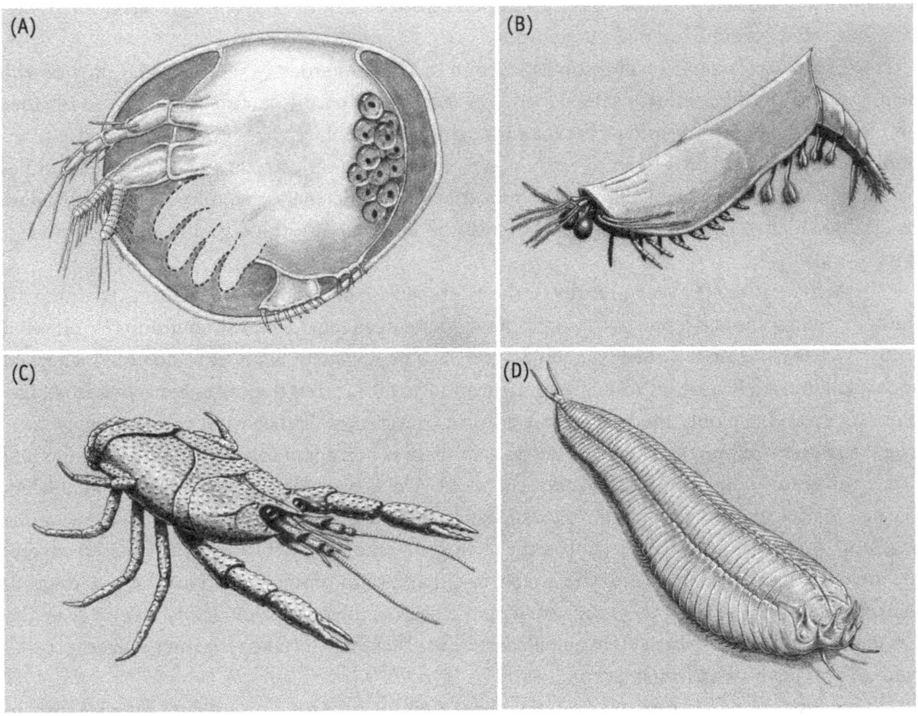

Fig. 2.3.
Hypothetical line drawing reconstructions of several extinct crustacean taxa. (A) The Late Ordovician *Luprisca incuba* from Beecher's Trilobite Bed (the earliest known ostracod with preserved limbs). Dashed lines indicate unknown anatomy. (B) The Middle Silurian (Wenlock) *Cinerocaris magnifica* from the Herefordshire Lagerstätten (stem-lineage leptostracan, and one of the earliest crown malacostracans). (C) The Late Triassic (Norian–Rhaethian) *Platykotta akaina* (possibly the earliest anomuran). (D) The Middle Silurian (Wenlock) *Tanazios dokeron* from the Herefordshire Lagerstätten (earliest stem-lineage remipede). A, from Siveter et al. (2014); B, from Briggs et al. (2003); C, from Chablais et al. (2011); D, from Siveter et al. (2007a). Artwork by R. E. Johnson.

limbs) (Pazinato et al. 2016). However, if mysids prove not to be peracarids, then pygocephalomorphs may follow them out of the peracarid tree (Jones et al. 2016, Wolfe et al. 2016).

Internal relationships among Peracarida are poorly resolved. The phylogeny has not been addressed in recent years with either morphological or molecular data. Furthermore, modern taxonomic groups are distinctive, and there is a dearth of transitional fossils to help understand the course of character evolution in the group. Thus, there have been many different suggested topologies for the peracarid tree (Poore 2005, Spears et al. 2005).

Ultimately, what one regards as the oldest peracarid depends on the phylogeny. If pygocephalomorphs prove to be crown-group peracarids, then the oldest peracarid would be *Tealliocaris walloniensis* (Gueriau et al. 2014b) from the Devonian (Fammenian, ~372–358 Ma). Next in line is the stem-group tanaidacean *Anthracocaris scotica* (Peach 1883), which is Visean in age (Early Carboniferous, ~346–330 Ma; Schram 1983, Schram et al. 1986b). Several peracarid groups have Carboniferous-aged (~360–300 Ma) stem- (Cumacea, Spelaeogriphacea, Tanaidacea) or crown-group (Isopoda) representatives.

Hoplocarida

Hoplocarids, or mantis shrimp, are predaceous crustaceans with distinctive rapacious second maxillipeds for spearing or smashing. The hoplocarids figured prominently in Schram's (1982) review of the crustacean fossil record. Our understanding of the relationships within Hoplocarida has become increasingly refined (Schram 2007, Haug et al. 2010a) since then. The earliest fossil attributed to the Hoplocarida is *Eopterum devonicum* from the Devonian (~420–360 Ma) of Russia (coincidentally, it is the only Devonian hoplocarid). The fossil was originally interpreted to represent insect wings (Rohdendorf 1961) but was subsequently reinterpreted to represent the tail of a palaeostomatopod (see Schram 2010). There are diverse early hoplocarids (Aeschronectida, "Palaeostomatopoda," "Archaeostomatopoda") known from the Carboniferous (~360–300 Ma) (Jenner et al. 1998), but none known from the Permian (~300–250 Ma) and Triassic (~250–200 Ma). The Permian and Triassic represent a significant gap in the record of hoplocarids, when they likely nearly went extinct. The hoplocarid clade Unipeltata (including Sculdidae, Pseduosculdidae and Verunipeltata; see Haug et al. 2010a) diversified in the post-Triassic Mesozoic. The family Sculdidae (Fig. 2.4C) ranges through the Jurassic (~200–145 Ma) and Cretaceous (~145–66 Ma) (Schram 2010). Pseudosculdidae is an exclusively Late Cretaceous group with few representatives (Schram 2010). Both families may have been casualties of the end-Cretaceous extinction event. Verunipeltatans (crown-group mantis shrimp) appear in the Late Cretaceous and radiated during the Cenozoic (~66–0 Ma).

Syncarida (Anaspidacea, Bathynellacea, and Palaeocaridacea)

Syncarids can be characterized as shrimp without carapaces. They were first discovered as fossils prior to the discovery of living specimens in Tasmania. They have traditionally been divided into three groups: Anaspidacea, Palaeocaridacea, and Bathynellacea. However, both molecular (Jenner et al. 2009) and morphological (Wills et al. 2009) evidence suggest that these three groups are not monophyletic (contra Schram 1986), i.e., Syncarida may be a paraphyletic grade at the base of the Eumalacostraca (Wills et al. 2009, Wolfe et al. 2016).

The bathynellaceans have no fossil record; as they are miniaturized and unmineralized, they are unlikely candidates for fossilization. The extinct Palaeocaridacea have a diverse record that extends from the Carboniferous (~360–300 Ma) to the Permian (~300–250 Ma). Fossils from the Devonian (~420–360 Ma) previously attributed to Palaeocaridacea have been reinterpreted as indeterminate basal eumalacostracans (i.e., non-palaeocaridacean; Brooks 1962). Existing phylogenies offer some support for palaeocaridacean monophyly (Schram 1984, Wills et al. 2009), but this hypothesis remains to be tested in the broader context of other early fossil eumalacostracans. The Anaspidacea have a limited fossil record, with only two fossils assigned (Schram 1984, Jell and Duncan 1986). The Jurassic (~200–145 Ma) syncarid fragments mentioned by Borns et al. (1971) are insect nymph fragments (TAH, pers. obser.). The fossil record of the Anaspidacea extends only from the Triassic to the Cretaceous (i.e., the Mesozoic, ~250–66 Ma), and no fossil anaspidaceans are known from the Cenozoic (~66–0 Ma), although the family Anaspididae might have radiated during the Paleogene (~66–23 Ma) (Jarman and Elliot 2000) or Early Neogene (~23–2.5 Ma) (Richter et al. 2018). Furthermore, no fossil anaspidaceans are known outside Australia.

Eucarida

Eucarida is currently envisioned as monophyletic, comprising the Euphausiacea and Decapoda (Wolfe et al. 2016). A presumed third order, the extant Amphionidacea, comprising one species, *Amphionides reynaudii*, is a larval caridean decapod (De Grave et al. 2015). Thus, Euphausiacea and

Decapoda are the only known fossil or extant Eucarida. No amphionidacean-like decapod fossil forms are known.

Euphausiacea

Euphausiacea (krill; represented by two families, 11 genera, and 87 species) are marine, usually planktonic filter-feeding eucarids distributed worldwide, and account for one of the largest biomasses known today (Spiridonov and Casanova 2010). Despite their abundance and broad distribution, their fossil record is dubious and no concrete fossil eucaridans can be attributed to crown-group Euphausiacea with certainty. Only a few Paleozoic forms such as *Anthracophausia* from the Mississippian (~360–320 Ma) of Scotland and the United States (Peach 1908) and *Archangeliphausia* from the Early Devonian (~400 Ma) of Russia (Dzik et al. 2004) may represent stem groups of euphausiaceans, eucarids, or even mysidaceans (Jarman 2001, Casanova 2003, Schram 2009, Spiridonov and Casanova 2010). Other Late Paleozoic fossils have also been identified as possible euphausiaceans or mysidaceans, but their relationships with other eumalacostracans remain poorly understood (Schram 1986, Maas and Waloszek 2001, Spiridonov and Casanova 2010).

Although the systematic affinities of *Anthracophausia* and *Archangeliphausia* with euphausiaceans are yet to be tested, the presence of total group (crown + stem) euphausiaceans during the Devonian and Carboniferous could be anticipated. The closest extant relation of Euphausiidae is the monotypic deep-water family Bentheuphausiidae (Colosi 1917), which noticeably differ from all other euphausiaceans in their ecology, bathymetry, and morphology. The extant *Bentheuphausia amblyops* has been considered as an "archaic" form due to its plesiomorphic well-developed eyes and thoracopods and male first pleopods being not modified (Schram 1986, Maas and Waloszek 2001, Spiridonov and Casanova 2010), traits not seen among euphausiid species, and which would be expected to have been present in some stem-group euphausiaceans.

Decapoda

The origins of Decapoda. Decapods (crabs, shrimps, lobsters, and allies) are one of the most diverse groups of non-hexapod pancrustaceas, characterized by a carapace covering all of the thoracomeres, the presence of three pairs of maxillipeds, and five pairs of thoracic limbs or pereopods (hence the name "Decapoda"). They fossilize well thanks to their usually larger sizes compared to other "crustaceans" (particularly crabs and lobsters) and more mineralized bodies and appendages. Despite their abundant fossil record in post-Paleozoic rocks, the early origins of Decapoda during the Paleozoic remain poorly understood (Schram 2009, Schweitzer and Feldmann 2015). No stem-group forms are currently recognized (Schram 2009, Jones et al. 2016). To date, the oldest confirmed crown-group Decapoda are macrurous forms representing both decapod suborders: Dendrobranchiata, i.e., the penaeoidean shrimp *Aciculopoda mapesi* from the Late Devonian (Famennian, ~370–360 Ma) of Oklahoma (Feldmann and Schweitzer 2010b), and Pleocyemata, i.e., the stenopodidean shrimp *Devonostenopus pennsylvaniensis* from the Late Devonian–early Mississippian (~360 Ma) of Pennsylvania (Jones, in Jones et al. 2014). A third taxon, the lobster-like *Palaeopalaemon newberryi*, is known from the Late Devonian (Famennian) of the United States (Whitfield 1880, Schram et al. 1978, Feldmann and Schweitzer 2010b, Jones et al. 2014). Although the affinities of *A. mapesi* and *D. pennsylvaniensis* are well constrained, the relationship of *P. newberryi* with other decapods is less clear. Karasawa et al. (2013) recovered *P. newberryi* in an unresolved trichotomy together with the Polychelida and a clade formed by the rest of the lobsters (i.e., Achelata + (Glypheida + Astacidea)), which prompted them to elevate it to its own infraordinal level. On the contrary, De Grave et al. (2009) considered palaeopalaemonids as astacidean lobsters, and Jones (2016) suggested proximity to Polychelida, a modern group confined to the deep sea. Regardless of the affinities of *P. newberryi*, both decapod suborders must have diverged earlier than 372 Mya, with their most recent common ancestor rooted in the Early–Middle Devonian or earlier (Schram 2009, Bracken-Grissom et al. 2013).

Another Late Devonian eumalacostracan clade with possible decapod affinities is the puzzling Angustidontidae first known from the Frasnian–Famennian (~380–360 Ma) of North America and Europe (Cooper 1936, Gueriau et al. 2014a, Zatoń et al. 2014, and references therein). Angustidontids, previously described as actinopterygian fish jaws, eurypterid raptorial appendages, or as crustaceans with stomatopod or uncertain affinities, may be eumalacostracans closer to Decapoda (Rolfe and Dzik 2006, Gueriau et al. 2014a).

The only other purported Late Paleozoic decapods known to date are *Imocaris tuberculata* (Fig. 2.4D) from the Mississippian (~360–320 Ma) of the United States, *I. colombiensis* from the Pennsylvanian (~20–300 Ma) of Colombia, and *Protoclytiopsis antiqua* from the Permian (~300–250 Ma) of Russia (Birshtein 1958, Schram and Mapes 1984, Racheboeuf and Villarroel 2003). Due to the incompleteness of the material representing *Imocaris* (i.e., unknown thoracic appendages, and pleonal elements), it has been considered as a dromiacean brachyuran crab (Schram and Mapes 1984, Schram 2009), a pygocephalomorph (Racheboeuf and Villarroel 2003; see comments in Guinot et al. 2007 and Jones et al. 2014), as Decapoda *incertae sedis* (Schram and Dixon 2004), or within the brachyuran Dakoticancroidea (Števčić 2005). The absence of orbits in *Imocaris*, the groove pattern of the dorsal carapace, and the regions delimited by them strongly advocate for a non-brachyuran relationship, with a likely proximity to lobster-like superfamilies. The systematic position of *P. antiqua*, however, seems to be better constrained, possibly related to glypheoid or erymoid lobsters (Schram 1980, Karasawa et al. 2013, Jones et al. 2014).

Shrimp- and lobster-like decapods were widespread during the Triassic but mostly restricted to stem penaeoid and caridean shrimps, polychelidan and astacidean lobsters, and even the purported earliest meiurans (discussed below). However, only a handful of crown families have confirmed fossils in the Triassic (~250–200 Ma), i.e., Penaeidae (prawns, shrimp) (Garassino and Pasini 2002), Palinuridae and Glypheidae (rock and glypheid lobsters) (Charbonnier et al. 2013), and Upogebiidae (mud shrimps) (Hyžný and Klompmaker 2015). The occurrence of several other decapod groups in the pre-Jurassic (before ~200 Ma) can be inferred by the traces of their burrows (e.g., several glypheidean lobsters and axiidean mud shrimps), but unfortunately the absence of preserved hard parts precludes a proper identification.

The Triassic "origins" of several stem and crown decapod infraorders may well be a sampling or preservation artifact, and it would not be surprising if in the years to come new findings push back several records into the Late Paleozoic.

The roots of Anomura and Brachyura (collectively Meiura). The morphological diversity seen across Triassic (~250–200 Ma) decapods is remarkable (Figs. 2.3C and 2.4E), especially considering the small number of Late Paleozoic taxa known. Curiously, all of these Triassic forms can be placed among the shrimp-like or lobster-like macrurous infraorders. More typical crab-looking decapods, such as the Anomura (hermit crabs, squat lobsters, false crabs, stone crabs) and Brachyura (true crabs), are mostly restricted to the post-Triassic, greatly obscuring our understanding of their origins and phylogenetic relationships with other decapod infraorders.

Anomurans and brachyurans are usually considered sister groups, together constituting the monophyletic Meiura (Scholtz and Richter 1995, Schram and Dixon 2004), which are characterized by the invariably second and third pereopods not being chelate (Scholtz and Richter 1995). To date, the only putative meiuran older than Jurassic (~200–145 Ma) is the anomuran-like *Platykotta akaina* (Fig. 2.3C), known from a single specimen of Late Triassic (Norian–Rhaetian) age (Chablais et al. 2011). Although relatively well preserved dorsally, its ventral side is badly weathered, precluding a detailed comparison with other lobster- and crab-like superfamilies. Despite this, several of its dorsal cephalothoracic, pleonal, and appendicular features are reminiscent of those seen among anomurans (Chablais et al. 2011), supporting the view of *P. akaina* as the oldest known anomuran.

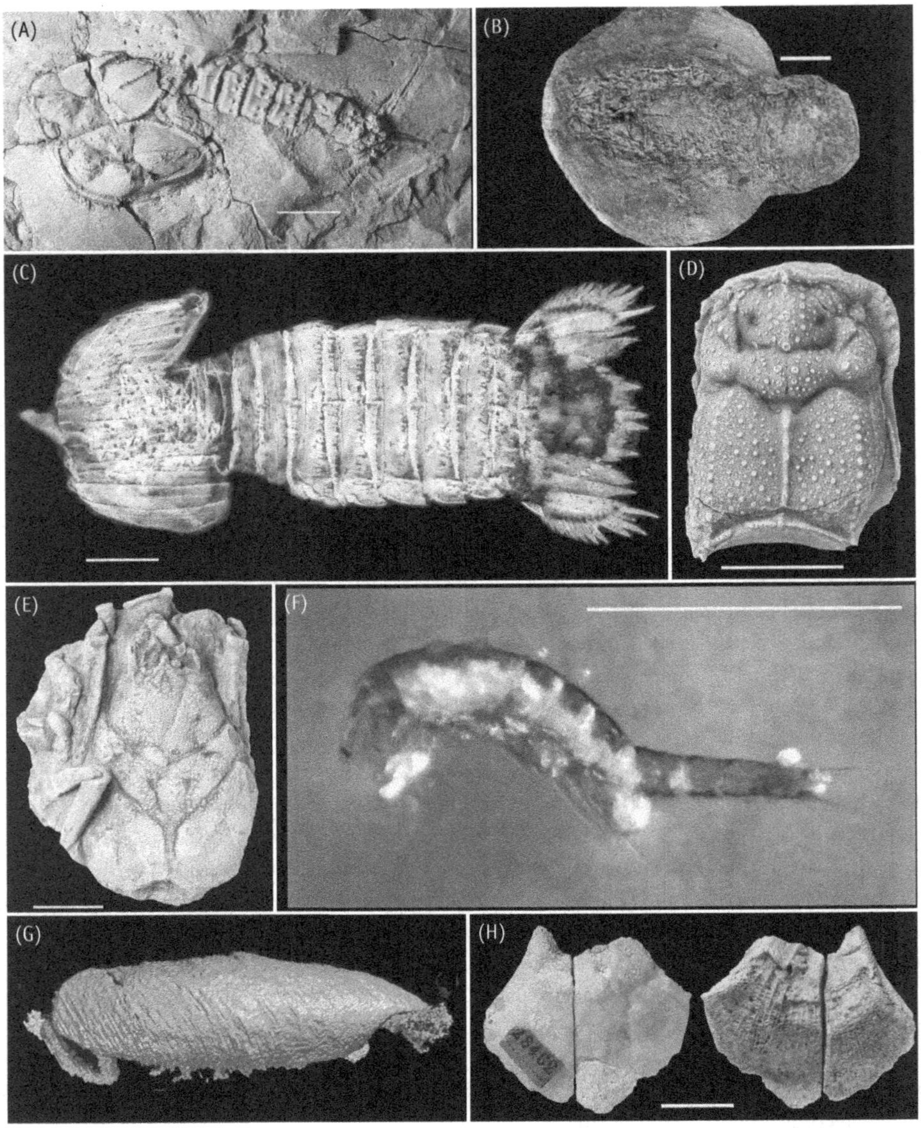

Fig. 2.4.
Multicrustaceans. (A) The phyllocarid *Echinocaris socialis* (Yale Peabody Museum Invertebrate Paleontology (YPMIP) 9674) from the Late Devonian of Pennsylvania. Scale bar is 1 cm. (B) The pygocephalomorph *Hoplita ginsburgi* (Collection of Fossil Invertebrates of the Facultad de Ciencias of Uruguay (FC-DPI) 4826) from the Early Permian of Uruguay. Scale bar is 5 mm. (C) The hoplocarid? *Sculda pennata/spinosa* (collection of Roger Frattigiani, Laichingen) from the Late Jurassic Solnhofen Plattenkalk of Germany. Scale bar is 3 mm. (D) The decapod *Imocaris tuberculata* (San Diego Society of Natural History 25139) from the Middle Carboniferous of Arkansas. Scale bar is 5 mm. Photo courtesy of R. M. Feldmann. (E) The meiuran *Eocarcinus praecursor* (British Museum of Natural History In 18425) from the Early Jurassic of Great Britain. Scale bar is 1 cm, courtesy of R. M. Feldmann. (F) An unidentified female ectinosomatid copepod (IHNFG—4899/Cop1) preserved in amber from the Miocene of Chiapas, Mexico. Scale bar is 200 μm. (G) A virtual reconstruction of the fossil cirripede cypris *Rhamphoverritor reduncus* from the Middle Silurian (Wenlock) of Herefordshire, England. (H) The putative turtle barnacle *Emersonius cybosyrinx* (American Museum of Natural History 28462) from the Eocene of Florida. Scale bar is 1 cm. A, from Beecher (1902) courtesy of the Yale Peabody Museum; B, from Piñeiro et al. (2012), courtesy of G. Piñeiro; C, from Haug et al. (2010a), courtesy of C. Haug; D, from Schram and Mapes (1984), courtesy of R. M. Feldmann; E, from Feldmann and Schweitzer (2010a), courtesy of R. M. Feldmann; F, from Huys et al. (2016) courtesy of F. J. Vega; G, from Briggs et al. (2005), courtesy of Yale Peabody Museum; H, from Ross and Newman (1967), courtesy of the American Museum of Natural History. See color version of this figure in centerfold.

However, the apparent occurrence of a second pair of chelate pereopods posterior to the first, typically chelate or claw-like in all anomurans and brachyurans (Chablais et al. 2011), strongly conflicts with the current definition of a monophyletic Meiura.

The early origins of anomurans and brachyurans become even murkier when discussing the phylogenetic relationships of *Eocarcinus praecursor* (Fig. 2.4E), a crab-like species from the Lower Jurassic of England (Withers 1932) considered either as a basal brachyuran (e.g., De Grave et al. 2009, Haug and Haug 2014), or as an early anomuran offshoot (Feldmann and Schweitzer 2010a). The presence of pleonal articular rings, the apparent lack of true orbits, and large antennae support an anomuran affinity. However, these may just be plesiomorphic conditions present in stem anomurans and/or brachyurans and are not conclusive in themselves (Haug and Haug 2014). Furthermore, *E. praecursor* has been described as also having truly chelate second pereopods (Feldmann and Schweitzer 2010a), which, just as for *P. akaina*, disagrees with the apparently apomorphic meiuran condition (as reconstructed in Fig. 2.4).

Regardless of the affinities of *P. akaina* and *E. praecursor*, given the known fossil and molecular information available, the origins of Anomura may be rooted in the Early Triassic or even the Permian (Bracken-Grissom et al. 2013) and of Brachyura in the earliest Jurassic (Tsang et al. 2014).

Copepoda

Copepods (Fig. 2.4F) are the most abundant metazoans in modern oceans, earning the title "insects of the sea." They are tiny (typically 0.5–5 mm; Huys and Boxshall 1991) crustaceans with typically teardrop-shaped bodies and a transparent, unmineralized exoskeleton. Despite their abundance, they have an exceedingly spotty fossil record, owing to their unmineralized exoskeleton and small size, limiting insights into when and how they diversified and conquered the oceans. Huys et al. (2016) recently reviewed the fossil record of copepods. The earliest fossils of copepods are highly fragmentary (Selden et al. 2010, Harvey and Pedder 2013) and include possible copepod mandibles from the Cambrian (~540–485 Ma) of Tennessee (Harvey and Pedder 2013). More certain body fossils are known from the Cretaceous (~145–66 Ma) of Brazil (Cressey and Boxshall 1989) and the Miocene (~23–5 Ma) of California (Palmer 1960, Hołyńska et al. 2015). Additional evidence of copepod activity is found in fossil echinoids that preserve the exocysts of parasitic copepods (Radwańska and Poirot 2010); curiously, these exocysts are known only from the Jurassic (~200–145 Ma) (Klompmaker and Boxshall 2015) and Recent time periods. The depauperate crown-group fossil record has challenged the ability to accurately estimate molecular clock divergences of crown copepods, with mean estimates ranging from Carboniferous (~360–300 Ma) (Oakley et al. 2013) to the Permian/Triassic boundary (~250 Ma) (Eyun 2017).

Thecostraca

The thecostracans are a group of sessile crustaceans with an often-overlooked fossil record. The relationships between the four major groups are stable and have the topology of (Ascothoracida, (Acrothoracica, (Rhizocephala, Thoracica))); Høeg et al. 2009). The thoracicans, barnacles, alone possess mineralized hard parts that stand a reasonable chance of making it into the fossil record. All other thecostracans have parasitic or epibiontic lifestyles, with many questions remaining about their evolution.

Ascothoracids are a small group parasitic on cnidarians and echinoderms. They lack the frontolateral horns in the nauplius that the Cirripedia (acrothoracicans, rhizocephalans and thoracicans) possess (Høeg et al. 2009). They have a very sparse, indirect fossil record of cysts and traces (Voigt 1967), with all records coming from the upper Cretaceous (~100–66 Ma).

Acrothoracicans are cirripedes that bore into calcareous substrata, producing slit-like traces (Schram 1986). Their fossil record is exclusively made up of these trace fossils. The oldest acrothoracican-like trace is from the Devonian (~420–360 Ma) (Rodriguez and Gutschick 1977). As with the ascothoracids, without the preserved body of the trace-maker, it is difficult to be certain of the identity of the maker.

Rhizocephalans are highly modified to the point of appearing non-arthropodan as adults. They are parasites of mainly decapod crustaceans and practice parasitic castration of their hosts (Schram 1986). It is through this parasitic castration that they modify the abdomen of their male hosts, creating a recognizable intersex morphology that can be identified in the fossil record (Klompmaker and Boxshall 2015). Fossil male crabs that are "feminized," suggestive of rhizocephalan parasitism, are known from the Miocene (~23–5 Ma) of New Zealand and the upper Cretaceous (~100–66 Ma) of South Dakota (Feldmann 1998). As noted by Klompmaker and Boxshall (2015), this attribution to rhizocephalan parasitism is not without ambiguity, as other parasites can also cause feminization of males.

Barnacles are what most people picture when they think of thoracicans. They are sessile and (primitively) plated cirripedes. Previously, barnacles were divided into the sessile Sessilia and the stalked Pedunculata. However, it seems that the stalked condition is primitive for the group and that the Pedunculata are paraphyletic (Pérez-Losada et al. 2008). The earliest supposed barnacle is *Priscansermarinus barnetti* from the Cambrian Burgess Shale (508 Ma) (Collins and Rudkin 1981), but it has been rejected as a barnacle (Briggs and Conway Morris 1986, Briggs et al. 2005). *Rhamphoverritor reduncus* (Fig. 2.4G) is a curious cyprid and juvenile fossil from the Silurian Herefordshire Lagerstätte (~425 Ma); the fossils are interpreted to represent the first stem-group representative of the cirripedian lineage (Briggs et al. 2005, Høeg et al. 2009). *Cyprilepas holmi* is known from organic fossils from the upper Silurian (~425 Ma) of Estonia recovered via acid etching of limestones (Wills 1962). Due to this method of recovery, it cannot be verified if it originally possessed calcareous valves (Briggs and Conway Morris 1986). Like *R. reduncus*, it is interpreted as a stem-group cirripedian (Høeg et al. 2009). The first crown-group thoracican is *Illilepas damrowi* from the Late Carboniferous (Middle Pennsylvanian, ~309 Ma) of Illinois (Schram 1975, Wolfe et al. 2016).

Some barnacles have adopted life as an epibiont. Being an epibiont requires modification of settling behavior (for larvae to settle on a moving target such as a marine turtle or whale) and evolution of new modes of attachment to living skin. This lifestyle seems to have evolved after the Mesozoic (after 66 Ma), suggested by the fact that there are no records of epibiontic barnacles colonizing Mesozoic marine reptiles. Coronuloidea is the clade of barnacles that are obligate epibionts. The oldest coronuloid barnacle is the extinct Late Eocene (~56–34 Ma) barnacle *Emersonius cybosyrinx* (Fig. 2.4H; Ross and Newman 1967). Although interpreted as a turtle barnacle (Ross and Newman 1967), it is preserved without a host association. The oldest evidence of a barnacle–host association is from skeletal scarring on a sea turtle interpreted as barnacle traces from the Miocene (~23–5 Ma) of Italy (Hayashi et al. 2013). Turtle barnacles seem to be the most primitive members of the coronuloid clade (Hayashi et al. 2013); perhaps the epibiontic lifestyle of coronuloid barnacles evolved first with turtles and then other hosts were acquired via host switching.

ALLOTRIOCARIDA

Branchiopoda

Modern branchiopods are a diverse set of almost exclusively freshwater organisms, with all large branchiopods except for cyclestherids specializing in ephemeral freshwater habitats. The group comprises the anostracans (fairy shrimp, brine shrimp), notostracans (tadpole shrimp), paraphyletic clam shrimp (collectively, "conchostracans": laevicaudatans, spinicaudatans, and cyclestherids), and the cladocerans (water fleas). Ephemeral freshwater ecology seems to be plesiomorphic for crown Branchiopoda and implies that the last common ancestor for Branchiopoda had already acquired the necessary adaptations to flourish in these unpredictable environments.

Rehbachiella kinnekullensis (Fig. 2.5A) is known from the upper Cambrian (~500–485 Ma) of Sweden and was previously proposed to be a anostracan (Walossek 1993). Recent phylogenetic analyses (Olesen 2009, Hegna 2012, Legg et al. 2013) indicate that it is more likely a stem-group branchiopod, or possibly even a stem-group cephalocarid (Wolfe and Hegna 2014). The first crown-group member of the Branchiopoda is *Lepidocaris rhyniensis* from the Early Devonian Rhynie Chert (~410 Ma) in Scotland (Scourfield 1926, Wolfe et al. 2016). All the major clades of the large branchiopods (anostracans, notostracans, spinicaudatans) appeared as stem groups by the Devonian (Lagebro et al. 2015, Gueriau et al. 2016). The spinicaudatan stem-group record may even extend into the Late Silurian (~425 Ma) (Péneau 1936, but these records need re-evaluation). Earlier supposed "conchostracan" records (Soot-Ryen 1960) are likely cephalopod diaptachii. Laevicaudatans and cyclestherids have a much poorer fossil record; despite this, the fossil record indicates that the laevicaudatan lineage may have been present by the Permian (~300–250 Ma) (Schneider et al. 2006), and certainly by the Late Jurassic (~163–145 Ma) (Shen and Chen 1984). The cyclestherid lineage may have appeared by the Devonian (Novozhilov 1961).

Related to the notostracans are the enigmatic kazacharthrans (Olesen 2009, Hegna 2012). Kazacharthrans (Fig. 2.5B) have generally spinier or wider carapaces than notostracans and an elongate telson (not anal plate). Traditionally, the group has been conceived of as a brief, morphologically diverse, geographically isolated radiation in the Triassic (~250–200 Ma) of central Asia, which contrasts strikingly with the conservative and widespread nature of the notostracans. Key parts of this concept are in question, including the age (*Jeholops hongi*; Hegna and Ren 2010) and geography (Martins-Neto et al. 2001, Wagner et al. 2018). Most importantly, the group has been only weakly supported in phylogenetic analyses (Hegna 2012).

Cladocerans are unlikely fossils; being both tiny and soft-bodied they are difficult to preserve. Despite this, they are known as compression fossils in lacustrine settings. They may be more ubiquitous as fossils, as few paleontologists examine lacustrine bedding planes with an eye for these tiny fossils. The oldest cladocerans are Jurassic (~200–145 Ma) in age and were reviewed thoroughly by Van Damme and Kotov (2016).

Remipedia

Remipedes are elongate crustaceans with a head shield rather than a carapace and a pair of biramous appendages on each segment (Schram 1986). They have no crown-group fossils, and their origin (both ecologically and phylogenetically) is enigmatic and controversial. Three fossils have been attributed to the remipede stem lineage. The upper Mississippian (Middle Carboniferous, ~330 Ma) *Tesnusocaris goldichi* (Fig. 2.5C), originally regarded as a branchiopod crustacean (Brooks 1955), was subsequently reinterpreted as a stem-group remipede (Schram et al. 1986a, Emerson and Schram 1991). Unfortunately, *T. goldichi* has not been included in any recent phylogenetic analyses. The slightly younger (Pennsylvanian, ~309 Ma) *Cryptocaris hootchi* has likewise been attributed to the remipede stem lineage (Emerson and Schram 1991) but has also not been included in phylogenetic analyses. More recently, Legg et al.'s (2013) phylogenetic analysis found the Silurian-aged *Tanazios dokeron* (Fig. 2.3D) from the Herefordshire Biota (Siveter et al. 2007a) to be the sister taxon to remipedes.

Hexapoda

Hexapods comprise insects and related wingless groups such as springtails. Molecular analyses have made abundantly clear that the group "Crustacea" is paraphyletic with regards to Hexapoda. Therefore, hexapods are nothing more than terrestrial crustaceans. The origin of hexapods can only

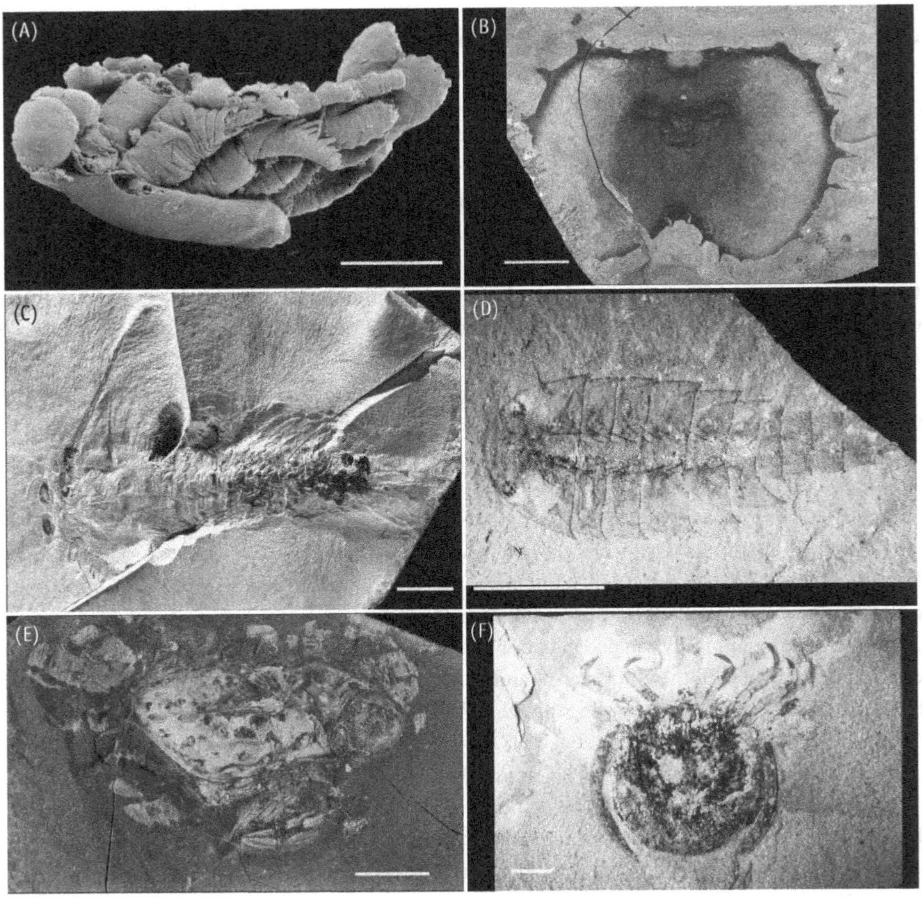

Fig. 2.5.
Allotriocarida and problematic groups. (A) The stem-group branchiopod *Rehbachiella kinnekullensis* from the Upper Cambrian of Sweden. Scale bar is 200 μm. Photo courtesy of D. Waloszek. (B) The kazacharthran branchiopod *Panacanthocaris ketmenica* (Nanjing Institute of Geology Paleontology and Stratigraphy K102, 84838) from the Late Triassic of Xinjiang, China. Scale bar is 5 mm. (C) Holotype of the stemgroup remipede *Tesnusocaris goldichi* (United States National Museum 124173a) from the Late Carboniferous of Texas. Scale bar is 1 cm. (D) The enigmatic crustacean-like euthycarcinoid arthropod *Euthycarcinus kessleri* (Louis Grauvogel Collection, Staatliches Museum für Naturkunde in Stuttgart) from the Middle Triassic of Bas-Rhin, France. Scale bar is 5 mm. Photo courtesy of L. Grauvogel-Stamm. (E) The enigmatic crustacean-like thylacocephalan arthropod *Dollocaris ingens* (Muséum National d'Histoire Naturelle, France 50939) from La Voulte sur Rhône, France. Scale bar is 1 cm. Photo courtesy of S. Charbonnier. (F) The enigmatic crustacean-like cyclid arthropod *Halicyne ornata* (Louis Grauvogel Collection, Staatliches Museum für Naturkunde in Stuttgart) from the Middle Triassic of Bas-Rhin, France. Scale bar is 3 mm. A, from Walossek (1993), courtesy of D. Waloszek; B, from Chen and Zhou (1985), courtesy of L. Gang, photographed by TAH; C, from Neiber et al. (2011); D, from Gall and Grauvogel (1964) and Gall et al. (2006), courtesy of L. Grauvogel-Stamm; E, from Secretan (1985), courtesy of S. Charbonnier—Investissements d'avenir (ANR-11-INBS-0004-RECOLNAT); F, from Gall and Grauvogel (1967) and Gall et al. (2006), courtesy of L. Grauvogel-Stamm. See color version of this figure in centerfold.

be understood by identifying the crustacean sister group of hexapods. This question has proved to be a challenging one: several different crustacean clades have been put forth as the sister group of hexapods (see Chapter 5).

PROBLEMATIC GROUPS

Cyclida

Cyclids (Fig. 2.5F) are a puzzling group of arthropods whose affinities have been debated for over 150 years. They have been given the common name "sea ticks" for their superficial resemblance to ticks; they are distinct from other crustaceans in that they have a shield-like carapace fused to all thoracic segments, uniramous antennae, subchelate maxillae and maxillipeds, among other characters (Schram et al. 2006). Recent work (Dzik 2008, Schram et al. 1997) placed them squarely in the Crustacea. Their affinities with the Crustacea are demonstrated by their possession of two sets of antennae, mandibles, two sets of maxillae and maxillipeds (Schram et al. 1997). Schram et al. (1997) found them to be a sister group to the copepods in a phylogenetic analysis. Other authors have observed the shield-like carapace, derived gills, and apparently serrate mandibles (Schram et al. 1997) and attributed them to the parasitic branchiurans (Dzik 2008, Schweigert et al. 2009). However, cyclids lack the apomorphies of branchiurans associated with a parasitic lifestyle and host attachment (Schram et al. 1997). A stem-group branchiuran position is probably more reasonable but is nonetheless in need of further phylogenetic testing.

Cyclids ranged from the Carboniferous (~360–300 Ma) to the Cretaceous (~145–66 Ma). Several species vie for the title of oldest cyclid, all being from Viséan age strata (Early Carboniferous, Mississippian, ~346–323 Ma) (Dzik 2008). The earliest cyclid with preserved limbs is *Schramine montanaensis* from the Late Mississippian Bear Gulch Limestone (~318 Ma) of central Montana (Schram et al. 2006). Their diversity is severely restricted after the Triassic (Dzik 2008). The youngest cyclid is *Maastrichtiocaris rostrata* from the Late Maastrichtian (~72–66 Ma, Cretaceous) of Maastricht, Netherlands (Fraaije et al. 2003). Dzik (2008) remarks that the Middle to Late Mesozoic forms are odd in a number of respects and do not seem to link to the Triassic forms. The Middle to Late Mesozoic forms lack preserved soft parts, so a definitive conclusion on their affinities must wait. It should be noted that there is a precedent for reinterpreting odd fossil cyclids as larval eumalacostracans (Hyžný et al. 2016); some of the odd Middle to Late Mesozoic forms may turn out to be something else entirely.

The ecology of cyclids is likewise a puzzle. Schram et al. (1997) pointed out the convergent similarities between cyclids and crabs. Given this morphological similarity, cyclids may have initiated the ecological niche that was taken over by true crabs during the Mid- to Late Mesozoic. Ecologically, cyclids occur in open marine limestones (Goldring 1967) to more transitional environments, and even in fresh water (Dzik 2008). The proposed relationship of cyclids to branchiurans raises some additional issues about their ecology. If cyclids are not parasitic but still related to branchiurans, then parasitism likely originated independently in the branchiuran-pentastomid lineage.

Euthycarcinoidea

Euthycarcinoids (Fig. 2.5D) are a distinctive group of crustacean-like animals. Their elongate bodies are divided into a distinctive cephalon, limb-bearing preabdomen and postabdomen (Ortega-Hernández et al. 2010). Their preabdomen is often made up of dorsally fused segments with two or three pairs of uniramous appendages attached to them. All euthycarcinoids inhabit either brackish or fresh water except for one of the earliest species, *Apankura machu*, which is marine (Vaccari et al. 2004).

The systematic placement of euthycarcinoids has long been poorly understood (Ortega-Hernández et al. 2010). Handlirsch (1914) originally allied them with copepods, and since then they have been hypothesized to fit all over the map of arthropod phylogeny (reviewed by Ortega-Hernández et al. 2010). Modern workers have placed them either as a stem lineage to the rest of the mandibulates (Vaccari et al. 2004) or as the immediate sister to the hexapods (Legg et al. 2013). Euthycarcinoid phylogeny has been analyzed at the genus level (Legg et al. 2013), but not at the species level.

Seventeen euthycarcinoid species are known (Collette and Hagadorn 2010a, Ortega-Hernández et al. 2010) ranging from the Middle to Late Cambrian (~509–497 Ma; Collette and Hagadorn 2010a) to the Middle Triassic (~247–237 Ma; Handlirsch 1914, Riek 1964). The highest diversity of euthycarcinoids occurred in the Carboniferous (~360–300 Ma). Several reports have attributed subaerial trackways to euthycarcinoid arthropods (MacNaughton et al. 2002), leading to the inference that euthycarcinoids may have been the earliest land-going animals (Collette and Hagadorn 2010a).

Like the other problematic groups discussed, the euthycarcinoids survived the end-Permian extinction (~250 Ma). Unlike those other two groups, however, they did not make it past the end-Triassic extinction event (~200 Ma).

Thylacocephala

Thylacocephalans (Fig. 2.5E) are moderately sized bivalved arthropods that resemble phyllocarids without an elongate abdomen. Many have prominent anterior raptorial appendages (Broda et al. 2015). The history of the thylacocephalans is intertwined with that of the phyllocarids for the simple fact that many thylacocephalans were originally interpreted as phyllocarids (Broda et al. 2015). The earliest bona fide thylacocephalans are those from the Silurian (~443–430 Ma; Haug et al. 2014), though there are thylacocephalan-like carapaces known from the Cambrian (Vannier et al. 2006). They seem to have become extinct at the end-Cretaceous (~66 Ma) extinction event, or slightly earlier (Hegna et al. 2014).

No consensus exists for their phylogenetic origin; recent analyses have found thylacocephalans to be either sister to the Eumalacostraca (Legg et al. 2013) or proposed to be the sister group to remipedes (Haug et al. 2014). Much remains unknown about the head appendages and the boundary of the head—for example, whether the raptorial appendages are cephalic or thoracic.

SUMMARY AND CONCLUSIONS

Old fossils can yield new insights into character evolution; new approaches to morphological description, such as whitening, autofluorescence, and computed tomography, reveal fossils' long-hidden morphological secrets (e.g., Hegna 2010, Haug and Haug 2011, Eriksson et al. 2016). These techniques are critical for identifying new characters in previously described fossils and improve their phylogenetic placement. Furthermore, taphonomic studies ranging from diagenesis of mineralization to cuticle microstructure have substantially improved understanding of how different anatomical features are preserved (McNamara et al. 2012, Gueriau et al. 2014c) to actualistic decay experiments (Krause et al. 2011, Hegna 2012, Butler et al. 2015). Taphonomy is critical for understanding the likelihood of preservation of phylogenetically informative characters; absence of a fossil record may be due to rapid decay rather than actual absence and could mislead phylogenetic inference (Sansom and Wills 2013).

Recent fossil discoveries from Konservat-Lagerstätten, Herefordshire, and Cenozoic amber deposits have been fruitful in expanding the age ranges and morphological disparity of pancrustacean fossils. Moreover, the framework of large-scale morphological matrices with hundreds of characters (e.g., Legg et al. 2013, Oakley et al. 2013, Briggs et al. 2016) permits robust testing of phylogenetic positions for old and new enigmatic fossils, which were previously limited by inappropriate understanding of outgroups. For example, past assumptions that cephalocarids and/or remipedes were "primitive" (see Chapter 5) caused significant biases in character polarization leading to erroneous topologies. Advances in models that combine molecular and fossil data (Bracken-Grissom et al. 2014, Donoghue and Yang 2016) permit simultaneous estimation of fossil topology and divergence times and allow the incorporation of fossils and their preservation history into diversification models. The unification of fossils with molecular phylogenetics signals a new era of comparative pancrustacean biology, improving investigations of topics such as trait evolution, inference of ancestral states, rates of evolution, biogeography and habitat colonization, coevolution of hosts and parasites, and conservation biology (Wolfe et al. 2016).

ACKNOWLEDGMENTS

Robert Johnson (WIU) created the four artistic reconstructions in this contribution. JL thanks the Izaak Walton Killam Memorial Scholarship (Canada), and JMW thanks the National Science Foundation NSF EAR-1615426 and G. Fournier.

REFERENCES

Beecher, C.E. 1902. Revision of the Phyllocarida from the Chemung and Waverly Groups of the Pennsylvanian. Quarterly Journal of the Geological Society of London 58:441–449.

Bergmann, A., and J. Rust. 2014. Morphology, palaeobiology and phylogeny of *Oryctocaris balssi* gen. nov. (Arthropoda), a phyllocarid from the Lower Devonian Hunsrück Slate (Germany). Journal of Systematic Palaeontology 12:427–444.

Birshtein, Y.A. 1958. Drevneishii predsta vitel otryada desyatinogikh rakoobraznikh, *Protoclytiopsis antiqua*, in Permakikh otlozhenii zapadnoi Sibiri. Doklady Akademii Nauk SSSR 122:477–480.

Boomer, I., D.J. Horne, and I.J. Slipper. 2003. The use of ostracods in palaeoenvironmental studies, or what can you do with an ostracod shell. Paleontological Society Papers 9:153–180.

Börner, N., B. De Baere, Q. Yang, K.P. Jochum, P. Frenzel, M.O. Andreae, and A. Schwalb. 2013. Ostracod shell chemistry as proxy for paleoenvironmental change. Quaternary International 313:17–37.

Borns, J.H.W., B.A. Hall, H.W. Ball, and H.K. Brooks. 1971. Mawson Tillite, Victoria Land, East Antarctica: reinvestigation continued. Antarctic Journal of the United States 7:106–107.

Boxshall, G.A. 2004. The evolution of arthropod limbs. Biological Reviews 79:253–300.

Boxshall, G.A. 2007. Crustacean classification: on-going controversies and unresolved problems. Zootaxa 1668:313–325.

Bracken-Grissom, H., M. Cannon, P. Cabezas, R. Feldmann, C. Schweitzer, S. Ahyong, D. Felder, R. Lemaitre, and K. Crandall. 2013. A comprehensive and integrative reconstruction of evolutionary history for Anomura (Crustacea: Decapoda). BMC Evolutionary Biology 13:128.

Bracken-Grissom, H.D., S.T. Ahyong, R.D. Wilkinson, R.M. Feldmann, C.E. Schweitzer, J.W. Breinholt, M. Bendall, F. Palero, T.Y. Chan, D.L. Felder, R. Robles, K.H. Chu, L. M. Tsang, D. Kim, J.W. Martin, and K.A. Crandall. 2014. The emergence of lobsters: phylogenetic relationships, morphological evolution and divergence time comparisons of an ancient group (Decapoda: Achelata, Astacidea, Glypheidea, Polychelida). Systematic Biology 63:457–479.

Briggs, D.E.G., and S. Conway Morris. 1986. Problematica from the Middle Cambrian Burgess Shale of British Columbia. Pages 167–183 *in* A. Hoffman and M.H. Nitecki, editors. Problematic Fossil Taxa (5). Oxford University Press, New York.

Briggs, D.E.G., M.D. Sutton, D.J. Siveter, and D.J. Siveter. 2003. A new phyllocarid (Crustacea: Malacostraca) from the Silurian Fossil-Lagerstätte of Herefordshire, UK. Proceedings of the Royal Society of London Series B 271:131–138.

Briggs, D.E.G., M.D. Sutton, D.J. Siveter, and D.J. Siveter. 2005. Metamorphosis in a Silurian barnacle. Proceedings of the Royal Society of London Series B 272:2365–2369.

Briggs, D.E.G., D.J. Siveter, D.J. Siveter, M.D. Sutton, and D. Legg. 2016. Tiny individuals attached to a new Silurian arthropod suggest a unique mode of brood care. Proceedings of the National Academy of Science USA 113:4410–4415.

Broda, K., T.A. Hegna, and M. Zatoń. 2015. Fossils Explained 65: Thylacocephala. Geology Today 31:116–120.

Brooks, H.K. 1955. A crustacean from the Tesnus Formation (Pennsylvanian) of Texas. Journal of Paleontology 29:852–856.

Brooks, H.K. 1962. Devonian Eumalacostraca. Arkiv för Zoologi 15:307–317.

Butler, A.D., J.A. Cunningham, G.E. Budd, and P.C. Donoghue. 2015. Experimental taphonomy of *Artemia* reveals the role of endogenous microbes in mediating decay and fossilization. Proceedings of the Royal Society B 282:20150476.

Casanova, B. 2003. Ordre des Euphausiacea Dana, 1852 (The order Euphausiacea Dana, 1852). Crustaceana 76:1083–1121.

Chablais, J., R.M. Feldmann, and C.E. Schweitzer. 2011. A new Triassic decapod, *Platykotta akaina*, from the Arabian shelf of the northern United Arab Emirates: earliest occurrence of the Anomura. Paläontologische Zeitschrift 85:93–102.

Charbonnier, S., A. Garassino, G. Schweigert, and M. Simpson. 2013. A Worldwide Review of Fossil and Extant Glypheid and Litogastrid Lobsters (Crustacea, Decapoda, Glypheoidea). Muséum National d'Histoire Naturelle, Paris.

Chen, P.J., and H. Zhou. 1985. A preliminary study on fossil Kazacharthra from Turpan Basin. Kexue Tongbao 30:950–954.

Christoffersen, M.L., and J.E. De Assis. 2013. A systematic monograph of the Recent Pentastomida, with a compilation of their hosts. Zoologische Mededelingen Leiden 87:1–206.

Collette, J.H., and J.W. Hagadorn. 2010a. Three-dimensionally preserved arthropods from Cambrian Lagerstätten of Quebec and Wisconsin. Journal of Paleontology 84:646–667.

Collette, J.H., and J.W. Hagadorn. 2010b. Early evolution of phyllocarid arthropods: phylogeny and systematics of Cambrian-Devonian archaeostracans. Journal of Paleontology 84:795–820.

Collins, D., and D.M. Rudkin. 1981. *Priscansermarinus barnetti*, a probable lepadomorph barnacle from the Middle Cambrian Burgess Shale of British Colombia. Journal of Paleontology 55:1006–1015.

Colosi, G. 1917. Crostacei—Parte II. Eufausiacei. Raccolte planctoniche fatte dalla R. Nave "Liguria," Vol. 2, Fasc. 7. Publicationi 1 st. del studi superiori practici e di perfezionamento in Firenze: Sezione di Scienze Fisiche e Naturali 29:165–205.

Cooper, C.L. 1936. Actinopterygian jaws from the Mississippian black shales of the Mississippi Valley. Journal of Paleontology 10:92–94.

Cressey, R., and G. Boxshall. 1989. *Kabatarina pattersoni*, a fossil parasitic copepod (Dichelesthiidae) from a Lower Cretaceous fish. Micropaleontology 35:150–167.

De Grave, S., N.D. Pentcheff, S.T. Ahyong, T.Y. Chan, K.A. Crandall, P.C. Dworschak, D.L. Felder, R.M. Feldmann, C.H.J.M. Fransen, L.Y.D. Goulding, R. Lemaitre, M.E.Y. Low, J.W. Martin, P.K.L. Ng, C.E. Schweitzer, S.H. Tan, D. Tshudy, and R. Wetzer. 2009. A classification of recent and fossil genera of decapod crustaceans. Raffles Bulletin of Zoology Supplement 21:1–109.

De Grave, S., T.Y. Chan, K.H. Chu, C.H. Yang, and J.M. Landeira. 2015. Phylogenetics reveals the crustacean order Amphionidacea to be larval shrimps (Decapoda: Caridea). Scientific Reports 5:17464.

Donoghue, P.C.J., and Z. Yang. 2016. The evolution of methods for establishing evolutionary timescales. Philosophical Transactions of the Royal Society B 371:20160020.

Doruk, N. 1973. On *Mutilus speyeri* (Brady). Stereo-Atlas of Ostracod Shells 1:137–140.

Dzik, J. 2008. Gill structure and relationships of the Triassic cycloid crustaceans. Journal of Morphology 269:1501–1519.

Dzik, J., A.Y. Ivantsov, and Y.V. Deulin. 2004. Oldest shrimp and associated phyllocarid from the Lower Devonian of northern Russia. Zoological Journal of the Linnean Society 142:83–90.

Emerson, M.J., and F.R. Schram. 1991. Remipedia, Part 2. Paleontology. Proceedings of the San Diego Society of Natural History 7:1–52.

Eriksson, M.E., F. Terfelt, R. Elofsson, A. Maas, F. Marone, A. Lindskog, D. Waloszek, B. Schmitz, and M. Stampanoni. 2016. Baring it all: undressing Cambrian "Orsten" phosphatocopine crustaceans using synchrotron radiation X-ray tomographic microscopy. Lethaia 49:312–326.

Eyun, S.-I. 2017. Phylogenomic analysis of Copepoda (Arthropoda, Crustacea) reveals unexpected similarities with earlier proposed morphological phylogenies. BMC Evolutionary Biology 17:23.

Feldmann, R.M. 1998. Parasitic castration of the crab, *Tumidocarcinus giganteus* Glaessner, from the Miocene of New Zealand: coevolution within the Crustacea. Journal of Paleontology 72:493–498.

Feldmann, R.M. 2003. The Decapoda: new initiatives and novel approaches. Journal of Paleontology 77:1021–1039.

Feldmann, R.M., and C.E. Schweitzer. 2010a. Is *Eocarcinus* Withers, 1932, a basal brachyuran? Journal of Crustacean Biology 30:241–250.

Feldmann, R.M., and C.E. Schweitzer. 2010b. The oldest shrimp (Devonian: Famennian) and remarkable preservation of soft tissue. Journal of Crustacean Biology 30:629–635.

Feldmann, R.M., A.A. Garassino, and C.E. Schweitzer. 2004. The presumed decapod, *Palaeopemphix* Gemmellaro, 1890, is a unique member of the Phyllocarida (Palaeopemphicida [Palaeopemphida]: Palaeopemphidae). Journal of Paleontology 78:340–348.

Fraaije, R.H.B., F.R. Schram, and R. Vonk. 2003. *Maastrichtiocaris rostratus* new genus and species, the first Cretaceous cycloid. Journal of Paleontology 77:386–389.

Gall, J.C., and L. Grauvogel. 1964. Un arthropode peu connu le genre *Euthycarcinus* Handlirsch. Annales de Paléontologie 50:1–18.

Gall, J.C., and L. Grauvogel. 1967. Faune du Buntsandstein II.—Les Halicynés. Annales de Paléontologie 53:1–14.

Gall, J.C., L. Grauvogel-Stamm, and F. Papier. 2006. Archive de Grès: L'Alsace il y a 240 Millions d'Années. Pierron, Sarreguemines, France.

Garassino, A., and G. Pasini. 2002. Studies on Permo-Trias of Madagascar. 5. *Ambilobeia karojoi* n. gen., n. sp. (Crustacea, Decapoda) from the Lower Triassic (Olenekian) of Ambilobè region (NW Madagascar). Atti della Società italiana di Scienze naturali e del Museo civico di Storia naturale in Milano 143:95–104.

Goldring, R. 1967. *Cyclus martinensis* sp. nov. (Crustacea) from the Upper Visean of the Mendip Hills, England. Palaeontology 10:317–321.

Gueriau, P., S. Charbonnier, and G. Clément. 2014a. Angustidontid crustaceans from the Late Devonian of Strud (Namur Province, Belgium): insights into the origin of Decapoda. Neues Jahrbuch für Geologie und Paläontologie, Abhandlungen 273:327–337.

Gueriau, P., S. Charbonnier, and G. Clément. 2014b. First decapod crustaceans in a Late Devonian continental ecosystem. Palaeontology 57:1203–1213.

Gueriau, P., C. Mocuta, D.B. Dutheil, S.X. Cohen, D. Thiaudière, OT1 Consortium, S. Charbonnier, G. Clément, and L. Bertrand. 2014c. Trace elemental imaging of rare earth elements discriminates tissues at microscale in flat fossils. PLoS One 9:e86946.

Gueriau, P., N. Rabet, G. Clément, L. Lagebro, J. Vannier, D.E.G. Briggs, S. Charbonnier, S. Olive, and O. Béthoux. 2016. A 365-million-year-old freshwater community reveals morphological and ecological stasis in branchiopod crustaceans. Current Biology 26: 383–390.

Guinot, D., A. De Angeli, and A. Garassino. 2007. *Herbertides jurassica* n. gen., n. sp. (Crustacea, Decapoda, Brachyura) from the Middle Jurassic (Bathonian) of Normandy (France). Atti della Società italiana di Scienze naturali e del Museo civico di Storia naturale in Milano 148:241–260.

Handlirsch, A. 1914. Eine interessante crustaceenform aus der trias der Vogesen. Verhandlungen der kaiserlich-koniglichen zoologisch-botanischen Gesellschaft in Wien 64:1–8.

Harvey, T.H.P., and B.E. Pedder. 2013. Copepod mandible palynomorphs from the Nolichucky Shale (Cambrian, Tennessee): implications for the taphonomy and recovery of small carbonaceous fossils. Palaios 28:278–284.

Harvey, T.H., M. I. Velez, and N. J. Butterfield. 2012. Exceptionally preserved crustaceans from western Canada reveal a cryptic Cambrian radiation. Proceedings of the National Academy of Science of the United States of America 109:1589–1594.

Haug, C., D.E.G. Briggs, D.G. Mikulic, J. Kluessendorf, and J.T. Haug. 2014. The implications of a Silurian and other thylacocephalan crustaceans for the functional morphology and systematic affinities of the group. BMC Evolutionary Biology 14:1–15.

Haug, J.T., and C. Haug. 2011. Fossilien unter langwelligem Licht: Grün-Orange-Fluoreszenz an makroskopischen Objekten [Fossils under long-wave light: green-orange fluorescence in macroscopic objects]. Archaeopteryx 29:20–23.

Haug, J.T., and C. Haug. 2014. *Eoprosopon klugi* (Brachyura)—the oldest unequivocal and most "primitive" crab reconsidered. Palaeodiversity 7:149–158.

Haug, J.T., C. Haug, A. Maas, V. Kutschera, and D. Waloszek. 2010a. Evolution of mantis shrimps (Stomatopoda, Malacostraca) in the light of new Mesozoic fossils. BMC Evolutionary Biology 10:290.

Haug, J.T., D. Waloszek, C. Haug, and A. Maas. 2010b. High-level phylogenetic analysis using developmental sequences: the Cambrian †*Martinssonia elongata*, †*Musacaris gerdgeyeri* gen. et sp. nov. and their position in early crustacean evolution. Arthropod Structure and Development 39:154–173.

Haug, J.T., A. Maas, C. Haug, and D. Waloszek. 2013. Evolution of crustacean appendages. Pages 34–73 *in* L. Watling and M. Thiel, editors. The Natural History of the Crustacea, Vol. 1: Functional Morphology and Diversity. Oxford University Press, Oxford.

Hayashi, R., B.K. Chan, N. Simon-Blecher, H. Watanabe, T. Guy-Haim, T. Yonezawa, Y. Levy, T. Shuto, and Y. Achituv. 2013. Phylogenetic position and evolutionary history of the turtle and whale barnacles (Cirripedia: Balanomorpha: Coronuloidea). Molecular Phylogenetics and Evolution 67:9–14.

Hegna, T.A. 2010. Photography of soft-bodied crustaceans via drying, whitening, and splicing. Journal of Crustacean Biology 30:351–356.

Hegna, T.A. 2012. Phylogeny and fossil record of branchiopod crustaceans: an integrative approach. Ph.D. dissertation. Yale University Department of Geology and Geophysics, New Haven, CT.

Hegna, T.A., and D. Ren. 2010. Two new "notostracans," *Chenops* gen. nov. and *Jeholops* gen. nov. (Crustacea: Branchiopoda: ?Notostraca) from the Yixian Formation, northeastern China. Acta Geologica Sinica (English Edition) 84: 886–894.

Hegna, T.A., F.J. Vega, and K.A. González-Rodríguez. 2014. First Mesozoic thylacocephalans (Arthropoda, ?Crustacea; Cretaceous) in the western hemisphere: new discoveries from the Muhi Quarry Lagerstätte. Journal of Paleontology 88:606–616.

Høeg, J.T., M. Pérez-Losada, H. Glenner, G.A. Kolbasov, and K.A. Crandall. 2009. Evolution of morphology, ontogeny and life cycles within the Crustacea Thecostraca. Arthropod Systematics and Phylogeny 67:199–217.

Hołyńska, M., L. Leggitt, and A. Kotov. 2015. Miocene cyclopid copepod from a saline paleolake in Mojave, California. Acta Palaeontologica Polonica 61:345–361.

Horne, D.J., A.C. Cohen, and K. Martens. 2002. Taxonomy, morphology and biology of quaternary and living Ostracoda. Pages 5–36 *in* J.A. Holmes and A.R. Chivas, editors. The Ostracoda: Applications in Quarternary Research. American Geophysical Union, Washington, DC.

Hou, X., M. Williams, D.J. Siveter, D.J. Siveter, R.J. Aldridge, and R.S. Sansom. 2010. Soft-part anatomy of the Early Cambrian bivalved arthropods *Kunyangella* and *Kunmingella*: significance for the phylogenetic relationships of Bradoriida. Proceedings of the Royal Society B 277:1835–1841.

Huys, R., and G. Boxshall. 1991. Copepod Evolution. The Ray Society, London.

Huys, R., E. Suárez-Morales, M.L. Serrano-Sánchez, E. Centeno-García, and F.J. Vega. 2016. Early Miocene amber inclusions from Mexico reveal antiquity of mangrove-associated copepods. Science Reports 6:34872.

Hyžný, M., and A.A. Klompmaker. 2015. Systematics, phylogeny, and taphonomy of ghost shrimps (Decapoda): a perspective from the fossil record. Arthropod Systematics and Phylogeny 73:401–437

Hyžný, M., C. Haug, and J.T. Haug. 2016. *Mesoprosopon triasinum* from the Triassic of Austria revisited: The oldest eumalacostracan larva known to date and its significance for interpreting fossil cycloids. Gondwana Research 37:86–97.

Jarman, S.N. 2001. The evolutionary history of krill inferred from nuclear large subunit rDNA sequence analysis. Biological Journal of the Linnean Society 73:199–212.

Jarman, S.N., and N.G. Elliott. 2000. DNA evidence for morphological and cryptic Cenozoic speciations in the Anaspididae, "living fossils" from the Triassic. Journal of Evolutionary Biology 13:624–633.

Jell, P.A., and P.M. Duncan. 1986. Invertebrates, mainly insects, from the freshwater, Lower Cretaceous, Koonwarra Fossil Bed (Korumburra Group), South Gippsland, Victoria. Memoirs of the Association of Australasian Palaeontologists 3:111–205.

Jenner, R.A., C.H.J. Hof, and F.R. Schram. 1998. Palaeo- and archaeostomatopods (Hoplocarida, Crustacea) from the Bear Gulch Limestone, Mississippian (Namurian), of central Montana. Contributions to Zoology 67:155–185.

Jenner, R.A., C. Ní Dhubhghaill, M.P. Ferla, and M.A. Wills. 2009. Eumalacostracan phylogeny and total evidence: limitations of the usual suspects. BMC Evolutionary Biology 9:1–20.

Jones, P.J., and E. Olempska. 2013. The ostracode suborder Eridostracina Adamczak, 1961. Neues Jahrbuch für Geologie und Paläontologie, Abhandlungen 267:341–352.

Jones, W.T. 2016. A taxonomic and morphologic assessment of some crustaceans and crustacean-like problematica. Ph.D Dissertation. Kent State University, Kent, OH.

Jones, W.T., R.M. Feldmann, C.E. Schweitzer, F.R. Schram, R.A. Behr, and K.L. Hand. 2014. The first Paleozoic stenopodidean from the Huntley Mountain Formation (Devonian–Carboniferous), north-central Pennsylvania. Journal of Paleontology 88:1251–1256.

Jones, W.T., D.G. Mikulic, and R.M. Feldmann. 2015. Archaeostracan (Phyllocarida: Archaeostraca) antennulae and antennae: sexual dimorphism in early malacostracans and *Ceratiocaris* M'Coy, 1849 as a possible stem eumalacostracan. Journal of Crustacean Biology 35:191–201.

Jones, W.T., R.M. Feldmann, F.R. Schram, C.E. Schweitzer, and E.P. Maguire. 2016. The proof is in the pouch: *Tealliocaris* is a peracarid. Palaeodiversity 9:75–88.

Karasawa, H., C.E. Schweitzer, and R.M. Feldmann. 2013. Phylogeny and systematics of extant and extinct lobsters. Journal of Crustacean Biology 33:78–123.

Keyser, D., and F. Friedrich. 2016. An exceptionally well-preserved new species of ostracod (Crustacea) with soft parts in Baltic amber. Historical Biology 29: 53–62.

Klompmaker, A.A., and G.A. Boxshall. 2015. Fossil crustaceans as parasites and hosts. Advances in Parasitology 90:233–289.

Kobayashi, T. 1937. An occurrence of a new Permian phyllocarid in South Chosen. Journal of the Geological Society of Japan 44:337–340.

Kornicker, L.S., B.W.M. Van Bakel, R.H.B. Fraaiji, and J.W.M. Jagt. 2006. Revision of Mesozoic Myodocopina (Ostracoda) and a new genus and species, *Mesoleberis hollandica*, from the Upper Cretaceous of Belgium and The Netherlands. Zootaxa 1246:15–54.

Krause, R.A., K. Parsons-Hubbard, and S.E. Walker. 2011. Experimental taphonomy of a decapod crustacean: long-term data and their implications. Palaeogeography, Palaeoclimatology, Palaeoecology 312:350–362.

Lagebro, L., P. Gueriau, T.A. Hegna, N. Rabet, A.D. Butler, and G.E. Budd. 2015. The oldest notostracan (Upper Devonian Strud locality, Belgium). Palaeontology 58:497–509.

Legg, D.A., M.D. Sutton, and G.D. Edgecombe. 2013. Arthropod fossil data increase congruence of morphological and molecular phylogenies. Nature Communications 4:2485.

Lozano-Fernandez, J., R. Carton, A.R. Tanner, M.N. Puttick, M. Blaxter, J. Vinther, J. Olesen, G. Giribet, G.D. Edgecombe, and D. Pisani. 2016. A molecular palaeobiological exploration of arthropod terrestrialization. Philosophical Transactions of the Royal Society B 371:1–12.

Luo, H., S. Hu, L. Chen, S. Zhang, and Y. Tao. 1999. Early Cambrian Chengjiang Fauna from Kunming Region, China. Yunnan Science and Technology Press, Kunming, China.

Maas, A., and D. Waloszek. 2001. Larval development of *Euphausia superba* Dana, 1852 and a phylogenetic analysis of the Euphausiacea. Hydrobiologia 448:143–169.

Maas, A., D. Waloszek, and K.J. Müller. 2003. Morphology, ontogeny and phylogeny of the Phosphatocopina (Crustacea) from the Upper Cambrian "Orsten" of Sweden. Fossils and Strata 49:1–238.

Maas, A., A. Braun, X.P. Dong, P.C.J. Donoghue, K.J. Müller, E. Olempska, J.E. Repetski, D. Siveter, M. Stein, and D. Waloszek. 2006. The "Orsten"—more than a Cambrian Konservat-Lagerstätte yielding exceptional preservation. Palaeoworld 15:266–282.

MacNaughton, R.B., J.M. Cole, R.W. Dalrymple, S.J. Braddy, D.E.G. Briggs, and T.D. Lukie. 2002. First steps on land; arthropod trackways in Cambrian-Ordovician eolian sandstone, southeastern Ontario, Canada. Geology 30:391–394.

Malzahn, E. 1958. Ein neuer jungpaläozoischer Kreb aus dem niederrheinischen Zechstein. Zeitschrift der Deutschen Geologischen Gesellschaft 110:352–359.

Martens, K., I. Schön, C. Meisch, and D.J. Horne. 2008. Global diversity of ostracods (Ostracoda, Crustacea) in freshwater. Hydrobiologia 595:185–193.

Martins-Neto, R.G., O.F. Gallego, and R.N. Melchor. 2001. The first record of kazacharthids (Branchiopoda) in the Triassic of South America. Ameghiniana 38: 36R-37R.

Matzke-Karasz, R., J.V. Neil, R.J. Smith, R. Symonová, L. Mořkovský, M. Archer, S.J. Hand, P. Cloetens, and P. Tafforeau. 2014. Subcellular preservation in giant ostracod sperm from an Early Miocene cave deposit in Australia. Proceedings of the Royal Society B 281:20140394.

McNamara, M.E., D.E.G. Briggs, P.J. Orr, H. Noh, and H. Cao. 2012. The original colours of fossil beetles. Proceedings of the Royal Society B 279:1114–1121.

Meland, K., and E. Willassen. 2007. The disunity of "Mysidacea" (Crustacea). Molecular Phylogenetics and Evolution 44:1083–1104.

Müller, K.J. 1983. Crustacea with preserved soft parts from the Upper Cambrian of Sweden. Lethaia 16:93–109.

Müller, K.J., and D. Walossek. 1986. *Martinssonia elongata* gen. et sp.n., a crustacean-like euarthropod from the Upper Cambrian "Orsten" of Sweden. Zoologica Scripta 15:73–92.

Nagler, C., C. Haug, U. Resch, and J.T. Haug. 2016. 150 million years old isopods on fishes: a possible case of palaeo-parasitism. Bulletin of Geosciences 91:1–12.

Neiber, M.T., T.R. Hartke, T. Stemme, A. Bergmann, J. Rust, T.M. Iliffe, and S. Koenemann. 2011. Global biodiversity and phylogenetic evaluation of Remipedia (Crustacea). PLoS ONE 6:e19627.

Novozhilov, N.I. 1961. Dvustvorcatye Listonogie Devona [Devonian Conchostracans]. Akademiâ Nauk SSSR Trudy Paleontogiceskogo Instituta 81:1–132.

Oakley, T.H., J.M. Wolfe, A.R. Lindgren, and A.K. Zaharoff. 2013. Phylotranscriptomics to bring the understudied into the fold: monophyletic Ostracoda, fossil placement, and pancrustacean phylogeny. Molecular Biology and Evolution 30:215–233.

Olempska, E. 2012. Morphology and affinities of Eridostracina: palaeozoic ostracods with moult retention. Hydrobiologia 688:139–165.

Olempska, E., D.J. Horne, and H. Szaniawski. 2012. First record of preserved soft parts in a Palaeozoic podocopid (Metacopina) ostracod, *Cytherellina submagna*: phylogenetic implications. Proceedings of the Royal Society B 279:564–570.

Olesen, J. 2009. Phylogeny of Branchiopoda (Crustacea)—character evolution and contribution of uniquely preserved fossils. Arthropod Systematics and Phylogeny 67:3–39.

Ortega-Hernández, J., D.A. Legg, J. Tremewan, and S.J. Braddy. 2010. Fossils explained 59: Euthycarcinoids. Geology Today 26:195–198.

Palmer, A.R. 1960. Miocene copepods from the Mojave Desert, California. Journal of Paleontology 34:447–452.

Pazinato, P.G., M.B. Soares, and K. Adami-Rodrigues. 2016. Systematic and palaeoecological significance of the first record of Pygocephalomorpha females bearing oöstegites (Malacostraca, Peracarida) from the lower Permian of southern Brazil. Palaeontology 59:817–826.

Peach, B.N. 1883. On some new crustaceans from the Lower Carboniferous rocks of Eskdale and Liddesdale. Transactions of the Royal Society of Edinburgh 30:73–91.

Peach, B.N. 1908. Monograph on the higher Crustacea of the Carboniferous rocks of Scotland. Memoirs of the Geological Survey of Great Britain. Palaeontology 1–82.

Péneau, J. 1936. Présence d'*Estheria* dans les schites attribués au Gothlandien du Synclinal de Laval. Bulletin de la Société des Sciences Naturelles de l'Ouest de la France, 5e Série 6:219–226.

Pérez-Losada, M., M. Harp, J. T. Høeg, Y. Achituv, D. Jones, H. Watanabe, and K. A. Crandall. 2008. The tempo and mode of barnacle evolution. Molecular Phylogenetics and Evolution 46:328–346.

Perrier, V., M. Williams, and D.J. Siveter. 2015. The fossil record and palaeoenvironmental significance of marine arthropod zooplankton. Earth-Science Reviews 146:146–162.

Piñeiro, G., E. Morosi, A. Ramos, and F. Scarabino. 2012. Pygocephalomorph crustaceans from the Early Permian of Uruguay: constraints on taxonomy. Revista Brasileira de Paleontologia 15:33–48.

Poore, G.C.B. 2005. Peracarida: monophyly, relationships and evolutionary success. Nauplius 13:1–27.

Racheboeuf, P.R., and C. Villarroel. 2003. *Imocaris colombiensis* n. sp. (Crustacea: Decapoda) from the Pennsylvanian of Colombia. Neues Jahrbuch für Geologie und Paläontologie, Monatshefte 10:577–590.
Radwańska, U., and E. Poirot. 2010. Copepod-infested Bathonian (Middle Jurassic) echinoids from northern France. Acta Geologica Polonica 60:549–555.
Richter, S., M. Schwentner, C.S. Wirkner, and S.T. Ahyong. 2018. Phylogeny and species diversity of Tasmanian mountain shrimps and their relatives (Crustacea, Anaspidesidae). Zoologica Scripta 47:84–105.
Riek, E.F. 1964. Merostomoidea (Arthropods, Trilobitomorpha) from the Australian Middle Triassic. Records of the Australian Museum 26:327–332.
Riley, J. 1986. The biology of pentastomids. Advances in Parasitology 25:45–128.
Rode, A.L., and B.S. Lieberman. 2002. Phylogenetic and biogeographic analysis of Devonian phyllocarid crustaceans. Journal of Paleontology 76:271–286.
Rodendorf, B.B. 1961. Description of the first winged insect from the Devonian beds of the Timan. Entomological Review 40:260–262.
Rodriguez, J., and R.C. Gutschick. 1977. Barnacle borings in live and dead hosts from the Louisiana Limestone (Famennian) of Missouri. Journal of Paleontology 51:718–724.
Rolfe, W.D.I. 1981. Phyllocarida and the origin of the Malacostraca. Geobios 14:17–27.
Rolfe, W.D.I., and J. Dzik. 2006. *Angustidontus*, a Late Devonian pelagic predatory crustacean. Transactions of the Royal Society of Edinburgh, Earth Sciences 97:75–96.
Ross, A., and W.A. Newman. 1967. Eocene Balanidae of Florida, including a new genus and species with a unique plan of "turtle-barnacle" organization. American Museum Novitates 1–21.
Sansom, R.S., and M.A. Wills. 2013. Fossilization causes organisms to appear erroneously primitive by distorting evolutionary trees. Scientific Reports 3:2545.
Schneider, J., R. Werneburg, G. Gand, M. Vianey-Liaud, and S. Steyer. 2006. First Permian playa fossil lagerstaette. Page 174 *in* Q. Yang, Y. Wang, and E.A. Weldon, editors. Ancient Life and Modern Approaches: Abstracts of the Second International Palaeontological Congress. University of Science and Technology of China Press, Beijing, China.
Scholtz, G., and S. Richter. 1995. Phylogenetic systematics of the reptantian Decapoda (Crustacea, Malacostraca). Zoological Journal of the Linnaean Society 113:289–328.
Schram, F.R. 1973. On some phyllocarids and the origin of the Hoplocarida. Fieldiana Geology 26:77–94.
Schram, F.R. 1975. A Pennsylvanian lepadomorph barnacle from the Mazon Creek Area, Illinois. Journal of Paleontology 49:928–930.
Schram, F.R. 1980. Miscelaneous Late Paleozoic Malacostraca of the Soviet Union. Journal of Paleontology 54:542–547.
Schram, F.R. 1982. The fossil record and evolution of Crustacea. Pages 93–147 *in* L.G. Abele, editor. Systematics, the Fossil Record, and Biogeography. Academic Press, New York.
Schram, F.R. 1983. Lower Carboniferous biota of Glencartholm, Eskdale, Dumfriesshire. Scottish Journal of Geology 19:1–15.
Schram, F.R. 1984. Fossil Syncarida. Transactions of the San Diego Society of Natural History 20:189–246.
Schram, F.R. 1986. Crustacea. Oxford University Press, New York.
Schram, F.R. 2007. Paleozoic proto-mantis shrimp revisited. Journal of Paleontology 81:895–916.
Schram, F.R. 2009. On the origin of Decapoda. Pages 3–13 *in* J.W. Martin, K.A. Crandall, and D.L. Felder, editors. Decapod Crustacean Phylogenetics. Crustacean Issues. Vol. 18. CRC Press, Taylor and Francis Group, Boca Raton, FL.
Schram, F.R. 2010. Catalog of the Fossil and Recent Stomatopoda. Bay Ridge Press, Langley, Washington.
Schram, F.R. 2013. Comments on crustacean biodiversity and disparity of body plans. Pages 1–33 *in* L. Watling and M. Thiel, editors. Functional Morphology and Diversity. Oxford University Press, New York.
Schram, F.R., and C.J. Dixon. 2004. Decapod phylogeny: addition of fossil evidence to a robust morphological cladistic data set. Bulletin of the Mizunami Fossil Museum 31:1–19.
Schram, F.R., and E. Malzahn. 1984. The fossil leptostracan *Rhabdouraea bentzi* (Malzahn, 1958). Transactions of the San Diego Society of Natural History 20:95–98.

Schram, F.R., and R.H. Mapes. 1984. *Imocaris tuberculata*, n. gen., n. sp. (Crustacea: Decapoda) from the Upper Mississippian Imo Formation, Arkansas. Transactions of the San Diego Society of Natural History 20:165–168.

Schram, F.R., R.M. Feldmann, and M. Copeland. 1978. The Late Devonian Palaeopalaemonidae and the earliest decapod crustaceans. Journal of Paleontology 52:1375–1387.

Schram, F.R., J. Yager, and M.J. Emerson. 1986a. Remipedia. Part I. Systematics. San Diego Society of Natural History Memoir 15:1–60.

Schram, F.R., J. Sieg, and E. Malzahn. 1986b. Fossil Tanaidacea. Transactions of the San Diego Society of Natural History 21:127–144.

Schram, F.R., R. Vonk, and C.H.J. Hof. 1997. Mazon Creek Cycloidea. Journal of Paleontology 71:261–283.

Schram, F.R., and S. Koenemann. 2004. Are crustaceans monophyletic? Pages 319–329 *in* J. Cracraft and M.J. Donaghue, editors. Assembling the Tree of Life. Oxford University Press, New York.

Schram, F.R., A.C. Boere, and N. Thomas. 2006. Cycloidea of the Mississippian Bear Gulch Limestone of central Montana. Contributions in Science from the Natural History Museum of Los Angeles County 504:1–8.

Schweigert, G., R. Fraaije, and M. Wulf. 2009. Die Cycliden: Eine rätselhafte Krebsgruppe. Fossilien 2009:17–22.

Schweitzer, C.E., and R.M. Feldmann. 2015. Faunal turnover and niche stability in marine Decapoda in the Phanerozoic. Journal of Crustacean Biology 35:633–649.

Scourfield, D.J. 1926. On a new type of crustacean from the Old Red Sandstone (Rhynie Chert Bed, Aberdeenshire)—*Lepidocaris rhyniensis*, gen et sp. nov. Philosophical Transactions of the Royal Society of London B 214:153–187.

Secrétan, S. 1985. Conchyliocarida, a class of fossil crustaceans: relationships to Malacostraca and postulated behaviour. Transactions of the Royal Society of Edinburgh: Earth Sciences 76:381–389.

Selden, P.A., R. Huys, M.H. Stephenson, A.P. Heward, and P.N. Taylor. 2010. Crustaceans from bitumen clast in Carboniferous glacial diamictite extend fossil record of copepods. Nature Communications 1:1–6.

Serrano-Sánchez, M.D.L., T.A. Hegna, P. Schaaf, L. Pérez, E. Centeno-García, and F.J. Vega. 2015. The aquatic and semiaquatic biota in Miocene amber from the Campo La Granja mine (Chiapas, Mexico): paleoenvironmental implications. Journal of South American Earth Sciences 62:243–256.

Shen, Y.B., and P.J. Chen. 1984. Late Middle Jurassic conchostracans from Tuchengzi Formation of W. Liaoning, NE China. Bulletin of the Nanjing Institute of Geology and Palaeontology, Academica Sinica 9:309–326.

Siveter, D.J. 1982. On *Homeokiesowia epicopa* Siveter sp. nov. Stereo-Atlas of Ostracod Shells 9:89–92.

Siveter, D.J. 2008a. The Silurian Herefordshire Konservat-Lagerstätte: a unique window on the evolution of life. Proceedings of the Shropshire Geological Society 13:58–61.

Siveter, D.J. 2008b. Ostracods in the Palaeozoic? Senckenbergiana Lethaea 88:1–9.

Siveter, D.J., M.D. Sutton, D.E.G. Briggs, and D.J. Siveter. 2003. An ostracode crustacean with soft parts from the Lower Silurian. Science 302:1749–1751.

Siveter, D.J., M.D. Sutton, D.E.G. Briggs, and D.J. Siveter. 2007a. A new probable stem lineage crustacean with three-dimensionally preserved soft-parts from the Herefordshire (Silurian) Lagerstätte, UK. Proceedings of the Royal Society of London Series B Biological Sciences 274:2099–2107.

Siveter, D.J., D.J. Siveter, M.D. Sutton, and D.E.G. Briggs. 2007b. Brood care in a Silurian ostracod. Proceedings of the Royal Society B 274:465–469.

Siveter, D.J., D.E.G. Briggs, D.J. Siveter, and M.D. Sutton. 2010. An exceptionally preserved myodocopid ostracod from the Silurian of Herefordshire, UK. Proceedings of the Royal Society B 277:1539–1544.

Siveter, D.J., D.E.G. Briggs, D.J. Siveter, M.D. Sutton, and S.C. Joomun. 2013. A Silurian myodocope with preserved soft-parts: cautioning the interpretation of the shell-based ostracod record. Proceedings of the Royal Society B 280:20122664.

Siveter, D.J., G. Tanaka, U.C. Farrell, M.J. Martin, D.J. Siveter, and D.E. Briggs. 2014. Exceptionally preserved 450-million-year-old Ordovician ostracods with brood care. Current Biology 24:801–806.

Siveter, D.J., D.E.G. Briggs, D.J. Siveter, and M.D. Sutton. 2015. A 425-million-year-old Silurian pentastomid parasitic on ostracods. Current Biology 25:1632–1637.

Siveter, D.J., D.E.G. Briggs, D.J. Siveter, M.D. Sutton, and D. Legg. 2017. A new crustacean from the Herefordshire (Silurian) Lagerstätte, UK, and its significance in malacostracan evolution. Proceedings of the Royal Society B 284:20170279.

Smith, R.J. 2000. Morphology and ontogeny of Cretaceous ostracods with preserved appendages from Brazil. Palaeontology 43:63–98.

Soot-Ryen, H. 1960. The Middle Ordovician of the Oslo region, Norway; 12, Notostraca and Conchostraca. Norsk Geologisk Tidsskrift 40: 123–131.

Spears, T., R.W. DeBry, L.G. Abele, and K. Chodyla. 2005. Peracarid monophyly and interordinal phylogeny inferred from nuclear small-subunit ribosomal DNA sequences (Crustacea: Malacostraca: Peracarida). Proceedings of the Biological Society of Washington 118:117–157.

Spiridonov, V., and B. Casanova. 2010. Order Euphausiacea Dana, 1852, treatise on zoology—anatomy taxonomy biology. Pages 5–82 *in* The Crustacea, Vol. 9 Part A. Brill.

Števčić, Z. 2005. The reclassification of brachyuran crabs (Crustacea: Decapoda: Brachyura). Natura Croatica 14:1159.

Swanson, K.M. 1989. *Manawa staceyi* n. sp. (Punciidae, Ostracoda): soft anatomy and ontogeny. Courier Forschungsinstitut Senckenberg 113:235–249.

Tinn, O., and T.H. Oakley. 2008. Erratic rates of molecular evolution and incongruence of fossil and molecular divergence time estimates in Ostracoda (Crustacea). Molecular Phylogenetics and Evolution 48:157–167.

Tsang, L. M., C.D. Schubart, S.T. Ahyong, J.C.Y. Lai, E.Y.C. Au, T.-Y. Chan, P,K.L. Ng, and K.H. Chu. 2014. Evolutionary history of true crabs (Crustacea: Decapoda: Brachyura) and the origin of freshwater crabs. Molecular Biology and Evolution 31:1173–1187.

Vaccari, N.E., G.D. Edgecombe, and C. Escudero. 2004. Cambrian origins and affinities of an enigmatic fossil group of arthropods. Nature 430:554–557.

Van Damme, K., and A.A. Kotov. 2016. The fossil record of the Cladocera (Crustacea: Branchiopoda): evidence and hypotheses. Earth-Science Reviews 163:162–189.

Vannier, J., J.Y. Chen, D.Y. Huang, S. Charbonnier, and X.Q. Wang. 2006. The Early Cambrian origin of thylacocephalan arthropods. Acta Palaeontologica Polonica 51:201–214.

Vannier, J., S.Q. Wang, and M. Coen. 2001. Leperditicopid arthropods (Ordovician-Late Devonian): functional morphology and ecology range. Journal of Paleontology 75:75–95.

Voigt, E. 1967. Ein vermutlicher Ascothoracide (*Endosacculus*? *najdini*) als Bewohner einer kretazisehen *Isis* aus der UdSSR. Paläontologische Zeitschrift 41:86–90.

Wagner, P., J.T. Haug, J. Sell, and C. Haug. 2018. A fossil crustacean from the Upper Triassic of southern Germany with kazacharthran affinities. Paleontological Research 22:57–63.

Walossek, D. 1993. The Upper Cambrian *Rehbachiella* and the phylogeny of Branchiopoda and Crustacea. Fossils and Strata 32:1–202.

Walossek, D., and K.J. Müller. 1990. Upper Cambrian stem-lineage crustaceans and their bearing upon the monophyletic origin of Crustacea and the position of *Agnostus*. Lethaia 23:409–427.

Walossek, D., and K.J. Müller. 1994. Pentastomid parasites from the lower Palaeozoic of Sweden. Transactions of the Royal Society of Edinburgh: Earth Sciences 85:1–37.

Walossek, D., J.E. Repetski, and K.J. Müller. 1994. An exceptionally preserved parasitic arthropod, *Heymonsicambria taylori* n.sp. (Arthropoda incertae sedis; Pentastomida), from Cambrian-Ordovician boundary beds of Newfoundland, Canada. Canadian Journal of Earth Sciences 31:1664–1671.

Waloszek, D., A. Maas, J.Y. Chen, and M. Stein. 2007. Evolution of cephalic feeding structures and the phylogeny of Arthropoda. Palaeogeography Palaeoclimatology Palaeoecology 254:273–287.

Weitschat, W. 1983. Ostracoden (O. Myodocopida) mit Weichkörper-Erhaltung aus der Unter-Trias von Spitzbergen. Paläontologische Zeitschrift 27:309–323.

Whitfield, R.P. 1880. Notice of new forms of fossil crustaceans from the Upper Devonian rocks of Ohio, with descriptions of new genera and species. American Journal of Science 19:33–42.

Wilkinson, I.P., P.R. Wilby, M. Williams, D.J. Siveter, A.A. Page, L. Leggitt, and D.A. Riley. 2010. Exceptionally preserved ostracodes from a Middle Miocene palaeolake, California, USA. Journal of the Geological Society 167:817–825.

Williams, M., D. J. Siveter, A.C. Ashworth, P.R. Wilby, D.J. Horne, A.R. Lewis, and D.R. Marchant. 2008. Exceptionally preserved lacustrine ostracods from the Middle Miocene of Antarctica: implications for high-latitude palaeoenvironment at 77° south. Proceedings of the Royal Society of London Series B Biological Sciences 275:2449–2454.

Wills, L.J. 1962. A pedunculate cirripede from the upper Silurian of Oesel, Estonia. Nature 194:567.

Wills, M.A., R.A. Jenner, and C. Ní Dhubhghaill. 2009. Eumalacostracan evolution: conflict between three sources of data. Arthropod Systematics and Phylogeny 67:71–90.

Withers, T.H. 1932. XXXIII.—A Liassic crab, and the origin of the Brachyura. Annals and Magazine of Natural History 9:313–323.

Wolfe, J.M., and T.A. Hegna. 2014. Testing the phylogenetic position of Cambrian pancrustacean larval fossils by coding ontogenetic stages. Cladistics 30:366–390.

Wolfe, J.M., A.C. Daley, D.A. Legg, and G.D. Edgecombe. 2016. Fossil calibrations for the arthropod Tree of Life. Earth-Science Reviews 160:43–110.

Zatoń, M., P. Filipiak, M. Rakociński, and W. Krawczyński. 2014. Kowala Lagerstätte: Late Devonian arthropods and non-biomineralized algae from Poland. Lethaia 47:352–364.

Zhai, D., J. Ortega-Hernández, J.M. Wolfe, X. Hou, C. Cao, and Y. Liu. 2019. Micro-computed tomography reveals three-dimensionally preserved appendages in an Early Cambrian stem-group pancrustacean. Current Biology 29:171–177.

Zhang, X.G., D.J. Siveter, D. Waloszek, and A. Maas. 2007. An epipodite-bearing crown-group crustacean from the Lower Cambrian. Nature 449:595–598.

Zhang, X.G., A. Maas, J.T. Haug, D.J. Siveter, and D. Waloszek. 2010. A eucrustacean metanauplius from the Lower Cambrian. Current Biology 20:1075–1079.

3

EVOLUTION AND RADIATION OF CRUSTACEA

Shane T. Ahyong

Abstract

The evolution of the Crustacea following their origins in the Cambrian is outlined, with an overview of their paleontological history and global distributions into modern times. Major recent developments in arthropod evolution include recognition that Hexapoda is nested within Crustacea. Perspectives also changed during the last decades of the 20th century on the form of the crustacean ancestor, from being a long-bodied, serially homonomous form (like a remipede or cephalocarid) to a short-bodied, possibly ostracod-like form similar to Cambrian stem and crown group fossil forms. These changes have come through a shift to formal methods of phylogenetic analysis combined with the much larger volume of both morphological and molecular data now available. The most extensive current phylogenies typically recover the short-bodied Oligostraca (containing ostracods and a few minor groups) as basal crustaceans; Malacostraca and Maxillopoda are high in the tree; and Cephalocarida and Remipedia are derived forms as sister to Branchiopoda and Hexapoda, respectively. Each of these major groups can be understood through variations in tagmatization (differentiation of body segments into regions). The early crustacean fossil record (especially the Ordovician) is dominated by ostracods. Malacostracans, although having Cambrian origins, did not significantly radiate until the Mesozoic. Eumalacostraca continued to actively radiate in the Cenozoic and are now the most ubiquitous and morphologically disparate crustaceans. The processes driving crustacean evolution remain to be fully evaluated. Contingency and external factors are undoubtedly important, but most deep lineages of the Crustacea show pervasive macroevolutionary trends toward increasing tagmatization. These trends are apparently driven, meaning the formation of new body plans is not merely a contingent outcome—intrinsic factors may contribute to increasing tagmatization. Further data are required from ontogeny and developmental genetics, paleontology, and phylogenetics in order to better understand how crustaceans have evolved.

Evolution and Biogeography. Edited by Gary C. B. Poore and Martin Thiel.

INTRODUCTION

Some 67,000 species of living Crustacea, of which more than 50,000 are marine, are currently known worldwide (Ahyong et al. 2011, Appeltans et al. 2012), occupying almost every habitat from the Himalayan mountain peaks to abyssal ocean depths. The crustaceans are among the most successful arthropods, with deep origins approximately 500 Ma in the Cambrian. Definitively, crustacean fossils have been identified in the upper Cambrian Orsten fauna of southern Sweden. All are small, often only millimeters in length, but known in exquisite detail owing to their phosphatic preservation that preserves extremely fine structures (Walossek and Müller 1998). Walossek and Müller (1990) distinguished both stem-lineage and crown-group crustaceans, recognizable by their differential acquisition of crustacean apomorphies (see Chapter 1 for the origins and early evolution of the Crustacea). Accordingly, the Orsten fauna is one of the most important lenses through which to observe the early evolution of crustaceans. Suffice it to say, however, evolution in the stem forms, particularly around limb arrangements and body segmentation, brought numerous changes in locomotion, reproduction, and feeding leading to "true" crustaceans (Walossek and Müller 1998, Haug et al. 2013). Subsequent morphological diversification, in addition to increased body size, most notably from the Silurian (420 Ma) onward, has produced the major branches of the crustacean tree we recognize today. The present survey outlines the evolution of the Crustacea following their origins in the Cambrian, including overviews of paleontological history and global distributions into modern times.

The modularity of the arthropod construction has enabled a disparate array of body plans in Crustacea, particularly through varying degrees of body regionalization and limb specialization. This differentiation of body regions and specialization of limbs along the body axis is known as tagmatization. In crustaceans, basic body tagma are the cephalon (head) and trunk. The trunk itself may be divided into the thorax and abdomen, and in malacostracans the thorax is further divided into a pereon and pleon. Most workers have thought the pereon and pleon of malacostracans to be equivalent to the respective thorax and abdomen of other crustacean groups. Hox gene expression patterns, however, suggest that both the pereon and pleon belong to the thoracic tagma (Abzhanov and Kaufman 2004, Schram and Koenemann 2004a), essentially corroborating the "thorax I" and "thorax II" of Walassek and Müller (1997).

CRUSTACEAN ANCESTORS

Widely recognized groups within the Crustacea are Malacostraca, Maxillopoda (usually including ostracods), Branchiopoda, Cephalocarida, and, most recently, Remipedia. Until recently, the high-level phylogeny of crustaceans generally saw Malacostraca and Maxillopoda at the "top" of the tree as sister to the Branchiopoda. The more recently discovered Remipedia and Cephalocarida have typically been seen in basal or near-basal positions (Schram and Hof 1998). Various hypotheses have been proposed. Sanders (1957), Hessler and Newman (1975), and Hessler et al. (1982), for instance, influenced by models of arthropod monophyly, favored an origin for Crustacea from a trilobite-like precursor, out of which the cephalocarid-like ancestral form could apparently be derived, i.e., "the cephalocarid theory." Schram (1982, 1983), on the other hand, influenced by hypotheses of arthropod polyphyly, favored an origin of crustaceans independent from other arthropods, i.e., the "biramous theory" (Cannon and Manton 1927). The "biramous theory" assumed a long, serially homonomous, ancestral form with biramous natatory limbs, so the discovery of remipedes seemed to nicely corroborate the theory (Schram 1983). Specific details aside, both models were influenced by the old paradigm of arthropod evolution that assumed a linear progression from simple to complex, starting with a long, serially homonomous, worm-like animal with anamorphic development, leading to a shortened form with increasingly diversified body regions (e.g., head, thorax, and abdomen; head and thorax; or head, pereon, and pleon) and limbs, and metamorphic development, i.e., simple to complex tagmatizaton. In addition, before the increasing application of cladistic analysis among carcinologists since the late 1980s, phylogenetic inference was typically focused on singular or "favorite" character suites, impeding effective testing of phylogenetic hypotheses. As a result, significant advances in crustacean phylogeny did not take place until the new methods (i.e., cladistic analysis) were adopted in the last decades of the 20th century. Currently, neither cephalocarids nor remipedes are considered "basal" crustaceans, indicating the apparently ancestral features that are better interpreted as specializations.

PANCRUSTACEA

A number of controversies have erupted over phylogenetic relationships within Crustacea, but its monophyly has historically not been doubted. A recent key development in arthropod phylogeny is the recognition that insects (Hexapoda) are nested within Crustacea. The unexpected position of hexapods among crustaceans was discovered largely through molecular phylogenetic studies, with results initially treated with suspicion but now consistently corroborated by increasing volumes of data. Pancrustacea, Tetraconata, and Crustaceomorpha have been proposed as names for the crustacean-hexapod clade. Insects can now be seen as terrestrial crustaceans, a perspective neatly "solving" the long-standing puzzle of why insects were unable to successfully diversify in the ocean and why crustaceans have had comparatively limited success on land. Evidently, crustaceans successfully invaded land *as insects* (Glenner et al. 2006). The crustacean sister group of the hexapods remains in contention, and how these results might translate into an acceptable formal classification is yet to be settled. Whatever the names, however, hexapods now seem firmly embedded within the crustaceans (Schwentner et al. 2017). The derivation of insects from Crustacea is treated in detail in Chapter 4 of this volume.

The phylogeny of Pancrustacea is a work in progress. The most comprehensive phylogenetic study to date for taxon coverage and multiple molecular and morphological data sources serves as our working hypothesis (Oakley et al. 2013; Fig. 3.1); it identifies Remipedia as sister to the hexapods. Hexapoda + Remipedia form the sister clade to Branchiopoda + Cephalocarida, altogether comprising of Allotriocarida. Allotriocarida is sister to the Multicrustacea of Regier et al. (2010), comprising the malacostracans and maxillopodans. Oligostraca, also proposed by Regier et al. (2010), is recovered as an assemblage of mostly short-bodied forms comprising former maxillopodans (Branchiura, Mystacocarida, Pentastomida, Ostracoda) as "basal" in Pancrustacea. Consensus is yet to be reached on pancrustacean relationships, but the most recent major studies (Regier et al. 2010, von Reumont et al. 2012a, Oakley et al. 2013, Rota-Stabelli 2013) recovered several common patterns: close proximity of Remipedia to Hexapoda, a close relationship between Maxillopoda and Malacostraca, and recovery of Oligostraca as usually basal in Pancrustacea. Oligostracans, rather than cephalocarids or remipedes, appear to be phenotypically nearest to the stem lineage crustaceans, consistent with assessments of the Cambrian Orsten Crustacea (Walossek and Müller 1990).

The major Crustacean groupings follow differences in tagmatization. Oligostracans have a short body with gonopores, usually on the fourth thoracomere (thoracic somite or segment), and a short abdomen (except in the vermiform mystacocarids). Although significantly divergent structurally, the maxillopodans, malacostracans, remipedes, and cephalocarids share gonopores positioned on thoracomeres 6–8. The maxillopodan body plan includes four abdominal somites, whereas malacostracans evidently lack the abdomen altogether, and instead the thorax is differentiated into a pereon and pleon (according to Hox gene expression; Schram and Koenemann 2004a). The remipedes lack trunk regionalization and have gonopores on the seventh trunk segment in females and the fourteenth in males. Cephalocarids and branchiopods both have an elongated thorax and

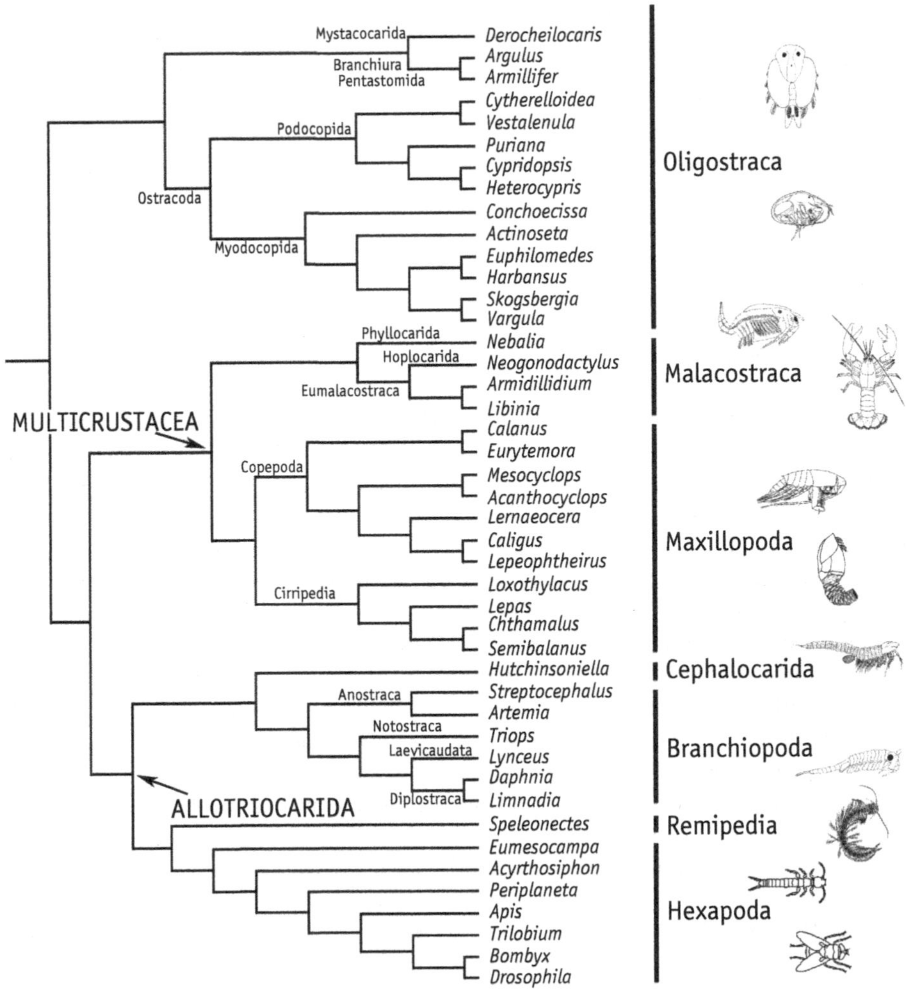

Fig. 3.1.
Phylogeny of the Pancrustacea based on combined morphological and transcriptomic data. Modified after Oakley et al. (2013), with permission from Oxford University Press.

abdomen, but have gonopores on the sixth and twelfth somites, respectively. Diversification of these major groups is further considered below.

MAJOR CRUSTACEAN CLADES

Oligostraca

Oligostraca comprises Ostracoda and the former maxillopodans, Mystacocarida, Branchiura, and Pentastomida (Regier et al. 2010, Oakley et al. 2013; Fig. 3.2). The body form of each of the component groups is very different, but all share a short thorax with the gonopores on the fourth somite where confirmed (noting that the segment homologies of the gonopore positions of ostracods remain difficult to interpret; Horne et al. 2005). The bivalved ostracods are highly speciose, with

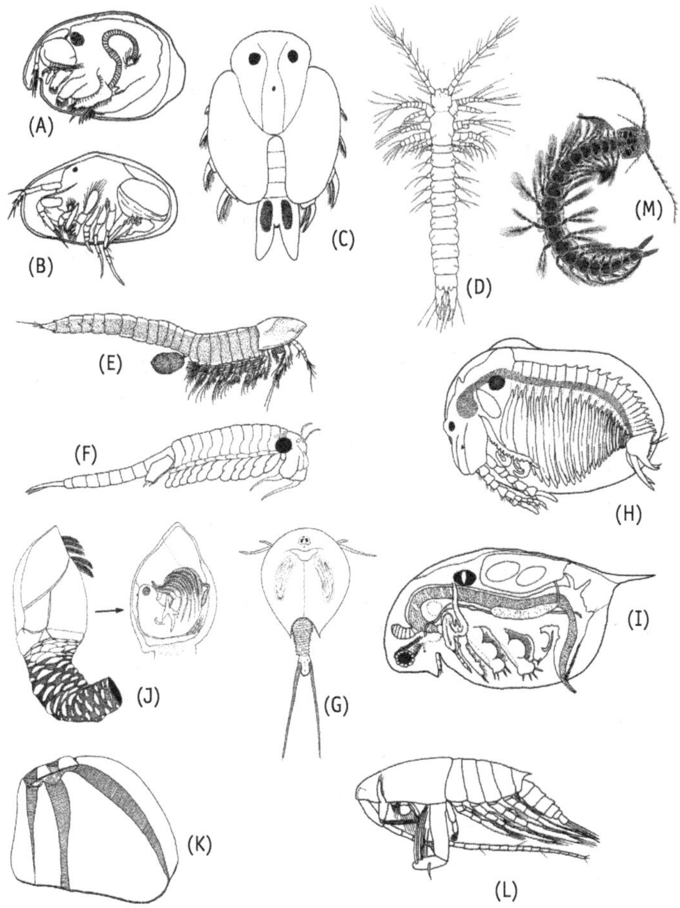

Fig. 3.2.
(A–D) Oligostraca: (A) myodocopid ostracod; (B) podocopid ostracod; (C) Branchiura; (D) Mystacocarida. (E) Cephalocarida. (F–I) Branchiopoda: (F) Anostraca; (G) Notostraca; (H) Laevicaudata (left valve removed); (I) Diplostraca (left valve removed). (J–L) Maxillopoda: (J) Thecostraca, stalked barnacle; (K) Thecostraca, sessile barnacle; (L) Copepoda, calanoid. (M) Remipedia. A–L, modified after McLaughlin (1980); E, modified from Martin et al. (2002); M, modified from Schram (1986), with permission from Oxford University Press.

some 7,600 Recent species, whereas mystacocarids with 13 species, branchiurans with 168, and pentastomidans with 130 have relatively small radiations (Ahyong et al. 2011). Mystacocarids are tiny, elongated interstitial specialists of beaches in the Atlantic basin. Branchiura and Pentastomida occur worldwide as highly derived ectoparasites of fishes and respiratory parasites of higher vertebrates, respectively. Ostracods occur globally in almost all habitats, ranging from damp soil to abyssal ocean depths, and comprise two major divergent clades, Podocopida and Myodocopida, along with a third group Paleocopida, represented by a single living species (Horne et al. 2005). Both major clades are well represented in marine habitats, but podocopids are the primary freshwater and terrestrial forms. The highest freshwater diversity is in the Palearctic region of the Northern Hemisphere. Ostracod monophyly has been questioned, but Oakley et al. (2013) comprehensively demonstrated monophyly, including reciprocal monophyly of Podocopida and Myodocopida.

Oligostraca was only recently formally recognized (Regier et al. 2010, Oakley et al. 2013), but some of the phylogenetic connections have been on view for two decades through sperm ultrastructure, somatic morphology, and molecular data (Zrzavy et al. 1998, Schram and Koenemann 2004b). Removing the oligostracans from Maxillopoda means that maxillopodans can now be better understood as a natural group, rather than as something of a "wastebasket" taxon. Perhaps of greater importance, however, is the phylogenetic position of Oligostraca as sister to the remaining Pancrustacea. A recurring theme in studies of the Orsten stem- and proto-crustaceans is the mismatch between the short-bodied forms actually found as fossils and the serially homonomous concept assumed to represent the primitive crustacean, whether in having the form of cephalocarids, branchiopods, or remipedes. Thus, recovery of oligostracans as basal Crustacea parsimoniously accords with the early crustacean fossil record.

The ostracod fossil record was once thought to extend to the Early Cambrian in the Phosphatocopina and Bradoriida, but these ostracod-like groups have since been shown to lie on the pancrustacean stem-line (Maas and Waloszek 2005, Hou et al. 2010). At present, the earliest ostracods, based on fossil mandibles closest to those of myodocopids, are known from the Late Cambrian of Canada (Harvey et al. 2012). Nevertheless, the ostracod fossil record is strongest in the Ordovician (~460 Ma), suffering several major declines toward the end of the Paleozoic (~250 Ma) followed by recovery in the Mesozoic where they have remained well represented in the fossil record ever since (Sepkoski 2000; Fig. 3.3). Of the remaining oligostracans, the mystacocarids and branchiurans lack a fossil record, but pentastomidans are recorded from the Cambrian-Ordovician Orsten faunas (Walossek and Müller 1994). Notably, fossil pentastomidans have small trunk limbs but lack an oral proboscis. Walossek and Müller (1994) interpreted these fossils as parasites, although they could equally represent free-living infaunal forms (Schram 2013).

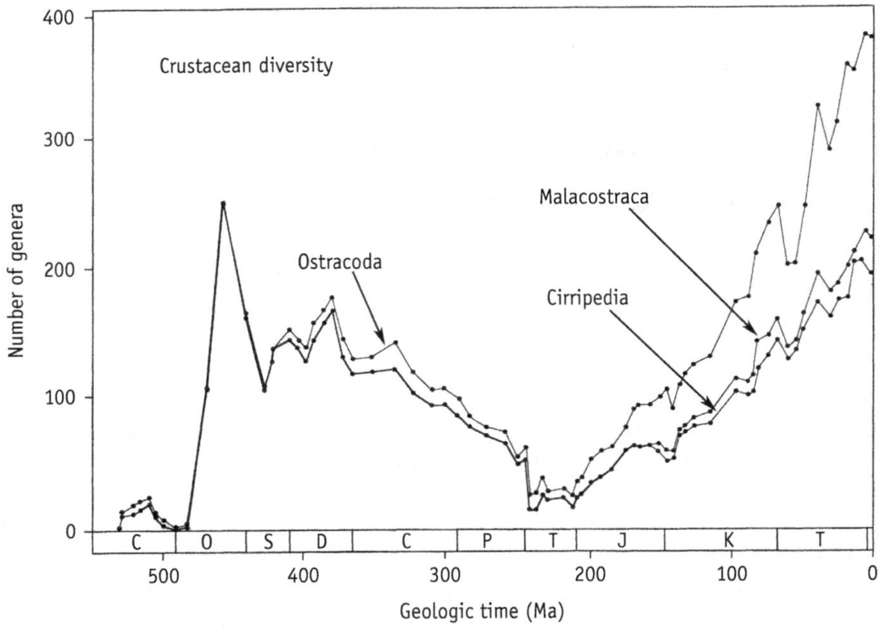

Fig. 3.3.
Diversity through the Phanerozoic of three major marine crustacean groups, Ostracoda, Cirripedia, and Malacostraca. Modified after Sepkoski (2000).

Cephalocarida

Cephalocarids occur worldwide, placed in a single family of 5 genera and 13 described species. Only a few millimeters in length, they are seldom encountered but occur in benthic sediments from shallow subtidal depths to more than 1,200 m (Hessler and Elofsson 2013).

Cephalocarids are morphologically uncomplicated (Fig. 3.2). Tagmata consist of a broad, shield-like cephalon, followed by a thorax of eight segments and an abdomen of twelve. The last abdominal segment bears paired caudal rami. Cephalocarids undergo anamorphic development and are hermaphroditic; male and female gonopores are located on the sixth thoracic segment. There are five pairs of cephalic appendages (antennule, antenna, maxillule, maxilla, and mandible) and seven or eight pairs of similar thoracic appendages. Notably, the maxillule is highly reduced and the maxilla has the same form as the thoracic appendages. They function together in metachronal pattern used in feeding and locomotion and resemble the morphology in some Cambrian Orsten groups (Walossek and Müller 1990). The antennae of adult cephalocarids are in front of the mouth as in larvae but unlike those of other adult Crustacea.

Following their discovery, cephalocarids were ascribed a central position in crustacean evolution (Sanders 1957). The small-size, larva-like features such as the placement of the antennules, unspecialized maxillae and thoracic limbs, anamorphic development, and elongated, simplified body plan of cephalocarids correspond to what might have been expected in the crustacean ancestor. Taken together, these features of cephalocarids have been interpreted as ancestral. Of course, other interpretations are possible. The "simplicity" of cephalocarids, rather than being ancestral, might be the result of progenetic pedomorphosis in which the developmental sequence is truncated by the early onset of sexual maturity (Gould 1977). Some of the so-called ancestral features might instead represent specializations (Schram 1982), a perspective corroborated by phylogenetic analysis (Oakley et al. 2013). Interestingly, although with little traction at the time, several earlier workers recognized a cephalocarid-branchiopod grouping (e.g., Dahl 1956, Mckenzie 1991), as did Schram and Hof (1998).

Despite their prominence in discussions of early crustacean evolution, and consequent assumption of great antiquity, cephalocarids presently have no confirmed fossil record. The Devonian *Lepidocaris rhyniensis,* suspected as a possible cephalocarid (Schram 1986), has since been demonstrated to have branchiopod affinities (Olesen 2009). The Cambrian *Rehbachiella kinnekullensis,* widely accepted as stem-lineage Branchiopoda (Sun et al. 2016), and Cambrian *Yicaris* were suggested as possible stem cephalocarids based on semaphoront analyses (Wolfe and Hegna 2014). The most recent molecular divergence estimate suggests a Late Cambrian to Early Ordovican age for the most recent common ancestor of Cephalocarida and Branchiopoda (Oakley et al. 2013). Cephalocarids lack pelagic dispersal stages, and given their antiquity, benthic habit, and wide geographic range of most genera, the present distributions probably owe much to plate tectonics (Hessler and Eloffson 2013).

Branchiopoda

Branchiopods occur worldwide, primarily in ephemeral surface waters; some cladocerans are marine. The approximately 1,650 Recent species are arranged in four orders: Anostraca (fairy shrimp), Notostraca (tadpole shrimps), Laevicaudata (smooth clam shrimps), and Diplostraca (water fleas, spiny clam shrimps, tropical clam shrimps; Cáceres and Rogers 2015). The orders Lipostraca and Kazacharthra are known only as fossils (Olesen 2009).

Branchiopods have diverse body forms (Fig. 3.2), ranging from the long-tailed notostracans with, and anostracans without, a carapace to laevicaudatans and diplostracans with bivalved carapaces that enclose the body. Immediate outward differences aside, however, branchiopods typically share a series of serially uniform, phyllopodous trunk limbs, reduced or absent maxillae, gonopores of both sexes usually located on the 12th thoracic segment, and a limbless abdomen (if present). Cladocerans (Diplostraca) have extreme body reduction, but other branchiopods may have almost 40 postcephalic segments.

Prior to the discovery of cephalocarids and remipedes, branchiopods were frequently regarded as best representing the ancestral crustacean (e.g., Tiegs and Manton 1958). The minimal tagmatization with multiramous, minimally diversified mouthparts and largely unsegmented phyllopodous limbs were the closest known approximation of the ancestor under the "biramous theory." Rather than being primitive, however, the unsegmented phyllopods of branchiopods are uniquely specialized for their mode of life and better interpreted as apomorphic (Schram and Hof 1998). Walossek and Müller (1998) regarded the endite arrangement of the maxillule and thoracic limbs of cephalocarids as a viable "precursor state" to the branchiopod apparatus. Not surprisingly, cephalocarids and branchiopods prove to be closely related.

Branchiopod phylogeny has been extensively studied (Walossek and Müller 1998, Stenderup et al. 2006, Olesen 2009, Schwentner et al. 2017). Although consensus is yet to be reached, the stem branchiopod probably had a subcylindrical carapace and natatory antennae (functioning together with the natatory trunk limbs), as exemplified by the Upper Cambrian *Rehbachiella kinnekullensis*. Within the crown group, the carapace-less anostracans are widely accepted as basal, the bivalved-carapace forms high in the tree, and simple-carapaced notostracans occupying the "middle ground." The anostracans and notostracans retained a natatory function for the trunk limbs, but not for the antennae, although these appear to be independent reductions (Olesen 2009). In contrast, the laevicaudatans and diplostracans retained a natatory function for the antennae but not the trunk limbs, now enclosed within the bivalved carapace.

The branchiopod fossil record, though not voluminous, represents the six recognized orders (Sun et al. 2016). *Rehbachiella* is known from the Swedish Orsten fauna and usually regarded as stem Branchiopoda (Oakley et al. 2013, Sun et al. 2016), although Wolfe and Hegna (2014) suggested a cephalocarid affinity instead. *Rehbachiella* has been extensively studied and clearly shows anamorphic development in some 30 stages (Walossek and Müller 1998). Mandibular fragments attributed to crown-group Branchiopoda have been found in Middle–Late Cambrian marine strata of Canada (Harvey et al. 2012). *Lepidocaris rhyniensis* (Lipostraca) has been identified as sister to Anostraca and *Castracollis wilsonae* as stem-lineage notostracan; both are of Devonian age, having lived around 410 Ma; a still older abdominal fragment (Silurian; >421 Ma) has been identified as stem Notostraca (Olesen 2009, Sun et al. 2016). Laevicaudatans are recorded from the Middle Jurassic of China (~165 Ma), diplostracans (Spinicaudata) the Middle Devonian (~390 Ma) of China, and Cladocera from the Early Jurassic, and possibly the Devonian (~390 Ma) of western-central Asia (Olesen 2009, Astrop and Hegna 2015). Thus, the presence of crown-group fossils from throughout the branchiopod tree, dating from early in the evolution of Crustacea, indicates the antiquity of the major branchiopod lineages. Fossil Branchiopoda are almost all from freshwater deposits, and these are probably the principal habitats of crown-group diversification. The marine or estuarine occurrence of some cladocerans is probably secondary, rather than reflecting the original habitat. Indeed, the lipostracan *Lepidocaris* and a probable cladoceran, both Devonian, were found in hot spring habitats (Scourfield 1926, Anderson et al. 2004). Moreover, recent molecular divergence analyses estimate a Late pre-Cambrian origin for Branchiopoda (~563 Ma) overall and a Middle Cambrian to Early Ordovician (478–512 Ma) age for the crown-group (and origin of the freshwater biota) (Sun et al. 2016). Limb structure has changed little during branchiopod evolution suggesting specialization from early on (Haug et al. 2013). As is, the unique mode of limb formation may constrain development of anything other than the thin foliaceous phyllopods, and along with the weak general body sclerotization, temporary or standing surface waters may be the only suitable habitats in which the group can thrive (Schram 2013).

Remipedia

Since their first report, remipedes have consistently attracted special attention (Schram 1983). This is partly because of their potential phylogenetic significance, though even basic knowledge of their biology is still unfolding. For instance, larval stages were described as recently as 2006, and confirmation of venom production was revealed only recently (von Reumont et al. 2015). Remipedes uniquely lack trunk tagmatization and have a long, serially homonomous body (or nearly so) with biramous natatory limbs; the mouthparts are specialized for holding prey, and the maxilla is fang-like, through which digestive venom is injected (Fig. 3.2). Remipedes have even been functionally compared to centipedes (von Reumont et al. 2015). More significantly, however, the strikingly "simple" morphology of remipedes spoke directly to cherished hypotheses about the ancestral crustacean form. The remipedes seemingly confirmed the "biramous-theory" of Cannon and Manton (1927) and were immediately identified in some quarters as "primitive," if not basal, crustaceans (e.g., Schram 1983, Wills 1997, Schram and Hof 1998). Current analyses, however, indicate a higher rather than basal position in the Crustacea. This higher position indicates that, like the cephalocarids, the assumed plesiomorphies of remipedes, according to previous concepts of the ur-crustacean, actually represent specializations—they are highly derived, and their uniqueness within the Crustacea suggests as much. Moreover, major recent studies identified remipedes as sister to the hexapods (von Reumont et al. 2012, Oakley et al. 2013). The position of Hexapoda within Crustacea is controversial, but a remipede–hexapod sister relationship is morphologically plausible given similarities in brain structure and neuroanatomy, haemocynanin chemistry, and aspects of somatic morphology (Koenemann and Iliffe 2013; see also Chapter 4).

To date, extant remipedes comprise 27 species in three families, placed in the order Nectiopoda. Two fossil species, both from the Upper Carboniferous of North America (~320 Ma), belong to the extinct order Enantiopoda: *Tesnusocaris goldichi* and *Cryptocaris hootchi* (Koenemann and Iliffe 2013). The modern distribution is highly disjunct and relictual, being known only from anchialine caves in the Canary Islands (one locality), Western Australia (one locality), and the Caribbean Sea (multiple localities; Koenemann and Iliffe 2013). This present-day distribution suggests a once-continuous range in the former Tethys Sea that separated Laurasia from Gondwana during the Mesozoic. With the breakup of the ancient continents, the surviving remipede populations were presumably carried to their present-day positions by plate tectonics.

Maxillopoda

Maxillopoda, typically regarded as comprising Branchiura, Copepoda, Mystacocarida, Ostracoda, Tantulocarida, and Thecostraca, has always been difficult to define (Boxshall 1983, Grygier 1983; Fig. 3.2). Indeed, these groups are not all closely related, and phylogenetic analyses have consistently revealed maxillopodan polyphyly. With formal removal of the oligostracan groups (Branchiura, Mystacocarida) and Ostracoda from Maxillopoda, the remaining taxa form a coherent group united by a common body plan typically of five cephalic, seven thoracic, and four abdominal somites, i.e., a 5-7-4 formula. This is equivalent to the 5-6-5 formula of Newman (1983) except that our knowledge of Hox gene expression in crustaceans indicates that the gonopore-bearing segment in maxillopodans, previously thought to be the first abdominal segment, is in fact the last, seventh, thoracic segment in both sexes (Schram and Koenemann 2004a).

The largest maxillopodan group, Copepoda, includes more than 16,000 species worldwide, occurring in all aquatic habitats. They are extremely diversified, and many are free-living, such as the benthic harpacticoids and typically planktonic calanoids and cyclopoids. Several groups, however, are parasitic and have varying degrees of specialization and structural reduction. For instance, poecilostomatoidans, ectoparasites of fish and invertebrates, may have a flattened, somewhat branchiuran-like habitus and highly modified appendages, whereas notodelphoidans have highly reduced limbs and appear little more than worm-like. Few copepod fossils are known, but fish-gill-parasites are recorded from the Lower Cretaceous (~120 Ma) (Cressy and Boxshall 1989), free-living forms from the Miocene (~5–20 Ma) (Palmer 1969), and mandibular fragments of a non-parasitic form were identified from the Middle-Late Cambrian of Canada (~500) (Harvey et al. 2012).

Tantulocarids, a cosmopolitan group of 36 species, are parasitic on other small marine crustaceans, including copepods, ostracods, and isopods. Tantulocarids are infective during the mobile benthic larval phase, which is presumed short, after which they remain attached to the integument of their host (Huys 1991).

The thecostracans comprise the Cirripedia (barnacles; ~1,300 species) and two smaller marine groups: the parasitic Ascothoracida (107 species) and enigmatic Facetotecta (y-larvae; 11 species). Adult facetotectans are yet to be recognized but are assumed parasitic (Hoeg et al. 2014).

Ascothoracidans are a small group of ecto- and endoparasites of echinoderms and anthozoan corals; they occur in all major oceans and are recorded from the Cretaceous as trace fossils (Schram 1982). Newman (1983) and Grygier (1983) considered ascothoracidans as potentially near the stem of the cirripedes or even the maxillopodans as a whole. In particular, the unfused head and first thoracic segment and similarity of the first thoracic limb to following pairs is considered plesiomorphic.

By far the most diverse thecostracans are the Cirripedia, which include both parasitic (Rhizocephala) and non-parasitic radiations (Acrothoracica, the burrowing barnacles; and Thoracica, the sessile barnacles). The fossil record is extensive. Acrothoracicans are known as far back as the upper Devonian and Thoracica the Upper Silurian, with balanomorphs first appearing in the Cretaceous (Schram 1982, 1986). Previous Middle Cambrian (*Priscanserimarinus*) and Carboniferous (*Pabulum*) records are now considered unreliable, the former a non-thoracican and the latter a bivalve mollusc (Buckeridge and Newman 2006). Rhizocephala, having soft externa less amenable to fossilization than "shelled" barnacles, are implicated in the feminization of a Miocene brachyuran crab (Feldmann 1998).

The shelled, sessile barnacles are widely accepted as the more derived forms, with most previous hypotheses of their evolution focusing on sequential addition of shell plates as a key indicator. Although the stem thoracican probably had only four shell plates, comprehensive molecular and morphological analyses do not find significant phylogenetic signal in tracing shell-plate numbers. Instead, soft-part morphology will probably prove a better lead. The thoracicans were initially benthic and, by the Late Triassic to Early Jurassic, had expanded into epiplanktonic habitats. This shift in mode of life was accompanied by a shift in shell chemistry from phosphatic to calcitic, which is energetically more easily deposited and thus selectively advantageous (Gale and Schweigert 2016). Molecular divergence estimates take Rhizocephala to the Late Triassic, Thoracica to the Upper Carboniferous, and balanomorphs to the Early Cretaceous (Pérez-Losada et al. 2008).

MALACOSTRACA

The malacostracans are the most diversified crustaceans (Figs. 3.4, 3.5). Like the remipedes, the malacostracan ground plan includes a complete or nearly complete set of biramous trunk limbs. Malacostracans, however, are the most highly tagmatized crustaceans. They typically have 19 or 20 body segments divided into a 5-segmented cephalon and a 14- or 15-segmented trunk. The anterior trunk has traditionally been termed thorax (eight segments), and the posterior trunk termed "abdomen" or pleon (six or seven segments). The cephalon carries the anterior sensory structures (eyes, antennules, antennae) and mouthparts, differentiated into food processing and handling structures (namely mandibles, maxillules, maxillae). The thoracic limbs have typically become modified as maxillipeds, chelipeds, and walking or swimming legs. The abdominal limbs have lost their epipods and are frequently used for swimming and brooding. Recent evidence from Hox gene expression in selected malacostracans, however, indicates that the malacostracan abdomen is not homologous with that of branchiopods and maxillopodans, for instance (Abzhanov and Kaufman 2004). Instead, the malacostracan "abdomen" is better termed pleon (Schram and Koenemann 2004a). Thus, the developmental pathways of the posterior trunk region of malacostracans differ fundamentally from those of other crustaceans and might be considered apomorphic. It should be

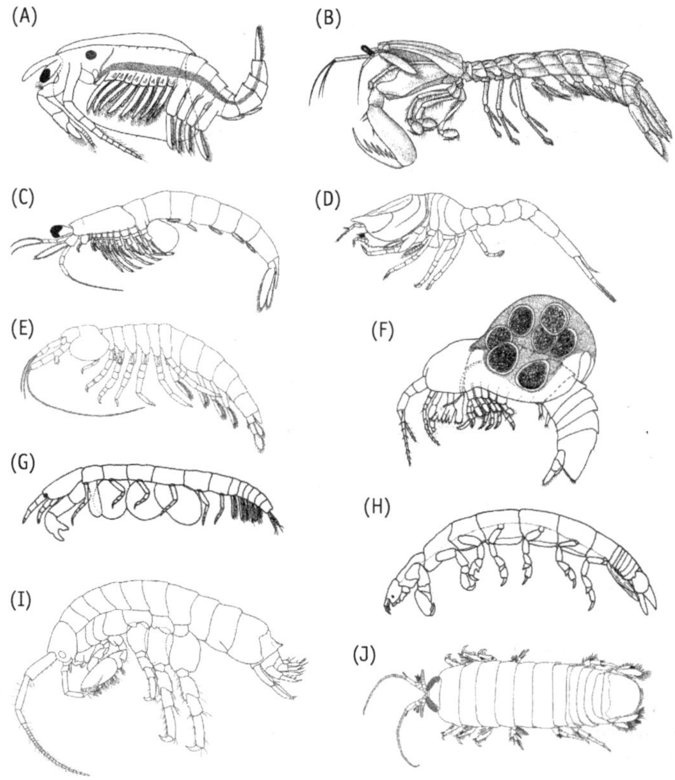

Fig. 3.4.
Malacostraca. (A) Phyllocarida, Leptostraca. (B–I) Eumalacostraca: (B) Hoplocarida, Stomatopoda; (C–I) Peracarida: (C) Mysida; (D) Cumacea; (E) Spelaeogriphacea; (F) Thermosbaenacea; (G) Tanaidacea; (H) Isopoda, Anthuridea; (I) Amphipoda, Gammaridea; (J) Isopoda, Flabellifera. A, C–H, modified after McLaughlin (1980); B, modified after Schram et al. (2013), with permission from Brill; I, modified after Kilgallen and Ahyong (2011), with permission from Magnolia Press©; J, modified after Paiva and Sousa-Filho (2015), with permission from Magnolia Press©.

noted, however, that Hox gene expression is known for only a few malacostracans, and additional groups should be analyzed.

Malacostraca are widely accepted to comprise two classes, Phyllocarida and Eumalacostraca. The Hoplocarida have variously been treated as a separate class or (more usually) included within Eumalacostraca. The malacostracan fossil record is extensive and spans the Cambrian through to the Recent.

Phyllocarida

Phyllocarids (Fig. 3.4) are considered the most "primitive" malacostracans based on retention of caudal rami and other plesiomorphies such as a simple antennal exopod, a long tubular heart that extends into the pleon. Distinctive features of phyllocarids among malacostracans are the bivalved carapace that encloses the foliaceous feeding limbs, an articulated rostrum (if present), and most significantly, a seven- rather than six-segmented pleon in addition to the telson usually with caudal ramae. The phyllocarids are a small group today; all 44 extant species are small (<10 mm long) and belong to the order Leptostraca (Permian to Recent). The phyllocarids, however, were much more diverse in the

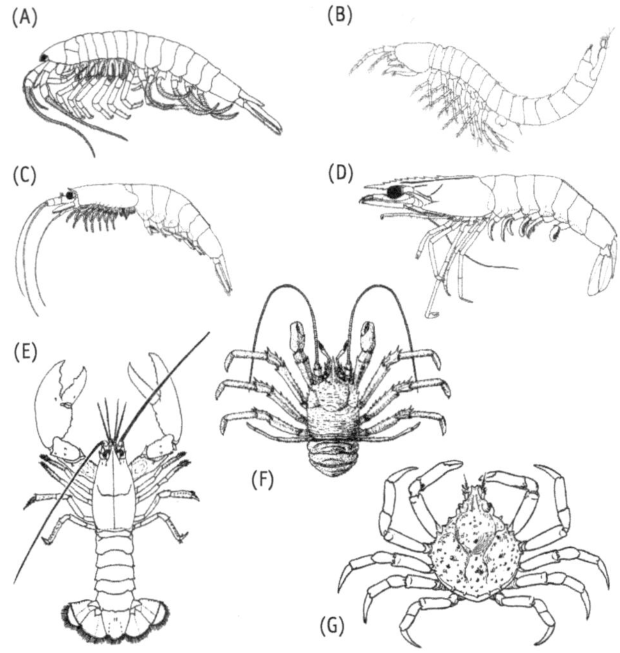

Fig. 3.5.
Malacostraca, Eumalacostraca. (A–B) Syncarida: (A) Anaspidacea; (B) Bathynellacea. (C–G) Eucarida: (C) Euphausiacea; (D–G) Decapoda: (D) Dendrobranchiata; (E) Astacidea; (F) Anomura; (G) Brachyura. A–D, modified after McLaughlin (1980); E, G, courtesy of the Smithsonian Libraries; F, modified after Tsang et al. (2011), with permission from Oxford University Press.

past. The leptostracan fossil record is limited to a single Permian species (*Rhabdouraea bentzi*), but the wholly extinct Archaeostraca has an extensive Mid-Cambrian to Late Triassic record (Collette and Hagadorn 2010). Although only one species of archaeostracan is known from Cambrian strata, significant diversification occurred in the Paleozoic through to the Silurian in a major lineage, Ceratiocaridina. By the Devonian, ceratiocaridids had declined and were replaced by other lineages, particularly, Rhinocaridina, which diversified in the Carboniferous. By the Mesozoic, however, phyllocarids had all but disappeared from the fossil record, leaving no further evidence for their existence apart from the extant leptostracans. Collette and Hagadorn (2010) implicated the Permian mass extinctions in the decline of archaeostracans leading to their ultimate extinction by the Triassic and suggested that leptostracans, which have survived into the present without a Mesozoic or Cenozoic fossil record, may have occupied nearshore high-energy nearshore habitats unfavorable to their fossilization.

Eumalacostraca

Eumalacostraca, the "higher crustaceans," includes the largest and most familiar crustaceans: shrimps, crabs, lobsters, slaters, krill, and mantis shrimps (Figs. 3.4, 3.5). They exhibit a "higher" level of structural organization than the phyllocarids and an extremely diverse range of limb specializations. Derived features of Eumalacostraca include loss of the caudal ramae, replaced by a tailfan formed by the telson and uropods, loss of the seventh pleonite through fusion with the sixth somite, and scale-like antennal exopod. Pleopod series are usually present, and the anterior pleopods in males are frequently modified for copulation and sperm transfer. The eumalacostracans comprise the Hoplocarida, Peracarida, Syncarida, and Eucarida, but see also Schwentner et al. (2017) for alternative renditions.

After more than a century of study, however, the interrelationships among these groups remain ambiguous. Central to recent concepts of eumalacostracan evolution has been the "caridoid facies" (Calman 1909). The "caridoid facies," described by Calman, was considered to be the malacostracan archetype, in this case, a pelagic, shrimp-like form. The morphology of mysid or lophogastrid shrimp essentially embodies the characters making up the "caridoid facies." Much of the late 20th-century debate over eumalacostracan phylogeny revolved around the merits or not of the facies (e.g., Dahl 1983, Hessler 1983), and although cladistic methods were beginning to take hold (e.g., Schram 1986), little real progress was made during this period in understanding how the Eumalacostraca are structured. Hessler et al. (1982) and Richter and Scholtz (2001) usefully summarized the history of competing hypotheses. Richter and Scholtz (2001) provided a firm basis for further research into eumalacostran phylogeny, which requires significant attention at all taxonomic levels.

The hoplocarids are usually included in the Eumalacostraca. A phyllocarid rather than eumalacostracan affinity, however, was also proposed on the basis of the number of arteries in the anterior pleon that could be interpreted as indicating that anterior rather than posterior pleonites were fused, convergently resulting in the six-somite pattern of other eumalacostracans (formed by fusion of posterior somites), and on the basis of cephalic kinesis in which the segments bearing the antennules and eyes are articulated rather than fused (Siewing 1963). However, the anterior pleonal arterial patterns proved inconsistent, and embryological development indicates fusion of the posterior pleonites as in other Eumalacostraca (Hessler 1983). Although hoplocarids have a number of plesiomorphic malacostracan features, such as an elongated heart that extends into the pleon and lack of transverse pleonal muscles, they also possess eumalacostracan synapomorphies such as the well-developed scale-like antennal exopod, coiled pleonal muscles, six pleonites, and well-developed uropods forming a tailfan. This indicates they should remain in Eumalacostraca.

Phylogenetic analyses based on both morphology (Richter and Scholtz 2001) and extensive molecular data sets (Regier et al. 2010, Oakley et al. 2013) recover Hoplocarida as sister to the remaining eumalacostracans, consistent with their retention of a number of malacostracan plesiomorphies.

Hoplocarida

The Hoplocarida includes the mantis shrimps (Order Stomatopoda) (Fig. 3.4) and their extinct forebears, the Order Aeshronectida. The almost 500 species of modern Stomatopoda are well known today for their finely tuned hunting apparatus—extraordinary vision combined with the powerful and rapid raptorial strike of the enlarged second maxilliped—making them among the most efficient of invertebrate predators (Ahyong and Jarman 2009). The remarkable hunting adaptations of the modern mantis shrimp, however, are not fundamental features of the Hoplocarida as a whole. Characteristic of hoplocarids are triflagellate antennules, a two-segmented antennal exopod in which the distal article is developed as a laminar scaphocerite, an articulated rostrum, and dendrobranchiate gills on the pleopods. Body tagmata consist of a cephalothorax (cephalon fused with pereonites 1–2 in Stomatopoda), pereon, and pleon including tailfan.

The aeschronectidans were shrimp-like, with unspecialized, achelate thoracic limbs and a carapace fully covering the pereon. Aeschronectidans are known only from the Lower Carboniferous of the former super-continent Laurentia (modern-day Europe and North America) and might be regarded as a kind of short-lived "experiment." It is from within the Stomatopoda, however, that the "true" mantis shrimp evolved through shortening of the carapace and differentiation of the thoracopods into subchelate maxillipeds (five pairs) and ambulatory pereopods (three pairs), of which the second maxilliped would become massively enlarged and specialized for ballistic strikes. Stomatopoda includes the Palaeostomatopodea (Late Devonian to Late Mississipian), Archaeostomatopodea (Lower Carboniferous to Upper Pennsylvanian), and Unipeltata (Jurassic to Recent; Ahyong and Harling 2000, Haug et al. 2010, Schram et al. 2013). The former two groups of proto-mantis shrimps essentially form a "transition series" that witnessed shortening of the carapace to expose the posterior pereonites and increasing differentiation of the second maxilliped as a raptorial claw. The maxillipeds of palaestomatopodeans were subequal in size, and probably served a general grasping function. The archaeostomatopodeans saw increasing size of the second maxilliped and a concomittant reduction in size of the remaining maxillipeds. The second maxilliped reaches maximum development as a powerful raptorial claw in the Unipeltata, the group to which all modern stomatopods belong, the "true" mantis shrimps. Although well developed as raptorial appendage in stem-linage unipelatatans, molecular-clock estimates indicate that specializations for highly optimized spearing or smashing strike modes, evident in modern stomatopods, evolved in the Late Jurassic to Early Cretaceous (126–155 Ma) (Van Der Wal 2017). Hoplocarid disparity was greatest in the Paleozoic, after which increasing refinement resulted in the single, highly specialized lineage, Unipeltata, surviving today. In recognition of changes in feeding mode and specialization of the maxillipeds between stem- and crown-group hoplocaridans, Haug et al. (2012) proposed additional cephalothoracic tagma for modern forms.

Unipeltata first appeared in the Jurassic–Cretaceous stem-lineage forms, Pseudosculdidae and Sculdidae. The fossil record indicates that the major modern stomatopod superfamilies diverged by the Late Cretaceous and have remained clearly recognizable since then (Ahyong and Harling 2000). Today, stomatopods are typically shallow-water, tropical to temperate species, although several deep-water species are known (Bathysquilloidea).

The families and superfamilies are largely cosmopolitan, but the global distributions of species and genera correspond to the two major warm-water faunal regions: the Indo-West Pacific region and Atlanto-East Pacific region. Thus, modern stomatopods exhibit the classic Tethyan distribution pattern, suggesting that the group was once widespread in the former Tethys Sea (Schram et al. 2013). The most recent phylogenetic analyses of modern stomatopods are those of Ahyong and Harling (2000), based on morphology, and molecular studies by Ahyong and Jarman (2009), Porter et al. (2010) and Van Der Wal et al. (2017). Among extant stomatopods, species of *Hemisquilla* (Hemisquillidae) are morphologically closest to the stem-lineage Pseudosculdidae (Ahyong and Jarman 2009).

Syncarida

The syncarids, with more than 260 extant species, comprise the orders Anaspidacea, Bathynellacea, and the extinct Palaeocaridacea (Fig. 3.5). Tagmata are similar to other eumalacostracans with a cephalon, pereon, and elongated pleon, each having differentiated limb types. Syncarids lack a carapace. The cephalon is separate from the first pereonite in bathynellaceans and palaeocaridaceans, and the first pereopod resembles the remaining pereopods. In anaspidaceans, however, the cephalon is apomorphically fused with the first pereonite, whose appendage is modified as a maxilliped. The remaining pereopods are biramous and modified as walking legs and have an almost full complement of exopods and epipods. Bathynellaceans are structurally reduced, apparently via progenetic pedomorphosis and as a result of their interstitial habitat. The surface-dwelling anaspidaceans, however, are otherwise widely acknowledged as retaining a suite of stem eumalacostracan features, such as biflagellate antennules, stalked eyes, and a well-developed antennal scale. For this reason, and because some living anaspidaceans are almost indistinguishable from Triassic fossil forms, anaspidaceans have been considered "living fossils."

Anaspidaceans may be relatively large. Some Anaspidesidae (formerly Anaspididae; Ahyong and Alonso-Zarazaga 2017) of Tasmania reach about 60 mm in body length (Ahyong 2016) and occur in surface waters or subterranean streams. Other anaspidaceans occur in groundwaters of mainland Australia, New Zealand, and South America (Chile and eastern Argentina). Bathynellaceans are all small species, less than 5 mm in length, and confined to interstital and groundwaters worldwide except for two benthic species in Lake Baikal and several euryhaline species occurring close to the seashore (Coineau and Camacho 2013).

Syncarids were known as fossils prior to discovery of living forms, with the fossil record beginning in the Paleozoic (Schram 1986). The palaeocaridaceans, known from the Permian-Carboniferous of the Northern Hemisphere, had a generalized shrimp-like morphology and were probably surface dwellers. The Anaspidacea have a disjunct Gondwanan distribution in the Southern Hemisphere. Fossil anaspidaceans are known only from the Triassic and Cretaceous of eastern Australia (*Anaspidites antiquus, Koonaspides indistinctus*), and living forms are now restricted to southeastern Australia, New Zealand, and South America. Bathynellacea presently lack a fossil record. However, on the basis of their worldwide distribution currently restricted to groundwaters, especially of the Northern Hemisphere, bathynellaceans are thought to have originated in ancient Laurentia during the Late Paleozoic (~350 Ma) and then dispersed north and south with the formation of Pangea (Schram 1986). The combined palaeontological record and Recent distributions of syncarids suggests they were widespread over Laurentia in the Carboniferous (300–360 Ma) and dispersed over Pangea in the Permian (250–300 Ma). Anaspidacea evolved separately on Gondwana, which separated from the remainder of Pangea in the Triassic (~200 Ma) (Coineau and Camacho 2013).

Peracarida

The 15,000 species of Peracarida are arrayed in the major extant orders Amphipoda, Isopoda, Tanaidacea, Cumacea, and Mysida, along with the minor groups, Lophogastrida, Spelaeogriphacea, Thermosbaenacea, Mictacea, and Bochusacea (Fig. 3.4). The major groups are hyperdiverse, especially Amphipoda and Isopoda, and are abundant in most marine habitats. Lophogastrids are pelagic, and the remaining groups are specialists of anchialine caves, interstitial habitats, or groundwaters. Although most peracarid orders are marine, with only minor radiations into freshwater or phreatic habitats, isopods have also extensively radiated on land in the Oniscidea, and amphipods have extensive subterranean radiations in the crangonyctoids.

Peracarids are morphologically diverse, but with a relatively consistent body plan. Peracarid tagmata include a cephalothorax (with one or more thoracomeres fused with the cephalon), pereon (comprising the remaining free thoracomeres), and pleon. The pereonal appendages comprise a maxilliped, followed by ambulatory or natatory pereopods, and a series of pleonal appendages, of which the anterior three to five pairs are pleopods and the posterior-most are uropods. Chief peracaridan derivations are based on the presence of an accessory articulated process on the mandible (lacinia mobilis) and the development in females of a thoracic brood chamber (marsupium) formed by wide plates (oostegites) originating from the pereopodal coxae (except in Thermosbaenacea, which uniquely have a dorsal brood pouch derived from the carapace; Olesen et al. 2015). Note that thermosbaenaceans have in the past been separated as another superorder, Pancarida, on account of the lack of a marsupium, but the most comprehensive phylogenies consistently align them with other peracarids (Poore 2005, Spears et al. 2005, Wilson 2009). Peracarids lack a dispersive larval stage. Instead, eggs hatch as juveniles or mancas and are carried in the marsupium prior to release. Unlike other malacostracan orders, the carapace is variably present or absent; it is well developed in the natatory, shrimp-like mysidans and lophogastridans, short in the epibenthic and benthic thermosbaenaceans, mictaceans and spelaeogriphaceans, cumaceans and tanaidaceans, and absent in the typically benthic amphipods and isopods (Oleson 2013). Similarly, the eyes may be stalked, sessile, or absent. Variability in what are usually considered fundamental features, such as carapace development, has raised questions over monophyly of peracarids more so than for other high-ranked malacostracan taxa (Watling 1983, Poore 2005).

Peracaridan monophyly is presently widely held, mostly on the basis of external morphology and internal ultrastructure (Richter and Scholtz 2001, Poore 2005, Wirkner and Richter 2010). Most molecular studies, however, are ambiguous, especially in the position of mysids and lophogastrids, and consequently also ambivalent on relationships among numbers of higher eumalacostracan taxa (e.g., Spears et al. 2005, Meland and Willassen 2007, Jenner et al. 2009). Additional molecular markers are required, given the limited resolution possible from the suite currently employed (Jenner et al. 2009, Wilson 2009). Not surprisingly, the interrelationships of the major peracarid groups remain far from resolved; the chief points of conflict are encompassed in how Amphipoda and Isopoda are related (or not), combined with differing interpretations of carapace development in different groups. Amphipods and isopods, both lacking the carapace and with sessile eyes (when present), have frequently been considered as sister groups (the old Edriophthalma; Poore 2005). Recent studies, however, find isopods more closely related to tanaids and cumaceans (Wilson 2009, Schwentner et al. 2017), taxa united by developing through manca stages, having the dorsally folded embryo and similar development of the midgut—the "mancoids" (Hessler 1983, Richter and Scholtz 2001, Wirkner and Richter 2010).

As a group, peracarids have a long fossil record extending back to the Paleozoic. The dominant Paleozoic (Carboniferous to Permian) forms are extinct shrimp- and sometimes crab-like Pygocephalomorpha from shallow marine habitats. Their usually shrimp-like body plan links them to other "mysidacean" groups, Lophogastrida, and Mysida. Lophogastridan fossils, closely resembling modern forms, are recorded from the Middle Triassic of North America and China (Feldmann et al. 2017) and Mysida from the Jurassic Solnhofen limestones of Germany (Taylor et al. 2001). Spelaeogriphaceans lack a Cenozoic fossil record, being known only from the Lower Carboniferous of Canada, Upper Jurassic of China, and Lower Cretaceous of Spain and Russia; all are regarded as stem-lineage forms (Jaume et al. 2013, Hegna et al. 2016). Cumaceans and tanaidaceans each have sparse fossil records starting in the Lower Carboniferous (Schram 1986, Schram et al. 2003). Mictaceans, bochusaceans, and thermosbaenaceans are yet to be recorded as fossils (Poore 2015). Isopods have an extensive fossil record extending from the Middle Carboniferous to Cenozoic. The earliest known isopod fossils are marine, with freshwater fossils recorded from the Upper Carboniferous (Lins et al. 2012, Etter 2014). Notably, molecular divergence estimates implicate an Ordovician origin for the Isopoda and suggest colonization of the deep sea as far back as the Early Permian (~290 Ma). Unlike many other deep-sea marine groups, the deep-sea isopods apparently survived major extinctions caused by widespread anoxic events during the Mesozoic and Cenozoic (Lins et al. 2012).

Of the major peracarid orders, Amphipoda are highly conspicuous in completely lacking a Paleozoic or Mesozoic fossil record. A Triassic fossil recently reported as an amphipod has since been shown to represent a partial decapod tailfan (Starr et al. 2016). The oldest confirmed fossil amphipods are from the Eocene Baltic Amber (Bousfield and Poinar 1995). Nevertheless, given the phylogenetic position of Amphipoda outside of the mancoid groups, and that these groups all have Paleozoic fossil records, the amphipods can be expected to have originated in the Carboniferous or earlier.

At an ordinal level, the fossil record and modern distributions of the major peracarid groups indicate ancient origins that predate the formation of the Pangea in the Permian. Most groups are essentially cosmopolitan, although modern spelaeogriphaceans have a relictual, largely Gondwanan distribution and modern thermosbaenaceans a relictual Tethyan pattern, suggesting a former wide range in the Tethys Sea during the Cretaceous (Jaume 2008).

Eucarida

Eucarida include the orders Euphausiacea (krill) and Decapoda (crabs, shrimps, lobsters), the most familiar of all Crustacea (Fig. 3.5). The monotypic Amphionidacea were once thought to represent a third eucarid order but were recently demonstrated to represent a larval caridean decapod (De Grave et al. 2015). Eucaridans are united by a well-developed carapace, tagmata consisting of a cephalothorax (eight thoracomeres fused with head), and pleon (of seven segments including telson); the gonopores are associated with the sixth (in females) or eighth (in males) thoracomere, usually on the pereopodal coxae. The mandible lacks the lacinia mobilis, the antennules are plesiomorphically bi-flagellate, antennal scale usually well developed, and gills derived from the thoracic epipods.

The pelagic shrimp-like euphausiaceans (krill), with some 87 extant species worldwide, are common in polar waters, where they are a major food source for whales. They usually have eight pairs of similar gill-bearing thoracic appendages, of which the first pair is not modified as a maxilliped, five pairs of pleopods, and a well-developed tailfan. The gills are fully exposed and not enclosed by the carapace. Their weakly sclerotized bodies in combination with the pelagic habitat probably minimize prospects for fossilization, although they apparently first appeared as fossils in the Carboniferous (Schram 1986).

The Decapoda, with some 15,000 species and a highly diverse array of body forms, are the most highly tagmatized malacostracans. The basic decapod body plan includes a cephalothorax with well-differentiated thoracopods, of which the first three pairs are modified as maxillipeds surrounding the oral field, and the posterior five pairs are modified as ambulatory pereopods, of which one or more pairs are usually adapted as pincers. The gills are associated with the thoracic appendages, and in addition to epipodal gills (podobranchs), others may arise from the arthrodial membranes between the limb and the body (arthrobranchs) and from the body wall itself (pleurobranchs). In decapods, the carapace encloses the gills in a branchial chamber, which is actively ventilated by movement of the lamellar exopod of the maxilla. The pleon typically has a series of biramous pleopods and a tailfan formed by the biramous uropods and telson. The appendages are further modified in almost every respect among different decapod lineages.

Although decapod phylogeny has attracted considerable interest for more than a century, the past decade or so has seen a surge in phylogenetic analyses using both morphological and molecular data (e.g., Scholtz and Richter 1995, Ahyong and O'Meally 2004, Tsang et al. 2008, Timm and Bracken-Grissom 2015, Wolfe et al. 2019). Whereas the hoplocarids have become increasingly specialized over time, the decapods have become increasingly diversified. Phylogenetic consensus is yet to be reached, but consistent patterns show dendrobranchiate, caridean, and stenopodidean shrimps (formerly grouped in Natantia) forming a paraphyletic grade leading to a monophyletic Reptantia containing all other decapods. Among the reptants, the long-tailed polychelidan and achelate lobsters are "basal" or "near basal" and the short-tailed crab-like Anomura + Brachyura (forming Meiura) feature the most highly derived forms. Ambiguity remains in the "middle ground" among the clawed lobsters and freshwater crayfish (Astacidea), finger lobsters (Glypheidea), and mud shrimps and ghost shrimps (Gebiidea and Axiidea, formerly Thalassinidea but now thought not to be sister taxa). A general trend toward attaining crab-like form has been observed in Decapoda, known as "carcinization," beginning with swimming forms with a long muscular pleon and more or less similar pereopods and leading to crawling, crab-like forms with a short pleon and body and well-differentiated pereopods variously specialized for prey handling, locomotion, and other functions such as swimming or carrying objects. The reduction of the muscular pleon reflects the shift from a pelagic habitat and primarily pleopodal locomotion to benthic habitat and pereopodal locomotion (Glaessner 1969). Phylogenetic relationships within the decapods are not easy to untangle. The classic example of carcinization, derivation of king crabs (Lithodidae) from hermit crabs (Paguridae), is, however, well corroborated (e.g., Tsang et al. 2011, Bracken-Grissom et al. 2013), and among the true crabs, strong evidence points to paraphyly of the "primitive" podotreme crabs with respect to the "higher" crabs (heterotremes and thoracotremes). How the remaining crabs are related remains a vexing issue (Tsang et al. 2014).

The decapod fossil record is substantial. Molecular estimates derive crown-group Decapoda in the Ordovician (Wolfe et al. 2019) but the shrimp-like Dendrobranchiata and Stenopodidea made their first fossil appearance in the Late Devonian (Jones et al. 2014), and Caridea the Late Triassic (Garassino and Teruzzi 1993). Reptantia first appeared in the Late Devonian clawed lobster, *Palaeopalaemon newberryi*, which is also notable as the oldest known decapod and placed in its own infraorder, Palaeopalaemonida. *Palaeopalaemon newberryi* represents a "derived" lobster-like reptant and appears to be the sister group to the Polychelida (Karasawa et al. 2013, Jones et al. 2018), indicating that most major decapod lineages had already diverged in the Paleozoic. Glypheidea is first recorded from the Permian (Karasawa et al. 2013). Other lobster groups, Polychelida and

Achelata, are first recorded in the Triassic. Clawed lobsters and freshwater crayfish are first recorded in the Jurassic, as are the "thalassinidean" mud shrimps. High in the decapod tree, short-bodied crab-like anomurans date from the Triassic, but the earliest known brachyurans date from the Early Jurassic (~90 Ma) (Schweitzer and Feldmann 2010). The Lower Carboniferous genus *Imocaris*, originally believed to be a brachyuran crab, has been shown to represent a pygocephalomorph peracarid (Racheboef and Villarroel 2003). Thus, most major decapod infraorders diverged in the Paleozoic, but major radiations began in the Mesozoic in more or less successive waves with distinct faunal turnover (Schweitzer and Feldmann 2015; Fig. 3.6).

The dendrobranchiate shrimps and glypheid lobsters peaked in the Late Triassic with rapid subsequent decline. Polychelidan lobsters showed the highest diversity in the Jurassic. Clawed lobsters peaked in the Mid-Jurassic to early Cretaceous and declined into the Cenozoic. Unfortunately, the freshwater crayfish have a poor fossil record, but on the basis of the few known fossils, modern distributions, palaeohistory, and estimates of molecular sequence divergence, a Triassic origin is likely. Northern Hemisphere (Astacoidea) and Southern Hemisphere (Parastacoidea) clades subsequently radiated following the separation of Gondwana from Pangea in the Permian (Toon et al. 2010). Anomurans and ghost shrimps have a relatively stable presence since their first appearance in the Triassic and Jurassic, respectively. The brachyuran crabs appeared in three successive

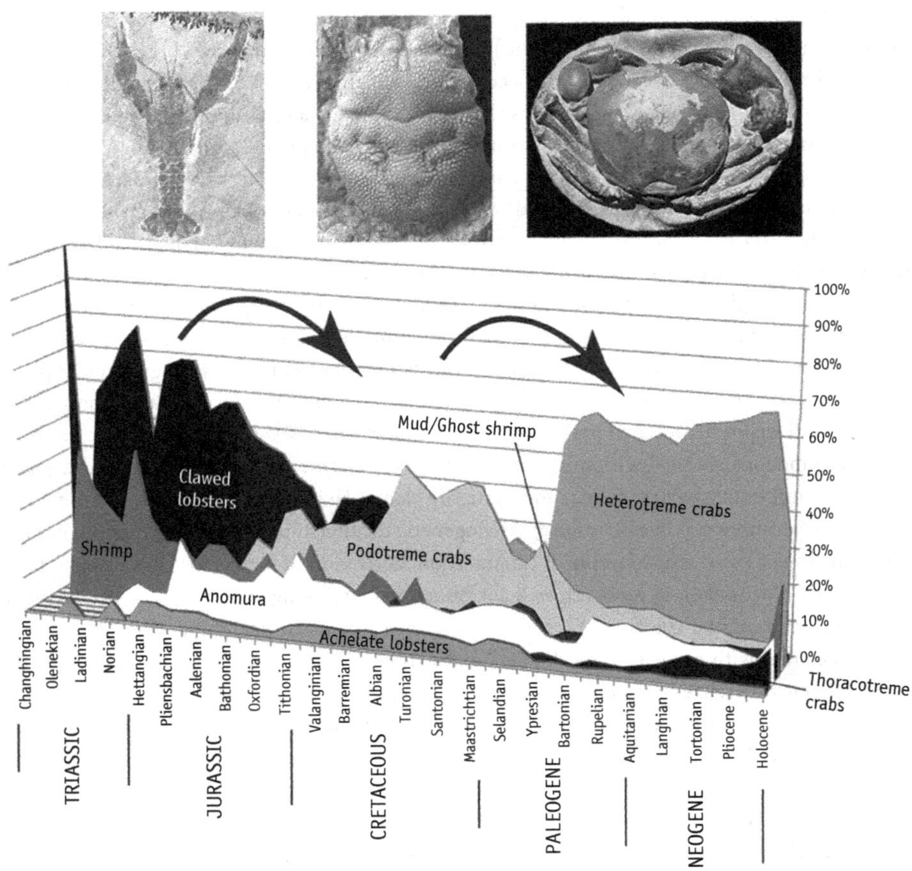

Fig. 3.6.
Proportional faunal turnover of major groups of clawed decapods throughout the Phanerozoic. Modified after Schweitzer and Feldmann (2015), with permission from Oxford University Press.

waves, with the so-called "primitive" podotremes (gonopores located on the coxae), followed by heterotremes (gonopores sternocoxal) and thoracotremes (gonopores sternal). The brachyuran fossil fauna is dominated by podotremes in the Jurassic and Cretaceous, giving way to heterotremes and thoracotremes, which radiated significantly in the Eocene and are the dominant groups today. Thus, following the presumed Devonian origin, the fossil record shows major Triassic radiations among shrimps and lobsters, moderate "activity" during the Jurassic, principally in the rise of anomurans and podotreme crabs (peaking in the Cretaceous), followed by extensive radiations of heterotreme and thoracotreme crabs in the Eocene that dominate the decapod record into the present. It is tempting to invoke a significant role for Cretaceous/Paleogene (K/P) extinctions in the tempo of decapod evolution, although it may have had less impact on decapods than for other taxon groups. Almost 80% of Late Cretaceous decapod families survived into the Paleogene, and 66% remain extant (Schweitzer and Feldmann 2005).

The oldest decapod fossils are from the Devonian of North America, suggesting a possible origin in ancient Laurentia, as is likely for other eumalacostracans (Schram 1977). Many questions remain, however, regarding decapod origins (and eumalacostracans as a whole). If the decapods and other eumalacostracans evolved in the Devonian, then diversification must have been rapid given that major infraordinal groups are present by the Late Paleozoic or soon after. Lange and Schram (1999) observed, however, that all major crustacean groups have fossil histories extending back to the Cambrian, except the Eumalacostraca, which first appear in the Devonian. Eumalacostracans are conspicuous for their absence from the fossil record for much of early crustacean history. Lange and Schram (1999) proposed a tantalizing hypothesis that decapods (and other eumalacostracans) may be much older than the fossil record shows and may have originated in deep-water environments, with subsequent movement into shallow waters only in the Late Palaezoic. This hypothesis warrants further investigation. Molecular sequence divergence estimates raise the possibility of an Ordovician or Silurian origin for Decapoda (Bracken-Grissom et al. 2014). Of possible significance is that the "basal" taxa in a number of decapod infraorders are from deep water, such as a number of lobster groups (Polychelida, Astacidea, Glypheidea) and the true crabs, Brachyura. Schweitzer and Feldmann (2015), however, suggested that groups such as clawed lobsters and podotreme crabs, although now primarily in deep water, moved there from onshore environments over evolutionary time. Much remains to be discovered about the early decapod origins.

Modern distributions of major decapod groups reflect the Mesozoic timing of major radiations, as would be expected from having formerly wide distributions in the Tethys Sea. With subsequent breakdown of the Tethys Sea and ultimate formation of the Atlantic and Pacific ocean basins during the Eocene, most warm-water marine decapod groups now have essentially discrete Indo-West Pacific and Atlanto-East Pacific distributions, with few shared species, some shared genera, and typically cosmopolitan families. The relatively few decapod species that occur in polar waters are mostly caridean shrimps. Exclusively freshwater groups exhibit different patterns as in the Aeglidae, restricted to South America, and the freshwater crayfish with independent Northern Hemisphere and Gondwanan Southern Hemisphere clades. The hyperdiverse freshwater crabs have a more recent origin, probably leaving the sea during the Cenozoic (Tsang et al. 2014).

SUMMARY AND FUTURE DIRECTIONS

The last two decades have brought significant advances in our understanding of the evolution of Crustacea, principally through the use of cladistic analysis and the introduction of large volumes of molecular sequence data. Molecular phylogenetic analyses in particular have driven several important developments. Chief among them is the derivation of insects (Hexapoda) from within the Crustacea, forming Tetraconata/Pancrustacea. Just where the Hexapoda inserts into the Crustacea is yet to be settled, but a position near the remipedes finds both molecular and morphological support. More recently, but perhaps less controversially, large-scale, primarily molecular phylogenies suggest alternatives to the long-held model of the ancestral crustacean as a long-bodied, serially homonomous form that preceded short-bodied, well-tagmatized forms. These views persisted until very recently, whether invested in a remipede- or cephalocarid-like form, according to the most popular models (Schram and Hof 1998). The most extensive current phylogenies, however, typically recover remipedes and cephalocarids as derived forms, with the short-bodied Oligostraca (containing ostracods and several minor groups) as sister to the remaining crustaceans. Thus, as originally proposed by Walossek and Müller (1990) on the basis of the Cambrian Orsten fauna, the ancestral crustacean may be a short-bodied form with relatively few body segments. Correspondingly, ostracods overwhelmingly dominate the crustacean fossil record throughout the Ordovician (Sepkoski 2000). Further large-scale analyses are required to corroborate the stability of such results, but multiple lines of evidence currently point to a rather different type of ancestor from that which was historically assumed. Historical burden, however, can be difficult to escape because the significance of the short-bodied structural plans of the Orsten crustaceans is "yet to be fully absorbed by most authorities" (Schram and Koenemann 2004b).

Crustaceans evolved in the Cambrian, the earliest derived being the Oligostraca, which dominates the Ordovician fossil record. Branchiopods are known from the Cambrian and maintain a somewhat consistent, although never expansive, presence in the fossil record. Cephalocarids appear to be closely related to the branchiopods, and together they form the sister clade to Remipedia + Hexpoda, all encompassed in the Allotriocarida. Allotriocarida is sister to Multicrustacea, the clade formed by Malacostraca and Maxillopoda. The multicrustaceans, although traced to the Cambrian, did not radiate significantly until the Mesozoic, after which they continued as major components of the crustacean fossil record. The higher malacostracans, Eumalacostraca, continued to actively radiate in the Cenozoic and are now the most ubiquitous and morphologically disparate crustaceans. The Eumalacostraca, however, are conspicuously absent from the Early Paleozoic fossil record, posing a significant palaeontological knowledge gap that warrants further investigation. Phylogenetic relationships at all taxonomic levels, particularly within the malacostracans and maxillopodans, should be priority areas for further research in concert with the palaeontological record.

Although the broad patterns of crustacean evolution may be clearer, underlying processes are much less understood. How intrinsic and extrinsic factors differentially contribute to evolutionary "success" remains speculative. Among Crustacea, progenetic pedomorphosis is recognized as a widespread phenomenon, but drivers are unknown. Schram's (1981, 1982) "stochastic mosaicism" emphasizes the strong role of chance in determining which body plans arise and which do not. Taking as an example the body-plan characters of the eumalacostracan higher-level groups and randomizing the possible permutations, a high proportion of resulting "mosaics" exist as known fossil or extant taxa. The "successful" character combinations (i.e., body plans) that go on to radiate are the result "as much of chance combination as actual selection for particular character suites" (Schram 1981). Taking a different approach, McKenzie (1991), influenced by the "new catastrophism," identified major environmental changes as primary drivers of evolution (and extinction). These changes or events interrupted periods of relative stasis and led to bursts of evolution that created ecological niches into which the new taxa expand. By correlating trends in earth history and crustacean phylogeny, McKenzie attributed the timing of major crustacean radiations to significant geological events in Earth history including the Carboniferous–Permian glaciations, the formation of the Tethys Sea in the Triassic, the formation and subsequent breakup of Pangea, the breakup of Gondwana, the formation of the major ocean basins, and other events continuing into the present. Renema et al. (2008) also identified the role of tectonic events in marine speciation. More recently, the role of ecological factors have drawn focus, particularly how herbivory promoted diversification in Crustacea (Poore et al. 2017).

Contingency and external factors undoubtedly affect the fate of lineages. Most deep lineages of the Crustacea, however, show pervasive macroevolutionary trends toward increasing complexity in the differentiation of limbs and fusion of body segments, i.e., tagmatization. Moreover, these trends appear to be driven, meaning that the formation of tagmata, and thus new body plans, is not merely a contingent outcome (Adamowicz et al. 2008). Further research is required to understand the mechanisms of tagmatization and whether lineages are intrinsically inclined as such or whether better differentiated forms are merely at a selective advantage (or both). These studies will require input from ontogeny and developmental genetics, paleontology, and phylogenetics. Either way, it seems that tagmatization remains the central organizing principle around which crustacean evolution will be understood.

ACKNOWLEDGMENTS

Thanks go to Gary Poore and Martin Thiel for inviting this contribution, and to Rod Feldmann and Carrie Schweitzer for kindly assisting with Figure 3.6.

REFERENCES

Abzhanov, A., and T.C. Kaufman. 2004. *Hox* genes and tagmatization of the higher Crustacea (Malacostraca). Pages 43–74 *in* G. Scholtz, editor. Evolutionary developmental biology of Crustacea. Crustacean Issues 15. A.A. Balkema, Lisse, The Netherlands.

Adamowicz, S.J., A. Purvis, and M.A. Wills. 2008. Increasing morphological complexity in multiple parallel lineages of the Crustacea. Proceedings of the National Academy of Sciences of the USA 105:4786–4791.

Ahyong, S.T. 2016. The Tasmanian mountain shrimp, *Anaspides* Thomson, 1894 (Crustacea: Syncarida: Anaspididae). Records of the Australian Museum 68:313–364.

Ahyong, S.T., and M.A. Alonso-Zarazaga. 2017. Anaspidesidae, a new family for syncarid crustaceans formerly placed in Anaspididae Thomson, 1893. Records of the Australian Museum 69: 257–258.

Ahyong, S.T., and C. Harling. 2000. The phylogeny of the stomatopod Crustacea. Australian Journal of Zoology 48:607–642.

Ahyong, S.T., and S.N. Jarman. 2009. Stomatopod interrelationships: preliminary results based on analysis of three molecular loci. Arthropod Systematics and Phylogeny 67:91–98.

Ahyong, S.T., and D. O'Meally. 2004. Phylogeny of the Decapoda Reptantia: resolution using three molecular loci and morphology. Raffles Bulletin of Zoology 52:673–693.

Ahyong, S.T., J.K. Lowry, M. Alonso, R.N. Bamber, G.A. Boxshall, P. Castro, S. Gerken, G.S. Karaman, J.W. Goy, D.S. Jones, K. Meland, D.C. Rogers, and J. Svavarsson. 2011. Subphylum Crustacea Brünnich, 1772. Animal biodiversity: an outline of higher-level classification and survey of taxonomic richness. Zootaxa 3148:165–191.

Anderson, L.I., W.R.B. Crighton, and H. Haas. 2004. A new univalve crustacean from Early Devonian Rhynie chert hot-spring complex. Transactions of the Royal Society of Edinburgh: Earth Sciences 94:355–369.

Appeltans, W., S.T. Ahyong, G. Anderson, M.V. Angel, T. Artois, N. Bailly, R.N. Bamber, A. Barber, I. Bartsch, A. Berta, M. Błażewicz-Paszkowycz, P. Bock, G.A. Boxshall, C.B. Boyko, S.N. Brandão, R.A. Bray, N.L. Bruce, S.D. Cairns, T.Y. Chan, L. Chan, A.G. Collins, T. Cribb, M. Curini-Galletti, F. Dahdouh-Guebas, P.J.F. Davie, M.N. Dawson, O. De Clerck, D. Wim, S. De Grave, N.J. de Voogd, D.P. Domning, C. Emig, C. Erséus, W. Eschmeyer, K. Fauchald, D.G. Fautin, S.W. Feist, C.H.J.M. Fransen, H. Furuya, O. Garcia-Alvarez, S. Gerken, D. Gibson, A. Gittenberger, S. Gofas, L. Gómez-Daglio, D.P. Gordon, M.D. Guiry, F. Hernandez, B.W. Hoeksema, R. Hopcroft, D. Jaume, P. Kirk, N. Koedam, S. Koenemann, J.B. Kolb, R.M. Kristensen, A. Kroh, G. Lambert, D.B. Lazarus, R. Lemaitre, M. Longshaw, J.K. Lowry, E. Macpherson, L.P. Madin, C. Mah, G. Mapstone, P.A. McLaughlin, J. Mees, K. Meland, C. Messing, C. Mills, T. Molodtsova, R. Mooi, B. Neuhaus, P.K.L. Ng, C. Nielsen, J. Norenburg, D.M.

Opresko, M. Osawa, G. Paulay, W. Perrin, J. Pilger, G.C.B. Poore, P. Pugh, G.B. Read, J.D. Reimer, M. Rius, R. Rocha, J. Saiz-Salinas, V. Scarabino, B. Schierwater, A. Schmidt-Rhaesa, K. Schnabel, M. Schotte, P. Schuchert, E. Schwabe, H. Segers, C. Self-Sullivan, N. Schenkar, V. Siegel, W. Sterrer, S. Stöhr, B. Swalla, M. Tasker, E.V. Thuesen, T. Timm, A. Todaro, X. Turon, S. Tyler, P. Uetz, J. van der Land, B. Vanhoorne, L. van Ofwegen, R. van Soest, J. Vanaverbeke, G.K. Walker-Smith, T.C. Walter, A. Warren, G.C. Williams, S. Wilson, and M. Costello. 2012. The magnitude of global marine species diversity. Current Biology 22:2189–2202.

Astrop, T.I., and T.A. Hegna. 2015. Phylogenetic relationships between living and fossil spinicaudatan taxa (Branchiopoda Spinicaudata): reconsidering the evidence. Journal of Crustacean Biology 35:339–354.

Bousfield, E.L., and G.O. Poinar. 1995. New terrestrial amphipod from Tertiary amber deposits of the Dominican Republic. Journal of Crustacean Biology 15:746–755.

Boxshall, G.A. 1983. A comparative functional analysis of major maxillopodan groups. Pages 121–143 *in* F.R. Schram, editor. Crustacean Phylogeny. Crustacean Issues 1. A.A. Balkema, Rotterdam, The Netherlands.

Bracken-Grissom, H.D., M.E. Cannon, P. Cabezas, R.M. Feldmann, C.E. Schweitzer, S.T. Ahyong, D.L. Felder, R. Lemaitre, and K.A. Crandall. 2013. A comprehensive and integrative reconstruction of evolutionary history for Anomura (Crustacea: Decapoda). BMC Evolutionary Biology 13:128.

Bracken-Grissom, H.D., S.T. Ahyong, R.D. Wilkinson, R.M. Feldmann, C.E. Schweitzer, J.W. Breinholt, M. Bendall, F. Palero, T.Y. Chan, D.L. Felder, R. Robles, K.H. Chu, L.M. Tsang, D. Kim, J.W. Martin, and K.A. Crandall. 2014. The emergence of the lobsters: phylogenetic relationships, morphological evolution and divergence time comparisons of an ancient group (Decapoda: Achelata, Astacidea, Glypheidea, Polychelida). Systematic Biology 63:457–479.

Buckeridge, J.S., and W.A. Newman, 2006. A revision of the Iblidae and the stalked barnacles (Crustacea: Cirripedia: Thoracica), including new ordinal, familial and generic taxa, and two new species from New Zealand and Tasmanian waters. Zootaxa 1136:1–38.

Cáceres, C.E., and D.C. Rogers. 2015. Class Branchiopoda. Pages 687–708 *in* J.H. Thorp and D.C. Rogers, editors. Thorp and Covich's Freshwater Invertebrates, Vol. 1. Ecology and General Biology, 4th ed. Academic Press, Cambridge, MA.

Calman, W.T. 1909. Crustacea *in* R. Lankester, editor. A Treatise on Zoology 7(3): i–viii, 1–346. Adam & Charles Black, London.

Cannon, H.G., and S. Manton. 1927. On the feeding mechanism of the mysid crustacean, *Hemimysis lamornae*. Transactions of the Royal Society of Edinburgh 55:219–254.

Coineau, N., and A.I. Camacho. 2013. Superorder Syncarida Packard, 1885. Pages 357–449 *in* F.R. Schram and J.C. von Vaupel Klein, editors. Treatise on Zoology—Anatomy, Taxonomy, Biology. The Crustacea. Vol. 4A. Brill, Leiden, The Netherlands.

Collette, J.H., and J.W. Hagadorn. 2010. Early evolution of phyllocarid arthropods: phylogeny and systematics of Cambrian-Devonian archaeostracans. Journal of Paleontology 84:795–820.

Cressy, R., and G. Boxshall. 1989. *Kabatarina pattersoni*, a fossil parasitic copepod (Dichelesthiidae) from a Lower Cretaceous fish. Micropaleontology 35:150–167.

Dahl, E. 1956. Some crustacean relationships. Pages 138–147 *in* K. Wingstrand, editor. Bertil Hanstrom: Zoological Papers in Honour of His Sixty-Fifth Birthday, November 20th, 1956. Lund Zoological Institute, Lund, Sweden.

Dahl, E. 1983. Malacostracan phylogeny and evolution. Pages 189–212 *in* F.R. Schram, editor. Crustacean Phylogeny. Crustacean Issues 1. A.A. Balkema, Rotterdam, The Netherlands.

De Grave, S., T.Y. Chan, K.H. Chu, C.H. Yang, and J.M. Landeira. 2015. Phylogenetics reveals the crustacean order Amphionidacea to be larval shrimps (Decapoda: Caridea). Scientific Reports 5:17464.

Etter, W. 2014. A well-preserved isopod from the Middle Jurassic of southern Germany and implications for the isopod fossil record. Palaeontology 57:931–949.

Feldmann, R.M. 1998. Parasitic castration of the crab, *Tumidocarcinus giganteus* Glaessner, from the Miocene of New Zealand: coevolution within the Crustacea. Journal of Paleontology 72:493–498.

Feldmann, R.M., C.E. Schweitzer, S. Hu, J. Huang, C. Zhou, Q. Zhang, W. Wen, T. Xie, F.R. Schram, and W.T. Jones. 2017. Earliest occurrence of lophogastrid mysidacean arthropods (Crustacea, Eucopiidae) from the Anisian Luoping Biota, Yunnan Province, China. Journal of Paleontology 9:100–115.

Gale, A., and G. Schweigert. 2016. A new phosphatic-shelled cirripede (Crustacea, Thoracica) from the Lower Jurassic (Toarcian) of Germany—the oldest epiplanktonic barnacle. Palaeontology 59:59–70.

Garassino, A., and G. Teruzzi. 1993. A new decapod crustacean assemblage from the Upper Triassic of Lombardy (N. Italy). Paleontologia Lombarda 1:1–27.

Glaessner, M.F. 1969. Decapoda. Pages 399–533 *in* R.C. Moore, editor. Arthropoda 4 Part R, Vol. 2. Treatise on invertebrate paleontology. Geological Society of America and University of Kansas Press, Lawrence.

Glenner, H., P.F. Thomsen, M.B. Hebsgaard, M.V. Sørensen, and E. Willerslev. 2006. The origin of insects. Science 314:1883–1884.

Gould, S.J. 1977. Ontogeny and phylogeny. Belknap Press, Cambridge, MA.

Grygier, M.J. 1983. Ascothoracida and the unity of Maxillopoda. Pages 73–104 *in* F.R. Schram, editor. Crustacean Phylogeny. Crustacean Issues 1. A.A. Balkema, Rotterdam, The Netherlands.

Harvey, T.H.P., M.I. Vélez, and N.J. Butterfield. 2012. Exceptionally preserved crustaceans from western Canada reveal a cryptic Cambrian radiation. Proceedings of the National Academy of Sciences of the USA 109:1589–1594.

Haug, J.T., C. Haug, A. Maas, V. Kutschera, and D. Waloszek. 2010. Evolution of mantis shrimps (Stomatopoda, Malacostraca) in the light of new Mesozoic fossils. BMC Evolutionary Biology 10:290.

Haug, C., W.S. Sallam, A. Maas, D. Waloszek, V. Kutschera, and J.T. Haug. 2012. Tagmatization in Stomatopoda—reconsidering functional units of modern-day mantis shrimps (Verunipeltata, Hoplocarida) and implications for the interpretation of fossils. Frontiers in Zoology 9:31.

Haug, J.T., A. Maas, C. Haug, and D. Waloszek. 2013. Evolution of crustacean appendages. Pages 34–73 *in* L. Watling and M. Thiel, editors. The Natural History of the Crustacea, Vol. 1: Functional Morphology and Diversity. Oxford University Press, New York.

Hegna, T.A., S.T. Ahyong, and G.D.F. Wilson. 2016. *Squilloides* Tchindonova and Tchesunov, 1993 is a spelaeogriphacean, not and isopod. Journal of Crustacean Biology 36:860–862.

Hessler, R.R. 1983. A defense of the caridoid facies; wherein the early evolution of the Eumalacostraca is discussed. Pages 145–164 *in* F.R. Schram, editor. Crustacean Phylogeny. Crustacean Issues 1. A.A. Balkema, Rotterdam, The Netherlands.

Hessler, R.R., and R. Elofsson. 2013. Class Cephalocarida Sanders, 1955. Pages 79–124 *in* F.R. Schram and J.C. von Vaupel Klein, editors. Treatise on Zoology—Anatomy, Taxonomy, Biology. The Crustacea. Vol. 4A. Brill, Leiden, The Netherlands.

Hessler, R.R., and W.A. Newman. 1975. A trilobitomorph origin for the Crustacea. Fossils and Strata 4:437–459.

Hessler, R.R., B.M. Marcotte, W.A. Newman, and R.F. Maddocks. 1982. Evolution within the Crustacea. Pages 150–239 *in* L.G. Abele, editor. The Biology of Crustacea. Vol. 1: Systematics, the Fossil Record, and Biogeography. Academic Press, New York.

Høeg, J.T., B.K.K. Chan, G.A. Kolbasov, and M.J. Grygier. 2014. Facetotecta. Pages 100–103 *in* J.W. Martin, J. Olesen, and J.T. Høeg, editors. An Atlas of Crustacean Larvae. Johns Hopkins University Press, Baltimore, MD.

Hou, X., M. Williams, D.J. Siveter, R.J. Aldridge, and R.S. Sansom. 2010. Soft-part anatomy of the Early Cambrian bivalved arthropods *Kunyangella* and *Kunmingella*: significance for the phylogenetic relationships of Bradoriida. Proceedings of the Royal Society of London B: Biological Sciences 277:1835–1841.

Horne, D.J., I. Schön, R.J. Smith, and K. Martens. 2005. What are Ostracoda? A cladistic analysis of the extant superfamilies of the subclass Myodocopa and Podocopa (Crustacea: Ostracoda). Pages 249–273 *in* S. Koenemann and R.A. Jenner, editors. Crustacea and Arthropod Relationships. Crustacean Issues 16. CRC Press, Boca Raton, FL.

Huys, R. 1991. Tantulocarida (Crustacea: Maxillopoda): A new taxon from the temporary meiobenthos. Pubblicazioni della Stazione zoologica di Napoli I: Marine Ecology 12: 1–34.

Jaume, D. 2008. Global diversity of spelaeogriphaceans and thermosbaenaceans (Crustacea; Spelaeogriphacea and Thermosbaenacea) in freshwater. Hydrobiologia 595:219–224.

Jaume, D., E.V.A. Pinardo-Moya, and G.A. Boxshall. 2013. A presumed spelaeogriphacean crustacean from an upper Barremian wetland (Las Hoyas; Lower Cretaceous; Central Spain). Palaeontology 56:15–28.

Jenner, R.A., C.N. Dhubhghaill, M.P. Ferla, and M.A. Wills. 2009. Eumalacostracan phylogeny and total evidence: limitations of the usual suspects. BMC Evolutionary Biology 9:1–20.

Jones, W.T., R.M. Feldmann, C.E. Schweitzer, F.R. Schram, R.A. Behr, and K.L. Hand. 2014. The first Paleozoic stenopodidean from the Huntley Mountain Formation (Devonian–Carboniferous), north-central Pennsylvania. Journal of Paleontology 88:1251–1256.

Jones, W.T., R.M. Feldmann, J.T. Hannibal, C.E. Schweitzer, M.C. Garland, E.P. Maguire, and J.N. Tashman. 2018. Morphology and paleoecology of the oldest lobster-like decapod, *Palaeopalaemon newberryi* Whitfield, 1880 (Decapoda: Malacostraca). Journal of Crustacean Biology 38:302–314.

Karasawa, H., C.E. Schweitzer, and R.M. Feldmann. 2013. Phylogeny and systematics of extant and extinct lobsters. Journal of Crustacean Biology 33:78–123.

Kilgallen, N.M., and S.T. Ahyong. 2011. The genus *Mallacoota* (Crustacea, Amphipoda, Maeridae) in New Zealand. Zootaxa 2929:22–36.

Koenemann, S., and T.M. Iliffe. 2013. Class Remipedia Yager, 1981. Pages 125–177 *in* F.R. Schram and J. C. von Vaupel Klein, editors. Treatise on Zoology—Anatomy, Taxonomy, Biology. The Crustacea. Vol. 4A. Brill, Leiden, The Netherlands.

Lange, S., and F.R. Schram. 1999. Evolución y filogenia de los Crustáceos. Boletin de la Sociedad Entomológica Aragonesa 26:235–254.

Lins, L.S.F., S.Y.W. Ho, G.D.F. Wilson, and N. Lo. 2012. Evidence for Permo-Triassic colonization of the deep sea by isopods. Biology Letters 8:979–982.

Maas, A., and D. Waloszek. 2005. Phosphatocopina—ostracode-like sister group of Eucrustacea. Hydrobiologia 538:139–152.

Martin, J.W., D.B. Cadien, and T.L. Zimmerman. 2002. First record and habitat notes for the genus *Lightiella* (Crustacea, Cephalocarida, Hutchinsoniellidae) from the British Virgin Islands. Gulf and Caribbean Research 14:75–79.

McKenzie, K.G. 1991. Crustacean evolutionary events: sequences and consequences. Memoirs of the Queensland Museum 31:19–38.

McLaughlin, P.A. 1980. Comparative Morphology of Recent Crustacea. W.H. Freeman, San Francisco.

Meland, K., and E. Willassen. 2007. The disunity of "Mysidacea" (Crustacea). Molecular Phylogenetics and Evolution 44:1083–1104.

Newman, W.A. 1983. Origin of the Maxillopoda; Urmalacostracan ontogeny and progenesis. Pages 105–119 *in* F.R. Schram, editor. Crustacean Phylogeny. Crustacean Issues 1. A.A. Balkema, Rotterdam, The Netherlands.

Oakley, T.H., J.M. Wolfe, A.R. Lindgren, and A.K. Zaharoff. 2013. Phylotranscriptomics to bring the understudied into the fold: monophyletic Ostracoda, fossil placement, and pancrustacean phylogeny. Molecular Biology and Evolution 30:215–233.

Olesen, J. 2009. Phylogeny of Branchiopoda (Crustacea)—character evolution and contribution of uniquely preserved fossils. Arthropod Systematics and Phylogeny 67:3–39.

Olesen, J. 2013. The crustacean carapace morphology, function, development, and phylogenetic history. Pages 103–139 *in* L. Watling and M. Thiel, editors. The Natural History of the Crustacea, Vol. 1: Functional Morphology and Diversity. Oxford University Press, New York.

Olesen, J., T. Boesgaard, and T.M. Iliffe. 2015. The unique dorsal brood pouch of Thermosbaenacea (Crustacea, Malacostraca) and description of an advanced developmental stage of *Tulumella unidens* from the Yucatan Peninsula (Mexico), with a discussion of mouth part homologies to other Malacostraca. PLoS ONE 10:e0122463.

Paiva, R.J.C., and J.F. Souza-Filho. 2015. A new species of *Dolicholana* Bruce, 1986 (Isopoda, Cymothoidea, Cirolanidae), the first record of the genus from the Atlantic Ocean. Zootaxa 4039:276–288.

Palmer, A.R. 1969. Copepoda. Pages R200–R203 *in* R.C. Moore, editor. Arthropoda 4, Part R, Vol. 1. Treatise On Invertebrate Paleontology. Geological Society of America and University of Kansas Press, Lawrence, Kansas.

Pérez-Losada, M., M. Harp, J.T. Høeg, Y. Achituv, D. Jones, H. Watanabe, and K.A. Crandall. 2008. The tempo and mode of barnacle evolution. Molecular Phylogenetics and Evolution 46:328–346.

Poore, A.G.B., S.T. Ahyong, J.K. Lowry, and E.E. Sotka. 2017. Plant feeding promotes diversification in the Crustacea. Proceedings of the National Academy of Sciences 114:8829–8834.

Poore, G.C.B. 2005. Peracarida: monophyly, relationships and evolutionary success. Nauplius 13:1–27.
Poore, G.C.B. 2015. Orders Bochusacea, Mictacea and Spelaeogriphacea. Pages 77–92 *in* F.R. Schram and J. C. von Vaupel Klein, editors. Treatise on Zoology—Anatomy, Taxonomy, Biology. The Crustacea, Vol. 5. Brill, Leiden, The Netherlands.
Porter, M., Y. Zhang, S. Desai, R.L. Caldwell, and T.W. Cronin. 2010. Evolution of anatomical and physiological specialization in the compound eyes of stomatopod crustaceans. Journal of Experimental Biology 213:3473–3486.
Racheboef, P.R., and C. Villarroel. 2003. *Imocaris colombiensis* from the Pennsylvanian of Colombia. Neues Jahrbuch für Geologie und Paläontologie Mitteilungen 2003:577–590.
Rathbun, M.J. 1925. The Spiders Crabs of America. Bulletin—United States National Museum (no. 129 1925). Smithsonian Institution Press, Washington, DC. https://library.si.edu/digital-library/book/bulletinunitedst1291925unit.
Regier, J.C., J.W. Schultz, A. Zwick, A. Hussey, B. Ball, R. Wetzer, J.W. Martin, and C.W. Cunningham. 2010. Arthropod relationships revealed by phylogenomic analysis of nuclear protein-coding sequences. Nature 463:1079–1083.
Renema, W., D.R. Bellwood, J.C. Braga, K. Bromfield, R. Hall, K.G. Johnson, P. Lunt, C.P. Meyer, L.B. McGonagle, R.J. Morley, A. O'Dea, J.A. Todd, F.P. Wesselingh, M.E.J. Wilson, and J.M. Pandolfi. 2008. Hopping hotspots: global shifts in marine biodiversity. Science 321:654–657.
Richter, S., and G. Scholtz. 2001. Phylogenetic analysis of the Malacostraca (Crustacea). Journal of Zoological Systematics and Evolutionary Research 39:113–136.
Rota-Stabelli, O., N. Lartillot, H. Philippe, and D. Pisani. 2013. Serine codon-usage bias in deep phylogenomics: pancrustacean relationships as a case study. Systematic Biology 62:121–133.
Sanders, H.L. 1957. The Cephalocarida and crustacean phylogeny. Systematic Zoology 6:112–129.
Scholtz, G., and S. Richter. 1995. Phylogenetic systematics of the reptantian Decapoda (Crustacea, Malacostraca). Zoological Journal of the Linnean Society 113:289–328.
Schram, F.R. 1977. Paleozoogeography of Late Paleozoic and Triassic Malacostraca. Systematic Zoology 26:367–379.
Schram, F.R. 1981. On the classification of Eumalacostraca. Journal of Crustacean Biology 1:1–10.
Schram, F.R. 1982. The fossil record and the evolution of Crustacea. Pages 93–147 *in* L.G. Abele, editor. The Biology of Crustacea, Vol. 1: Systematics, the Fossil Record, and Biogeography. Academic Press, New York.
Schram, F.R. 1983. Remipedia and crustacean phylogeny. Pages 23–28 *in* F.R. Schram, editor. Crustacean Phylogeny. Crustacean Issues 1. A.A. Balkema, Rotterdam, The Netherlands.
Schram, F.R. 1986. Crustacea. Oxford University Press, New York.
Schram, F.R. 2013. Comments on crustacean biodiversity and disparity of body plans. Pages 1–33 *in* L. Watling and M. Thiel, editors. The Natural History of the Crustacea, Vol. 1: Functional Morphology and Diversity. Oxford University Press, New York.
Schram, F.R., and C.H.J. Hof. 1998. Fossils and the interrelationships of major crustacean groups. Pages 233–302 *in* G.D. Edgecombe, editor. Arthropod Fossils and Phylogeny. Columbia University Press, New York.
Schram, F.R., and S. Koenemann. 2004a. Developmental genetics and arthropod evolution: on body regions of Crustacea. Pages 75–92 *in* G. Scholtz, editor. Evolutionary Developmental Biology of Crustacea. Crustacean Issues 15. A.A. Balkema, Lisse, The Netherlands.
Schram, F R., and S. Koenemann. 2004b. Are the crustaceans monophyletic? Pages 319–329 *in* J. Cracraft and M.J. Donoghue, editors. Assembling the Tree of Life. Oxford University Press, New York.
Schram, F.R., C.H.J. Hof, R.H. Mapes, and P. Snowdon. 2003. Paleozoic cumaceans (Crustacea, Malacostraca, Peracarida) from North America. Contributions to Zoology 72:1–16.
Schram, F.R., S.T. Ahyong, S.N. Patek, P.A. Green, M.V. Rosario, M.J. Bok, T.W. Cronin, K.S. Mead Vetter, R.L. Caldwell, G. Scholtz, K.D. Feller, and P. Abelló. 2013. Subclass Hoplocarida Calman, 1904: Order Stomatopoda Latreille, 1817. Pages 197–355 *in* F.R. Schram and J. C. von Vaupel Klein, editors. Treatise on Zoology—Anatomy, Taxonomy, Biology. The Crustacea, Vol. 4A. Brill, Leiden, The Netherlands.
Schweitzer, C.E., and R.M. Feldmann. 2005. Decapod crustaceans, the K/P event, and paleocene recovery. Pages 17–53 *in* S. Koenemann and R.A. Jenner, editors. Crustacea and Arthropod Relationships. Crustacean Issues 16. CRC Press, Boca Raton, FL.

Schweitzer, C.E., and R.M. Feldmann. 2010. The oldest Brachyura (Decapoda: Homolodromioidea: Glaessneropsoidea) known to date (Jurassic). Journal of Crustacean Biology 30:251–256.

Schweitzer, C.E., and R.M. Feldmann. 2015. Faunal turnover and niche stability in marine Decapoda in the Phanerozoic. Journal of Crustacean Biology 35:633–649.

Schwentner, M., S. Richter, D.C. Rogers, and G. Giribet. 2018. Tetraconatan phylogeny with special focus on Malacostraca and Branchiopoda: highlighting the strength of taxon-specific matrices in phylogenomics. Proceedings of the Royal Society B: Biological Sciences 285: 20181524.

Schwentner, M., D. J. Combosch, J. Pakes Nelson, and G. Giribet. 2017. A phylogenomic solution to the origin of insects by resolving crustacean-hexapod relationships. Current Biology 27:1818–1824.

Scourfield, D.J. 1926. On a new type of crustacean from the Old Red Sandstone (Rhynie Chert Bed, Aberdeenshire)—*Lepidocaris rhyniensis*. Philosophical Transactions of the Royal Society of London B: Biological Sciences 214:153–187.

Sepkoski, J.J., Jr. 2000. Crustacean biodiversity through the marine fossil record. Contributions to Zoology 69:213–222.

Siewing, R. 1963. Studies in malacostracan morphology: results and problems. Pages 85–103 *in* H.B. Whittington and W.D.I. Rolfe, editors. Phylogeny and Evolution of Crustacea. Museum of Comparative Zoology Special Publication, Cambridge, MA.

Spears, T., R.W. DeBry, L.G. Abele, and K. Chodyla. 2005. Peracarid monophyly and interordinal phylogeny inferred from nuclear small-subunit ribosomal DNA sequences (Crustacea: Malacostraca: Peracarida). Proceedings of the Biological Society of Washington 118:117–157.

Starr, H., T.A. Hegna, and M.A.S. McMenamin. 2016. Epilogue to the tale of the Triassic amphipod: *Rosagammarus* McMenamin, Zapata and Hussey, 2013 is a decapod tail (Luning Formation, Nevada). Journal of Crustacean Biology 36:525–529.

Stenderup, J.T., J. Oleson, and H. Glenner. 2006. Molecular phylogeny of the Branchiopoda (Crustacea)—multiple approaches suggest a "diplostracan" ancestry of the Notostraca. Molecular Phylogenetics and Evolution 41:182–194.

Sun, X.Y., X. Xia, and Q. Yang. 2016. Dating the origin of the major lineages of Branchiopoda. Palaeoworld 25:303–317.

Taylor, R.S., F.R. Schram, and Y.B. Shen. 2001. A new Upper Middle Triassic shrimp (Crustacea: Lophogastrida) from Guizhou, China, with discussion regarding other fossil "mysidaceans." Journal of Paleontology 75:310–318.

Tiegs, O.W., and S.M. Manton. 1958. The evolution of the Arthropoda. Biological Reviews of the Cambridge Philosophical Society 33:255–337.

Timm, L., and H.D. Bracken-Grissom. 2015. The forest for the trees: evaluating molecular phylogenies with an emphasis on higher-level Decapoda. Journal of Crustacean Biology 35:577–592.

Toon, A., M. Pérez-Losada, C.E. Schweitzer, R.M. Feldmann, M. Carlson, and K.A. Crandall. 2010. Gondwanan radiation of the Southern Hemisphere crayfishes (Decapoda: Parastacidae): evidence from fossils and molecules. Journal of Biogeography 37:2275–2290.

Tsang, L.M., K.Y. Ma, S.T. Ahyong, T.Y. Chan, and K.H. Chu. 2008. Phylogeny of Decapoda using two nuclear protein-coding genes: origin and evolution of the Reptantia. Molecular Phylogenetics and Evolution 48:359–368.

Tsang, L.M., T.Y. Chan, S.T. Ahyong, and K.H. Chu. 2011. Hermit to king, or hermit to all: multiple transitions to crab-like forms from hermit crab ancestors. Systematic Biology 60:1–14.

Tsang, L.M., C.D. Schubart, S.T. Ahyong, J.C.Y. Lai, E.Y.C. Au, T.Y. Chan, P.K.L. Ng, and K.H. Chu. 2014. Evolutionary history of true crabs (Crustacea: Decapoda: Brachyura) and the origins of freshwater crabs. Molecular Biology and Evolution 31:1173–1187.

van Der Wal, C., S.T. Ahyong, S.Y.W. Ho, and N. Lo. 2017. The evolutionary history of Stomatopoda (Crustacea: Malacostraca) inferred from molecular data. PeerJ 5:e3844.

von Reumont, B.M., R.A. Jenner, M.A. Wills, E. Dell'Ampio, G. Pass, I. Ebersberger, B. Meyer, S. Koenemann, T.M. Iliffe, A. Stamatakis, O. Niehuis, K. Meusemann, and B. Misof. 2012. Pancrustacean phylogeny in the light of new phylogenomic data: support for Remipedia as the possible sister group of Hexapoda. Molecular Biology and Evolution 29:1031–1045.

von Reumont, B.M., A. Blanke, S. Richter, F. Alvarez, C. Bleidorn, and R.A. Jenner. 2015. The first venomous crustacean revealed by transcriptomics and functional morphology: remipede venom glands express a unique toxin cocktail dominated by enzymes and a neurotoxin. Molecular Biology and Evolution 31:48–58.

Walossek, D., and K.J. Müller. 1990. Upper Cambrian stem-lineage crustaceans and their bearing upon the monophyletic origin of Crustacea and the position of *Agnostus*. Lethaia 23:409–427.

Walossek, D., and K.J. Müller. 1994. Pentastomid parasites from the Lower Paleozoic of Sweden. Transactions of the Royal Society of Edinburgh: Earth Sciences 85:1–37.

Walossek, D., and K.J. Müller. 1997. Cambrian "Orsten"-type arthropods and the phylogeny of Crustacea. Pages 139–153 *in* R.A. Fortey and R.H. Thomas, editors. Arthropod Relationships. Chapman and Hall, London.

Walossek, D., and K.J. Müller. 1998. Early arthropod phylogeny in the light of the Cambrian "Orsten" fossils. Pages 185–231 *in* G.D. Edgecombe, editor. Arthropod Fossils and Phylogeny. Columbia University Press, New York.

Watling, L. 1983. Peracaridan disunity and its bearing on eumalacostracan phylogeny with a redefinition of eumalacostracan superorders. Pages 213–228 *in* F.R. Schram, editor. Crustacean Phylogeny. Crustacean Issues 1. A.A. Balkema, Rotterdam, The Netherlands.

Wills, M. 1997. A phylogeny of recent and fossil Crustacea derived from morphological characters. Pages 189–209 *in* R.A. Fortey and R.H. Thomas, editors. Arthropod Relationships. Chapman and Hall, London.

Wilson, G.D.F. 2009. The phylogenetic position of the Isopoda in the Peracarida (Crustacea: Malacostraca). Arthropod Systematics and Phylogeny 67:159–198.

Wirkner, C.S., and S. Richter. 2010. Evolutionary morphology of the circulatory system in Peracarida (Malacostraca; Crustacea). Cladistics 26:143–167.

Wolfe, J.M., and T.A. Hegna. 2014. Testing the phylogenetic position of Cambrian pancrustacean larval fossils by coding ontogenetic stages. Cladistics 30:366–390.

Wolfe, J.M., J.W. Breinholt, K.A. Crandall, A.R. Lemmon, E.M. Lemmon, L.E. Timm, M.E. Siddall, and H.D. Bracken-Grissom. 2019. A phylogenomic framework, evolutionary timeline and genomic resources for comparative studies of decapod crustaceans. Proceedings of the Royal Society B: Biological Sciences 286: 20190079.

Zrzavy, J., S. Mihulka, P. Kepka, A. Bezdek, and D. Tietz. 1998. Phylogeny of the metazoa based on morphological and 18S ribosomal DNA evidence. Cladistics 14:249–285.

4

THE PANCRUSTACEAN CONUNDRUM: A CONFLICTED PHYLOGENY WITH EMPHASIS ON CRUSTACEA

Heather Bracken-Grissom and Joanna M. Wolfe

Abstract

Pancrustacea represents a diverse clade of arthropods and includes the crustaceans and insects. The morphological and ecological diversity is among the most impressive of all invertebrates, with representatives including barnacles, shrimps, crabs, copepods, ostracods, brine shrimp, dragonflies, grasshoppers, true bugs, butterflies, and beetles, among others. For decades, morphological and molecular studies have attempted to reconstruct the evolutionary relationships between the major lineages, but a consensus on pancrustacean phylogeny is still under debate. Much of the uncertainty surrounding the evolutionary history of this ancient group stems from uneven and unrepresentative sampling, conflicting data, and different analytical approaches. First, we will introduce the major lineages of Pancrustacea and discuss the evidence for monophyly within these groups. Next, we will review the pioneering morphological and molecular studies that have contributed to current understanding of higher-level (class/subclass) relationships within Pancrustacea, with an emphasis on Crustacea. Within these sections, we will start with a series of early studies and describe how the development of methods and technology allowed for the advancement of morphological and molecular phylogenetics. A summary of current knowledge of pancrustacean phylogeny will be followed by a brief discussion of new emergent methods in the field that can be applied to future phylogenetic and phylogenomic studies.

INTRODUCTION

Pancrustaceans are a morphologically diverse clade of arthropods with a rich fossil record dating back approximately 514 million years into the Cambrian (Zhang et al. 2007, Wolfe et al. 2016). With the recent inclusion of Hexapoda, the group represents a monophyletic taxon with over 1,000,000 described species, accounting for more than half of the described metazoan species (Brusca

Evolution and Biogeography. Edited by Gary C. B. Poore and Martin Thiel.

et al. 2016). These include a variety of familiar and unfamiliar forms, including the seed shrimp, fish lice, and tongue worms of Oligostraca; the crabs, lobsters, shrimps, copepods, and barnacles of Multicrustacea; and the insects, remipedes, and cephalocarids of Allotriocarida (Fig. 4.1). Morphological disparity across pancrustaceans is the most astonishing of all invertebrates, with body sizes ranging from the order of micrometers to meters. The ecological and associated physiological diversity is equally impressive, with representatives found from the open ocean to a single isolated anchialine cave, from abyssal oceanic plains to coral reefs, and from terrestrial to aquatic ecosystems. Many species are vital for ecosystem health and survival through their role in pollination, trophic food webs, and community structure, while others play a fundamental role in the economy through commercial fisheries, aquaculture, and the aquarium and terrarium trade. Many pose a threat because they represent invasive species or vectors for disease, while others are targets of significant conservation efforts as they are threatened by anthropogenic encroachment, habitat loss, and climate change. Pancrustaceans are often used as model species for studies in evolution, adaptation, biodiversity, developmental biology, physiology, and ecosystem health and resilience. Work on pancrustaceans has contributed to a critical understanding of fundamental ecological and evolutionary processes and patterns across the planet.

Due to the biological, economic, and ecological importance of Pancrustacea, the need for a resolved phylogeny is pressing. An understanding of evolutionary relationships is the foundation for future studies in ecology, biogeography, diversification, speciation, and conservation biology, among other disciplines. With the inclusion of Hexapoda, monophyly of the Pancrustacea is well established. This is supported by numerous phylogenetic analyses, including nuclear protein-coding genes (Regier et al. 2005, 2010), transcriptomes (Meusemann et al. 2010, von Reumont et al. 2012), morphology (Schram and Koenemann 2004, Strausfeld and Andrew 2011, Legg et al. 2013), and combined morphological and molecular data (Oakley et al. 2013). However, internal relationships among the 90 orders remain only partially resolved and, for the most part, unknown. Only ~45 of the 90 orders within Pancrustacea have been included in past molecular analyses (most studies contain far fewer), thus, ~50% of ordinal diversity has never been sampled. Likewise, the most comprehensive morphological phylogeny to date included 35 orders (753 characters), omitting 60% of ordinal diversity (Legg et al. 2013). Unsurprisingly, many underrepresented groups have been excluded due to the difficulties in collection methods, rarity in nature, or, in the case of molecular work, the inability to obtain high-quality molecular grade material. For these reasons, some of the most functionally diverse and taxonomically challenging groups have remained a phylogenetic mystery with respect to their position in the Tree of Life.

Even in the face of sampling challenges, substantial advancements have been made in pancrustacean phylogenetics. Herein, we focus on the higher-level relationships (class or subclass and above) within Pancrustacea, with an emphasis on the major lineages within Crustacea. First, we briefly introduce the major lineages within Crustacea and succinctly discuss the evidence for the monophyly or polyphyly of these groups. Next, we review the morphological and molecular phylogenetic studies that have contributed to our current understanding of higher-level pancrustacean relationships. We conclude with a synthesis of these findings and the future of phylogenetics and phylogenomics within the group.

INTRODUCTION TO THE MAJOR GROUPS

The pancrustacean clade is comprised of three major groups: Oligostraca (Ostracoda, Branchiura, Pentastomida, and Mystacocarida), Multicrustacea (Malacostraca, Copepoda, Thecostraca, and Tantulocarida), and Allotriocarida (Hexapoda, Branchiopoda, Cephalocarida, and Remipedia) (Oakley et. al 2013). While this clade has also been called Tetraconata (introduced by Dohle

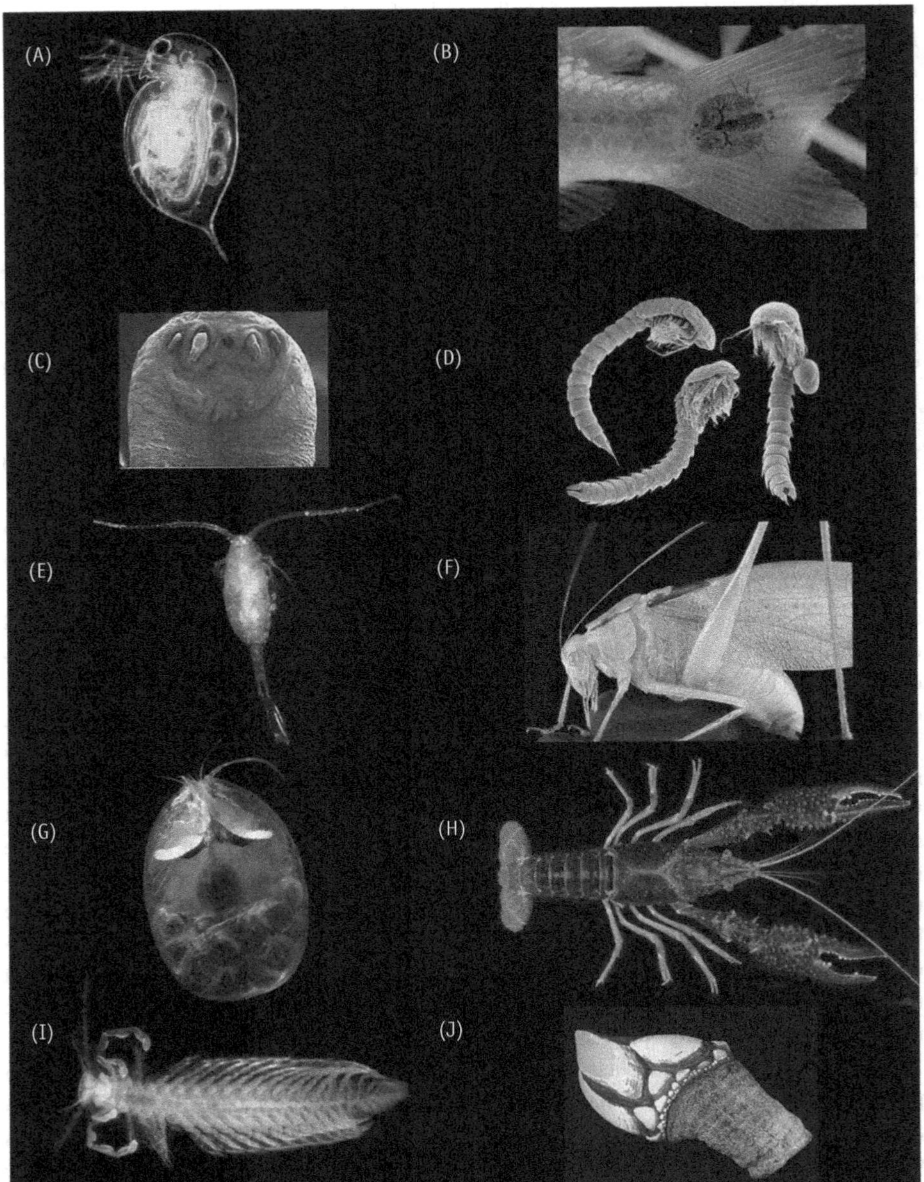

Fig. 4.1.

Examples demonstrating the morphological diversity across Pancrustacea. (A) *Daphnia pulex*, Branchiopoda (from Gewin 2005, under License CC-BY); (B) *Argulus foliaceus*, Branchiura (© G. Boxshall, under License CC-BY-NC-SA); (C) *Porocephalus crotali*, Pentastomida (from Abele et al. 1989, with permission from Oxford University Press); (D) *Lightiella monniotae*, Cephalocarida (© D. Waloszek and J. Olesen); (E) *Ridgewayia* sp., Copepoda (© T. Iliffe); (F) *Amblycorypha oblongifolia*, Hexapoda (from USGS Bee Inventory and Monitoring Lab); (G) *Gigantocypris* sp., Ostracoda (© D. Fenolio/ DEEPEND); (H) *Procambarus clarkii*, Malacostraca (© D. Felder); (I) *Cryptocorynetes* sp., Remipedia (© T. Iliffe); (J) *Pollicipes cornucopia*, Thecostraca (© H. Hillewaert, under License CC-BY-SA). See color version of this figure in centerfold.

2001) based on the shared ommatidia in the compound eyes, the name Pancrustacea (introduced by Zrzavý and Stys 1997) will be used herein. Across Pancrustacea there are 2 subphyla, 7 classes, 16 subclasses, and 90 orders, of which 59 orders belong to Crustacea (Table 4.1). Here we will introduce the major lineages (class or subclass) across Pancrustacea and briefly summarize studies that have provided evidence for these groupings.

Branchiopoda

This class of crustaceans includes the fairy/brine shrimp, tadpole shrimp, clam shrimp, and water fleas (Fig. 4.1A). The monophyly of Branchiopoda has been widely established based on phylogenetic analysis of molecules (Koenemann et al. 2010, Regier et al. 2010, von Reumont et al. 2012), morphology (Legg et al. 2013), and combined molecular and morphological data (Oakley et al. 2013). This group contains the only fully annotated non-insect crustacean genome to date, *Daphnia pulex*, as of 2016 (see GIGA Community of Scientists 2014 for list of genomic resources).

Branchiura and Pentastomida

These two subclasses include some of the most infamous parasitic crustaceans. Branchiurans, also known as fish lice, are obligate ectoparasites of fish and amphibians (Fig. 4.1B). They are economically important, as they can decimate fish aquacultures. Pentastomids are a small group of vermiform animals commonly known as tongue worms, living as obligate endoparasites of vertebrate respiratory tracts (though larvae infect intermediate host species; Fig. 4.1C). The unusual morphology of pentastomids (adults are so degenerated that they lack any recognizable arthropod characters) have led to uncertain placement within the metazoan Tree of Life with allies such as Arthropoda, Tardigrada, Annelida, Platyhelminthes, and Nematoda (Riley et al. 1978, de Oliveira Almeida and Christoffersen 1999). However, placement within the arthropods and, more specifically, Pancrustacea, is now widely accepted (see section "Where Are We Now?"). Branchiura include one order, Arguloida, whereas Pentastomida include two orders, Cephalobaenida and Porocephalida.

Cephalocarida

This small group of crustaceans (Fig. 4.1D) was discovered in 1955 in benthic marine sediments, living as part of the interstitial meiofauna. They can be found up to depths of 1,500 m and feed on detrital material within the sediment. Cephalocarids are certainly monophyletic but have been included in few molecular and morphological trees, with uncertain phylogenetic placement.

Copepoda

Copepods are a well-known, diverse group of tiny crustaceans found in nearly all aquatic environments from fresh to hypersaline waters, submerged caves, and terrestrial ponds (Fig. 4.1E). They comprise up to 91% of the ocean's zooplankton biomass (Kosobokova and Hopcraft 2010) and are thus an integral part of the marine food chain. Past studies have recovered copepods as monophyletic (Huys et al. 2007, Oakley et al. 2013), and, historically, they have been divided into 10 orders, namely Calanoida, Cyclopoida, Gelyelloida, Harpacticoida (benthic copepods), Misophrioida, Monstrilloida, Mormonilloida, Platycopioida, Poecilostomatoida, and Siphonostomatoida (fish parasites). However, a very recent study based on molecular data provided an updated position of copepods within Pancrustacea, tested the major subdivisions of Copepoda, and resolved deep internal relationships (Khodami et al. 2017). This paper found 7 of the 10 orders to be monophyletic (Platycopioida, Calanoida, Misophrioida, Monstrilloida, Siphonostomatoida, Gelyelloida, and Mormonilloida), redefined two orders (Cyclopoida and Harpacticoida), and proposed a new

Table 4.1. Current Classification of Extant Pancrustacea*

Subphylum	Clade	Class	Subclass	Order
Crustacea				
	Oligostraca			
			Mystacocarida	
				Mystacocaridida
		Ichthyostraca		
			Branchiura	
				Arguloida
			Pentastomida	
				Cephalobaenida
				Porocephalida
		Ostracoda		
			Myodocopa	
				Halocyprida
				Myodocopida
			Palaeocopa	
			Podocopa	
				Podocopida
			Platycopa	
				Platycopina
	Multicrustacea			
		Hexanauplia		
			Copepoda	
				Calanoida
				Canuelloida
				Cyclopoida
				Gelyelloida
				Harpacticoida
				Misophrioida
				Monstrilloida
				Mormonilloida
				Platycopioida
				Siphonostomatoida
			Tantulocarida	
				Tantulocarida
			Thecostraca	
				Akentrogonida
				Cryptophialida
				Cyprilepadiformes
				Dendrogastrida
				Facetotecta (Infraclass)
				Ibliformes
				Kentrogonida
				Laurida
				Lepadiformes

Table 4.1. ***(Continued)***

Subphylum	Clade	Class	Subclass	Order
				Lithoglyptida
				Scalpelliformes
				Sessilia
		Malacostraca		
			Eumalacostraca	
				Amphipoda
				Anaspidacea
				Bathynellacea
				Bochusacea
				Cumacea
				Decapoda
				Euphausiacea
				Isopoda
				Lophogastrida
				Mictacea
				Mysida
				Spelaeogriphacea
				Stygiomysida
				Tanaidacea
				Thermosbaenacea
			Hoplocarida	
				Stomatopoda
			Phyllocarida	
				Leptostraca
	Allotriocarida			
		Branchiopoda		
			Calmanostraca	
				Notostraca
			Diplostraca	
				Anomopoda
				Ctenopoda
				Cyclestherida
				Haplopoda
				Laevicaudata
				Onychopoda
				Spinicaudata
			Sarsostraca	
				Anostraca
		Cephalocarida		
				Brachypoda
		Remipedia		
				Nectiopoda
Hexapoda				~31 orders

*Following Recent Studies and World Register of Marine Species (crustacean taxa) and Oakley et al. (2013).

order, Canuelloida (Table 4.1). Even with these advancements, shallow-level relationships are still in dire need of phylogenetic research.

Hexapoda

Tracing the phylogenetic relationships of insects, with their tremendous diversity, remains a challenge in evolutionary biology (Fig. 4.1F). Currently, Hexapoda is divided into three lineages including the insects (Insecta), coneheads (Protura), and springtails (Collembola; Misof et al. 2014). The monophyly of hexapods is well established (Regier et al. 2010, Sasaki et al. 2013, Dell'Ampio et al. 2014, Misof et al. 2014), and recent phylogenomic methods have been applied to resolve the evolutionary relationships and origins among major lineages (Misof et al. 2014). As with any group with such extensive diversity, familial, generic, and species relationships within the 31 orders continue to be targets for phylogenetic research, and many questions still remain. Extensive discussion on hexapod relationships, origins, and phylogenetic placement within Pancrustacea can be found in Chapter 5.

Ostracoda

This diverse class of aquatic crustaceans is commonly referred to as seed shrimp (Fig. 4.1G). They can be easily recognized due to their bivalved, calcified carapace, with the limbs often fully enclosed. These carapaces are prone to fossilization; hence ostracods have an excellent carapace-only fossil record dating over 478 mya (Wolfe et al. 2016). Ostracod monophyly is supported by phylogenetic analyses using nuclear protein-coding genes, transcriptomes, and/or morphology (Legg et al. 2013, Oakley et al. 2013, Wolfe and Hegna 2014). However, relationships within the two major lineages, Myodocopa and Podocopa, remain unresolved.

Malacostraca

This hyperdiverse class of crustaceans includes approximately 17 orders (with the recent inclusion of Amphionidacea into Decapoda; De Grave et al. 2015) and ~20,000 species. This group contains well-known representatives such as amphipods, isopods, crabs, lobsters, shrimps, stomatopods, as well as less familiar forms (Fig. 4.1H). Malacostraca is a well-supported clade within Pancrustacea, with consistent evidence from analyses of morphology (Wills et al. 1998, Legg et al. 2013), molecules (Regier et al. 2005, 2010, von Reumont et al. 2012), and total evidence (Oakley et al. 2013). Although several nested groups within Malacostraca have also been recovered consistently as monophyletic (i.e., Eumalacostraca, Eucarida, Decapoda), many taxonomic and phylogenetic questions still exist within and across major lineages. For example, peracarids, including isopods, amphipods, and mysids, among others, have been neglected in phylogenetic studies in spite of their tremendous biodiversity. Although some studies have recognized peracarids as a monophyletic group (Richter and Scholtz 2001, Poore 2005, Wills et al. 2009, Wirkner and Richter 2010), lack of sampling has left internal relationships unresolved (Jenner et al. 2009). To date, no higher-level molecular studies are available for peracarids. The monophyly and placement of the Superorder Syncarida within Crustacea remains questionable, with previous studies finding Syncarida as para- or polyphyletic (Camacho 2003, Camacho et al. 2002, Jenner et al. 2009, Wills et al. 2009).

Mystacocarida

The Mystacocarida is a subclass of minute (<1 mm), elongate, and cylindrical crustaceans originally described as a new order by Pennak and Zinn (1943). All extant species (~13) are found living in between sand grains as part of the meiofauna, feeding upon microalgae and bacteria that grow in these environments. Although inconspicuous, mystacocarids have been included in higher-level molecular and morphological phylogenies.

Remipedia

The remipedes are a fascinating and enigmatic group of blind crustaceans with a global distribution restricted to anchialine caves and oceanic blue holes (Fig. 4.1I). The class is composed of one extant order, Nectiopoda, containing 8 families, 12 genera, and 29 species (Olesen et al. 2017, Hoenemann et al. 2013). Although remipedes may appear morphologically primitive (Yager 1994), they possess a highly specialized nervous system (Stemme et al. 2012, 2013) and feeding mouthparts and are voracious predators (von Reumont et al. 2014). The monophyly of Remipedia has been established based on molecular (Neiber et al. 2011) and robust morphological analyses (Koenemann et al. 2007), although internal relationships remain contentious.

Thecostraca

This charismatic group includes sessile and parasitic barnacles (Fig. 4.1J). Twelve or thirteen (if tanulocarids are included) orders make up the thecostracans, and these include bizarre representatives such as the Facetotecta (y-larvae), with no known adult form, and rhizocephalans that parasitize decapods, causing a series of behavioral, reproductive, and morphological changes in the host. Superficially, adults do not seem to resemble pancrustaceans or even arthropods (though dissection of a barnacle reveals that they use a series of biramous thoracic limbs for feeding). Their affinities were discovered by early 20th-century observation of their larvae, which share many characters with other pancrustaceans. Thecostraca is considered to be monophyletic based on total evidence (Pérez-Losada et al. 2014); however, this hypothesis has not been formally addressed in a robust evolutionary framework. In a recent study the parasitic group Tantulocarida was suggested to be part of Thecostraca based on one ribosomal gene (Petrunina et al. 2013), but more evidence is needed to resolve this issue. Although some studies have been done at the ordinal level (e.g., Sessilia, Pedunculata; Pérez-Losada et al. 2004), many basic phylogenetic and taxonomic questions remain across and within Thecostraca.

THE HISTORY OF PANCRUSTACEAN PHYLOGENETICS WITH AN EMPHASIS ON HIGHER-LEVEL RELATIONSHIPS

The higher-level phylogenetic relationships within Pancrustacea have been fraught with controversy. Several phylogenetic hypotheses, both morphological and molecular, have been proposed with little consensus in sight (Figs. 4.2 and 4.3). For the purpose of this review, we will introduce some of the monumental studies that have contributed to our current understanding of pancrustacean relationships. Due to the high volume of morphological and molecular studies that have emerged over the past few decades, we will not have the space to discuss them all. For this reason, we will limit our discussion to some of the most profound contributions in pancrustacean phylogenetics, with emphasis on subclass and class-level crustacean relationships (for detailed discussion on Hexapoda-Crustacea relationships, see Chapter 5).

Morphological Insights: from Observation to Rigorous Analysis

The earliest hypotheses of crustacean relationships were, of course, based broadly on morphology (Fig. 4.2). Linnaeus classified six genera containing 87 species we now recognize as crustaceans (as

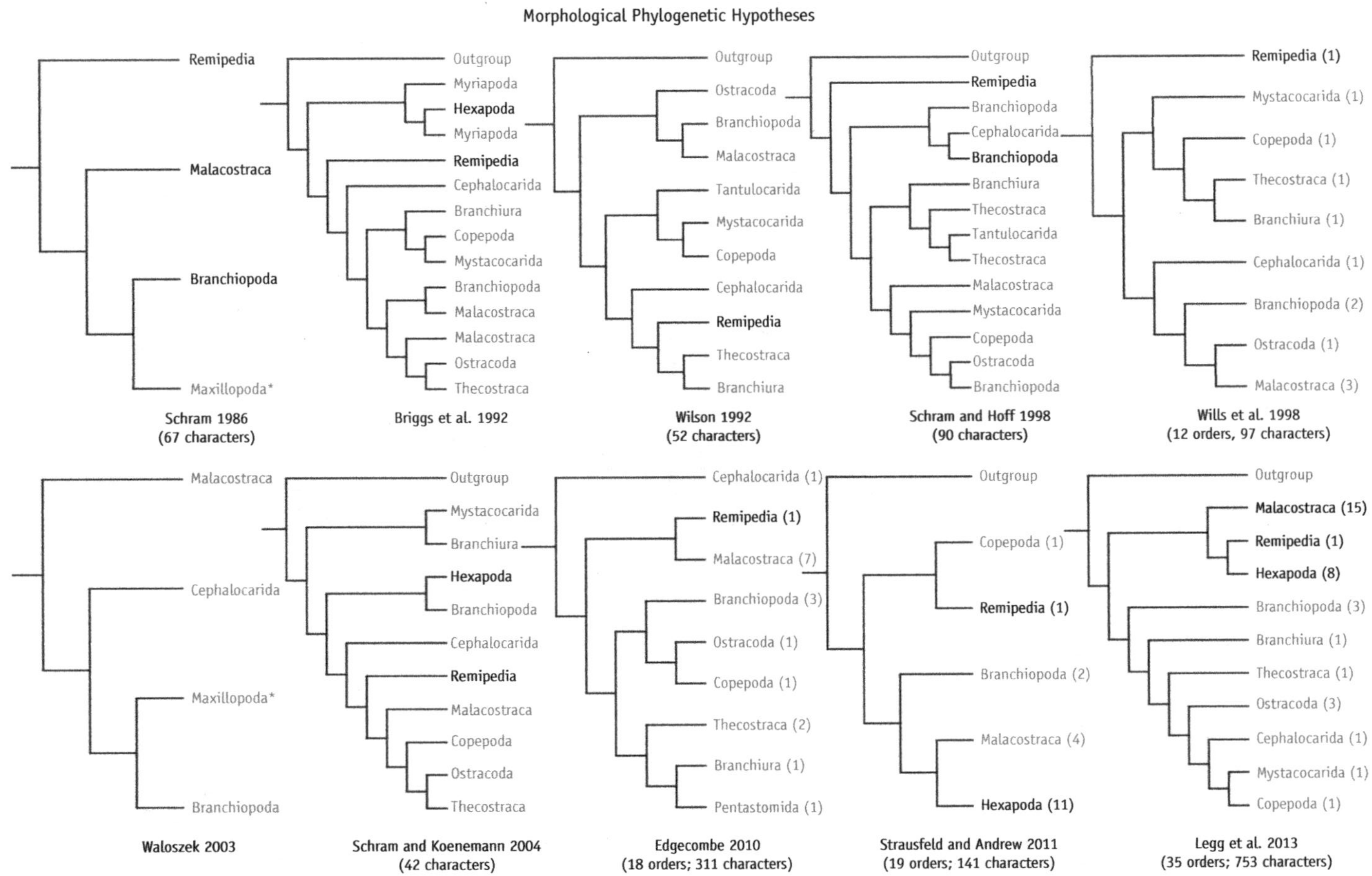

Fig. 4.2.
Alternative hypotheses of the Pancrustacean phylogeny based on morphological data. Numbers adjacent to taxon names in the tree represent how many orders for that taxon were included in the analyses (when available). Caption below citation indicates the total number of orders and characters used in the analyses. Fossil taxa excluded from phylogram for clarity. Class Maxillopoda, denoted by an asterisk, no longer constitutes a valid taxon. However, the source literature did not specify the maxillopodan taxa employed in said analyses. See color version of this figure in centerfold.

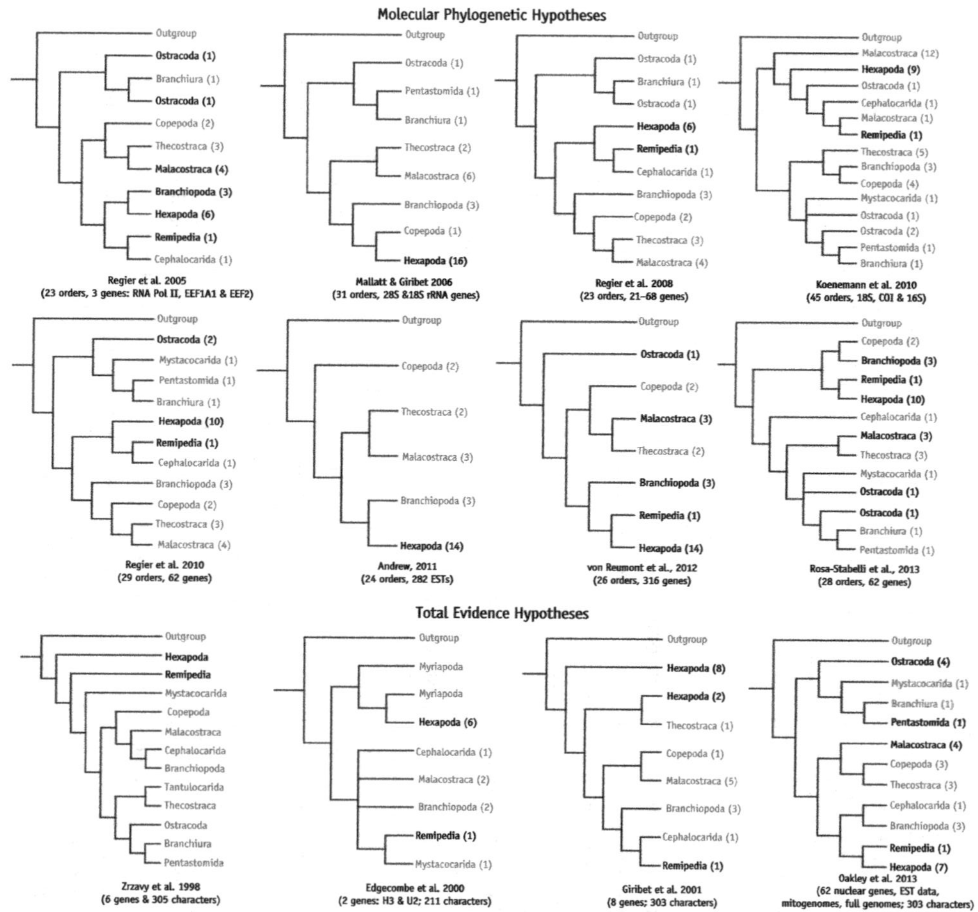

Fig. 4.3.

Alternative hypotheses of the Pancrustacean phylogeny based on genetic data (rows 1 and 2) and combined morphological and genetic data (row 3). Numbers adjacent to taxon names in the tree represent how many orders for that taxon were included in the analyses (when available). Caption below citation indicates the total number of orders, genes, and/or characters used in the analyses. Fossil taxa excluded from phylogram for clarity. Abbreviations: RNA Pol II, RNA polymerase II; EEF1A1, elongation factor alpha subunit 1, EEF2, elongation factor 2; 28S, 28S ribosomal mitochondrial gene; 18S, 18S ribosomal mitochondrial gene; COI, cytochrome oxidase I; 16S, 16S ribosomal mitochondrial gene; EST, expressed sequence tags; U2, small nuclear RNA. See color version of this figure in centerfold.

well as a xiphosuran and several molluscs; Boxshall 2007). For over 200 years, hexapod relationships with crustaceans were rarely assessed, as they were assumed to be more closely related to myriapods (Snodgrass 1938, Tiegs and Manton 1958, Manton 1977; also see Chapter 5). Early research on crustacean systematics focused on reconstructing characteristics of the hypothetical crustacean ancestor. For this endeavor, gross morphology of the exoskeleton and differentiation of the limbs and body regions were emphasized, followed by internal anatomy and functional morphology (Manton 1977). Crustaceans were assumed to plesiomorphically share a five-segmented head bearing two pairs of antennae, gnathobasic mandibles, and early development characterized as a nauplius (either free-living larva or egg-nauplius; Tiegs and Manton 1958, Cisne 1982). Walossek's (1999) concept of the crustacean "groundplan" added several possible synapomorphies for the group: the presence of a proximal endite on post-antennular limbs, multi-annulated exopods on post-antennular limbs, and post-antennular limbs with no more than five podomeres on the endopod.

Work in the later 20th century frequently compared crustaceans to trilobites, with an assumption that the newly discovered Cephalocarida were "primitive" (Hessler and Newman 1975, Cisne 1982). The notostracan and anostracan Branchiopoda also received special attention, as they shared with Cephalocarida a long series of undifferentiated thoracic limbs and the flattened, leaf-like limb morphology (Fig. 4.4A). Thus, Branchiopoda and Cephalocarida were united in the literature as a clade called Phyllopoda; however, the same name has been used to refer to other clade compositions, sometimes including Ostracoda, excluding Cephalocarida and Leptostraca, or referring to non-anostracan Branchiopoda; thus the name Phyllopoda is now seldom used (Martin and Christiansen 1995, Martin and Davis 2001). Once discovered, remipedes were also viewed as a model for ancestral crustaceans, on the basis that their uniramous thoracic limbs must be plesiomorphic (e.g., Emerson and Schram 1990). Assumptions of remipede plesiomorphy have colored polarization of morphological characters in several early analyses (see below), where they were by default assumed to be the outgroup to other crustaceans.

Within Crustacea, one of the earliest described taxonomic divisions was that between Malacostraca and Entomostraca. The Entomostraca comprised most other crustaceans (Branchiopoda, Branchiura, Copepoda, Mystacocarida, Ostracoda, Thecostraca, and later Cephalocarida), based on the shared lack of a mandibular palp in the adult, presence of a maxillule with four median endites, and abdomen of at least four limb-less somites (Walossek 1999, Boxshall 2007). In Walossek's studies of developmental sequences, Cephalocarida are viewed as secondarily specialized and unrelated to Branchiopoda (Walossek 1999).

Taxonomy was revolutionized by the paradigm of phylogenetic systematics (Hennig 1966) and swiftly thereafter by the development of software to calculate the most parsimonious or most likely topology. The first cladistic analyses of any crustacean clades used matrices of morphological characters, scored as shared (homologous) between taxa at the familial or ordinal level, and analyzed using maximum parsimony (Tanaidacea: Sieg 1983; Eumalacostraca: Schram 1984). Shortly thereafter, the first comprehensive cladistic analysis of crustacean morphology was conducted by Schram (1986; Fig. 4.2). The morphological matrix comprised 23 extant taxa, coded for 67 characters (Schram 1986). Hexapoda were not included. The most parsimonious topology supported Remipedia as the earliest branching crustaceans, with Malacostraca as sister to a clade of Phyllopoda + Maxillopoda (Schram 1986). Note that in this case, Phyllopoda refers specifically to the clade containing Branchiopoda, Cephalocarida, and Leptostraca, united by the presence of multiramous, leaflike thoracic limbs (Schram 1986). Monophyly of Maxillopoda (Branchiura, Copepoda, Mystacocarida, Ostracoda, and Thecostraca) was supported by the presence of 11 trunk segments, fewer than six thoracic segments, a short bulbous heart, and a naupliar eye (Schram 1986). The clade of Maxillopoda and Phyllopoda was characterized by lack of a mandibular palp in the adult (Fig. 4.4B), a character that has recently been deconstructed on developmental grounds (Fig. 4.4C; Wolfe and Hegna 2014).

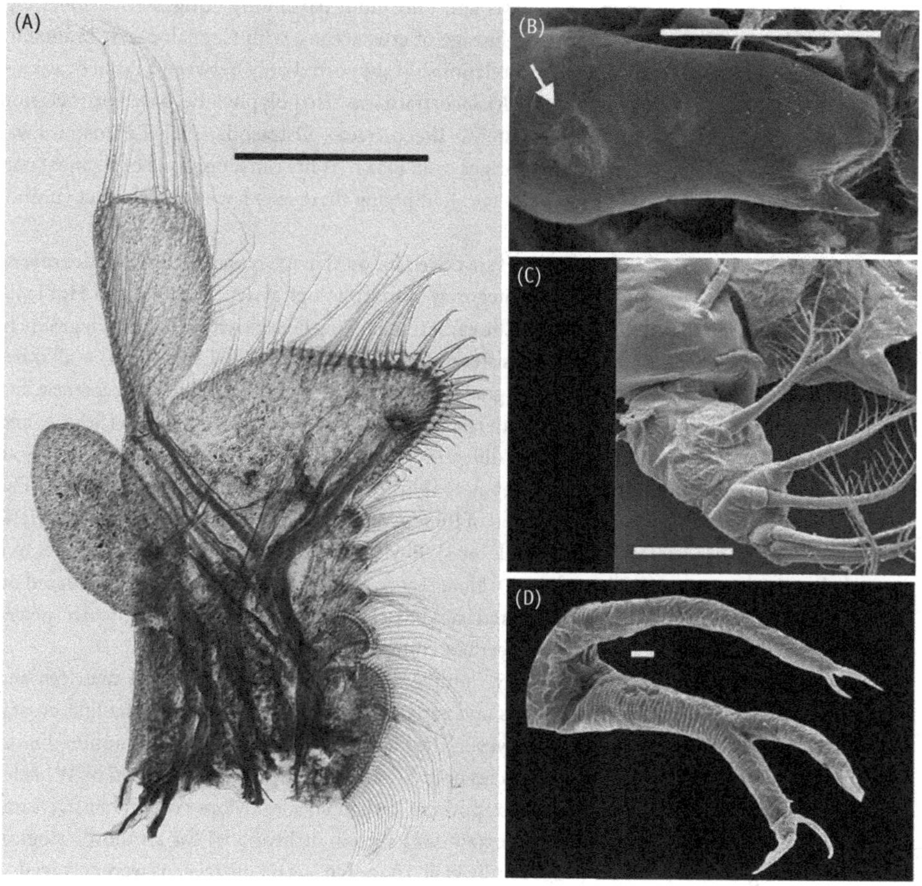

Fig. 4.4.
Morphological characters from limbs and body organization. (A) The phyllopodous eighth trunk limb of an anostracan branchiopod, *Branchinencta brushi*, with large epipodite visibly lacking intrinsic musculature (YPM IZ 024277, photo courtesy of E. Lazo-Wasem). Scale bar = 500 μm. (B) Adult mandible of the anostracan *Branchinectus raptor*, with a scar (arrow) indicating the position where the mandibular palp was lost during ontogeny (modified from Olesen 2009, with permission from Museum für Tierkunde Dresden, Senckenberg Naturhistorische Sammlungen Dresden). Scale bar = 1 mm. (C) Well-developed larval mandibular palp in an anostracan, *Eubranchipus grubii*, a character shared across most crustaceans (modified from Olesen 2009). Scale bar = 50 μm. (D) The deutocerebral antennule and tritocerebral antenna of the remipede *Pleomothra apletocheles* represent a traditional synapomorphy for all non-hexapod crustaceans (modified from Olesen et al. 2014, with permission from Johns Hopkins University Press). The antennule is always uniramous (above, with a distal secondary branch, or flagellum), while many taxa have truly biramous post-antennular limbs (below) (Boxshall 2004). Scale bar = 50 μm.

Exceptionally preserved fossils (such as those from the Cambrian Burgess Shale and Orsten deposits, and the Devonian Rhynie Chert) and their ability to polarize character evolution did not escape notice. In the 1990s, a series of morphological matrices that included fossils were constructed and analyzed. An early analysis (17 fossil and extant taxa coded for 59 characters) divided crustaceans into a clade comprising Branchiopoda + Malacostraca and another clade rendering Maxillopoda as polyphyletic (i.e., containing Cephalocarida and Remipedia; Fig. 4.2; Wilson 1992). The same year, 46 fossil and extant taxa (across all arthropods) coded for 134 characters were analyzed using principal component analysis and maximum parsimony (Briggs et al. 1992; Fig. 4.2). Remipedia was the earliest branching lineage of crustaceans, with Cephalocarida branching next. Maxillopoda was polyphyletic, with a relationship supported only between Mystacocarida + Branchiura + Copepoda (Briggs et al. 1992). Malacostraca was also polyphyletic, based on inclusion of two representative taxa. Decapoda was sister to Thecostraca + Ostracoda, while Leptostraca was sister to Branchiopoda and several fossils (Briggs et al. 1992). At the time, only one of six most parsimonious trees was presented (Briggs et al. 1992), implying there may have been conflict (homoplasy) unreported in this matrix.

An updated matrix of 64 fossil and extant taxa coded for 90 characters was analyzed under several parameters, removing fossils and different categories of morphological data (Schram and Hof 1998; Fig. 4.2). All results supported Remipedia as the earliest branching lineage within the crown group. In the most complete analysis, Remipedia was sister to Phyllopoda (excluding Leptostraca) + all other crustaceans. Maxillopoda was a paraphyletic grade (separate clades of Branchiura + Thecostraca and Mystacocarida + Copepoda + Ostracoda) with respect to a monophyletic Malacostraca (Schram and Hof 1998). Meanwhile, in the same book, a different matrix of 64 fossil and extant taxa (across all arthropods) coded for 97 characters (Wills et al. 1998) produced dramatically different results (Fig. 4.2). The main similarity was the basal position of Remipedia. Maxillopoda was largely monophyletic (with the exception of Ostracoda; Wills et al. 1998). Phyllopoda was paraphyletic with respect to a monophyletic Malacostraca (Wills et al. 1998). Note that codings in these analyses were still based on "groundplans," i.e., the terminals were not scored based on characters observed from particular species or specimens but from expert knowledge of features common to orders or families.

Total evidence analyses (simultaneously optimizing data from morphological matrices and DNA sequences) also emerged in the 1990s and early 2000s, at the same time as the first crustacean phylogenies incorporating fossils. However, the first total evidence analyses combined morphology and molecular data from extant species only. The initial analysis was presented by Wheeler et al. (1993) but only included two non-hexapod crustacean species. More comprehensive sampling came several years later, sampling 36 extant taxa (across arthropods) for 211 morphological characters and 2 genes, H3 and U2 (Edgecombe et al. 2000; Fig. 4.3). Crustaceans were resolved as monophyletic, with little internal resolution. Interestingly, the presence of a second antenna (Fig. 4.4D) and nauplius were symplesiomorphic, although in this early analysis, morphological signal is strongly observed in the topology (Edgecombe et al. 2000). Cephalocarida or Cephalocarida + Branchiopoda (i.e., partial Phyllocarida) was the most basal clade, with Remipedia + Mystacocarida as sister to monophyletic Malacostraca (Edgecombe et al. 2000).

The next and most seminal total evidence analysis sequenced 8 genes in addition to coding 303 morphological characters for 54 extant taxa across arthropods (Giribet et al. 2001; Fig. 4.3). An important advance was the adoption of species-based morphological coding, where the matrix was constructed of character states relevant to the same species (rather than an assumed groundplan) that were represented by molecular sequences. Although crustacean paraphyly was recovered with respect to hexapods in the optimal topology, a thecostracan barnacle was nested within a clade also containing *Drosophila* and a dipluran (apterygote hexapod). Other analyses exploring this issue found that *Drosophila* had an aberrant gene sequence; thus the significance of this result was restricted to hexapods as the sister group of crustaceans, to the exclusion of myriapods (Giribet et al. 2001). Within non-hexapod crustaceans, Copepoda + Malacostraca comprised the sister group to (Cephalocarida + Remipedia) + Branchiopoda (Giribet et al. 2001). In hindsight, the relationship between Cephalocarida and Remipedia is strikingly similar to an early molecular analysis with extensive taxon sampling, discussed below (Regier et al. 2010).

A new morphological analysis incorporated 31 fossil and extant taxa coded for 42 characters, but based on groundplan codings (Schram and Koenemann 2004; Fig. 4.2). The earliest branching crustaceans were Branchiura + Mystacocarida, resulting in polyphyly of Maxillopoda. Another

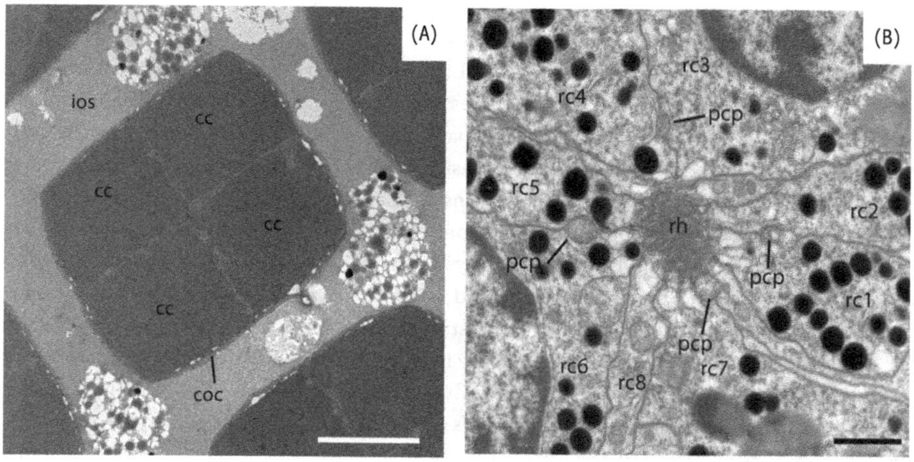

Fig. 4.5.
Morphological characters from the visual system, defining the alternate clade name of Tetraconata (shared by malacostracans and hexapods). (A) Cross section through the square crystalline cone of an ommatidium in the decapod *Palaemon elegans*, clearly showing the tetracone wrapped in a sleeve of the two corneagenous cells. Scale bar = 5 μm. (B) Cross section of the distal part of the retinula of the decapod *Albunea carabus*, containing eight retinula cells surrounding the rhabdom in the center. The irregular, unpigmented eighth retinula cell is a tetralobed apparatus in the interspace of the seven regularly pigmented retinula cells. Between the retinula cells are four proximal processes of the cone cells. Scale bar = 1 μm. Abbreviations: cc, crystalline cone cell or Semper cell; coc, corneagenous cell (circumconical sheath); ios, interommatidial space; pcp, proximal process of cone cell; rc, retinula cell; rh, rhabdom. Photos courtesy of C. Müller.

supported clade comprised Cephalocarida, Remipedia, monophyletic Malacostraca, and the remainder of Maxillopoda (Schram and Koenemann 2004). This paper was notable as it was the first significant morphological cladistic result depicting crustacean paraphyly with respect to Hexapoda (i.e., with Branchiopoda as the hexapod sister group); however, it did not present a substantial discussion of character support apart from some possible similarities from the then-emerging field of developmental genetics (Schram and Koenemann 2004).

Meanwhile, many conserved morphological characters supported crustacean paraphyly with respect to Hexapoda but had been overlooked by systematists. Shared architecture of the nervous system was indeed recognized since the late 19th century, without direct phylogenetic context (reviewed by Strausfeld 2012). The presence of four crystalline cone cells in the eye (Fig. 4.5) is eponymous for the 'Tetraconata' (Dohle 2001, Richter 2002). Neural morphology supported a sister relationship not only between crustaceans and hexapods but also between specific crustacean groups (i.e., paraphyly of crustaceans). Neurophylogeny was then developed, which identified morphological characters from the nervous system to argue in support of a sister relationship between Malacostraca and Hexapoda (Harzsch 2006). These comparisons, however, were merely mapped onto an assumed tree topology. Meanwhile, similar results were also discovered in a series of neural cladistic analyses, where neural characters were used as the basis of a morphological matrix (the most recent of which is Strausfeld and Andrew 2011; Fig. 4.2). This matrix included 42 extant taxa (across arthropods) coded for 141 characters only from the nervous system. Only one species each of Maxillopoda and Remipedia were included; they formed the most basal clade within paraphyletic Pancrustacea (Strausfeld and Andrew 2011). Branchiopoda was sister to the clade of Malacostraca + Hexapoda, a result that suggests the secondary simplification of the branchiopod brain (Strausfeld and Andrew 2011, Strausfeld 2012). Neural characters continue to increase in importance, and data are even available from select fossils (Ma et al. 2015).

Seminal studies by Edgecombe (2010) and Legg et al. (2013) demonstrated congruence between molecular and morphological phylogenies of arthropods when fossils are included in the analysis (Fig. 4.2). Each of these papers built paired morphological phylogenies of their given data matrix, with and without fossils. The inclusion of 391 characters coded for 16 fossil and 68 extant taxa added fossil character combinations that conflict with Pentastomida falling outside Euarthropoda, instead supporting their grouping within the Crustacea (Zrzavý et al. 1998, Edgecombe 2010). A much more extensive analysis of 753 morphological characters for 215 fossil and 96 extant arthropod taxa further increased congruence between molecular and morphological phylogenies (Legg et al. 2013). The most significant result was recovery of paraphyletic Crustacea with respect to Hexapoda, found only in analyses including fossils. Analyses excluding fossils produced monophyletic sister groups of Crustacea and Hexapoda, likely driven by convergence in terrestrial characters between hexapods and myriapods (Legg et al. 2013). Futhermore, the extant sister group of Hexapoda was Remipedia in the fossil analysis, a relationship first proposed from transcriptome data (von Reumont et al. 2012, Legg et al. 2013, Oakley et al. 2013). Note, however, that with or without fossils, this matrix recovered monophyletic Maxillopoda, including Ostracoda (Legg et al. 2013).

Oakley et al. (2013) conducted the most extensive total evidence analysis of pancrustacean phylogeny, with a morphological data partition coding 16 fossil and 93 extant taxa for 183 characters. A massive molecular partition was added, with 1,001 genes represented, the results of which are discussed below (see section "Molecular Insights"). Briefly, monophyly of Ostracoda was supported within Oligostraca (including support for Pentastomida with Branchiura, i.e., Ichthyostraca). There was weak support for a new clade, Hexanauplia, comprising Copepoda and Thecostraca (also supported by larva-based analysis by Wolfe and Hegna 2014). Remipedia was again supported as the sister group of Hexapoda (Oakley et al. 2013). Simultaneous analysis of fossil and extant morphology with extant molecular data promises to improve divergence time estimates by adding uncertainty in fossil ages directly to the model ("tip dating"; e.g., O'Reilly et al. 2015), an approach that has not yet been applied to non-hexapod pancrustaceans.

Most recent morphological studies continue to support crustacean paraphyly, recently showing Remipedia as the sister group to Hexapoda, congruent with phylogenomic work (von Reumont et al. 2012, Legg et al. 2013, Oakley et al. 2013). Mounting morphological evidence for crustacean paraphyly casts doubt on the previously designated synapomorphies of Crustacea (none of which have truly persisted in the modern molecular era). Sperm morphology and subsequent phylogenies support a position for Pentastomida within crustaceans, usually grouped with Branchiura as the clade Ichthyostraca (Zrzavý et al. 1998, Edgecombe 2010, Oakley et al. 2013). There is potential support for the Hexanauplia clade (Oakley et al. 2013), but widespread support for Maxillopoda still persists in morphology-only studies and is a significant challenge to reconcile with molecular data (see section "Molecular Insights"). None of the higher-level clades continually found in recent molecular analyses (Oligostraca, Multicrustacea, Allotriocarida) have ever been supported by morphology alone, driving the search for new morphological characters and fossil data to make sense of possible character transformations upon robust molecular branches.

Molecular Insights: from Single Gene Trees to the Genomics Revolution

The monophyly of Crustacea has long been questioned from both a morphological and a molecular perspective. Few apomorphies have been proposed to unite the group (Lauterbach 1983, Richter 2002, Muller et al. 2003), and all recent molecular studies show hexapods nested within paraphyletic Crustacea (see below). Although this relationship is now largely accepted, the internal relationships including the sister group to Hexapoda and internal relationships among Crustacea are still widely debated. Below, we will review some of the pioneering studies that investigated pancrustacean relationships using molecular data over the past two decades. It is important to note that several monumental phylogenetic studies that focus on hexapod-crustacean relationships or arthropod phylogeny have been omitted due to limited sampling across Pancrustacea (see Chapter 5; see also section "Morphological Insights").

Prior to the 21st century, molecular investigations of crustacean phylogeny included limited taxon sampling and markers. Some of the earliest studies to investigate crustacean relationships used 18S, a nuclear small-subunit ribosomal DNA gene, to generate single gene trees (Abele et al. 1989, Abele et al. 1992, Spears and Abele 1999). One of these studies included 10 crustacean species and 3 other arthropod outgroups (Spears and Abele 1999). Findings of this study rejected the hypothesis that crustaceans possessing foliaceous limbs (branchiopods, cephalocarids, phyllocarids) form a monophyletic clade. They recovered three major groupings, some of which are still supported today (i.e., Malacostraca), but scarce sampling of taxa and markers did not allow for robust testing of Pancrustacea relationships (for a discussion on the use of single markers in phylogenetic reconstruction, see Timm and Bracken-Grissom 2015).

In the early 2000s, the ability to generate larger amounts of sequence data allowed for the investigation of evolutionary relationships with more robust datasets. One outstanding question in the early 21st century was the uncertain phylogenetic position of Pentastomida in relation to other arthropod groups. Using complete mtDNA sequences and gene arrangements, Lavrov et al. (2004) concluded that pentastomids are crustaceans with a close affinity to branchiurans. Although this work was done over a decade ago, the monophyly of Branchiura + Pentastomida is still commonly accepted. This relationship is supported based on sperm morphology (Wingstrand 1972, Storch and Jamieson 1992, Giribet and Ribera 2000), morphology + protein coding genes (Zrzavý et al. 1998), transcriptomes (Oakley et al. 2013), and molecules only (Møller et al. 2008, Regier et al. 2010).

Around the same time, researchers began to substantially increase taxon sampling to investigate phylogenetic relationship across a much broader representation of crustaceans. Although molecular studies had provided support for Pancrustacea, the internal relationships within Crustacea were contentious. In addition to increased sampling, the use of nuclear protein-coding genes allowed Regier et al. (2005) to test the monophyly of Crustacea across 62 arthropod lineages and 2 outgroups. Using three nuclear markers, i.e., elongation factor-1α, RNA polymerase II, and elongation factor-2, Regier et al. (2005) reconfirmed previous studies that united hexapods and crustaceans, but with uncertain interrelationships (Fig. 4.3). In addition to recovering a pancrustacean clade, Regier et al. (2005) recovered four major groups within Crustacea. First, Branchiopoda were recovered as a monophyletic group sister to Hexapoda, but with low support. Second, Cephalocarida + Remipedia were recovered as a monophyletic group. Third was the recovery of Malacostraca + Cirripedia (Thecostraca includes Cirripedia) + Copepoda, with Malacostraca + Cirripedia strongly supported across all analyses. The final major grouping was Ostracoda + Branchiura, abeit with low bootstrap support values. Findings from this study rendered the traditional Maxillopoda (Thecostraca, Copepoda, Mystacocarida, Branchiura, Ostracoda) para- or polyphyletic. The non-monophyly of Maxillopoda had long been suspected based on rampant morphological variation across the group (Martin and Davis 2001) and is supported by subsequent phylogenetic analyses (Mallatt and Giribet 2006, Regier et al. 2008, Regier et al. 2010). Other deep relationships were recovered with uncertain or conflicting support (i.e., high posterior probabilities but low bootstrap).

The following year, Mallatt and Giribet (2006) used complete ribosomal RNA sequences, 28S and 18S, to investigate the relationships across Ecdysozoa (i.e., arthropods, nematodes, tardigrades, onychophorans, nematomorphs, priapulans, kinorhynchs, and loriciferans; Fig. 4.3). They included 20 crustaceans and 25 hexapods for a total of 45 pancrustaceans and included newly generated and published sequences. Many overarching findings concurred with those of Regier et al. (2005) but differed slightly due to the taxa and molecular markers included in the study. Branchiura + Pentastomida were recovered with high support. Ostracoda were included as sister to this clade, but with no support. Similar to Regier et al. (2005), Malacostraca and Cirripedia formed a monophyletic group, however the current study found Copepoda to be sister to Hexapoda. This sister relationship between copepods and hexapods was later found to be an analytical artifact based on a sensitivity analysis (Mallatt and Giribet 2006). As in Regier at al. (2005), Branchiopoda were ultimately deemed to be the sister group to Hexapoda, and Maxillopoda were recovered as non-monophyletic. Several deeper relationships were left unanswered, and the authors noted that the phylogenetic positions of ostracods and copepods were extremely unstable.

A non-trivial issue to consider in any morphological or molecular study is the selection of analysis parameters in phylogenetic reconstruction. An important study by Koenemann et al. (2010) examined the conflict in arthropod phylogenetic relationships under different analysis parameters. This study was based on two mitochondrial markers, 16S rDNA and COI, and the nuclear ribosomal marker, 18S. In conclusion, several different phylogenetic reconstructions were recovered, highlighting the confounding effects of secondary structure on tree building. The major findings from this study recovered several monophyletic groups within Pancrustacea, namely Malacostraca and Insecta. As seen in previous studies (Regier et al. 2005, Mallatt and Giribet 2006) Maxillopoda was recovered as paraphyletic (Fig. 4.3). More surprisingly, Hexapoda emerged as a paraphyletic group. Resolving arthropod relationships was not a primary goal of this research, rather this study brought into light the careful consideration that needs to be given to alignment methods, nucleotide coding, and model selection during phylogenetic reconstruction (also see Timm and Bracken-Grissom 2015).

For the past decade, most molecular phylogenies relied on a limited number of nuclear genes or mitochondrial sequences to infer pancrustacean relationships. However, in two consecutive monumental studies, Regier et al. (2008, 2010) investigated arthropod relationships again based on the largest dataset to date: 62 single-copy nuclear protein-coding genes representing over 41kb of sequence data. Here, we will focus on the later study, Regier et al. (2010), which included greater taxon sampling across Pancrustacea (see Fig. 4.3). In this study, Regier et al. included 75 arthropod species and 5 outgroups, with 25 and 19 species represented by crustaceans and hexapods, respectively. For the first time, Remipedia were included in a molecular phylogeny. This resulted in a monophyletic clade including Remepedia + Cephalocarida (=Xenocarida, as seen in previous studies; Giribet et al. 2001) and sister to Hexapoda (=Miracrustacea). As seen in previous studies, Thecostraca + Malacostraca + Copepoda were recovered with high bootstrap support and herein referred to as Multicrustacea. Branchiopoda were recovered as sister to Multicrustacea, a relationship that had not been recovered in previous studies (Regier et al. 2005, Mallatt and Giribet 2006). Pentastomida and Branchiura (=Ichthyostraca) formed a strongly supported clade, as seen in other molecular phylogenies that included both groups (Lavrov 2004, Mallatt and Giribet 2006). For the first time, Mystacocarida were included in a comprehensive molecular analysis and showed an affinity to Ichthyostraca, but with no support. Ostracoda formed a statistically significant clade with Ichthyostraca + Mystacocarida (=Oligostraca). Importantly, this was the first phylogenomic study that recovered high support for several deep relationships across Pancrustacea and supported the notion that increased taxon and gene sampling was needed to improve resolution across the arthropod Tree of Life.

A year later, Andrew (2011) used expressed sequence tags (ESTs) mined from GenBank to infer insect–crustacean relationships. ESTs are short fragments of mRNA sequenced from cDNA libraries that can be used in phylogenetic inference. This phylogenomic approach allowed Andrew (2011) to identify ~282 orthologs (=67,726 amino acids) across 156 taxa (23 species of Crustacea and 63 of Hexapoda for a total of 86 pancrustaceans). The major finding from this study confirmed the

Tetraconata (=Pancrustacea) hypothesis with strong support (Fig. 4.3). As seen in other studies (Regier et al. 2005, 2008, 2010, Mallatt and Giribet 2006) Malacostraca were recovered as a monophyletic group; however, many orders within Malacostraca were excluded in the tree (only Decapoda, Amphipoda, and Euphausicea included). Thecostraca (represented as Cirripedia) were recovered as sister to Malacostraca, and Copepoda were recovered as sister to Thecostraca + Malacostraca (=Multicrustacea). With the recovery of Multicrustacea, Maxillopoda were found to be paraphyletic with respect to Malacostraca. Findings from the EST datasets were in accordance with overarching phylogenetic patterns recovered with protein-coding (Regier et al. 2005, 2008, 2010), ribosomal and/or mitochondrial DNA datasets (Mallatt and Giribet 2006, Koenemann et al. 2010); however, the lack of sampling by Andrew (2011) prohibited a comprehensive evaluation of pancrustacean phylogeny.

Several consistent, yet general patterns of Pancrustacea phylogeny were emerging. However, the limited sampling across the group hindered robust testing of fundamental questions such as "What is the sister group to Hexapoda?" A group of researchers set out to answer this question using EST data in a series of analyses to evaluate pancrustacean relationships (von Reumont et al. 2012; Fig. 4.3). Although the number of pancrustacean species differed across analyses, Ostracoda, Malacostraca, Cirripedia, Copepoda, Branchiopoda, Remipedia, and Hexapoda were included in all. Von Reumont et al. (2012) consistently found Remipedia to be the sister to Hexapoda, similar to that of Regier et al. (2010), which found Remipedia-Cephalocarida (=Xenocarida) as sister to Hexapoda. These findings challenged previous thoughts that suspected remipedes were an early-branching lineage on the crustacean Tree of Life (Schram 1986); however, in recent years, mounting evidence based on morphological, developmental, molecular, and neuroanatomical data has suggested a more derived placement (see section "Remipedia"). Other findings included the monophyly of Pancrustacea, Malacostraca, Multicrustacea, and Branchiopoda. As seen in most other studies, the phylogenetic relationships of deep lineages remain uncertain.

Further studies continued to test the effects of analytical approaches (Rota-Stabelli et al. 2013), similar to those performed a few years earlier (Koenemann et al. 2010). In a set of varied data analyses and parameters, Rota-Stabelli et al. (2013) reused the Regier et al. (2010) dataset to investigate the effects of nucleotide or amino acid alignments in deep phylogenomic reconstruction. More specifically, they performed a series of analyses to test if the same relationships would emerge using different models and different starting alignments (nucleotides vs. amino acids). Rota-Stabelli et al. (2013) found that the two different alignments resulted in strikingly different pancrustacean relationships (Fig. 4.3). The authors concluded that the nucleotide dataset that recovered a Branchiopoda + Malacostraca + Thecostraca + Copepoda clade was inaccurate based on codon (serine) biases in the alignment. Alternatively, the amino acid dataset recovered a Branchiopoda + Copepoda + Remipedia + Hexapoda clade, and sometimes (depending on the model used) Cephalocarida. They also recovered Malacostraca + Oligostraca + Thecostraca. The authors argued that the relationships recovered using amino acids resulted in relationships more similar to EST datasets (Meusemann et al. 2010, von Reumont et al. 2012) and that taxa with similar codon composition tended to group together artificially, implying that amino acid–based studies may be more robust.

The same year, Oakley et al. (2013) investigated the phylogenetic position of Ostracoda, a group often neglected in many higher-level phylogenetic studies. In addition to studying the position of ostracods across the pancrustacean Tree of Life, this study sought to answer several outstanding questions concerning higher-level relationships across Pancrustacea. Using a robust total evidence dataset comprising morphology, EST, mitochondrial genome, nuclear genome, and rDNA data, Oakley et al. (2013) found support for several pancrustacean clades. This study found Ostracoda to be a monophyletic group (but see Oakley et al. 2013 for a detailed discussion) closely related to a clade comprising Mystacocarida, Branchiura, and Pentastomida (= Oligostraca). Oligostraca were recovered to be the sister group to the remaining Pancrustacea. As seen in several other studies, Multicrustacea were recovered with high support (Regier et al. 2010); however, internal relationships between Malacostraca, Cirripedia, and Copepoda remained debatable. As recovered by von Reumont et al. (2012), Remipedia were found as sister to Hexapoda, albeit with low support. A new clade, Allotriocarida, was recovered that included Cephalocarida, Branchiopoda, Remipedia, and Hexapoda, which contradicted the findings of Regier et al. (2010; i.e., with respect to position of Branchiopoda). It was noted that the recovery of Allotriocarida was most probably attributed to the data selection (proteomic and transcriptomic datasets). It is also important to note that data from the remipede and cephalocarid were recycled from the Regier et al. (2010) dataset, possibly perpetuating relationships recovered from previous studies (for a discussion on data recycling, see Timm and Bracken-Grissom 2015). Even so, this was the most comprehensive molecular dataset to date, recovering several well-supported clades with Pancrustacea.

WHERE ARE WE NOW? A SYNTHESIS OF HIGHER-LEVEL SUPPORT AND OUTSTANDING QUESTIONS

Although several outstanding questions still remain across Pancrustacea phylogeny, many higher-level relationships are consistently recovered in molecular studies. The three major clades that include all pancrustacean lineages include Multicrustacea, Oligostraca, and, based on recent evidence, Allotriocarida. It is important to note that none of these higher-level clades (Oligostraca, Multicrustacea, Allotriocarida) are supported by morphology alone; however, all recent total evidence approaches concur with the relationships discussed below.

Multicrustacea are consistently recovered as a monophyletic group with high support and include Malacostraca, Copepoda, and Thecostraca. Monophyly was first demonstrated by nuclear protein-coding genes (Regier et al. 2010) and later supported by EST sequences (von Reumont et al. 2012) and combined analysis of molecular and morphological data (Lee et al. 2013, Oakley et al. 2013). This clade has, however, not been recovered in one recent molecular-only study (Rota-Stabelli et al. 2013) nor any morphology-only phylogenetic analyses, presumably owing to widespread support for Malacostraca as sister to the rest of Pancrustacea (the Entomostraca hypothesis; see Walossek 1999, and Wolfe and Hegna 2014 for a morphological deconstruction of Entomostraca). The internal relationships of Multicrustacea are less well resolved. Although Malacostraca continue to be seen as monophyletic, many rare or understudied groups (e.g., Anaspidacea, Bochusacea, Mictacea, Spelaeogriphacea) have not yet been available for molecular phylogenetic analyses. Additionally, the phylogenetic position of Malacostraca relative to Copepoda and Thecostraca is unresolved, with competing hypotheses suggesting (Malacostraca, (Copepoda+Thecostraca)) (Oakley et al. 2013) or ((Malacostraca + Thecostraca), Copepoda) (Regier et al. 2010, von Reumont et al. 2012).

The superclass Oligostraca also continues to be supported in many molecular phylogenetic studies. The clade, composed of Mystacocarida, Branchiura, Pentastomida, and Ostracoda, has been recovered with nuclear protein coding genes (Regier et al. 2010, Zwick et al. 2012) and total evidence approaches that combine morphology and molecules (Zrzavý et al. 1998, Lee et al. 2013, Oakley et al. 2013). Some internal relationships are well accepted, while others need much attention. Currently, the monophyly of Branchiura + Pentastomida (=Ichthyostraca) is commonly accepted. This relationship is supported by sperm morphology (Wingstrand 1972, Storch and Jamieson 1992, Giribet and Ribera 2000). Additional lines of evidence based on total evidence (Zrzavý et al. 1998, Oakley et al. 2013) and molecules only (Møller et al. 2008, Regier et al. 2010) also support this group. Due to recent total evidence approaches (Oakley et al. 2013), the Ostracoda are monophyletic and fall as a sister clade to Ichthyostraca + Mystacocarida.

Increased taxon sampling from across ostracods and mystacocarids would help test these findings.

The last and perhaps most contentious clade, Allotriocarida, includes Branchiopoda, Cephalocarida, Remipedia, and Hexapoda. Monophyly of this clade was proposed by a combined phylogenetic analysis of transcriptomes, nuclear protein-coding genes, and morphology (Oakley et al. 2013). Internal relationships are in complete flux. The sister lineage to Hexapoda is hotly debated and has been argued to be Remipedia (von Reumont et al. 2012, Oakley et al. 2013), Remipedia + Cephalocarida (=Xenocarida; Regier et al. 2010), or possibly Branchiopoda (Regier et al. 2005, Mallatt and Giribet 2006). Future phylogenetic studies need to target lineages within this group to resolve these important outstanding questions.

Finally, there is growing evidence for Altocrustacea, a group that unites Multicrustacea and Allotriocarida. Monophyly has been supported by phylogenetic analysis of nuclear protein-coding genes (Regier et al. 2010, Zwick et al. 2012), EST data (von Reumont et al. 2012), and combined analysis of morphology and nuclear protein-coding genes (Lee et al. 2013) or morphology and transcriptomes (Oakley et al. 2013). However, this clade has been challenged as paraphyletic (containing Oligostraca) by Rota-Stabelli et al. (2013) and has not been supported by morphological data alone.

THE FUTURE OF PANCRUSTACEAN PHYLOGENY AND CONCLUDING REMARKS

The field of phylogenetics has undergone an exciting revolution in recent years, and emergent methods have the ability to transform morphological and molecular pancrustacean phylogenetics. New technologies for morphological description, such as micro-computed and synchrotron radiation X-ray tomography, autofluorescence imaging, whitening, immunolabeling, spatiotemporal gene expression, and geochemical and taphonomic analysis of fossils, continue to reveal insights into new characters (e.g., Hegna 2010, Haug et al. 2011, Eriksson et al. 2015, Ma et al. 2015, Stegner et al. 2015). These techniques are critical for expanding morphological sampling to taxa that are only available as fixed museum specimens (i.e., difficult-to-collect material which may be unavailable for nucleic acid extraction) and fossils. Besides morphological methods, recent advancements in sequencing techniques have yet to be applied widely across pancrustaceans. Such advancements include anchored hybrid enrichment (AHE) methods for high-throughput sequencing (Lemmon et al. 2012, Lemmon and Lemmon 2013). AHE is capable of targeting hundreds of loci, allowing for resolution from deep to shallow phylogenetic scales (Lemmon et al. 2012, Lemmon and Lemmon 2012). The starting material for this method is genomic DNA, allowing the use of ethanol-preserved specimens. This approach can therefore leverage the thousands of ethanol-preserved tissues housed in museum collections, eliminating challenges associated with tissue collection, preservation, and handling of RNA-preserved or "fresh" specimens. This presents an advantage over alternative "phylotranscriptomic" approaches. A recent study has demonstrated the potential of AHE methods, successfully recovering ~410 loci and ~86 kb (kilobases) of genomic data across the order Decapoda (Wolfe et al. 2019). With heterogeneous sequencing methods and even genome sizes, 100% coverage of all genes for all species may be intractable; optimizing sampling in computer analyses for phylogenetically informative genes may prove fruitful (Fernández et al. 2016). As new morphological and molecular characters are added to taxon sampling, pancrustacean systematists must be mindful of other developments in large-scale phylogenetic analysis, such as methods to mitigate conflict where the topologies of individual gene trees differ from the summary "species tree" when many genes are included (Edwards 2009).

The future of pancrustacean phylogeny is promising, and many lessons have been learned from a conflicted past. It is imperative that concerted efforts be made to include rare and understudied groups to validate or refute assumptions of monophyly within and among major pancrustacean lineages. Increased taxon sampling in combination with new molecular and morphological characters need to be incorporated into future studies to increase resolution at deep splits and avoid the perpetuation of artifacts being included in the same data across multiple studies (i.e., taxon recycling; see Timm and Bracken-Grissom 2015). As learned from previous studies (Koenemann et al. 2010, Rota-Stabelli et al. 2013), future datasets must undergo a series of robust analyses with different parameters to account for potential biases in models, alignments (nucleotides vs. amino acids), and data (i.e., codon usage, gene tree conflict). Emergent technologies in morphological and molecular phylogenetics have the ability to overcome many of these obstacles. Past studies have contributed greatly to our understanding of pancrustacean relationships, and future endeavors must build upon this knowledge to resolve the pancrustacean Tree of Life.

REFERENCES

Abele, L.G., W. Kim, and B.E. Felgenhauer. 1989. Molecular evidence for inclusion of the phylum Pentastomida in the Crustacea. Molecular Biology and Evolution 6:685–691.

Abele, L.G., T. Spears, W. Kim, and M. Applegate. 1992. Phylogeny of selected maxillopodan and other crustacean taxa based on 18S ribosomal nucleotide sequences: a preliminary analysis. Acta Zoologica 73:373–382.

Andrew, D.R. 2011. A new view of insect-crustacean relationships. II. Inferences from expressed sequence tags and comparisons with neural cladistics. Arthropod Structure and Development 40:289–302.

Boxshall, G.A. 2004. The evolution of arthropod limbs. Biological Reviews 79:253–300.

Boxshall, G.A. 2007. Crustacean classification: on-going controversies and unresolved problems. Zootaxa 1668:313–325.

Briggs, D.E., R.A. Fortey, and M.A. Wills. 1992. Morphological disparity in the Cambrian. Science 256:1670–1673.

Brusca, R., W. Moore, and S. Shuster. 2016. Invertebrates. Sinauer Associates, Sunderland, MA.

Camacho, A.I. 2003. Historical biogeography of Hexabathynella, a cosmopolitan genus of groundwater Syncarida (Crustacea, Bathynellacea, Parabathynellidae). Biological Journal of the Linnean Society 78:457–466.

Camacho, A.I., I. Rey, B.A. Dorda, A. Machordom, and A.G. Valdecasas. 2002. A note on the systematic position of the Bathynellacea (Crustacea, Malacostraca) using molecular evidence. Contributions to Zoology 71:123–130.

Cisne, J.L. 1982. Origin of the Crustacea. Pages 65–93 *in* L.G. Abele, editor. The Biology of Crustacea. Vol 1. Systematics, the Fossil Record, and Biogeography. Academic Press, New York.

De Grave, S., T.Y. Chan, K.H. Chu, and J.M. Landeira. 2015. Phylogenetics reveals the crustacean order Amphionidacea to be larval shrimps (Decapoda: Caridea). Scientific Reports 5:17464.

De Oliveira, W., and M.L. Christoffersen. 1999. A cladistic approach to relationships in Pentastomida. Journal of Parasitology 85:695–704.

Dell'Ampio, E., K. Meusemann, N.U. Szucsich, R.S. Peters, B. Meyer, J. Borner, M. Petersen, A.J. Aberer, A. Stamatakis, M.G. Walzl, B.Q. Minh, A. von Haeseler, I. Edesberger, G. Pass, and B. Misof. 2014. Decisive data sets in phylogenomics: lessons from studies on the phylogenetic relationships of primarily wingless insects. Molecular Biology and Evolution 31:239–249.

Dohle, W. 2001. Are the insects terrestrial crustaceans? A discussion of some new facts and arguments and the proposal of the proper name 'Tetraconata' for the monophyletic unit Crustacea+Hexapoda. Annales De La Societe Entomologique De France 37:85–103.

Edgecombe, G.D. 2010. Palaeomorphology: fossils and the inference of cladistic relationships. Acta Zoologica 91:72–80.

Edgecombe, G.D., G.D.F. Wilson, D.J. Colgan, M.R. Gray, and G. Cassis. 2000. Arthropod cladistics: combined analysis of histone H3 and U2 snRNA sequences and morphology. Cladistics 16:155–203.

Edwards, S.V. 2009. Is a new and general theory of molecular systematics emerging? Evolution 63:1–19.

Emerson, M.J., and F.R. Schram. 1990. The origin of crustacean biramous appendages and the evolution of arthropoda. Science 250:667–669.

Eriksson, M.E., F. Terfelt, R. Elofsson, A. Maas, F. Marone, A. Lindskog, D. Waloszek, B. Schmitz, and M. Stampanoni. 2015. Baring it all: undressing Cambrian 'Orsten' phosphatocopine crustaceans using synchrotron radiation X-ray tomographic microscopy. Lethaia 49:312–326.

Fernández, R., G.D. Edgecombe, and G. Giribet. 2016. Exploring phylogenomic relationships within Myriapoda and the effects of matrix composition and occupancy on phylogenomic reconstruction. Systematic Biology 65:871–889.

Gewin, V. 2005. Functional genomics thickens the biological plot. PLoS Biology 3:e219.

GIGA Community of Scientists. 2014. The Global Invertebrate Genomics Alliance (GIGA): developing community resources to study diverse invertebrate genomes. Journal of Heredity 105:1–18.

Giribet, G., and C. Ribera. 2000. A review of arthropod phylogeny: new data based on ribosomal DNA sequences and direct character optimization. Cladistics 16:204–231.

Giribet, G., G.D. Edgecombe, and W.C. Wheeler. 2001. Arthropod phylogeny based on eight molecular loci and morphology. Nature 413:157–160.

Harzsch, S. 2006. Neurophylogeny: architecture of the nervous system and a fresh view on arthropod phyologeny. Integrative and Comparative Biology 46:162–194.

Haug, J.T., C. Haug, V. Kutschera, G. Mayer, A. Maas, S. Liebau, C. Castellani, U. Wolfram, E.N. Clarkson, and D. Waloszek. 2011. Autofluorescence imaging, an excellent tool for comparative morphology. Journal of Microscopy 244:259–272.

Hegna, T.A. 2010. Photography of soft-bodied crustaceans via drying, whitening, and splicing. Journal of Crustacea Biology 30:351–356.

Hennig, W. 1966. Phylogenetic Systematics. University of Illinois Press, Urbana.

Hessler, R.R., and W.A. Newman. 1975. A trilobitomorph origin for the Crustacea. Fossils and Strata 4:437–459.

Hoenemann, M., M.T. Neiber, W.F. Humphreys, T.M. Iliffe, L. Difei,, F.R. Shram, and S. Koenemann. 2013. Phylogenetic analysis and systematic revision of Remipedia (Nectiopoda) from Bayesian analysis of molecular data. Journal of Crustacean Biology 33(5):603–619.

Huys, R., J. Llewellyn-Hughes, S. Conroy-Dalton, P.D. Olson, J.N. Spinks, and D.A. Johnston. 2007. Extraordinary host switching in siphonostomatoid copepods and the demise of the Monstrilloida: integrating molecular data, ontogeny and antennulary morphology. Molecular Phylogenetetics and Evolution 43:368–378.

Jenner, R., C. Dhubhghaill, M. Ferla, and M. Wills. 2009. Eumalacostracan phylogeny and total evidence: limitations of the usual suspects. BMC Evolutionary Biology 9:21.

Khodami, S., J.V. McArthur, L. Blanco-Bercial and P.M. Arbizu. 2017. Molecular phylogeny and revision of copepod orders (Crustacea: Copepoda). Scientific Reports 7:9164.

Koenemann, S., F.R. Schram, M. Honemann, and T.M. Iliffe. 2007. Phylogenetic analysis of Remipedia (Crustacea). Organisms Diversity and Evolution 7:33–51.

Koenemann, S., R.A. Jenner, M. Hoenemann, T. Stemme, and B.M. von Reumont. 2010. Arthropod phylogeny revisited, with a focus on crustacean relationships. Arthropod Structure and Development 39:88–110.

Kosobokova, K.N., and R.R. Hopcraft. 2010. Diversity and vertical distribution of mesozooplankton in the Arctic's Canada Basin. Deep Sea Research II 57:96–110.

Lauterbach, K.E. 1983. Zum Problem der Monophylie der Crustacea. Abhanglungen und Verhandlungen des Naturwissenschaftlichen Vereins in Hamburg 26:293–320.

Lavrov, D.V., W.M. Brown, and J.L. Boore. 2004. Phylogenetic position of the Pentastomida and (pan) crustacean relationships. Proceedings of the Royal Society B: Biological Sciences 271:537–544.

Lee, M.S.Y., J. Soubrier, and G.D. Edgecombe. 2013. Rates of phenotypic and genomic evolution during the Cambrian explosion. Current Biology 23:1889–1895.

Legg, D.A., M.D. Sutton, and G.D. Edgecombe. 2013. Arthropod fossil data increase congruence of morphological and molecular phylogenies. Nature Communications 4:2485.

Lemmon, E.M., and A.R. Lemmon. 2013. High-throughput genomic data in systematics and phylogenetics. Annual Review of Ecology, Evolution, and Systematics 44:99–121.

Lemmon, A.R., S.A. Emme, and E.M. Lemmon. 2012. Anchored hybrid enrichment for massively high-throughput phylogenomics. Systematic Biology 61:727–744.

Ma, X., G.D. Edgecombe, X. Hou, T. Goral, and N.J. Strausfeld. 2015. Preservational pathways of corresponding brains of a Cambrian euarthropod. Current Biology 25:2969–2975.

Mallatt, J., and G. Giribet. 2006. Further use of nearly complete, 28S and 18S rRNA genes to classify Ecdysozoa: 37 more arthropods and a kinorhynch. Molecular Phylogenetics and Evolution 40:772–794.

Manton, S.M. 1977. The Arthropoda: Habits, Functional Morphology, and Evolution. Clarendon Press, Oxford, UK.

Martin, J.W., and J.C. Christiansen. 1995. A morphological comparison of the phyllopodous thoracic limbs of a leptostracan (*Nebalia* sp.) and a spinicaudate conchostracan (*Leptestheria* sp.), with comments on the use of Phyllopoda as a taxonomic categor. Canadian Journal of Zoology 73:2283–2291.

Martin, J.W., and G.E. Davis. 2001. An updated classification of the Recent Crustacea. Natural History Museum of Los Angeles County: Science Series 39:1–124.

Meusemann, K., B.M. von Reumont, S. Simon, F. Roeding, S. Strauss, P. Kuck, I. Ebersberger, M. Walzl, G. Pass, S. Breuers, V. Achter, A. von Haeseler, T. Burmester, H. Hadrys, J.W. Wägele, and B. Misof. 2010. A phylogenomic approach to resolve the arthropod Tree of Life. Molecular Biology and Evolution 27:2451–2464.

Misof, B., S. Liu, K. Meusemann, et al. 2014. Phylogenomics resolves the timing and pattern of insect evolution. Science 346:763–767.

Møller, O.S., J. Olesen, A. Avenant-Oldewage, P.F. Thomsen, and H. Glenner. 2008. First maxillae suction discs in Branchiura (Crustacea): development and evolution in light of the first molecular phylogeny of Branchiura, Pentastomida, and other "Maxillopoda." Arthropod Structure and Development 37:333–346.

Muller, C.H.G., J.L. Rosenberg, S. Richter, and V.B. Meyer-Rochow. 2003. The compound eye of *Scutigera coleoptera* (Linneaus, 1758) (Chilopoda: Notostigmophora): an ultrastrutural reinvestigation that adds support to the Mandiulata concept. Zoomorphology 122:191–209.

Neiber, M.T., T.R. Hartke, T. Stemme, A. Bergmann, J. Rust, T.M. Iliffe, and S. Koenemann. 2011. Global biodiversity and phylogenetic evaluation of Remipedia (Crustacea). PLoS ONE 6:e19627.

O'Reilly, J.E., M. dos Reis, and P.C. Donoghue. 2015. Dating tips for divergence-time estimation. Trends in Genetics 31:637–650.

Oakley, T.H., J.M. Wolfe, A.R. Lindgren, and A.K. Zaharoff. 2013. Phylotranscriptomics to bring the understudied into the fold: monophyletic ostracoda, fossil placement, and pancrustacean phylogeny. Molecular Biology and Evolution 30:215–233.

Olesen, J. 2009. Phylogeny of Branchiopoda (Crustacea)—character evolution and contribution of uniquely preserved fossils. Arthropod Systematics and Phylogeny 67:3–39.

Olesen, J., S.V. Martinsen, T.M. Iliffe, and S. Koenemann. 2014. Remipedia. Pages 84–89 *in* J.W. Martin, J. Olesen, and J.T. Hoeg, editors. Atlas of Crustacean Larvae. Johns Hopkins University Press, Baltimore.

Olesen, J., K. Meland, H. Glenner, and T.M. Iliffe. 2017. *Xibalbanus cozumelensis*, a new species of Remipedia (Crustacea) from Cozumel, Mexico, and a molecular phylogeny of *Xibalbanus* on the Yucatan Peninsula. European Journal of Taxonomy 316:1–27.

Pennak, R.W., and D.J. Zinn. 1943. Mystacocarida, a new order of Crustacea from intertidal beaches in Massachusetts and Connecticut. Smithsonian Miscellaneous Collections 103:1–11.

Pérez-Losada, M., J.T. Hoeg, and K.A. Crandall. 2004. Unraveling the evolutionary radiation of the thoracian barnacles using molecular and morphological evidence: A comparison of several divergence time estimation approaches. Systematic Biology 53:244–264.

Pérez-Losada, M., J.T. Hoeg, and K.A. Crandall. 2009. Remarkable convergent evolution in specialized parasitic Thecostraca (Crustacea). BMC Biology 7:15.

Petrunina, A.S., T.V. Neretina, N.S. Mugue, and G.A. Kolbasov. 2013. Tantulocarida versus Thecostraca: inside or outside? First attempts to resolve phylogenetic position of Tantulocarida using gene sequences. Journal of Zoological Systematics and Evolutionary Research 52:100–108.

Poore, G.C.B. 2005. Peracarida: monophyly, relationships, and evolutionary success. Nauplius 13:1–27.

Regier, J.C., J.W. Shultz, and R.E. Kambic. 2005. Pancrustacean phylogeny: hexapods are terrestrial crustaceans and maxillopods are not monophyletic. Proceedings of the Royal Society B: Biologial Sciences 272:395–401.

Regier, J.C., J.W. Shultz, A.R.D. Ganley, A. Hussey, D. Shi, B. Ball, A. Zwick, J.E. Stajich, M.P. Cummings, J.W. Martin, and C.W. Cunningham. 2008. Resolving arthropod phylogeny: exploring phylogenetic signal within 41 kb of protein-coding nuclear gene sequence. Systematic Biology 57:920–938.

Regier, J.C., J.W. Shultz, A. Zwick, A. Hussey, B. Ball, R. Wetzer, J.W. Martin, and C.W. Cunningham. 2010. Arthropod relationships revealed by phylogenomic analysis of nuclear protein-coding sequences. Nature 463:1079–1083.

Richter, S. 2002. The Tetraconata concept: hexapod-crustacean relationships and the phylogeny of Crustacea. Organisms, Diversity and Evolution 2:217–237.

Richter, S., and G. Scholtz. 2001. Phylogenetic analysis of the Malacostraca (Crustacea). Journal of Zoological Systematics and Evolutionary Research 39:113–136.

Riley, J., A.A. Banja, and J.L. James. 1978. The phylogenetic relationships of the Pentastomida: the case for their inclusion within the Crustacea. International Journal of Parasitology 8:245–254.

Rota-Stabelli, O., N. Lartillot, H. Philippe, and D. Pisani. 2013. Serine codon-usage bias in deep phylogenomics: pancrustacean relationships as a case study. Systematic Biology 62:121–133.

Sasaki, G., K. Ishiwata, R. Machida, T. Miyata, and Z.H. Su. 2013. Molecular phylogenetic analyses support the monophyly of Hexapoda and suggest the paraphyly of Entognatha. BMC Evolutionary Biology 13:236–236.

Schram, F.R. 1984. Relationships within eumalacostracan Crustacea. Transactions of the San Diego Society of Natural History 20:301–312.

Schram, F.R. 1986. Crustacea. Oxford University Press, New York.

Schram, F.R., and C.H.J. Hof. 1998. Fossils and the interrelationships of major crustacean groups. Pages 223–302 *in* G.D. Edgecombe, editor. Arthropod Fossils and Phylogeny. Columbia University Press, New York.

Schram, F.R., and S. Koenemann. 2004. Are the crustaceans monophyletic? Pages 319–329 *in* J. Cracraft and M.J. Donoghue, editors. Assembling the Tree of Life: Where We Stand at the Beginning of the 21st Century. Oxford University Press, New York.

Sieg, J. 1983. Evolution of Tanaidacea. Pages 229–256 *in* Crustacean Phylogeny. Crustacean Issues, Vol. 1. A.A. Balkema, Rotterdam, Netherlands.

Snodgrass, R.E. 1938. Evolution of the Annelida, Onychophora and Arthropoda. Smithsonian Miscellaneous Collections 97:1–159.

Spears, T., and L.G. Abele. 1999. Phylogenetic relationships of crustaceans with foliaceous limbs: an 18S rDNA study of Branchiopoda, Cephalocarida, and Phyllocarida. Journal of Crustacean Biology 19:825–843.

Stegner, M.E., T. Stemme, T.M. Iliffe, S. Richter, and C.S. Wirkner. 2015. The brain in three crustaceans from cavernous darkness. BMC Neuroscience 16:19–19.

Stemme, T., T.M. Iliffe, G. Bicker, S. Harzsch, and S. Koenemann. 2012. Serotonin immunoreactive interneurons in the brain of the Remipedia: new insights into the phylogenetic affinities of an enigmatic crustacean taxon. BMC Evolutionary Biology 12:168–168.

Stemme, T., T.M. Iliffe, B.M. von Reumont, S. Koenemann, S. Harzsch, and G. Bicker. 2013. Serotonin-immunoreactive neurons in the ventral nerve cord of Remipedia (Crustacea): support for a sister group relationship of Remipedia and Hexapoda? BMC Evolutionary Biology 13:119–119.

Storch, V., and B.G.M. Jamieson. 1992. Further spermatological evidence for including the Pentastomida (tongue worms) in the Crustacea. International Journal for Parasitology 22:95–108.

Strausfeld, N.J. 2012. Arthropod Brains: Evolution, Functional Elegance, and Historical Significance. Belknap Press, Cambridge, MA.

Strausfeld, N.J., and D.R. Andrew. 2011. A new view of insect-crustacean relationships. I. Inferences from neural cladistics and comparative neuroanatomy. Arthropod Structure and Development 40:276–288.

Tiegs, O.W., and S.M. Manton. 1958. The evolution of the Arthropoda. Biological Reviews 33:255–333.

Timm, L., and H.D. Bracken-Grissom. 2015. The forest for the trees: evaluating molecular phylogenies with an emphasis on higher-level Decapoda. Journal of Crustacean Biology 35:577–592.

von Reumont, B.M., R.A. Jenner, M.A. Wills, E. Dell'Ampio, G. Pass, I. Ebersberger, B. Meyer, S. Koenemann, T.M. Iliffe, A. Stamatakis, O. Niehuis, K. Meusemann, and B. Misof. 2012. Pancrustacean phylogeny in the light of new phylogenomic data: support for Remipedia as the possible sister group of Hexapoda. Molecular Biology and Evolution 29:1031–1045.

von Reumont, B.M., A. Blanke, S. Richter, F. Alvarez, C. Bleidorn, and R.A. Jenner. 2014. The first venomous crustacean revealed by transcriptomics and functional morphology: remipede venom glands express a unique toxin cocktail dominated by enzymes and a neurotoxin. Molecular Biology and Evolution 31:48–58.

Walossek, D. 1999. On the Cambrian diversity of Crustacea. Pages 3–27 *in* F.R. Schram and J.C. von Vaupel Klein, editors. Crustaceans and the Biodiversity Crisis: Proceedings of the Fourth International Crustacean Congress, Amsterdam. Brill Academic Publishers, Leiden, The Netherlands.

Wheeler, W.C., P. Cartwright, and C.Y. Hayashi. 1993. Arthropod phylogeny: a combined approach. Cladistics 9:1–39.

Wills, M.A., D.E.G. Briggs, R.A. Fortey, M. Wilkinson, and P.H. Sneath. 1998. An arthropod phylogeny based on fossil and Recent taxa. Pages 33–105 *in* G.D. Edgecombe, editor. Arthropod Fossils and Phylogeny. Columbia University Press, New York.

Wills, M.A., R.A. Jenner, and C.N. Dhubhghaill. 2009. Eumalacostracan evolution: conflict between three sources of data. Arthropod Systematics and Phylogeny 67:71–90.

Wilson, G.D.F. 1992. Computerized analysis of crustacean relationships. Acta Zoologica 73:383–389.

Wingstrand, K.G. 1972. Comparative spermatology of a pentastomid, *Raillietiella hemidactyli*, and a branchiuran crustacean, *Argulus foliaceus*, with a discussion of pentastomid relationships. Det Kongelige Danske Videnskabernes Selskab Biologiske Skrifter 19:1–72.

Wirkner, C.S., and S. Richter. 2010. Evolutionary morphology of the circulatory system in Peracarida (Malacostraca; Crustacea). Cladistics 26:143–167.

Wolfe, J.M., and T.A. Hegna. 2014. Testing the phylogenetic position of Cambrian pancrustacean larval fossils by coding ontogenetic stages. Cladistics 30:366–390.

Wolfe, J.M., A.C. Daley, D.A. Legg, and G.D. Edgecombe. 2016. Fossil calibrations for the arthropod Tree of Life. Earth Science Reviews 160:43–110.

Wolfe, J.M., J.W. Breinholt, K.A. Crandall, A.R. Lemmon, E.M. Lemmon, L.E. Timm, M.E. Siddall, and H.D. Bracken-Grissom. 2019. A phylogenomic framework, evolutionary timeline, and genomic resources for comparative studies of decapod crustaceans. Proceedings of the Royal Society B 286:20190079.

Yager, J. 1994. Speleonectes gironensis, new species of remipede (Crustacea: Remipedia: Speleonectidae) from anchialine caves in Cuba, with remarks on biogeography and ecology. Journal of Crustacean Biology 14:752–762.

Zhang, X.G., D.J. Siveter, D. Waloszek, and A. Maas. 2007. An epipodite-bearing crown-group crustacean from the Lower Cambrian. Nature 449:595–598.

Zrzavý, J., and P. Stys. 1997. The basic body plan of arthropods: insights from evolutionary morphology and developmental biology. Journal of Evolutionary Biology 10:353–367.

Zrzavý, J., V. Hypsa, and M. Vlaskova. 1998. Arthropod phylogeny: taxonomic congruence, total evidence and conditional combination approaches to morphological and molecular data sets. Pages 97–107 *in* R.A. Fortey and R.H. Thomas, editors. Arthropod Relationships. Chapman and Hall, London.

Zwick, A., J.C. Regier, and D.J. Zwickl. 2012. Resolving discrepancy between nucleotides and amino acids in deep-level arthropod phylogenomics: differentiating serine codons in 21-amino-acid models. PLoS ONE 7:e47450.

5

CRUSTACEANS AND INSECT ORIGINS

Björn M. von Reumont and Gregory D. Edgecombe

Abstract

Although Crustacea has a long history of being recognized as a formal taxonomic group in arthropod classification, the past 30 years have witnessed repeated challenges to crustacean monophyly. Few unambiguous autapomorphic characters for crustaceans have been proposed by morphologists, and many diagnostic characters can be interpreted as symplesiomorphies of Mandibulata. More serious challenges arise from molecular phylogenetics: irrespective of the scope of taxonomic and/or character sampling or analytical methods, a pancrustacean clade in which "Crustacea" is paraphyletic with respect to Hexapoda is retrieved. However, most traditional single to multigene studies infer phylogenies that display considerable mutual conflict. Although hexapod monophyly is robust and its deep branchings have recently been recovered using large-scale transcriptomic datasets, its crustacean sister group has been contentious. To some extent, a conclusive result is still hindered by uneven taxonomic coverage, with some key groups still being undersampled in phylogenomic studies. Nonetheless, phylogenomic analyses provide some robust results: notably, Hexapoda is part of a pancrustacean clade named Allotriocarida, which includes Cephalocarida and Branchiopoda as a grade or each other's sister group, and Remipedia as the closest relatives to Hexapoda. Neuroanatomical support for a rival malacostracan-remipede-hexapod clade is incongruent with molecular datasets, which instead group Malacostraca, Copepoda, and Cirripedia as a clade. However, cirripedes resolve either as a sister group to copepods or to malacostracans, and this instability casts doubt on the typical pattern in molecular analyses that position malacostracans unexpectedly deep within the crustacean lineage. Pancrustacean phylogeny requires critical interpretation of phylogenomic data to reveal conflict in the data and ambiguous signals within the selected set of orthologous genes.

Evolution and Biogeography. Edited by Gary C. B. Poore and Martin Thiel.

INTRODUCTION

Twenty-five years ago, the notion that *The Natural History of the Crustacea* would treat insects as crustaceans was all but unthinkable. Arthropod phylogenies had long positioned insects and their primitively flightless hexapod relatives as allied to myriapods rather than to crustaceans, a framework in which terrestrial and marine members of the jawed arthropods, Mandibulata, were systematically separated. Diverse suites of molecular data have nonetheless provided a very compelling case for "Crustacea" being a paraphyletic assemblage within a broader clade named Pancrustacea (Zrzavý and Štys 1997) or Tetraconata (Dohle 2001). The most species-rich clade within Pancrustacea is Hexapoda (=Insecta and the flightless, entognathous Protura, Collembola, and Diplura), known from an estimated 1.2 million species, most of which are insects sensu stricto. Crustacean phylogeny has evolved from a search for a hypothetical ancestral crustacean based on morphological comparisons to analyses based on hundreds of orthologous genes that aim to resolve the sequence of crustacean sister groups to Hexapoda. Herein we review recent hypotheses about crustacean–hexapod interrelationships, focusing on recent phylogenomic analyses based on transcriptomic approaches rather than early efforts based on targeted sequencing of a few loci.

CRUSTACEAN MONOPHYLY OR PARAPHYLY: MORPHOLOGICAL ARGUMENTS

Crustacea has a long history of taxonomic usage, and for the most part the membership of the group had been widely agreed (Fig. 5.1). Some taxa were recognized as crustaceans comparatively recently, such as the proposal that Pentastomida (tongue worms; Fig. 5.1A) are allied to Branchiura (fish lice; Fig. 5.1B) within a "maxillopodan" group, originally based on sperm ultrastructure (Wingstrand 1972) and later robustly amplified by molecular data (Møller et al. 2008). Mitochondrial gene order rearrangements provide further support for placement of pentastomids within Pancrustacea (Lavrov et al. 2004). In more recent times, a multigene analysis by Regier et al. (2010) provided strong support for a clade named Ichthyostraca (Branchiura + Pentastomida) within a larger clade of Oligostraca that also includes Ostracoda (Fig. 5.1C) and Mystacocarida (Fig. 5.1F), hypotheses that first emerged in the 1990s based on analyses of just a few molecular markers (Zrzavý et al. 1998). These findings were corroborated by Oakley et al. (2013), who analyzed a mixed dataset that included the Regier et al. (2010) data and the first transcriptomic data for Branchiura and Ostracoda. These hypotheses are in conflict with the relegation of pentastomids to a "protarthropod" grade as many morphologists have advocated (Almeida et al. 2008, Castellani et al. 2011).

In the case of Crustacea, the group's taxonomic recognition was perpetuated by systematic inertia as much as by cogent phylogenetic arguments. As long ago as 1983, K.-E. Lauterbach made the observation that crustacean monophyly can be defended by a (then) surprisingly few putative autapomorphic characters. These amounted to the naupliar eye and the restriction of nephridial derivatives to two cephalic segments, those of the second antenna and maxilla (Lauterbach 1983). Additional characters have, however, been proposed to support crustacean monophyly by including information from exceptionally preserved Cambrian fossils (Walossek 1999, Haug et al. 2012). In this scheme, several characters support Crustacea s.l. (which includes stem-group forms), additional characters support a more restricted clade named Labrophora composed of the fossil Phosphatocopina and the crustacean crown group, and other characters support the crustacean crown group (=Eucrustacea of Waloszek and collaborators). In some formulations, Crustacea s.l. or "crustaceans" are equivalent to Mandibulata (Zhang et al. 2007), i.e., a monophyletic group united by the post-tritocerebral appendages being modified as mandibles, but this framework has sometimes excluded hexapods and myriapods from consideration or explicitly argued that they are not members of Eucrustacea (e.g., Haug et al. 2012).

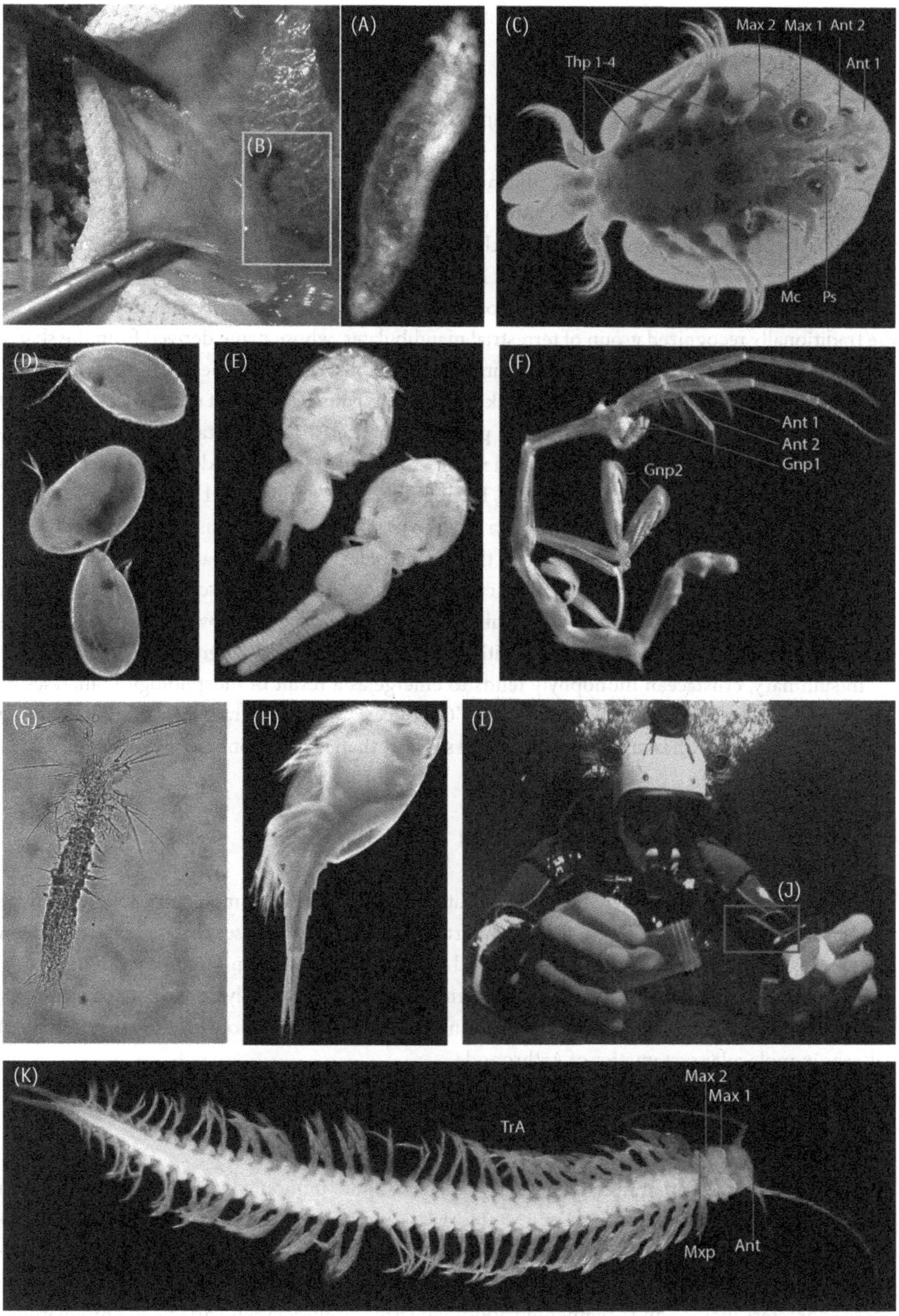

Fig. 5.1.
Diversity and morphological disparity of crustaceans. Many crustaceans are parasites and show derived body shapes adapted to their hosts. An extreme example is pentastomids (tongue worms), such as (A) *Raillietiella* sp., which lives in the lung tracts of vertebrates. (B) Two pentastomids in vivo in the right lung wing of a house gekko. The branchiuran fishlouse *Argulus foliaceaus* (C) parasitizes fish; interestingly the sealouse *Caligus* also parasitizes the gills of salmonids, and its parasitic, blood-sucking feeding mode is convergent to branchiurans. Many taxa are extremely modified and miniaturized. These include (D) ostracods (*Heterocypris* sp.), (E) copepods (*Caligus rogeresseryi*), and (G) mystacocarids (*Derocheilocaris typicus*). (F, H) Examples of two relatively unknown malacostracans that contradict the typical "lobster-like" image of the group. (F) *Caprella causa* (male), a caprellid or skeleton shrimp, a rather derived group of small malacostracans that inhabit sea-grass beds. (H) *Nebalia antarctica*, a leptostracan species representing an older malacostracan lineage with seven pleonite segments. The remipede *Xibalbanus tulumensis* (J, K), collected by the first author in a Mexican anchialine cave system (I). Abbreviations: Ant, antenna; Gnp, gnathopod; Max, maxilla; Mc, mouth cone; Mxp, maxilliped; Ps, preoral spine; Thp, thoracopod; TrA, trunk appendages. All images copyright B. von Reumont. See color version of this figure in centerfold.

Crustacean monophyly has not been universally endorsed by morphologists (Richter et al. 2009). Available morphological support for Crustacea was critiqued by Schram and Koenemann (2004), who produced a tree based on morphological characters that depicted "Crustacea" as a grade with hexapods part of the ingroup (as sister group to Branchiopoda). Earlier, a morphology-based analysis using Hennigian argumentation had likewise depicted "Crustacea" as paraphyletic within Mandibulata but disputed the monophyly of Pancrustacea (Moura and Christoffersen 1996). In this scheme, remipedes were singled out as the closest relatives of Tracheata (= Myriapoda + Hexapoda), the traditionally recognized group of terrestrial mandibulates, whose name derived from the shared presence of a tracheal respiratory system. A modified version of this hypothesis has been advocated and formalized taxonomically by Fanenbruck (2009), who named a series of nodes that deconstruct "Crustacea" as a paraphyletic assemblage of primitively marine Mandibulata. In a numerical parsimony analysis including the Cambrian fossil members of "Crustacea s.l." and most of the characters used to defend the monophyly of Crustacea, Labrophora, and Eucrustacea, Legg et al. (2013) recovered the fossils as stem group Mandibulata (rather than stem group Crustacea), monophyly of Pancrustacea, and "Eucrustacea" as a grade that include Hexapoda. Neuroanatomists have repeatedly argued for crustacean paraphyly in the context of Pancrustacea (Harzsch 2006, Strausfeld and Andrew 2011), based on characters of the brain and optic neuropils shared by hexapods and subsets of "Crustacea." These are discussed below with reference to potential sister groups to Hexapoda.

In summary, crustacean monophyly tends to emerge as a result of morphological analyses by excluding hexapods, and the hypothesis that "Crustacea" may be a grade of pancrustaceans or mandibulates, or both, has more morphological support than is commonly acknowledged.

CRUSTACEAN PARAPHYLY: MOLECULAR ARGUMENTS

Whereas crustacean monophyly has been debated by morphologists, molecular data lean much more in favor of paraphyly (Fig. 5.2, Fig 5.3). Apart from some early analyses that included only a few crustaceans and a few genes (e.g., Friedrich and Tautz 1995), crustacean monophyly is not recovered with molecular datasets. The detailed pattern of crustacean paraphyly differs between analyses, but the ingroup positioning of Hexapoda within "Crustacea" has been one of the most persistent patterns in molecular systematics of Arthropoda.

Hexapod Monophyly and the Deep Nodes in Hexapoda

Morphologists have sometimes commented that hexapod monophyly is supported by fewer characters than might be expected. For example, Klass and Kristensen (2001) cited the specific position of the thoracic-abdominal boundary as the principal autapomorphy of Hexapoda and considered that some other characters previously used to support hexapod monophyly were ambiguous. These characters include an 11-segmented abdomen (fixed in Insecta but variable in entognathous hexapods), 6-segmented trunk appendages, and a lack of *Distal-less* expression in the mandible (not yet examined in all entognathous groups).

Hexapod monophyly came under fire in a series of papers based on analyses of sequence data for complete mitochrondrial genomes ("mitogenomics"). These separated Collembola from other hexapods and typically yielded reciprocal paraphyly of Crustacea and Hexapoda (Nardi et al. 2003, Carapelli et al. 2007), though the efficacy of mitochondrial sequences for resolving these questions has been disputed (Cameron et al. 2004, Hassanin 2006). Likewise, other early molecular analyses based on just a few genes and sometimes combined with morphology (Giribet et al. 2004, Koenemann et al. 2010) occasionally failed to retrieve hexapod monophyly. However, this is a footnote in history because as gene samples grew through the late 2000s, hexapod monophyly was

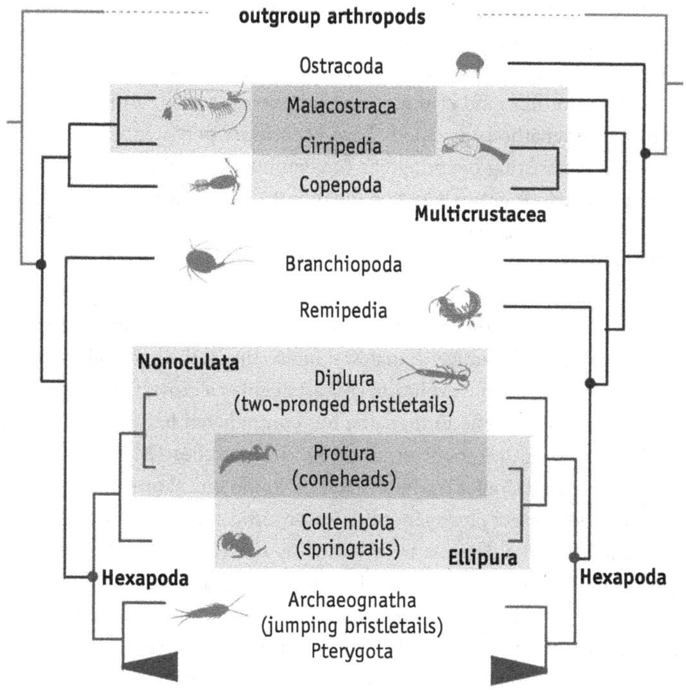

Fig. 5.2.
Summary of phylogenomic hypotheses for Pancrustacea from the first well-sampled analyses of expressed sequence tag (EST) data. (Left) Topology from Meusemann et al. (2010). (Right) Topology from von Reumont et al. (2012).

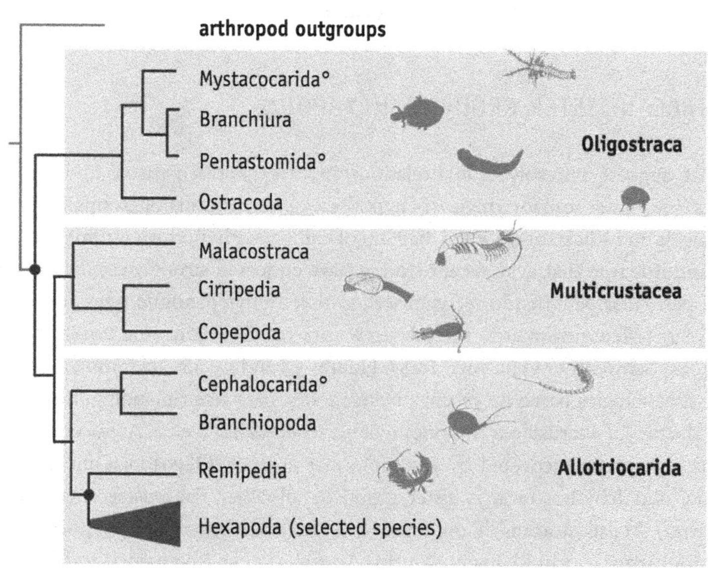

Fig. 5.3.
Summary of a phylogenomic study on pancrustaceans by Oakley et al. (2013), including also morphological and multi-gene data. Hollow circles behind taxon names indicate missing transcriptomic data at that time.

repeatedly and strongly recovered (Timmermans et al. 2008, Meusemann et al. 2010, Regier et al. 2010, Andrew 2011, von Reumont et al. 2012).

While Ectognatha (including jumping bristletails, silverfish and firebrats, and winged insects) as one of the two larger major hexapod clades is well supported in many molecular studies, phylogenetic relationships of entognathous hexapods have been more persistently debated. Some authors considered Entognatha, consisting of springtails (Collembola), coneheads (Protura) and diplurans (Diplura), to be a monophyletic group based on the mouthparts being concealed in gnathal pouches. Within a monophyletic Entognatha, two conflicting hypotheses were supported: Ellipura unites Protura and Collembola, whereas Nonoculata groups Protura and Diplura. Ellipura has long been endorsed by morphologists based on, for example, the style of entognathy (the cranial folds meeting at a *linea ventralis*). Nonoculata is the grouping that was so far recovered in most molecular studies, particularly based on rRNA and multigene analyses. Interestingly, the first major phylogenomic analyses on arthropod and pancrustacean evolution (Meusemann et al. 2010, von Reumont et al. 2012; see Fig. 5.2) recovered the monophyletic Entognatha but contradicted both Nonoculata (Meusemann et al. 2010) and Ellipura (von Reumont et al. 2012). Both studies (Meusemann et al. 2010, von Reumont et al. 2012) had used similar taxon sets and methodological procedures. This puzzling outcome on primarily wingless insect phylogenomics was investigated more deeply by Dell'Ampio et al. (2014), who concluded that the datasets were biased by misleading signals and missing data. The persistent question of monophyly versus paraphyly of Entognatha is currently resolved in favor of paraphyly: Collembola and Protura unite as Ellipura, and Ellipura is sister group to a clade composed of Diplura and Ectognatha/Insecta (Misof et al. 2014). The latest analyses in hexapod phylogenomics (Fig. 5.4) robustly support Hexapoda and resolve its deepest divergences in a manner that is highly congruent with morphology (Misof et al. 2014). A dipluran-insect clade (named Cercophora, for the shared presence of cerci), for example, has compelling morphological support, including characters from embryology (Tomizuka and Machida 2015) and accessory microtubules in the sperm (Dallai et al. 2011). Within Insecta, the pattern of paraphyly of primitively flightless taxa (Archaeognatha and Zygentoma, including *Tricholepidion* as successive sister groups to Pterygota) is precisely as advocated by most morphologists. These results mean that when considering the status of Hexapoda within Pancrustacea, hexapods can be taken as a single (astoundingly speciose) lineage.

THE CRUSTACEAN SISTER GROUP OF HEXAPODA

Diagrams that suggest unresolved interrelationships are routinely used to depict crustacean phylogenetics (see Jenner 2010 for a review of hypotheses about how major groups are related; and see Fig. 12.3 in Wägele and Kück 2014, a Venn diagram of conflicting hypotheses from molecular datasets in general), and it is true that even recent studies have endorsed strongly conflicting relationships. Nonetheless, particular patterns do recur in well-sampled phylogenomic trees constructed during the past few years (Meusemann et al. 2010, Andrew 2011, von Reumont et al. 2012, Oakley et al. 2013, Misof et al. 2014; Schwentner et al. 2017, 2018). Figures 5.5 and 5.6 aim to summarize the congruent or incongruent topologies based on recent phylogenomic data sets (including the multigene study by Regier et al. 2010). Pancrustacea is monophyletic, whereas Tracheata/Atelocerata is not. Indeed, Tracheata has never been recovered by any molecular analysis. Oligostraca, including Ostracoda, Mystacocarida, and Ichythostraca, is sister group to all other Pancrustacea (=Altocrustacea in Regier et al. 2010). Malacostracans do not unite with hexapods, but rather group with copepods and cirripedes (=Multicrustacea in Regier et al. 2010). Branchiopoda, Remipedia, and Hexapoda form a clade together with Cephalocarida (=Allotriocarida in Oakley et al. 2013).

The following section addresses the case for and against current candidates for the closest relatives of Hexapoda in the context of crustacean paraphyly and the Allotriocarida hypothesis.

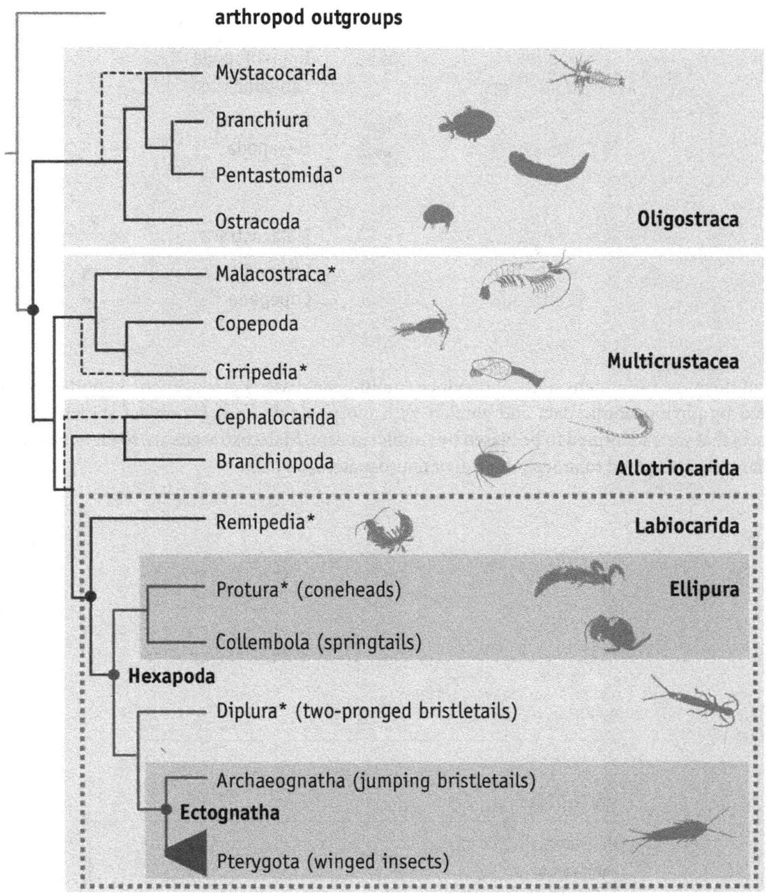

Fig. 5.4.
Summary cladogram of a phylogenomic study on pancrustaceans, including results of Misof et al. (2014) on insect phylogenomics. Hollow circles behind taxon names indicate missing transcriptomic data. Asterisks indicate important subgroups that still need to be covered more thoroughly. Schwentner et al. (2017) contributed new data on cephalocarids and mystacocarids. Dashed lines indicate inferred topologies favored by Schwentner et al. (2017), but the ambiguous position of cirripedes—discussed by Meusemann et al. (2010), von Reumont et al. (2012)—was not addressed in their study.

Malacostraca

The hypotheses that Crustacea are paraphyletic and that Malacostraca are more closely related to Hexapoda than to any "entomostracans" was advanced by Hanström (1926) drawing on neuroanatomical similarities. Modern neuroanatomical comparisons have reinforced this interpretation (Strausfeld 2009, Strausfeld and Andrew 2011): putative shared derived characters of Malacostraca and Insecta include three nested optic neuropils (lamina, medulla, and lobula), the adjoining axons in the form of chiasmata, a glomerular olfactory lobe connected to the lateral protocerebrum, and a stratified, spindle-shaped central complex. Arthropod phylogenies based on discrete neuroanatomical characters retrieve Malacostraca as sister group of Insecta (Strausfeld and Andrew 2011), although it is worth noting that remipedes and entognathous hexapods have sometimes not been coded as terminals. Similarities in the brains of remipedes and malacostracans, such as the presence

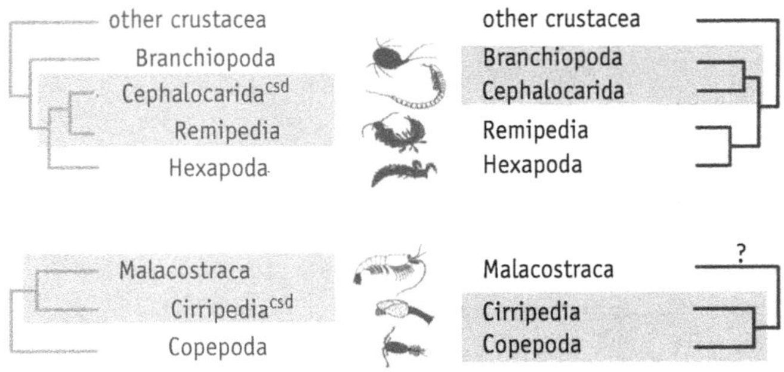

Fig. 5.5.
Summary of the most prominent shifts of previous (multi-gene–based) phylogenetic hypotheses, which were contradicted by phylogenomic data and conflict with morphology. "csd" (=conflict shown and discussed) indicates taxa that were identified to be biased by conflict in data. Malacostracans are so far anomalously placed in molecular trees compared to morphological or neuroanatomical data.

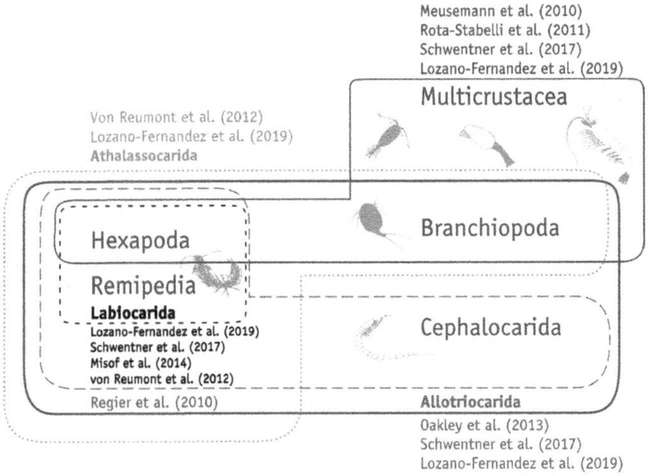

Fig. 5.6.
Venn diagram summarizing recent phylogenomic studies and inferred clades for Hexapoda and closely related crustacean lineages.

of hemiellipsoid bodies and a chiasm in the olfactory globular tract, have however been cited as evidence that they are each other's closest relative (Stemme et al. 2012). The presence of mushroom bodies that resemble those of insects has been established within Malacostraca, in stomatopods (Wolff et al. 2017). It can reasonably be summarized that ((Remipedia + Malacostraca) Hexapoda) is the neuroanatomical solution to the "insects as crustaceans" question.

An alliance between remipedes and hexapods has considerable molecular support (as discussed below), but a close alliance of malacostracans to either does not. The position of malacostracans relative to hexapods and remipedes is a case of character conflict (Fig. 5.5), with several possible explanations. One explanation could be that malacostracans are spuriously attracted to copepods and thecostracans as a result of systematic error in molecular datasets.

Alternatively, the brains and optic lobes of malacostracans and insects would have to share convergent rather than homologous similarities, or the brains of branchiopods have undergone secondary character loss from an ancestor that possessed characters shared by malacostracans (and remipedes) and insects (Strausfeld and Andrew 2011; Wolff et al. 2017). It is noteworthy that several of the rather ancient malacostracan groups such as Leptostraca (Fig. 5.1H) are still underrepresented in phylogenomic analyses (but see Schwentner et al. 2018). Representatives of such lineages might share and harbor more ancient molecular characters than the relatively deeply nested malacostracan species that so far predominate in analyses. Inclusion of older lineages might lead to a more stable position of Malacostraca relative to other pancrustacean taxa by adding more positions that are not autapomorphic characters of the derived groups.

Allotriocarida 1: Branchiopoda

A sister group relationship between Branchiopoda and Hexapoda was first found in various analyses of a few nuclear protein-coding genes, with a Cephalocarida + Remipedia clade being their sister group (summarized by Regier et al. 2005). This hypothesis is thus compatible with the broader clade Allotriocarida. A branchiopod-hexapod clade has also been retrieved in some analyses using transcriptomes, although in most cases these were lacking data for remipedes and cephalocarids (Meusemann et al. 2010, Andrew 2011, Rota-Stabelli et al. 2011) or included just one remipede (Borner et al. 2014, Lozano-Fernandez et al. 2016). In the latter cases, pancrustacean interrelationships were resolved as (Malacostraca (Remipedia (Branchiopoda + Hexapoda))).

The possibility of a branchiopod sister group to hexapods led to a theory of freshwater origins for Hexapoda (Glenner et al. 2006). A challenge to this hypothesis, however, is that Cambrian fossils suggest a marine origin of total-group Branchiopoda (i.e., the branchiopod crown group together with its fossil stem group). The late Cambrian *Rehbachiella kinekullensis*, which is unquestionably marine, is generally regarded as a stem group branchiopod (Olesen 2009), although some analyses group it more closely with Cephalocarida (Oakley et al. 2013). Cambrian "Small Carbonaceous Fossils" that have been identified as mandibles and filter plates of Branchiopoda (Harvey and Butterfield 2008, Harvey et al. 2012) have likewise been extracted from marine sediments.

Allotriocarida 2: Remipedia

The first molecular evidence in favor of a remipede-hexapod sister group relationship was presented by Ertas et al. (2009) based on three highly similar subunits of hemocyanin. Regier et al. (2010), who conducted a multigene study on arthropod evolution using specifically designed primer-based amplification of 62 genes, inferred a clade of Remipedia + Cephalocarida (named Xenocarida) to be sister group to Hexapoda. The remipede-cephalocarid grouping was contradicted in reanalyses of the original dataset that formalized it taxonomically (see Regier et al. 2010 vs, Rota-Stabelli et al. 2013b). One should also note that the number of overlapping genes for cephalocarids and remipedes in the Regier et al. (2010) analyses does not cover all genes; in general, only 41 of 62 genes were present for cephalocarids. However, since its first occurrence and description by Spears and Abele (1998), the grouping of Cephalocarida + Remipedia has been discussed critically because both taxa show typically long branches in molecular phylogenies, suggestive of artifactual attraction. Novel neuronatomical descriptions of cephalocarids (Stegner and Richter 2011, Stegner et al. 2014) and remipedes (Stemme et al. 2012, 2013) have not yielded support for an especially close relationship between the two groups, nor a derived position of cephalocarids. An alternative alliance between branchiopods and cephalocarids, retrieved in other multi-gene analyses (von Reumont et al. 2009, Oakley et al. 2013), is more compatible with most morphological data (see Fig. 5.5), although a few characters of ovary structure have been cited as possible synapomorphies of remipedes, cephalocarids, and hexapods (Kubrakiewicz et al. 2012). Based on rRNA data, von Reumont et al. (2009) first showed that the systematic placement of cephalocarids is biased by substitution models that do not implement base composition changes along the tree.

The first phylogenomic study on pancrustaceans to include novel transcriptome data for remipedes and other crustaceans (von Reumont et al. 2012) yielded strong support for remipedes as closest relatives to Hexapoda. More broadly, this study also recovered Allotriocarida with its internal relationships resolved as (((Branchiopoda + Cephalocarida), Remipedia), Hexapoda), a set of results presented later in Oakley et al. (2013; see Fig. 5.1 and Fig. 5.2). However, von Reumont et al. (2012) also demonstrated the importance and possible impact of orthology inference; using slightly differing settings or including slightly different genome taxa to compile ortholog sets resulted in different topologies.

In their combined study, Oakley et al. (2013) added further transcriptomes of ostracods and branchiurans and also included existing multigene data and morphological characters. This study corroborated remipedes as the closest relatives to Hexapoda, with branchiopods and cephalocarids as each other's sister group, the four clades together constituting Allotriocarida.

In terms of molecular sampling, an obvious deficiency in current datasets was the lack of transcriptomic data for Cephalocarida, which means that a putative remipede-cephalocarid grouping (Xenocarida) had not been strongly tested. Until recently, published phylogenomic data for remipedes had been based on just a single species, *Xibalbanus tulumensis* (formerly *Speleonectes tulumensis*; Fig. 5.1K). Addition of Illumina transcriptomes for a second remipede, *Lasionectes entrichoma*, as well as two cephalocarids (*Hurchinsoniella macracantha* and *Lightiella incisa*) has, on the one hand, provided additional support for monophyly of a remipede-hexapod clade and, on the other hand, provided evidence that cephalocarids and remipedes attract spuriously, biased by long branch attraction and/or compositional heterogeneity (Schwentner et al. 2017). The remipede-hexapod grouping has been formalized taxonomically as Labiocarida. In most analyses by Schwentner et al (2017), cephalocarids are sister group to remaining Allotriocarida (see also Lozano-Fernandez et al. 2019). Whether this resulting clade is robust remains to be evaluated using hypotheses-based tests for conflicts in the data and increased taxon sampling of crustaceans that are known to induce conflicting signal (such as cirripedes and malacostracans).

Ecological Consequences of Competing Sister Groups of Hexapoda

Hexapoda is fundamentally a terrestrial clade. A crustacean, rather than myriapod, sister group of Hexapoda opens up the possibility that a suite of characters traditionally interpreted as apomorphies of Tracheata may have independently evolved in hexapods and myriapods. Some of these are correlated with terrestrial life habits, such as Malpighian tubules as the excretory organs, a tracheal respiratory system, and uniramous thoracic appendages (inferred loss of an exopod that may have served an ancestral role in aquatic locomotion or respiration). The reoptimization of some of these characters in the hexapod ground pattern is partly dependent on how internal relationships of entognathous Hexapoda are resolved, given that Malpighian tubules and tracheae are not universally present in Ellipura. Irrespective of this uncertainty about how particular characters evolved at or near the base of Hexapoda, there remains little question that the last common ancestor of the hexapod crown group was terrestrial. The question of the hexapod stem group, however, is wide open: no existing fossils (either marine, freshwater, or terrestrial) have been convincingly assigned to the hexapod stem group. One candidate, *Devonohexapodus bocksbergensis* (=*Wingertshellicus backesi*), a marine fossil from the Lower Devonian, proved upon subsequent restudy to be excluded from the mandibulate crown group, let alone the Hexapoda (Kühl and Rust 2009).

As discussed above under Branchiopoda, hypotheses about hexapod origins from freshwater crustaceans are challenged by relevant fossils being recovered from marine sediments. A possible

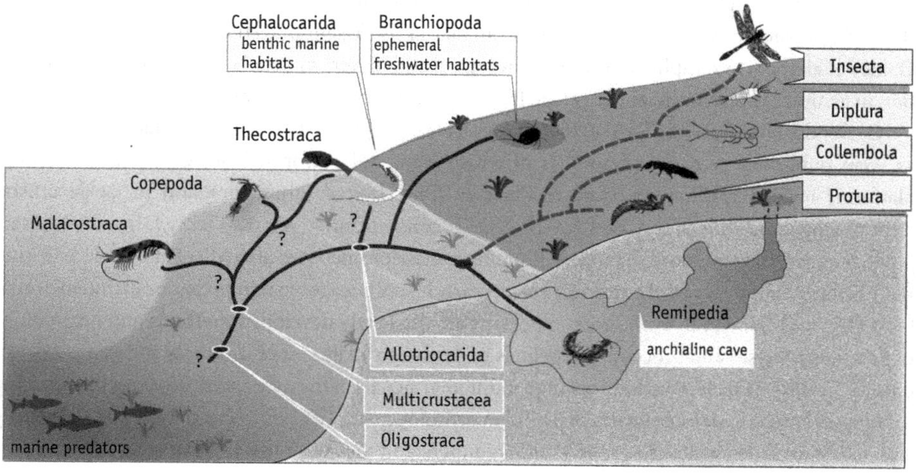

Fig. 5.7.
Evolutionary scenario depicting the transition from marine to terrestrial habitats by pancrustaceans. Shown is the most likely process supported by current data. Question marks indicate lineages that are still not robustly resolved. Hexapod lineages are illustrated by dotted branches, crustaceans by continuous lines. This figure is simplified (e.g., malacostracans such as hermit crabs also conquered land later). Modified from von Reumont et al. (2012), with permission from Oxford University Press. See color version of this figure in centerfold.

evolutionary scenario for terrestrialization in hexapods in the context of pancrustacean phylogeny is shown in Fig. 5.7. A sister group relationship between remipedes and hexapods, as increasingly finds support, suggests that their most recent common ancestor likely inhabited an environment permitting invasion of, on the one hand, anchialine caves and, on the other, the nearshore terrestrial realm. This would favor a shallow marine setting for the stem species of Remipedia + Hexapoda. An estimated divergence time for remipedes and hexapods is in the mid- to Late Cambrian (Rota-Stabelli et al. 2013a; Schwentner et al. 2017), when the trace fossil record begins to show evidence for arthropods making subaerial excursions onto tidal flats (Collette et al. 2010).

FUTURE DIRECTIONS

We have insufficient genomes for crustacean and early insect lineages. Genome sequencing initiatives like the 5000 arthropod genome sequencing consortium (i5K, http://i5k.github.io) are in most cases directed at questions of applied science, e.g., organization of genomes of species that are important pests or kept in aquaculture. Consequently, the handful of ongoing genome projects of crustaceans are overrepresented by aquaculture species or pests (e.g., *Litopaeneus, Eurythemora, Caligus, Lepeoptheirus*) and the rather derived genomes of the evolutionary developmental biology model *Daphnia* and of the amphipod *Parhyale hawaiensis* (still ongoing). To train adequate search strategies like Hidden Markov Models and to compile ortholog gene sets, however, not only depends on sophisticated ortholog prediction but also requires well-sampled genome data. An additional hurdle will be that crustaceans show an incredible diversity of genome sizes (Jeffery and Gregory 2014); sizes can vary within groups in the range of several gigabases (see animal genome size database, http://genomesize.com/).

We need to also consider the extent to which transcriptomic studies are biased by the inclusion of multigene or "older" EST data when combined with Illumina sequenced transcripts that generally show a high coverage of all expressed genes of sampled tissues. Dell'Ampio et al. (2014) showed that such missing data can have an impact in phylogenomic studies. On the other hand, Oakley et al. (2013) exemplified how morphological and "older" molecular data can provide resolution in analyses where sufficient coverage from transcriptomes is not available.

Currently, NCBI's SRA, TSA archive shows 93 transcriptome projects of crustaceans. Of those, 45 were published in 2015 and 2016, 13 of which were still performed on "older" 454 Titanium platforms instead of Illumina-based technology. One desired future direction to tackle crustacean phylogenomics is showcased by the 1KITE consortium (www.1KITE.org), in which insect phylogenetics draw on more than 1,200 Illumina transcriptomes (see also Misof et al. 2014). Within 1KITE, representatives of the putatively closest early insect lineages to remipedes are included in analyses (Fig. 5.4). However, even more important are the newly developed methods and adaptations of previous analyses and processing pipelines to be able to analyze such amounts and complexity of data. Necessary next steps include testing for different signals in the data and comparing the resulting topologies based on different parts of included ortholog genes (e.g., domains only) or on differently selected or excluded genes (i.e., to read and to understand the data and its structure).

SUMMARY AND CONCLUSIONS

The Pancrustacea/Tetraconata hypothesis is one of the most stable results of molecular phylogenetics of Arthropoda, withstanding the transition from analyses of single or a few genes to current phylogenomic approaches. Throughout the past two decades of research, "Crustacea" has consistently been found to be paraphyletic with respect to Hexapoda, a resolution in which neurophylogeny/neural cladistics and molecular phylogeny are in agreement. Morphological arguments for crustacean monophyly have been influenced by the exclusion of hexapods from consideration. Hexapod monophyly is firmly established; early studies that yielded reciprocal paraphyly with Crustacea reflected the limits of mitogenomics to resolve deep branchings. Some key questions in pancrustacean phylogeny remain open, notably the position of Malacostraca, for which neuroanatomy and multi-gene or transcriptomic approaches are conspicuously incongruent. In spite of this, some patterns are repeatedly recovered across well-sampled molecular trees, irrespective of substantial differences in analytical methods. These include a basal division of Pancrustacea into Oligostraca and Altocrustacea and, within the latter, Hexapoda nesting in a clade that includes cephalocarids, branchiopods, and remipedes. The latter are the strongest candidate for the sister group of Hexapoda.

The next frontier will be the methodological transfer from transcriptomic to genomic data. Adaptation of software and tools to supercomputer systems given the increased size of phylogenomic datasets is already happening. This should be followed by a collaborative effort to sample transcriptome data equally for all crustacean groups on a lower level. A final goal will be to combine 1KITE data with a complementary dataset and methodological analyses of crustaceans.

Compared to genomic approaches, solving the question of the crustacean-hexapod transition paleontologically is less predictable in terms of devising a methodology. Molecular dating predicts that stem group Hexapoda should have diverged from their closest crustacean relatives by the mid-Cambrian to Early Ordovician (Rota-Stabelli et al. 2013a, Misof et al. 2014, Schwentner et al. 2017), but hexapods first appear in the fossil record in the Lower Devonian as crown group members of Collembola and Insecta. Now that the internal phylogeny of entognathous hexapods is stabilizing in phylogenomic trees, it becomes imperative that morphological characters are mapped onto a stable topology to elucidate the hexapod ground pattern. This will facilitate a "search image" for the most recent common ancestor of extant Hexapoda that can be used to infer the morphological consequences of potential stem group fossils under different allotriocarid sister group hypotheses.

REFERENCES

Almeida, W.O., M.L. Christoffersen, D.S. Amorim, and E.C. Costa Eloy. 2008. Morphological support for the phylogenetic position of Pentastomida and allied fossils. Biotemas 21:81–90.

Andrew, D.R. 2011. A new view of insect-crustacean relationships II. Inferences from expressed sequence tags and comparisons with neural cladistics. Arthropod Structure and Development 40:289–302.

Borner, J., P. Rehm, R.O. Schill, I. Ebersberger, and T. Burmester. 2014. A transcriptome approach to ecdysozoan phylogeny. Molecular Phylogenetics and Evolution 80:79–87.

Cameron, S.L., K.B. Miller, C. D'Haese, M.F. Whiting, and S.C. Barker. 2004. Mitochondrial genome data alone are not enough to unambiguously resolve the relationships of Entognatha, Insecta and Crustacea *sensu lato* (Arthropoda). Cladistics 20:534–557.

Carapelli, A., P. Lió, F. Nardi, E. van der Wath, and F. Frati. 2007. Phylogenetic analysis of mitochondrial protein coding genes confirms the reciprocal paraphyly of Hexapoda and Crustacea. BMC Evolutionary Biology 7:S8.

Castellani, C., A. Maas, D. Waloszek, and J.T. Haug. 2011. New pentastomids from the Late Cambrian of Sweden—deeper insight of the ontogeny of fossil tongue worms. Palaeontographica A 293:95–145.

Collette, J.H., J.W. Hagadorn, and M.A. LaCelle. 2010. Dead in their tracks: Cambrian arthropods and their traces from intertidal sandstones of Qúebec and Wisconsin. Palaios 25:475–486.

Dallai, R., D. Mercati, A. Carapelli, F. Nardi, R. Machida, K. Sekiya, and F. Frati. 2011. Sperm accessory microtubules suggest the placement of Diplura as the sister-group of Insecta s.s. Arthropod Structure and Development 40:77–92.

Dell'Ampio, E., K. Meusemann, N.U. Szucsich, R.S. Peters, B. Meyer, J. Borner, M. Petersen, A.J. Aberer, A. Stamatakis, M.G. Walzl, B. Quang Minh, A. von Haeseler, I. Ebersberger, G. Pass, and B. Misof. 2014. Decisive data sets in phylogenomics: lessons from studies on the phylogenetic relationships of primarily wingless insects. Molecular Biology and Evolution 31:239–249.

Dohle, W. 2001. Are the insects terrestrial crustaceans? A discussion of some new facts and arguments and the proposal of the proper name Tetraconata for the monophyletic unit Crustacea + Hexapoda. Pages 85–103 *in* T. Deuve, editor. Origin of the Hexapoda. Annales de la Sociétè entomologique de France (N.S.), 37.

Ertas, B., B.M. von Reumont, J.-W. Wägele, B. Misof, and T. Burmester. 2009. Hemocyanin suggests a close relationship of Remipedia and Hexapoda. Molecular Biology and Evolution 26:2711–2718.

Fanenbruck, M. 2009. Die Kopfanatomie der Mandibulata. Eine anatomische Studie des Skelettmuskel-systems zur Klaärung der Verwandtschaftsverhältnisse von Krebsen, Tausendfüssern und Insekten. Südwestdeutscher Verlag für Hockschulschriften, Saarbrücken, Germany.

Friedrich, M., and D. Tautz. 1995. Ribosomal DNA phylogeny of the major extant arthropod classes and the evolution of myriapods. Nature 376:165–167.

Giribet, G., G.D. Edgecombe, J.M. Carpenter, C. D'Haese, and W.C. Wheeler. 2004. Is Ellipura monophyletic? A combined analysis of basal hexapod relationships with emphasis on the origin of insects. Organisms, Diversity and Evolution, 4:319–340.

Glenner, H., P.F. Thomsen, M.B. Hebsgaard, M.V. Sørensen, and E. Willerslev. 2006. The origin of insects. Science 314:1883–1884.

Hanström, B. 1926. Eine genetische Studie über die Augen und Sehzentren von Turbellarien, Anneliden und Arthropoden (Trilobiten, Xiphosuren, Eurypeteriden, Arachniden, Myriopoden, Crustaceen und Insekten). Kungliga. Svenska Vetenskapsakademiens Handlingar, Tredje Serien 4:1–176.

Harvey, T.H.P., and N.J. Butterfield. 2008. Sophisticated particle-feeding in a large Early Cambrian crustacean. Nature 452:868–871.

Harvey, T.H.P., M.I. Vélez, and N.J. Butterfield. 2012. Exceptionally preserved crustaceans from western Canada reveal a cryptic Cambrian radiation. Proceedings of the National Academy of Sciences of the USA 109:1589–1594.

Harzsch, S. 2006. Neurophylogeny: architecture of the nervous system and a fresh view on arthropod phylogeny. Integrative and Comparative Biology 46:162–194.

Hassanin, A. 2006. Phylogeny of Arthropoda inferred from mitochondrial sequences: strategies for limiting the misleading effects of multiple changes in pattern and rates of substitution. Molecular Phylogenetics and Evolution 38:100–116.

Haug, J.T., A. Maas, C. Haug, and D. Waloszek. 2012. Evolution of crustacean appendages. Pages 34–73 *in* L. Watling and M. Thiel, editors. The Natural History of the Crustacea. Vol. 1: Functional Morphology and Diversity. Oxford University Press, New York.

Jeffery, N.W., and T.R. Gregory. 2014. Genome size estimates for crustaceans using Feulgen image analysis densitometry of ethanol-preserved tissues. Cytometry 85:862–868.

Jenner, R.A. 2010. Higher-level crustacean phylogeny: conflict and consensus. Arthropod Structure and Development 39:143–153.

Klass, K.-D., and N. P. Kristensen. 2001. The ground plan and affinities of hexapods: recent progress and open problems. Pages 265–298 *in* T. Deuve, editor. Origin of the Hexapoda. Annales de la Société entomologique de France (N.S.), 37.

Koenemann, S., R.A. Jenner, M. Hoenemann, T. Stemme, and B.M. von Reumont. 2010. Arthropod phylogeny revisited, with a focus on crustacean relationships. Arthropod Structure and Development 39:88–110.

Kubrakiewicz, J., M.K. Jaglarz, T.M. Iliffe, S.M. Bilinski, and S. Koenemann. 2012. Ovary structure and early oogenesis in the remipede *Godzilliognomus frondosus* (Crustacea, Remipedia): phylogenetic implications. Zoology 115:261–269.

Kühl, G, and J. Rust. 2009. *Devonohexapodus bocksbergensis* is a synonym of *Wingertshellicus backesi* (Euarthropoda)—no evidence for marine hexapods living in the Devonian Hunsrück Sea. Organisms Diversity and Evolution 9:215–231.

Lavrov, D.V., W.M. Brown, and J.L. Boore. 2004. Phylogenetic position of the Pentastomida and (pan) crustacean relationships. Proceedings of the Royal Society B: Biological Sciences 271:537–544.

Lauterbach, K.-E. 1983. Zum Problem der Monophylie der Crustacea. Verhandlungen des Naturwissenschaftlichen Vereins in Hamburg 26:293–320.

Legg, D.A., M.D. Sutton, and G.D. Edgecombe. 2013. Arthropod fossil data increase congruence of morphological and molecular phylogenies. Nature Communications 4:2485.

Lozano-Fernández, J., R. Carton, A.R. Tanner, M.N. Puttick, M. Blaxter, J. Vinther, J. Olesen, G. Giribet, G.D. Edgecombe, and D. Pisani. 2016. A molecular palaeobiological exploration of arthropod terrestrialisation. Philosophical Transactions of the Royal Society B: Biological Sciences 371:20150133.

Lozano-Fernandez, J., M. Giacomelli, J.F. Fleming, A. Chen, J. Vinther, P.F. Thomsen, H. Glenner, F. Palero, D.A. Legg, T.M. Iliffe, D. Pisani, and J. Olesen. 2019. Pancrustacean evolution illuminated by taxon-rich genomic-scale data sets with an expanded remipede sampling. Genome Biology & Evolution 11:2055–2070.

Meusemann, K., B.M. von Reumont, S. Simon, F. Roeding, S. Strauss, P. Kück, I. Ebersberger, M. Walzl, G. Pass, S. Breuers, V. Achter, A. von Haesler, T. Burmester, H. Hadrys, J.-W. Wägele, and B. Misof. 2010. A phylogenomic approach to resolve the arthropod tree of life. Molecular Biology and Evolution 27:2451–2464.

Misof, B., S. Liu, K. Meusemann, R.S. Peters, A. Donath, C. Mayer, P.B. Frandsen, J. Ware, T. Flouri, R.G. Beutel, O. Niehuis, M. Petersen, F. Izquierdo-Carrasco, T. Wappler, J. Rust, A.J. Aberer, U. Aspöck, H. Aspöck, D. Bartel, A. Blanke, S. Berger, A. Böhm, T.R. Buckley, B. Calcott, J. Chen, F. Friedrich, M. Fukui, M. Fujita, C. Greve, P. Grobe, S. Gu, Y. Huang, L.S. Jermiin, A.Y. Kawahara, L. Krogmann, M. Kubiak, R. Lanfear, H. Letsch, Y. Li, Z. Li, J. Li, H. Lu, R. Machida, Y. Mashimo, P. Kapli, D.D. McKenna, G. Meng, Y. Nakagaki, J.L. Navarrete-Heredia, M. Ott, Y. Ou, G. Pass, L. Podsiadlowski, H. Pohl, B.M. von Reumont, K. Schütte, K. Sekiya, S. Shimizu, A. Slipinski, A. Stamatakis, W. Song, X. Su, N.U. Szucsich, M. Tan, X. Tan, M. Tang, J. Tang, G. Timelthaler, S. Tomizuka, M. Trautwein, X. Tong, T. Uchifune, M.G. Walzl, B.M. Wiegmann, J. Wilbrandt, B. Wipfler, T.K. Wong, Q. Wu, G. Wu, Y. Xie, S. Yang, Q. Yang, D.K. Yeates, K. Yoshizawa, Q. Zhang, R. Zhang, W. Zhang, Y. Zhang, J. Zhao, C. Zhou, L. Zhou, T. Ziesmann, S. Zou, Y. Li, X. Xu, Y. Zhang, H. Yang, J. Wang, J. Wang, K.M. Kjer, and X. Zhou. 2014. Phylogenomics resolves the timing and pattern of insect evolution. Science 346:763–767.

Møller, O.S., J. Olesen, A. Avenant-Oldewage, P.F. Thomsen, and H. Glenner. 2008. First maxillae suction discs in Branchiura (Crustacea): development and evolution in light of the first molecular phylogeny of Branchiura, Pentastomida, and other "Maxillopoda." Arthropod Structure and Development 37:333–346.

Moura, G., and M.L. Christoffersen. 1996. The system of the mandibulate arthropods: Tracheata and Remipedia as sister groups, "Crustacea" non-monophyletic. Journal of Comparative Biology 1:95–113.

Nardi, F., G. Spinsanti, J.L. Boore, A. Carapelli, R. Dallai, and F. Frati. 2003. Hexapod origins: monophyletic or paraphyletic? Science 299:1887–1889.

Oakley, T.H., J.M. Wolfe, A.R. Lindgren, and A.K. Zaharoff. 2013. Phylotranscriptomics to bring the understudied into the fold: monophyletic Ostracoda, fossil placement, and pancrustacean phylogeny. Molecular Biology and Evolution 30:215–233.

Olesen, J. 2009. Phylogeny of Brachiopoda (Crustacea)—character evolution and contribution of uniquely preserved fossils. Arthropod Systematics and Phylogeny 67:3–39.

Regier, J.C., J.W. Shultz, and R.E. Kambic. 2005. Pancrustacean phylogeny: hexapods are terrestrial crustaceans and maxillopods are not monophyletic. Proceedings of the Royal Society B: Biological Sciences 272:395–401.

Regier, J.C., J.W. Shultz, A. Zwick, A. Hussey, B. Ball, R. Wetzer, J.W. Martin, and C.W. Cunningham. 2010. Arthropod relationships revealed by phylogenomic analysis of nuclear protein-coding sequences. Nature 463:1079–1083.

Richter, S., O.S. Møller, and C.S. Wirkner. 2009. Advances in crustacean phylogenetics. Arthropod Systematics and Phylogeny 67:275–286.

Rota-Stabelli, O., L. Campbell., H. Brinkmann., G.D. Edgecombe, S.J. Longhorn, K.J. Peterson, D. Pisani, H. Philippe, and M.J. Telford. 2011. A congruent solution to arthropod phylogeny: phylogenomics, microRNAs and morphology support monophyletic Mandibulata. Proceedings of the Royal Society B: Biological Sciences 278:298–306.

Rota-Stabelli, O., A.C. Daley, and D. Pisani. 2013a. Molecular timetrees reveal a Cambrian colonization of land and a new scenario for ecdysozoan evolution. Current Biology 23:292–298.

Rota-Stabelli, O., N. Lartillot, H. Philippe, and D. Pisani. 2013b. Serine codon-usage bias in deep phylogenomics: pancrustacean relationships as a case study. Systematic Biology 62:121–133.

Schram, F.R., and S. Koenemann. 2004. Are the crustaceans monophyletic? Pages 319–329 *in* J. Cracraft and M.J. Donoghue, editors. Assembling the Tree of Life. Oxford University Press, New York.

Schwentner, M., D.J. Combosch, J.P. Nelson, and G. Giribet. 2017. A phylogenomic solution to the origin of insects by resolving crustacean–hexapod relationships. Current Biology 27:1818–1824.

Schwentner, M., S. Richter, D.C. Rogers, and G. Giribet, G. 2018. Tetraconatan phylogeny with special focus on Malacostraca and Branchiopoda: Highlighting the strength of taxon-specific matrices in phylogenomics. Proceedings of the Royal Society B: Biological Sciences 285:20181524.

Spears, T., and L.G. Abele. 1998. Crustacean phylogeny inferred from 18S rRNA. Pages 169–187 *in* R.A. Fortey and R.H. Thomas, editors. Arthropod Relationships. Chapman and Hall, London.

Stegner, M.E.J., and S. Richter. 2011. Morphology of the brain in *Hutchinsoniella macracantha* (Cephalocarida, Crustacea). Arthropod Structure and Development 40:221–243.

Stegner, M.E.J., G. Brenneis, and S. Richter. 2014. The ventral nerve cord in Cephalocarida (Crustacea): new insights into the ground pattern of Tetraconata. Journal of Morphology 275:269–294.

Stemme, T., T.M. Iliffe, G. Bicker, S. Harzsch, and S. Koenemann. 2012. Serotonin immunoreactive interneurons in the brain of the Remipedia: new insights into the phylogenetic affinities of an enigmatic crustacean taxon. BMC Evolutionary Biology 12:168.

Stemme, T., T.M. Iliffe, B.M. von Reumont, S. Koenemann, S. Harzsch, and G. Bicker. 2013. Serotonin immunoreactive neurons in the ventral nerve cord of Remipedia (Crustacea): support for a sister group relationship of Remipedia and Hexapoda? BMC Evolutionary Biology 13:119.

Strausfeld, N.J. 2009. Brain organization and the origin of insects: an assessment. Proceedings of the Royal Society B: Biological Sciences 276:1929–1937.

Strausfeld, N.J., and D.R. Andrew. 2011. A new view of insect-crustacean relationships. I. Inferences from neural cladistics and comparative neuroanatomy. Arthropod Structure and Development 20:276–288.

Timmermans, M.J.T.N., D. Roelofs, J. Mariën, and N.M. van Straalen. 2008. Revealing pancrustacean relationships: phylogenetic analysis of ribosomal protein genes places Collembola (springtails) in a monophyletic Hexapoda and reinforces the discrepancy between mitochondrial and nuclear DNA markers. BMC Evolutionary Biology 8:83.

Tomizuka, S., and R. Machida. 2015. Embryonic development of a collembolan, *Tomocerus cuspidatus* Börner, 1909: with special reference to the development and developmental potential of serosa (Hexapoda: Collembola: Tomoceridae). Arthropod Structure and Development 44:157–172.

von Reumont, B.M., K. Meusemann, N.U. Szucsich, E. Dell'Ampio, V. Gowri-Shankar, D. Bartel, S. Simon, H.O. Letsch, R.R. Stocsits, L.-X. Luan, J.W. Wägele, G. Pass, H. Hadrys, and B. Misof. 2009. Can comprehensive background knowledge be incorporated into substitution models to improve phylogenetic analyses? A case study on major arthropod relationships. BMC Evolutionary Biology 9:1–19.

von Reumont, B.M., R.A. Jenner, M.A. Wills, E. Dell'Ampio, G. Pass, I. Ebersberger, B. Meyer, S. Koenemann, T.M. Iliffe, A. Stamatakis, O. Niehuis, O., K. Meusemann, and B. Misof. 2012. Pancrustacean phylogeny in the light of new phylogenomic data: support for Remipedia as the possible sister group of Hexapoda. Molecular Biology and Evolution 29:1031–1045.

Wägele, J.-W., and P. Kück. 2014. Arthropod phylogeny and the origin of Tracheata (=Atelocerata) from Remipedia-like ancestors. Pages 285–341 *in* J.-W. Wägele and T. Bartolomaeus, editors. Deep Metazoan Phylogeny: The Backbone of the Tree of Life. Walter De Gruyter, Berlin.

Walossek, D. 1999. On the Cambrian diversity of Crustacea. Pages 3–27 *in* F.R. Schram and J.C. von Vaupel Klein, editors. Crustaceans and the Biodiversity Crisis. Brill, Leiden, The Netherlands.

Wingstrand, K.G. 1972. Comparative spermatology of a pentastomid *Raillietiella hemidactyli*, and a branchiuran crustacean, *Argulus foliaceus*, with a discussion of pentastomid relationships. Det Kongelige Danske Videnskabernes Selsklab, Biologiske Skrifter 19:1–72.

Wolff, G.H., H.H. Thoen, J. Marshall, M.E. Sayre, and N.J. Strausfeld. 2017. An insect-like mushroom body in a crustacean brain. eLIFE 6:e29889.

Zhang, X., D.J. Siveter, D. Waloszek, and A. Maas. 2007. An epipodite-bearing crown-group crustacean from the Lower Cambrian. Nature 449:595–598.

Zrzavý, J., and P. Štys. 1997. The basic body plans of arthropods: insights from evolutionary morphology and developmental biology. Journal of Evolutionary Biology 10:357–367.

Zrzavý, J., V. Hypša, and M. Vlášková. 1998. Arthropod phylogeny: taxonomic congruence, total evidence and conditional combination approaches to morphological and molecular data sets. Pages 97–107 *in* R.A. Fortey and R.H. Thomas, editors. Arthropod Relationships. Chapman and Hall, London.

6

BIOGEOGRAPHY OF THE OCEANS

Claudia Halsband, Shane T. Ahyong, Angelika Brandt, Ksenia Kosobokova, Peter Ward, Will P. Goodall-Copestake, and Enrique Macpherson

Abstract

This chapter summarizes global patterns and mechanisms of both ecological and historical crustacean biogeography resulting in the contemporary species distributions described over the past decades. In the pelagic realm, hydrographic features such as ocean currents, physical depth profiles, and latitudinal temperature gradients are major structuring elements, as well as selection pressure exerted by the environment and species interactions, which have resulted in speciation over evolutionary time. Benthic crustacean distributions are additionally constrained longitudinally by continental barriers and submarine features such as ridges and seamounts. The main biogeographic patterns of both benthic and pelagic crustaceans are described for all ocean basins and the polar regions, of which the Indian Ocean is the least well studied. The Copepoda and Decapoda are generally represented with the highest number of described species, followed by Amphipoda and Isopoda. Life cycles with pelagic larvae (e.g., decapods and stomatopods) increase dispersal and enable wide distributions, while a lack of dispersive larvae promotes endemism in benthic forms (e.g., amphipods). Restricted regions with high species richness and endemism, such as the "coral triangle" (the Indo-Australian Archipelago), the Red Sea, and the Mediterranean, represent important biodiversity hotspots. Endemics are also suitable markers for past earth history events. Only a few studies cover the biogeography of crustacean taxa in all of the world's oceans, but a few exceptions exist for decapods, amphipods, and isopods. Although the world's oceans have been reasonably well studied for crustacean distribution and diversity, species complexes and cryptic species lacking morphological diagnostic features leave us with a large number of unconsolidated taxa. Emerging molecular tools may be able to assist with refinement of nomenclature and hence increase the resolution of crustacean biogeography in the future.

Evolution and Biogeography. Edited by Gary C. B. Poore and Martin Thiel.

INTRODUCTION

Some 67,000 species of crustaceans are known worldwide (Ahyong et al. 2011), which comprise more than 20% of all marine species. Copepods, with more than 16,000 species worldwide, are among the most versatile groups of marine crustaceans, are the most abundant, and contribute the highest biomass to the food web (Longhurst 1985, Mauchline 1998). Their environmental preferences range from pelagic to benthic, estuarine to oceanic, shallow to deep-sea, tropical to polar, and free-living to parasitic (Boxshall 2004). Decapoda, represented with over 12,000 described species, also live in all parts of the world's oceans (Appeltans et al. 2012, Table 6.1). They exhibit a considerable diversity of modes of reproduction, life cycles, and capacities for dispersal. Amphipod and isopod crustaceans live primarily on the sea floor and occur in almost all environments. A few species are widespread, but most are rare. Some thrive in the deep sea and occur down to >10,000 m hadal depths (e.g., Lysianassidae, Macrostylidae). They cover a broad size range from less than 1 mm (suborder Asellota) to about 40 cm (*Bathynomus giganteus*). Also widespread but exclusively pelagic are Euphausiacea (krill) and some Caridea (shrimps) and Sergestioidea (dendrobranchiate shrimps). Worldwide, some 86 species of euphausiids have been recorded, being present in all of the world's oceans, where they are good indicators of water masses (Brinton 1962, Baker et al. 1990, Gibbons et al. 1999).

Modern distributions of marine invertebrates are the result of both contemporary physical-ecological processes and ancient events in geological history, the former being the domain of ecological biogeography, also referred to as macroecology, and the latter of historical biogeography. Although modern distributions are ultimately constrained by present-day environmental variables, modern distributions of most marine crustaceans have their roots in the Mesozoic. During the Cretaceous, the northern supercontinent Laurasia and southern supercontinent Gondwana were separated by the equatorial Tethys Sea, within which a widespread tropical fauna evolved. Subsequent tectonic events, especially with the suturing of Gondwana from Laurasia, the subsequent breakup of Gondwana, and the formation of the separation of the Pacific, Indian, and Atlantic oceans in the Eocene, separated the formerly widespread Tethyan fauna, facilitating lineage divergence. The eastern Pacific fauna remained connected to that of the Atlantic until relatively "recently," i.e., the formation of the Isthmus of Panama in the Pliocene (O'Dea et al. 2016).

Ekman's (1953) *Zoogeography of the Sea* was the first global-scale marine biogeographic survey based on the then-known distributions of all marine species. He identified two major warm-water shelf regions: the Atlanto-East Pacific, for the Atlantic Ocean and eastern Pacific; and the Indo-West Pacific, spanning the Indian Ocean and western half of the Pacific Ocean (Hawaii and French Polynesia to Australia). The Indo-West Pacific is the largest and most speciose warm-water shelf region in the world, with Australia and the Indo-Australian Archipelago lying almost at its geographic center and separating the Pacific and Indian Ocean basins. The highest taxonomic diversity in the Indo-West Pacific is concentrated in the area roughly bounded by the Philippines, Indonesia, and New Guinea, the so-called Indo-Australian Archipelago or Coral Triangle. Although the major biogeographic regions recognized by Ekman are consistently corroborated, especially for benthic species, recent global biogeographic classifications recognize further fine-scale subdivisions of the world's oceans including pelagic and benthic provinces. Fundamental biogeographic breaks along continents and across oceans are consistent but differ somewhat between authors. Spalding et al.'s (2007) Marine Ecoregions of the World (MEOW) is a nested system of 12 realms, 62 provinces, and 232 ecoregions for coastal and shelf marine systems (Fig. 6.1). Longhurst's (2007) system covers pelagic ecosystems and proposes, in contrast, 4 principal biomes and 51 provinces, showing that major oceanographic discontinuities are generally responsible for producing different biogeographic provinces in the open oceans. In the mesopelagic zone (200–1000 m deep), an iterative Delphi method was applied to describe ecoregions and their boundaries based on biological and physical proxies (Sutton et al. 2017). The deep sea is much less well mapped, in terms of both biological data and delineation of water masses and currents, rendering the current biogeographic provinces somewhat hypothetical. With the available historical

Table 6.1. Number of Known and Unknown Marine Crustacean Species

Estimates of Known and Unknown Marine Species Diversity

		Total Known	Described (Accepted)	% Syn	Undescribed (Collected)		Undiscovered (Morpho)		Undiscovered (Molecular Cryptic)		Total Unknown (Experts)		Total Unknown (Model)		Total Estimated		% Known		New spp. (1999–2008)
					min	max	min	max	min	max	min	max	min	max	min	max	min	max	
Decapoda		12029											4500	5100	21073	24204	50	57	1611
	Dendrobranchiata		551	31	50		100		NB		150		—				79		
	Achelata		142	38	10		30	70	5	10	45	90	—				61	76	
	Chirostyloidea		206	2	250		580		10	55	840	885	—				19	20	
	Galatheoidea		715	8	300		830		19	97	1149	1227	—				37	38	
	Hippoidea		81	19	3		10		NB		13		—				86		
	Lithodoidea		129	20	10		40		—		50		—				72		
	Lomisoidea		1	—	—		—		—		—		—				100		
	Paguroidea		1106	17	150	200	400		NB		550	600	—				65	67	
	Enoplometopoidea		12	20	—		2	7	42372		3	10	—				55	80	
	Glypheoidea		2	—	—		1	2	—		1	2	—				50	67	
	Nephropoidea		54	24	1		10	28	42405		13	34	—				61	81	
	Brachyura		5688	30	300		3550	6400	—		3850	6700	—				46	60	
	Procarididea		6	—	—		2		NB		2		—				75		
	Caridea		2572	25	400		1500		NB		1900		—				58		
	Polychelida		38	27	—		42566		42372		8	18	—				68	83	
	Stenopodidea		68	16	10		50		NB		60		—				53		
	Gebiidea		203	10	50		100		—		150		—				58		
	Axiidea		455	10	50		200		—		250		—				65		
Peracarida		17115											**		132297	228231	131		2275
	Amphipoda		6947	—	?		20000		—		20000		4000	4300			26		
	Bochusacea		5	—	—		10		NB		10		—				33		
	Cumacea		1444	2	45		6000		—		6045		**				19		
	Isopoda		6345	2	3400		60000	120000	—		63400	123400	**				5	9	
	Lophogastrida		56	24	10		120		42374		131	135	—				29	30	
	Mictacea		1	—	—		—		—		—		—				100		
	Mysida		1180	32	80	100	2000	4000	10	20	2090	4120	340	450			22	36	
	Tanaidacea		1130	6	900		22600	56500	NB		23500	57400	**				2	5	
	Thermosbaenacea		7	—	1		5		—		6		—				54		

(continued)

Table 6.1. ***(Continued)***

Estimates of Known and Unknown Marine Species Diversity

	Total Known	Described (Accepted)	% Syn	Undescribed (Collected)		Undiscovered (Morpho)		Undiscovered (Molecular Cryptic)		Total Unknown (Experts)		Total Unknown (Model)		Total Estimated		% Known		New spp. (1999–2008)
Other	21086													55604	107594	20	38	
Crustacea																		
Branchiopoda		90	3	—		—		—		—		—				100		
Cephalocarida		12	—	—		—		NB		10		—				55		
Amphionidacea		1	—	—		—		—		—		—				100		
Euphausiacea		86	42	—		—		—		—		—				100		
Stomatopoda		468	19	52		200		—		252		—				65		
Leptostraca		49	2	50	100	200	600	—		250	700	—				7	16	
Branchiura		44	12	2	3	50	80	NB		528	83	—				35	46	
Copepoda		10000	17	1500	2000	28500	48000	125		30125	50125	**				17	25	
Mystacocarida		13	—	1		10		NB		11		—				54		
Pentastomida		10	—	?		?		—		—		—				—		
Tantulocarida		36	—	60		1000		NB		1060		—				3		
Thecostraca		1400	7	?		100	200	NB		100	200	—				88	93	
Ostracoda		8853	7	1000	2000	1625	32000	NB		2625	34000	**				21	77	
Remipedia		24	4	8		20	50	5	9	33-67		—				26	42	
Hexapoda (Insecta, Collembola)	2037	2037	15	30	60	30	100	NB		60	160	110	250	2097	2197	93	97	30
Myriapoda	61	61	58	?		190		—		190		—		251		24		2
Total	**52328**	**52328**	**549**	**8,723**	**4,463**	**1,92,641**	**2,68,037**	**1,69,697**	**191**	**1,58,984**	**2,79,956**	**8,950**	**10,100**	**2,11,322**	**3,62,226**	**2,263**	**1,156**	**3,918**

Described = number of currently described and taxonomically accepted species; % Syn = percent of all nominal species names considered subjective synonyms; Undescribed = undescribed species in specimen collections; Undiscovered (Morpho) = unsampled and undiscovered morphospecies; Undiscovered (Molecular Cryptic) = cryptic species only distinguishable by molecular methods; Total Unknown (Experts) = undescribed + undiscovered based on expert opinions; Total Unknown (Model) = unknown species based on the statistical model; Total Estimated = number of species (expert-based); %Known = estimated percent of all existing species that are currently described; New spp. = number of new species published in the last decade (1999–2008; data from WoRMS). Symbols used: ?, not estimated; -, no data; NB, no basis for judgment; **, rate of discovery still rising, so no meaningful estimate of total species numbers can be made using the statistical model.

Source: From Appeltans et al. (2012), with permission from Elsevier.

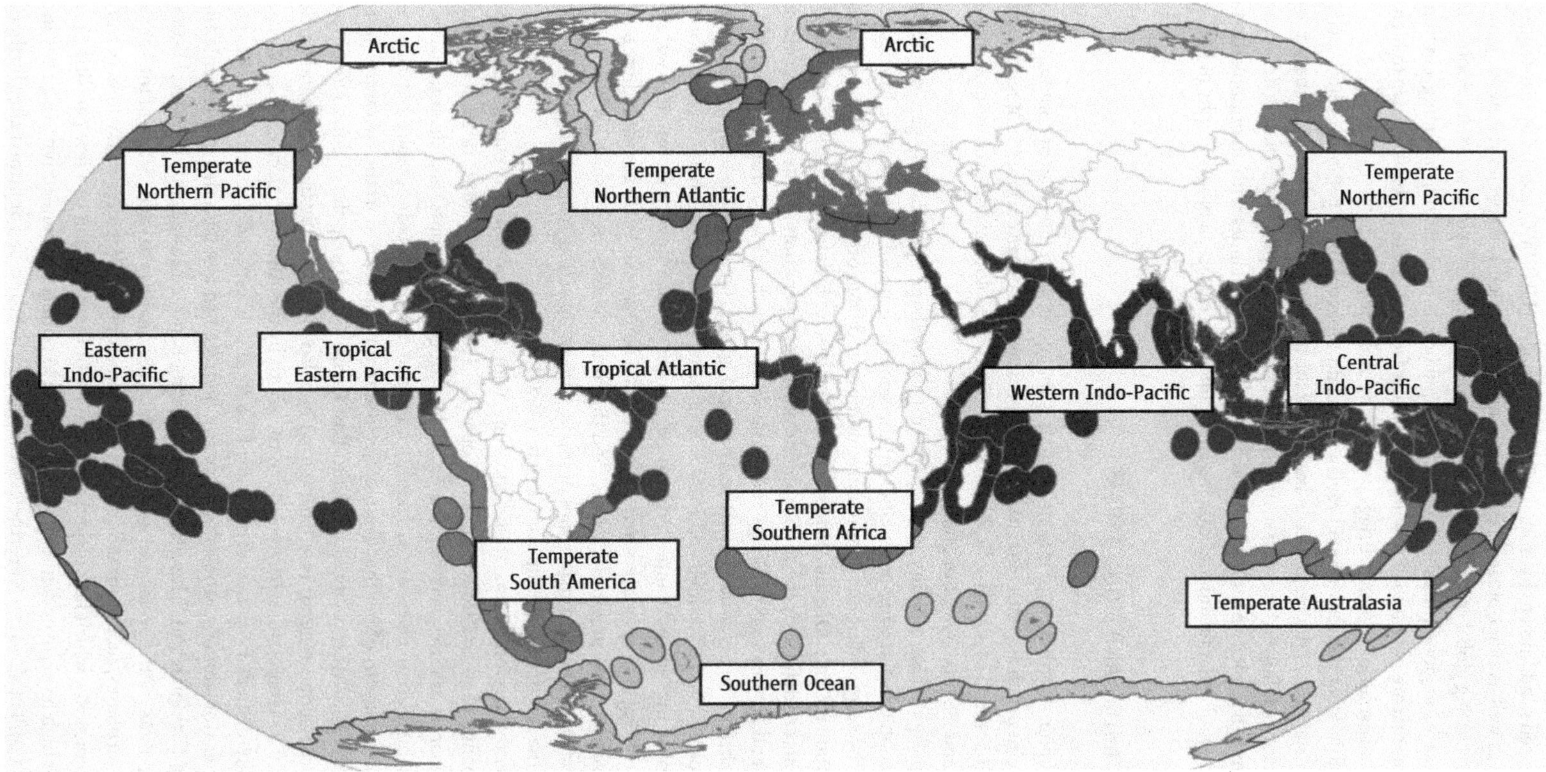

Fig. 6.1.
Biogeographic realms with ecoregion boundaries according to Spalding et al. 2007 (their Fig. 2a). See color version of this figure in centerfold.

and recent (partly unpublished) information at hand, lower bathyal, abyssal, and hadal provinces have been suggested in addition to hydrothermal vent provinces (UNESCO 2009). The existence of these provinces, as well as other biogeographic patterns (e.g., gradients in species richness), have proved useful in discussing global species patterns (e.g., Błażewicz-Paszkowycz et al. 2012m Poore and Bruce 2012). These ecoregions probably have a non-equilibrium explanation, including environmental mechanisms that are not mutually exclusive and other factors that are probably related to high speciation rates and evolutionary time (Briggs 2003, Willig et al. 2003).

In the following sections we will review the main biogeographic patterns in both pelagic and benthic crustaceans in the major ocean basins and the polar regions, including an analysis of biodiversity hotspots, followed by a section on global species richness patterns including a comparison of diversity and endemicity patterns across taxa. The chapter concludes with an outlook on the value of emerging molecular techniques that may help advance crustacean biogeography in the future.

OCEAN BASINS AND BOUNDARIES

The geographic distribution of marine animals is the result of an interplay of contemporary physical constraints, such as continental barriers, temperature and ocean currents, biological constraints limiting the survival of populations, and the past effects of geological and evolutionary history (van der Spoel and Heyman 1983, Longhurst 2007).

In the pelagic realm, mobility and dispersal capability shape a populations' geographic extent, as do hydrographic features and boundaries. At the same time, populations respond to selection pressure from the surrounding environmental conditions and from other species, leading to speciation over evolutionary time. Oceanic dispersal of marine species is defined in part by latitudinal zones, which are in turn defined by the surface and subsurface temperatures of different water masses (Van der Spoel and Heyman 1983). Warm and saline waters flow northward from the Southern Ocean into the Labrador and Nordic Seas of the Atlantic (Fig. 6.2). Small, localized deep-water formation areas are located in the Weddell and Ross Seas of Antarctica in the south and in the Labrador and Nordic Seas of the North Atlantic in the north. The density of Southern Ocean deep waters results in their spreading to deeper layers than those from the North Atlantic. In contrast, there is no deep-water formation occurring in the Pacific, and the surface waters are fresher. Pacific mid-latitudes are characterized by widespread zones of mixing-driven upwelling, compared to wind-driven upwelling in the Antarctic Circumpolar Current (ACC).

The benthic region begins at the shoreline with the intertidal and extends offshore along the continental shelf and slope out to the open sea (Lincoln et al. 1998). In addition, numerous other submarine features may be present, such as ridges, seamounts, and deep trenches, which increase habitat complexity. These features influence not only oceanography but also biodiversity. The distributions of benthic crustaceans essentially correspond to the two warm-water shelf regions—the Indo-West Pacific and Atlanto-East Pacific—and two polar regions. Distributions are constrained by bottom depth (see Chapter 16) but also longitudinally by continental barriers and latitudinally by temperature, particularly in the south by the Antarctic Convergence. Polar faunas are typically distinct from the warm-water faunas. This is especially so for polar decapods and Subantarctic stomatopods, whose ancestry can be traced to low latitudes in Mesozoic time: the Tethys Sea (Feldman and Schweitzer 2006). Among benthic crustaceans, high-level taxa tend to be cosmopolitan, with species and genera typically regionally restricted. Many of the species with pelagic larval phases, such as most decapods and stomatopods, are widespread within ocean basins, whereas peracarids, which lack dispersive larvae (with the exception of a few parasitic taxa), usually exhibit high levels of endemicity and a decrease in population connectivity (e.g., Bryan-Brown et al. 2017, Cahil et al. 2017). Nevertheless, this pattern needs to be confirmed after further studies using molecular data and species with different geographical distributions (Havermans et al. 2013).

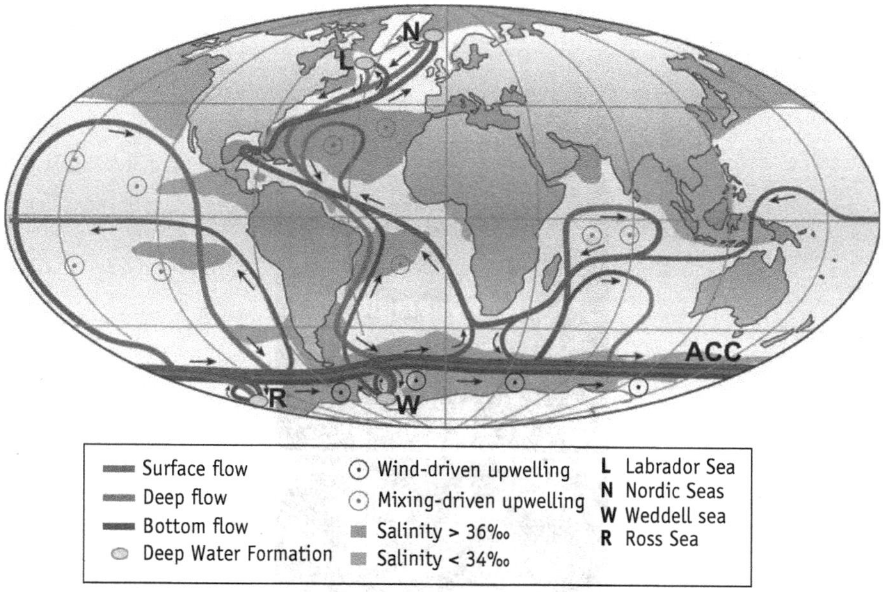

Fig. 6.2.
The global overturning circulation system. From Kuhlbrodt et al. (2007), with permission from the American Geophysical Union. See color version of this figure in centerfold.

Similar to the benthic crustaceans, the primary global distribution of pelagic groups can also be described along latitudinal and longitudinal gradients, and the third axis, water depth, plays an additional crucial role in habitat partitioning and transport patterns (Fig. 6.3): a continuous bathypelagic layer reaches from 1 to 4 km in depth, the mesopelagic layer (200–1000 m deep) divides into subpolar and equatorial-central compartments, and the epipelagial structures the upper 200 m into subpolar, transitional, central, and equatorial regions. Reid et al. (1978) found biological provinces in the seas congruent with the gross patterns of ocean circulation and wind-driven convergence and divergence of surface water (Fig. 6.2). Norris (2000) postulated that many pelagic groups are much more limited by their ecological niche than by dispersal barriers preventing exchange between populations. Thus, sympatric or parapatric speciation mechanisms may dominate in pelagic crustaceans. Seemingly gradual morphological changes may obscure relatively rapid speciation events involving cladogenesis and multiple, cryptic speciation events caused by paleoceanographic and climatic changes and associated emergence of new habitats and environmental gradients. Norris (2000) considered changes in vertical structure and seasonality of water masses as particularly important determinants of the depth and timing of reproduction. Dispersal to distant locations, in contrast, may facilitate speciation by regularly transporting individuals across major oceanic fronts and exposing them to very different selection pressures than those in their native habitat (Norris 2000).

Within the world's oceans, the species richness of many if not most pelagic taxa exhibit latitudinal gradients decreasing from the tropics to the poles (Angel 1997), although within some groups, a peak in the mid-latitudes is apparent (Boltovskoy 1999, see Chapter 15). The reasons for this cline are still far from understood, despite no shortage of causative ideas, such as the role of a number of correlates of latitude, for example temperature (Rutherford et al. 1999, Rombouts et al. 2009), primary productivity and energy availability (Wright 1983), and seasonality (Woodd-Walker et al. 2002). The latitudinal cline of many groups is, however, rarely smooth. In the South Atlantic, within many taxa, step changes are seen when passing from the

Fig. 6.3.
Depth zonation of the pelagos. Creative commons license—Wikipedia. See color version of this figure in centerfold.

transitional temperate regions to the cooler Subantarctic and Antarctic waters, and indeed for many groups, increases in the number of species occur at this point where different faunas are mixed together (Angel 1993, Boltovskoy 1999). In the tropics and subtropics, production cycles are generally of low amplitude, continue through the year, and primary and secondary production are tightly coupled. This leads to a highly retentive system where diversity is generally high (Conover 1979, Longhurst and Pauly 1987, Angel 1993). Polewards of 40–45°S, seasonal overturning of the upper water column takes place and environmental factors such as ice cover, irradiance, allied to variable day length, iron and silicic acid availability are among the key determinants for phytoplankton growth (Boyd 2002). Thus, production and consumption in the pelagial become increasingly uncoupled, and diversity within the epipelagic zone is governed by temporal patterns of primary and secondary production, rather than by temperature as such (Woodd-Walker et al. 2002). Pelagic diversity is low at both poles, despite the very different evolutionary histories, which suggests that there is no single or simple explanation for large-scale patterns of zooplankton richness.

Atlantic Ocean

In the Atlantic Ocean, the boundaries of ranges of pelagic crustaceans tend to coincide with transition zones between oceanic domains (around 50°N and 40°S) that have distinct biogeochemical properties and plankton communities, domains identified by Longhurst et al. (1995) (see also Macpherson 2002). In several taxa, e.g., decapods and euphausiids, the number of species increases from the poles to ~50°N (Northern Hemisphere) and ~40°S (Southern Hemisphere); between these two parallels, the increase is very small or non-existent.

Planktonic copepods are dominated by calanoids. Razouls et al. (2005–2016) reported 1,348 species in the Atlantic basin, of which 1,046 are calanoids. Between 60°N and 63°S, copepod diversity exhibits a smooth latitudinal cline in taxonomic richness from the tropics to the poles, with abrupt changes in evenness and distinctness at both 40°N and 40°S (Woodd-Walker et al. 2002), resulting in four primary biomes driven by the temporal patterns of primary productivity and seven zoogeographic regions (Woodd-Walker 2001). High abundances of small calanoids versus low abundance of *Lucicutia* and *Corycaeus* defined the northern temperate region; the northern and southern subtropical regions had high numbers of *Acartia*. The equatorial and southern tropical regions had no diagnostic genera to discriminate them from the other regions. The southern temperate region consisted of high *Oithona* abundance but low numbers of *Oncaea* and *Corycaeus*. The Benguela upwelling region was lower in copepod richness than the adjacent open ocean.

Beaugrand et al. (2002a) proposed nine major species associations for the North Atlantic, integrating distributions from neritic to oceanic, seasonalities and biological compositions. They emphasize a major contrast between low diversity of calanoids in the Subarctic waters south of Greenland and high diversity in the waters influenced by the Gulf Stream extension (~40°N to 50°N, ~40°W to 50°W), the Bay of Biscay, and the southeast North Sea. A clear east–west distinction originates from the difference between gradual diversity changes in the eastern oceanic North Atlantic and the western part, where the contrast is sharper, mainly due to the effects of the Gulf Stream extension. The American and European continental shelves also differ in taxonomic richness, with few species coming south with the Labrador Current over the American shelf, whereas the North Atlantic Current, and probably the shelf edge current, transport more species northward along the European side. Beaugrand et al. (2002b) demonstrated that these communities respond rapidly to changes in sea surface temperatures brought about by climate change. They calculated a northward movement of these species assemblages at a rate of >23 km yr^{-1}, which is far greater than responses in the terrestrial realm. The ecologically important *Calanus* and *Pseudocalanus* species complexes are represented with morphologically similar species throughout the Atlantic from north to south: *C. hyperboreus, C. glacialis, C. finmarchicus,* and *C. helgolandicus*, as well as *P. minutus, P. elongatus,* and *P. acuspes*, where the continuity in the distribution of *P. acuspes* between the Baltic Sea and the North Atlantic and Arctic remains obscure (Grabbert et al. 2010, Aarbakke et al. 2011). The relationships between the degree of biogeographic separation and thermal preferences were studied by Halsband-Lenk et al. (2002, 2004) for congeneric pairs of calanoids in the North Sea and the Mediterranean. The congeners *Temora longicornis* and *T. stylifera* showed distinct ranges of temperature tolerance of reproduction and development and thus little spatial overlap. The species pair *Centropages hamatus* and *C. typicus*, in contrast, varied in their life history approaches. While *C. hamatus* had a defined boreal life cycle and cold-temperate distribution pattern, *C. typicus* showed higher flexibility by shifting its reproductive period between seasons in order to keep recruitment within the tolerable thermal window, enabling the species to populate a wider latitudinal range from the equatorial to the Subarctic province (van der Spoel and Heyman 1983).

Frontal hydrography influences copepod assemblages and distribution patterns in the Southern Hemisphere. In the Brazil–Malvinas Confluence area between 35°S and 55°S, two zones with a total of 35 species of copepods from 23 genera and 13 families were distinguished based on their copepod composition, the subtropical waters of the Brazil current, and the subantarctic waters of the Malvinas Current (Berasategui et al. 2005). Six biogeographic provinces were postulated from euphausiid presence/absence data, corresponding to Antarctic, Subantarctic, cold-temperate, warm-temperate (subtropical), and tropical waters, as well as the Agulhas Current. Congruence with the biogeochemistry of the region is good in the south where temperature plays a major role in defining species assemblages, while coastal and tropical biogeochemical provinces have many species in common and differed in quantitative composition (Gibbons 1997). The *Calanus* complex is represented by *C. australis*, which is widely distributed in shelf waters off Argentina, and decreasing abundances offshore, where it is gradually replaced by *C. simillimus*.

Benthic species show a different pattern, being strongly influenced by hydrographic events (especially upwelling processes and river discharges), and, consequently, the gradients are not symmetric along both sides of the Atlantic. Eastern and western Atlantic faunas are usually distinct at species level, with higher diversity in the west. For instance, the western Atlantic stomatopod fauna (85 species) is almost triple that of the eastern Atlantic (30 species) despite having similar numbers of genera and families (Schram et al. 2013). The distributions of range end-points are mainly influenced by the Sahara and Benguela upwellings in the eastern Atlantic, and the boundaries of the Labrador and Falkland currents and the Amazon/Orinoco river discharge in the western Atlantic. The zones around these boundaries coincide with marked changes in species richness and with a progressive decrease in the abundance of many species (Brown et al. 1996). The effect of the oceanographic events marking the biogeographic boundaries on the shelf and slope is much lower on the broad slope/rise plains and, in consequence, the boundaries are poorly defined, which corroborates the low level of zonation in the slope/rise region (Carney 2005). Furthermore, a clear increase in depth range with maximum depth was observable for benthic species, and they displayed a discernible latitudinal pattern, right-skewed at high latitudes and left-skewed at low latitudes (Macpherson 2003). However, the geographic distribution of many deep-sea species has numerous uncertainties, and any conclusion is speculative (Rowden et al. 2010, Schnabel et al. 2011).

Most studies divide the eastern Atlantic into northern and southern provinces (e.g., Spalding 2007, Biggs and Bowen 2012), but a biogeographic break occurs at 10°N (Macpherson 2002). The northern province includes Iceland and the British Isles, Mediterranean Sea, northwest Africa and the oceanic islands of the Azores, Cape Verde and Canary Islands, while the southern province runs from the Gulf of Guinea around the Cape of Good Hope and as far as Durban along the east coast of South Africa. The break between the temperate and tropical Atlantic fauna (the "West African Transition") is consistent between studies across a range of taxa including stomatopods and decapods (Manning 1977, Macpherson 2002).

The western Atlantic is significantly subdivided, where the Caribbean and the Brazilian region constitute distinct regional assemblages that are significantly different from the temperate northern waters off the Atlantic United States and the Columbia/Panama area. Boschi's (2000) analysis of decapod species across the Americas separated the West Atlantic into a northern temperate fauna north of Cape Canaveral, a tropical Caribbean and a tropical Brazilian fauna south of the Delta Rio Orinoco (see also Astorga et al. 2003).

Pacific Ocean

The origin of biodiversity in the Pacific Ocean, as in other oceans, is the consequence of a combination of historical and geological processes that, together with the past and present oceanographic and climate regimes, have shaped the present distribution of species (Poore and O'Hara 2007, Waters 2008).

An extensive study of pelagic crustacean biogeography in the Pacific is the classic work by Brinton (1962) documenting the vertical and horizontal distribution of 59 euphausiid species (Fig. 6.4). *Bentheuphausia amblyops* and *Thysanopoda* spp. are typical deep-sea taxa. *Bentheuphausia*

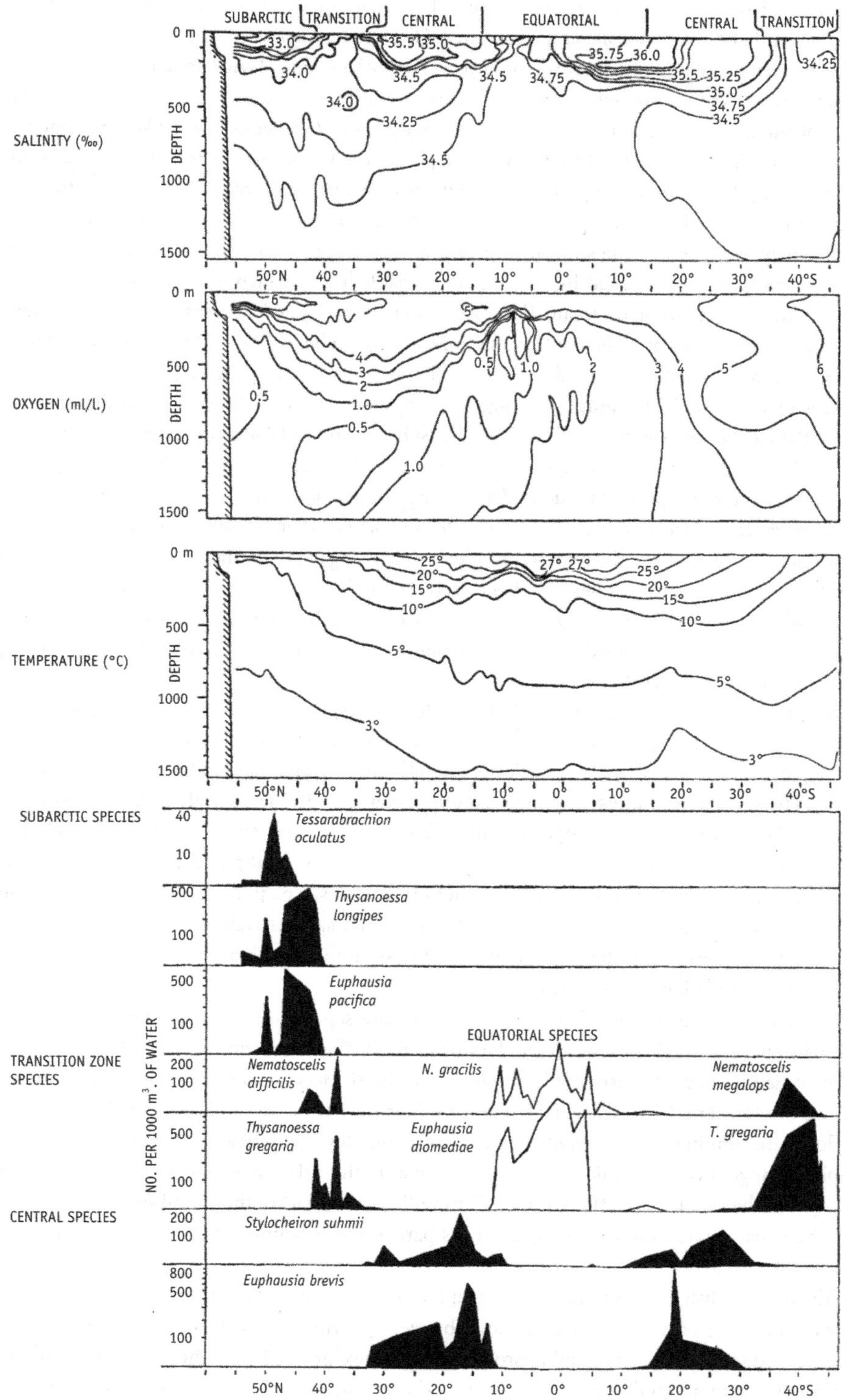

Fig. 6.4.
North–south abundance of euphausiid species representing Pacific faunal groups compared with salinity, oxygen, and temperature profiles based on the same stations at which the plankton was sampled (Brinton 1962)

ambloyops, *T. cornuta*, and *T. egregia* have wide bathypelagic ranges, where the adults are found below 1,500 m depth. Other *Thysanopoda* species and the five genera *Euphausia*, *Thysanoessa*, *Nematoscelis*, and *Stylocheiron* are primarily meso- (500–1,500 m depth) and epipelagic (<700 m depth). The epipelagic species show congruence with four Pacific water masses, featuring typical diagnostic species: the Subarctic domain (1) with *Thysanoessa longipes* and *Tessarabrachion oculatus* ranging from north of the North Pacific Drift to the northern part of the California Current; transitional waters (2) extending from the cooler part of the California Current south of 40°N to the North Pacific Drift and Japan in the west, typically with *Thysanoessa gregaria* and *Nematoscelis difficilis* (in the north) or *N. megalops* (in the south). Two central regions (3) around the mid-latitude gyres between 15 and 40° are separated by a broad equatorial belt (4), which narrows from east to west and extends into the Indian Ocean. The warm water *E. distinguenda*, an eastern tropical endemic along a westward axis at 10°N, contrasts with *N. microps*, which shows a broad, warm-water distribution throughout the west and central Pacific but with a clear hiatus in the east. In addition, Brinton (1962) described a composite range for a tropical western Pacific group extending eastward across the Pacific between 10°S and 20°S. Finally, Subantarctic and Antarctic species are distributed south of the Subantarctic Convergence, with examples both among the euphausiids (*Euphausia longirostris*) and the copepods (*Clausocalanus brevipes*, *Calanoides acutus*).

Among pelagic decapods, species of *Sergestes* and *Sergia* are most numerous. These genera are meso- or bathypelagic, and the numbers of species occurring in the Pacific are quite similar to those in other ocean basins: 38 species have been recorded in the Central and West Pacific versus 32 in the Atlantic and 37 in the Indian Ocean. The East Pacific is inhabited by a relatively low number of species that may be diminished because of oxygen-depleted conditions recorded in many areas of this region (Vereshchaka et al. 2014).

Studies on the biogeography of benthic decapod species are mostly geographically restricted to a single region, centered on the Eastern Pacific (Wicksten 1989, Boschi 2000, Astorga et al. 2003), central Pacific (Poupin 2008), and southern Australia (Poore and O'Hara 2007). However, few studies have been conducted at a Pacific-wide scale, e.g., on pontoniine shrimps (De Grave 2001), trapeziid crabs (Castro 1999), and squat lobsters (Macpherson et al. 2010, Schnabel et al. 2011; see also Reaka and Manning 1987 and Reaka et al. 2008 on mantis shrimps from the Indo-Pacific). Other crustaceans, e.g., peracarids, have been studied along the Pacific coast of South America, showing that species richness increased poleward, following an inverse latitudinal gradient of richness coupled with an increase in bathymetric distribution (Rivadeneira et al. 2011).

The Western/Central and Eastern Pacific faunas are separated by approximately 5,000–8,000 km, a distance that has been considered the world's most effective marine barrier to larval dispersal, the so-called East Pacific Barrier (Ekman 1953). This barrier has operated for as long as 65 My, separating the Western/Central Pacific fauna from Eastern Pacific assemblages, which show closer phylogenetic affinities to the Atlantic fauna (Grigg and Hey 1992). Examination of this wide geographical range reveals a similar west–east pattern as in the Atlantic, with two clear biogeographical faunas: the rich fauna of the Western-Central Pacific (as part of the Indo-West Pacific region) and the depauperate Eastern Pacific fauna (as part of the Atlanto-East Pacific region; Schnabel et al. 2011).

The present distribution of the oceanographic currents can explain, at least partially, the presence of numerous provinces. The boundaries between provinces along the coasts of America tend to be related to major oceanographic processes, e.g., upwelling of Peru/Chile and California, and conform to previous studies on other decapod crustaceans (i.e., Boschi 2000, Macpherson et al. 2010, Schnabel et al. 2011; see also Sudre and Morrow 2008). A major faunistic break along the central eastern Pacific margin is apparent off of Ecuador at about 10°S, separating a distinct northern assemblage that includes the central American region and the Galapagos Islands from a southern assemblage that runs as far south as Patagonia. This faunal discontinuity coincides with changes in ocean current regimes and lies where the South Pacific Humboldt Current meets the North Pacific California Current.

The large biogeographic breaks in the Western Pacific can be partially attributed to northern and southern equatorial currents. Currents around Japan can also explain the separation of this province from adjacent ones (Sudre and Morrow 2008). The separation of the provinces along the east coast of Australia and New Zealand, dating from the fragmentation of Gondwana and drifting of continental fragments, may be maintained by the deep Tasman Sea and East Australian Current, a warm current that flows southward along the east coast of Australia, turning eastward and forming the Tasman Front towards New Zealand (Godfrey et al. 1980). Relatively few decapods and stomatopods are shared by Australia and New Zealand (Dell 1968, Ahyong 2012, Ahyong 2017).

Along the western Pacific, the fauna north of Indonesia is usually separated from the fauna of Melanesia and also clearly differentiated from the eastern coasts of Australia and New Zealand. Biogeographic studies on western Pacific fauna, and especially those on the Indo-Malay-Philippine archipelago, indicate that this area hosts a higher number of marine species than any other oceanic region (Roberts et al. 2002). This hotspot is much older than previously thought (at least 50 My old), suggesting vicariance distribution with separately derived faunas from pan-Tethyan and Western Atlantic species (Renema et al. 2008). The present high diversity, however, seems to have been established with the Early Miocene closure between the Mediterranean and Indian Ocean, the formation of the Antarctic Circumpolar Current, and the final closure of the Isthmus of Panama in the Pliocene (Floeter et al. 2008).

Indian Ocean

In the Indian Ocean, as in other basins, the regional oceanographic and other environmental characteristics align well with observed biogeographic patterns. The barriers among provinces are also influenced by the global thermohaline circulation that pushes cold Antarctic waters into the eastern Indian Ocean. This is the primary inflow, although some waters enter over the shallow Indonesian sill, and the outflow in the west is through the southern Agulhas Current into the Atlantic (Schiller et al. 2008). Studies suggest the existence of four large ecoregions: the easternmost part of the Temperate Southern Africa; the Western Indo-Pacific; the westernmost portions of the Central Indo-Pacific; and temperate Australasia (Spalding et al. 2007, Fig. 6.1). These ecoregions can be subdivided into nine provinces: four along the western basin (Agulhas Current, Somali Coastal Current, Red Sea, Arabian Sea), two in the central ocean (Bay of Bengal and Gulf of Thailand), and three along the Eastern part (West Central Australian Shelf, Northwest Australian Shelf, and Southwest Australian Shelf; Sherman and Hempel 2009).

Little information is available about the distribution and diversity of pelagic crustaceans in the Indian Ocean. Various data sets of zooplankton abundance and distribution originate from the International Indian Ocean Expedition (IIOW) in the 1960s (UNESCO 1965–1972). Croce and Venugopal (1972) reviewed species distribution of pelagic cladocerans, which mainly include neritic forms. Abundances were estimated for a number of zooplankton groups in the Bay of Bengal in 1978 (Nair et al. 1981), including copepods, ostracods, euphausiids, amphipods, mysids, and decapods, in addition to diversity indices for all taxa combined, followed by a review of epipelagic calanoid copepods (Madhuratap and Haridas 1986). The latter study reports 183 frequent epipelagic copepod species, another 15 rare species, and 45 deep-water forms sporadically encountered in the upper layers due to vertical migration. The authors conclude that the Indian Ocean community is more homogeneous with the Pacific than with the Atlantic fauna. Rakhesh et al. (2006) conducted multivariate analyses of zooplankton assemblages of the northeast coast of India, dominated by copepods, and defined inshore, shelf, and oceanic communities.

For the benthic taxa, Schnabel et al. (2011) studied the distribution of squat lobsters in the Indian Ocean and showed a distinct East African region that includes the Red Sea, a second province containing the Arabian Basin (west of India and including the Maldives), and a third zone including the Bay of Bengal/Andaman Basin. The Western Australian and Indonesian faunas are clearly differentiated from the previous provinces, being more closely related to the Western Pacific fauna. Generally, however, this ocean basin is the least studied.

Polar Regions

Arctic

The term "Arctic" is applied here to the marine pelagic areas in which unmixed water from the upper layers of the Arctic Ocean (polar water) is found in the surface layers. This definition includes deep Arctic Ocean basins: the Nansen, Amundsen, Makarov, and Canada basins and their marginal seas—the East Greenland shelf and slope, the narrow shelf off northern Greenland and the Canadian Arctic Archipelago (excluding the archipelago itself), the eastern Labrador Sea, the northern Barents and Kara Seas, the White, Laptev, East-Siberian, Chukchi, Beaufort, and Lincoln Seas (Jakobsson et al. 2008).

During the history of Arctic biogeography, numerous schemes of zonal regionalization have been produced based on the distribution of benthos and plankton, with contradictory hierarchical and zonal characteristics (Mironov 2013; see Chapter 9). Here we use a generalized scheme subdividing the large biogeographic region of the Arctic Ocean into the Arctic Deep Basin Subregion and the Shelf-Upper Slope Subregion based on numerous previous works (Mironov 2013) and the large differences in the environmental conditions in the pelagic neritic (shelf) zone relative to the conditions at greater depths. In addition to differences between the Arctic deep basins and shelves, one more important feature of the Arctic Ocean is a strong freshwater input. The estuaries of Arctic rivers and immediate shelf regions differ from open marine shelf regions. Based on biocenotic and landscape criteria, Zenkevitch (1947, 1963) suggested delineating the Brackish-Water Province forming a continuous belt along the Eurasian coast from Yamal to the Bering Strait and along the American coast.

The faunal composition of pelagic Crustacea in the Arctic has been studied in detail (Sirenko 2001, Petrjashov 2004, 2009, Vinogradov and Semenova 2009, Kosobokova et al. 2011). According to the extensive account of the metazoan plankton fauna of the deep Arctic Ocean (Kosobokova et al. 2011), at least 125 species of pelagic crustaceans are known from the Arctic deep basins, and a similar number (about one-half different) is known from immediately adjacent shelf seas (Shelf-Upper Slope Subregion; Sirenko 2001).

Pelagic crustaceans strongly dominate in the Arctic Ocean among other taxonomic groups in terms of species number (~70%). The copepods are the most diverse group among them, both in deep basins and shelf regions. In the deep basins, they account for over 90 species and >74% of the total crustacean species number. In contrast, other crustacean groups are represented by very few species. Pelagic Amphipoda of the deep basins include 16 species, Ostracoda 5 species, Mysidacea 4 species, Euphausiacea 4 species, Isopoda 2 species (both parasitic on meso- and macroplanktonic crustaceans), and Decapoda one species (Kosobokova et al. 2011). In the shelf regions, the diversity of pelagic copepods is lower than in the deep Arctic basins, and the diversity of pelagic amphipods is also markedly higher offshore compared to the shelf (Sirenko 2001).

The Arctic Ocean patterns in diversity and biogeography are closely tied to the geological history and semi-isolation of the area as well as to the connections with the Pacific and Atlantic Oceans

(Golikov and Scarlato 1989, Bluhm et al. 2015). Originally an embayment of the North Pacific, the Arctic deep sea was influenced by Pacific fauna until ~80 million years ago when the deep-water connection closed (Marincovich et al. 1990). Exchange with the deep Atlantic began ~40 million years ago, coinciding with a strong cooling period (Savin et al. 1975). The deep-sea pelagic crustacean fauna is believed to originate primarily from other areas of the deep World oceans. Almost half of all resident (species able to reproduce and maintain their independent populations in the Arctic) meso- and macroplanktonic crustaceans of the Arctic Ocean have wide distribution ranges with the majority of those being cosmopolitan (~25%), bipolar (~10%), or species in common with the North Atlantic and the North Pacific (20–25%) (Kosobokova 2012, Bluhm et al. 2015). The modern North Atlantic fauna of pelagic crustaceans contributes ~20%, although it was long believed that at least half of the species in the deep Arctic Ocean were of North Atlantic origin due to extensive deep-water exchange through the Fram Strait (Brodsky 1956, Brodsky and Pavshtiks 1977, Grainger 1989). The elements in common with the North Pacific are very few in the deep basins (~2%). The shelf crustacean fauna also includes cosmopolitan and bipolar species, Arctic endemics, and boreal elements originating from both the North Atlantic and North Pacific (Sirenko 2001, Ershova et al. 2015). A phylogenetic analysis of Pan-Arctic *Pseudocalanus* population connectivity based on the mitochondrial cytochrome oxidase I (COI) gene revealed that dispersal patterns and pathways differed among three congeners: the North Atlantic and Pacific currents are most important for *P. acuspes*, *P. minutus* shows Pan-Arctic connectivity, and *P. newmani* shows panmixia (Questel et al. 2018).

A striking feature of the resident crustacean fauna of the deep Arctic Ocean basin is the respectable degree of endemism (Dunbar and Harding 1968, Kosobokova 2012). About one-third of the Arctic endemics are ice-associated (Kosobokova 2012, Bluhm et al. 2015), while the remaining two-thirds are the bathypelagic species, with a clear preference for depths below 1,000 m (Kosobokova and Hopcroft 2010, Kosobokova et al. 2011, Kosobokova 2012). The deep-dwelling Arctic endemics include the pelagic copepods *Mimocalanus damkaeri, Spinocalanus elongatus, S. horridus, Chiridiella reductella, Paraeuchaeta polaris, Pseudaugaptilus polaris, Scaphocalanus polaris,* and *Lucicutia pseudopolaris* (Fig. 6.5). The evolution of these deep-water species is without doubt related to the geological history of the Arctic Ocean, whose deep-water habitat was cut off from other

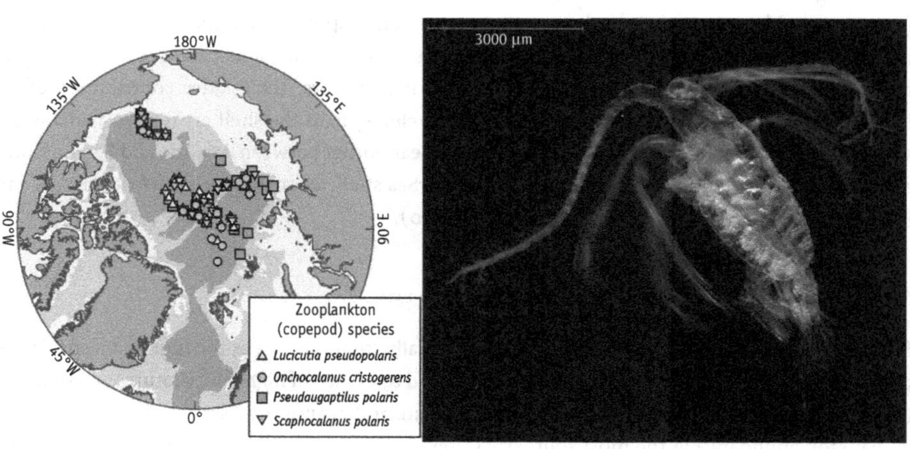

Fig. 6.5.
Distribution of Arctic pelagic crustaceans, including the endemic bathypelagic copepod *Pseudaugaptilus polaris* (photo: Russ Hopcroft). From Bluhm et al. (2015). See color version of this figure in centerfold.

deep oceans until the opening of Fram Strait 17.5 million years ago (Jakobsson et al. 2007). The ice-associated endemics include the copepods *Jaschnovia tolli, J. brevis,* and *Eurytemora richingsi* and the amphipods *Onisimus glacialis, O. nanseni, Eusirus holmi, Eusirogenes arctica, Apherusa glacialis,* and *Gammarus wilkitzkii.*

Although several crustacean species inhabiting the shelf province and associated with brackish water are Arctic endemics (e.g., the copepods *Pseudocalanus major, Drepanopus bungei,* and *Senecella sibirica,* the amphipods *Monoporea affinis* and *Onisimus botkini,* the isopod *Saduria entomon*; Deubel et al. 2003, Lyubina and Sayapin 2008, Galkin et al. 2010), there are no known planktonic endemics from estuarine zones. The brakish water fauna has a high tolerance for salinity and temperature variations and inhabits predominantly brakish waters of the Arctic shelf seas.

Depth is the major structuring element for crustacean distributions in the deep Arctic basins. The overall diversity of pelagic Crustacea increases with depth (Kosobokova 1989, Auel 1999, Kosobokova and Hirche 2000, Kosobokova et al. 2011). The epipelagic Crustacea of the deep basins are low in diversity and consist of a few ubiquitous forms and seasonally migrating Arctic species, such as the copepods *Calanus hyperboreus, C. glacialis,* and *Metridia longa.* The maximum crustacean diversity is found between 300 and 1,000 m, in the Atlantic water layer. Below this layer, both diversity and abundance of crustaceans decrease (Kosobokova et al. 2011, Kosobokova 2012, Bluhm et al. 2015). The important point is that most Arctic deep basins residents have characteristic but wide vertical distribution ranges (Dunbar and Harding 1968, Grainger 1989, Kosobokova and Hopcroft 2010, Kosobokova et al. 2011), inhabiting two or more of the conventionally defined water masses of the Arctic Ocean (Coachman and Aagaard 1974).

Despite the prominent underwater ridge systems, there appears to be no obvious general zoogeographical barrier for crustacean meso- and macroplankton between the Eurasian and Amerasian Basins of the Arctic Ocean (Kosobokova et al. 2011, Bluhm et al. 2015). Deep-water and endemic species are found on both sides of the Lomonosov Ridge, and the vertical structures of epipelagic, mesopelagic, and bathypelagic communities were found to be consistent between basins (Kosobokova et al. 2011).

Substantial differences in the physical environment of Arctic shelves, not only in their mean depth and relative width but also in the degree to which they are connected to adjoining waters, are reflected in the faunal composition and structure of their pelagic communities. The most distinct pelagic communities with a strong contribution of expatriates from the adjoining shelf regions (species not able to reproduce and maintain independent populations in the Arctic; e.g., Ekmann 1953, Beklemishev 1969) occur in the inflow regions of the Barents and Chukchi Seas as well as brackish-water regions in the vicinity of the Arctic's major rivers. The relative contribution of the North Atlantic, North Pacific, or widespread faunal elements to the shelf crustacean fauna is not studied in detail on a circumpolar scale; however, a clear border between Atlantic and Pacific faunal elements is supposed to exist over the East-Siberian Sea shelf, an analogy with distribution of various groups of benthos (Gurjanova 1951, Golikov 1980).

Antarctica

In the Southern Ocean, the plankton is characteristically stenothermal as the functional specialization to permanently low temperatures implies reduced tolerance of high temperatures as a trade-off (Pörtner et al. 2007). The net result of low temperature and pronounced seasonality is a reduction in plankton abundance in the most southerly waters where extended development and multi-year life cycles are apparent (Chiba et al. 2001, Mackey et al. 2012).

There is evidence to suggest that many plankton species, even the rarer ones, have a broadly continuous circumpolar distribution (Baker 1954). Genetic studies on *Euphausia superba* have not provided any evidence for genetic structure in Antarctic krill at large geographical scales (Fevolden and Schneppenheim 1989, Bortolotto et al. 2011, Deagle et al. 2015), suggesting a single undifferentiated breeding population throughout its range, and the same may be true for many other species distributed within the ACC. Furthermore, molecular data has suggested that the speciation process in some genera, e.g., *Euphausia*, started around 20 mya with the separation of the Antarctic clade and a Subantarctic one from an ancestral lineage (Patarnello et al. 1996, Bargelloni et al. 2000).

The total number of copepod species reported for the Southern Ocean appears to vary somewhat depending upon which orders are included and how the limits of the Southern Ocean are defined (Razouls et al. 2000; Park and Ferrari 2009; Kouwenberg et al. 2014). In common with many other taxonomic groups, endemism within the Southern Ocean Copepoda is high. For instance, of 184 deep-water species, 50 originally described from the Southern Ocean have not been found elsewhere (Park and Ferrari 2009).

For many, if not the majority of, copepod species in the Southern Ocean, a circumpolar distribution exists or can be inferred. There may, however, be exceptions (e.g., *Paraeuchaeta erebi* to date is only found in the Ross Sea; Fontaine 1988). Whether such distributions are the result of undersampling and rarity remains to be seen. Epi- to mesopelagic species are sometimes found in southern temperate, subtropical, and tropical latitudes, while deeper dwelling species may be encountered from Antarctic to boreal seas.

Pelagic ostracods are a relatively understudied group, at least in the Southern Ocean. While details of their life histories are insufficiently known and they have a somewhat confusing taxonomy, general patterns of distribution have been inferred from patterns occurring in the North Atlantic (Angel 1999). Ostracods have been reported from the surface to the abyssal floor in open water and under pack ice (Angel and Blachowiak–Samolyk 2014). Circumpolar distributions are assumed for the majority of the 47 species described by Blachowiak–Samolyk and Angel (2003) belonging to 31 genera from south of 52°S.

For the Lophogastrida and Mysida, an extensive biogeographical review in Russian was carried out by Petryashov (2004, 2007). Regional faunal endemism is low, with only three of 24 genera endemic to the Antarctic and Subantarctic regions. The pelagic fauna comprises 28 species, the rest being benthopelagic. Of the pelagic species, 50% are panoceanic and distributed through the world ocean.

Within the Southern Ocean, pelagic mysids are characteristically distributed in two provinces, first, from the Antarctic coast to the Southern Antarctic Circumpolar Current Front (SACCF), and second, from the SACCF to the Subtropical Convergence. Other, more specific distributions are apparent for the bentho-pelagic species, which include the region close to the Antarctic continent (including subregions) such as the major gyres and the Antarctic Peninsula and beyond along the Scotia Arc to coastal South America (Petryashov 2014).

The Antarctic decapod fauna is characterized by an overall low diversity and a high percentage of endemism (Thatje and Arntz 2004; see Chapter 9). This is, however, not the case for other crustacean taxa in the same waters. For groups such as Amphipoda and Isopoda, Antarctic diversity is similar to that described for tropical areas (De Broyer et al. 2014). Moreover, mid- to high-latitude diversity may be asymmetrical, with higher absolute diversity in the Southern Hemisphere (Platnick 1991, Poore and Wilson 1993, Poore and Bruce 2012). Frequent fluctuations in habitat availability originating from oscillations in the extent of the continental ice sheet, together with the effects of the long-term cooling of seawater (Clarke and Crame 1997), may have been the main causes for the absence of many decapod families. Only about a dozen caridean shrimp species are known from the Southern Ocean south of the Antarctic Convergence, with only three species left on the high Antarctic continental shelves, where temperatures are below zero all year. Although they are low in species number, in terms of abundance these species represent a major component of the mobile benthic fauna on the continental shelf (Thatje and Arntz 2004). Brachyuran crab species south of the Polar Front are very few, and only lithodid crabs (Anomura) can tolerate environmental and physiological constraints imposed by the low temperatures and short periods of food availability at high Antarctic latitudes (Clarke and Crame 1997), a pattern that appears to be very similar in the Arctic >70° N (Zaklan 2002). Stomatopoda are absent from Antarctic waters (Manning 1977).

In contrast to Decapoda, isopod diversity in Antarctica is high (Brandt 1991). To date, 744 isopod species have been reported for the Southern Ocean, 14.9% of all Isopoda described worldwide (Kaiser 2014). Molecular data have provided evidence for multiple colonizations of the deep sea by at least four major lineages of isopods (Raupach et al. 2009). Isopods are well represented at all depths and even increase in numbers of species with increasing depth, reaching the highest numbers of species at bathyal/abyssal depths (Brandt et al. 2007a, Ellingsen et al. 2007).

Despite their direct development and brooding, several Antarctic isopod species seem to be widely distributed, e.g., *Notopais quadrispinosa* and *Ilyarachna antarctica* (Munnopsidae). The evolutionary history based on plate tectonics and Quarternary glacial and warm periods probably influenced the eurybathy of many Southern Ocean taxa including the Isopoda (Brandt et al. 2007a–c, Kaiser et al. 2013, Kaiser 2014). These processes might also explain the multiple origins of the deep-sea isopods (Raupach et al. 2009) as well as emergence and submergence processes (Brandt 1991, 1992).

Other peracarid taxa follow similar trends. The Tanaidacea are represented by 160 species from a total of currently known >1,100 tanaidaceans worldwide (Błażewicz-Paszkowycz 2014). There is also a strong decrease in the knowledge of species richness with depths. Within the Cumacea, a total of 98 species are known from the subantarctic and Antarctic of the Southern Ocean, with 66 of these (67%) occurring downwards to 900 m depth, 13 species occurring down to 2,000 m, and 19 species found only deeper than 2,000 m (Mühlenhardt-Siegel 2014).

DISTRIBUTION OF HOTSPOTS

Marine and terrestrial biodiversity hotspots have been defined as restricted regions with high species richness and high endemism (Myers et al. 2000). The presence of hotspots in the different oceans depends not only on species' biological characteristics (e.g., habitat preferences, dispersal capabilities) but also on the historical processes of speciation, extinction, and range fragmentation via vicariance events (Connolly et al. 2003). Most hotspots are distributed in tropical zones, probably associated with historical processes and temperature. Temperature has been shown to have a significant impact on rates of genetic divergence and speciation facilitating the genesis of new species (Allen et al. 2006), although Held (2001) could not detect latitudinal differences in molecular substitution rates and concluded that temperature is not the key determinant of mutation rates. Briggs and Bowen (2013) synthesized the evolutionary implications of marine speciation and the relationships among the biogeographic regions and provinces of the oceans. They showed that endemism reflected the degree of evolutionary innovation in each hotspot, although dominant species appear to arise within smaller centers of high species diversity and maximum interspecies competition. These diversity centers increase or maintain species diversity within their areas of influence.

Unfortunately, studies analyzing the global diversity of all crustacean taxa are not numerous, preventing a complete view of the distributional patterns. Nevertheless, the data hosted at WoRMS (World Register of Marine Species) are facilitating these overviews, and some studies have been published in recent years (e.g., Schnabel et al. 2011, Błażewicz -Paszkowycz et al. 2012, Poore and Bruce 2012).

Some of these hotspots have operated over a period of 50 million years (Renema et al. 2008). Numerous studies have contributed to the widely accepted paradigm that posits the Indo-Australian Archipelago ("coral triangle") as a major marine biodiversity hotspot, with the western Indian Ocean and the Caribbean Sea being at most secondary centers of diversity (Roberts et al. 2002, Briggs 2003, Malay and Paulay 2010). Nevertheless, the non-random distribution patterns observed in the different taxa also support the existence of additional centers of diversity (e.g., Coral Sea, French Polynesia, Mediterranean Sea provinces). The Western basin of the Indian Ocean contains two areas with a high proportion of endemism, and they can also be considered as hotspots: the Red Sea and the Madagascar area. The Red Sea is characterized by a unique fauna and historical periods of desiccation, hypersalinity, and intermittent isolation (DiBattista et al. 2016). The isolation of the Red Sea fauna is probably reinforced by cold-water upwelling off the northeast African and southern Arabian coasts. Although global maps for species distributions and "hotspots" are primarily based on shallow-water tropical taxa, it is evident that many of these patterns may hold across an extended depth range (reviewed by Barber 2009).

The centers of diversity for the Atlantic and eastern Pacific seem to have originated during the Miocene, associated with the tectonic events and/or climate shifts, e.g., the final closure between the Mediterranean and the Indian Ocean and the final closure of the Isthmus of Panama (Floeter et al. 2008, Lessios 2008). Feldmann and Schweitzer (2006) also suggested that the temperate faunas off South America and Australia/New Zealand originated after the formation of the Antarctic Circumpolar Current during the Miocene. In the Pacific and Indian Oceans, some authors (Renema et al. 2008, Williams and Duda 2008) have suggested that tectonic activity along plate boundaries during the Oligocene-Miocene significantly increased habitat complexity in the area and promoted rapid speciation. The results obtained in some crustacean decapods confirm a common period of diversification in both shallow and deepwater environments during the Late Oligocene-Miocene (Cabezas et al. 2012).

The collision of the plates in all oceans also increased volcanism, resulting in the formation of new island-arc systems, the formation of new seamounts, and the uplift (and subduction) of existing seamounts and islands (Neall and Trewick 2008). These processes likely provided new habitats for colonization and stepping stone dispersal (Williams and Duda 2008) and also disrupted connectivity in other areas by loss of habitat at various water depths.

GLOBAL PATTERN OF SPECIES RICHNESS

The geographical pattern of increasing biodiversity from the poles to the equator (Fig. 6.6) is one of the most pervasive features of life on Earth (Gaston and Blackburn 2000, Macpherson 2002, Willig et al. 2003; see Chapter 15). Although some exceptions exist that are associated with depth or habitat characteristics (e.g., Rohde 1999), this gradient is one of the most widely recognized ecological patterns (Brown et al. 1996). Recently, Chaudhary et al. (2017) documented, however, that species richness in the marine realm is bimodal with latitude.

A latitudinal gradient holds nonetheless for many crustaceans, including pelagic and benthic species not only on each side of the Atlantic (Macpherson 2002) but also in the Indian and Pacific Oceans (Boschi 2000, Astorga et al. 2003, Macpherson et al. 2010, Schnabel et al. 2011). In contrast, species assemblages in some habitats such as the continental slope and abyss show a more homogeneous distribution of species richness with respect to latitude (Macpherson 2002, Schnabel et al. 2011). In the Atlantic, species richness patterns are strongly influenced by major local hydrographic features, including large upwelling areas, such as those off Benguela and the Sahara coast, and discharges of very large rivers (e.g., Congo, Orinoco, and Amazon). Consequently, richness of

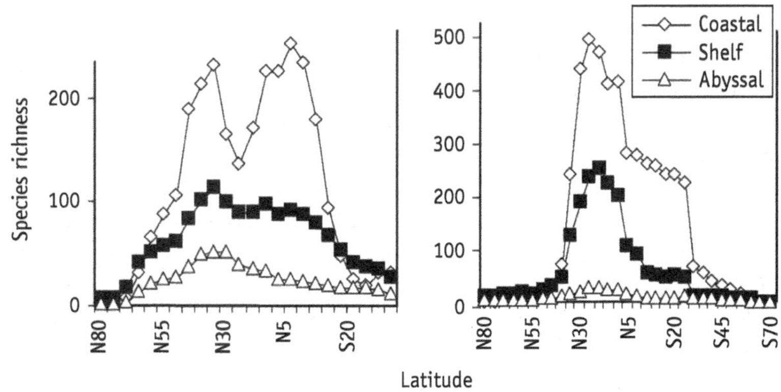

Fig. 6.6.
Latitudinal patterns of decapod species richness. From Macpherson (2002).

benthic decapods does not peak at the Equator itself but does so northerly, between 9°N and 20°N (Macpherson 2002, Chaudhary et al. 2017).

Pelagic crustaceans, e.g., decapods and euphausiids, are usually widely distributed. In the Atlantic Ocean, there are strong poleward decreases in species richness among pelagic species at around 50°N and 40°S (Macpherson 2002). In the eastern Pacific, a convex diversity pattern with latitude holds for some decapods (Boschi 2000, Schnabel et al. 2011).

Strong longitudinal variations in species richness are also evident in some marine taxa (Mora et al. 2003). Within the Indo-Pacific, species richness of stomatopods and some groups of decapods and most peracarids declines strongly with distance from the Indo-Australian Archipelago (Reaka et al. 2008, Macpherson et al. 2010, Schnabel et al. 2011). This decrease can be explained by the reduction of habitats as one moves eastward onto the Pacific or westward onto the Indian plates.

The Indian and Pacific oceans both have the greatest concentrations of decapod, stomatopod, and peracarid diversity on their western boundaries (Schnabel et al. 2011). A similar pattern exists in the Atlantic basin, where benthic crustacean diversity is greater on the western than the eastern side (Manning 1969, 1977, Macpherson 2002). The eastern Pacific is isolated from the rest of the Indo-Pacific by the East Pacific Barrier and is faunistically allied to the Atlantic as a result of the marine connection that existed prior to the Pliocene formation of the Isthmus of Panama. The East Pacific Barrier is evidently a significant barrier to gene flow. Normal currents across the barrier are sufficiently slow that transit times are probably beyond the larval durations of most species. Consequently, few decapod, stomatopod, and other benthic species cross the barrier, although many reach the outer oceanic islands in the eastern Pacific, which have habitats similar to the islands from which they originated in the central Pacific. Only three species of Stomatopoda span the East Pacific Barrier (Ahyong 2005).

The Indo-Malayan triangle harbors the greatest marine species diversity in the world, declining eastward across the Pacific and westward across the Indian Ocean (Bellwood and Hughes 2001). Several factors, including changes in sea level and oceanic circulation, high tectonic activity, and varying temperature regimes, were likely involved in shaping the distribution and diversification of the highly diverse fauna in this region (Williams and Duda 2008).

COMPARISON OF TAXA

Most studies on the general distribution of Crustacea are based on Decapoda. However, Schram et al. (2013) reviewed the biogeography and patterns of species richness in Stomatopoda, and Abele

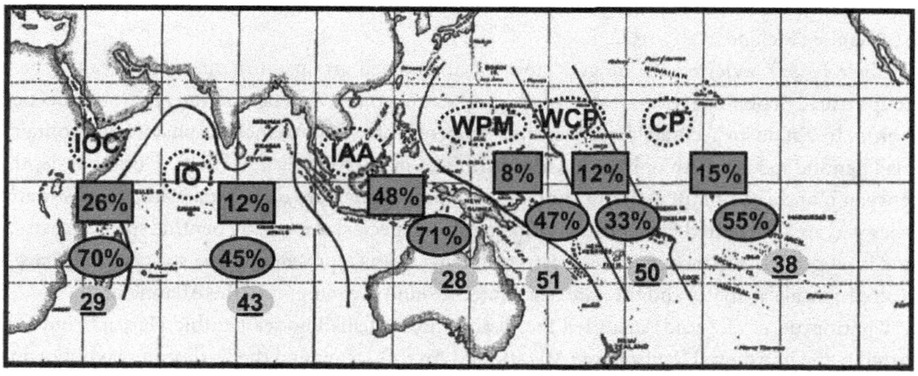

Fig. 6.7.
Endemism and body size of reef stomatopods in six different regions (dashed circles) consisting of the IOC, IO, IAA (the western margin of the IAA abuts the IOC off Burma), Western Pacific margin (WPM), West Central Pacific (WCP), and CP. Numbers in each region represent (1) percentage of species in each region that are endemic (square, top number in each region; percentage is used to avoid confounding diversity with endemism and because the regions are not equal in area), (2) percentage of species in each region <40 mm in body length (oval, middle number in each region; 40 mm is the median body size across all regions), and (3) median body size (millimeters total length) among species in each region (underline bar, bottom number in each region). From Reaka et al. (2008) ©National Academy of Sciences, USA.

(1982) reviewed other crustacean taxa. Stomatopoda are by far the most diverse in the Indo-West Pacific (331 species), being twice as diverse as those of the Atlanto-East Pacific (Schram et al. 2013). The species richness and percentage endemism of Stomatopoda peaks in the Indo-Australian Archipelago, or "coral-triangle," and declines in the central Indian Ocean and Central Pacific (Reaka et al. 2008, Schram et al. 2013). Unfortunately, studies on most crustacean groups usually lack complete overviews covering all oceans, although some exceptions provide interesting results (Fig. 6.7). For instance, the distribution of Isopoda shows that about half of all coastal, shelf, and upper bathyal species have been recorded in the temperate realms, 40% in tropical regions, and the remainder in the polar seas (Poore and Bruce 2012). The main hotspots are found in the Indo-Australian Archipelago for tropical species, being home to more species than any other region, and Australasia for temperate species. This may in part be due to the efforts of these two authors and others before them describing numerous species from the region. The Tanaidacea are understudied, and many species in most ocean regions remain to be discovered. However, most genera have their distributions in one ocean basin, usually the Indo-Pacific region (Błażewicz-Paszkowycz et al. 2012; Fig. 6.6). The world distribution of the Amphipoda shows some interesting patterns due to the lack of pelagic dispersal phases (as do Isopoda and Tanaidacea), so the present distributions can be considered good "markers" of past earth history (Myers and Lowry 2009). In fact, some authors have suggested that the patterns of amphipod endemicity are quite similar to that of terrestrial plants. The highest level of endemicity is recorded on the intensely studied Great Barrier Reef (Lowry and Myers 2009, Myers and Lowry 2009). Other crustacean taxa, containing few species, exhibit a disjointed global distribution pattern. For instance, Remipedia (with only 24 species living in anchialine caves) has the highest abundance and diversity in the Caribbean region and isolated species in the Canary Islands and in Western Australia (Neiber et al. 2011). Ostracoda are numerous, containing ~7,600 species and showing a clear increase in species richness with decreasing latitude. A similar latitudinal pattern in species richness is observed in Cirripedia, Cumacea, Euphausiacea, and Stomatopoda, with the greatest numbers of species occurring between 20°N and 5°S (see review by Abele 1982; Mühlenhardt-Siegel 1999, Macpherson 2002). Mysidacea seem to be more abundant in the northern than in the Southern Hemisphere, without a strong latitudinal gradient in species richness (Abele 1982), with species richness generally higher in the Indo-Pacific than in the Atlantic (Meland et al. 2015).

Abele (1982) reviewed the biogeography of other small groups of marine crustaceans. These groups usually contain a low number of species, and in some cases taxonomic revision is needed in order to obtain an accurate picture of their biogeography. For instance, Cephalocarida contains small benthic species, with only 13 species distributed in the Atlantic and Pacific Oceans at depths between 0 and 1,550 m. Branchiura contains 168 species; they are usually parasitic on freshwater fishes and are distributed in all areas. Leptostraca (44 species) are mostly benthic species, distributed in all oceans. Mystacocarida are very small crustaceans represented by 13 species. They have a small dispersal capability and they are distributed around the margins of the Atlantic basin.

Chertoprud et al. (2010) analyzed the distributions of shallow-sea benthic Harpacticoida reported in the Northern Hemisphere (Atlantic and Arctic Oceans and the European seas), showing a significant latitudinal trend, with a few endemics present in polar waters. Furthermore, the richest faunas were observed in the Northeast Atlantic and the North and Mediterranean Seas. Most families are widely distributed (probably cosmopolitan), though others are restricted to boreo-subtropical or even tropical zones, and no purely Arctic families were discovered. The benthopelagic species are the most widely distributed, followed by phytal species, whereas benthic species have a more restricted distribution. A similar decreasing latitudinal diversity gradient in benthopelagic calanoid copepods (Renz and Markhaseva 2015) was also observed from the Southwest Atlantic to the Southern Ocean.

Deep-sea species show different distributional patterns and larger range sizes than shallower species. For instance, Brandt et al. (2012) analyzed the global distribution of peracarid species found below 2,000 m. They observed that to date, only 328 benthic Amphipoda among ~7,000 marine species have been found below 2,000 m, and among these only 29 species appear to be distributed in two or more oceans. A similar pattern was observed in Cumacea; among c. 1,500 species and 140 genera, only 26 genera have a wide distribution (represented in three or four oceans). The Isopoda contain ~6,500 species, and only about 1,300 species occur below 2,000 m depth, over 320 species widely distributed in the deep sea. Recent Tanaidacea are represented by ~1,100 species, and 30% (324 species) occur below 2,000 m. In summary, only 11 peracarid species have been recorded in all oceans, are geographically the most widespread and potentially cosmopolitan deep-sea species, but represent a tiny fraction of the more than 3,000 peracarid species recorded from below 2,000 m (Brandt et al. 2012). Peracarids also show an inverse latitudinal gradient of species richness, increasing poleward in South America, coupled with an increase in bathymetric distribution (Rivadeneira et al. 2011). They suggest that the southern region may be a source of evolutionary novelties and/or exhibit higher diversification rates, and this might in part account for the asymmetrical pattern of mid-high-latitude species richness between hemispheres. These high-latitude species, as it has been observed in Decapoda, could have originated in the Southern Hemisphere during the Jurassic through Eocene, when the temperate climate was favorable for decapods. The rich fossil record indicated that this was the area of origin of many crustacean taxa (Feldmann and Schweitzer 2006).

THE ROLE OF GENETICS IN CRUSTACEAN BIOGEOGRAPHY

The application of genetic methods to identify, delineate, and infer the evolutionary history of species is well established and was summarized within a crustacean context by Held et al. (2011). Since then, the field, which is primarily technology-driven, has developed in ways that offer increased opportunities for crustacean biogeography. These largely stem from the application of high-throughput DNA sequencing technologies, e.g., Illumina HiSeq, MiSeq, and NextSeq.

At the finest spatial and temporal scale, population genetics, these new sequencing technologies allow researchers to more readily develop multi-locus datasets that address the limitations of single-locus mtDNA genetics outlined by Held et al. (2011). However, the increased descriptive power of these technologies has also revealed limitations, a case in point being Antarctic krill. In an admirable study, Deagle et al. (2015) analyzed over a billion krill DNA sequences to assess population structure in the species, but due to its highly duplicated genome and immense population size, even a dataset of this magnitude was greatly reduced in its efficacy to detect genetic structure. This finding has ramifications for other highly abundant crustaceans such as copepods. New sequencing technologies hold more promise for crustacean biogeography when it comes to inferences above the population level. Traditional sequencing of the mitochondrial DNA barcode (COI, cox1) continues to provide an effective tool with which the diversity and taxonomic complexity of marine crustaceans can be quantified in a standardized manner, with particular advances in the DNA barcoding of decapods (reviewed in Raupach and Radulovici 2015). For example, combined morphological and molecular approaches have helped to discern cryptic species of squat lobsters of the genera *Agononida* and *Uroptychus* in the Indo-Pacific (Poore and Andreakis 2011, 2014). Coupled with the high throughput of new sequencing approaches, COI DNA barcoding can be applied to complex multi-taxon samples to provide a comprehensive "metabarcoding" measure of species presence, which would be time-consuming to acquire using traditional approaches (e.g., Leray et al. 2013). Examples of such multi-taxon samples that are of significance to marine crustaceans include benthic sediment, gut contents, and seawater samples. The latter comprises the "marine snow" left behind by organisms, which is known to degrade within days and thus can be used to provide an environmental DNA (eDNA) snapshot of marine communities (Thomsen et al. 2012). These snapshots have great potential to advance the field of marine biogeography by virtue of the relative ease with which marine eDNA samples can be collected over vast spatial scales, e.g., from the underway water supplies of ships, from specific depths via a CTD carousel, or autonomously via moorings or remotely operated or autonomous underwater vehicles (ROVs and AUVs). Due to the technical limitations with the use of primers for the amplification of the COI barcode, the amplification of other more amenable DNA regions such as 18S ribosomal DNA has dominated high-throughput sequencing of multi-taxon samples to date (e.g., in copepods; Hirai et al. 2017). Ribosomal DNA (18S) is limited to discerning taxonomic groups higher than species, which places some constraints on the application of 18S data for biogeography. However, it is worth bearing in mind that, as the genetic profiling of natural diversity outstrips the number of morphological groups identified, assigning species names to genetic profiles becomes less important, because comparisons are made between molecular operational taxonomic units (MOTUs). These are likely to play an increasingly important role in refining taxonomic comparisons for biogeography.

SUMMARY AND CONCLUSIONS

The geographical pattern of increasing biodiversity of crustaceans from the poles to the equator, or some variant of this pattern, is one of the most pervasive features of life in the oceans. An ongoing challenge for oceanic biogeographic research is, however, the vastness of the oceans versus the high cost of collecting samples, a frequent lack of regional isolating features and thus high level of dispersal, especially of planktonic forms, and the difficulty of observing populations that vary in all three dimensions and time (Longhurst 2007). In addition, some apparently widespread species have proven to be species complexes based on new molecular analysis tools, requiring refinement of the taxonomic baseline for biogeography and creating a large body of unconsolidated nomenclature.

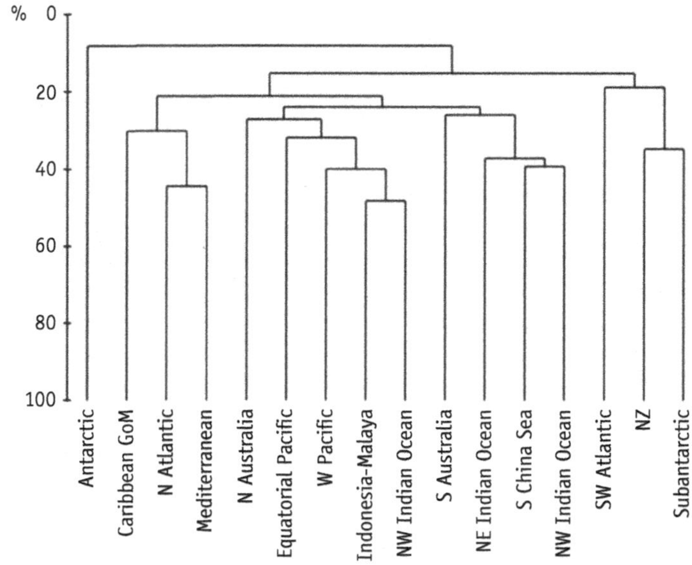

Fig. 6.8.
Cluster analysis of Bray-Curtis similarities (%) between various sea areas, based on the numbers of shallow water species of tanaidacean genera per region. From Blazewicz-Paszkowycz et al. (2012).

In the pelagic realm, most taxa exhibit latitudinal gradients in diversity, decreasing from the tropics to the poles, although within some groups a peak in the mid-latitudes is apparent. Oceanic dispersal of pelagic species is related to latitudinal zones, which are in turn defined by the surface and subsurface temperatures of the different water masses. As different taxa have very different evolutionary histories, however, no single or simple mechanism provides an explanation for the large-scale patterns of crustacean zooplankton richness.

The distributions of most benthic crustaceans are constrained longitudinally by continental barriers and latitudinally by temperature or one of its correlates. Polar faunas are typically distinct from the warm-water faunas, even if they may have originated in warm waters. In general, deep-sea species show different distributional patterns and larger distributional range sizes than shallower species.

The presence of hotspots in the different oceans depends not only on the biological characteristics of species but also on the historical processes of speciation, extinction, and range fragmentation via vicariance events. Most hotspots are in tropical zones, probably associated with historical processes such as changes in sea level and oceanic circulation, high tectonic activity, and global warming (Renema et al. 2008, Cabezas et al. 2012). Some of these hotspots have operated over a period of 50 million years—e.g., the Indo-Malayan triangle. This triangle harbors the highest marine species diversity in the world, declining eastward, northward, and southward across the Pacific and westward across the Indian Ocean. The western Indian Ocean and the Caribbean Sea, in the Atlantic Ocean, can be also considered as secondary centers of diversity.

The Atlantic, Indian, and Pacific oceans have the greatest concentrations of decapod, stomatopod, and peracarid diversity on their western boundaries. This east–west gradient can be explained by a longitudinal reduction of habitats. The eastern Pacific is isolated from the rest of the Indo-Pacific by the East Pacific Barrier and is faunistically allied to the Atlantic as a result of the marine connection that existed prior to the Pliocene formation of the Isthmus of Panama. The East Pacific Barrier is evidently a significant barrier to gene flow.

These patterns are not homogeneous in all crustacean taxa. For instance, peracarids show an inverse latitudinal gradient of species richness, increasing poleward in South America, coupled with an increase in bathymetric distribution. This supports the hypothesis that the Southern Hemisphere may be a source of evolutionary novelties and/or exhibits higher diversification rates (Rivadeneira et al. 2011). This may in part account for the asymmetrical pattern of mid-high-latitude species richness between the two hemispheres.

As the differentiation of species and populations no longer solely relies on the visibility of phenotypic variation, thanks to new molecular tools a combination of population genetics and phylogenetic approaches will increase the availability of molecular data sets and help resolve some of the remaining questions in crustacean biogeography.

ACKNOWLEDGMENTS

The research of KK was performed in the framework of the state assignment of IO RAS (theme No. 0149-2018-0035).

REFERENCES

Aarbakke, O.N.S., A. Bucklin, C. Halsband, and F. Norrbin. 2011. Discovery of *Pseudocalanus moultoni* (Frost, 1989) in Northeast Atlantic waters based on mitochondrial COI sequence variation. Journal of Plankton Research 33:1487–1495.

Abele, L.G. 1982. Biogeography. Pages 242–304 *in* L.G.Abele, editor. The Biology of Crustacea, Vol. 1. Systematics, The Fossil Records, and Biogeography. Academic Press, New York.

Ahyong, S.T. 2005. Phylogenetic analysis of the Squilloidea (Crustacea: Stomatopoda). Invertebrate Systematics 19:189–208.

Ahyong, S.T. 2012. The marine fauna of New Zealand: mantis shrimps (Crustacea: Stomatopoda). NIWA Biodiversity Memoir 125:1–111.

Ahyong, S.T., J.K. Lowry, M. Alonso, R.N. Bamber, G.A. Boxshall, P. Castro, S. Gerken, G.S. Karaman, J.W. Goy, D.S. Jones, K. Meland, D.C. Rogers, and J. Svavarsson. 2011. Subphylum Crustacea Brünnich, 1772. Animal biodiversity: an outline of higher-level classification and survey of taxonomic richness. Zootaxa 3148:165–191.

Ahyong, S.T. 2017. Biogeography of Australian Marine Invertebrates. Pages 81–99 *in* M.C. Ebach, editor. Handbook of Australasian Biogeography. CRC Press and Taylor & Francis, Boca Raton.

Allen, A.P., J.F. Gillooly, V.M. Savage, and J.H. Brown. 2006. Kinetic effects of temperature on rates of genetic divergence and speciation. Proceedings of the National Academy of Sciences 103:9130–9135.

Angel, M.V. 1993. Biodiversity of the Pelagic Ocean. Conservation Biology 7:760–772.

Angel, M.V. 1997. Pelagic diversity. Pages 35–68 *in* R.F.G. Ormond, J.D. Gage, and M.V. Angel, editors. Marine Biodiversity, Patterns and Processes. Cambridge University Press, Cambridge, UK.

Angel, M.V. 1999. Ostracoda. Pages 815–868 *in* D. Boltovskoy, editor. South Atlantic Zooplankton. Backhuys, Leiden, The Netherlands.

Angel, M.V., and K. Blachowiak-Samolyk. 2014. Halocyprid Ostracoda of the Southern Ocean. Pages 297–302 *in* C. De Broyer, P. Koubbi, H.J. Griffiths, B. Raymond, C. d'Udekem d'Acoz, A.P. Van De Putte, B. Danis, B. David, S. Grant, J. Gutt, C. Held, G. Hosie, F. Huettmann, A. Post, and Y. Ropert-Coudert, editors. Biogeographic Atlas of the Southern Ocean. Scientific Committee on Antarctic Research, Cambridge, UK.

Appeltans, W., S.T. Ahyong, G. Anderson, M.V. Angel, T. Artois, N. Bailly, R.N. Bamber, A. Barber, I. Bartsch, A. Berta, M. Błażewicz-Paszkowycz, P. Bock, G.A. Boxshall, C.B. Boyko, S.N. Brandão, R.A. Bray, N.L. Bruce, S.D. Cairns, T.Y. Chan, L. Chan, A.G. Collins, T. Cribb, M. Curini-Galletti, F. Dahdouh-Guebas, P.J.F. Davie, M.N. Dawson, O. De Clerck, D. Wim, S. De Grave, N.J. de Voogd, D.P. Domning,

C. Emig, C. Erséus, W. Eschmeyer, K. Fauchald, D.G. Fautin, S.W. Feist, C.H.J.M. Fransen, H. Furuya, O. Garcia-Alvarez, S. Gerken, D. Gibson, A. Gittenberger, S. Gofas, L. Gómez-Daglio, D.P. Gordon, M.D. Guiry, F. Hernandez, B.W. Hoeksema, R. Hopcroft, D. Jaume, P. Kirk, N. Koedam, S. Koenemann, J.B. Kolb, R.M. Kristensen, A. Kroh, G. Lambert, D.B. Lazarus, R. Lemaitre, M. Longshaw, J.K. Lowry, E. Macpherson, L.P. Madin, C. Mah, G. Mapstone, P.A. McLaughlin, J. Mees, K. Meland, C. Messing, C. Mills, T. Molodtsova, R. Mooi, B. Neuhaus, P.K.L. Ng, C. Nielsen, J. Norenburg, D.M. Opresko, M. Osawa, G. Paulay, W. Perrin, J. Pilger, G.C.B. Poore, P. Pugh, G.B. Read, J.D. Reimer, M. Rius, R. Rocha, J. Saiz-Salinas, V. Scarabino, B. Schierwater, A. Schmidt-Rhaesa, K. Schnabel, M. Schotte, P. Schuchert, E. Schwabe, H. Segers, C. Self-Sullivan, N. Schenkar, V. Siegel, W. Sterrer, S. Stöhr, B. Swalla, M. Tasker, E.V. Thuesen, T. Timm, A. Todaro, X. Turon, S. Tyler, P. Uetz, J. van der Land, B. Vanhoorne, L. van Ofwegen, R. van Soest, J. Vanaverbeke, G.K. Walker-Smith, T.C. Walter, A. Warren, G.C. Williams, S. Wilson, and M. Costello. 2012. The magnitude of global marine species diversity. Current Biology 22:2189–2202.

Astorga, A., M. Fernandez, E.E. Boschi, and N. Lagos. 2003. Two oceans, two taxa and one mode of development: latitudinal diversity patterns of South American crabs and test for possible causal processes. Ecology Letters 6:420–427.

Auel, H. 1999. The ecology of Arctic deep-sea copepods (Euchaetidae and Aetideidae). Aspects of their distribution, trophodynamics and effect on the carbon flux. Berichte für Polarforschung 319:1–97.

Baker, A. de C. 1954. The circumpolar continuity of Antarctic plankton species. Discovery Reports 28:201–218.

Baker, A. de C., B.P. Boden, and E. Brinton. 1990. A Practical Guide to the Euphausiids of the World. Natural History Museum Publications, London, UK.

Barber, P. H. 2009. The challenge of understanding the Coral Triangle biodiversity hotspot. Journal of Biogeography 36:1845–1846.

Bargelloni, L., Z. Lorenzo, N. Derome, G. Lecointre, and T. Patarnello. 2000. Molecular zoogeography of Antarctic euphausiids and notothenioids: from species phylogenies to intraspecific patterns of genetic variation. Antarctic Science 12:259–268.

Beaugrand, G., F. Ibañez, J.A. Lindley, and P.C. Reid. 2002a. Diversity of calanoid copepods in the North Atlantic and adjacent seas: species associations and biogeography. Marine Ecology Progress Series 232:179–195.

Beaugrand, G., P.C. Reid, F. Ibanez, J.A. Lindley, and M. Edwards. 2002b. Reorganization of North Atlantic marine copepod biodiversity and climate. Science 296:1692–1694.

Beklemishev, K.V. 1969. [Ecology and biogeography of the pelagial]. Nauka Press, Moscow. 291 pp. [in Russian].

Bellwood, D. R., and T.P. Hughes, 2001. Regional-scale assembly rules and biodiversity of coral reefs. Science 292:1532–1535.

Berasategui, A. D., F.C. Ramírez, and A. Schiariti. 2005. Patterns in diversity and community structure of epipelagic copepods from the Brazil–Malvinas Confluence area, south-western Atlantic. Journal of Marine Systems 56:309–316.

Blachowiak-Samolyk, K., and M.V. Angel. 2003. An Atlas of Southern Ocean Planktonic Ostracoda. http://ocean.iopan.gda.pl/ostracoda/.

Błażewicz-Paszkowycz, M. 2014. Tanaidacea. Pages 173–180 *in* C. De Broyer, P. Koubbi, H.J. Griffiths, B. Raymond, C. d'Udekem d'Acoz, A.P. Van De Putte, B. Danis, B. David, S. Grant, J. Gutt, C. Held, G. Hosie, F. Huettmann, A. Post, and Y. Ropert-Coudert, editors. Biogeographic Atlas of the Southern Ocean. Scientific Committee on Antarctic Research Cambridge, UK.

Błażewicz-Paszkowycz, M., R. Bamber, and G. Anderson.2012. Diversity of Tanaidacea (Crustacea: Peracarida) in the world's oceans—How far have we come? Plos One 7:e33068.

Bluhm, B.A., K.N. Kosobokova, and E.C. Carmack. 2015. A tale of two basins: an integrated physical and biological perspective of the deep Arctic Ocean. Progress in Oceanography 139:89–121.

Boltovskoy, D., M.J. Gibbons, L. Hutchings, and D. Binet. 1999. General biological features of the South Atlantic. Pages 1–42 *in* D. Boltovskoy, editor. South Atlantic Zooplankton. Backhuys, Leiden, The Netherlands.

Bortolotto, E., A. Bucklin, M. Mezzavilla, L. Zane, and T. Patarnello. 2011 Gone with the currents: lack of genetic differentiation at the circum-continental scale in the Antarctic krill *Euphausia superba*. BMC Genetics 12:1.

Boschi, E.E. 2000. Species of decapod crustaceans and their distribution in the American marine zoogeographic provinces. Revista de Investigación y Desarrollo Pesquero 13:1–136.

Boxshall, G.A., and S.H. Halsey. 2004. An Introduction to Copepod Diversity. Ray Society Andover, UK.

Boyd, P.W. 2002. Environmental factors controlling phytoplankton processes in the Southern Ocean. Journal of Phycology 38:844–86.

Brandt, A. 1991. Zur Besiedlungsgeschichte des antarktischen Schelfes am Beispiel der Isopoda (Crustacea, Malacostraca). Berichte zur Polarforschung 98:1–240.

Brandt, A. 1992. Origin of Antarctic Isopoda (Crustacea, Malacostraca). Marine Biology 113:415–423.

Brandt, A., S. Brix, W. Brökeland, M. Choudhury, S. Kaiser, and M. Malyutina. 2007a. Deep-sea isopod biodiversity, distribution and endemism in the Atlantic sector of the Southern Ocean—results from the ANDEEP I–III expeditions. Deep Sea Research II 54:1760–1775.

Brandt, A., A.J. Gooday, S.N. Brandão, S. Brix, W. Brökeland, T. Cedhagen, M. Choudhury, N. Cornelius, B. Danis, I. De Mesel, R.J. Diaz, D.C. Gillan, B. Hilbig, J. Howe, D. Janussen, S. Kaiser, K. Linse, M. Malyutina, J. Pawlowski, M. Raupach, and A. Vanreusel. 2007b. First insights into the biodiversity and biogeography of the Southern Ocean deep sea. Nature 447:307–311.

Brandt, A., C. De Broyer, I. De Mesel, K.E. Ellingsen, A. Gooday, B. Hilbig, K. Linse, M. Thomson, and P. Tyler. 2007c. The deep benthos. Pages 39–66 *in* A. Rogers, editor. Antarctic Ecology: From Genes to Ecosystems, Royal Society, London. Philosophical Transactions of the Royal Society B 362:39–66.

Brandt, A., M. Blazewicz-Paszkowycz, R.N. Bamber, U. Mühlenhardt-Siegel, M.V. Malyutina, S. Kaiser, C. De Broyer, and Ch. Havermans. 2012. Are there widespread peracarid species in the deep sea (Crustacea: Malacostraca)? Polish Polar Research 33:139–162.

Briggs, J.C. 2003. Marine centres of origin as evolutionary engines. Journal of Biogeography 30:1–18.

Briggs, J. C., and B.W. Bowen. 2012. A realignment of marine biogeographic provinces with particular reference to fish distributions. Journal of Biogeography, 39: 12–30.

Briggs, J.C., and B.W. Bowen. 2013. Marine shelf habitat: biogeography and evolution. Journal of Biogeography 40:1023–1035.

Brinton, E. 1962. The Distribution of Pacific Euphausiids. Scripps Institution of Oceanography, Berkeley & Los Angeles, CA.

Brodsky, K.A. 1956. Deep-sea life in the Arctic Basin. Priroda 5:41–48.

Brodsky, K.A., and E.A. Pavshtiks. 1977. Plankton of the central part of the Arctic Basin. Polar Geography 1:143–161.

Brown, J. H., G.C. Stevens, and D.M. Kaufman. 1996. The geographic range: size, shape, boundaries and internal structure. Annual Review of Ecology and Systematics 27:597–623.

Bryan-Brown, D.N., C.J. Brown, J.M. Hughes, and R.M. Connolly. 2017. Patterns and trends in marine population connectivity research. Marine Ecology Progress Series 585: 243–256.

Cabezas, P., I. Sanmartín, G. Paulay, E. Macpherson, and A. Machordom. 2012. Deep under the sea: unraveling the evolutionary history of the deep-sea squat lobster *Paramunida* (Decapoda, Munididae). Evolution 66:1878–1896.

Cahill, A.E., A. De Jode, S. Dubois, Z. Bouzaza, D. Aurelle, E. Boissin, O. Chabrol. R. David, E. Egea, J.-B. Ledoux, B. Mérigot, A. Weber, and A. Chenuil. 2017. A multispecies approach reveals hot spots and cold spots of diversity and connectivity in invertebrate species with contrasting dispersal modes. Molecular Ecology 26:6563–6577.

Carney, R.S. 2005. Zonation of deep biota on continental margins. Pages 211–278 *in* R. N. Gibson, R. J. A. Atkinson, and J. D. Gordon, editors. Oceanography and Marine Biology: An Annual Review 43. Taylor and Francis, London, UK.

Castro, P. 1999. Biogeography of trapeziid crabs (Brachyura, Trapeziidae) symbiotic with corals and other cnidarians. The biodiversity crisis and Crustacea. Crustacean Issues 12:65–75.

Chaudhary, C., H. Saeedi, and M.J. Costello. 2017. Marine species richness is bimodal with latitude: a reply to Fernandez and Marques. Trends in Ecology & Evolution 32:234–237.

Chertoprud, E. S., L.A. Garlitska, and A.I. Azovsky. 2010. Large-scale patterns in marine harpacticoid (Crustacea, Copepoda) diversity and distribution. Marine Biology 40:301–315.

Chiba S, T. Ishimaru, G.W. Hosie, and M. Fuckuchi. 2001. Spatio-temporal variability of zooplankton community structure off east Antarctica (90°E to 160°E). Marine Ecology Progress Series 216:95–108.

Clarke, A., and J.A. Crame. 1997. Diversity, latitude and time: patterns in shallow sea. Pages 122–147 *in* R.F.G. Ormond, J.D. Gage, and M.V. Angel, editors. Marine Biodiversity: Patterns and Processes. Cambridge University Press, Cambridge, UK.

Coachman, L.K., and K. Aagard. 1974. Physical oceanography of Arctic and Subarctic Seas. Pages 1–72 *in* Y. Herman, editor. Marine Geology and Oceanography of Arctic Seas. Springer Verlag, New York.

Connolly, S.R., D.R. Bellwood, and T. Hughes. 2003. Indo-Pacific biodiversity of coral reefs: deviations from a mid-domain model. Ecology 84:2178–2190.

Conover, R.J. 1979. Secondary production as an ecological phenomenon. Pages 50–86 *in* S. van der Spoel and A.C. Pierrot-Bults, editors. Zoogeography and Diversity of Plankton. Bunge Scientific Publishers, London.

Croce, N.D., and Venugopal, P. 1972. Distribution of marine cladocerans in the Indian Ocean. Marine Biology 15:132–138.

Deagle, B.E., C. Faux, S. Kawaguchi, B. Meyer, and S.N. Jarman. 2015. Antarctic krill population genomics: apparent panmixia, but genome complexity and large population size muddy the water. Molecular Ecology 24:4943–4959.

De Broyer, C., P. Koubbi, H.J. Griffiths, B. Raymond, C. d'Udekem d'Acoz, A.P. Van de Putte, B. Danis, B. David, S. Grant, J. Gutt, C. Held, G. Hosie, F. Huettmann, A. Post, and Y. Ropert-Coudert, editors. 2014. Biogeographic Atlas of the Southern Ocean. Scientific Committee on Antarctic Research, Cambridge, UK.

De Grave, S. 2001. Biogeography of Indo-Pacific Pontoniinae (Crustacea, Decapoda): a PAE analysis. Journal of Biogeography 28:1239–1253.

Dell, R.K. 1968. Composition and distribution of the New Zealand brachyuran fauna. Transactions of the Royal Society of New Zealand, Zoology 10:225–240.

Deubel, H., I. Fetzer, S. Gagayev, H.J. Hirche, M. Klages, V. Larionov, P. Lubin, O. Lubina, E.M. Noethig, Y. Odkolodkov, and E. Rachor. 2003. The Kara Sea ecosystem: phytoplankton, zooplankton and benthos communities influenced by river run-off. Pages 237–265 *in* R. Stein, K. Fahl, D.K. Fuetterer, E. Galimov, and O. Stepanets, editors. Siberian River Run-off in the Kara Sea: Characterisation, Quantification, Variability, and Environmental Significance. Proceedings in Marine Sciences. Elsevier, Amsterdam.

DiBattista, J.D., M.B. Roberts, J. Bouwmeester, B.W. Bowen, D.J. Coker, D.F. Lozano-Cortés, J.H. Choat, M.R. Gaither, J-P.A. Hobbs, M.T. Khalil, M. Kochzius, R.F. Myers, G. Paulay, V.S.N. Robitzch, P. Saenz-Agudelo, E. Salas, T.H. Sinclair-Taylor, R.J. Toonen, M.W. Westneat, S.T. Williams, and M.L. Berumen. 2016. A review of contemporary patterns of endemism for shallow water reef fauna in the Red Sea. Journal of Biogeography 43:423–439.

Dunbar, M.J., and G. Harding. 1968. Arctic Ocean watermasses and plankton—a reappraisal. Pages 315–326 *in* J.E. Sater, editor. Arctic Drifting Stations: A Report on Activities Supported by the Office of Naval Research. Arctic Institute of North America, Washington, DC.

Ekman, S. 1953. Zoogeography of the Sea. Sidgwick and Jackson, London.

Ellingsen, K., A. Brandt, B. Hilbig, and K. Linse. 2007. The diversity and spatial distribution of polychaetes, isopods and bivalves in the Atlantic sector of the deep Southern Ocean. Polar Biology 30:1265–1273.

Ershova, E.A., R.R. Hopcroft, and K.N. Kosobokova. 2015. Inter-annual variability of summer mesozooplankton communities of the western Chukchi Sea: 2004–2012. Polar Biology 38:1461–1481.

Feldmann, R.M., and C.E. Schweitzer. 2006. Paleobiogeography of Southern Hemisphere decapod Crustacea. Journal of Paleontology 80:83–103.

Fevolden, S.E., and R. Schneppenheim. 1989. Genetic homogeneity of krill (*Euphausia superba* Dana) in the Southern Ocean. Polar Biology 9:533–539.

Floeter, S.R., L.A. Rocha, D.R. Robertson, J.C. Joyeux, W.F. Smith-Vaniz, P. Wirtz, A.J. Edwards, J.P. Barreiros, C.E.L. Ferreira, J.L. Gasparini, A. Brito, J.M. Falcón, B.W. Bowen, and G. Bernardi. 2008. Atlantic reef fish biogeography and evolution. Journal of Biogeography 35:22–47.

Fontaine, M. 1988. Taxonomy and distribution of the Antarctic species group of the genus *Euchaeta* (Copepoda, Calanoida). Biology of the Antarctic Seas 19:27–57.

Galkin, S.V., N.V. Kucheruk, K.V. Minin, A.K. Rayskiy, and E.I. Goroslavskaya. 2010. Macrobenthos of the Ob River estuarine zone and of the adjacent regions of the Kara Sea. Okeanologia 50:793–797 [In Russian, English abstract].

Gaston, K.J., and T.M. Blackburn. 2000. Pattern and Process in Macroecology. Blackwell Science, London.

Gibbons, M. J. 1997. Pelagic biogeography of the south Atlantic Ocean. Marine Biology 129:757–768.

Gibbons, M.J., V.A. Spiridonov, and G.A. Tarling. 1999. Euphausiacea. Pages 1241–1280 *in* D. Boltovskoy, editor. South Atlantic Zooplankton. Backhuys, Leiden, The Netherlands.

Godfrey, J.S., G.R. Cresswell, T.J. Golding, and A.F. Pearce. 1980. The separation of the East Australian current. Journal of Physical Oceanography 10:430–440.

Golikov, A.N. 1980. [Mollusca Buccininae of the World Ocean]. Fauna SSSR [Fauna of the USSR]. Novaya Seriya Is. 85. Mollusca 5. No.2. 508 p. [in Russian].

Golikov, A.N., and O.A. Scarlato. 1989. Evolution of arctic ecosystems during the Neogene period. Pages 257–259 *in* Y. Herman, editor. The Arctic Seas: Climatology, Oceanography and Biology. Van Nostrand Reinhold, New York.

Grabbert, S., J. Renz, H.J. Hirche, and A. Bucklin. 2010. Species-specific PCR discrimination of species of the calanoid copepod *Pseudocalanus, P. acuspes* and *P. elongatus*, in the Baltic and North Seas. Hydrobiologia 652:289–297.

Grainger, E.H. 1989. Vertical distribution of zooplankton in the central Arctic Ocean. Pages 48–60 *in* L. Rey and V. Alexander, editors. Proceedings of the 6th Conference Comité Arctique International, E.J. Brill, Leiden, The Netherlands.

Grigg, R.W., and R. Hey. 1992. Paleoceanography of the tropical eastern Pacific Ocean. Science 255:172–178.

Gurjanova, E.F. 1951. Side-swimmers of the seas of the USSR and adjacent waters (Amphipoda, Gammaridea). Opredeliteli po faune SSSR [Keys to the fauna of the USSR] 41:1–1029 [in Russian].

Halsband-Lenk, C., H.J. Hirche, and F. Carlotti. 2002. Temperature impact on reproduction and development of congener copepod populations. Journal of Experimental Marine Biology and Ecology 271:121–153.

Halsband-Lenk, C., F. Carlotti, and W. Greve. 2004. Life-history strategies of calanoid congeners under two different climate regimes: a comparison. ICES Journal of Marine Science 61:709–720.

Havermans, Ch., G. Sonet, C. d'Udekem d'Acoz, Z.T. Nagy, P. Martin, S. Brix, T. Riehl, S. Agrawal, and C. Held. 2003. Genetic and morphological divergences in the cosmopolitan deep-sea amphipod *Eurythenes gryllus* reveal a diverse abyss and a bipolar species. PLoS ONE 8:e74218.

Held, C. 2001. No evidence for slow-down of molecular substitution rates at subzero temperatures in Antarctic serolid isopods (Crustacea, Isopoda, Serolidae). Polar Biology 24: 497–501.

Held, C., S. Koenemann, and C.D. Schubart. 2011. Phylogeography and Population Genetics in Crustacea. CRC Press, Boca Raton, FL.

Hirai, J., S. Nagai, and K. Hidaka. 2017. Evaluation of metagenetic community analysis of planktonic copepods using Illumina MiSeq: Comparisons with morphological classification and metagenetic analysis using Roche 454. PLoS ONE 12:e0181452.

Jakobsson, M., J. Backmann, B. Rudels, J. Nikander, M. Frank, L. Mayer, W. Jokat, F. Sangiorgi, M. D'Oregan, H. Brinkhuis, J. King, and K. Moran. 2007. The Early Miocene onset of a ventilated circulation regime in the Arctic Ocean. Nature 447:986–990.

Jakobsson, M., R. Macnab, L. Mayer, R. Anderson, M. Edwards, J. Hatzky, H.W. Schenke, and P. Johnson. 2008. An improved bathymetric portrayal of the Arctic Ocean: implications for ocean modeling and geological, geophysical and oceanographic analyses. Geophysical Research Letters 35:L07602.

Kaiser, S., S.N. Brandão, S. Brix, D.K.A. Barnes, D. Bowden, J. Ingels, F. Leese, K. Linse, S. Schiaparelli, C. Arango, N. Bax, M. Blazewicz-Paszkowycz, A. Brandt, A.I. Catarino, B. Danis, B. David, C. De Ridder, P. Dubois, K.E. Ellingsen, A. Glover, H.J. Griffiths, J. Gutt, K. Halanych, C. Havermans, C. Held, D. Janussen, A.N. Lörz, D. Pearce, B. Pierrat, T. Riehl, A. Rose, C.J. Sands, A. Soler i Membrives, M. Schüller, J. Strugnell, A. Vanreusel, G. Veit-Köhler, N. Wilson, and M. Yasuhara. 2013. Pattern, process and vulnerability of Southern Ocean benthos—a decadal leap in knowledge and understanding. Marine Biology 160:2295–2317.

Kaiser, S. 2014. Southern Hemisphere isopod crustaceans. Pages 166–172 *in* C. De Broyer, P. Koubbi, H.J. Griffiths, B. Raymond, C. d'Udekem d'Acoz, A.P. Van de Putte, B. Danis, B. David, S. Grant, J. Gutt, C. Held, G. Hosie, F. Huettmann, A. Post, and Y. Ropert-Coudert, editors. Biogeographic Atlas of the Southern Ocean. Scientific Committee on Antarctic Research, Cambridge, UK.

Kosobokova, K.N. 1989. Vertical distribution of plankton animals in the eastern part of the central Arctic Basin. Marine Plankton. Explorations of the Fauna of the Seas 41:24–31 [in Russian, English abstract].

Kosobokova, K.N. 2012. Zooplankton of the Arctic Ocean: Community structure, Ecology, Spatial distribution. GEOS, Moscow, Russia. [in Russian, English abstract].

Kosobokova, K., and H.J. Hirche. 2000. Zooplankton distribution across the Lomonosov Ridge, Arctic Ocean: species inventory, biomass and vertical structure. Deep-Sea Research I 47:2029–2060.

Kosobokova, K.N., and R.R. Hopcroft. 2010. Diversity and vertical distribution of mesozooplankton in the Arctic's Canada Basin. Deep-Sea Research II 57:96–110.

Kosobokova, K.N., R.R. Hopcroft, and H.J. Hirche. 2011. Patterns of zooplankton diversity through the depths of the Arctic's central basins. Marine Biodiversity 41: 29–50.

Kouwenberg, J.H.M., C. Razouls, and N. Desreumaux. 2014. Southern Ocean pelagic copepods. Pages 290–296 *in* C. De Broyer, P. Koubbi, H.J. Griffiths, B. Raymond, C. d'Udekem d'Acoz, A.P. Van de Putte, B. Danis, B. David, S. Grant, J. Gutt, C. Held, G. Hosie, F. Huettmann, A. Post, and Y. Ropert-Coudert, editors. Biogeographic Atlas of the Southern Ocean. Scientific Committee on Antarctic Research, Cambridge, UK.

Kuhlbrodt, T., A. Griesel, M. Montoya, A. Levermann, M. Hofmann, and S. Rahmstorf. 2007. On the driving processes of the Atlantic meridional overturning circulation. Review of Geophysics 45:RG2001.

Leray, M., J.Y. Yang, C.P. Meyer, S.C. Mills, N. Agudelo, V. Ranwez, J.T. Boehm, and R. Machida. 2013. A new versatile primer set targeting a short fragment of the mitochondrial COI region for metabarcoding metazoan diversity: application for characterizing coral reef fish gut contents. Frontiers in Zoology 10:34.

Lessios, H.A. 2008. The great American schism: divergence of marine organisms after the rise of the Central American Isthmus. Annual Review of Ecology, Evolution, and Systematics 39:63–91.

Lincoln, R., G. Boxshall, and P. Clark. 1998. A Dictionary of Ecology, Evolution and Systematics, 2nd ed. Cambridge University Press, Cambridge, UK.

Longhurst, A.R. 1985. The structure and evolution of plankton communities. Progress in Oceanography 15: 1–35.

Longhurst, A.R. 2007. Ecological Geography of the Sea, 2nd ed. Elsevier, Amsterdam.

Longhurst, A.R., and D. Pauly. 1987. Ecology of Tropical Oceans. Academic Press, Orlando, FL.

Longhurst, A.R., S. Sathyendranath, T. Platt, and C. Caverhill. 1995. An estimate of global primary production in the ocean from satellite radiometer data. Journal of Plankton Research 17:1245–1271.

Lowry, J.K., and A.A. Myers. 2009. Foreword. *In* J.K. Lowry and A.A. Myers, editors, Benthic Amphipoda (Crustacea: Peracarida) of the Great Barrier Reef, Australia. Zootaxa 2260:17–108.

Lyubina, O.S., and V.V. Sayapin. 2008. [Amphipods (Amphipoda, Gammaridea) from various geographical areas. Species composition, distribution, ecology]. Kola Science Centre RAS, Apatity [in Russian].

Mackey, A.P., A. Atkinson, S.L. Hill, P. Ward, N.J. Cunningham, N.M. Johnston, and E.J. Murphy. 2012. Antarctic macroplankton of the southwest Atlantic sector and Bellingshausen Sea: baseline historical distributions (Discovery Investigations, 1928–1935) related to temperature and food, with projections for subsequent ocean warming. Deep-Sea Research II 59–60:130–146.

Macpherson, E. 2002. Large-scale scies-richness gradients in the Atlantic Ocean. Proceedings of the Royal Society of London, Series B 269:1715–1720.

Macpherson, E. 2003. Species range size distributions for some marine taxa in the Atlantic Ocean. Effect of latitude and depth. Biological Journal of the Linnean Society 80:437–455.

Macpherson, E., B. Richer de Forges, K.E. Schnabel, S. Samadi, M.C. Boiselierand, A. Garcia-Rubies. 2010. Biogeography of the deep-sea galatheid squat lobsters of the Pacific. Deep-Sea Research I 57:228–238.

Madhupratap, M., and P. Haridas. 1986. Epipelagic calanoid copepods of the northern Indian Ocean. Oceanologica Acta 9:105–117.

Malay, M.C., and G.P. Paulay. 2010. Peripatric speciation drives diversification and distributional pattern of reef hermit crabs (Decapoda: Diogenidae: Calcinus). Evolution 64:634–662.

Manning, R.B. 1969. Stomatopod Crustacea of the western Atlantic. Studies in Tropical Oceanography, Miami 8:1–380.

Manning, R.B. 1977. A monograph of the West African stomatopod Crustacea. Atlantide Report 12: 25–181.

Marincovich Jr, L., E.M. Brouwers, D.M. Hopkins, and M.C. McKenna. 1990. Late Mesozoic and Cenozoic paleogeographic and paleoclimatic history of the Arctic Ocean Basin, based on shallow-water marine faunas and terrestrial vertebrates. Geology of North America 50:403–426.

Mauchline, J. 1998. The biology of calanoid copepods. *In* J.H. Blaxter, B. Douglas, and P.A. Tyler, editors. Advances in Marine Biology, Vol. 33. Academic Press, San Diego, CA.

Meland, K., J. Mees, M. Porter, and K.J. Wittmann. 2015. Taxonomic review of the orders Mysida and Stygiomysida (Crustacea, Peracarida). PLoS ONE 10:e0124656.

Mironov, A.N., A.B. Dilman, and E.M. Krylova. 2013. Global distribution patterns of genera occurring in the Arctic Ocean deeper than 2000 m. Invertebrate Zoology 10:167–194.

Mora, C., P.M. Chittaro, P.F. Sale, J.P. Kritzer, and S.A. Ludsin. 2003. Patterns and processes in reef fish diversity. Nature 421:933–936.

Mühlenhardt-Siegel, U. 1999. On the biogeography of Cumacea (Crustacea, Malacostraca). A comparison between South America, the Subantarctic Islands and Antarctica: present state of the art. Pages 295–302 *in* W.E. Arntz and C. Rios, editors. Magellan-Antarctic: Ecosystems that Drifted Apart. Scientia Marina 63 (suppl. 1).

Mühlenhardt-Siegel, U. 2014. Southern Ocean Cumacea. Pages 181–184 *in* C. De Broyer, P. Koubbi, H.J. Griffiths, B. Raymond, C. d'Udekem d'Acoz, A.P. Van de Putte, B. Danis, B. David, S. Grant, J. Gutt, C. Held, G. Hosie, F. Huettmann, A. Post, and Y. Ropert-Coudert, editors. Biogeographic Atlas of the Southern Ocean. Scientific Committee on Antarctic Research, Cambridge, UK.

Myers, A.A., and J. K. Lowry. 2009. The biogeography of Indo-West Pacific tropical amphipods with particular reference to Australia. Zootaxa 2260:109–127.

Myers, N., R.A. Mittermeier, C.G. Mittermeier, G.B. da Fonseca, and J. Kent. 2000. Biodiversity hotspots for conservation priorities. Nature 403:853–858.

Nair, S.R.S., V.R. Nair, C. Achuthankutty, and M. Madhupratap. 1981. Zooplankton composition and diversity in western Bay of Bengal. Journal of Plankton Research 3:493–508.

Neall, V. E., and S.A. Trewick. 2008. The age and origin of the Pacific islands: a geological overview. Philosophical Transactions of the Royal Society B 363:3293–3308.

Neiber, M.T., T. R. Hartke, T. Stemme, A. Bergmann, J. Rust, T. M. Iliffe, and S. Koenemann. 2011. Global biodiversity and phylogenetic evaluation of Remipedia (Crustacea). PLoS ONE 6:e19627.

Norris, R.D. 2000. Pelagic species diversity, biogeography, and evolution. Paleobiology 26:236–258.

O'Dea, A., H.A. Lessios, A.G. Coates, R.I. Eytan, S.A. Restrepo-Moreno, A.L. Cione, L.S. Collins, A. de Queiroz, D.W. Farris, R.D. Norris, R.F. Stallard, M.O. Woodburne, O. Aguilera, M.P. Aubry,, W.A. Berggren, A.F. Budd, M.A. Cozzuol, S.E. Coppard, H. Duque-Caro, S. Finnegan, G.M. Gasparini, E.L. Grossman, K.G. Johnson, L.D. Keigwin, N. Knowlton, E.G., Leigh, J.S. Leonard-Pingel, P.B. Marko, N.D. Pyenson, P.G. Rachello-Dolmen, E. Soibelzon, L. Soibelzon, J.A. Todd, G.J. Vermeij, and J.B. Jackson. 2016. Formation of the Isthmus of Panama. Science Advances 2:e1600883.

Park, E.T., and F.D. Ferrari. 2009. Species diversity and distributions of pelagic Calanoid copepods from the Southern Ocean. Pages 143–180 *in* I. Krupnik, M.A. Lang, and S.E. Miller, editors. A Selection from Smithsonian at the Poles. Contributions to International Polar Year Science, Smithsonian Institution Scholarly Press, Washington, DC.

Patarnello, T., L. Bargelloni, V. Varotto, and B. Battaglia. 1996. Krill evolution and the Antarctic ocean currents. Marine Biology 126:603–608.

Petryashov, V.V. 2004. [Mysids (Crustacea, Mysidacea) of the Eurasiatic subbasin of the Arctic basin and the adjacent seas: The Barents, Kara and Laptev Seas]. Explorations of the Fauna of the Seas 54:124–145 [in Russian].

Petryashov, V.V. 2007. Biogeographical division of Antarctic and subantarctic by Mysid (Crustacea: Mysidacea) Fauna. Russian Journal of Marine Biology 33:1–16 (in Russian).

Petryashov, V.V. 2009. [Biogeographical division of the Arctic and North Atlantic based on the fauna of mysids (Crustacea: Mysidacea)]. Russian Journal of Marine Biology 35:87–106 [in Russian, English abstract].

Petryashov, V.V. 2014. Lophogastrida and Mysida (Crustacea: Malocostraca: Peracarida) of the Southern Ocean. Pages 149–154 *in* C. De Broyer, P. Koubbi, H.J. Griffiths, B. Raymond, C. d'Udekem d'Acoz, A.P. Van de Putte, B. Danis, B. David, S. Grant, J. Gutt, C. Held, G. Hosie, F. Huettmann, A. Post, and Y. Ropert-Coudert, editors. Biogeographic Atlas of the Southern Ocean. Scientific Committee on Antarctic Research, Cambridge, UK.

Platnick, N.I. 1991. Patterns of biodiversity: tropical vs temperate. Journal of Natural History 25:1083–1088.

Poore, G.C.B., and N. Andreakis. 2011. Morphological, molecular and biogeographic evidence support two new species in the *Uroptychus naso* complex (Crustacea: Decapoda: Chirostylidae). Molecular Phylogenetics and Evolution 60:152–169.

Poore, G.C.B., and N. Andreakis. 2014. More species of the *Agononida incerta* complex revealed by molecules and morphology (Crustacea: Decapoda: Anomura: Munididae). Zootaxa 3860:201–225.

Poore, G.C.B., and N.L. Bruce. 2012. Global diversity of marine Isopods (except Asellota and Crustacean Symbionts). PLoS ONE 7:e43529.

Poore, G.C.B., and T.D. O'Hara. 2007. Marine biogeography and biodiversity of Australia. Pages 177–198 *in* S.D. Connell and B.M. Gillanders, editors. Marine Ecology. Oxford University Press, Oxford.

Poore, G.C.B., and G.D.F. Wilson. 1993. Scientific correspondence: marine species richness (with reply from R. M. May). Nature 361:597–598.

Pörtner, H.O., L. Peck, and G. Somero. 2007. Thermal limits and adaptation in marine Antarctic ectotherms: an integrative view. Philosophical Transactions of the Royal Society of London B, Biological Sciences 362:2233–2258.

Poupin, J. 2008. Biogeography of the decapod and stomatopod Crustacea of the tropical Pacific: Issues and prospects. Pacific Science 62:377–383.

Questel, J., O.N.S. Aarbakke, L. Blanco-Bercial, R.R. Hopcroft, F. Norrbin, C. Halsband, A. Weydmann, E.A. Ershova, K.N. Kosobokova, and A. Bucklin. Pan-Arctic phylogeography and connectivity of *Pseudocalanus* (Copepoda). POLAR2018 Open Science Conference, June 19–23, 2018, Davos, Switzerland.

Rakhesh, M., A.V. Raman, and D. Sudarsan. 2006. Discriminating zooplankton assemblages in neritic and oceanic waters: a case for the northeast coast of India, Bay of Bengal. Marine Environmental Research 61: 93–109.

Raupach, M. J., and A.E. Radulovici. 2015. Looking back on a decade of barcoding crustaceans. Zookeys 539:53–81.

Raupach, M.J., C. Mayer, M. Malyutina, and J.W. Wägele. 2009. Multiple origins of deep-sea Asellota (Crustacea: Isopoda) from shallow waters revealed by molecular data. Proceedings of the Royal Society B 276:799–808.

Razouls, C., F. de Bovée, J. Kouwenberg, and N. Desreumaux. 2005–2016. Diversity and Geographic Distribution of Marine Planktonic Copepods. https://copepodes.obs-banyuls.fr/en/.

Razouls, S., C. Razouls, and F. De Bovée. 2000. Biodiversity and biogeography of Antarctic copepods. Antarctic Science 12:343–362.

Reaka, M.L., and R.B. Manning. 1987. The significance of body size, dispersal potential, and habitat for rates of morphological evolution in stomatopod Crustacea. Smithsonian Contributions to Zoology 448:1–46.

Reaka, M.L., P. J. Rodgers, and A.U. Kudla. 2008. Patterns of biodiversity and endemism on Indo-West Pacific coral reefs. Proceedings of the National Academy of Sciences 105:11474–11481.

Reid, J.L., E. Brinton, A. Fleminger, E.L. Venrick, and J.A. McGowan. 1978. Ocean circulation and marine life. Pages 65–130 *in* Advances in Oceanography. Springer.

Renema, W., D. R. Bellwood, J.C. Braga, K. Bromfield, R. Hall, K.G. Johnson, P. Lunt, C.P. Meyer, L.B. McGonagle, R.J. Morley, A. O'Dea, J.A. Todd, F.P. Wesselingh, M. E. J. Wilson, and J. M. Pandolfi. 2008. Hopping hotspots: global shifts in marine biodiversity. Science 321:654–657.

Renz, J., and E.L. Markhaseva. 2015. First insights into genus level diversity and biogeography of deep sea benthopelagic calanoid copepods in the South Atlantic and Southern Ocean. Deep-Sea Research I 105: 96–110.

Rivadeneira, M.M., M. Thiel, E.R. Gonzalez, and P.A. Haye. 2011. An inverse latitudinal gradient of diversity of peracarid crustaceans along the Pacific Coast of South America: out of the deep south. Global Ecology and Biogeography 20:437–448.

Roberts, C.M., C.J. McClean, J.E.N. Veron, J.P. Hawkins, G.R. Allen, D.E. McAllister, C.G. Mittermeier, F.W. Schueler, M. Spalding, F. Wells, C. Vynne, and T.B. Werner. 2002. Marine biodiversity hotspots and conservation priorities for tropical reefs. Science 295:1280–1284.

Rohde, K. 1999. Latitudinal gradients in species diversity and Rapoport's rule revisited: a review of recent work and what can parasites teach us about the causes of the gradients? Ecography 22:593–613.

Rombouts, I., G. Beaugrand, F. Ibañez, S. Gasparini, S. Chiba, and L. Legendre. 2009. Global latitudinal variations in marine copepod diversity and environmental factors. Proceedings of the Royal Society of London B: Biological Sciences :rspb20090742.

Rowden, A.A., Schnabel, K.E., Schlacher, T.A., Macpherson, E., Ahyong, S.T., and Richer de Forges, B. 2010. Squat lobster assemblages on seamounts differ from some, but not all, deep-sea habitats of comparable depth. Marine Ecology 31 (suppl. 1):63–83.

Rutherford, S., S. D'Hondt, and W. Prell. 1999. Environmental controls on the geographic distribution of zooplankton diversity. Nature 400:749–753.

Savin, S.M., R.C. Douglas, and F.G. Stehli. 1975. Tertiary marine paleo-temperatures. Geological Society American Bulletin 86:1499–1510.

Schiller, A., P.R. Oke, G. Brassington, M. Entel, R. Fiedler, D.A. Griffin, and J.V. Mansbridge. 2008. Eddy-resolving ocean circulation in the Asian-Australian region inferred from an ocean reanalysis effort. Progress in Oceanography 76:334–365.

Schnabel, K.E., P. Cabezas, A. McCallum, E. Macpherson, S.T. Ahyong, and K. Baba. 2011. World-wide distribution patterns of squat lobsters. Pages 149–182 *in* G.C.B. Poore, S.T. Ahyong, and J. Taylor, editors. The Biology of Squat Lobsters. CSIRO Publishing, Melbourne, Australia.

Schram, F.R., S.T. Ahyong, S.N. Patek, P.A. Green, M.V. Rosario, M.J. Bok, T.W. Cronin, K.S. Mead Vetter, R.L. Caldwell, G. Scholtz, K.D. Feller, and P. Abelló. 2013. Subclass Hoplocarida Calman, 1904: Order Stomatopoda Latreille, 1817. Pages 197–355 *in* F.R. Schram and J. C. Von Vaupel Klein, editors. Treatise on Zoology—Anatomy, Taxonomy, Biology. The Crustacea, Vol. 4A. Brill, Leiden, The Netherlands.

Sherman, K., and G. Hempel, editors. 2009. The UNEP Large Marine Ecosystem Report. A Perspective on Changing Conditions in LMEs of the world's regional seas. UNEP Regional Seas Report and Studies No. 182. United Nations Environment Program 852, Nairobi, Kenya.

Sirenko, B.I. 2001. List of species of free-living invertebrates of Eurasian Arctic seas and adjacent deep waters. Explorations of the Fauna of the Seas 51:131.

Spalding, M.D., H.E. Fox, G.R. Allen, N. Davidson, Z.A. Ferdaña, M. Finlayson, B.S. Halpern, M.A. Jorge, A. Lombana, S.A. Lourie, K.D. Martin, E. McManus, J. Molnar, C.A. Recchia, and J. Robertson. 2007. Marine ecoregions of the world: a bioregionalization of coastal and shelf areas. Bioscience 57:573–583.

Sudre, J., and R.A. Morrow. 2008. Global surface currents: a high-resolution product for investigating ocean dynamics. Ocean Dynamics 58:101–118.

Sutton, T.T., M.R. Clark, D.C. Dunn, P.N. Halpin, A.D. Rogers, J. Guinotte, S.J. Bograd, M.V. Angel, J.A.A. Perez, K. Wishner, R.L. Haedrich, D.J. Lindsay, J.C. Drazen, A. Vereshchaka, U. Piatkowski, T. Morato, K. Błachowiak-Samołyk, B.H. Robison, K.M. Gjerde, A. Pierrot-Bults, P. Bernal, G. Reygondeau, and M. Heino. 2017. A global biogeographic classification of the mesopelagic zone. Deep Sea Research Part I: Oceanographic Research Papers:j.dsr.2017.05.006.

Thatje, S., and W.E. Arntz. 2004. Antarctic reptant decapods: more than a myth? Polar Biology 27:195–201.

Thomsen, P. F., J. Kielgast, L. Iversen, P.R. Møller, M. Rasmussen, and E. Willerslev. 2012. Detection of a diverse marine fish fauna using environmental DNA from seawater samples. PLoS ONE 7:e41732.

UNESCO. 1965–1972. International Indian Ocean Expedition. Collected Reprints, 1–8. UNESCO, Paris.

UNESCO. 2009. Global Open Oceans and Deep Seabed (GOODS) biogeographic classification. UNESCO-IOC. IOC Technical Series 84:1–87.

Van der Spoel, S., and R.P. Heyman. 1983. A Comparative Atlas of Zooplankton. Bunge, Utrecht, The Netherlands.

Vereshchaka, A.L., O. Olesen, and A.L. Lunina. 2014. Global diversity and phylogeny of pelagic shrimps of the former genera *Sergestes* and *Sergia* (Crustacea, Dendrobranchiata, Sergestidae), with definition of eight new genera. PLoS ONE 9: e112057.

Vinogradov, M.E., and T.N. Semenova. 2009. Suborder Hyperiidea (Class Crustacea, Order Amphipoda). Pages 85–98 *in* S.V. Vassilenko and V.V. Petryashov, editors. Illustrated Keys to Free-Living Invertebrates of Eurasian Arctic Seas and Adjacent Deep Waters. Rotifera, Pycnogonida, Cirripedia, Leptostraca, Mysidacea, Hyperiidea, Caprellidea, Euphausiacea, Dendrobranchiata, Pleocyemata, Anomura, and Brachyura. Alaska Sea Grant, University of Alaska Fairbanks.

Waters, J.M. 2008. Marine biogeographical disjunction in temperate Australia: historical landbridge, contemporary currents, or both? Diversity and Distributions 14:692–700.

Wicksten, M. 1989. Ranges of offshore decapod crustaceans in the eastern Pacific Ocean. Transations of the San Diego Society of Natural History 21:291–316.

Williams, S. T., and T. F. Duda, Jr. 2008. Did tectonic activity stimulate Oligo-Miocene speciation in the Indo-West Pacific? Evolution 62:1618–1634.

Willig, M.R., D.M. Kaufman, and R.D. Stevens. 2003. Latitudinal gradients of biodiversity: pattern, process, scale and synthesis. Annual Review of Ecology and Systematics 34:273–309.

Woodd-Walker, R.S. 2001. Spatial distributions of copepod genera along the Atlantic Meridional Transect. Hydrobiologia 453–454:161–170.

Woodd-Walker, R.S., P. Ward, and A. Clarke. 2002. Large-scale patterns in diversity and community structure of surface water copepods from the Atlantic Ocean. Marine Ecology Progress Series 236:189–203.

Wright, D.H. 1983. Species-energy theory: an extension of species-area theory. Oikos 41:496–506.

Zaklan, S. 2002. Review of the family Lithodidae (Crustacea: Anomura: Paguroidea): distribution, biology and fisheries. Pages 751–845 *in* A.J. Paul, E.G., Dawe, R. Elner, G.S. Jamieson, G.H. Kruse, R.S. Otto, B. Sainte-Marie, T.C. Shirley, and D. Woodby, editors. Crabs in Cold Water Regions: Biology, Management, and Economics. University of Alaska Sea Grant Program, Fairbanks, Alaska.

Zenkevitch, L.A. 1947. [Fauna and biological productivity of the sea]. Izdatelstvo Sovetskaya Nauka, Moscow. [in Russian].

Zenkevitch, L. 1963. Biology of the seas of the USSR. Allen & Unwin, London.

7

CONQUERING THE OCEAN DEPTHS OVER THREE GEOLOGICAL ERAS

Torben Riehl, Simone Nunes Brandão, and Angelika Brandt

Abstract

The deep sea is the largest habitat on Earth, but it is the least accessible and comprehensible. This apparently harsh environment is inhabited by Crustacea of diverse evolutionary lineages that have, to various degrees, evolved uniquely specialized morphologies and lifestyles. Following a century of debate about the antiquity of the deep-sea fauna, studies of Crustacea reveal that the faunas of the deep and shallow oceans have been continuously and repeatedly exchanged, probably since the Mid-Paleozoic. Deep-sea colonization and subsequent diversification has occurred across many crustacean lineages, during several periods, and may still be underway. Despite a commonly held view that shallow–deep phylogenetic relationships are unidirectional, there is also evidence for evolutionary emergence from the abyss into shallower zones. As a result, the present-day fauna represents an amalgamation of clades of various ages. Environmental factors such as pressure, temperature, and energy supply differ substantially between shallow- and deep-water layers, creating gradients that pose important ramifications for crustacean physiology. Consequently, depth-range expansions require adaptations, which may lead to peculiar phenomena such as gigantism and dwarfism, as well as diverse crustacean radiations.

INTRODUCTION

This chapter summarizes evidence for and hypotheses on the origins of the benthic deep-sea Crustacea and their evolutionary linkages to faunas dwelling at shallower depths. We provide a review of what is known about potential deep-sea extinctions and how crustaceans play a role in the reconstruction of paleoenvironments and global environmental change. We also highlight the survivors of historic global change as well as recurring colonization and radiations, which together form the diverse crustacean deep-sea fauna comprising ancient and young elements.

Evolution and Biogeography. Edited by Gary C. B. Poore and Martin Thiel.

The oceans below 300 m deep are commonly referred to as deep sea. The seafloor is dominated by the abyssal zone stretching out between the bathyal zone, reaching down to 3,500 m, and the hadal zone, deeper than 6,500 m (Watling et al. 2013) (Fig. 7.1). Consequently, the focus of this chapter is on the abyss, but we also consider crustaceans of the structurally diverse continental slopes and those inhabiting the deepest spots in the oceans, the hadal trenches. Its enormous dimensions, relatively difficult accessibility, and environmental characteristics render the abyss an unusual habitat from a human point of view: solar light is absent; temperatures are low; diurnal, seasonal, and annual cycles are relatively slight (Johnson et al. 2007) and have been maintained relatively stable over geological time scales. Under the effects of these conditions and other factors, some crustaceans such as the mostly meiofaunal (<1.10 mm body length) harpacticoid Copepoda as well as macrofaunal (1.10–10.0 mm) Isopoda have evolved specialized morphologies and lifestyles unique to the deep sea. Decapods and other megafaunal Crustacea are rather rare (Wilson and Ahyong 2015). An account of the lifestyles of deep-sea crustaceans was recently given in another volume of this series (Wilson and Ahyong 2015); bathymetric gradients and the adaptations required for range expansions in crustaceans are discussed in Chapter 16 in this volume; in this chapter we focus on evolutionary patterns and invasions.

BATHYMETRIC AND PHYLOGENETIC PATTERNS

The benthos (i.e., the organisms inhabiting the seafloor) does not often reflect the clear geomorphological delineation into depth zones. Rather, taxon composition changes slowly with continuous species replacement along a bathymetric gradient until the benthos of shelf and abyss have little in common (Rex and Etter 2010, Wei et al. 2010), at least on a species level. As we will demonstrate, crustaceans thrive at all depths, but the degree of specialization and diversity depends partially on the period each taxon has existed there. In this respect, phylogenetic links between abyssal and shallow-water Crustacea can be found at all levels of the systematic hierarchy. This is reflected in the distinct, taxon-specific patterns of evolutionary histories shown by the Recent (taxonomically narrowly defined as opposed to extinct, fossil taxa which are more broadly defined) deep-sea crustaceans.

Decapods and stomatopods are irregularly encountered in the abyssal environment (Hessler and Wilson 1983), and their diversity decreases with depth (Rosa et al. 2012). Yet blind lobsters, family Polychelidae, and the squat lobster (Galatheoidea) genera *Munidopsis* and *Galacantha* have the greatest diversity on the continental slopes (Galil 2000, Macpherson et al. 2010). Multiple decapod lineages have colonized abyssal and hadal waters, but mostly with few species. Prawns (Dendrobranchiata) and several families of caridean shrimps (Fig. 7.2C) (e.g., Alvinocarididae, Crangonidae, Glyphocrangonidae, Nematocarcinidae, Pandalidae, Bresiliidae (vents), and Hippolytidae) have penetrated the deep oceans. Moreover, some species range into the hadal zone, such as the benthesicymid prawn *Benthesicymus crenatus* and species of *Acanthephyra* (Caridea: Acanthephyridae) (Jamieson et al. 2009).

Like decapods, most peracarid orders have relatively few representatives at abyssal and even fewer at hadal depths. Particularly among the isopods, the Antarcturidae, Anthuroidea, Asellota, and Serolidae have true abyssal or, in case of the Antarcturidae and Asellota, even hadal representatives. Yet the core distributions of most Isopoda are the subtidal and continental shelves (Poore and Bruce 2012). An exception is the isopod suborder Asellota, which has its greatest diversity in the deep sea (Hessler et al. 1979). Many of its families have eurybathic distributions ranging from shelf to the abyss (Brandt et al. 2009) and, in some cases, from subtidal to the deepest hadal, such as the Macrostylidae (Riehl and Brandt 2010).

While cirripeds occur predominantly in shallow waters, sessile and stalked barnacles are common inhabitants of eastern and western Pacific hydrothermal vent fields (Yamaguchi and Newman 1990) and

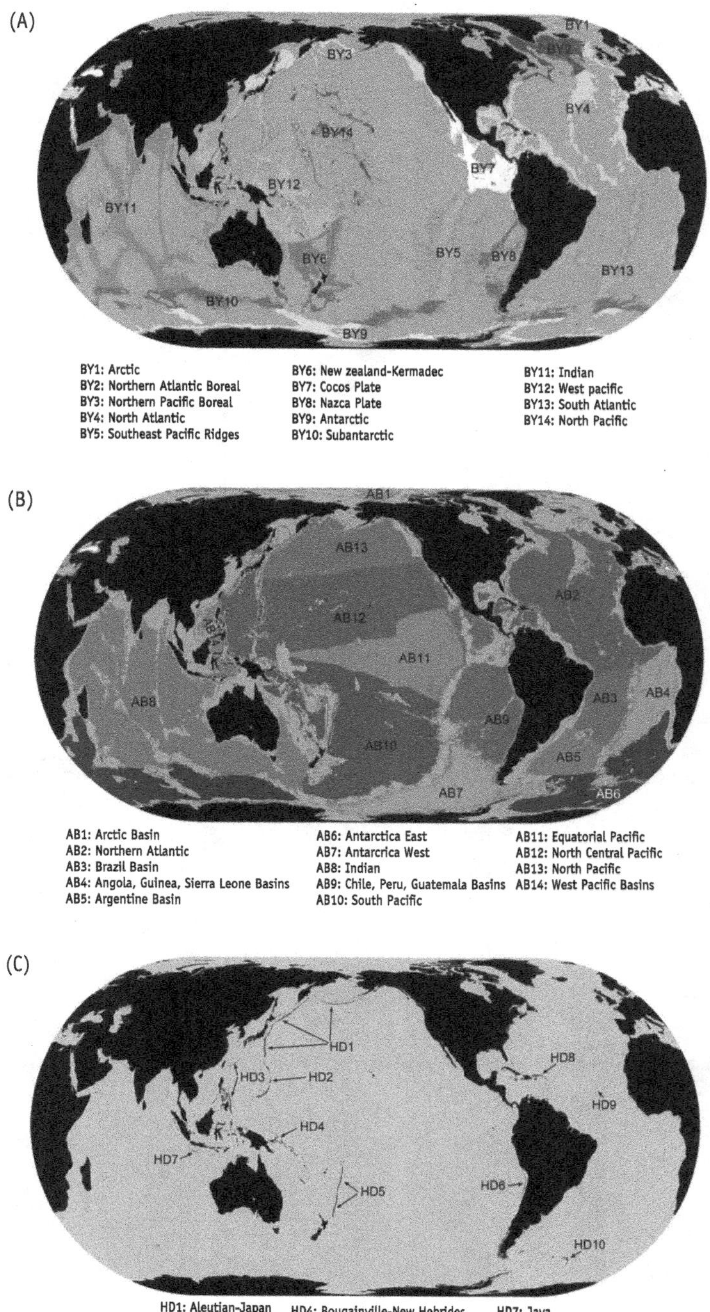

Fig. 7.1.
Biogeographic regions of the deep sea. (A) Lower bathyal provinces including continental slopes and ridges, as well as island slopes, and several thousand seamounts. (B) Abyssal provinces separated by mid-ocean ridges and other bathyal topography creating deep basins as well as by varying levels of POC flux to the seafloor. (C) Hadal provinces include all areas deep areas below 6,500 m. Modified after Watling et al. (2013), with permission from Elsevier. See color version of this figure in centerfold.

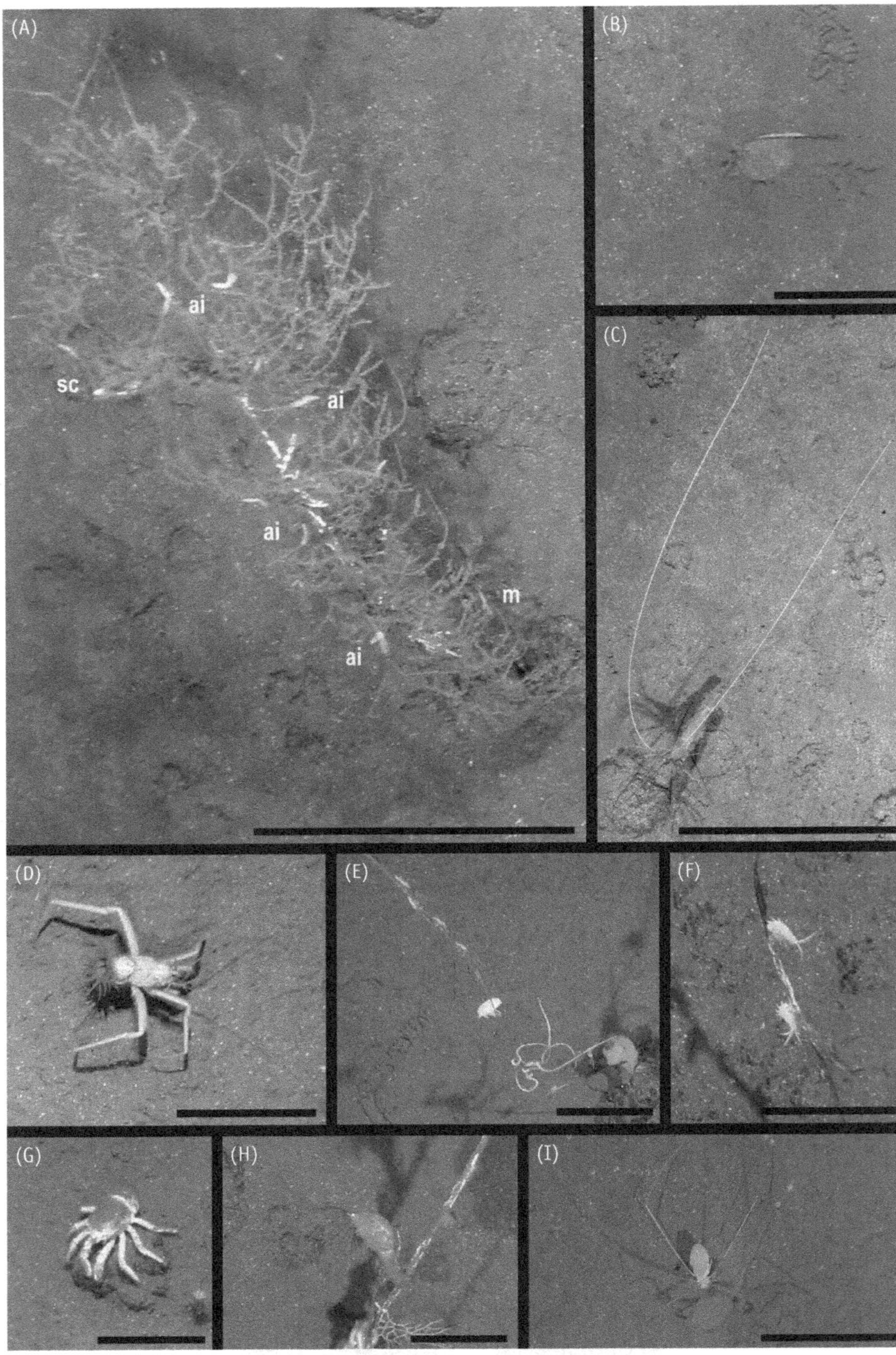

Fig. 7.2.
Megafaunal Crustacea from the Peru Basin taken at >4,000 m depth in the DISCOL manganese nodule area. Many megafaunal Crustacea seem to rely on some form of hard substrate for a habitat, such as nodules or structures provided by living as well as dead sessile animals. (A) Arcturid isopods (ai), scalpelliform cirriped (sc), and mysid (m) associated on a gorgonian octocoral. Scale bar = 25 cm. (B) Polychelid decapod dug into the sediment. Scale bar = 10 cm. (C) Caridean shrimp, Stylodactiloidea. Scale bar = 25 cm. (D) Parapagurid decapod (*Probeebei mirabilis*) with anemone on pleon. (E) Association on manganese nodule: coral stalk, isopods (i), lyssianassoid amphipod (la), ophiuroid (o), mysid (m), sponge (s), and munnopsid isopod (mi). Scale bar = 10 cm. (F) Gammaridean amphipods in a manganese nodule field. Scale bar = 10 cm. (G) *Munidopsis* sp. on manganese nodule. Scale bar = 10 cm. (H) scalpelliform barnacle. Scale bar = 10 cm. (I) *Paropsurus* sp. (Isopoda, Munnopsidae) feeding on salp. Scale bar = 10 cm. Photos courtesy of Alfred-Wegener-Institute/ SONNE SO242-2, project JPIO / MIDAS; Ocean Floor Observation System (OFOS) imaging, used with permission under the CC4-BY-NC-SA license. See color version of this figure in centerfold.

where hard substrata such as in manganese nodule areas exist in abyssal plains (Fig. 7.2A, H). The cosmopolitan Cephalocarida are distributed differently. They occur discontinuously on flocculent, organic-rich surface sediments from the intertidal zone to the abyss. Their morphology seems basal and uniform. Accordingly, they have been interpreted as Paleozoic relics and generalists that, despite an ancient natural history, never underwent a process of adaptive radiation (Hessler and Sanders 1973).

Most families of the omnipresent Copepoda Harpacticoida occur over wide depth ranges, but several genera are considered "deep-sea" specialists, for instance, those of the basal oligoarthran families Aegisthidae and Neobradyidae. Deep-sea origins and diversification have been hypothesized for these groups (Seifried 2004). It has even been suggested that the origin of the whole and super-diverse Harpacticoida may lie in lower bathyal or abyssal depths from where several subsequent processes of colonization lead them to shallower zones and secondary migrations back into the abyss (Seifried 2004).

Except for the meiofaunal harpacticoids with great dispersal potential (Easton and Thistle 2016), the colonization history between abyss and shallower zones has mostly been seen as unidirectional: from shallow to deep. There is, however, no reason to reject a priori the possibility of the opposite process, the evolutionary emergence from the abyss into shallower zones (Hessler and Wilson 1983, Brandt 1992). It seems unlikely that one or a few time-constrained and unidirectional events of deep-sea colonization are the underlying explanation for the present deep-sea fauna. If that were the case, the differences and overlap between biogeographic and phylogenetic patterns of shallow-water and deep-sea faunas would require less parsimonious explanations. Ancestral depth-occurrence reconstruction on Macrostylidae showed that this isopod family diversified in depth greater than 3,000 m (Riehl 2014). Subsequently and multiple times independently, macrostylid species or lineages colonized shallower depth zones. The phylogenetic relationships, furthermore, indicated that, secondarily, some macrostylids have extended their depth range from shallow to deep (Riehl 2014). This data thus point toward a shallow-water-refuge and deep-sea-recolonization scenario (Fig. 7.3). The existence of clades with distinct bathymetric distribution patterns illustrates that depth barriers occur and the underlying bathymetric gradients require physiological adaptations (see Chapter 16).

Radiation or Invasion?

Popular concepts of the origins of the deep-sea fauna (elaborated below) suggest that deep-sea colonization and in situ radiation are alternatives. The following examples, however, show that these evolutionary trajectories are not mutually exclusive. We treat the terms invasion and colonization as synonyms here, referring to depth-range expansion of taxa that originally inhabited shallow water depth and subsequently propagated into the deep sea.

The asellote isopods of the superfamily Janiroidea are one of the most diverse crustacean groups in the deep sea, and the lineage referred to as the munnopsoid radiation (Raupach et al. 2004) is thought to be the oldest of the Recent deep-sea radiations (Lins et al. 2012) (Fig. 7.4). It contains highly modified taxa, such as Macrostylidae and Munnopsidae, and abyss-endemic families, like Urstylidae (Riehl et al. 2014). However, deep-sea Asellota comprise lineages that have independently colonized the abyss (Raupach et al. 2009). The distribution of their diversity with most species occurring at great depths (bathyal and abyssal, OBIS 2017) and phylogenetic relationships indicate that some groups originated and diversified at great depth (e.g., Hessler and Thistle 1975, Wilson 1999, Riehl and Brandt 2010, Riehl 2014), while rather recent colonization can be observed in Janiridae, Paramunnidae, and Munnidae (Wilson 1980, Linse et al. 2014). This diversity may have depended upon unlocking deep-sea protists (e.g., Komokiacea) as a food source (Svavarsson et al. 1993, Brökeland 2010, Riehl et al. 2018). Some munnopsids have secondarily evolved the ability to swim backward using paddle-shaped posterior pereopods to propel actively and, furthermore, to use anterior pereopods to "walk" up the water column and decrease sinking. These special features, as opposed to other Asellota, allow them both to swim, even against water movement, and to make increased use of currents through floatation. Allowing for extended individual ranges, these

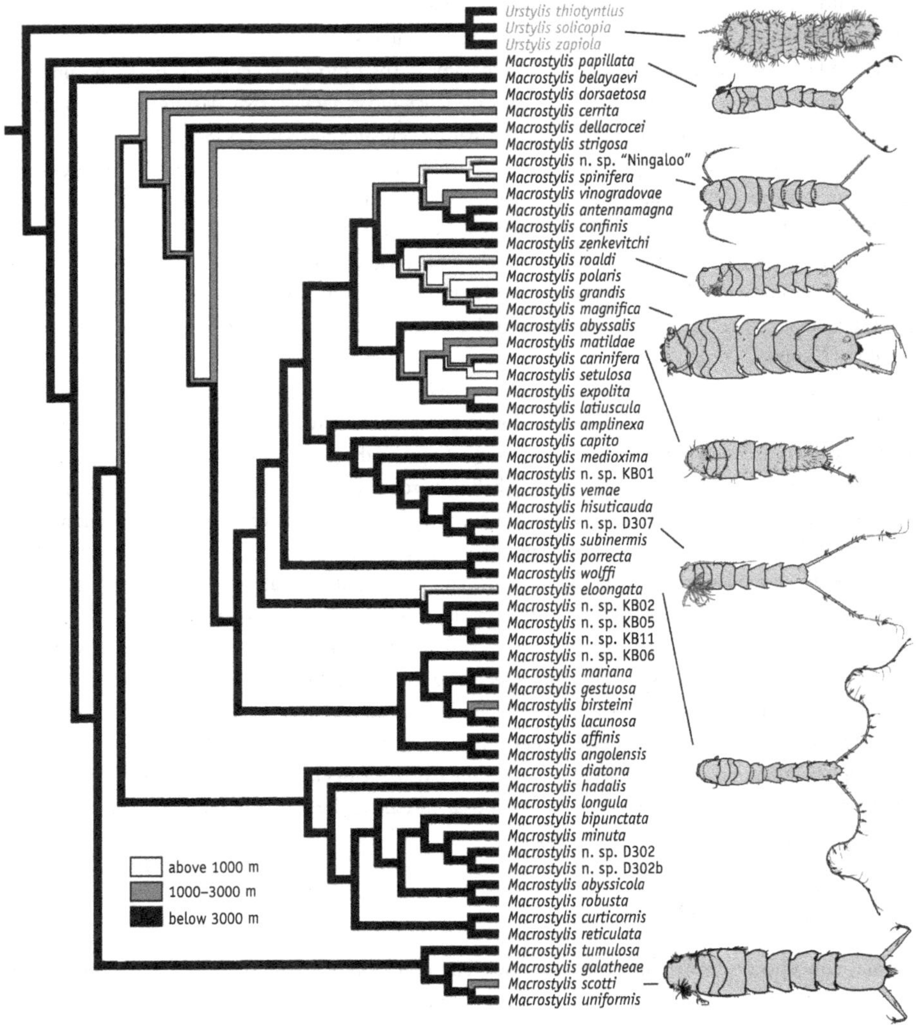

Fig. 7.3.
The abyss-shelf colonization history of Macrostylidae (Isopoda: Asellota). An ancestral state reconstruction was conducted based on a morphological phylogeny using depth zones "sublittoral" (0–1,000 m), "bathyal" (1,000–3,000 m), and "abyssal" + "hadal" (>3,000 m depth) as discrete characters. A parsimony reconstruction scheme as implemented in Mesquite was applied. White branches indicate sublittoral, grey branches indicate bathyal, black bars represent abyssal and hadal. Species (and predecessors) occurring across more than one depth zone have mixed coloration. An abyssal origin of Macrostylidae is most parsimonious. Within Macrostylidae, six emergence events into the bathyal became apparent, of which four led to shelf emergence. Potential reversals (submergence) have occurred, and one species (*Macrostylis magnifica*) has a eurybathic distribution across all three depth zones. Three species of Urstylidae formed the outgroup. Modified from Riehl (2014), with permission from the University of Hamburg. See color version of this figure in centerfold.

characteristics of Munnopsidae can be considered a key to understanding this important adaptive radiation in the deep sea (Thistle and Hessler 1976, Wilson 1989, Osborn 2009). The isopod taxa discussed above represent examples of diverse groups that have evolved in situ, although with continued bidirectional exchange between depth zones (e.g., Riehl 2014; Fig. 7.3).

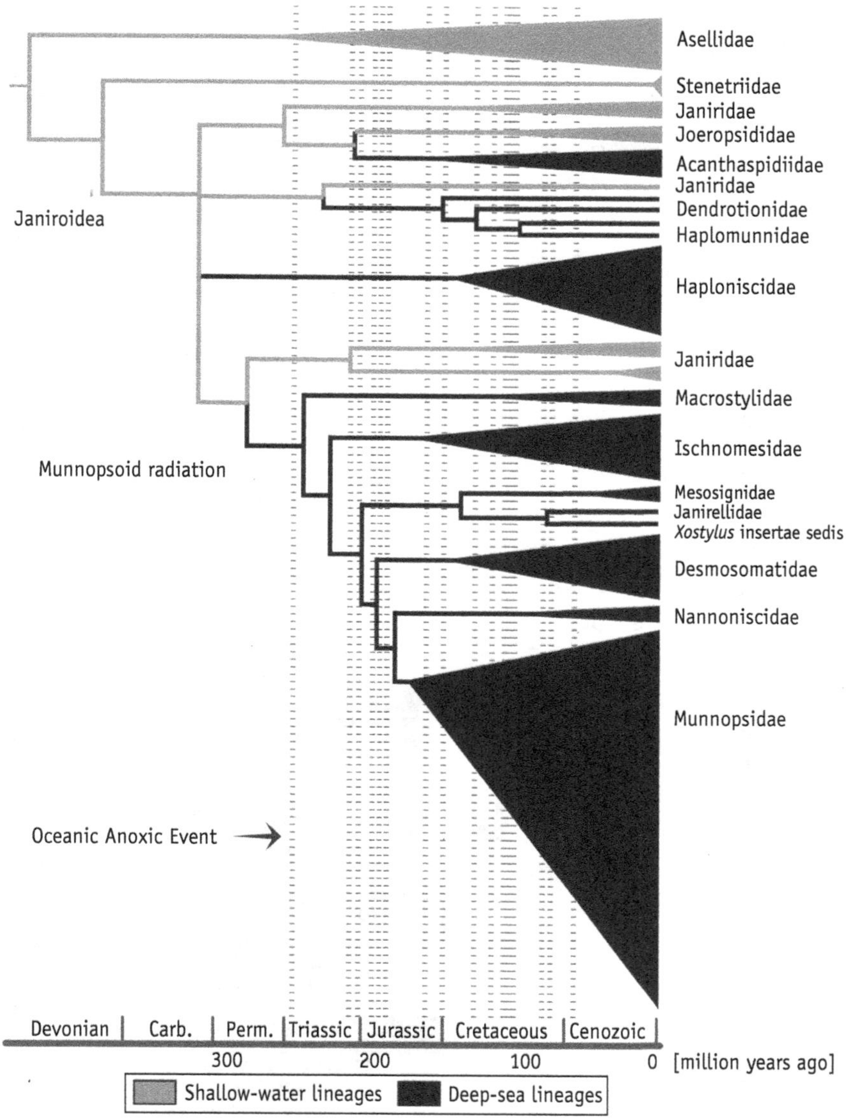

Fig. 7.4.
Graphical representation of a dated phylogenetic reconstruction of the Janiroidea (Crustacea, Isopoda). On family level, the dendrogram is collapsed for simplicity. The occurrences of proposed global or ocean-spanning oceanic anoxic events (OAEs) are indicated by dashed vertical bars. Lineages have been categorized as deep-sea or shallow-water lineages based on biogeographic patterns as well as presence or absence of eyes. This figure has been interpreted as counter-evidence for the global extents and fatal consequences of OAEs as some of the specialized deep-sea clades have evolutionary origins distinctly older than the OAEs themselves. Modified after Lins et al. (2012).

Like the Isopoda, Tanaidacea is a peracarid order representing a high and probably old diversity in the deep-sea benthos. The apseudomorphs and especially the Neotanaidomorpha have gone through extensive radiation and adaptation in the deep sea, which harbors a large diversity of these groups (Błażewicz-Paszkowycz et al. 2012; Fig. 7.5)

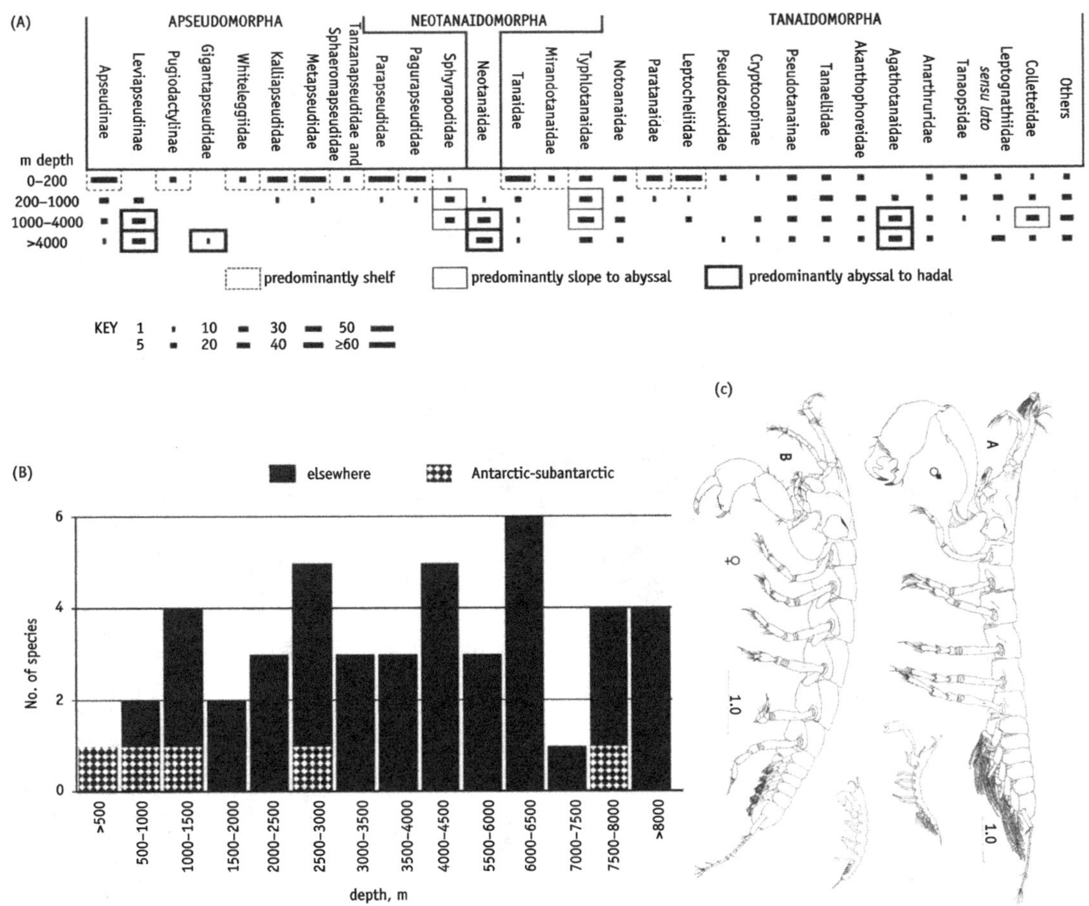

Fig. 7.5.
Tanaidacea depth distribution and species numbers per depth layer. (A) Numbers of species subdivided by four depth bands for most families of the Tanaidacea (some monotypic families excluded for clarity), arranged by suborder. The four depth bands represent sublittoral to the shelf edge, upper bathyal, lower bathyal to upper abyssal, and lower abyssal to hadal. (B) Number of species of Neotanaidomorpha in various depth bands. (C) Lateral views of male and female *Neotanais micrompher* as an example for the deep-sea family Neotanaidae. Scale lines = 1 mm. A, B, from Blazewicz-Paszkowycz et al. (2012) under Creative Commons license (BY); C, from Gardiner (1975).

Diversification of lysianassoid amphipods may be linked to feeding preferences. Progressive specialization from non-scavenging behavior to specialized scavengers around 66 million years ago (Ma) drove their diversification at the Eocene–Oligocene transition 33.9 Ma. Ancestral state reconstruction of taxonomically important mouthpart characters that furthermore indicate feeding-guild memberships showed evolutionary progression from a non-scavenging ancestor to opportunistic, obligate, and specialized scavengers. This transition was associated with increasing diversification over time in these amphipods (Corrigan et al. 2014) (Fig. 7.6).

Two major clades of Penaeoidea (prawns) diverged in the Triassic (>200 Ma), and their subsequent radiations may be associated with differential preferences for bathyal versus shelf habitats. The predominantly bathyal clade composed of Aristeidae, Benthesicymidae, and Solenoceridae seems to have a Jurassic origin. However, it remains difficult to reconstruct the depth range of their common ancestor (Ma et al. 2009). The situation is different for deep-sea Cumacea. Based on biogeographic and bathymetric patterns of its most basal genera, diversification of the deep-sea family Bodotriidae is linked to the opening of the South Atlantic during the Cretaceous (146–66 Ma). From this center of origin, bodotriids subsequently colonized both the North Atlantic (including Mediterranean) as well as the other oceans via the Southern Ocean (Haye 2007).

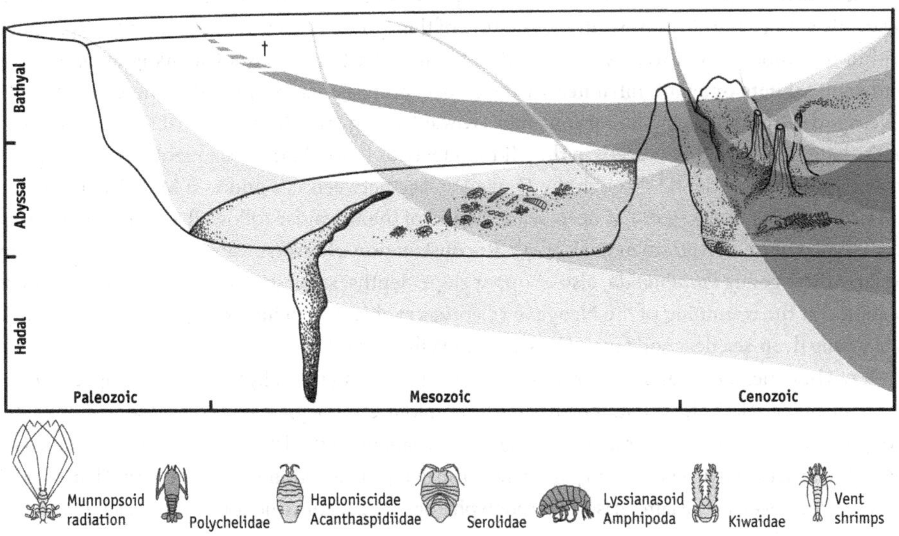

Fig. 7.6.
Overview of selected deep-sea crustacean lineages showing approximate times of origin, radiation, and depth distribution in a depth-time scheme from Paleozoic (541–252.2 Ma) until today. The time scale is vague on purpose, acknowledging the high uncertainties involved. Each lineage is represented by a colored beam. The height of each beam at the right margin indicates the approximate core depth distribution today. Phylogenetic links to potential shallow-water ancestors are indicated at the top left of each beam. Sea-floor structures are not correlated to the time line. The isopods of the munnopsoid radiation as well as the Haploniscidae/Acanthaspidiidae and Serolidae have colonized the abyss independently (Raupach et al. 2009). While the first originated before 250 Ma, the latter invaded the deep sea during the Mesozoic (252–66 Ma) (Lins et al. 2012). The first three groups are diverse on soft sediments, which may have been the key to their success. Subsequently, they colonized shallower depth zones and the hadal. The Serolidae has colonized abyssal depth without radiation at great depth. Polychelidae got extinct on continental shelves but inhabits intermediate depths today (Galil 2000). Among the lyssianassoid amphipods an end-Mesozoic–early-Cenozoic adaptive radiation resulted in a clade of specialized scavengers (Corrigan et al. 2014). Kiwaidae and other seep and vent decapods radiated in extreme habitats after the Paleocene (66–56 Ma) of the Cenozoic (66–0 Ma) (Roterman et al. 2013). With permission from Nele Johannsen. See color version of this figure in centerfold.

Already adapted to oligotrophy, stenothermy, and dark environments, cave inhabitants in shallow water may have been well suited for transition into deep-ocean environments. Ostracods (Thaumatoconcha) and copepods (Misophrioida, Novocrinidomorpha) have derived representatives in deep and "open" environments, whereas plesiomorphic species inhabit caves. Boxshall and Jaume (2000) hypothesized that secondary colonization from shallow anchialine caves into the deep sea happened <40 Ma in late through post-Eocene glacial periods (Benson et al. 1984, Corgosinho et al. 2016).

Colonization of meiofauna is hard to infer because a high capacity for dispersal seems associated with their small size. In Harpacticoida, "extreme" dispersal capabilities can be assumed, promoted by their small size and a high probability of passive or semi-passive drift (Menzel et al. 2011, Easton and Thistle 2016). This may explain how some genera can achieve a distribution ranging from the intertidal to the abyss (Veit-Köhler 2005). While no deep-sea endemic harpacticoid lineages on deep systematic levels are known, most genera of Aegisthidae and Neobradyidae evolved at great depths and achieved cosmopolitan distributions there (Seifried 2004). Moreover, a deep-sea origin of Oligoarthra and possibly of all Harpacticoida has been hypothesized, and some lineages within Syngnatharthra have apparently made multiple deep-to-shallow transitions, and, much like the Macrostylidae, several lineages subsequently re-invaded greater depths (Seifried 2004).

Molecular genetic analyses suggest that deep-sea vent shrimps (Caridea: Alvinocarididae) are relatively young. The cosmopolitan distribution of this vent- and seep-endemic group was achieved within the Miocene and was hypothesized to be no older than ~12 Ma (Shank et al. 1999). Seep and vent habitats often are inhabited by dense decapod populations, notably the Chirostyloidea (Decapoda: Anomura), including yeti crabs (Kiwaidae; Fig. 7.6). Molecular data suggest an origin of the latter family at Pacific seeps in the Mid-Cretaceous (~100 Ma), subsequently colonizing both vent sites and the Atlantic through the Drake Passage between 13.4 and 25.9 Ma (Roterman et al. 2013). It has been suggested that deep-sea radiation of the Kiwaidae followed a Palaeocene-Eocene low-oxygen period (Roterman et al. 2013). Reconstruction of diversification rates of the common squat lobster genus *Paramunida,* also at upper slope depths, suggests that it went through a major radiation at the beginning of the Neogene (Cabezas et al. 2012), in line with previous assumptions of a young deep-sea decapod fauna (Hessler and Wilson 1983).

It can be concluded that while certain deep-sea crustaceans, especially the decapods, apparently are descendants of relatively recent, mostly Cenozoic colonizers, the great diversity observed in asellote isopods originated in situ over a much longer time span, possibly dating back to the Paleozoic (Fig. 7.6). Similar ages and diversification patterns are also likely for tanaidomorphs and several harpacticoid lineages. However, in the latter cases, thorough phylogenetic reconstructions are still lacking.

The Role of the High Latitudes

Two opposing theories attribute the origin of the deep-sea fauna either to relatively recent invasion from the Antarctic shelf (submergence) or deep-sea diversification and subsequent colonization of shallow environments (emergence) (Hessler and Wilson 1983). High latitudes, areas of deep-water formation, and upwelling are likely to favor the evolution of large depth ranges in species because of the isothermal water column (Hessler and Thistle 1975, Hessler et al. 1979, Brandt 1992). Deep-water formation today occurs particularly in the Weddell Sea. The Southern Ocean may have been the place of deep-water formation since the Early Cretaceous (146–100 Ma) and has therefore probably facilitated faunal exchange between shallow and deep (Murphy and Thomas 2012).

The submergence theory (e.g., Menzies and Imbrie 1958, Kussakin 1973, Menzies et al. 1973) relies on the assumption that Miocene (23–5.3 Ma) glaciations lead to cooling and major extinctions in the abyss and hadal environment that could be only survived by eurybathic species (Bruun 1956).

Accordingly, the tropical shallow-water fauna should be the most ancient, followed by high-latitude shallow-water fauna, and deep-sea fauna being the youngest (Kussakin 1973).

The emergence theory hypothesizes abyssal in situ diversification and origin of many taxa that presumably survived unfavorable environmental changes occurring in shallow seas by occupying a more stable deep ocean. Subsequently deep-water clades propagated into shallower zones (Hessler and Thistle 1975, Hessler et al. 1979). The theory of ancient deep-sea lineages is based mostly on high morphological diversification and biogeographic and bathymetric patterns of isopods. Indeed, representatives of typical deep-sea families of Asellota are also found on the Antarctic shelf and Arctic Fjords (Hessler and Wilson 1983, Brandt 1992).

One scenario or the other, however, may be an oversimplification (Wilson 1999). While several families such as Desmosomatidae, Macrostylidae, and Munnopsidae evidently originated in the deep sea where they are most diverse, they subsequently colonized continental shelves of predominantly cold-water regions (Hessler et al. 1979, Brandt 1999). Some species can thus be found even in shallow near-shore habitats (Hessler and Wilson 1983, Riehl 2014). Several cymothoid, valviferan and sphaeromatidean isopod families, on the other hand, likely colonized abyssal depths recently from the shelf of Southern Hemisphere high latitudes promoted by deep-water formation (Wilson 1998, Brandt 1999).

The eye-bearing species of the usually blind ostracod genera *Poseidonamicus* (Fig. 7.7) and *Dutoitella* recorded by Dingle (2003, 2009) from Marion Island are further examples of emergence in cold shallow habitats of the Southern Ocean. Dingle (2009) proposed that reactivation of genes caused the reappearance of eyes where sightedness had been lost during the Cretaceous, a pattern not observed in isopods or other deep-sea crustaceans, indicating their even earlier deep-sea colonization.

Other examples of emergence and submergence, respectively, are the ostracod genera *Abyssocythere* (Fig. 7.7) and *Dutoitella* (Brandão et al. 2016). Both originated in the Cretaceous Southern Ocean/South Atlantic and are now widespread in the deep sea. While the former has likely evolved in the deep sea, the other first appeared in shallow environments. *Dutoitella* had colonized the Indian Ocean by the end of the Eocene (33.9 Ma). It continued to spread after the establishment of cold water masses around 40 Ma and diversified into numerous Miocene species. *Abyssocythere* also appeared in the South Atlantic and, although with fewer species, is widespread in the modern deep sea.

Consequently, evolutionary emergence and submergence are both suitable explanations for taxon-specific bathymetric patterns, especially in Southern Ocean glacial and warm periods. Besides representing a connection between shallow and deep zones, the unique glaciological history of the isolated Antarctic continent was hypothesized to generate an "evolutionary incubator" and a "diversity pump" (Fig. 7.8). Repeated glacial expansions caused pulses of species range contraction into refugia leading to vicariance (Clarke and Crame 1989, Watling and Thurston 1989, Rogers 2007). Apparently, this process facilitated diversification in crustacean groups such as Munnopsidae (Malyutina and Brandt 2007) and Eusiridae (Amphipoda) (Baird et al. 2011) and may have had an effect on the diversity of many other groups in the deep sea.

Antiquity of the Deep-Sea Fauna

The deep sea was first imagined as azoic (Forbes 1844) and then as a refuge for archaic forms of life that went extinct in the shallow seas (Thomson 1873). Since the mid-20th century, the abyss has been known to be inhabited predominantly by a fauna of at most late Mesozioc age (max. 68 Ma), phylogenetically younger than bathyal or littoral faunas (Menzies et al. 1961, Gage 2004, Danovaro et al. 2014). This view is founded partially on an oversimplification of a primarily sedimentological concept, the so-called Oceanic Anoxic Events (Schlanger and Jenkyns 1976).

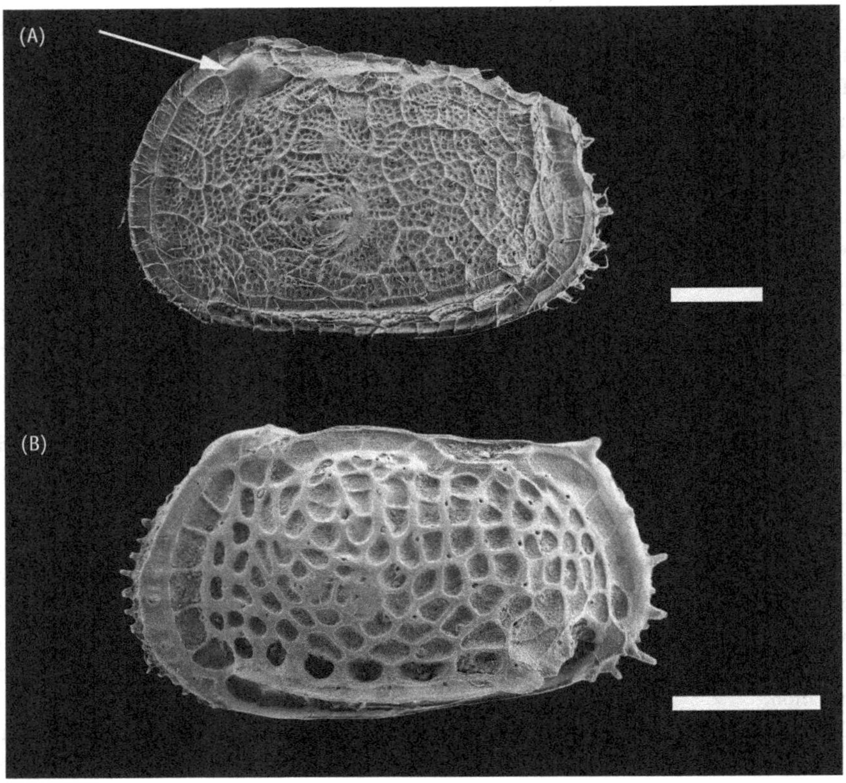

Fig. 7.7.
(A) *Australicythere polylyca* is typical of the Antarctic continental shelf, with published occurrence records ranging from 17 to 495 m depth (compilation by Brandão 2012). It has eyes (arrow pointing to the eye tubercles of the shell), which are typical for shallow-water inhabitants. However, this specimen has been collected alive at 1,048 m depth in the Weddell Sea during the ANDEEP III expedition (station 74-6; Brandão 2012). (B) *Poseidonamicus* is abundant and widely distributed in the modern deep sea. Its appearance in the Palaeocene fossil record was used by Benson (1975) as indicator for the development of a "psychrosphere" (the deep ocean layer, typical of todays' oceans, which is considerably colder than the shallow layer). Scale bars: A = 0.2 mm; B = 0.3 mm. Images from Brandão et al. (2017) under Creative Commons license (BY-NC-SA).

Recolonization after Catastrophic Events?

Oceanic anoxic events (OAEs) were likely a consequence of climate warming (Hay 2008) and occurred especially during the Cretaceous (Föllmi 2012). Jacobs and Lindberg (1998) suggested that throughout the Phanerozoic Eon (541–0 Ma) during recurring OAEs, global anoxia in the oceans eradicated the deep-sea fauna entirely to be replaced by survivors from the shallow subtidal. Other significant changes in the deep-sea fauna composition have been supposedly caused at the onset of the psychrosphere, the stratified cold-water–dominated deep sea, maintained by the ocean current system commencing around 40 Ma (Benson 1975, Benson et al. 1984). The fauna has, according to these theories, repeatedly reinvaded the deep sea after mass extinctions (Menzies et al. 1961, Jacobs and Lindberg 1998).

The deposition of black shales, sedimentary rock rich in organic carbon, is evidence for conditions with low dissolved oxygen in the prehistoric water column. Yet it has been argued that most of the ocean circulation, and thus oxygen supply to the deep seabed, was not very different

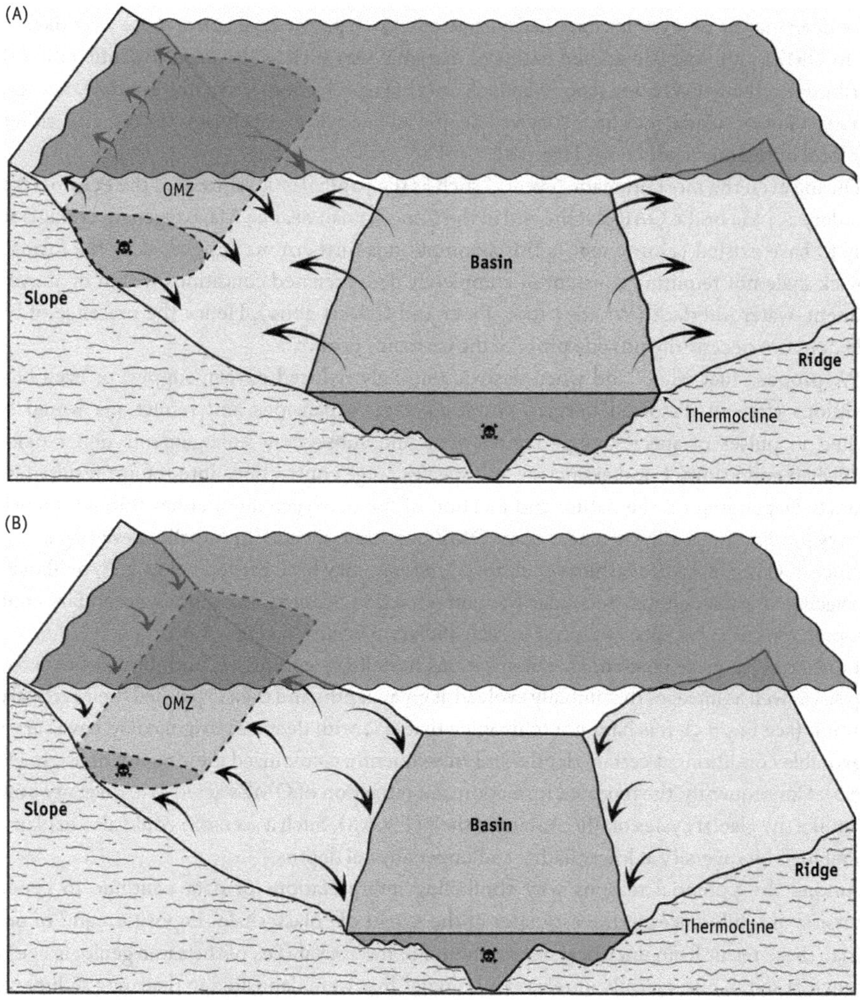

Fig. 7.8.
Schematic diagram showing pulses of species range contraction and expansion during periods of (A) deoxygenation and (B) ventilation and oxygenation of oceanic basins and/or bathyal environments. Both show anoxic zones covering both an expanded oxygen minimum layer model of an anoxic/dysoxic ocean as well as a stagnant-ocean scenario (Takashima et al. 2006); both are not necessarily related in causation or time of occurrence. Both forms of dysoxia/anoxia are characterized by local extinction and range contraction into refugia in depth zones outside the unfavorable environment (A). By pulses of repeated depth expansion (A) and retraction (B) of both forms of anoxia, linked to climate change as well as plate tectonics, species may have locally died out or experienced range contraction and subsequent expansion with secondary contact of populations or sister species as well as radiation of clades. Modified from Takashima et al. (2006), with permission from Oceanography.

from today during the times of black shale deposition (Hay 2008, Murphy and Thomas 2012). Oxygen-minimum zones (OMZs) should have accordingly existed rather on lower shelves and along the slopes than stretching across whole ocean basins (Diaz and Rosenberg 1995), especially in the Panthalassa (i.e., the superocean surrounding the supercontinent Pangaea, covering ~70% of Earth's surface 250 Ma). Therefore, while there is evidence that large-scale low-oxygen conditions in the deep marine ecosystems have had austere effects on parts of the fauna, these were likely limited to OMZs and (semi-)enclosed basins of marginal seas such as the Proto-Atlantic rather than distributed globally (Whatley 1990, Wagreich 2012) (Fig. 7.8). Some taxa, for example, lineages of Isopoda and Ostracoda, may have survived deep-water anoxic events in situ (or refugia), suffering only local or regional extinctions (Fig. 7.8).

During even the most dramatic "events," such as the panthalassic "anoxia" at the Permo-Triassic boundary 252 Ma or the OAE-2 at the end of the Cenomanian near 94 Ma, oxygenated refugia were likely to have existed (Horne 1999). This assumption is furthermore supported by the formation of black shale not requiring constant or completely deoxygenated conditions within or above the sediment–water interface (Whatley 1990, Piper and Calvert 2009). Hence the presence of black shale does not necessarily provide proof of the existence of anoxia.

We propose that where and when dysoxic (strongly reduced oxygen supply) or even anoxic conditions (free of dissolved oxygen) occurred, their expansions and retractions would have resulted in pulses of species range contractions and subsequent enlargements under oxic re-established conditions. Local extinction and species range contractions into refugia resulted in vicariance. Depending on the nature and location of the deoxygenation, either through increased primary production leading to a near-coast OMZ or a strengthened thermocline resulting in a stagnant ocean (Fig. 7.8 A) (Takashima et al. 2006), refuges may have existed below and/or above the deoxygenated water column. Secondarily, upon retraction of anoxic conditions, secondary contact of populations or sister species as well as clade radiation occurred (Fig. 7.8 B).

Furthermore, these post-anoxia colonizations have likely comprised both lineages new to the deep sea as well as lineages that initially evolved at great depths and then expanded their vertical distribution (see Fig. 7.5). It is hard not to imagine that taxa with deep-sea origin could have survived unfavorable conditions at certain depths and subsequently recolonized their region of origin (Figs. 7.3, 7.6). Consequently, the repeated formation and retraction of OMZs acted as a "diversity pump," not unlike the glacial cycles on the Antarctic shelf (Fig. 7.8). Such a scenario could also explain the general high biodiversity at lower bathal and upper abyssal depths.

Among the potential reasons why conflicting interpretations of data continue to generate discussion are differences in the estimates of the extent of "black shale" deposition and of ocean anoxia. These result from uncritical interpretation of the coeval (i.e., of the same geologic age) occurrence of sedimentologically unrelated lithologies that were deposited in disparate sedimentary environments (Trabucho-Alexandre et al. 2012).

The Fossil Record

Identifying ages and origins of the deep-sea fauna is hampered by a lack of fossils of actual deep-sea organisms (Smith and Stockley 2005), with the exceptions of ostracods and other microfossils (Benson 1975; see, e.g., http://deepseadrilling.org/about.htm for the results of the Deep-Sea Drilling Project [DSDP], Ocean Drilling Program [ODP], and Integrated Ocean Drilling Program [IODP]). Due to their calcified shells, ostracods have an extensive fossil record from strata of the last 400 Ma. They have inhabited the deep sea since the Palaeozoic (>251 Ma), but the modern fauna is mostly composed of direct descendants of Cretaceous forms (Benson 1988).

The evolution, ecological adaptations, as well as submergence and emergence of several taxa can be traced for millions of years through the fossil record. Thus, ostracods can be used as proxies in studies on historical biodiversity and biogeography and especially phylogeny covering the past tens to hundreds of million years (Benson 1988, Hunt 2007). Based on ostracod fossils in deep-sea sediment cores, Benson (1975) proposed that the beginning of modern deep-sea circulation strongly impacted the ostracod faunas (Benson 1975). From 75 to 50 Ma, the deep-sea ostracod fauna supposedly adapted to a warm deep sea. A major change occurred around 40 Ma (Benson 1975), when more genera appeared (Benson et al. 1984) than during earlier and successive periods. According to Benson's reconstructions, the genus *Atlanticythere*, common before this "event" of global change, was largely replaced by *Poseidonamicus* and *Abyssocythere*. These faunal changes were possibly linked to the origin of the present cold, deep-water masses (Benson 1975). We provide an updated, comprehensive literature review, which clarifies that these extinctions were affecting species and species richness but that all three above-mentioned genera continued to flourish in the deep sea well into the Neogene (Fig. 7.9).

The poor fossil record of isopods, especially the typical deep-sea taxon Asellota, is likely due to their thin cuticle (Selden et al. 2016). Discoveries of isopod fossils are thus far limited to terrestrial and shallow-water sediments. However, recent discoveries of several isopod suborders and families (Asellota: Stenetriidae; Cymothoida: Cirolanidae; Valvifera: Chaetiliidae, Idoteidae) from Cenomanian strata show that since around 100 Ma a virtually modern isopod fauna has existed, at least in shallow water (Wilson and Morel 2016). This is in line with the previously discovered serolid-like *Schweglerella* (Sphaeromatidea) from the late Jurassic (Brandt et al. 1999). Other fossil data set the date for well-derived Asellota (Paramunnidae, a distally derived branch of Janiroidea; T. Riehl and G. D. F. Wilson, unpublished data) to the Late Triassic (237–201 Ma), also showing a clear Paleozoic occurrence for the now wholely limnic Phreatoicidea (*Hesslerella*; Schram 1970), considering the possible sister group relationship between Asellota and Phreatoicidea (Wilson 2009). This indirectly supports Paleozoic origins of many deep-sea isopods.

Among Cirripedia, some taxa have been found to be remarkably basally derived, or "primitive." *Neolepas* (Eolepadinae), for instance, as well as *Verruca* and *Scillaelepas*, have fossil records dating back to the Early Jurassic (201–174 Ma), while extant species occur frequently at vents. This suggests evolutionary origins at least dating back to that period. The well-above-average generic age attributed to those taxa may justify the use of the term "living fossil" (Newman 1985). Absence of extant non-vent close relatives additionally supports considering *Neolepas* a relic of a Mesozoic (252–66 Ma) radiation (Newman 1985). The discovery of a living representative of the Brachylepadomorpha, a group known from fossils dating back to the Jurassic (201–145 Ma), indicated that the fauna of chemosynthetic oases encompassed mid-Mesozoic marine relics (Yamaguchi et al. 2004).

Recent representatives of Polychelidae, blind lobsters, occur exclusively below 600 m down to abyssal depths (Galil 2000). Fossil data support their Late Triassic (237–201.3 Ma) shallow water origin (Fig. 7.6). The sightedness of their extinct relatives suggests a secondary deep-sea colonization (Ahyong 2009), for which dates are currently wanting.

Molecular Data

The use of DNA-sequence data has shed light on multiple facets of deep-sea crustacean evolution. For instance, molecular phylogenies allow age estimations on phylogenetic trees, thus allowing one to interpret radiations and colonizations from new angles. Dated molecular phylogenies inferred the approximate date of radiation of the major vent barnacle lineages in the Cenozoic (Herrera et al. 2015), similar to the relatively young decapod lineages at vents and seeps coinciding with the end of the Palaeocene/Eocene Thermal Maximum some 55.5 Ma (Roterman et al. 2013). In this case, recent advances in molecular dating contradict earlier assumptions about the antiquity of vent barnacles and decapods. Similar to vent and seep decapods, the spiny lobsters (Palinuridae) have colonized the deep sea from shallow depths and then radiated (Tsang et al. 2009). These decapod lineages mentioned above, however, are rather bathyal than abyssal representatives. Molecular data for truly abyssal decapods do not yet exist in public databases.

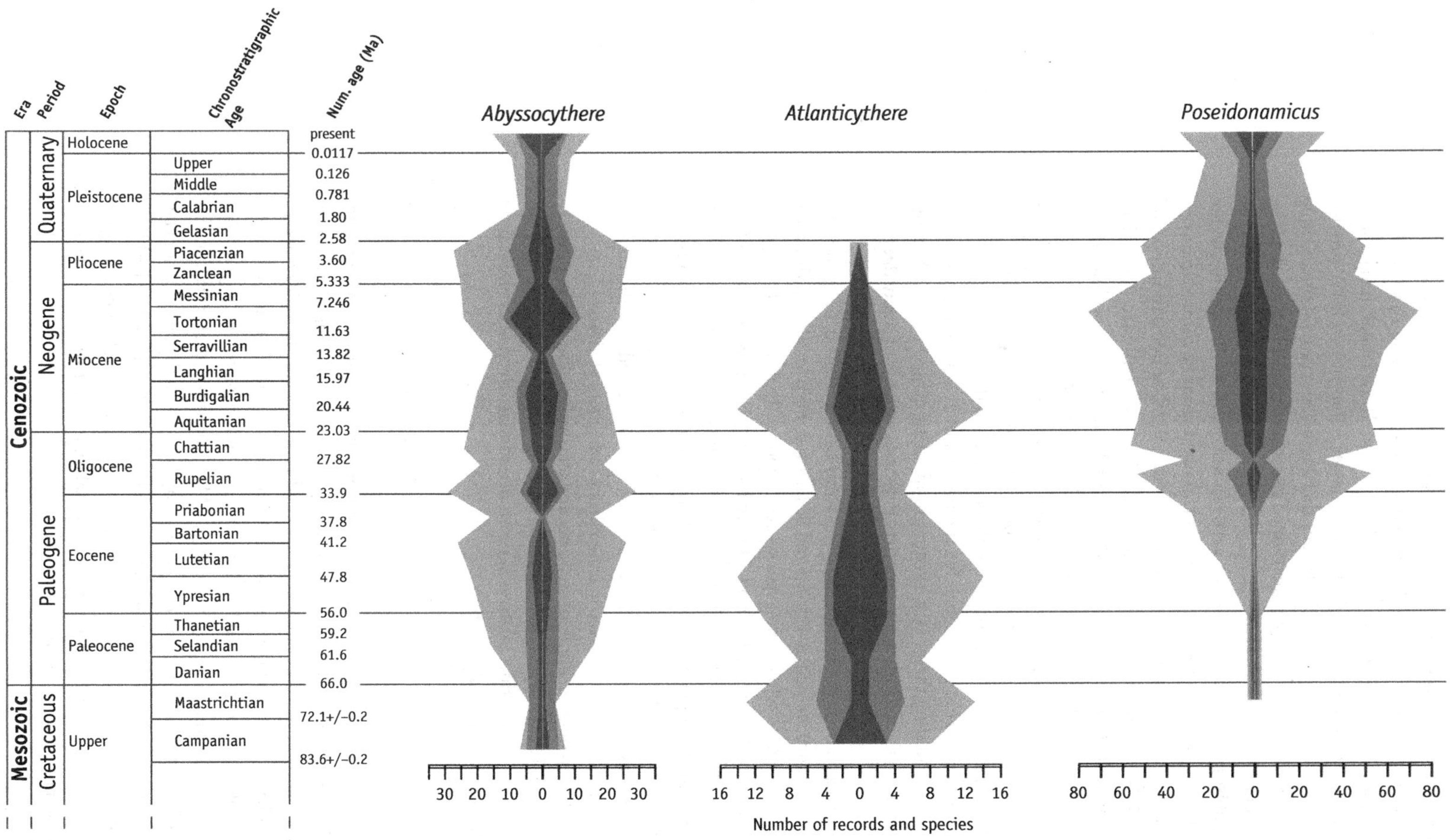

Fig. 7.9.
The first appearances, records, and species diversity (i.e., number of species) of the ostracod genera *Abyssocythere, Atlanticythere*, and *Poseidonamicus*, assembled from records in the literature. Dark gray represents the most conservative estimation of species richness (numbers of species reported from every time period) only taking into account records that have been assigned to a defined chronostratigraphic stage* (e.g., Maastrichtian, Danian, Selandian, Thanetian). In the case of ostracods identified only to genus level, all unidentified records of a certain genus in a given chronostratigraphic stage were counted as one. Medium gray further takes into account records that were assigned only to a certain chronostratigraphic Epoch* (i.e., Upper Cretaceous, Paleocene, Eocene, Oligocene), but not to a specific chronostratigraphic stage*. Light gray represents the total number of records including a sum of repeated, coeval reports of the same species and also accounting for identifications restricted to the genus level.

* As defined by the International Commission on Stratigraphy (Cohen, K.M., Finney, S.C., Gibbard, P.L. and Fan, J.-X. (2013; updated) The ICS International Chronostratigraphic Chart. Episodes 36: 199–204. http://www.stratigraphy.org/ICSchart/ChronostratChart2017-02.pdf.

For deep-sea isopods, on the other hand, molecular data are in line with fossil data. They colonized the deep sea multiple times from shallow waters (Raupach et al. 2009), with the earliest intrusions dating back to the late Paleozoic and subsequent radiations (Lins et al. 2012; Fig. 7.4). While the munnopsoid radiation (Raupach et al. 2004) may have occurred during colonization of the deep sea more than 250 Ma (Fig. 7.4), in some families, such as Janiridae, gradual adaptations to the deep-sea environment, such as loss of eyes, can be observed when comparing species (Linse et al. 2014).

Molecular studies on deep-sea crustaceans contribute to the understanding of depth-limited species distribution and the recognition of genetically distinct but morphologically "cryptic" species. This highlights recent or continuing diversification (e.g., France and Kocher 1996, Raupach et al. 2007, Poore and Andreakis 2011, 2014, Havermans et al. 2013, Brix et al. 2015, Bober et al. 2018). Moreover, species belonging to typical deep-sea families show distributions and a genetic structure that indicate recent colonization following the Last Glacial Maximum, such as by *Macrostylis roaldi* in the Amundsen Sea, Southern Ocean (Riehl and Kaiser 2012).

Molecular data on macrostylids furthermore contributed to our understanding of connectivity among abyssal as well as between abyssal and hadal populations of this common component of the macrobenthos. Population-genetic analyses revealed that while species may have ranges extending beyond oceanic ridges, across 3,000-m-depth gradients, as well as over 2,000 km geographic distance, connectivity is restricted (Riehl et al. 2018). All three factors—depth gradients, physical obstacles, and the merely enormous extent of oceanic abyssal basins—thus represent barriers to dispersal and contribute to diversification in the deep sea.

Rooted phylogenies have the potential to trace the evolution of certain characters or the direction of colonization patterns. The phylogenetic trees inferred by Held (2000) for serolid isopods support the theory of the Antarctic shelf as a "diversity generator" from which submergence through the deep sea was colonized (see above). Similar assumptions have been published for the spiny lobsters of the family Palinuridae, of which the majority of species inhabit shallow reef habitats and few lineages have adopted deep-sea sediments as habitat (Tsang et al. 2009).

Molecular data for other groups, such as the most diverse of all Crustacea in the abyss, the Harpacticoida, or their related Siphonostomatoida, are just beginning to emerge and have still limited applicability in answering the question of the age of the deep-sea fauna. They confirm, however, that harpacticoids have remarkable dispersal capabilities across 1,000-km-scale geographic ranges and several hundred meters depth (Gollner et al. 2010a, Easton and Thistle 2016).

Comprehensive molecular phylogenies often lack sufficient taxon sampling for deep-sea representatives of the clade under investigation. Inference of the phylogenetic position of deep-sea clades among their much better sampled shallow-water relatives thus commonly remains insufficiently supported. That is the case, for instance, among the Stomatopoda (e.g., Van Der Wal et al. 2017).

Depth Endemism

Endemism refers to a taxon being unique to a defined geographic location. This could be a location, a defined zone, or a habitat. Here we discuss the endemicity with regard to the deep sea and especially its three above-defined depth zones. Possibly as a consequence of the relative isolation of the shallow and deep faunas caused by strong environmental gradients (see Chapter 16), peculiar faunal patterns can be observed at great depths (Rex and Etter 2010). These are characterized by deep-sea endemism, predominantly on the species level, but in some Crustacea, for instance, isopod crustaceans, at higher taxonomic levels as well (Hessler et al. 1979, Hessler and Wilson 1983, Brandt et al. 2009).

Decapods and stomatopods are irregularly encountered and can be seen as "newcomers" in the abyssal environment (Hessler and Wilson 1983). Among the Decapoda, multiple lineages have colonized abyssal and hadal waters, most of which have few representative species and thus a low degree of endemism there and only on low taxonomic levels. The family Polychelidae, once inhabiting shallow marine zones (see above), is now confined to depths below 600 m, with a mainly bathyal distribution and few occurrences also in the abyss (OBIS 2017; Fig. 7.2). Among the caridean shrimps, especially the family Alvinocarididae, is a vent and seep specialist and a notorious deep-sea endemic with a depth range of 252 to 4960 m (Komai and Chan 2010, Nye et al. 2012).

Among the Anomura, the Munidopsidae contains the deepest-dwelling squat lobsters (Galatheoidea), with some occurring at abyssal depths. Many species of this family have reduced their eyes as an adaptation for life in low light on the outer continental shelf, slope, or abyssal plain (Ahyong et al. 2011). Genera like *Munidopsis* and *Galacantha* are considered lower-bathyal or abyssal genera as their main diversity is found at great depth (despite also occurring in shallow water) (Baba et al. 2008). In general, however, galatheoid species have restricted depth ranges, and the order is most diverse at upper bathyal depths (Schnabel et al. 2011). The hermit-crab family Parapaguridae, including the two likely deepest occurring benthic decapods *Probeebei mirabilis* (Fig. 7.2D) and *Tylaspis anomala,* is one of few truly abyssal endemics (Tsang et al. 2011). The stomatopod family Bathysquillidae appears to be confined to the outer shelf and upper slope covering roughly 200–1,500 m depth and rendering this group endemic to the upper bathyal zone (Manning et al. 1990). This group has also adapted to living below the photic zone by reducing ommatidial midbands, as well as the ommatidia themselves (Van Der Wal et al. 2017).

Recent genera of ostracods tend to dominate in either shallow or deep water, but most have some representatives across a wide bathymetric range (Maddocks 1990, Dingle 2003). To our knowledge, all marine families of ostracods occur in both shallow and deep environments, evidence of relatively recent and good dispersibility. Similar to other predominantly meiofauna groups, like harpacticoid copepods as well as larger amphipods and cumaceans, Cephalocarida show low endemism on higher taxonomic levels despite some of these meiofaunal groups being omnipresent in soft abyssal seafloor sediments (Hessler and Sanders 1973). The deep-sea hydrothermal vent copepods (Dirivultidae), however, seem to be an exception. They represent true endemics to the deep sea, despite large dispersal capacity (Gollner et al. 2010a), probably due to specialization for vent systems (Gollner et al. 2010b). A molecular phylogeny of the copepod orders suggested a basal position of Dirivultidae among the Siphonostomatoida (Khodami et al. 2017), highlighting a potential old age for this family as well as an origination linked to vent habitats.

Isopods of the superfamily Janiroidea (Fig. 7.10), however, are represented by numerous genera, subfamilies, and even families that are endemic to great depths or only sparsely occurring in the shallows, thus indicating early colonization and radiations by some isopod lineages (Hessler and Wilson 1983). Among the deep-sea endemics are the families Haplomunnidae, Janirellidae, and Mesosignidae. Among the subfamilies of the diverse Munnopsidae, the Lipomerinae is a deep-sea endemic example (Wilson 1989). The munnopsid genera *Bathyopsurus* and *Paropsurus* as well as *Urstylis* are confined to abyssal depth, rendering their parent taxa (subfamily Bathyopsurinae and family Urstylidae, respectively) abyssoendemic (OBIS 2017). Accordingly, isopod representatives comprise a variety of abyssal-endemic and specialized families, such as Mesosignidae and Urstylidae, to singleton species closely related to typical shallow-water taxa, such as Janiridae (Linse et al. 2014, Riehl et al. 2014). Deep-sea endemism and the wide distribution of some other deep-sea taxa conflict with the hypothesis of a generally young deep-sea fauna.

COMPARISON WITH OTHER TAXA

Other than microfossils and cirriped fragments, there is no fossil record of deep-sea Crustacea to help unravel long-term adaptive radiation and the global spread of higher taxa. Phylogenetic evidence must instead be inferred from molecular genetic data. To better delineate the processes

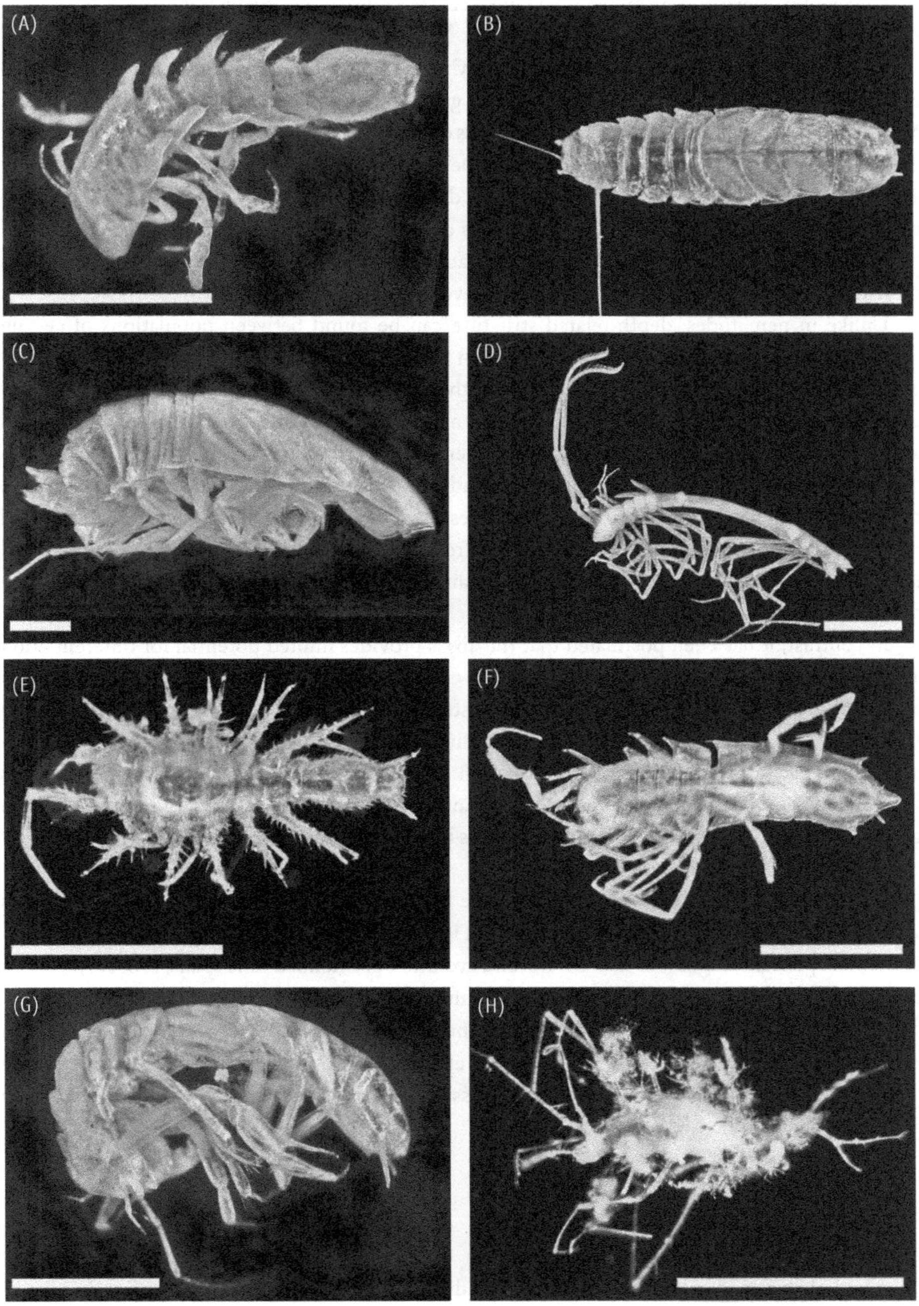

Fig. 7.10.
Common families of abyssal isopoda collected during the Vema-TRANSIT expedition in the North Atlantic. These families and genera are considered to be specialized deep-sea taxa because of their main distribution and diversity at great depths as well as the reduction of eyes. (A) Macrostylidae, *Macrostylis* aff. *magnifica*; (B) Munnopsidae, Eurycopinae, *Eurycope* sp.; (C) Munnopsidae, Betamorphinae, *Betamorpha* sp.; (D) Ischnomesidae, *Gracilimesus* sp.; (E) Mesosignidae, *Mesosignum* sp.; (F) Nannoniscidae, *Nannoniscus* sp.; (G) Desmosomatidae, *Chelator* sp.; (H) Dendrotionidae, *Acanthomunna* sp. All scale bars = 1 mm. From Devey et al. (2015), with permission from GEOMAR. See color version of this figure in centerfold.

shaping evolutionary origins of deep-sea fauna, comprehensive phylogenies are required for most deep-sea taxa.

Molecular evidence has revealed that, as in Crustacea, evolutionary origins across the whole deep-sea fauna are complex mixtures of age and geography with varying, taxon-specific evidence. Molecular data for polychaetes indicate numerous cryptic species (Wiklund et al. 2012, Janssen et al. 2015), suggesting that rather than multiple independent colonization events, speciation occurring at abyssal depth must have led to the observed diversity. Ubiquitous Nematoda show high connectivity with cosmopolitan species and species complexes. The frequent interchanges observed between shallow and abyssal depths (Bik et al. 2010) make identification of evolutionary origins difficult if not impossible on higher taxonomic levels.

Unlike in nematodes, depth-related structure can be found between populations of certain molluscs. As in amphipod crustaceans, they seem to be restricted to certain depth ranges, and species turnover has been detected, for example, at the transition between bathyal and abyssal depths at around 3,000 m (Rex and Etter 2010). Etter et al. (2005) placed this observation on molluscs in a theoretical framework called the Depth-Differentiation Hypothesis and postulated that, due to strong environmental gradients, evolution and adaptive radiation occur primarily on continental margins. Several mollusc groups have their highest diversity at intermediate (bathyal) depths and recent and continuous colonization of the abyss from shallower marine habitats. These, as opposed to in situ diversification, are thought to be the main sources of their great diversity observed in the deep (Etter et al. 2005).

In contrast, it has been postulated that the abyss provides limited potential for differentiation and speciation by isolation due to relatively low environmental variability (Etter et al. 2011). Low abyssal population densities of some molluscs distributed in both the bathyal and abyssal have promoted the concept of the abyss as a sink for biodiversity (Rex et al. 2005): abyssal populations are accordingly facing a high risk of local extinction and depend on recruitment from the slopes as abyssal densities are too low to allow reproduction. While this concept may hold true for certain groups, evidence suggests the contrary for others like isopod crustaceans (Riehl et al. 2014, Hardy et al. 2015).

When compared to shallow-water forms, the high distinctness in some deep-sea crustaceans such as Janiroidea and polychelid decapods (Fig. 7.2B), as well as in co-occurring (but rare) monoplacophorans, suggest in situ survival and evolution of extended time spans, probably in cases originating in the Paleozoic (>252 Ma). For Ophiuroidea, species richness peaks near highly productive continental margins. This indicates that the deep-sea fauna at least partially originates from range expansion of upper-bathyal taxa (Woolley et al. 2016). This ophiuroid deep-sea origin of the present-day diversity is supported by a microfossil record from the Early Jurassic onward (Thuy et al. 2012, Thuy 2013).

FUTURE DIRECTIONS

The main obstacles to gaining a good understanding of the evolution and colonization history of deep-sea Crustacea are technological and related to sampling (Brandt et al. 2014). Despite more than a century of deep-sea sampling, the largest landscape on Earth has been explored at few distributed sites (Ramirez-Llodra et al. 2010). Consequently, most of the fauna remains undescribed Typically, >90% of the abyssal macrofaunal species sampled at any single site in the abyssal Atlantic and Pacific are new to science (Ebbe et al. 2010, Kaiser et al. 2013, Poore et al. 2015). Given this extremely high rate of new discoveries, taxonomic descriptions of new taxa and their morphologies as well as inference of their phylogenies remains a primary objective. Increased sampling, for instance, by new autonomous samplers (Brandt et al. 2016) and strengthened taxonomic effort would allow for enhanced taxon sampling and phylogenetic reconstruction, eventually leading to a better understanding of deep-shallow and interocean-basin relationships.

Furthermore, improved genetic taxon and character sampling as well as genome-level analysis of the deep nodes within dominant deep-sea crustacean groups are promising approaches for filling present gaps, such as resolving the deep nodes of the munnopsoid radiation or inferring shallow-deep connections within Harpacticoida. A few pioneer studies have shown that DNA information is highly valuable for the inference of speciation and the factors promoting diversification in deep-sea crustaceans (France and Kocher 1996, Raupach et al. 2007, Brix et al. 2011, Havermans et al. 2013, Herrera et al. 2015, Easton and Thistle 2016). They highlight the enormous potential for new discoveries based on DNA concerning factors promoting speciation or representing barriers to dispersal at abyssal depth.

Another problem inhibiting the understanding of the adaptive value of certain traits (e.g., morphology, behavior) is the lack of in situ observations and laboratory experiments on living abyssal crustaceans. Moreover, understanding the evolutionary histories of deep-sea crustaceans requires more detailed reconstruction of oceanic paleoenvironments. With the current state of exploration, global and fatal "anoxic events" in the deep oceans should not be assumed to explain patterns of diversification and extinction, since various data suggest these events were probably not as global or as fatal as often stated (e.g., Lins et al. 2012).

Finally, it remains to be emphasized that a phylogenetic background is not yet available for many of the crustacean groups inhabiting great ocean depths, like the super-diverse Harpacticoida, or others, for example, the less diverse but omnipresent Cumacea and Mysidacea. Such a phylogenetic backbone is essential to conclude any evolutionary aspect of these taxa.

CONCLUSIONS

Today the deep sea is inhabited by crustaceans of various phylogenetic ages and origins. This variety probably began evolving at the beginning of the Paleozoic some 500 Ma. Depending on the age of their clades, deep-sea crustaceans show a wide range of adaptation levels from highly modified isopods to ostracods resembling shallow-water counterparts. Some species may even occur across wide depth ranges from the sublittoral down to the abyss.

Dramatic environmental changes occurred in the deep oceans throughout the Phanerozoic and significantly impacted the deep-sea fauna. However, there is ample evidence suggesting that the often-proclaimed global extents and catastrophic consequences of these "events" are exaggerated and too frequently cited uncritically.

The abyss thus cannot be seen solely as a sink for crustacean biodiversity generated at shallower depths. It is rather multifaceted, and while for some clades it may be considered an evolutionary dead end, for others the deep oceans may provide a refuge from temporal rigors in the shallows.

ACKNOWLEDGMENTS

During the preparation of this chapter, TR and AB were supported by and worked in the Vema-TRANSIT (Bathymetry of the Vema-Fracture Zone and Puerto Rico TRench and Abyssal AtlaNtic BiodiverSITy Study) project funded by the German Federal Ministry (03G0237A). CNPq provided a postdoctoral fellowship to SNB (process nr 374397/2013-9) and further funding (processes nr 400116/2013-8 442550/2014-6). Bente Stransky, Anne-Nina Lörz, Sabine Holst, and Inma Frutos are thanked for fruitful discussions. Yann Marcon is thanked for the OFOS images. Edelgard Baumann kindly provided the raw data of fossil ostracod appearances from literature research. The kind assistance of reviewers and editors that led to a much improved chapter is much appreciated.

REFERENCES

Ahyong, S.T. 2009. The polychelidan lobsters: phylogeny and systematics (Polychelida: Polychelidae). Pages 369–396 *in* J. W. Martin, K. A. Crandall, and D. L. Felder, editors. Decapod Crustacean Phylogenetics. CRC Press.

Ahyong, S.T., N. Andreakis, and J. Taylor. 2011. Mitochondrial phylogeny of the deep-sea squat lobsters, Munidopsidae (Galatheoidea). Zoologischer Anzeiger—A Journal of Comparative Zoology 250:367–377.

Baba, K., E. Macpherson, G.C.B. Poore, S.T. Ahyong, A. Bermudez, P. Cabezas, C.W. Lin, M. Nizinski, C. Rodrigues, and K.E. Schnabel. 2008. Catalogue of squat lobsters of the world (Crustacea: Decapoda: Anomura–families Chriostylidae, Galatheidae and Kiwaidae). Zootaxa 1905:1–220.

Baird, H.P., K.J. Miller, and J.S. Stark. 2011. Evidence of hidden biodiversity, ongoing speciation and diverse patterns of genetic structure in giant Antarctic amphipods. Molecular Ecology 20:3439–3454.

Benson, R.H. 1975. The origin of the psychrosphere as recorded in changes of deep-sea ostracode assemblages. Lethaia 8:69–83.

Benson, R.H. 1988. Ostracods and palaeoceanography. Pages 1–26 *in* P. De Deckker, J.P. Colin, and J.P. Peypouquet, editors. Ostracoda in the Earth Sciences. Elsevier, Amsterdam

Benson, R.H., R.E. Chapman, and L.T. Deck. 1984. Paleoceanographic events and deep-sea ostracodes. Science 224:1334–1336.

Bik, H.M., W.K. Thomas, D.H. Lunt, and P.J.D. Lambshead. 2010. Low endemism, continued deep-shallow interchanges, and evidence for cosmopolitan distributions in free-living marine nematodes (order Enoplida). BMC Evolutionary Biology 10:389.

Błażewicz-Paszkowycz, M., R. Bamber, and G. Anderson. 2012. Diversity of Tanaidacea (Crustacea: Peracarida) in the world's oceans—How far have we come? PloS ONE 7:p.e33068.

Bober, S., T. Riehl, S. Henne, and A. Brandt. 2018. New Macrostylidae (Isopoda) from the Northwest Pacific Basin described by means of integrative taxonomy with reference to geographical barriers in the abyss. Zoological Journal of the Linnean Society 182:549–603.

Boxshall, G.A., and D. Jaume. 2000. Discoveries of cave misophrioids (Crustacea: Copepoda) shed new light on the origin of anchialine faunas. Zoologischer Anzeiger 239:1–19.

Brandão, S.N. 2012. Southern Ocean Ostracoda (except Halocypridina). http://www.scarmarbin.be/imis.php?module=datasetanddasid=2443.

Brandão, S.N., A. Stuhlmann, H. Vital, and A. Brandt. 2016. Biogeography of *Abyssocythere* and *Dutoitella* (Ostracoda), with descriptions of three new species. Zootaxa 4139:391–418.

Brandão, S.N., M.V. Angel, I. Karanovic, V. Perrier, and M. Yasuhara. 2017. World Ostracoda Database. http://www.marinespecies.org/ostracoda.

Brandt, A. 1992. Origin of Antarctic Isopoda (Crustacea, Malacostraca). Marine Biology 113:415–423.

Brandt, A. 1999. On the origin and evolution of Antarctic Peracarida (Crustacea, Malacostraca). Scientia Marina 63:261–274.

Brandt, A., J.A. Crame, H. Polz, and M.R.A. Thomson. 1999. Late Jurassic tethyan ancestry of Recent southern high-latitude marine isopods (Crustacea, Malacostraca). Palaeontology 42:663–675.

Brandt, A., K. Linse, and M. Schüller. 2009. Bathymetric distribution patterns of Southern Ocean macrofaunal taxa: Bivalvia, Gastropoda, Isopoda and Polychaeta. Deep Sea Research Part I: Oceanographic Research Papers 56:2013–2025.

Brandt, A., H. Griffiths, J. Gutt, K. Linse, S. Schiaparelli, T. Ballerini, B. Danis, and O. Pfannkuche. 2014. Challenges of deep-sea biodiversity assessments in the Southern Ocean. Advances in Polar Science 25:204–212.

Brandt, A., J. Gutt, M. Hildebrandt, J. Pawlowski, J. Schwendner, T. Soltwedel, and L. Thomsen. 2016. Cutting the umbilical: new technological perspectives in benthic deep-sea research. Journal of Marine Science and Engineering 4:36.

Brix, S., T. Riehl, and F. Leese. 2011. First genetic data for species of the genus *Haploniscus* Richardson, 1908 (Isopoda: Asellota: Haploniscidae) from neighbouring deep-sea basins in the South Atlantic. Zootaxa 2838:79–84.

Brix, S., F. Leese, T. Riehl, and T. C. Kihara. 2015. A new genus and new species of Desmosomatidae Sars, 1897 (Isopoda) from the eastern South Atlantic abyss described by means of integrative taxonomy. Marine Biodiversity 45:7–61.

Brökeland, W. 2010. Redescription of *Haploniscus rostratus* (Menzies, 1962) (Crustacea: Peracarida: Isopoda) with observations on the postmarsupial development, size ranges and distribution. Zootaxa 2521:1–25.

Bruun, A. F. 1956. The abyssal fauna: its ecology, distribution and origin. Nature 177:1105–1108.

Cabezas, P., I. Sanmartín, G. Paulay, E. Macpherson, and A. Machordom. 2012. Deep under the sea: unraveling the evolutionary history of the deep-sea squat lobster *Paramunida* (Decapoda, Munididae). Evolution 66:1878–1896.

Clarke, A., and J.A. Crame. 1989. The origin of the Southern Ocean marine fauna. Geological Society, London, Special Publications 47:253–268.

Corgosinho, P.H., N.V. Schizas, and M.A. Lozano. 2016. A new species of *Atergopedia* (Copepoda: Harpacticoida: Novocriniidae) from a Caribbean mesophotic reef. Marine Biodiversity 46:841–852.

Corrigan, L.J., T. Horton, H. Fotherby, T.A. White, and A.R. Hoelzel. 2014. Adaptive evolution of deep-sea amphipods from the superfamily Lysiassanoidea in the North Atlantic. Evolutionary Biology 41:154–165.

Danovaro, R., P.V.R. Snelgrove, and P. Tyler. 2014. Challenging the paradigms of deep-sea ecology. Trends in Ecology and Evolution 29:465–475.

Devey, C.W., editor. 2015. RV SONNE Fahrtbericht / Cruise Report SO237 Vema-TRANSIT. Geomar Report 23:130.

Diaz, R.J., and R. Rosenberg. 1995. Marine benthic hypoxia: a review of its ecological effects and the behavioural responses of benthic macrofauna. Oceanography and Marine Biology, An Annual Review 33:245–303.

Dingle, R.V. 2003. Recent subantarctic benthic ostracod faunas from the Marion and Prince Edward Islands archipielago, Southern Ocean. Revista Española de Micropaleontología 35:119–155.

Dingle, R.V. 2009. Further visions of Dollo's Law through ostracods' eyes: an essay. Journal of Micropalaeontology 28:87–89.

Easton, E.E., and D. Thistle. 2016. Do some deep-sea, sediment-dwelling species of harpacticoid copepods have 1,000-km-scale range sizes? Molecular Ecology 25:4301–4318.

Ebbe, B., D.S. M. Billett, K.E. Ellingsen, A.G. Glover, S. Keller, M.V. Malyutina, P. Martinez Arbizu, T. Melodtsova, M.A. Rex, C.R. Smith, and A. Tselepides. 2010. Diversity of abyssal marine life. Pages 139–160 *in* A. D. McIntyre, editor. Life in the World's Oceans: Diversity, Distribution, and Abundance. Wiley-Blackwell, Chichester, UK.

Etter, R.J., M.A. Rex, M.R. Chase, and J.M. Quattro. 2005. Population differentiation decreases with depth in deep-sea bivalves. Evolution 59:1479–1491.

Etter, R.J., E.E. Boyle, A. Glazier, R.M. Jennings, E. Dutra, and M.R. Chase. 2011. Phylogeography of a pan-Atlantic abyssal protobranch bivalve: implications for evolution in the deep Atlantic. Molecular Ecology 20:829–843.

Föllmi, K.B. 2012. Early Cretaceous life, climate and anoxia. Cretaceous Research 35:230–257.

Forbes, E. 1844. Report on the Mollusca and Radiata of the Aegean Sea: and on their distribution, considered as bearing on geology. Pages 129–193 *in* Report of the British Association for the Advancement of Science for 1843. John Murray, London.

France, S.C., and T.D. Kocher. 1996. Geographic and bathymetric patterns of mitochondrial 16S rRNA sequence divergence among deep-sea amphipods, *Eurythenes gryllus*. Marine Biology 126:633–643.

Gage, J.D. 2004. Diversity in deep-sea benthic macrofauna: the importance of local ecology, the larger scale, history and the Antarctic. Deep Sea Research Part II: Topical Studies in Oceanography 51:1689–1708.

Galil, B.S. 2000. Crustacea Decapoda: review of the genera and species of the family Polychelidae Wood-Mason, 1874. Pages 285–387 *in* A. Crosnier, editor. Résultats des campagnes MUSORSTOM, Vol. 21. Muséum national d'Histoire naturelle, Paris.

Gardiner, L.F. 1975. The systematics, postmarsupial development, and ecology of the deep-sea family Neotanaidae (Crustacea: Tanaidacea). Smithsonian Contributions to Zoology 170:1–265.

Gollner, S., D. Fontaneto, and P.M. Arbizu. 2010a. Molecular taxonomy confirms morphological classification of deep-sea hydrothermal vent copepods (Dirivultidae) and suggests broad physiological tolerance of species and frequent dispersal along ridges. Marine Biology 158:221–231.

Gollner, S., V.N. Ivanenko, P.M. Arbizu, and M. Bright. 2010b. Advances in taxonomy, ecology, and biogeography of Dirivultidae (Copepoda) associated with chemosynthetic environments in the deep sea. PLoS ONE 5:e9801.

Hardy, S.M., C.R. Smith, and A.M. Thurnherr. 2015. Can the source–sink hypothesis explain macrofaunal abundance patterns in the abyss? A modelling test. Proceedings of the Royal Society of London B: Biological Sciences 282:20150193.

Havermans, C., G. Sonet, C. d'Udekem d'Acoz, Z.T. Nagy, P. Martin, S. Brix, T. Riehl, S. Agrawal, and C. Held. 2013. Genetic and morphological divergences in the cosmopolitan deep-sea amphipod *Eurythenes gryllus* reveal a diverse abyss and a bipolar species. PLoS ONE 8:e74218.

Hay, W.W. 2008. Evolving ideas about the Cretaceous climate and ocean circulation. Cretaceous Research 29:725–753.

Haye, P.A. 2007. Systematics of the genera of Bodotriidae (Crustacea: Cumacea). Zoological Journal of the Linnean Society 151:1–58.

Held, C. 2000. Phylogeny and biogeography of serolid isopods (Crustacea, Isopoda, Serolidae) and the use of ribosomal expansion segments in molecular systematics. Molecular Phylogenetics and Evolution 15:165–178.

Herrera, S., H. Watanabe, and T.M. Shank. 2015. Evolutionary and biogeographical patterns of barnacles from deep-sea hydrothermal vents. Molecular Ecology 24:673–689.

Hessler, R.R., and H.L. Sanders. 1973. Two new species of *Sandersiella* (Cephalocarida), including one from the deep sea. Crustaceana 24:181–196.

Hessler, R.R., and D. Thistle. 1975. On the place of origin of deep-sea isopods. Marine Biology 32:155–165.

Hessler, R.R., and G.D.F. Wilson. 1983. The origin and biogeography of malacostracan crustaceans in the deep sea. Pages 227–254 *in* R.W. Sims, J.H. Price, and P.E. S. Whalley, editors. Evolution, Time, and Space: The Emergence of the Biosphere. Academic Press, London.

Hessler, R.R., G.D.F. Wilson, and D. Thistle. 1979. The deep-sea isopods: a biogeographic and phylogenetic overview. Sarsia 64:67–75.

Horne, D.J. 1999. Ocean circulation modes of the Phanerozoic: implications for the antiquity of deep-sea benthonic invertebrates. Crustaceana 72:999–1018.

Hunt, G. 2007. Evolutionary divergence in directions of high phenotypic variance in the ostracode genus *Poseidonamicus*. Evolution 61:1560–1576.

Jacobs, D.K., and D.R. Lindberg. 1998. Oxygen and evolutionary patterns in the sea: onshore/offshore trends and recent recruitment of deep-sea faunas. Proceedings of the National Academy of Sciences 95:9396–9401.

Jamieson, A.J., T. Fujii, M. Solan, A.K. Matsumoto, P.M. Bagley, and I.G. Priede. 2009. First findings of decapod crustacea in the hadal zone. Deep Sea Research Part I: Oceanographic Research Papers 56:641–647.

Janssen, A., S. Kaiser, K. Meißner, N. Brenke, L. Menot, and P. Martínez Arbizu. 2015. A reverse taxonomic approach to assess macrofaunal distribution patterns in abyssal Pacific polymetallic nodule fields. PLoS ONE 10:e0117790.

Johnson, G.C., S. Mecking, B.M. Sloyan, and S.E. Wijffels. 2007. Recent bottom water warming in the Pacific Ocean. Journal of Climate 20:5365–5375.

Kaiser, S., S.N. Brandão, S. Brix, D.K.A. Barnes, D.A. Bowden, J. Ingels, F. Leese, S. Schiaparelli, C.P. Arango, R. Badhe, N. Bax, M. Blazewicz-Paszkowycz, A. Brandt, N. Brenke, A.I. Catarino, B. David, C.D. Ridder, P. Dubois, K.E. Ellingsen, A.G. Glover, H.J. Griffiths, J. Gutt, K.M. Halanych, C. Havermans, C. Held, D. Janussen, A.N. Lörz, D.A. Pearce, B. Pierrat, T. Riehl, A. Rose, C.J. Sands, A. Soler-Membrives, M. Schüller, J.M. Strugnell, A. Vanreusel, G. Veit-Köhler, N.G. Wilson, and M. Yasuhara. 2013. Patterns, processes and vulnerability of Southern Ocean benthos: a decadal leap in knowledge and understanding. Marine Biology 160:2295–2317.

Khodami, S., J.V. McArthur, L. Blanco-Bercial, and P.M. Arbizu. 2017. Molecular phylogeny and revision of copepod orders (Crustacea: Copepoda). Scientific Reports 7: 9164.

Komai, T., and T.Y. Chan. 2010. A new genus and two new species of alvinocaridid shrimps (Crustacea: Decapoda: Caridea) from a hydrothermal vent field off northeastern Taiwan. Zootaxa 2372:15–32.

Kussakin, O.G. 1973. Peculiarities of the geographical and vertical distribution of marine isopods and the problem of deep-sea fauna origin. Marine Biology 23:19–34.

Lins, L.S.F., S.Y.W. Ho, G.D.F. Wilson, and N. Lo. 2012. Evidence for Permo-Triassic colonization of the deep sea by isopods. Biology Letters 8:979–982.

Linse, K., J.A. Jackson, M.V. Malyutina, and A. Brandt. 2014. Shallow-water northern hemisphere *Jaera* (Crustacea, Isopoda, Janiridae) found on whale bones in the Southern Ocean deep sea: ecology and description of *Jaera tyleri* sp. nov. PLoS ONE 9:e93018.

Ma, K.Y., T.Y. Chan, and K.H. Chu. 2009. Phylogeny of penaeoid shrimps (Decapoda: Penaeoidea) inferred from nuclear protein-coding genes. Molecular Phylogenetics and Evolution 53:45–55.

Macpherson, E., B.R. de Forges, K. Schnabel, S. Samadi, M.C. Boisselier, and A. Garcia-Rubies. 2010. Biogeography of the deep-sea galatheid squat lobsters of the Pacific Ocean. Deep Sea Research Part I: Oceanographic Research Papers 57:228–238.

Maddocks, R.F. 1990. Living and Fossil Macrocyprididae. The University of Kansas Paleontological Institute, Lawrence.

Malyutina, M.V., and A. Brandt. 2007. Diversity and zoogeography of Antarctic deep-sea Munnopsidae (Crustacea, Isopoda, Asellota). Deep-Sea Research Part II 54:1790–1805.

Manning, R.B., R.K. Kropp, and J. Dominguez. 1990. Biogeography of deep-sea stomatopod Crustacea, family Bathysquillidae. Progress in Oceanography 24:311–316.

Menzel, L., K.H. George, and P.M. Arbizu. 2011. Submarine ridges do not prevent large-scale dispersal of abyssal fauna: a case study of *Mesocletodes* (Crustacea, Copepoda, Harpacticoida). Deep Sea Research Part I: Oceanographic Research Papers 58:839–864.

Menzies, R.J., and J. Imbrie. 1958. On the antiquity of the deep sea bottom fauna. Oikos 9:192.

Menzies, R.J., J. Imbrie, and B.C. Heezen. 1961. Further considerations regarding the antiquity of the abyssal fauna with evidence for a changing abyssal environment. Deep Sea Research 8:79–90.

Menzies, R.J., R.Y. George, and G.T. Rowe. 1973. Abyssal Environment and Ecology of the World Oceans. John Wiley, New York.

Murphy, D.P., and D.J. Thomas. 2012. Cretaceous deep-water formation in the Indian sector of the Southern Ocean. Paleoceanography 27:PA1211.

Newman, W.A. 1985. The abyssal hydrothermal vent invertebrate fauna: a glimpse of antiquity? Bulletin of the Biological Society of Washington 6:231–242.

Nye, V., J. Copley, and S. Plouviez. 2012. A new species of Rimicaris (Crustacea: Decapoda: Caridea: Alvinocarididae) from hydrothermal vent fields on the Mid-Cayman spreading centre, Caribbean. Journal of the Marine Biological Association of the United Kingdom 92:1057–1072.

OBIS. 2017. Ocean Biogeographic Information System. Intergovernmental Oceanographic Commission of UNESCO. www.iobis.org.

Osborn, K.J. 2009. Relationships within the Munnopsidae (Crustacea, Isopoda, Asellota) based on three genes. Zoologica Scripta 38:617–635.

Piper, D.Z., and S.E. Calvert. 2009. A marine biogeochemical perspective on black shale deposition. Earth-Science Reviews 95:63–96.

Poore, G.C., and N. Andreakis. 2011. Morphological, molecular and biogeographic evidence support two new species in the *Uroptychus naso* complex (Crustacea: Decapoda: Chirostylidae). Molecular Phylogenetics and Evolution 60: 152–169.

Poore, G.C., and N. Andreakis. 2014. More species of the *Agononida incerta* complex revealed by molecules and morphology (Crustacea: Decapoda: Anomura: Munididae). Zootaxa 3860:201–225.

Poore, G.C., and N.L. Bruce. 2012. Global diversity of marine isopods (except Asellota and crustacean symbionts). PLoS ONE 7:e43529.

Poore, G.C.B., L. Avery, M. Błażewicz-Paszkowycz, J. Browne, N.L. Bruce, S. Gerken, C. Glasby, E. Greaves, A.W. McCallum, D. Staples, A. Syme, J. Taylor, G. Walker-Smith, M. Warne, C. Watson, A. Williams, R.S. Wilson, and S. Woolley. 2015. Invertebrate diversity of the unexplored marine western margin of Australia: taxonomy and implications for global biodiversity. Marine Biodiversity 45:271–286.

Ramirez-Llodra, E., A. Brandt, R. Danovaro, B. De Mol, E. Escobar, C.R. German, L. A. Levin, P. Martinez Arbizu, L. Menot, P. Buhl-Mortensen, B.E. Narayanaswamy, C.R. Smith, D.P. Tittensor, P.A. Tyler, A. Vanreusel, and M. Vecchione. 2010. Deep, diverse and definitely different: unique attributes of the world's largest ecosystem. Biogeosciences 7:2851–2899.

Raupach, M.J., C. Held, and J.W. Wägele. 2004. Multiple colonization of the deep sea by the Asellota (Crustacea: Peracarida: Isopoda). Deep Sea Research Part II: Topical Studies in Oceanography 51:1787–1795.

Raupach, M.J., M.V. Malyutina, A. Brandt, and J.W. Wägele. 2007. Molecular data reveal a highly diverse species flock within the munnopsoid deep-sea isopod *Betamorpha fusiformis* (Barnard, 1920) (Crustacea: Isopoda: Asellota) in the Southern Ocean. Deep Sea Research Part II: Topical Studies in Oceanography 54:1820–1830.

Raupach, M.J., C. Mayer, M.V. Malyutina, and J.W. Wägele. 2009. Multiple origins of deep-sea Asellota (Crustacea: Isopoda) from shallow waters revealed by molecular data. Proceedings of the Royal Society B: Biological Sciences 276:799–808.

Rex, M.A., and R.J. Etter. 2010. Deep-Sea Biodiversity: Pattern and Scale. Harvard University Press, Cambridge, MA.

Rex, M.A., C.R. McClain, N.A. Johnson, R.J. Etter, J.A. Allen, P. Bouchet, and A. Warén. 2005. A source-sink hypothesis for abyssal biodiversity. The American Naturalist 165:163–178.

Riehl, T. 2014. A Phylogenetic Approach to the Classification of Macrostylid Isopods and Faunal Linkages Between the Deep Sea and Shallow-Water Environments. Dissertation, University of Hamburg, Hamburg, Germany.

Riehl, T., and A. Brandt. 2010. Descriptions of two new species in the genus *Macrostylis* Sars, 1864 (Isopoda, Asellota, Macrostylidae) from the Weddell Sea (Southern Ocean), with a synonymisation of the genus *Desmostylis* Brandt, 1992 with *Macrostylis*. Zookeys 57:9–49.

Riehl, T., and S. Kaiser. 2012. Conquered from the deep sea? A new deep-sea isopod species from the Antarctic shelf shows pattern of recent colonization. PLoS ONE 7:e49354.

Riehl, T., G.D.F. Wilson, and M.V. Malyutina. 2014. Urstylidae—a new family of deep-sea isopods and its phylogenetic implications. Zoological Journal of the Linnean Society 170:245–296.

Riehl, T., S. Bober, I. Voltski, M.V. Malyutina, and A. Brandt. 2018. Caught in the act: an abyssal isopod collected while feeding on Komokiaceae. Marine Biodiversity 48:157–158.

Riehl, T., L. Lins, and A. Brandt. 2018. The effects of depth, distance, and the Mid-Atlantic Ridge on genetic differentiation of abyssal and hadal isopods (Macrostylidae). Deep Sea Research Part II: Topical Studies in Oceanography 148:74–90.

Rogers, A.D. 2007. Evolution and biodiversity of Antarctic organisms: a molecular perspective. Philosophical Transactions of the Royal Society B: Biological Sciences 362:2191–2214.

Rosa, R., J. Boavida-Portugal, K. Trübenbach, M. Baptista, R. Araújo, and R. Calado. 2012. Descending into the abyss: bathymetric patterns of diversity in decapod crustaceans shift with taxonomic level and life strategies. Deep Sea Research Part I: Oceanographic Research Papers 64:9–21.

Roterman, C.N., J.T. Copley, K.T. Linse, P.A. Tyler, and A.D. Rogers. 2013. The biogeography of the yeti crabs (Kiwaidae) with notes on the phylogeny of the Chirostyloidea (Decapoda: Anomura). Proceedings of the Royal Society of London B: Biological Sciences 280:20130718.

Schlanger, S.O., and H.C. Jenkyns. 1976. Cretaceous oceanic anoxic events: causes and consequences. Netherlands Journal of Geosciences/Geologie en Mijnbouw 55:179–184.

Schnabel, K.E., P. Cabezas, A. McCallum, E.Macpherson, S.T. Ahyong, and K. Baba. 2011. Worldwide distribution patterns of squat lobsters. Pages 149–182 *in* G.C.B. Poore, S.T. Ahyong, and J. Taylor, editors. The Biology of Squat Lobsters. CSIRO Publishing, Collingwood, Australia.

Schram, F.R. 1970. Isopod from the Pennsylvanian of Illinois. Science 169:854–855.

Seifried, S. 2004. The importance of a phylogenetic system for the study of deep-sea harpacticoid diversity. Zoological Studies 43:435–445.

Selden, P.A., G.D.F. Wilson, L. Simonetto, and F.M.D. Vecchia. 2016. First fossil asellote (Isopoda: Asellota), from the Upper Triassic (Norian) of the Carnic Prealps (Friuli, northeastern Italy). Journal of Crustacean Biology 36:68–86.

Shank, T.M., M.B. Black, K.M. Halanych, R.A. Lutz, and R.C. Vrijenhoek. 1999. Miocene radiation of deep-sea hydrothermal vent shrimp (Caridea: Bresiliidae): evidence from mitochondrial cytochrome oxidase subunit I. Molecular Phylogenetics and Evolution 13:244–254.

Smith, A.B., and B. Stockley. 2005. The geological history of deep-sea colonization by echinoids: roles of surface productivity and deep-water ventilation. Proceedings of the Royal Society B: Biological Sciences 272:865–869.

Svavarsson, J., G. Gudmundsson, and T. Brattegard. 1993. Feeding by asellote isopods (Crustacea) on foraminifers (Protozoa) in the deep sea. Deep Sea Research Part I: Oceanographic Research 40:1225–1239.

Takashima, R., H. Nishi, B.T. Huber, and R.M. Leckie. 2006. Greenhouse world and the mesozoic ocean. Oceanography 19:64–74.

Thistle, D., and R.R. Hessler. 1976. Origin of a deep-sea family, the Ilyarachnidae (Crustacea: Isopoda). Systematic Biology 25:110–116.

Thomson, C.W. 1873. The Depths of the Sea. An Account of the General Results of the Dredging Cruises of H.M.S.S. "Porcupine" and "Lightning" during the summers of 1868, 1869, and 1870, under the Scientific Direction of Dr. Carpenter, F.R.S, J. Gwyn Jeffreys, F.R.S., and Dr. Whville Thompson, F.R.S. Macmillan, London.

Thuy, B. 2013. Temporary expansion to shelf depths rather than an onshore-offshore trend: the shallow-water rise and demise of the modern deep-sea brittle star family Ophiacanthidae (Echinodermata: Ophiuroidea). European Journal of Taxonomy 48:1–242.

Thuy, B., A.S. Gale, A. Kroh, M. Kucera, L.D. Numberger-Thuy, M. Reich, and S. Stöhr. 2012. Ancient origin of the modern deep-sea fauna. PloS ONE 710:e46913.

Trabucho-Alexandre, J., W.W. Hay, and P.L. De Boer. 2012. Phanerozoic environments of black shale deposition and the Wilson Cycle. Solid Earth 3:29.

Tsang, L.M., T.Y. Chan, M.K. Cheung, and K.H. Chu. 2009. Molecular evidence for the Southern Hemisphere origin and deep-sea diversification of spiny lobsters (Crustacea: Decapoda: Palinuridae). Molecular Phylogenetics and Evolution 51:304–311.

Tsang, L.M., T.Y. Chan, S.T. Ahyong, and K.H. Chu. 2011. Hermit to king, or hermit to all: multiple transitions to crab-like forms from hermit crab ancestors. Systematic Biology 60:616–629.

Van Der Wal, C., S.T. Ahyong, S.Y.W. Ho, and N. Lo. 2017. The evolutionary history of Stomatopoda (Crustacea: Malacostraca) inferred from molecular data. PeerJ 5:e3844.

Veit-Köhler, G. 2005. First deep-sea record of the genus Kliopsyllus Kunz, 1962 (Copepoda: Harpacticoida) with the description of Kliopsyllus diva sp. n.—the most abundant member of Paramesochridae at two different sites of the Angola Basin. Organisms Diversity and Evolution 5(1 Suppl.):29–41.

Wagreich, M. 2012. "OAE 3"—a low-to mid-latitude Atlantic oceanic event during the Coniacian-Santonian. Climate of the Past Discussions 8:1209–1227.

Watling, L., and M.H. Thurston. 1989. Antarctica as an evolutionary incubator: evidence from the cladistic biogeography of the amphipod family Iphimediidae. Geological Society, London, Special Publications 47:297–313.

Watling, L., J. Guinotte, M.R. Clark, and C.R. Smith. 2013. A proposed biogeography of the deep ocean floor. Progress in Oceanography 111:91–112.

Wei, C.-L., G.T. Rowe, E. Escobar-Briones, A. Boetius, T. Soltwedel, M.J. Caley, Y. Soliman, F. Huettmann, F. Qu, Z. Yu, C.R. Pitcher, R.L. Haedrich, M.K. Wicksten, M.A. Rex, J.G. Baguley, J. Sharma, R. Danovaro, I.R. MacDonald, C.C. Nunnally, J.W. Deming, P.A. Montagna, M. Lévesque, J.M. Weslawski, M. Wlodarska-Kowalczuk, B.S. Ingole, B.J. Bett, D.S.M. Billett, A. Yool, B.A. Bluhm, K. Iken, and B.E. Narayanaswamy. 2010. Global patterns and predictions of seafloor biomass using random forests. PLoS ONE 5:e15323.

Whatley, R.C. 1990. Ostracoda and global events. Pages 3–24 *in* R. Whatley and C. Maybury, editors. Ostracoda and Global Events. Chapman and Hall, London.

Wiklund, H., I.V. Altamira, A.G. Glover, C.R. Smith, A.R. Baco, and T.G. Dahlgren. 2012. Systematics and biodiversity of Ophryotrocha (Annelida, Dorvilleidae) with descriptions of six new species from deep-sea whale-fall and wood-fall habitats in the north-east Pacific. Systematics and Biodiversity 10:243–259.

Wilson, G.D.F. 1980. New insights into the colonization of the deep sea: systematics and zoogeography of the Munnidae and the Pleurogoniidae comb. nov. (Isopoda; Janiroidea). Journal of Natural History 14:215–236.

Wilson, G.D.F. 1989. A systematic revision of the deep-sea subfamily Lipomerinae of the isopod crustacean family Munnopsidae. Pages 1-138 in C.S. Cox, A. Fleminger, G.L. Kooyman, and R.H. Rosenblatt, editors. Bulletin of the Scripps Institution of Oceanography, Volu. 27. University of California Press, Berkley, Los Angeles, London.

Wilson, G.D.F. 1998. Historical influences on deep-sea isopod diversity in the Atlantic Ocean. Deep Sea Research Part II: Topical Studies in Oceanography 45:279–301.

Wilson, G.D.F. 1999. Some of the deep-sea fauna is ancient. Crustaceana 72:1019–1030.

Wilson, G.D.F. 2009. The phylogenetic position of the Isopoda in the Peracarida (Crustacea: Malacostraca). Arthropod Systematics and Phylogeny 67:159–198.

Wilson, G.D.F., and S.T. Ahyong. 2015. Lifestyles of the species-rich and fabulous: the deep-sea crustaceans. Pages 279–297 *in* M. Thiel and L. Watling, editors. The Natural History of the Crustacea, Volu. 2: Lifestyles and Feeding Biology. Oxford University Press, New York.

Wilson, G.D.F., and N. Morel. 2016. Isopod crustacean fossils from the Cenomanian stratotype: five new species in suborders Cymothoida, Asellota and Valvifera. Musée Vert, Muséum d'Histoire Naturelle du Mans, Le Mans, France.

Woolley, S.N.C., D.P. Tittensor, P.K. Dunstan, G. Guillera-Arroita, J.J. Lahoz-Monfort, B.A. Wintle, B. Worm, and T.D. O'Hara. 2016. Deep-sea diversity patterns are shaped by energy availability. Nature 533:393–396.

Yamaguchi, T., and W.A. Newman. 1990. A new and primitive barnacle (Cirripedia: Balanomorpha) from the North Fiji Basin abyssal hydrothermal field, and its evolutionary implications. Pacific Science 44:135–155.

Yamaguchi, T., W.A. Newman, and J. Hashimoto. 2004. A cold seep barnacle (Cirripedia: Neolepadinae) from Japan and the age of the vent/seep fauna. Journal of the Marine Biological Association of the United Kingdom 84:111–120.

8

BIOGEOGRAPHY AND HOST USAGE OF CORAL-ASSOCIATED CRUSTACEANS: BARNACLES, COPEPODS, AND GALL CRABS AS MODEL ORGANISMS

Benny K. K. Chan, Kingsley J. H. Wong, and Yu-Rong Cheng

Abstract

Most of the diverse groups of crustaceans associated with scleractinian and fire corals form symbiotic relationship with their coral hosts. Coral-associated barnacles include species from the orders Acrothoracica and Thoracica. Most of the coral-associated barnacles belong to the family Pyrgomatidae in Thoracica. Within Pyrgomatidae, the subfamily Ceratoconchinae contains mostly extant species and is present from Florida through the Caribbean to Brazilian waters. The subfamily Megatrematinae has lower species diversity and has a cosmopolitan distribution (except the Eastern Pacific). The Pyrgomatinae are the most species-rich subfamily and distributed only in Indo-West Pacific waters. Host usage of pyrgomatinid barnacles varies spatially, probably related to coral host diversity. Copepods are the most common and most abundant coral-associated crustaceans, often associated with scleractinian, gorgonian, and alcyonacean corals. More than 90% of coral-associated copepods are endemic to the Indo-West Pacific. In contrast, only a few species (<10%) have been discovered from the Atlantic due to several historical perturbations reducing the diversity of their coral hosts. The communities of coral-associated copepods thus show dramatic differences between geographic regions, notably between the Indian, Pacific, and Atlantic Oceans. Brachyurans of the family Cryptochiridae (gall crabs) are obligate associates or parasites, of scleractinian coral hosts in tropical and subtropical seas, being a monophyletic group of only 52 species, from the intertidal to the deep sea (to 512 m) habitats with most (46) recorded in the seas of the tropical Indo-West Pacific and none being cosmopolitan. Atlantic species of Cryptochiridae, apparently not phylogenetically related, display less strict host specificity than their Indo-West Pacific counterparts. Current phylogenetic understandings of the group remain preliminary, while one consistent Indo-West Pacific clade reflects rapid species diversification during the last ~15 million years.

Evolution and Biogeography. Edited by Gary C. B. Poore and Martin Thiel.

INTRODUCTION

Coral reefs are important habitats in marine biodiversity hotspots and are often referred to as "tropical rainforests of the sea" due to their structural complexity, high primary productivity, abundant biomass, and great species richness. Coral reefs encompass the highest biodiversity of any marine ecosystem, with estimates of the number of species ranging from 172,000 to over 9,000,000 (Reaka-Kudla 1997, Ruppert et al. 2004), and have been associated with the evolution of the largest diversity of symbiotic associations in the marine environment and possibly the biosphere (Castro 1988). According to a recent review of the diversity of coral-associated invertebrates, at least 869 species have been described as associates of corals, of which 636 crustacean species might be the largest group of coral-associated animals (Stella et al. 2011). Crustaceans associated with scleractinian and fire corals include barnacles, copepods, and decapods (palaemonid and alpheid shrimps, brachyuran crabs, and hermit crabs) (Goldberg 2013). Each group is dealt with here in turn.

Barnacles are sessile, and many species are specifically symbiotic with corals. Species in the family Pyrgomatidae in Thoracica are obligate commensals with corals, present in most coral reefs worldwide, exhibiting various degrees of host-specificity. $\delta^{13}C$ stable isotope studies on a few pyrgomatid species in the Red Sea suggested that organic matter produced by the coral-zooxanthellae do contribute part of their carbon source (Achituv and Mizrahi 1996, Achituv et al. 1997). In return, the ammonium released from the coral barnacles are absorbed by their zooxanthellae.

Copepods often associate with scleractinian, gorgonian, and alcyonacean corals. More than 90% of coral-associated copepods are endemic to the Indo-West Pacific (Stock 1988). Coral-associated copepods are either ectoparasites or endoparasites in the gastrovascular cavity of scleractinian corals. Some are obligatory and rely on the coral hosts for nutrients (Humes 1985a), while others form galls, dome-like surface enclosures (Kim and Yamashiro 2007).

Several groups of decapods associate with scleractinian corals. The snapping shrimp *Alpheus lottini* (Alpheidae), for example, associates with branching corals of the family Pocilloporidae, in which the shrimp defends the corals from predators, and in return the shrimp feeds on the mucus secreted from the hosts (Goldberg 2013). At least 13 genera of palaemonid shrimps are associated with scleractinian corals. Members of the palaemonid subfamily Pontoniinae are specialists, some genera and species being specific to their host (Goldberg 2013). Gall crabs (Cryptochiridae) live inside the coral skeleton and form galls on the surface.

Understanding the processes that determine the distribution of coral symbionts and factors affecting their host specificity is essential for understanding the diversity of coral reefs and protecting this important habitat. The present chapter reviews the biogeography and host usage of coral-associated barnacles, copepods, and gall crabs, one family of Decapoda.

CORAL-ASSOCIATED BARNACLES

Coral-associated barnacles are obligatory symbionts of stony corals (Scleractinia) and fire corals (Hydrocoralia) (Ross and Newman 1973). Barnacles include species from two superorders, Thoracica and Acrothoracica (Fig. 8.1). Thoracican barnacles have diverse lifestyles and inhabit almost every marine environment, from the intertidal zone to the deep sea (Chan et al. 2009). Most thoracican coral-inhabiting barnacles belong to the family Pyrgomatidae of the order Sessilia. In addition to pyrgomatid barnacles, a few species of *Megabalanus* (Balanidae, e.g., *M. ajax* and *M. stultus*) also live in living stony corals and hydrocorals. *Armatobalanus allium*, previously thought to belong to Archaeobalanidae, was shown to be nested within the molecular clade of Pyrgomatidae (Tsang et al. 2014a). The pyrgomatid barnacle *Wanella*, which inhabits fire corals, should be

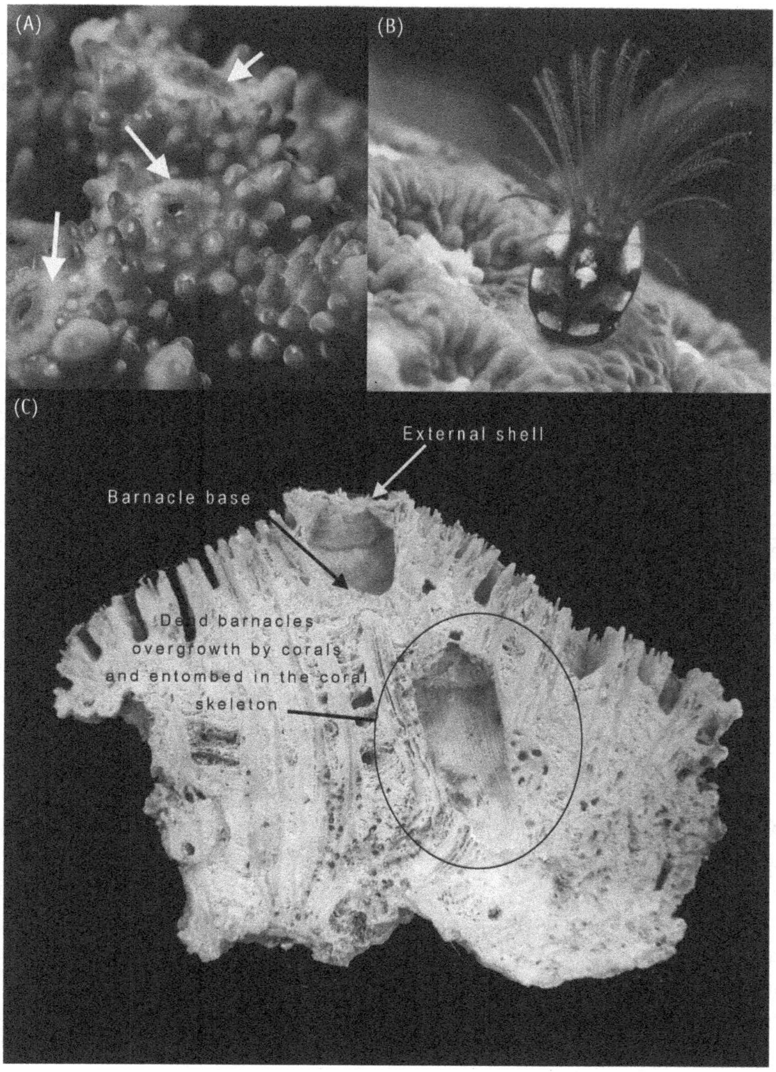

Fig. 8.1.
(A) The Prygomatid barnacles (Throacica) *Cantellius secundus* on *Acropora* corals (indicated by white arrows). (B) The Acrothoraican barnacle *Berndtia purpurea* bore in *Lepastrea* corals. (C) Cross section of a coral skeleton (*Alveopora japonica*) showing the barnacles *Nobia grandis*. On top, the live *Nobia* has its mantle cavity and base embedded in the coral skeleton and the shell is on the surface of corals. There is a dead *Nobia* entombed by coral. Once the barnacle is dead, the coral will gradually overgrow the surface and entomb the barnacle inside the skeleton. See color version of this figure in centerfold.

classified under a new family, as suggested from molecular evidence (Malay and Michonneau 2014, Tsang et al. 2014a).

Compared to free-living barnacles, pyrgomatid barnacles have a modified shell, in that the bases are usually cup-shaped or tubular and embedded in the coral skeleton. On the bases are longitudinal ribs linking to the internal ribs or septa of the shells. Concentric growth lines are also found on the bases (Ross and Newman 1973). Species of *Pyrgoma* have perforations in the base that are believed to enable chemical exchange with the host coral (Ross and Newman 2002). Most pyrgomatids have flat discoidal or elliptic shells. In some coral barnacles, the shells are low and conical. Shells of pyrgomatids are either solid or tubiferous. Tubiferous shells have primary or secondary longitudinal canals separated by longitudinal septa believed to improve shell strength. The spaces or gaps between the barnacle shell perimeter and the coral appear to allow for physiological interaction with the coral host, possibly by receiving metabolic substances (Ross and Newman 1973).

A major problem facing coral-inhabiting barnacles is the growth of coral tissue over the barnacle's orifice. There is therefore a "growth race" between the coral host and the barnacles. After the initial growth of the shell, most growth takes place at the perimeter of the base, crucially growing at the same pace as the coral in order to avoid being entombed, thus pushing the shell plate to the surface and, in some cases, even creating chimneys on the coral host surface.

Another mechanism pyrgomatids use to avoid coral tissue growing over the orifice is an aperture frill which expands over the orifice rim during cirral activity. The aperture frill, a glandular fold, and its secretions inhibit coral growth chemically (Anderson 1992, 1994). In addition to aperture frills, Ross and Newman (1973) have suggested that barnacles may exert a physiological control on the growth of the coral host since, once the barnacle dies, it is overgrown and entombed by the coral skeleton.

Under the present taxonomic classification, the number of shell plates in living species of Pyrgomatidae is four (carina, rostrum and paired laterals) that may be fused. The Pyrgomatidae are divided into three subfamilies, based on the shell and opercular plate structures. Species of Pyrgomatinae have separated or fused shells and separated or fused opercular plates. Megatrematinae have a conical-fused shell with separated balanoid-type opercular plates and inhabit both zooxanthellate and non-zooxanthellate corals. The subfamily Ceratoconchinae have a six- (in fossil species) or four-plated shell with balanoid-type opercular plates and differ from other subfamilies in having a single large crest for the depressor muscle on the tergum. Species of this subfamily inhabit corals in the Atlantic Ocean.

Barnacles of the superorder Acrothoracica have no external mineral shells and bore into calcareous substrata, such as molluscan and thoracican shells, corals, bryozoans and calcareous rocks (Tomlinson 1969, Chan et al. 2014a). Acrothoracican barnacles are dioecious, with a dwarf male and a much larger female. The somatic body of females is embedded in a mantle sac attached to the substratum by the base (attachment disk).

Paleobiogeography, Recent Biogeography, and Host Usage

Among the three subfamilies of Pyrgomatidae, the Certatoconchinae contain the greatest number of extinct species, with the oldest fossil record in the Oligocene. In contrast, the Pyrgomatinae contain no extinct species but extant species traced back to the Miocene.

Ceratoconchinae

The Ceratoconchinae contain two genera, the exclusively fossil genus *Eoceratochoncha* (three species) and *Ceratoconcha* (27 species, 23 extinct and 4 living species) (Ross and Newman 2000). The oldest pyrgomatid fossil was recorded as *Eoceratoconcha* sp. from the early Miocene in Jamaica (Newman and Ladd 1974). Its shell is of six plates, differing from that of *Ceratoconcha* with a four-plated shell. *Eoceratoconcha kugleri and E.renzi* were found in the middle Miocene of Trinidad (Newman and Ladd 1974), and subsequently *E. weisbordi* was found in the Pleistocene of Florida (Zullo and Portell 1990, 1991).

Fossils of *Ceratoconcha* are recorded from the Miocene in the Caribbean, including Jamaica, Puerto Rico, and Trinidad, and the Pliocene in Cuba, Barbados, and Florida (Withers 1926, Withers 1953, Brooks and Ross 1960, Weisbord 1972, Newman and Ladd 1974, Bacon 1976, Zullo and Portell 1991, 1992) (Fig. 8.2). The greatest diversity of fossil records of *Ceratoconcha* is found in the Miocene of the Paratethys (15 extinct species; Baluk and Radwanski 1967a, b, Ross and Newman 2000) and also in the Miocene of Spain (Moisette and Saint Martin 1990) and the Mediterranean, including Greece and Algeria (Baluk and Radwanski 1967a, b) (Fig. 8.2). However, no living species of *Ceratoconcha* is found in European and Mediterranean waters. Ceratoconchinae probably originated in Caribbean waters. The presence of European and Mediterranean *Ceratoconcha* fossils suggests that ceratoconchines may have been distributed from the Caribbean to the Mediterranean (Ross and Newman, 2000) during the Miocene and subsequently became extinct probably due to the interactions of sea levels, climatic change, and the drying of the Mediterranean Sea during the Messinian salinity events in the Miocene (Ross and Newman 2000). Four living species of *Ceratoconcha* inhabit the west Atlantic, including Gulf of Mexico, the Caribbean and from Brazil to Bermuda (Wells 1966, see review in Ross and Newman 2000) (Fig. 8.2).

Megatermatinae

Unlike Ceratoconchinae, megatrematines have almost worldwide distribution, absent only from the eastern Pacific. Megatrematinae are composed of two tribes, the Pyrgomini (*Adna* and *Pyrgomina*) and Megatrematini (*Megatrema* and *Memagreta*) (Ross and Pitombo 2002) of seven living and six extant species (Ross and Pitombo 2002).

The oldest megatrematine fossil was recorded as *Pyrgomina* in the early Pliocene of Sicily (Seguenza 1873, Ross and Newman 2000). Additional fossils were found in the late Pliocene in Crete, Greece (Baluk and Rawanski 1967a) and the Pliocene-Pleistocene of Italy (Withers 1953) (Fig. 8.3). Fossils of *Pyrgomina* were also recorded in the Pliocene in the Coralline Crag at Suffolk in England (Tilbrook 1997) and as Pleistocene to Holocene fossils in Japan (Asami and Yamaguchi 1997) (Fig. 8.3). The closure of the Betic corridor and the drying of the Mediterranean Sea during the Messinian events may have resulted in the extinction of the Mediterranean species of *Pyrgomina* (Ross and Newman 2000). With the refilling of the Mediterranean Sea in the Pleistocene, *Adna anglica* re-invaded the Mediterranean. *Adna anglica* is the only living species recorded in the Atlantic, including the British Isles, Madeira, and the Canary Islands, Nigeria and Angola in West Africa, and in the west Mediterranean including Oran, Algeria, and northeastern Tunisia (Stubbing 1967, Ross and Newman 2000, Ross and Pitombo 2002) (Fig. 8.3). *Megatrema madreporum* is currently present in the Caribbean (Tobago, Bonaire, Jamaica, Barbados, and Florida and south of Venezuela) (Scott 1987, Pilsbry 1931, Southward 1986, Ross and Pitombo 2002). The other living species, *M. youngi*, was recorded in Fortaleza, Brazil (Ross and Pitombo 2002).

The Indo-West Pacific region houses several living species of megatrematines. *Memagreta* and *Pyrgomina* occur in Australian waters of the Indian Ocean (Ross and Pitombo 2002). *Pyrgomina oulastreae* is recorded in Japan, Korea, Hong Kong, and southern Chinese waters (Sakakura 1934, Ren 1986, Asami and Yamaguchi 1997, Liu and Ren 2007, Kim 2011). Megatrematinae are absent from the Eastern Pacific, both as fossil and as Recent species.

Host usage of the Megatrematinae received less attention than that of Pyrgomatinae. Unlike pyrgomatines, which associate with zooxanthellate coral hosts, species of the Megatrematinae can inhabit both non-zooxanthellate and zooxanthellate corals (Ross and Pitombo 2002). In the tribe Pyrgomini, the fossil species *Pyrgomina culcatum* inhabited the non-zooxanthellate coral *Oculia* in the Pliocene of Sicily (Ross and Newman 2000). In the Atlantic Ocean, the living species *Adna anglica* is reported to inhabit the non-zooxanthellate corals *Caryophyllia smithii* (Rees 1962, 1966).

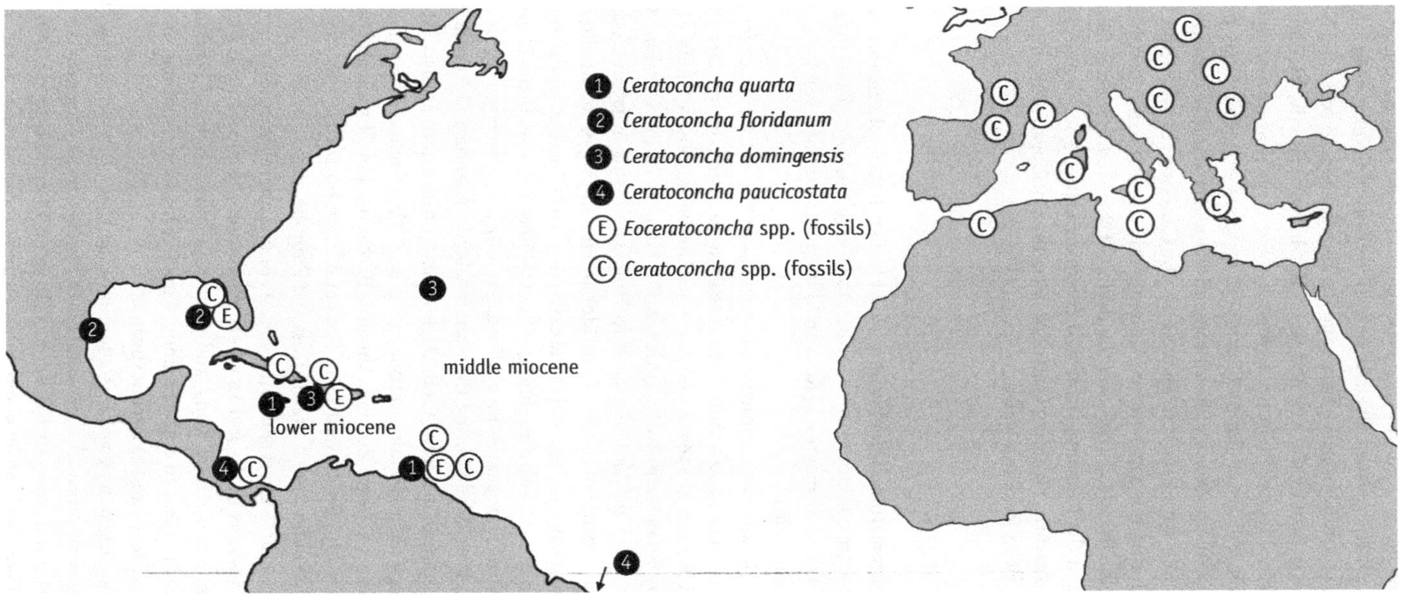

Fig. 8.2.
Paleobiogeography and recent biogeography of species in the subfamily Ceratoconchinae. Black circles indicate recent living species, numerical numbers indicate different species, open circles indicate fossil species. E is fossil *Eoceratoconcha,* which was found in the lower Miocene in Jamaica and the middle Miocene in Trinidad. C is fossil *Certatoconcha,* which was found in Miocene. Note that the Mediterranean had a high diversity of *Ceratoconcha* fossils but contains no living species nowadays. This suggests that *Ceratoconcha* sp. in the Miocene was driven to extinction in the Mediterrancean during the Messinian events and no species recolonized the Mediterranean Sea after the sea levels rose.

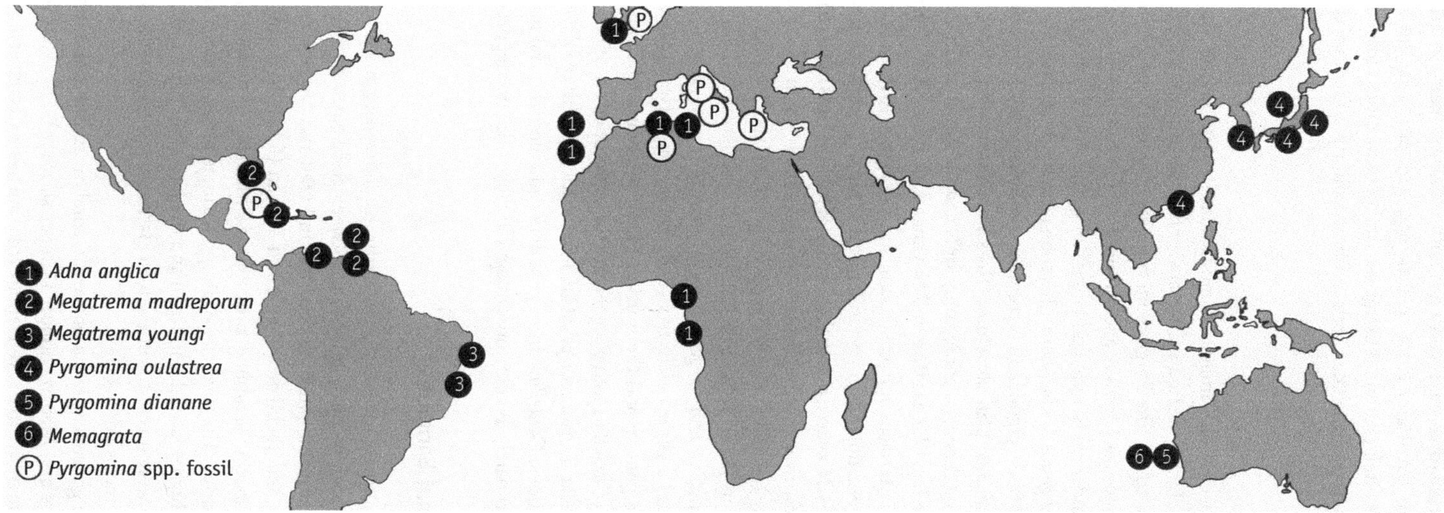

Fig. 8.3.
Paleobiogeography and recent biogeography of species in the subfamily Megatrematinae. Species in Megatrematinae had almost cosmopolitan distribution except that they are absent from the Eastern Pacific. Black circles indicate recent living species, numerical numbers indicate different species, open circles with P indicate fossil *Pyrgomina* species. Note that *Pyrgomina* fossils were present in the Mediterranean Sea but they did not survive through the Messinian events. After the sea levels rose, the Mediterranean Sea was reinvaded by *Adna anglica* and thus were only present in the Western waters of the Mediterranean.

Species of Megatrematini are all reported to live in zooxanthellate corals (*Agaricia agaricites*) in Caribbean and Brazilian waters (Young 1988, 1998).

In the West Pacific, *Pyrgomina oulastrea* was recorded in both zooxanthellate (*Oulastrea crispata*) and non-zooxanthellate corals in Japan and Korea (Hiro 1935, Kim 2011). In the West Pacific, *Pyrgomina oulastrea* is the species often present in marginal coral communities close to the edge of the coral distribution, where non-zooxanthellate corals dominate and Pyrgomatinae do less well.

Pyrgomatinae

The Pyrgomatinae are the most speciose subfamily of Pyrgomatidae (>20 genera, >80 species), being present in the Indo-West Pacific, including the Red Sea, but absent from the Atlantic and Mediterranean. The Pyrgomatinae comprise three tribes. The Pyrgosellini contains *Pyrgopsella*, which inhabit corals, and *Pyrgospongia*, which inhabit sponges (Achituv and Simon-Blecher 2006). *Pyrgospongia* has been recorded in the Philippines, the Andaman Islands, and Hong Kong. The Hoekiini include coral-eating barnacles residing only in the coral genus *Hydnophora*. This tribe is presently recorded from the Great Barrier Reef and Japan. The third tribe Pyrgomatini is composed of 20 genera, >80 species, all extant, and has diverse shell forms and opercular plate structures (Ross and Newman 1995, Ross 2000). Fossil records of Pyrgomatini are fragmented, being recorded only for *Nobia* and *Savignium* in the Miocene of Fiji and the Pleistocene of the Marshall Islands (Ross and Newman 2000). *Cantellius, Darwiniella*, and *Galkinius* are found in the Holocene terraces in Japan (Asami and Yamaguchi 1997, Ross and Newman 2000). Such great diversity of Pyrgomatini today may result from the rapid diversification and extensive radiation in former geological times. However, due to the scarcity of fossil records of this tribe, it is difficult to estimate when and how such diversification happened (Ross and Newman 2000).

Asami and Yamaguchi (1997) studied the biogeography of 23 species and 11 genera of Pyrgomatidae from 18 locations in Japanese waters. They found four latitudinal zones: southern waters of Okinawa, where 25 species and 10 genera were recorded; waters between Okinawa and Shikoku, where 18 species and 8 genera were recorded; the region between Shikoku and Kii Peninsula with 18 species and 78 genera; and the northern limit of coral at the Boso Peninsula, where 4 species from 2 genera (*Cantellius* and *Pyrgomina*) were recorded. It appears that most genera, including *Nobia, Darwiniella, Trevathana*, and *Galkinius*, associate with corals in subtropical and tropical waters at lower latitudes. *Cantellius* and *Pyrgomina* can colonize coral marginal communities at high latitudes. Unfortunately, Asami and Yamaguchi (1997) did not report coral hosts.

Host Specificity of Pyrgomatini Barnacles

The diversity of coral hosts and latitude appears to affect patterns of diversity and biogeography of pyrgomatid barnacles. Here, we examine the relationship between coral barnacle diversity and latitude using data from 14 sites in Honshu, Shukoku, Kyushu, and Okinawa (Asami and Yamaguchi 1997), Ogasawara Island (Ogawa and Tachikawa 2009), and Taiwan (Chan et al. 2013b), and plot the number of coral species at specific sampling sites (Japan: Ministry of Environment, Japan, 2004; Taiwan: Dai and Horng 2009). The diversity of Pyrgomatini is influenced by both latitude and coral diversity (Fig. 8.4). At higher latitudes in Japan and the Japan Sea region, where coral diversity is lowest, only two species were present, while at lower latitudes the highest diversity of coral barnacles in the West Pacific was recorded.

Pyrgomatid barnacles include generalists living on several coral hosts and specialists living with a limited number of hosts (Ogawa and Matsuzaki 1992, Tsang et al. 2014a). For example, species of *Armatobalanus* were recorded from more than seven families of coral hosts (Tsang et al. 2014a), while *Galkinius, Savignium* and *Trevathana* are more host-specific, found only on the family Faviidae (Tsang et al. 2014a). Generalists are located at the more basal position of the phylogenetic tree of Pyrgomatini

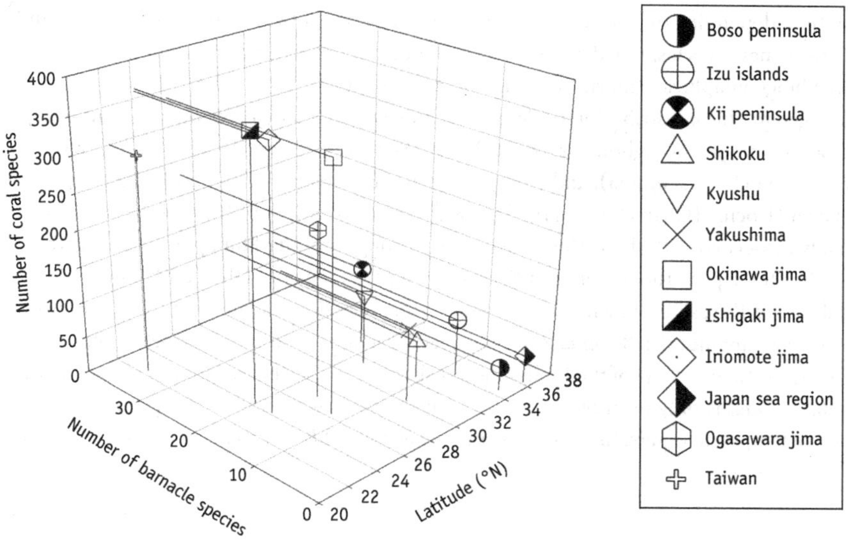

Fig. 8.4.
Relationship of Pyrgomatini species with latitudes and the number of coral species at the study sites. Barnacle diversity data collected from Asami and Yamaguchi (1997), Ogawa and Tachikawa (2009), and Chan et al. (2013a).

and specialists near the tip of the clades (Tsang et al. 2014a). Ogawa and Matzuzaki (1992) summarized their own work on host usage of Pyrgomatini. Records of coral hosts for each species were pooled from multiple locations from the whole Indo-West Pacific region. In the present study, we compared the spatial variation in host usage of Pyrgomatini species across the West Pacific and show that some exhibit a host-shifting pattern across regions. We compared data collected by Ogawa and Tachikawa (2009) in Ogasawara Island, Chan et al. (2013a) in Taiwan, and Poltaruka and Dutova (2007) from Nhatrang Bay, Vietnam. These three studies had extensive sampling efforts so that the variation in host usage across sites is unlikely to be affected by inadequate data. For example, *Galkinius indica* was recorded in *Acanthastrea, Goniastrea, Favia, Favites,* and *Platygyra* in Ogasawara Island, but only in *Hydnophora* in Taiwan (Chan et al. 2013b) and was recorded in corals of Fungiidae in Nhatrang Bay, Vietnam (Poltarukha and Dutova 2007). In Taiwan, corals of Fungiidae are not infested by any barnacles (Chan pers. obs). *Savignium crenatum* was recorded in *Goniastrea, Montastrea, Merulina,* and *Oxypora* in Ogasawara Island but recorded in *Favites* in Taiwan and *Cyphastrea* and *Favites* in Nhatrang Bay, Vietnam. In contrast, some species are relative specialists and do not exhibit host-shifting patterns among the three compared regions, including *Hiroa stubbingsi* (recorded in *Astreopora* in Ogasawara, Taiwan, and Vietnam), *Cantellius septimus* (only recorded in *Montipora*), and *C. secundus* (only recorded in *Acropora*). Further studies should examine the genetic differentiations of the same barnacle species across latitudinal gradients and hosts, to investigate cryptic diversity.

Acrothoracican Barnacles

In the superorder Acrothoracica, the genus *Berndtia* lives in only two genera of live corals, *Lepastrea* and *Psammocora* in west Pacific waters. Unlike thoracican barnacles, acrothoracicans do not have external shells; instead, they bore in live corals and live within burrows their whole life. The burrow opening of *Berndtia* species are oval, and the opercular bars are striped with black and blue coloration, depending on species. Chan et al. (2014b) revised the taxonomy of *Berndtia* and identified three new species, but there are now six species in West Pacific waters where they exhibit distinct biogeographical patterns with high host specificity.

The geographical distribution of *Berndtia* is mostly affected by oceanographic currents as their hosts are common throughout the West Pacific. *Berndtia purpurea* is reported in Taiwan, Japan, (Okinawa, Kochi, Wakayama), and Hong Kong, and *B. utinomii* in Taiwan, Hong Kong, Vietnam, and Japan (Kochi) (Chan et al. 2014b) (Fig. 8.5). The distribution of *B. purpurea* and *B. utinomii* is probably associated with the Kuroshio Current, flowing from the Philippines, passing Taiwan and on to Honshu, Japan. The presence of *B. purpurea* and *B. utinomii* in the South China Sea could be a result of the Kuroshio Branch Current, which enters the South China Sea from the Luzon Strait during the winter months. *Berndtia denticulata*, recorded in high abundance only in Okinawa and occasionally in Pacific waters of the Philippines, is absent from Kochi and Taiwan (Chan et al. 2014b). *Berndtia haradai* is only recorded in Honshu, Japan. This suggests that this species is a cold-water species of temperate waters. In contrast, *Berndtia nodosa* and *B. fossata* are tropical species (Fig. 8.5).

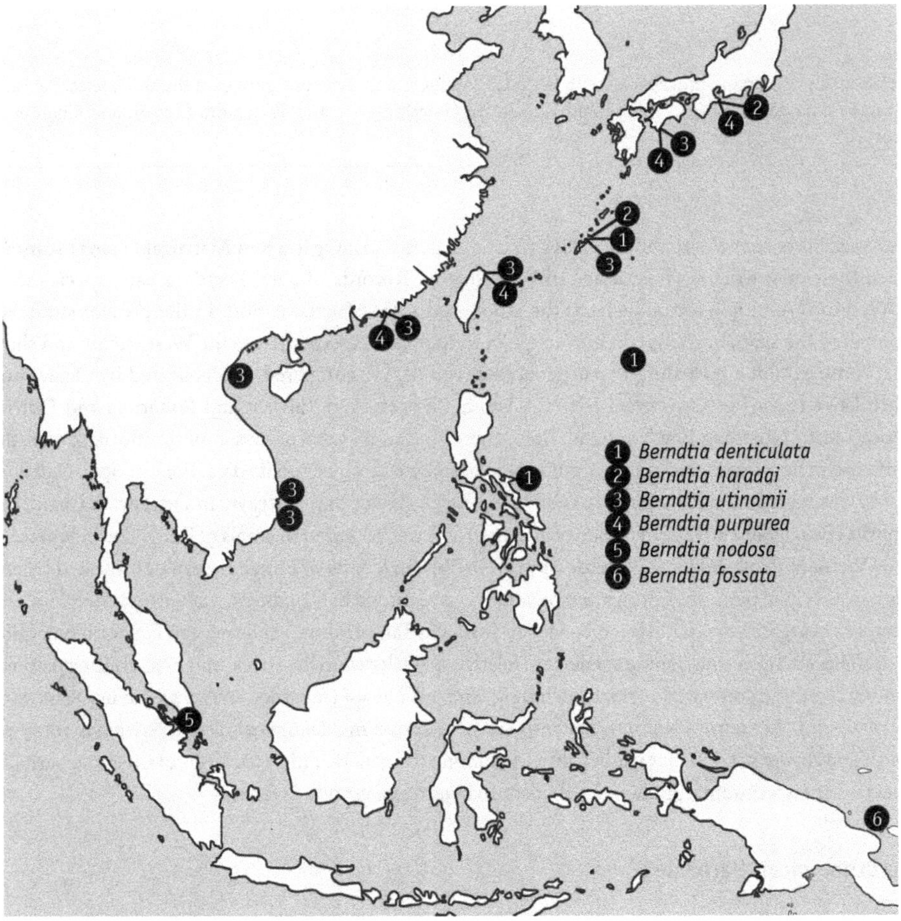

Fig. 8.5.
Biogeography of the Acrothoracican coral barnacle *Berndtia* species in the West Pacific. Figure modified from Figure 1 in Chan et al. (2014b).

CORAL-ASSOCIATED COPEPODS

Copepods have been highly successful in forming associations with many marine invertebrates, and the association between copepods and cnidarian hosts is common (Humes 1985a). More than 552 species, 37% of 1,500 known species from invertebrate hosts, are associated with cnidarians. A review of the scattered literature on the diversity and taxonomic composition of scleractinian-associated copepods revealed that the order Scleractinia contains 1,579 extant coral species in 243 genera, yet only 148 species (9.4%) representing 66 genera have so far been examined for copepods and serve as hosts to 363 copepod species (Cheng et al. 2016). Besides, Humes (1985a) concluded that approximately 227 species (172 species in 41 genera known from 118 species of shallow-water scleractinians) were associated with hexacorals and 161 copepod species with octocorals. He further proposed that shallow-water corals belonging to Scleractinia seemed more vulnerable to copepods than any other group of animals and one host coral may frequently support more than one species of copepod. Since then, the number of new copepod species associated with hard corals has substantially increased, and during the last 30 years new species have been described at a rate of 6.5 species per year. In recent decades, the number of copepod species and genera found to be associated with scleractinians has more than doubled, although only 30 additional hosts had been examined. It is apparent that the true diversity of coral-associated copepods is underestimated, and a vast number of new taxa are expected to be discovered in the future.

Cheng et al. (2012) performed a nonparametric estimation technique (using the idea of sample coverage) to estimate the number of potential undiscovered species in the mainly coral-associated copepod genus, *Xarifia*. Even though only one genus (of 99 genera) was included in this nonparametric analysis, this contains a projected undescribed ~136 copepod species. Recently, Mora et al. (2011) and Poore et al. (2015) proposed that 91% of marine species are yet to be described. This is especially true for the number of undiscovered species of coral-associated copepods. Taking those findings from our previous studies together, Copepoda, undoubtedly, is by far the most diverse group of coral-associated crustaceans.

The subclass Copepoda comprises one of the most diverse and abundant animal groups in marine ecosystems and is also the most common and widespread crustacean taxon living in symbiotic relationship with other organisms. Based on traditionally morphological characters, the group currently comprises 10 orders (Calanoida, Cyclopoida, Gelyelloida, Harpacticoida, Misophrioida, Monstrilloida, Mormonilloida, Platycopioida, Poecilostomatoida, and Siphonostomatoida) accommodating a total of 14,571 valid species (Walter and Boxshall 2017). Among the various orders of Copepoda, the Poecilostomatoida (to be encompassed in Cyclopoida based on molecular evidence; Huys et al. 2006, 2012) contain the greatest number of species living in association with hard corals (Fig. 8.6). In contrast, the orders Harpacticoida and Siphonostomatoida (Fig. 8.6), with relatively fewer species, are mainly ectosymbiotic and have a looser relationship with their scleractinian hosts (Humes 1985a). Until now, 363 copepod species (representing 99 genera and 19 families) have been recorded as associates of scleractinian corals (representing 66 genera and 15 families), including 288 cyclopoids, 68 siphonostomatoids, and 7 harpacticoids.

Within Cyclopoida (including Poecilostomatoida), 13 families are considered to utilize hard corals as hosts. The Anchimolgidae, Xarifiidae, and Rhynchomolgidae are the main groups and comprise approximately 141 (49% of known coral-associated copepods in Poecilostomatoida), 92 (32%), and 31 (11%) species, respectively. The remaining 10 families contain fewer than 10 species (<3%) symbiotic with a scleractinian host. The Siphonostomatoida are represented by four families, of which Asterocheridae (55 species, approximately 81% of known coral-associated copepods in Siphonostomatoida) and Coralliomyzontidae (8 species, 12%) are the most speciose. Although the hosts of most Artotrogidae and Entomolepididae are unknown, four and one species respectively, have been confirmed as associates of scleractinian corals. In the Harpacticoida, only a few species are associates of scleractinian corals.

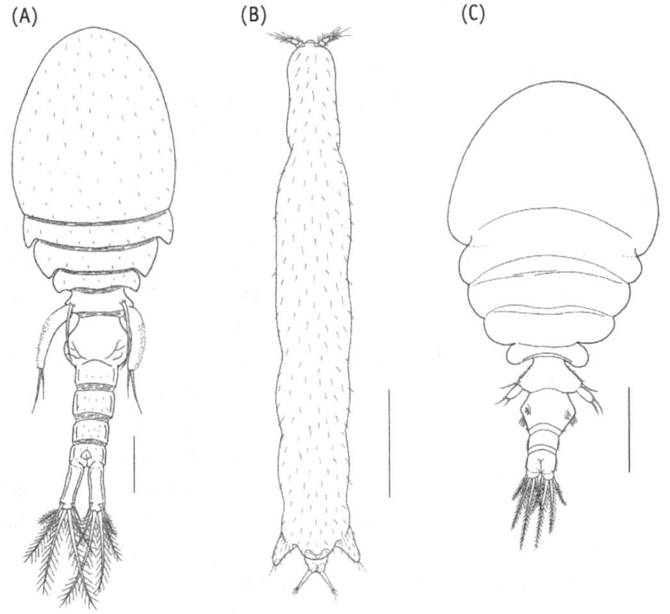

Fig. 8.6.
Coral-associated copepods in the tropical Indo-Pacific. (a) Anchimolgidae (Poecilostomatoida); (b) Xarifiidae (Poecilostomatoida); and (c) Asterocheridae (Siphonostomatoida). Scale bar: a, 500 μm; b, c, 200 μm.

Diversity in the Indo-West Pacific

The greatest number of copepod species (41%) comprise associates of scleractinian hosts and, in particular, more than 90% are distributed in Indo-West Pacific waters (Humes 1985a, Stock 1988). In contrast, only a few species (<10%) have been discovered from the Atlantic due to several historical perturbations (fluctuation of sea levels, low water temperatures, and isolation during geological time) reducing the diversity of their coral hosts. The copepod communities therefore differ between the Indian, Pacific, and Atlantic Oceans. For example, a great majority in the Indo-West Pacific are from the families Lichomolgidae (ecto-associated) and Xarifiidae (endo-parasitic), while Asterocheridae (ecto-associated) and Corallovexiidae (highly transformed endo-parasites) dominate the Atlantic (Stock 1988). The close association between copepods and scleractinian corals suggests that they may have interacted throughout their evolution. The occurrence of lichomolgid and xarifiid copepods on many scleractinian corals is common, particularly in the tropical Indo-West Pacific, but they are absent in the eastern Pacific Ocean and the Caribbean Sea where few species of several Indo-West Pacific coral genera (*Acropora, Favia, Porites,* and *Tubastrea*) appear. It has been speculated that these lichomolgids and xarifiids did not evolve or spread eastward from the Indo-West Pacific to the Caribbean and the Eastern Pacific when the Isthmus of Panama was still open in the pre-Pliocene (Humes 1985b). Although the main reason for differences in the faunas between major oceans is still unclear, they must be related to the diversity/species richness of coral hosts influenced by the fluctuation of sea levels, low water temperatures, and the isolation during geological time (Stock 1988).

Host Specificity

The scattered literature reveals that 148 scleractinian species serve as hosts for copepods (Figure 8.7). Without relative abundance of host coral species from each sampling locations, it is difficult

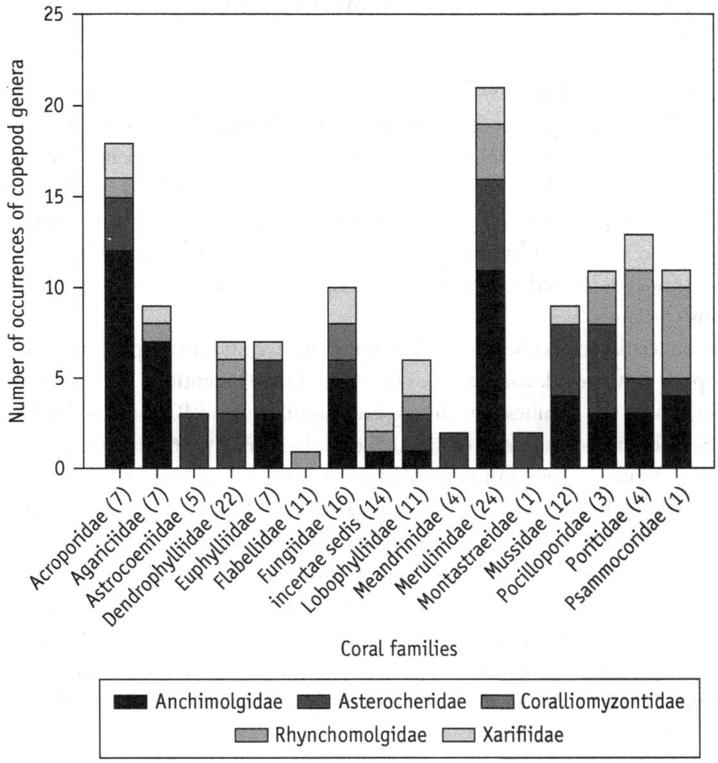

Fig. 8.7.
Distribution of five copepod families among 16 scleractinian coral families. Bar height indicates the total number of occurrences of copepod genera within each coral family; bar width with various colors is proportional to the number of genera of different copepod families occurring from each coral family. The number of parentheses indicates the total number of coral genera within each coral family.

to reveal the susceptibility of corals to copepod infestation. However, of the 16 families (including incertae sedis) of scleractinian corals, 8 families of copepods occur on Merulinidae, Acroporidae, and Pocilloporidae. The Merulinidae host the greatest diversity of copepods (26 genera), while Acroporidae and Pocilloporidae support 22 and 15 genera, respectively. However, at the species level, Acroporidae serve as host to 93 species, followed by Merulinidae (58), Pocilloporidae (48), Agariciidae (37), and Poritidae (34). These five coral families harbor nearly 75% of all described species, and one-third of copepods can be found from acroporid hosts. Flabellidae and Lobophylliidae (with 11 genera) and Mussidae (12 genera) support only one species of coral-associated copepod. Lobophylliidae and Mussidae are hermatypic and live in shallow water, while the Flabellidae are ahermatypic with representatives ranging in the less explored deep sea.

Even with a disproportionate number of species of Poecilostomatoida and Siphonostomatoida, it is clear that host specificity varies. The two most speciose families of Poecilostomatoida, Anchimolgidae and Xarifiidae, can be found in more than half of coral families (69–75%), but 74% species of Anchimolgidae and 58% species of Xarifiidae appear to utilize only one coral host. In contrast, the most speciose family of Siphonostomatoida, Asterocheridae, invades 75% of coral families and nearly 78% of species are known from a single host. In these cases, however, it is still difficult to determine if the association between corals and copepods represents true host specificity at the species level since only a small proportion of scleractinian corals have been examined.

Evolutionary Relationship between Symbiotic Copepods and Scleractinian Corals

The phylogenetic relationship between coral-associated copepods and their coral hosts has rarely been explored. Humes (1985b) was the first to apply morphological similarities and host preference, categorizing 75 species of *Xarifia* into 12 groups. He further carried out a phenetic study of *Xarifia* species based on the overall similarity of 26 selected characters, and the results were somewhat similar to the 12 groups classified by morphological similarities and host preferences. However, since the study did not include an outgroup and the evolutionary relationships of morphological characters were not considered, the exact relationship between coral-associated copepods and coral hosts remains obscure.

Recently, molecular studies have provided important insights into the phylogenetic relationships within copepods (Huys et al. 2006, 2012). Huys et al. (2006) identified a scleractinian-associated clade encompassing the families Anchimolgidae, Xarifiidae, and Rhynchomolgidae (AXR) and suggested that the common ancestor of the AXR clade had already established a symbiotic relationship with scleractinians. Cheng (2011) pointed out that the Xarifiidae were monophyletic and the most speciose genus, *Xarifia* (>85 species), originated in the Indian Ocean and gradually dispersed to the Pacific Ocean and formed a coevolutionary relationship with its hosts. The wide variety of host preference in the *Xarifia* species may involve morphological alterations and derived characters such as a higher number of aesthetascs on the antennule to detect hosts and mates and more spines on legs to hold their host. These characters might also increase the survival rate and reproductive success and thus promote their diversification. Since xarifiids are a recently derived family in the evolution of Copepoda (Huys et al. 2006), the extremely high species diversity in the *Xarifia* lineage implies a rapid evolutionary radiation.

Ecological Relationship between Copepods and Coral Hosts

Although plenty of copepods have been found living in various degrees of association with scleractinian corals, few studies have uncovered the ecological relationship between copepods and their hosts. Some findings described below provide insight.

Gall Formation

Copepods have evolved numerous ways to accommodate on or in their coral hosts. Most species live on the coral surface or inside the gastrovascular cavities of the polyps, but a few species inhabit gall structures (Fig. 8.8) and are referred as gallicolous or gall-inducing copepods. These copepods attach to the soft tissues of their hosts and elicit a defensive reaction, deposition of a calcareous barrier (gall) through their movement, or attaching/swimming appendages such as antenna and legs 1–4. Only five species have been referred to as gall-inducing copepods, living in galls of four families of scleractinian corals (Acroporidae, Mussidae, Pocilloporidae, Poritidae). Dojiri (1988) was the first to describe a gallicolous copepod, *Isomolgus desmotes* from *Seriatopora hystrix* in Indonesia. Dojiri and Grygier (1990) reported *Pionomolgus gallicolus* from *Echinopora lamellosa* collected in Australia, and Kim and Yamashiro (2007) observed two types of skeletal structure caused by two copepods: *Allopodion ryukyuensis* from the galls of *Montipora informis* and *Xenomolgus varius* from the crypts of a species of *Porites* in Japan. Recently, *A. ryukyuensis and Haplomolgus montiporae* were collected from a scleractinian coral *Montipora aequituberculata* in Taiwan (Cheng et al. 2010; Fig. 8.8). The mechanism of gall formation induced by these copepods and their influence on the coral skeleton are still unknown, but such skeletal overgrowth/depression probably reduces the

Fig. 8.8.
Photographs of *Montipora aequituberculata* with numbers of dome-shaped galls. (a) A colony showing several galls each inhabited by *Allopodion ryukyuensis* and *Hoplomolgus montiporae*. (b) A closer view of dome-shaped gall. White arrows indicate gall structure. See color version of this figure in centerfold.

tolerance of hosts to physical damage caused by typhoons, storm waves, and predators. It is interesting to note that all gall-inducing copepods discovered so far belong to Poecilostomatoida.

Chemical Release

Coral-associated copepods exhibit a marked diversity in morphology (body form and appendage structure) as well as feeding strategies and behavior in accordance with their ecological niches. For example, xarifiid copepods are endoparasites that inhabit the gastrovascular cavities of coral polyps and may digest and retain symbiotic algae *Symbiodinium* (zooanthellae) cells in their body temporarily as a nutrient source (Cheng and Dai 2010). In contrast, siphonostomatoids are ectoparasites and consume coral mucus via their tube-like mouths, and harpacticoids encompass planktonic, free-living, benthic, and invertebrate-associated species that generally consume microalgae, bacteria, particulate organic matter, and even animal tissues. In any case, their preference for residing in/on host coral colonies increases their chances of being attacked or killed by the host's nematocysts. According to the Red Queen hypothesis, however, it is highly likely that copepods may obtain some abilities to overcome this situation through their evolutionary history. Interestingly, based on the live observations of a reintroduction study, a new hypothesis of chemical release that induces the relaxation of coral polyps and enables immunization against the coral's nematocysts has been proposed (Cheng and Dai 2009; Fig. 8.9). These characteristics indeed enable the copepods to establish close associations with their host, and the mechanism and evolutionary significance of coral-parasite associations invite further studies.

A Three-Part Symbiosis (the Copepod-Zooxanthellae-Coral Relationship)

Histological and microscopic studies have demonstrated that coral-associated copepods may retain zooxanthellae cells in their guts for photosynthesis, and this photosymbiotic relationship may benefit their survival. Cheng and Dai (2010) found an abundance of unicellular algae displaying typical characteristics of *Symbiodinium* in the guts of xarifiid copepods. The presence of degraded algal cells with the chloroplast lamellae disintegrating (Fig. 8.10) suggested that the copepods may, to some extent, utilize these unicellular algae. Moreover, the copepod might retain the intact unicellular algae in their guts, and the algal cells possibly remained viable and photosynthetically active for a short period (two weeks) in laboratory-reared, starved copepods. This association between

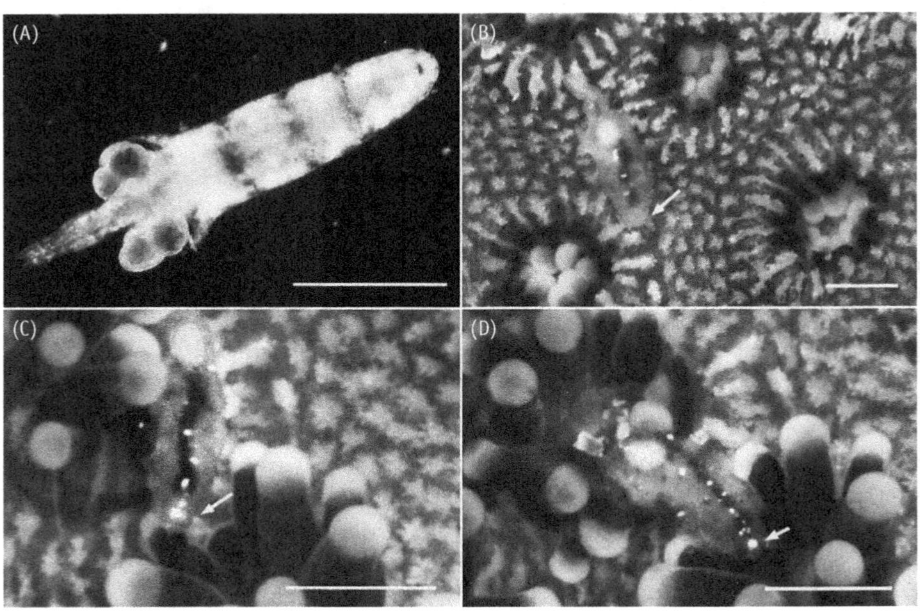

Fig. 8.9.
Live observations of reintroduced coral-associated copepod, *Xarifia obesa*, to its coral hosts. (a) An individual of female copepod carrying six eggs in dorsal view. (b) Copepod was seeking polyps after reintroduction to a coral colony. (c, d) Coral polyps started to extend their tentacles when in contact with copepod (probably antennule or antenna), and allowed copepod to enter its gastrovascular cavity. Scale bars = 0.5 mm. Arrows indicate the head of copepod. See color version of this figure in centerfold.

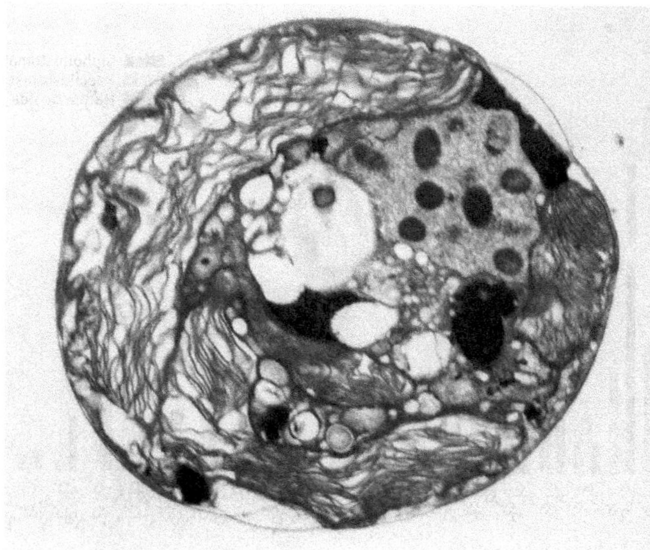

Fig. 8.10.
Transmission electron microscopic (TEM) images of *Symbiodinium* within a coral-associated copepod, showing scattered chloroplasts with disintegrating lamellae.

zooxanthellae and copepod, therefore, was referred to as a case of temporary retention or unstable symbiosis. In addition, cultivation experiments with various treatments (light/dark, with and without photosynthesis-inhibiting DCMU solution) were conducted to test whether unicellular algae in their guts might benefit copepods' survival. Mortality rate was significantly higher in groups in the dark/with DCMU than in groups in the light/without DCMU, suggesting unicellular algae may benefit survival, possibly through releasing photosynthetic products to the hosts. It is well known that zooxanthellae in corals fix carbon through photosynthesis and translocate as much as 95% of the amount fixed to corals, and this symbiotic relationship plays a key role in coral physiology and the primary productivity of coral reefs. The occurrence of *Symbiodinium* in corals and copepods may be beneficial to both hosts. This close association of three groups of organisms may represent a special case of a three-part symbiosis contributing to high species diversity and primary productivity of coral reefs.

A Potential Bioindicator of Coral Bleaching and Coral Disease Reservoir

It has been suggested that coral-associated copepods may consume and retain endosymbiotic zooxanthellae (*Symbiodinium* spp.) from their coral host for further photosynthesis since the existence of zooxanthellae in their gut is beneficial to their survival (see above). Thus, the occurrence of copepods might be controlled by zooxanthellae density or by the health of their coral hosts (zooxanthellae density is used to evaluate the health of corals). Additionally, as in other parasites, the virulence (infection ability) of copepods may also respond to the physiology of their hosts as well as to the abiotic milieu. The first study to document the dynamics of copepod infection over space and time in the model reef-building coral *Pocillopora damicornis* was carried out to estimate potential roles of copepods during coral bleaching (Cheng et al. 2016). Infestation parameters (prevalence, intensity, abundance, and density) and biodiversity of infected copepods of many coral colonies (n = 480) were documented in southern Taiwan from July 2007 (with a big coral bleaching event)

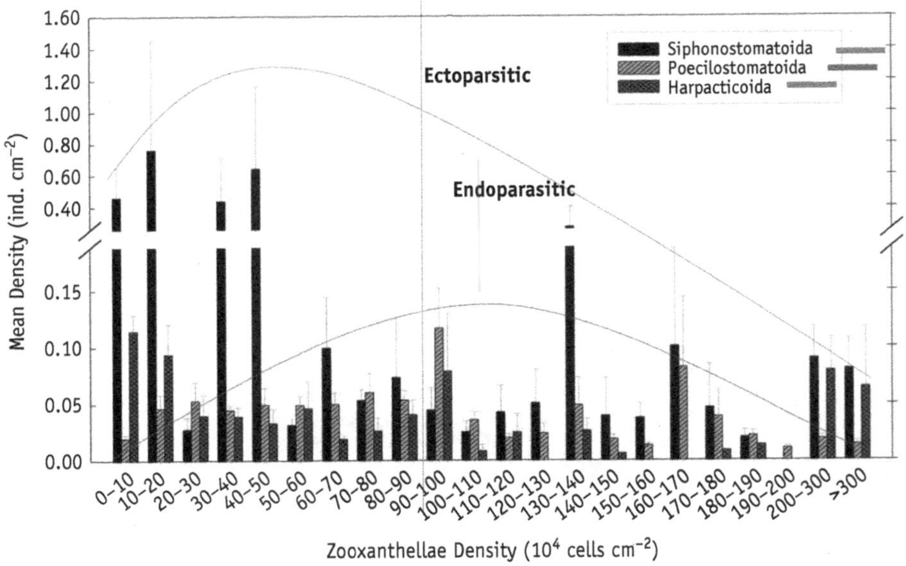

Fig. 8.11.
The mean density of three orders of coral-associated copepods (Harpacticoida, Poecilostomatoida, and Siphonostomatoida) collected from 480 *Pocillopora damicornis* colonies with various zooxanthellae density. The dotted lines indicate the threshold of coral bleaching (zooxanthellae density approximately 100 × 10^4 cells cm^{-2}).

to November 2008. The results provided a clear picture of the response of copepods to bleaching (Fig. 8.11). The highest infestation occurred in the summer of 2007 when the corals suffered from a severe bleaching event, and this was possibly related to the weakening of the defense mechanisms of the corals due to prolonged exposure to thermal stress. Although the results were inconsistent between different copepod taxa, virulence may have been related to feeding strategies (inferred from mandible structures) and behavior (ectoparasitic, endoparasitic, or benthic), as well as to surface area (capacity) of the host coral colonies, and concentration of nitrate and chlorophyll-*a* in the surrounding seawater. Thus, the potential to use coral-associated copepods as bioindicators for predicting the physiological performance of coral hosts in response to environmental change may be further explored and elaborated. Moreover, two *Vibrio*-like isolates (*Vibrio harveyi* and *Photobacterium rosenbergii*) were obtained from copepods parasitic in bleached corals (unpublished data), suggesting copepods might play the role as a disease reservoir and their impacts on corals should not be ignored. Further studies on the pathology and epidemiology are necessary for the conservation and management of coral reef ecosystems.

Dispersal and Possible Impacts on Coral Larvae

Dispersal strategies of coral-associated copepods remain uncertain. They can be found even at early life stages of the corals such as on bail-out polyps (individual polyps detaching from a coral colony under environmental stress) and primary polyps (Fig. 8.12). The infected bail-out polyp appears as an odd, outside shape with a swelling in lateral view and can be killed suddenly by these copepods. Undoubtedly, the ability of long-distance dispersal by larvae has played a major role in determining distribution. In the field, planulae and bail-out polpys from coral colonies typically settle and

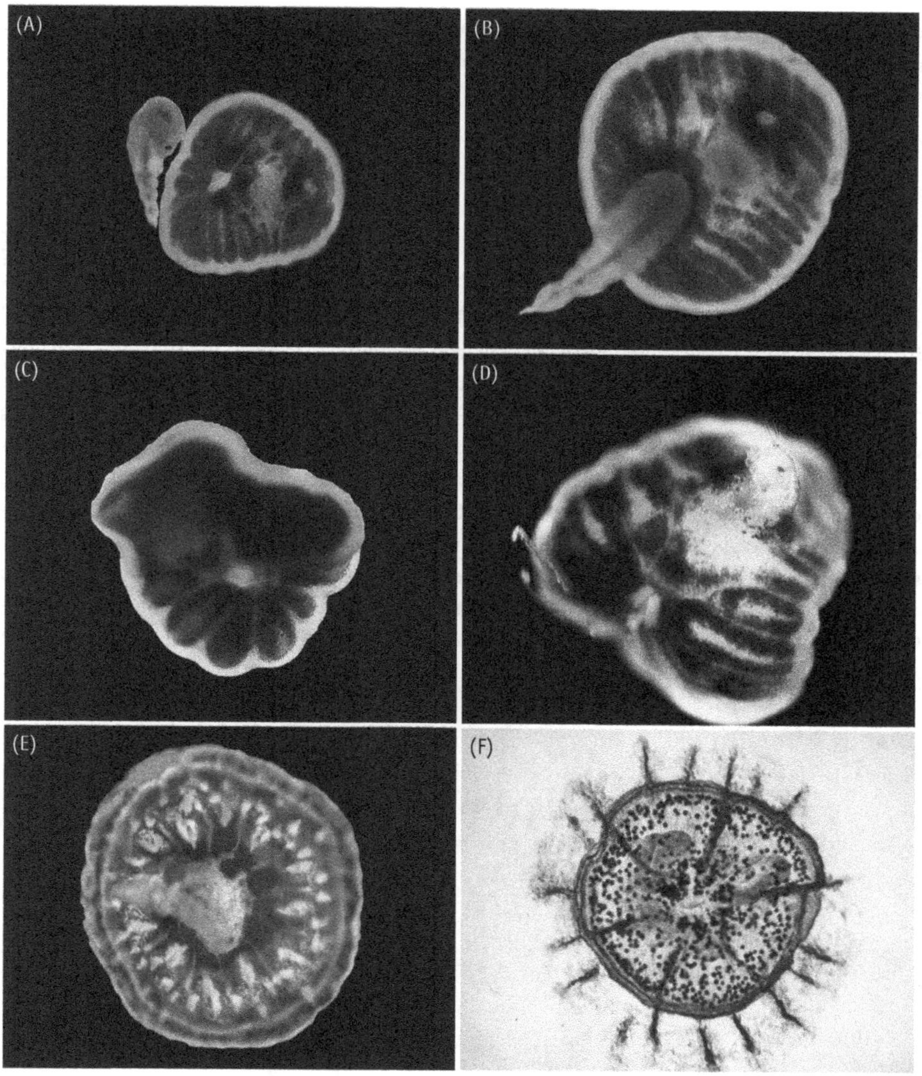

Fig. 8.12.
The reintroducing process of coral-associated copepods to bail-out polyps (a–d) and primary polyp (e, f). (a) Copepod was starting to look for polyps. (b) It tries to hide in the gastrovascular cavity of the polyp via an oral opening when approaching a coral polyp. (c) The infected bail-out polyp appeared an odd, outside shape with a swell in the lateral view. (d) Copepod was tearing the polyp suddenly and wanting to leave when we added a drop of 95% alcohol. (e) Copepod was starting to hide in the newly primary polyp. (f) Copepod infection might influence the development and skeleton forming of primary polyp of coral host. See color version of this figure in centerfold.

metamorphose within a relatively brief period, approximately two to nine days (Sammarco 1982, Glynn et al. 2008). In some cases, however, long-term dispersal planulae have occurred. Planulae can metamorphose and aggregate in clusters of several polyps without previous settlement on benthic substrata and remain alive for at least six months (Mizrahi et al. 2014). If so, in addition to the typical dispersal strategy of copepods (nauplius and copepodid larvae), infection of bail-out polyps or aggregated planulae might be a potential way for wide copepod dispersal. The copepod, however, might damage coral planulae and primary polyps, or at least influence their development, settlement, metamorphosis, and skeleton formation.

GALL CRABS

The number of species of decapod crustaceans found in coral-reef habitats (Patton 1966, Haig 1976) is fewer than that of barnacles and copepods. Among brachyuran crabs (Garth 1964, Serène 1972, Castro 1976) are gall crabs, the Cryptochiridae. They number 52 species in 21 genera (updated from Ng et al. 2008). Gall crabs, like their coral hosts, are found in tropical and subtropical shallow coral reefs in both the Indo-West Pacific and Atlantic, plus in the deep sea. The common name "gall crab" owes its origin to the domiciles of *Hapalocarcinus marsupialis* (Fig. 8.13A), which is composed of branches of *Pocillopora* corals modified into a chamber by two flattened hemispheres (Potts 1915, Hiro 1937). Domiciles of gall crabs vary in size according to that of its occupant, depending on the species and on its coral host. It takes the form of shallow depressions with circular or crescentic openings (Figs 8.14A, D), cylindrical chambers excavated perpendicular to the surface of the host (Fig. 8.14C), or tunnels (Wei et al. 2013). Females are more or less sedentary in excavated pits or extended galls with their abdomens considerably extended as a pouch-like structure for retaining eggs (Figs. 8.13A–E) and are often much larger than males (Figs. 8.13E, F), which are free-living and visit females for mating (Asakura 2009, Van der Meij 2014b).

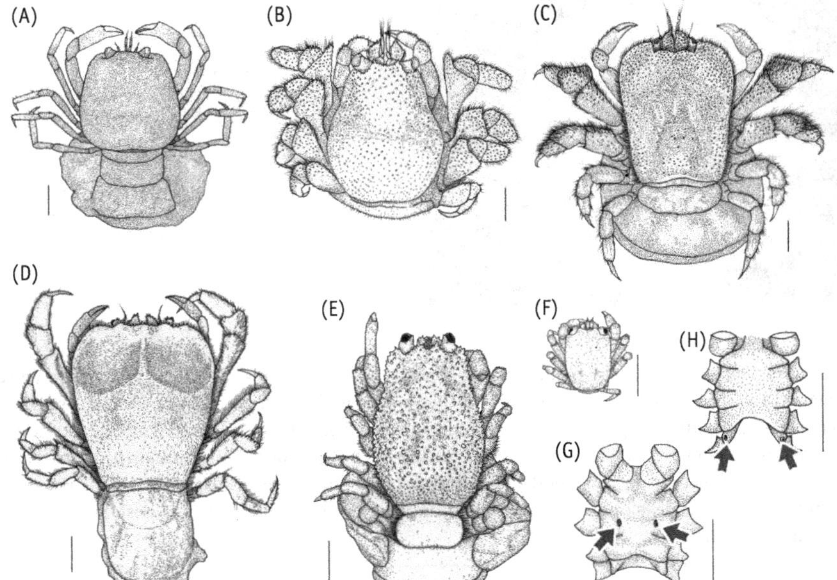

Fig. 8.13.
Diversity of gall crabs. (A) *Hapalocarcinus marsupialis.* (B) *Pseudocryptochirus viridis.* (C) *Fungicola fagei.* (D) *Fizesereneia heimi.* (E–H) *Utinomiella dimorpha.* A–E, female; F, male; G, female thoracic sternum; H, male thoracic sternum. Scale bars: A–C, E–G, 1 mm; D, 2 mm; H, 0.5 mm. In G and H, position of gonopores are indicated in grey arrows. A, C, D redrawn from Fize and Serène (1957); B from Utinomi (1944); E and F from Takeda and Tamura (1981); G and H from Henderson (1906).

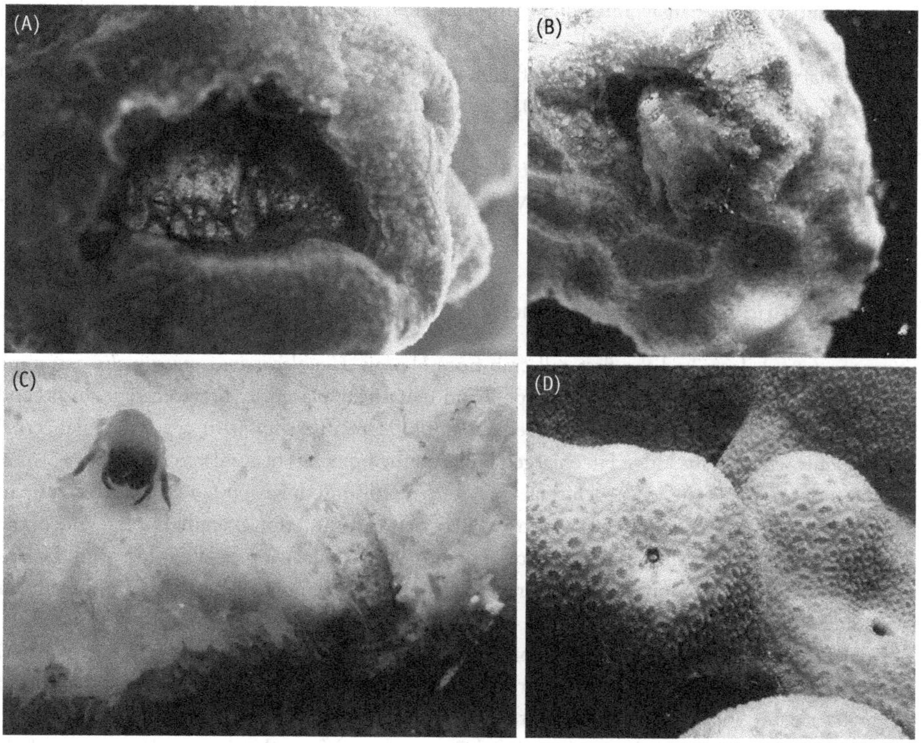

Fig. 8.14.
Gall crabs in domicile. (A) *Pseudocryptochirus viridis* dwelling in a shallow crescent-shaped crevice. (B) *Dacryomaia edmonsonsi* in coral *Psammocora* sp. (C) Vertical cylindrical domicile and its occupant, *Cryptochirus* or *Lithoscaptus* sp. (D) Round domicile openings of the previous in *Favia* sp. See color version of this figure in centerfold.

The third maxillipeds of cryptochirids are separated and barely cover the buccal cavity when closed, having a pronounced pronouncedly produced ischium with inner lobe convex, and merus slender as distal segments (Potts 1915), forming a four-segmented "palp". This has led to the conclusion that these buccal appendages together act as a sieve to filter plankton from the respiratory current (Hiro 1937, Marshall and Orr 1960, Castro 1976, McLaughlin 1982). However, Kropp (1986) found that *Hapalocarcinus marsupialis, Utinomia dimorpha,* and *Favicola rugosa* (formerly *Cryptochirus coralliodytes*) pick debris, coral tissue, and mucus flocks with their chelipeds from the surfaces of their coral hosts and do not filter feed. Scleractinian corals continuously secrete mucus for repelling sediments (Abe 1938, Dallmeyer et al. 1982), and the ingestion of mucus does not constitute a metabolic drain or detrimental effect on the host (Kropp 1986). The relationship between gall crabs and the host is therefore considered a mutualistic symbiosis. Understanding the association is complicated by some gall crabs (*Troglocarcinus corallicola*) maintaining and feeding on "pruned gardens" of filamentous algae around their burrow opening (Carricart-Ganivet et al. 2004). Conversely, Dallmeyer et al. (1982) concluded that the release of mucus as a clearing behavior inflicts costs by the reduction of chlorophyll content in coral tissue and the loss of zooxanththellae. Gall crabs modify the morphology of their hosts (Abelson et al. 1991), inhibiting growth rates (Simon-Blecher and Achituv 1997, Nogueira et al. 2014), perforating and weakening the skeletal structure, and fostering settlement of algae and fungi (Simon-Blencher et al. 1999). As such, a possible parasitic nature of the association can be inferred.

Regional intensive surveys of coral reef habitats for gall crabs have been limited to reports from the Red Sea, around Nhatrang (Vietnam), Orchid Island (Taiwan), the Ryukyu Archipelago and Japan, the Central Pacific, and the Atlantic. Numerous current nominal species are probable species complexes (e.g., *Hapalocarcinus marsupialis*), and some genera include known undescribed species (e.g., *Lithoscaptus, Opecarcinus*) (Van der Meij et al. 2015a). The diversity of the group therefore remains underreported.

Phylogeny, Origins, and Host Specificity

Stimpson (1859) described the first gall crab from Hawaii, *Hapalocarcinus marsupialis*, assuming that they "probably feed upon coral-polypes" and tentatively placed it "between *Pinnotheres* and *Hymenosoma*." Verrill (1867) first reported a symbiotic relationship with a scleractinian coral (*Pocillopora*) where crabs lodged in enlarged bulbs on the branches of corals.

The monophyly of the Cryptochiridae has not been disputed. Kropp (1988b) listed a suite of morphological features, including the gastric mill ossicles, the anterior-extended lateral lobe of the antennule, the anterior extension on the coxae of the last two pairs of ambulatory legs, and the lack of mandibular palp. Monophyly has been recently confirmed by molecular investigations (Van der Meij and Schubart 2014).

The most recent common ancestor of gall crabs emerged in the Cretaceous (108 ± 11 ma) (Tsang et al. 2014b). However, given their small size and poorly calcified bodies, fossil records of gall crabs are few (Glaessner 1969, De Grave et al. 2009, Jagt et al. 2015). Recent descriptions of *Montemagrechirus tethysianus* by De Angeli and Ceccon (2015) from the Eocene (Ypersian) in Italy reported to represent the first cryptochirid fossil record, which was later placed in a new family, Montemagrechiridae, by Klompmaker et al. (2016). These authors reported trace fossil evidence of gall crabs in fossils of five species of scleractinian corals from the upper Pliocene to the upper Pleistocene from Cuba and Florida, all bearing crescent-shaped pits, resembling those of extant *Troglocarcinus corallicola* and other species in the Indo-West Pacific.

The systematic position of the gall crabs has been contentious. Any resemblance to the anomuran crab genus *Porcellana* would appear to be a convergent adaption for a filter-feeding diet (but see Kropp 1986 and the discussion above). Various placements of the family within the Brachyura have been suggested (Borradaile 1907, Utinomi 1944, Balss 1957, Števčić 1971, Sakai 1976).

On the evidence of zoea and megalopa anatomy, Fize (1956) found similarities with pinnotherids but also grapsids, leucosiids, and *Calappa*. Scotto and Gore (1981) found zoea to resemble those of Pinnotheridae, Leucosiidae, and Hymenosomatidae. These families are strongly separated in the most widely accepted classifications of Brachyura (Guinot 1979, 2011). Spermatozoal evidence is equivocal (Jamieson and Tudge 2000, Tudge et al. 2014).

Molecular phylogenetics showed *Hapalocarcinus marsupialis* to have affinities to genera of Grapsoidea (Wetzer et al. 2009). The affiliation with thoracotreme crabs was confirmed by Tsang et al. (2014b), using only one species, and by Van de Meij and Schubart (2014), who included 10 species from 9 genera of gall crabs.

Kropp (1988b) recognized four groups of genera based on morphological cladistic analyses: (1) *Hapalocarcinus*, (2) *Pseudohapalocarcinus*, (3) the clade "detocarcini" (*Cecidocarcinus, Detocarcinus, Neotroglocarcinus, Pseudocryptochirus*, and *Utinomia*), and (4) "cryptochirini" (*Cryptochirus, Dacryomaia, Fizesereneia, Fungicola, Hiroia, Lithoscaptus, Opecarcinus, Pelycomaia, Troglocarcinus*, and *Xynomaia*). Recent molecular studies derived by mtDNA markers 16S and COI showed both "detocarcini" and "cryptochirini" to be artificial subdivisions (Van der Meij and Reijnen 2014).

In recent investigations on phylogenetic relationships and diversification of the group, a chronogram derived from the mtDNA 16S and COI, and nDNA H3 (Van der Meij and Klaus 2015) provided insights, as did a phylogenetic reconstruction based on COI, with the addition of "ancient DNA" obtained from museum specimens (Van der Meij and Nieman 2016) (reproduced as Fig. 8.15). Both studies showed *Fungicola, Lithoscaptus,* and *Xynomaia* to be non-monophyletic genera (Fig. 8.15), and the following three subclades of somewhat consistent host specificity are better supported: (1) *Opecarcinus* and *Pseudohapalocarcinus,* both symbionts of the Agariciidae and Siderastreidae; (2) Pacific genera *Neotroglocarcinus* and *Pseudocryptochirus,* both symbionts of *Turbinaria* species (the Dendrophylliidae); and (3) association of multiple Pacific genera, all diverging and intensely radiating during the Miocene (Van der Meij and Klaus 2015), all symbiotic with scleractinian corals of the "Robust Clade" families (Romano and Palumbi 1996, 1997, Romano and Cairns 2000).

Several genera appear to occupy basal positions to these three subclades—namely (1) *Troglocarcinus,* with trans-Atlantic *corallicola* displaying an extremely broad range of hosts (Kropp and Manning 1987), (2) currently monospecific trans-Indo-Pacific *Hapalocarcinus,* symbiotic with

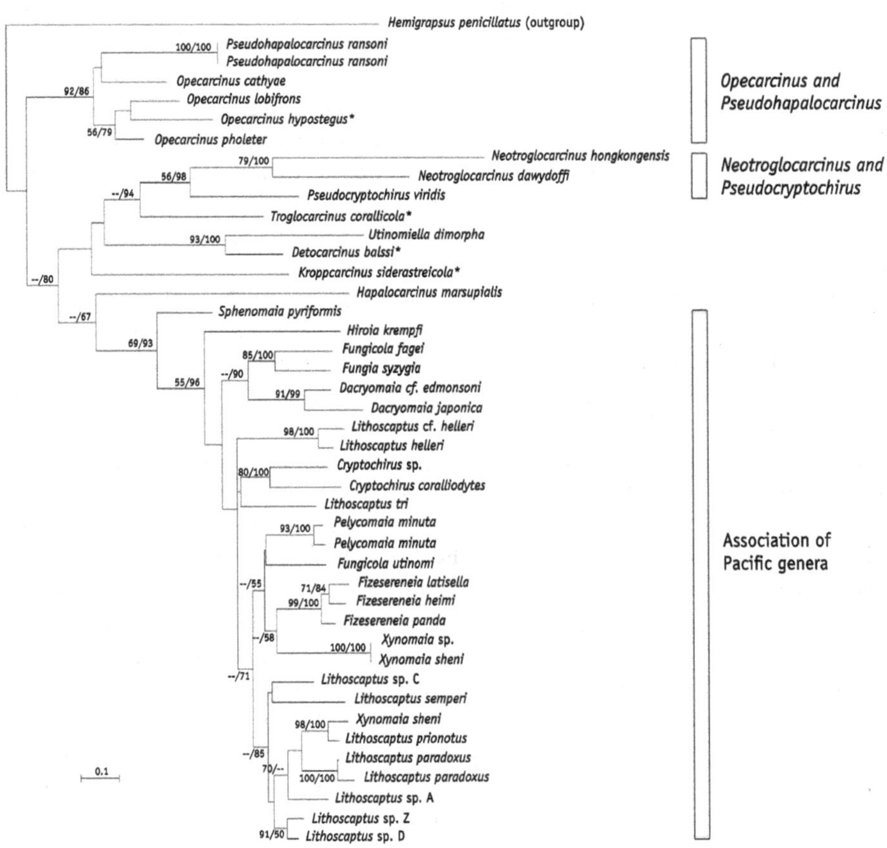

Fig. 8.15.
Phylogenetic reconstruction of the Cryptochiridae, based on mtDNA COI sequences, reproduced and slightly modified from Fig. 1 in Van der Meij and Nieman (2016). Atlantic species are indicated with an asterisk (*). Support values represent maximum likelihood/Baysian Inferences. Indications of sequence number and "ancient DNA" (sequences obtained from museum material) in the original work are omitted for simplicity.

the Pocilloporidae, and (3) western Atlantic *Kroppcarcinus*, symbiotic with *Siderastrea*—yet exact phylogenetic placements remain inconclusive (Van der Meij and Klaus 2015, Meij and Nieman 2016, Fig. 8.15).

General Observations on Biogeography and Vertical Distribution

All gall crab species are restricted to tropical and warm subtropical waters, with most (46) species distributed in the Indo-West Pacific. None of the 52 recognized species co-occurs in both regions. At a generic level, only *Opecarcinus* (see below) and *Cecidocarcinus* occur in both the Indo-West Pacific and the Atlantic-East Pacific. Among the nine known species of *Opecarcinus*, otherwise Pacific, only one (*hypostegus*) occurs in the Gulf of Mexico and Caribbean, very probably as a result of allopatric separation during closure of the Isthmus of Panama c. 2.8 mya (Van der Meij and Klaus 2015). *Cecidocarcinus* comprises two deep-sea species (see below) and morphologically resembles *Detocarcinus, Neotroglocarcinus,* and *Pseudocryptochirus* (Kropp and Manning 1987), but until molecular evidence is presented, its true affinities remains obscure.

Patterns follow that of coral reefs with a noticeable hiatus in the Bay of Bengal and the Arabian Sea, where extensive coral reefs are scarce (Veron 2000). Several species appear to be exceptionally widespread, spanning the western Indian Ocean (including the Red Sea) to the Central Pacific, e.g., *Lithoscaptus paradoxus* (Kropp 1988a), *Opecarcinus lobifrons* (Van der Meij et al.2015a; Fig. 8.16), and *Hapalocarcinus marsupialis.* The latter may well be a species complex (Van der Meij et al. 2015a). A few of the Indo-West Pacific species reach the East Pacific, namely *O. crescentus, O. lobifrons* (Fig. 8.16), and *H. marsupialis.* A number of species present in the northern portion of the tropical West Pacific are also found along the Japanese main islands, whereas two species, namely *Cryptochirus planus* and *Lithoscaptus grandis,* however, appear to be restricted to the Pacific coast of Honshu, Japan (Takeda and Tamura 1983, Kropp 1990). Among the remaining six Atlantic species with no apparent phylogenetic affinities, three, *Detocarcinus balssi, Opecarcinus hypostegus* and *Troglocarcinus corallicola,* are distributed across the Atlantic, while the remainder are restricted to confined distributions. Atlantic species, particularly *T. corallicola* and *D. balssi,* display a much broader or less strict host specificity than those of the Indo-West Pacific (Kropp and Manning 1987, Van der Meij 2014b).

Apart from some deep-water species, namely *Cecidocarcinus brychius* (512 m; host Dendrophylliidae), *C. zibrowii* (425–440 m; host Dendrophylliidae), *Luciades agana* (34–137 m; host Agariciidae), and *Zibrovia galea* (81–100 m; host Caryophylliidae) (Kropp and Manning 1987, 1996, Manning 1991), most species are restricted to the photic zone with their scleractinian hosts, typically from intertidal habitats to less than 30 m depth. Several species have been recorded as exceeding this depth: *Detocarcinus balssi* up to depths of 98 m (Kropp and Manning 1987), *Opecarcinus granulatus* to 82 m (Kropp 1989), and *O. hypostegus* to 60 m (Van der Meij et al. 2015b). A recent observation of *O. hypostegus* off Curaçao, southern Caribbean, reaching 60 m to the mesophotic zone by a manned submersible (Van der Meij et al. 2015b), suggests that reported depth limits might be artificial, possibly biased by maximum depth of SCUBA observations. Mesophotic reefs very likely accommodate more gall crabs than presently realized (Bongaerts et al. 2010, Van der Meij et al. 2015b). Moreover, considering 92% of the species of coral of the Caryophylliidae and 78% of Dendrophylliidae have been recorded deeper than 50 m (Roberts et al. 2009), the gall crab fauna associated with deep-water hosts may be unexplored.

FUTURE DIRECTIONS

At present there are no detailed studies on variations in the host usage of coral-associated invertebrates such as barnacles, copepods, and gall crabs along geographical regions in the

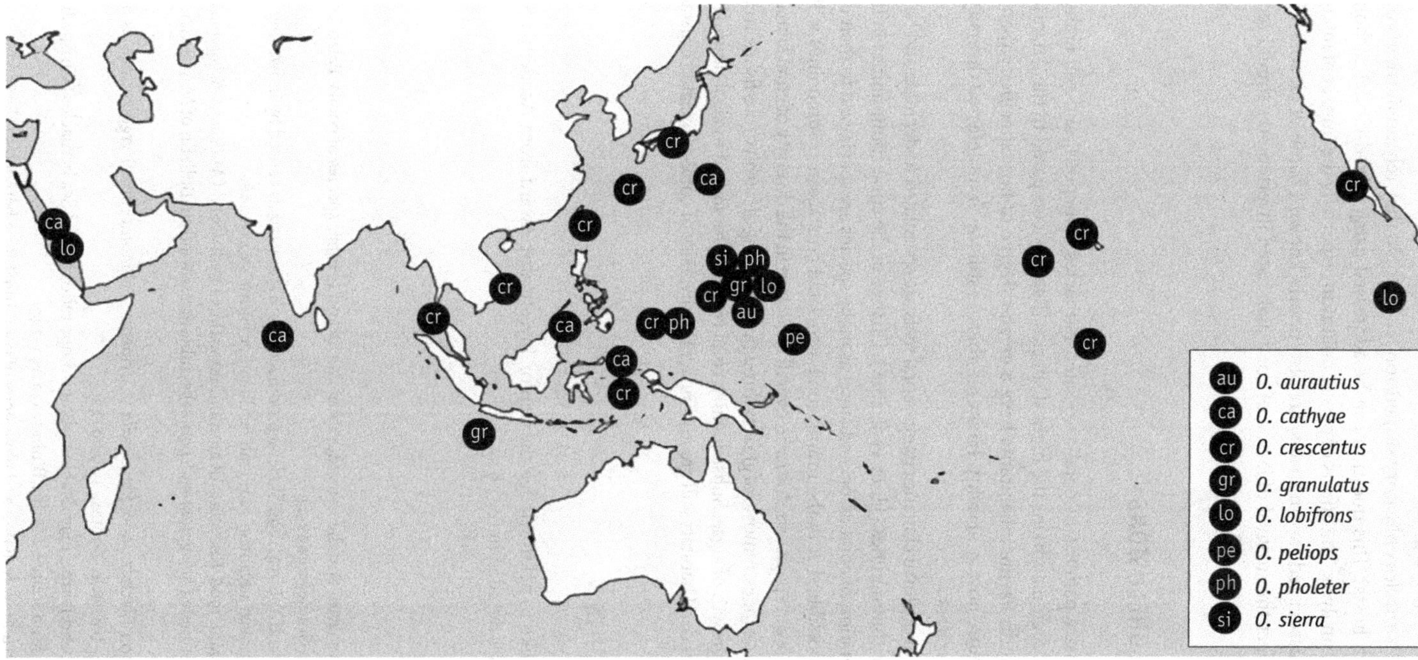

Fig. 8.16.
Distribution of Indo-Pacific *Opecarcinus* species. Note variations in biogeographical patterns: across the Indo-Pacific (*lobifrons*), Central Indian Ocean to West Pacific (*cathyae*), West to East Pacific (*crescentus*), and those so far from restricted regions (*aurautius, granulatus, peliops, pholeter*, and *sierra*). From Kropp 1989, 1990, Van der Meij 2014a.

Indo-Pacific. Given different choices of coral hosts, do these invertebrates use different hosts in different region? The feeding ecology of coral-associated invertebrates has received very little attension. Are coral-associated barnacles obligatory suspension feeders? Do they rely on nutrients or organic matter supplied from hosts? This question can be addressed using mixed-model carbon and nitrogen stable isotope analysis. The effects of plastic pollution, especially the impact of microplastics on corals and their associated inverterbates, are also poorly known (Lamb et al. 2018). Whether microplastics can affect the survival of larvae of coral-associated invertebrates should receive further attention.

SUMMARY AND CONCLUSIONS

Reviewing the diversity patterns of coral-associated barnacles, copepods, and gall crabs, we see that the Indo-Pacific region contains the highest species diversity compared to the Atlantic Ocean. It appears that the Indo-Pacific inhabitants have a greater degree of host specificity than Atlantic populations, given more host availability. However, there is still much undetected diversity in both Pacific and Atlantic waters.

Under the scenario of global climatic change, coral reefs are speculated to decrease and be replaced by other warm-water communities (Hoey et al, 2016). Change in the coral communities in the future may result in the alteration of coral-associated fauna as many species are specialized to their coral host species. It could be possible that in the future, coral-associated crustacean communities will include more generalists and fewer specialists due to an overall reduction in the diversity of coral fauna. In addition to global climatic changes, removal of grazers through overfishing ofherbivorous fishes in coral reefs may result in an overgrowth of algae. Subsequently, coral reefs will be shifted to algal-dominated reefs, and the destruction of coral habitats will greatly impacr the diversity of coral-associated crustaceans.

ACKNOWLEDGMENTS

The authors would like to acknowledge the Senior Investigator Award from Academia Sinica for supporting the studies detailed here.

REFERENCES

Abe, N. 1938. Feeding behavior and the nematocysts of *Fungia* and 15 other species of corals. Palao Tropical Biological Station Studies 1:469–521.

Abelson, A., B.S. Galil, and Y. Loya. 1991. Skeletal modifications in stony corals caused by indwelling crabs: hydrodynamical advantages for crab feeding. Symbiosis 10:233–248.

Achituv, Y., and L. Mizrahi. 1996. Recycling of ammonium within a hydrocoral (*Millepora dichotoma*)—zooxanthellae-cirripede (*Savignium milleporum*) symbiotic association. Bulletin of Marine Science 58:856–860.

Achituv, Y., and N. Simon-Blecher. 2006. *Pyrgopsella* (Cirripedia: Balanomorpha: Pyrgomatidae) is not a sponge-inhabiting barnacle. Zootaxa 1319:29–42.

Achituv, Y., I. Brickner, and J. Erez. 1997. Stable carbon isotope ratios in Red Sea barnacles (Cirripedia) as an indicator of their food source. Marine Biology 130:243–247.

Anderson, D.T. 1992. Structure, function and phylogeny of coral-inhabiting barnacles (Cirripedia, Balanoidea). Zoological Journal of the Linnean Society 106:277–339.

Anderson, D.T. 1994. Barnacles: Structure, Function, Development and Evolution. Chapman and Hall, London.

Asakura, A. 2009. The evolution of mating systems in decapod crustaceans. Pages 121–182 *in* J.W. Martin, K.A. Crandall, and D.L. Felder, editors. Decapod Crustacean Phylogenetics. Crustacean Issues. CRC Press, Boca Raton, FL.

Asami, K., and T. Yamaguchi. 1997. Distribution of living and fossil coral barnacles (Cirripedia; Pyrgomatidae) in Japan. Sessile Organisms 14:9–16.

Bacon, P.R. 1976. The Cirripedia of Trinidad. Studies on the Fauna of Curaçao and other Caribbean Islands 50:1–55.

Balss, H. 1957. Decapoda. Pages 1505–1672 *in* H.G. Bronns, editor. Klassen und Ordnungen des Tierreichs, Band 5, Abteilung 1, Buch 7. Leipzig and Heidelberg, Winter.

Baluk, W., and A. Radwanski. 1967a. Genus *Creusia* Leach, 1817, (Cirripedia) in the Croatian Miocene. Geoloski Vjesnik Beograd 20:137–152.

Bongaerts, P., T. Ridgeway, E.M. Sampayo, and O. Hoegh-Guldberg. 2010. Assessing the "deep reef refugia" hypothesis: focus on Caribbean reefs. Coral Reefs 29:309–327.

Borradaile, L.A. 1907. On the classification of the decapod crustaceans. Annals and Magazine of Natural History, Series 7 19:457–486.

Brooks, H.K., and A. Ross. 1960. *Pyrgoma prefloridanum*, a new species of cirriped from the Caloosahatchee Marl (Pleistocene) of Florida. Crustaceana 1:353–365.

Carricart-Ganivet, J.P., L.F. Carrea-Parra, L.I. Quan-Young, and M.S. García-Madrigal. 2004. Ecological note on *Troglocarcinus corallicola* (Brachyura: Cryptocchiridae) living in symbiosis with *Manicina areolata* (Cnidaria: Scleractinia) in the Mexico Caribbean. Coral Reefs 23:215–217.

Castro, P. 1976. Brachyuran crabs symbiotic with scleractinian corals. Marine Biology 46:237–245.

Castro, P. 1988. Animal symbioses in coral reef communities: a review. Symbiosis 5:161–184.

Chan, B.K.K., R.E. Prabowo, and K.S. Lee. 2009. Crustacean Fauna of Taiwan: Barnacles, Vol. 1—Cirripedia: Thoracica Excluding the Pyrgomatidae and Acastinae. National Taiwan Ocean University Press, Taiwan.

Chan, B.K.K., Y.Y. Chen, and Y. Achituv. 2013a. Crustacean Fauna of Taiwan: Barnacles, Vol. 2—Cirripedia: Thoracica: Pyrgomatidae. Biodiversity Research Center, Academia Sinica Press, Taiwan.

Chan, B. K. K., Y.Y. Chen, and H.C. Lin. 2013b. Biodiversity and host specificity of coral barnacles of *Galkinia* (Cirripedia: Pyrgomatidae) in Taiwan, with descriptions of six new species. Journal of Crustacean Biology 33:392–431.

Chan, B.K.K., W.P. Hsieh, and G.A. Kolbasov. 2014a. Crustacean Fauna of Taiwan: Barnacles, Vol. 3—Cirripedia: Acrothoracica. Biodiversity Research Center, Academia Sinica Press, Taiwan.

Chan, B.K.K., G.A. Kolbasov, M. Hirose, T. Mezaki, and R. Suwae. 2014b. Biodiversity and biogeography of the coral boring barnacles of the genus *Berndtia* (Cirripedia: Acrothoracica) in the West Pacific, with description of three new species. Journal of Natural History 48:1503–1541.

Cheng, Y.R. 2011. Systematics and ecology of Xarifiid Copepods associated with Scleractinian corals in the surrounding seas of Taiwan. Doctoral dissertation, Institute of Oceanography, National Taiwan University, Taiwan.

Cheng, Y.R., and C.F. Dai. 2009. The infection process of a parasitic copepod, *Xarifia obesa*, with *Stylophora pistillata*. Coral Reefs 28:681.

Cheng, Y.R., and C.F. Dai. 2010. Endosymbiotic copepods may feed on zooxanthellae from their coral host, *Pocillopora damicronis*. Coral Reefs 29:13–18.

Cheng, Y.R., C.F. Dai, and W.B. Chang. 2010. Two species of gall-inducing poecilostomatoid copepods on the scleractinian coral, *Montipora aequituberculata*. Journal of Fisheries Society of Taiwan 37:273–284.

Cheng, Y.R., J.S. Ho, and C.F. Dai. 2012. Species diversity of xarifiid copepods associated with scleractinian corals from Taiwan. Pages 113–121 *in* Proceedings of the 1st Mainland and Taiwan Symposium of Marine Biodiversity Studies. Ocean Press, Beijing, China.

Cheng, Y.R., B.M. Anderson, P.J. Meng, C.F. Dai, and R. Huys. 2016. Copepods associated with scleractinian corals: a worldwide checklist and a case study of their impact on the reef-building coral *Pocillopora damicornis* (Linnaeus, 1758) (Pocilloporidae). Zootaxa 4174:291–345.

Dai, C.F., and S. Horng. 2009. Scleractinia of Taiwan I. The Complex Group. National Taiwan University, Taiwan.

Dallmeyer, D.G., J.W. Porter, and G.J. Smith. 1982. Effects of particulate peat on the behavior and physiology of the Jamaican reef-building coral *Montastrea annularis*. Marine Biology 68:229–233.

De Angeli, A., and L. Ceccon. 2015. Nuovi crostacei brachiuri dell'Eocene di Monte Magrè (Vicenza, Italia settentrionale). Società Veneziana di Scienze Naturali 40:119–138.

De Grave, S., N.D. Pentcheff, S.T. Ahyong, T.Y. Chan, K.A. Crandall, P.C. Dworschak, D.L. Felder, R.M. Feldmann, C.H.J.M. Fransen, L.Y.D. Goulding, R. Lemaitre, M.E.Y. Low, J.W. Martin, P.K.L. Ng, C.E. Schweitzer, S.H. Tan, D. Tshudy, and R. Wetzer. 2009. A classification of living and fossil genera of decapod crustaceans. The Raffles Bulletin of Zoology Supplement 21:1–109.

Dojiri, M. 1988. Isomolgus desmotes, new genus, new species (Lichomolgidae), a gallicolous Poecilostome copepod from the scleractinian coral *Seriatopora hystrix* Dana in Indonesia, with a review of gall-inhabiting crustaceans of anthozoans. Journal of Crustacean Biology 8:99–109.

Dojiri, M., and M.J. Grygier.1990. *Pionomolgus gallicolus*, gen et sp. nov. (Poecilostomatoida, Lichomolgidae), a gall inducing copepod of a scleractinian coral from Australia. Australian Journal of Zoology 37:695–703.

Fize, A. 1956. Observations biologiques sur les hapalocarcinidés. Annales de la Faculté de Science Université National du Viêt Nam, Institut Océanographique de Nhatrang 22:1–30.

Fize, A., and R. Serène. 1957. Les Hapalocarcinidés du Việt-Nam. Mémoires de l'Institut Océanographique de Nhatrang 10:1–202.

Garth, J.S. 1964. The crustacea decapoda (Brachyura and Anomura) of Eniwetok Atoll, Marshall Islands, with special reference to the oblique commensals of branching corals. Micronesica 1:137–144.

Glaessner, M.F. 1969. Decapoda. Pages 217–340 *in* R.C. Moore, editor. Treatise on Invertebrate Paleontology, Part R. Arthropoda 4(2). The University of Kansas Press and The Geological Society of America, Lawrence.

Glynn, P.W., S.B. Colley, J.L. Mate, J. Cortes, H.M. Guzman, R.L. Bailey, J.S. Feingold, and I.C. Enochs. 2008. Reproductive ecology of the azooxanthellate coral *Tubastraea coccinea* in the Equatorial Eastern Pacific: Part V. Dentrophylliidae. Marine Biology 153:529–544.

Goldberg, W.M. 2013. Biology of Reefs and Reef Organisms. The University of Chicago Press, Chicago.

Guinot, D. 1979. Données nouvelles sur la morphologie, la phylogenèse et la taxonomie des Crustacés Décapodes Brachyoures. Mémoires du Muséum national d'Histoire naturelle. Nouvelle Série. Série A, Zoologie 112:3–354.

Guinot, D. 2011. The position of the Hymenosomatidae MacLeay, 1838, within the Brachyura (Crustacea, Decapoda). Zootaxa 2890:40–52.

Haig, J. 1976. Decapod Crustacea of Pacific coral reefs. Micronesica 12:183–185.

Henderson, J.R. 1906. On a new species of coral-infesting crab taken by the R.I.M.S. "Investigator" at the Andaman Islands. The Annals and Magazine of Natural History, Series 7 18:211–219.

Hiro, F. 1935. A study of cirripeds associated with corals occurring in Tanabe Bay. Records of oceanographic works in Japan 7:1–28.

Hiro, F. 1937. Studies on the animals inhabiting reef corals. I. *Hapalocarcinus* and *Cryptochirus*. Palao Tropical Biological Station Studies 1:137–154.

Hoey, A.S., E. Howells, J.L., Johansen, J.-P. A. Hobbs, V. Messmer, D. M. McCowan, S. K. Wilson, and M. S. Pratchett. 2016. Recent advances in understanding the effects of climate change on coral reefs. Diversity 8:12.

Humes, A.G. 1985a. Cnidarians and copepods: a success story. Transactions of the American Microscopical Society 104:313–320.

Humes, A.G. 1985b. A review of the Xarifiidae (Copepoda, Poecilostomatoida), parasites of scleractinian corals in the Indo-Pacific. Bulletin of Marine Science 36:467–632.

Huys, R., J. Llewellyn-Hughes, P.D. Olson, and K. Nagasawa. 2006. Small subunit rDNA and Bayesian inference reveal *Pectenophilus ornatus* (Copepoda incertae sedis) as highly transformed Mytilicolidae, and support assignment of Chondracanthidae and Xarifiidae to Lichomolgoidea (Cyclopoida). Biological Journal of the Linnean Society 87:403–425.

Huys, R., J. Mackenzie-Dodds, and J. Llewellyn-Hughes. 2009. Cancrincolidae (Copepoda, Harpacticoida) associated with land crabs: a semiterrestrial leaf of the ameirid tree. Molecular Phylogenetics and Evolution 51:143–156.

Huys, R., F. Fatih, S. Ohtsuka, and J. Llewellyn-Hughes. 2012. Evolution of the bomolochiform superfamily complex (Copepoda: Cyclopoida): new insights from ssrDNA and morphology, and origin of umazuracolids from polychaete-infesting ancestors rejected. International Journal for Parasitology 42:71–92.

Jagt, J.W.M., B.W.M. Bakel van, D. Guinot, R.H.B. Fraaije, and P. Artal. 2015. Fossil brachyura. Pages 847–920 *in* P. Castro, P.J.F. Davie, D. Guinot, F.R. Schram, and J.C. von Vaupel Klein, editors. Treatise on Zoology—Anatomy, Taxonomy, Biology. The Crustacea, Vol. 9, Part C-II. Brill, Leiden, The Netherlands.

Jamieson, B.G.M., and C.C. Tudge. 2000. Crustacea—Decapoda. Pages 1–95 *in* B.G.M. Jamieson, editor. Reproductive Biology of Invertebrates. Vol. 9, Part C. Progress in Male Gamete Ultrastructure and Phylogeny. John Wiley, Chichester, UK.

Kim, I.H. 2011. Invertebrate Fauna of Korea, Vol. 21: Barnacles. National Institute of Biological Resources, Ministry of Environment, The Republic of Korea.

Kim, I.H., and H. Yamashiro. 2007. Two species of poecilostomatoid copepods inhabiting galls on scleractinian corals in Okinawa, Japan. Journal of Crustacean Biology 27:319–326.

Klompmaker, A.A., R.W. Portell, and S.E.T. van der Meij. 2016. Trace fossil evidence of coral-inhabiting crabs (Cryptochiridae) and its implications for growth and palaeobiogeography. Scientific Reports 6:23443.

Kropp, R.K. 1986. Feeding biology and mouthpart morphology of three species of coral gall crabs (Decapoda: Cryptochiridae). Journal of Crustacean Biology 6:377–384.

Kropp, R.K. 1988a. The status of *Cryptochirus coralliodytes* Heller and *Lithoscapthus paradoxus* Milne Edwards (Brachyura: Cryptochiridae). Proceedings of the Biological Society of Washington 101:872–882.

Kropp, R.K. 1988b. Biology and systematics of coral gall crabs (Drustacea: Cryptochiridae). Ph.D. thesis, University of Maryland, Maryland.

Kropp, R.K. 1989. A revision of the Pacific species of gall crabs, genus *Opecarcinus* (Crustacea: Cryptochiridae). Bulletin of Marine Science 45:98–129.

Kropp, R.K. 1990. Revision of the genera of gall crabs (Crustacea: Cryptochiridae) occurring in the Pacific Ocean. Pacific Science 44:417–448.

Kropp, R.K., and R.B. Manning. 1987. The Atlantic gall crabs, family Cryptochiridae (Crustacea: Decapoda: Brachyura). Smithsonian Contributions to Zoology 462:1–21.

Kropp, R.K., and R.B. Manning. 1996. Crustacea Decapoda: two new genera and species of deep water gall crabs from the Indo-west Pacific (Cryptochiridae). Pages 531–539 *in* A. Crosnier, editor. Résultats des Campagnes MUSORSTOM, Vol. 15. Mémoires du Muséum National d'Histoire Naturelle, Paris.

Lamb, J. B., B.L. Willis, E.A. Fiorenza, C.S. Couch, R. Howard, R., D.N. Rader, and C.D. Harvell. 2018. Plastic waste associated with disease on coral reefs. Science 359:460–446.

Liu, R.Y., and X.Q. Ren. 2007. Fauna Sinica. Invertebrata. Vol. 42. Crustacea Cirripedia Thoracica. Science Press, Beijing, China.

Malay, M.C.M.D., and F. Michonneau. 2014. Phylogenetics and morphological evolution of coral-dwelling barnacles (Balanomorpha: Pyrgomatidae). Biological Journal of the Linnean Society, 113:162–179.

Manning, R.B. 1991. Crustacea Decapoda: *Cecidocarcinus zibrowii*, a new deep water gall crab (Cryptochiridae) from New Caledonia. Pages 515–520 *in* A. Crosnier, editor. Résultats des Campagnes MUSORSTOM, Vol. 9. Mémoires du Muséum National d'Histoire Naturelle, Paris.

Marshall, S.M., and A.P. Orr. 1960. Feeding and nutrition. Pages 227–258 *in* T. H. Waterman, editor. The Physiology of Crustacea, Vol. I. Metabolism and Growth. Academic Press, New York.

McLaughlin, P.A. 1982. Comparative morphology of crustacean appendages. Pages 197–256 *in* L.G. Abele, editor. The Biology of Crustacea, Vol. 2. Embryology, Morphology, and Genetics. Academic Press, New York.

Meij, S.E.T. van der. 2014a. A new species of *Opecarcinus* Kropp and Manning, 1987 (Crustacea: Brachyura: Cryptochiridae) association with the stony corals *Pavona clavus* (Dana, 1846) and *P. biparta* Nemenzo, 1980 (Scleractinia: Agariciidae). Zootaxa 3869:44–52.

Meij, S.E.T. van der. 2014b Host species, range extensions, and an observation of the mating system of Atlantic shallow-water gall crabs (Decapoda: Cryptoochiridae). Bulletin of Marine Science 90:1001–1010.

Meij, S.E.T. van der, and S. Klaus. 2015. Evolutionary diversification of coral-dwelling gall crabs (Cryptochiridae). Ph.D. thesis, Leiden University, The Netherlands.

Meij, S.E.T. van der, and A.M. Nieman. 2016. Old and new DNA unweave the phylogenetic position of the eastern Atlantic gall crab *Detocarcinus balssi* (Monod, 1956) (Decapoda: Cryptochiridae). Journal of Zoological Systematics and Evolutionary Research 54:189–196.

Meij, S.E.T. van der, and B.T. Reijnen. 2014. The curious case of *Neotroglocarcinus dawydoffi* (Decapoda, Cryptochiridae): unforeseen biogeographic patterns resulting from isolation. Systematics and Biodiversity 12:503–512.

Meij, S.E.T. van der, and C.D. Schubart. 2014. Monophyly and phylogenetic origin of the gall crab family Cryptochiridae (Decapoda: Brachyura). Invertebrate Systematics 28:491–500.

Meij, S.E.T. van der, M.L. Berumen, and G. Paulay. 2015a. The Red Sea and Arabia are a diversity and endemism hotspot for coral-dwelling gall crabs (Cryptochiridae). Pages 143–152 *in* S.E.T. van der Meij, Evolutionary diversification of coral-dwelling gall crabs (Cryptochiridae). Ph.D. thesis, Leiden University, The Netherlands.

Meij, S.E.T. van der, K.M. Tienderen van, and B.W. Hoeksema. 2015b. A mesophotic record of the gall crab *Opecarcinus hypostegus* from a Curaçaoan reef. Bulletin of Marine Science 91:205–206.

Mizrahi, D., S.A. Navarrete, and A.A.V. Flores. 2014. Groups travel further: pelagic metamorphosis and polyp clustering allow higher dispersal potential in sun coral propagules. Coral Reefs 33:443–448.

Moisette, P., and J.P. Saint Martin. 1990. Cirripedes *Pyrgomatidae actuels* et fossils d'Oranie (Algeria). Revue de Paleobiologie 9:37–47.

Mora, C., D.P. Tittensor, S. Adl, A.G.B. Simpson, and B. Worm. 2011. How many species are there on Earth and in the ocean? PLoS Biology 9:e1001127.

Newman, W.A., and H.S. Ladd. 1974. Origin of coral-inhabiting balanids (Cirripedia, Thoracica). Verhandlungen der Naturforschenden Gesellschaft in Basel 84:381–396.

Ng, P.K.L., D. Guinot, and P.J.F. Davie. 2008. Systema Brachyurorum: Part I. An annotated checklist of extant brachyuran crabs of the world. Raffles Bulletin of Zoology 17(Suppl.):1–286.

Nogueira, M.M., N.M. Menezes, R. Johnsson, and E. Neves. 2014. The adverse effects of cryptochirid crabs (Decapoda: Brachyura) on *Siderastrea stellata* Verril, 1868 (Anthozoa: Scleractinia): causes and consequences of cavity establishment. Cahiers de Biologie Marine 55:155–162.

Ogawa, K., and K. Matsuzkai. 1992. An essay on host specificity, systematic taxonomy, and evolution of the coral-barnacles. Bulletin of the Biogeographical Society of Japan 47:87–101.

Ogawa, K., and H. Tachikawa. 2009. Coral-inhabiting barnacles from the Ogasawara Islands. Bulletin of the Biogeographical Society of Japan 64:89–96.

Patton, W.K. 1966. Decapod crustacea commensal with Queensland branching corals. Crustaceana 10:271–295.

Pilsbry, H.A. 1931. The cirriped genus *Pyrgoma* in American waters. Proceedings of the Academy of Natural Sciences of Philadelphia 83:81–83.

Poltarukha, O.P., and T.N. Dautova. 2007. Barnacles (Cirripedia, Thoracica) of Nhatrang Bay. Pages 89–123 *in* T.A. Britayev and D.S. Pavlov, editors. Benthic Fauna of the Bay of Nhatrang, Southern Vietnam. KMK Scientific Press, Moscow.

Poore, G.C.B, L. Avery, M. Błażewicz-Paszkowycz, J. Browne, N.L. Bruce, S. Gerken, C. Glasby, E. Greaves, A.W. McCallum, D. Staples, A. Syme, J. Taylor, G. Walker-Smith, M. Warne, C. Watson, A. Williams, R.S. Wilson, and S. Woolley. 2015. Invertebrate diversity of the unexplored marine western margin of Australia: taxonomy and implications for global biodiversity. Marine Biodiversity 45: 271–286.

Potts, F.A. 1915. *Hapalocarcinus*, the gall-forming crab, with some notes on the related genus Cryptochirus. Papers from the Tortugas Laboratory of the Carnegie Institution of Washington 8:33–69.

Reaka-Kudla, M.L. 1997. The global biodiversity of coral reefs: A comparison with rain forests. Pages 83–108 *in* M.L. Reaka-Kudla, D.E. Wilson, and E.O. Wilson, editors. Biodiversity II: Understanding and Protecting our Natural Resources. Joseph Henry Press, Washington, DC.

Rees, W.J. 1962. The distribution of the coral, *Caryophyllia smithii* and the barnacle *Pyrgoma anglicum* in British waters. Bulletin of the British Museum (Natural History), Zoology 8:403–418.

Rees, W.J. 1966. Further notes on the distribution of *Caryophyllia smithii* Stokes and Broderip and *Pyrgoma anglicum*. Annals and Magazine of Natural History, Series 13 9:289–292.

Ren, X. 1986. Studies on Chinese Cirripedia (Crustacea), VII. Family Pyrgomatidae. Studia Marina Sinica 7:129–158.

Roberts, J.M., A. Wheeler, A. Freiwald, and S.D. Cairns. 2009. Cold-Water Corals. The Biology and Geology of Deep-Sea Coral Habitats. Cambridge University Press, Cambridge, UK.

Romano, S.L., and S.D. Cairns. 2000. Molecular phylogenetic hypotheses for the evolution of scleractinian corals. Bulletin of Marine Science 67:1043–1068.

Romano, S.L., and S.R. Palumbi. 1996. Evolution of Scleractinian corals inferred from molecular systematics. Science 271:640–642.

Romano, S.L., and S.R. Palumbi. 1997. Molecular evolution of a portion of the mithchondrial 16S ribosomal gene region in scleractinian corals. Journal of Molecular Evolution 45:397–411.

Ross, A. 2000. Coral-eating barnacles: wall morphology and the descriptions of two new species. Sessile Organisms 17:45–56.

Ross, A., and W.A. Newman. 1973. Revision of the coral inhabiting barnacles (Cirripedia: Balanidae). Transactions of San Diego Society of Natural History 17:137–174.

Ross, A., and W.A. Newman. 1995. A coral-eating barnacle, revisited (Cirripedia, Pyrgomatidae). Contributions to Zoology 65:129–175.

Ross, A., and W.A. Newman. 2000. A new coral-eating barnacle: the first record from the Great Barrier Reef, Australia. Memoirs of the Queensland Museum 45:585–591.

Ross, A., and W.A. Newman. 2002. A review of the *Pyrgoma cancellatum* species complex (Cirripedia: Pyrgomatidae). Journal of Natural History 36:407–421.

Ross, A., and F.B. Pitombo. 2002. Notes on the coral-inhabiting Megatrematinae and the description of a new tribe, new genus and three new species (Cirripedia: Sessilia: Pyrgomatidae). Sessile Organisms 19:58–68.

Ruppert, E.E., R.S. Fox, and R.D. Barnes. 2004. Invertebrate Zoology. A Functional Evolutionary Approach, 7th ed. Brooks/Cole Thompson Learning, Belmont, CA.

Sakai, T. 1976. Crabs of Japan and the Adjacent Seas. Kodansha, Tokyo.

Sakakura, K. 1934. Pleistocene Pyrgoma from Tiba Prefecture. Journal of the Geological Society of Japan 41:575–581.

Sammarco, P.W. 1982. Polyp bail-out: an escape response to environmental stress and a new means of reproduction in corals. Marine Ecology Progress Series 10:57–65.

Scott, P.J.B. 1987. Associations between corals and macroinfaunal invertebrates in Jamaica, with a list of Caribbean and Atlantic coral associates. Bulletin of Marine Science 40:271–286.

Scotto, L.E., and R.H. Gore. 1981. Studies on decapod Crustacea from the Indian River region of Florida. 23. The laboratory cultured zoel stages of the coral gall-forming crab *Troglocarcinus corallicola* Verrill, 1908 (Brachyura, Haplocarcinidae) and its familial position. Journal of Crustacean Biology 1:486–505.

Seguenza, G. 1873. Richerche paleontologiche intorno ai Cirripedi Terziarii della provincia di Messina. Con appendice intorno ai Cirripedi viventi nel Mediterraneo, e sui fossili Terziarii fossili dell'Italia meridionale. Pt. I, Fam. Balanidi e Verrucidi. Atti dell'Accademia Pontaniana, Napoli 10:266–366.

Serène, R. 1972. On the brachyuran fauna of the Indo-Pacific coral reefs. Pages 419–424 *in* C. Mukundan and C.S. Gopindha Pillai, editors. Proceedings of the First International Symposium on Corals and Coral Reefs. Marine Biological Association of India, Mandapam Camp, India.

Simon-Blecher, N., and Y. Achituv. 1997. Relationship between the coral pit crab *Cryptochirus coralliodytes* Heller and its host coral. Journal of Experimental Marine Biology and Ecology 215:93–102.

Simon-Blencher, N., A. Chemedanov, N. Eden, and Y. Achituv. 1999. Pit structure and trophic relationship of the coral pit crab *Cryptochirus coralliodytes*. Marine Biology 134:711–717.

Southward, A.J. 1986. Class Cirripedia (barnacles). Pages 299–305 *in* W. Sterrer, editor. Marine Fauna and Flora of Bermuda. A Systematic Guide to the Identification of Marine Organisms. John Wiley, New York.

Stella, J.S., M.S. Pratchett, P.A. Hutchings, and G.P. Jones. 2011. Coral-associated invertebrates: diversity, ecological importance and vulnerability to disturbance. Oceanography and Marine Biology: An Annual Review 49:43–104.

Števčić, Z. 1971. The main features of brachyuran evolution. Systematic Zoology 20: 331–340.

Stimpson, W. 1859. Communication: *Hapalocarcinus marsupialis*. Proceedings of the Boston Society of Natural History 6:412–413.

Stock, J.H. 1988. Copepods associated with reef corals: a comparison between the Atlantic and the Pacific. Hydrobiologia 167:545–547.

Stubbings, H.G. 1967. The cirriped fauna of tropical west Africa. Bulletin of the British Museum (Natural History). Zoology 15:227–319.

Takeda, M., and Y. Tamura. 1981. Coral-inhabiting crabs of the family Hapalocarcinidae from Japan. VIII. Genus *Pseudocryptochirus* and two new genera. Bulletin of the Biogeographical Society of Japan: 13–27, pls. 1–4.

Takeda, M., and Y. Tamura. 1983 Coral-inhabiting crabs of the Family Hapalocarcinidae from Japan. A Small Collection made at Kushimoto and Koza, Kii Peninsula IX. Bulletin of the National Science Museum, Tokyo, series A (Zoology) 9:1–12.

Tilbrook, K.J. 1997. Barnacle and bivalve associates of bryozoan-coral symbiosis from the Coralline Crag (Pliocene) of England. Tertiary Research 18:7–22.

Tomlinson, J. 1969. The burrowing barnacles: (Cirripedia: Order Acrothoracica). United States National Museum Bulletin 296:162.

Tsang, L.M., K.H. Chu, Y. Nozawa, and B.K.K. Chan. 2014a. Morphological and host specificity evolution in coral symbiont barnacles (Balanomorpha: Pyrgomatidae) inferred from a multi-locus phylogeny. Molecular Phylogenetics and Evolution 77:11–22.

Tsang, L.M., C.D. Schubart, S.T. Ahyong, J.C.Y. Lai, E.Y.C. Au, T.Y. Chan, P.K.L. Ng, and K.H. Chu. 2014b. Evolutionary history of true crabs (Crustacea: Decapoda: Brachyura) and the origin of freshwater crabs. Molecular Biology and Evolution 31:1173–1187.

Tudge, C.C., D.M. Scheltinga, B.G.M. Jamieson, D. Guinot, and B. Richer de Forges. 2014. Comparative ultrastructure of the spermatozoa of the Majoidea (Crustacea, Decapoda, Brachyura) with new data on six species in five genera. Acta Zoologica 95:1–20.

Utinomi, H. 1944. Studies on the animals inhabiting reef corals. III. A revision of the family Hapalocarcinidae (Brachyura) with some remarks on their morphological peculiarities. Palao Tropical Biological Station Studies 2:687–731.

Veron, J.E.N. 2000. Corals of the World. Australian Institute of Marine Science, Queensland, Australia.

Verrill, A.E. 1867. Remarkable instances of crustacean parasitism. American Journal of Science 44:126.

Walter, T.C., and G. Boxshall 2017. World of Copepods database. http://www.marinespecies.org/copepoda.

Wei, T.P., H.C. Chen, Y.C. Lee, M.L. Tsai, J.S. Hwang, S.H. Peng, and Y.W. Chiu. 2013. Gall polymorphism of coral-inhabiting crabs (Decapoda, Cryptochiridae): a new perspective. Journal of Marine Science and Technology 21 (Suppl.):304–307.

Weisbord, N.E. 1972. *Creusia neogenica,* a new species of coral-inhabiting barnacle from Florida. Tulane Studies in Geology and Paleontology 10:59–64.

Wells, H.W. 1966. Barnacles of the northwestern Gulf of Mexico. Quarterly Journal of the Florida Academy of Sciences 29:81–95.

Wetzer, R., J.W. Martin, and S.L. Boyce. 2009. Evolutionary origins of the gall crabs (family Cryptochiridae) based on 16S rDNA sequence data. Pages 475–490 *in* J.W. Martin, K. A. Crandall, and D. L. Felder, editors. Decapod Crustacean Phylogenetics. Crustacean Issues, Vol. 18. CRC Press, Boca Raton, FL.

Withers, T.H. 1926. Barnacles of the *Creusia-Pyrgoma* type from the Pleistocene of Barbados. Annals and Magazine of Natural History, Series 9 17:1–6.

Withers, T.H. 1953. Catalogue of fossil Cirripedia in the Department of Geology. Vol. 3. Tertiary. British Museum (Natural History), London.

Young, P.S. 1988. Recent cnidarian-associated barnacles (Cirripedia, Balanomorpha) from the Brazilian coast. Revista Brasileira de Zoologia 5:353–369.

Young, P.S. 1998. The Cirripedia (Crustacea) collected by the "Fisheries Steamer Meteor" in the eastern Atlantic. Arquivos do Museu Nacional Rio de Janeiro 58:1–54.

Zullo, V.A., and R.W. Portell. 1990. The coral barnacle *Eoceratoconcha* Newman and Ladd, 1974 in the Pliocene of southern Florida. Abstracts with Programs. Bulletin of the Geological Society of America 22:71.

Zullo, V.A., and R.W. Portell. 1991. A new species of *Eoceratoconcha* Newman and Ladd, 1974 (Cirripedia, Archaeobalanidae) from the Pliocene of Florida. Journal of Paleontology 65:271–276.

Zullo, V.A., and R.W. Portell. 1992. Paleobiogeography of Florida Neogene barnacles. The 3rd Bald Head Island Conference on Coastal Plains Geology, Hilton Head Island, South Carolina.

9

INVASION OF THE POLES

Joanna Legeżyńska, Claude De Broyer, and
Jan Marcin Węsławski

Abstract

Polar Crustacea show high taxonomic and functional diversity and hold crucial roles within regional food webs. Despite the differences in the evolutionary history of the two Polar regions, present data suggest rather similar species richness, with over 2,250 taxa recorded in the Antarctic and over 1,930 noted in the Arctic. A longer duration of isolated evolution resulted in a high percentage of endemic species in the Antarctic, while the relatively young Arctic ecosystem, subjected to advection from adjacent seas, shows a very low level of endemism. Low temperatures and seasonal changes of food availability have a strong impact on polar crustacean life histories, resulting in their slow growth and development, extended life cycles, and reproduction well synchronized with annual peaks of primary production. Many species, Antarctic amphipods in particular, exhibit a clear tendency to attain large size. In both regions, abundant populations of pelagic grazers play a pivotal role in the transport of energy and nutrients to higher trophic levels. The sea-ice habitat unique to polar seas supports a wide range of species, with euphausiids and amphipods being the most important in terms of biomass in the Antarctic and Arctic, respectively. Deep sea fauna remains poorly studied, with new species being collected on a regular basis. Ongoing processes, namely a decline of sea-ice cover, increasing levels of ultraviolet radiation, and invasions of sub-polar species, are likely to reshape crustacean communities in both Polar regions.

INTRODUCTION

The Antarctic Ocean (34.8 million km^2) and Arctic Ocean (14 million km^2) are significant parts of the world's oceans. Their cold waters (<4°C) resemble the largest marine domain, the deep ocean, but are peculiar in being strongly seasonal with long periods of low primary production and starvation

Evolution and Biogeography. Edited by Gary C. B. Poore and Martin Thiel.

for consumers. This cycle has been linked to slow growth rates, enhanced resistance to starvation, and reduced ecological competition. The Polar regions are responding the fastest to global warming and therefore stimulate public concern and increasing research efforts in recent decades. Crustacea are the most speciose taxon in the Polar regions and are the subject of large research programs and data management initiatives such as the Census of Antarctic Marine Life (CAML), Arctic Ocean Diversity (ArcOD), ANDEEP, OBIS, SCAR-MarBIN/AntaBIF, etc., which have contributed to the knowledge of diversity, evolution, and biogeography of the Crustacea.

ARCTIC VS. ANTARCTIC: SIMILARITIES AND DIFFERENCES

Whereas the high latitude of the Arctic Ocean permits it to experience polar day and polar night cycles with its associated contrasting productivity regime, the highest latitudes of the Antarctic are continental, and marine areas are distant from extreme latitudes (the southernmost is 78°S) (Table 9.1). Low water temperatures are shared but tend to be stable in the Antarctic, never exceeding −1.8°C in some places, and are more variable even in the high Arctic (80°N) where summer sea surface temperature may reach 3°C. The Arctic comprises an extensive, shallow shelf washed by one of the largest riverine system on the globe and a deep central basin divided by submarine ridges in four sub-basins. Antarctica has an exceptionally deep shelf and vast areas of deep ocean basins with few major isolating structures (ridges, plateaus) (Post et al. 2014).

The Arctic system is connected with the adjoining oceans via the deep (>2000 m) and wide (450 km) Fram Strait in the Atlantic sector and the shallow (50m) and narrow (85 km) Bering Strait in the Pacific sector. Main transport of heat and salt occurs from the Atlantic Ocean, and it is compensated by the outflow of cold, freshened waters along Greenland coasts (Fig. 9.1). Arctic near-bottom, dense saline waters are exported south and are of key importance to global thermohaline circulation (ACIA 2004).

The Antarctic system, on the other hand, is largely isolated from the warmer subtropical waters from the Pacific, Atlantic, and Indian oceans by the east-flowing Antarctic Circumpolar Current (ACC) and its hydrological fronts. The well-defined Antarctic Polar Front (APF, or Antarctic Convergence) marks the northernmost extent of cold surface Antarctic waters, which at these

Table 9.1. Arctic and Antarctic Specific Features of Importance for Crustacean Life

Factor	**Arctic**	**Antarctic**
Evolutionary time of cold climate (ice on the sea)	700 Ka	>30 Ma
Sea ice	Perennial	Annual
Land discharge	Large rivers	Ice sheet
Seabed disturbance by ice scours	Rare and shallow (<100 m)	Common to 250 m
Major pelagic predators	Birds, fish	Birds, sea mammals
Major benthic predators	Sea mammals	Demersal fish
Primary production distribution	Oligotrophic center, rich shallow shelf	Rich extensive frontal zones on the open sea

Source: Modified partly from Dayton et al. 1994.

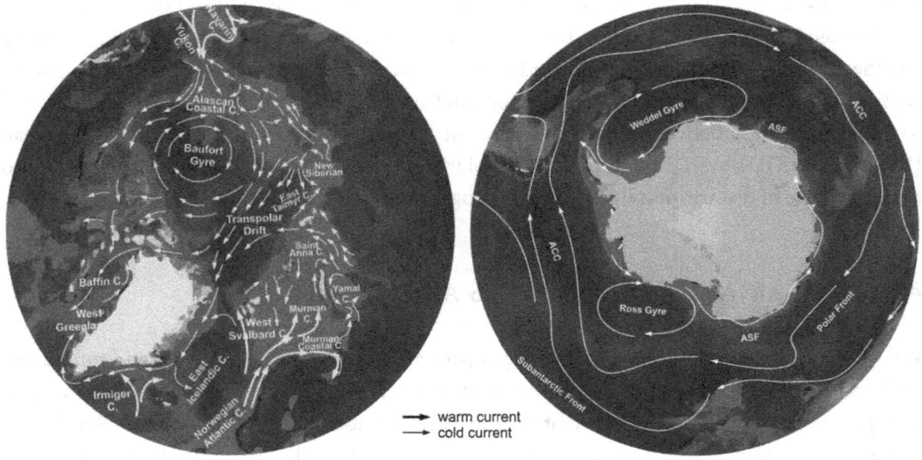

Fig. 9.1.
Major Arctic and Southern Ocean circulation features. ACC, Antarctic Circumpolar Current; ASF, Antarctic Slope Front. Adapted from Post et al. 2014, Rintoul et al. 2001.

latitudes plunges below the warmer subantarctic water masses and indicates the northern boundary of the Southern Ocean (SO) *sensu stricto*. Oceanographers and biogeographers often consider the Southern Ocean to extend to the Subtropical Front (STF) (between 41°S and 48°S; Orsi et al. 1995), and the term in this sense is adopted here (Fig. 9.1). At several locations the Southern Ocean shelf water is a source of deep-sea cold, saline, dense water (Antarctic Bottom Water) that spreads northward to the ocean basins.

These physical features shape the functioning of both systems, mainly primary production, which is highly seasonal, short, and intense in the Arctic with its rich supply of land-derived nutrients but is, conversely, more productive for longer in the Antarctic with a permanent deficit of silica and iron. Sea ice is perennial in the Arctic, with ice floes several meters thick, circulating for several years, and supporting diverse microflora and fauna, whereas in the Antarctic the seasonal/annual pack ice rarely exceeds 2 m in thickness (Arndt and Swadling 2006).

BIODIVERSITY OF POLAR CRUSTACEANS

Soft bottom areas markedly prevail in the Arctic where massive river runoff and coast washout lead to continuous accumulation of fine-grained sediments both on the wide shelf and in the deep basins where they are transported with the sea ice. In the Antarctic terrigenous material input results mostly from creeping down of the continental ice, which transports high amounts of coarse sediments (stones, gravel, and sand) on the narrow and deep shelf. Dropstones densely distributed around the whole Antarctic continent provide a substratum for a variety of sessile and epibenthic species. In the Arctic, dropstones are fairly abundant, but often not exposed, buried in soft sediments, and coarser sediments cover relatively limited areas where wave action or sea currents are strong enough to winnow fine particles (Sirenko 2009, Taylor et al. 2017). As a consequence, hard substratum epifauna, including numerous crustaceans, is rich and diverse in the Antarctic but much poorer in the Arctic, where soft-bottom species constitute a majority of taxa. The pelagic domain is uniform in the Antarctic, due to strong circum-Antarctic currents mixing the upper pelagial around the continent and impeding the penetration of northern species into the Antarctic seas.

On the other hand, the pelagic domain of the Arctic is strongly influenced by advection from the Atlantic and, to a lesser extent, from the Pacific. Thus, pelagic species are more numerous in the Arctic than in the Antarctic. The vast deep-sea Antarctic domain is ice-covered seasonally (allowing tight pelago-benthic coupling), while the Arctic Ocean consists of four deep (3,000–4,000 m) basins (Canada, Makarov, Nansen, and Amundsen) that are largely covered by multi-year ice and receive relatively low input of organic matter from limited primary production. As a consequence, the deep Antarctic fauna is highly diverse and endemic (Brandt et al. 2007b), while that of the Arctic is mostly a subsample of the North Atlantic fauna (Bluhm et al. 2011) (Fig. 9.2).

Peracarida (in particular Amphipoda and Isopoda) dominate the crustacean fauna of the Southern Ocean, while Cirripedia and Decapoda are scarce. Pelagic crustaceans (ostracods, copepods, euphausiids, hyperiid amphipods, natant decapods) number about 550 species, and the benthic and bentho-pelagic ones about 1,750 species (Table 9.2), but many more benthic species, in particular deep-sea isopods and amphipods, await description (De Broyer et al. 2004, Brandt et al. 2007b, Udekem d'Acoz and Verheye 2017). Despite substantial progress during the last two decades, important gaps remain in the faunal survey of Antarctica (Griffiths 2010). On the other hand, as elsewhere, molecular studies are detecting more and more cryptic or pseudocryptic species, challenging our current estimations of SO species richness (Riesgo et al. 2015).

The high species richness of benthic amphipods and isopods in the Antarctic seas has been attributed to their brooding habit, which limits the dispersal of juveniles and enhances reproductive isolation. This leads to reduced gene flow between populations and facilitates speciation. Many Antarctic amphipod species are known to be bottom crawlers, burrowers, nestlers, tube-dwellers, or clingers on or associates of algae or benthic sessile suspensivores (De Broyer et al. 2003). In addition, the Antarctic shelf offers diverse potential habitats, with macroalgal belts (Huang et al. 2007), biogenic sediments such as sponge spicule mats and bryozoan debris, and abundant and diverse sessile suspension-feeding communities, which provide three-dimensional substrata, shelters, food resources, and opportunities for symbioses (De Broyer et al. 2001, Gutt et al. 2015). Extinction events (in particular that of reptant decapods) during the Tertiary cooling of the Southern Ocean (Clarke and Crame 1989, Brandt 2000, De Broyer et al. 2003) may also have promoted amphipod and isopod diversification.

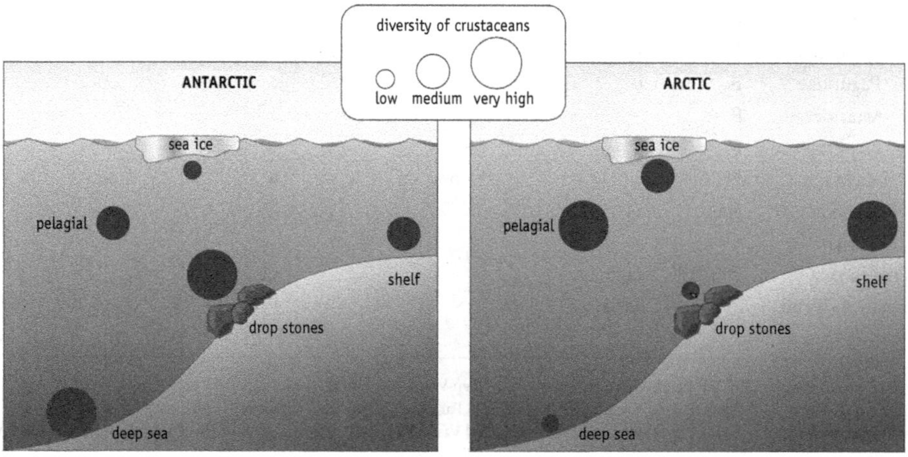

Fig. 9.2.
Diversity of Crustacea in different Antarctic and Arctic habitats.

Table 9.2. Biodiversity of Polar Crustaceans

		Antarctic				Arctic			
	Habitat	N. spp.	N. spp. Southern Ocean s.l.	Endemism Rate	Ref.	N. spp.	N. spp. Deep-Water	Endemism Rate	Ref.
Cladocera	P	0				6			[4]
Ostracoda	B	234	357		[1,2]	248	40		[4,5]
Ostracoda	P	47		21%	[2]	9	8		[4]
Copepoda		484			[2]	362			
Copepoda	P	388		14%	[2]	156	97	10%	[4,6]
Calanoidea	P	30	205	14%	[3]	130			[7]
Harpacticoida	B	96	261	78%	[2]	206	56		[4,5]
Ascothoracida	B, S	5			[1]	0	0		
Cirripedia	B	44			[1]	16	7	12.5%	[8]
Leptostraca	B	5			[1]	1	0	0%	[7]
Mysidacea	P, BP, B	44	72		[2]	35	13	17%	[8]
Amphipoda	P, BP, B	630	927			932	140		
Gammaridea + Senticaudata	B, BP	530	801	67%	[2]	920	132		[4,5]
Hyperiidea	P	66	74	12%	[2]	12	8	0%	[4,8]
Isopoda	B, P	441	740	87%	[2]	112	66 (B)		[5,9]
Tanaidacea	B	160		81%	[2]	26	17		[5,9]
Cumacea	B	93	98	80%	[2]	115	24		[2,5]
Euphausiacea	P	8	23	14%	[2]	9	2	0%	[4, 8]
Decapoda	P, B	55			[2]	68	40	7%	[8]
Decapoda (reptant)	B	22			[2]	19	6	0%	[8]
Anomura	B	16			[2]	13	4		[8]
Galatheidae	B	1			[2]	2	1	0%	[8]
Kiwaidae	B	1			[2]				
Lithodidae	B	13			[2]	5	1	0%	[8]
Paguridae	B	0				6	2	0%	[8]
Astacidea	B								
Nephropidae	B	3	4		[2]				
Brachyura	B	2 (+1)		0%	[2]	6	2	0%	[8]
Decapoda (natant)	P, B	23			[2]	49	34	10%	[8]
Malacostraca	P, B, BP	1436				1298			
Crustacea Total		**2250**				**1939**			

Habitat: P, pelagic; B, benthic; BP, benthopelagic; S, symbiotic; Deep-Water, >500 m.

Sources: [1] De Broyer and Danis 2011; [2] De Broyer et al. 2014; [3] Park and Ferrari 2009; [4] Sirenko et al. 2016) [5] Bluhm et al. 2011; [6] Kosobokova et al. 2011; [7] Sirenko 2001 (Eurasian Arctic only); [8] Vassilenko and Petryashov 2009 (Eurasian Arctic only); [9] Piepenburg et al. 2011.

The scarcity of Cirripedia, notably sessile species, has been correlated with the lack of suitable littoral habitats (disturbed by ice) and to the Cenozoic glaciations that eradicated the former balaniform species (Newman and Ross 1971).

Antarctic Decapoda are represented by few pelagic and benthic natant and reptant species (Table 9.2), in contrast to the richer Cenozoic fauna of macrurans and brachyurans (Feldmann et al. 1993). Like other invertebrates and fish, this Cenozoic temperate fauna was probably eradicated or forced to migrate by the cooling ocean and periodic extensions of the Antarctic ice cap (Clarke and Crame 1989, 1992). But curiously enough, decapods, in particular reptant species, were apparently unable to re-invade the high Antarctic shelf, although a relatively high diversity of decapod crustaceans is recorded in the subantarctic region (Gorny 1999).

Hypotheses explaining the depauperate decapod fauna in the high Antarctic were based on geological/historical, larval, and/or ecological patterns (De Broyer et al. 2003, Thatje et al. 2005). But physiology appears to be more significant here. Brachyuran and anomuran crabs, excluding caridean shrimps, have a low ability to regulate the concentration of hemolymph magnesium at temperatures <0°C, a threshold below which cold-induced cardiac and ventilator performance fails (Frederich et al. 2001). Wittmann et al. (2010) showed that Antarctic caridean decapods and amphipods maintained low hemolymph magnesium levels (6–20% ambient sea-water magnesium concentration), whereas subantarctic brachyuran and anomuran crabs and Antarctic isopods maintained high levels (54–96% ambient sea-water magnesium concentration). They concluded that magnesium regulation may explain the biogeographic pattern of decapods but not that of isopods.

Crustacea are the most species-rich invertebrate group of the pelagic, benthic, and sea-ice realms in the Arctic Ocean (Josefson et al. 2013). Copepods are the most speciose (>150 species) and, with hyperiid amphipods, dominate the metazoan biomass in the pelagic zone. Euphausiids and mysids are abundant but are represented by few species (Buchholz et al. 2010). The Arctic sea-ice community is dominated by four species of amphipods (*Gammarus wilkitzkii, Apherusa glacialis, Onisimus glacialis*, and *O. nanseni*) and by copepods dwelling inside the ice (harpacticoids, cyclopoids) or grazing on algae under the ice (calanoids). Pelagic (*Themisto* spp.) and benthic (*Gammarus, Weyprechtia, Gammaracanthus, Anonyx*, and *Onisimus*) amphipods may temporarily exploit sea-ice habitat as a feeding ground (Arndt and Swadling 2006).

Malacostraca are the most diverse class of benthic Crustacea, with 805 species recorded on the circum-Arctic shelf (Piepenburg et al. 2011), declining from >500 species in the Barents Sea to >250 in the East Siberian Sea (Sirenko 2001). The western Greenland Sea is also diverse (Brandt 1997). The diverse shallow-water substrata (<50 m), which provide good foraging opportunities and shelter from predators and environmental stressors, often host rich and functionally diverse malacostracan communities dominated by amphipods (Voronkov et al. 2013). The muddy bottom below 50 m supports relatively fewer species and is dominated by burrowing deposit-feeders or carnivores (Legeżyńska et al. 2017). In the Bering and Chukchi Seas, influenced by nutrient-rich Pacific water, strong pelagic–benthic coupling supports dense populations of tube-dwelling, filter-feeding ampeliscid amphipods which aggregate in dense mats over the bottom (up to 21,000 ind. m^{-1}; Conlan et al. 2013). Littoral communities, characterized by harsh environmental conditions, support the fewest species; mobile amphipods (*Onisimus* spp. and *Gammarus* spp.) and the barnacle *Semibalanus balanoides* associated with the kelp *Fucus* may be abundant where moderate ice scouring occurs (Węsławski et al. 1993).

The deep central Arctic basin has fewer species than the shelf, with only 261 malacostracan species recorded (Bluhm et al. 2011). However, depth-related trends differ between taxa; numbers of tanaidacean and isopod species tend to rise with increasing depth (Brandt 1997, Svavarsson 1997). Meiofaunal diversity is less well known: ostracods and harpacticoids number about 200 species each, but more can be expected (Bluhm et al. 2011, Josefson et al. 2013).

ECOFUNCTIONAL DIVERSITY OF POLAR CRUSTACEANS

Functional Groups and Life History Traits

The sea-ice zone of the Antarctic marine ecosystem is dominated by the Antarctic krill (*Euphausia superba*). Its circumpolar biomass has been estimated most recently at between 67 and 297 Mt, with the mean of 170 Mt (Siegel 2005). This biomass, its circumpolar distribution (Atkinson et al. 2004), and its nutritive value (Nicol et al. 2000) make it the key species of the Antarctic trophic web, providing a staple food resource for many pelagic and benthic (macro)invertebrates (including squid), fish, penguins, seabirds, seals, and whales. The whole SO (*s.l.*) hosts 13 euphausiid species unevenly distributed latitudinally from the coast of the continent to the Subtropical Front zone (Fig. 9.3, Cuzin-Roudy et al. 2014). Most species form swarms, which for *E. superba* can be 250 m thick and extend over several kilometers, with densities of up to 60,000 individuals and 15 kg m^{-3} (Everson 2000).

Postlarval krill is typically concentrated in the epipelagic zone in the summer (Siegel 2005), while in the early winter a significant fraction of the adult population goes to deeper water (150–450 m) (Lawson et al. 1994). Clarke and Tyler (2008) reported ROV observations of adult krill, including gravid females, actively feeding on fresh phytodetritus at the foot of the continental slope (3000 m) and on the abyssal plain (3500 m) in the western Antarctic Peninsula. Krill were observed digging in the sedimented phytodetritus and, just above the bottom, filtering the resuspended material. Abundant and large fast-sinking krill fecal pellets contribute significantly to the pelago-benthic organic flux (Smetacek et al. 2004), which sustains communities of suspension feeders and deposit feeders on the Antarctic shelf (Smith et al. 2006). At subantarctic latitudes (between APF and STF), hyperiid amphipods play a key trophic role in pelagic ecosystems. On the Southern Patagonian Shelf, around the Prince Edward and Marion Islands and around the Kerguelen Islands and South Georgia, *Themisto gaudichaudii* in particular are abundant and play trophic roles similar to that of *Euphausia superba* farther south (Bocher et al. 2001, Padovani et al. 2012, Watts and Tarling 2012). Antarctic benthic crustaceans present a large array of feeding roles, with a dominance of scavengers, micro-predators, and detritophages (Table 9.3).

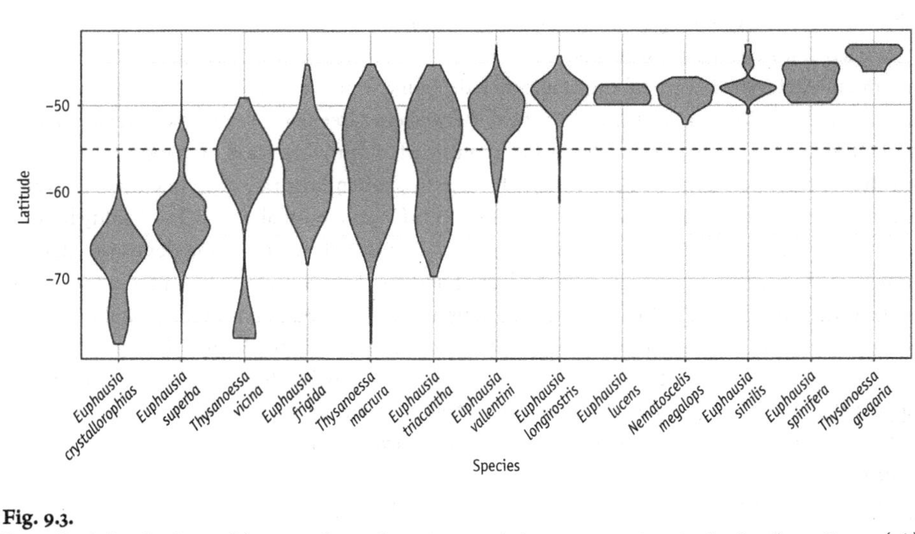

Fig. 9.3.
Latitudinal distributions of the 13 euphausiid species sampled at 14,397 stations in the Southern Ocean (s.l.), represented as violins (the wider the violin, the more presence at this latitude). The dashed line indicates the average position of the Antarctic Polar Front zone. From Cuzin-Roudy et al. 2014.

Table 9.3. Main Ecofunctional Groups of Marine Crustaceans in the Southern Ocean and Arctic Shelves with Examples of Key Species

Functional group	Examples from SO (s.s.)	Importance in SO ecosystems	Examples from Arctic	Importance in Arctic ecosystems
Small pelagic herbivores	Calanoid and cyclopoid copepods (*Calanus simillimus, Rhincalanus gigas, Oithona similis*) *Calanoides acutus*	Key elements of pelagic food web	Copepods (surface waters: *Calanus* spp., *Oithona similis*)	Key element of pelagic food web
Large pelagic herbivores	Six species of euphausiids	Dominant herbivores *E. superba*: keystone species in the SO (s.s.) ecosystem	Euphausids (*Thysanoessa* spp.)	Prey for top predators
Small pelagic carnivores	Hyperiids (*Themisto gaudichaudii*) Deep-water copepods (*Paraeuchaeta antarctica, P. rasa*)	Key element in subantarctic waters	Hyperiids (*Themisto* spp.), copepods (*Paraeucheta spp.*)	Prey for top predators
Supra-benthic omnivores	Mysids (*Antarctomysis maxima*)	May contribute significantly to bottom biogenic modification	Mysids (*Mysis oculata*)	May contribute to bottom biogenic modification
Benthic deposit feeders	Ostracods amphipods (*Epimeria* spp., *Paraceradocus* spp.) Shrimps (*Notocrangon antarcticus*)		Ostracods, amphipods, cumaceans (Leuconidae) shrimps (*Sabinea*)	
Benthic seston feeders	Amphipods (Ampeliscidae, Melphidippidae). cumaceans	Minor role in suspension-feeding communities	Amphipods (Ampeliscidae) and cumaceans (Diastylidae)	Ampeliscids are of key importance as food for grey whales
Benthic carrion feeders	Lysianassoid amphipods, cirolanid isopods	May occur in large numbers on food falls	Lysianassoid amphipods	May occur in large number feeding on animal carcasses
Benthic (micro) predators	Amphipods (*Epimeria* spp., Iphimediidae)	Usually specialized	Decapods (*Pandalus borealis*, crabs), shrimps (*Spirontocaris*), amphipods (Oedicerotidae, Phoxocephalidae)	

In the Arctic pelagic realm, copepods (especially Calanidae) and euphausids are widespread and abundant, holding crucial positions within food webs, directly linking primary production to higher trophic levels. Hyperiid amphipods are epipelagic predators feeding preferentially on *Calanus* spp. Benthic crustaceans have versatile feeding repertoires and are important prey for seabirds, fish, and bottom-feeding sea mammals. Abundant deep-water prawns (*Pandalus borealis*) are fed upon by demersal fish (e.g., cod) and are commercially exploited (Table 9.3).

Adaptations of Polar Crustaceans

Polar Gigantism

Several ecological, physiological, and biogeographic hypotheses have been put forward to explain polar gigantism (Moran and Woods 2012). Antarctic amphipods in particular exhibit a clear tendency to attain large size (Arnaud 1977, De Broyer 1977, Chapelle 2001). Chapelle and Peck (1999, 2004) and Peck and Chapelle (1999) analyzed and compared the size spectra of 2,892 species of benthic amphipods in 15 localities from the tropics to the poles, in fresh and marine waters (but excluding the deep sea). They showed a narrow relationship between the size of gammaridean amphipods and mean absolute oxygen concentration of the surface waters. At higher oxygen concentrations, the increase in size is slight in small animals, more pronounced in the middle of the spectrum, and greatest in the largest animals. This was explained by two physiological factors: the essentially passive nature of respiration in amphipods and the decrease of the surface-volume ratio with increasing size, setting an upper threshold beyond which the respiratory surface becomes insufficient for oxygen to supply the tissues. They concluded that oxygen availability does not act as a selection pressure in favor of large size but rather as a physicochemical ceiling limiting maximum potential size. They also showed that the minimum adult size is probably limited by the minimum possible egg size.

Exploring the links between the physical, chemical, and biological environments of polar seas and the evolution of gigantism, Moran and Woods (2012) distinguished eight major explanatory hypotheses, noting that even if the "oxygen theory" seems the most widely accepted, each deserves additional theoretical and experimental work.

Coping with Variable Food Conditions

Life history strategies at high latitudes are shaped by extreme environmental and resource seasonality. Several strategies to cope with these fluctuations are observed (Clarke and Peck, 1991, Torres et al. 1994, Huenerlage et al. 2015). With primary production restricted to two to four months per year, herbivores are more affected than other consumers (Obermüller et al. 2010). They often build up large lipid reserves during the summer, which sustain them throughout periods of low food supply and permit reproduction. Total lipid content in herbivorous zooplankters may surpass 70% dry weight, allowing survival without food uptake such as 210 days in Antarctic krill and 270 days in the Arctic copepod *Calanus glacialis* (Lee et al. 2006). Lipid-rich zooplankton display different overwintering strategies: herbivorous copepods migrate into deep waters and enter a period of reduced metabolism (diapause), while other species that remain active in surface waters, such as euphausiids, may survive due to reduced metabolism, stagnated growth, sexual regression, catabolism of stored lipids and tissue proteins coupled with occasional feeding (Lee et al. 2006, Huenerlage et al. 2015). In contrast, benthic crustaceans generally do not accumulate lipids and metabolize in the winter as in the summer (Graeve et al. 1997, Legeżyńska et al. 2012). Having access to a variety of year-round food sources, they have developed opportunistic feeding strategies, making them less dependent on fluctuations in primary productivity. A common, though often only facultative, feeding strategy among Arctic and Antarctic crustaceans is necrophagy, regarded as an important adaptive trait of polar benthos (Arnaud 1977). Mortality of small organisms due to ice scouring, wave action or mixing of different water masses, and large carrion falls (dead fish, mammals) provides food for scavengers year round. A key role in dispersing organic material in both Polar regions is played by lysianassoid amphipods (Britton and Morton 1994). Shallow-water genera, such as *Anonyx, Onisimus,* and *Orchomenella* in the Arctic and *Cheirimedon, Hippomedon, Pseudorchomene,* and *Waldeckia* in the Antarctic have more flexible feeding habits than deep-sea species (e.g., cosmopolitan and bipolar *Eurythenes gryllus*) that rely on infrequent large food falls.

Low temperature and seasonal pulses of food supply strongly influence life histories of polar crustaceans: typically, they demonstrate slow growth and development and extended life cycles (biannual or perennial; Table 9.4) with a single annual breeding period (Peck et al. 2006). Annual cohorts within the same species differ considerably in size, mobility, diet, and microhabitat and therefore may play different roles in the ecosystem (Węsławski et al. 2010).

Since successful recruitment of zooplankton herbivores depends on their ability to synchronize their seasonal migration, reproduction, and growth with local primary production, populations under differing environmental conditions show variable developmental patterns. In regions with long-lasting ice cover where phytoplankton blooms are preceded by ice algae blooms, the Arctic copepod *Calanus glacialis* utilizes ice algae to fuel spawning, while growth and development of the new generation is primarily supported by the later bloom of pelagic microalgae. In ice-free areas, this species uses accumulated lipid reserves to start reproduction early in the season, allowing early life stages to benefit from the phytoplankton bloom (Daase et al. 2013). High-latitude benthic crustaceans are typically characterized by long life spans, small brood sizes, large eggs, and semelparity (one reproduction event in a lifetime) (Sainte-Marie 1991), resulting in their modest reproductive potential. Therefore, reproduction must be well timed to maximum food availability. Success of offspring is ensured by parental care (in peracarids with a marsupium) until the start of the productive season. In most species, release of juveniles coincides with spring phytoplankton/ice algae blooms (Węsławski and Legeżyńska 2002), but in some it may be synchronized with the availability of more specific food sources. The scavenging Arctic amphipod *Onisimus caricus*

Table 9.4. Life Span of some Polar Crustaceans

Taxon	**Region, Domain**	**Life Span**	**Ref.**
Calanus hyperboreus (Copepoda)	Arctic, pelagic	Up to 5 years	[1]
Gammarus wilkitzkii (Amphipoda)	Arctic, sympagic	Up to 5 years	[2]
Sclerocrangon boreas (Decapoda)	Boreo-Arctic, benthic	Over 6 years	[3]
Antarctomysis maxima (Mysida)	Antarctic, suprabenthic	Up to 7 years	[4]
Eurythenes gryllus (Amphipoda)	Bipolar, deep water bentho-pelagic	Over 9 years	[5]
Bovallia gigantea (Amphipoda)	Antarctic, shallow benthic	Up to 4 years	[6]

Sources: [1] Hirche 1997; [2] Poltermann 2000; [3] Sainte-Marie et al. 2006; [4] San Vicente 2010; [5] Ingram and Hessler 1987; [6] Thurston 1972.

releases its juveniles at the same time there is mass mortality of zooplankton during the melting season (Nygård et al. 2009).

Coping with Thermal and Osmotic Stress

Unlike sublittoral and deep-sea polar species that experience little fluctuations of environmental conditions, organisms living in the intertidal zone are faced with temperature extremes and strong fluctuations of salinity resulting from freshwater runoff (Węsławski et al. 1993, Waller 2013). Most inhabitants of this zone migrate from the subtidal to graze on macroalgal detritus, epiphytes, or animal carcasses and bury in sediment or leave the area when conditions become less hospitable. Some species, however, adapt physiologically and stay. For example, some taxa prevent internal ice formation by lowering the freezing point of their body fluids (e.g., the Antarctic amphipod *Prostebbingia gracilis*; Waller et al. 2006). Further, littoral species are usually euryhaline. The Arctic amphipod *Onisimus litoralis*, which is an efficient osmoregulator, can survive at salinities from 5 to 55, enabling its colonization of the littoral zone and the sea-ice underside (Shea and Percy 1990). Species that gradually adjust their hemolymph concentration of sodium and chloride ions to the salinity of seawater (osmoconformers) have narrower tolerance that may limit their occurrence in osmotically stressful habitats (Shea and Percy 1990).

Strong salinity gradients characterize the shallow areas of the Siberian coast that is extensively influenced by riverine waters. Their evolutionary age is very modest, and its colonization probably occurred as a consequence of repeated invasions from the sea. In such habitats, sequences of closely related species of varying salinity tolerance may be found occupying different depth zones along salinity gradients. For example, in the Kara Sea *Onisimus litoralis* and *O. affinis* occur in the littoral and shallow sublittoral tolerating salinity <20, while *O. caricus* and *O. brevicaudatus* require deeper and more saline waters (Vader et. al. 2005).

ORIGIN OF POLAR CRUSTACEANS

Antarctic

The origin and evolution of the Antarctic fauna are closely linked to the tectonic, climatic. and glacial history of the region. The rich and diverse Antarctic continental shelf benthos consists of a mixture of taxa with different evolutionary histories and biogeographic affinities and is considered to have originated from three sources (Dell 1972, Clarke and Crame 1989, 1992):

1. From a relict autochthonous fauna of Gondwanan origin, which evolved during the cooling of the polar waters after the opening of the Drake Passage, the onset of the ACC, and thermal isolation of the Antarctic continent. Part of this Gondwanan fauna, such as brachyuran crabs, did not succeed and disappeared, leaving some fossil evidence. The isopod families Serolidae and Antarcturidae (Brandt 1992), the former possibly dating back to the Late Cretaceous (80 Ma), are Gondwanan (Wägele 1994).
2. From the adjacent deep-sea basins, for example, the isopod families Munnopsididae, Haploniscidae, Desmosomatidae, Nannoniscidae, and Ischnomesidae (Brandt et al. 2007a). An abyssal origin was also suggested for Antarctic stenothermic eurybathic tanaidaceans (Sieg 1992).
3. From the southern South American continental shelf by migration along the Scotia Arc. The islands of the Arc may serve as "stepping stones" for species able to cross an expanse of deep water. The configuration of the Scotia Arc has changed and deep water expanses found today

between the Arc islands didn't always exist in the past (Thomson 2004). Serolid isopods (Held 2000) and some notothenoid fishes (Bargelloni et al. 2000) colonized the Antarctic shelf from South America before the opening of the Drake Passage, which occurred about 33 Ma. According to Sieg (1986), some Antarctic shelf Tanaidacea migrated from the Magellan region, and Watling and Thurston (1989) suggested a similar origin for some iphimediid amphipods. This migration may still be ongoing on the Scotia Ridge, as suggested, for example, for Isopoda Asellota (Winkler 1994).

Whatever their origin(s), it is widely recognized that some benthic lineages have successfully radiated in the cold but constant environmental conditions of the Antarctic shelf (Clarke and Crame 1989). Examples are the isopod families Serolidae (Held 2000) and Antarcturidae (Brandt 2000) and the amphipod families Iphimediidae (Lörz and Held 2004) and Epimeriidae (Verheye et al. 2016).

The high diversity in Antarctic waters is also attributed to the "Antarctic Climatic Biodiversity Pump," which refers to several scenarios favoring persistence and speciation of Antarctic biota. During the Pliocene-Pleistocene glacial periods, the continental ice over the shelf extended and contracted on Milankovitch frequencies of about 100,000 years (Clarke and Crame 1989, 1992). The advance of grounded ice sheets across the continental shelf may have eradicated sessile or poorly mobile benthos over large parts of their ranges (e.g., Thatje et al. 2005, Allcock and Strugnell 2012). More motile animals may have been displaced to refugia: to the slope and deep sea, to ice-free areas on the shelf, to the Scotia Arc, northern Antarctic Peninsula and peri-Antarctic islands, or to oceanic ridges. Arguments against complete eradication of the shelf fauna during glacial maxima refer to the persistence of taxa present before the onset of glaciations (Clarke and Crame 1989, 1992), survival of benthic species under floating ice shelves (e.g., Riddle et al. 2007), and the occurrence of ice-free areas on the shelf. Grounded ice did not advance to the shelf edge during the Last Glacial Maximum, e.g., off Adélie Land, Prydz Bay, and George V Land (Anderson et al. 2002), and ice advances did not cover all the shelf at the same time. The isolation of fragmented populations during glacial periods may have resulted in some level of genetic differentiation. During interglacial periods, coalescence of fragmented populations during recolonization of the shelf from transient mosaic refuges may have led to allopatric speciation, explaining the apparently common occurrence of complexes of sibling or cryptic species (Allcock and Strugnell 2012, Rogers 2012).

Arctic

The Arctic Ocean was a cool embayment of the Pacific Ocean until ~80 Ma when this deep connection closed. Exchange with the Atlantic started ~40 Ma, with deep (>1000 m) and wide (>500 km) connection existing since the Eocene (27 Ma). Shallow-water (<100 m) and narrow (50 km) contact with the Pacific opened again in the Pliocene (3.5 Ma). The Arctic Ocean was initially cool-temperate (10°C) and gradually started to get colder, reaching negative temperatures and permanent pack ice cover in mid-Pleistocene (0.7 Ma). Cold-water, brackish species associated with sea ice and Siberian river mouths evolved less than 100,000 years ago. The Wisconsin/Würm shield developed 18,000 years ago, causing widespread extinctions. Deglaciation started 14,000 years ago, and the Pacific contributed to an inflow of shelf species, while the Atlantic contributed both deep water and littoral fauna (Dunton 1992). As a result of geological history and existing dispersal barriers, the majority of crustaceans occurring nowadays in the Arctic are boreo-Arctic forms of Atlantic origin, while species of Pacific origin and true Arctic taxa account for around 10% each (Josefson et al. 2013).

BIOGEOGRAPHY OF POLAR CRUSTACEANS

Antarctic

The general distribution pattern of pelagic crustaceans is mostly driven by the highly structured water masses and current system of the Southern Ocean, dominated by the ACC. Most epi- and mesopelagic species are circumpolar, but some, such as the Antarctic krill (e.g., Atkinson et al. 2004), show distinct longitudinal patterns in biomass, or in occurrence, such as some other euphausiids (Cuzin-Roudy et al. 2014). Oceanographic fronts are the main boundaries in the latitudinal ranges of pelagic species. Epi- and mesopelagic ostracods, copepods, amphipods, and euphausiids are distributed in somewhat consistent biogeographic bands around Antarctica; Antarctic species range from the seasonal sea-ice zone to the Antarctic Polar Front Zone (APFZ), or toward the subantarctic Front Zone (SAFZ), the Subtropical Front Zone (STFZ), or beyond (Kouwenberg et al. 2014).

Deep-sea zooplankton species are not restricted by hydrographic fronts and may be widely dispersed by deep-water currents (Park and Ferrari 2009). Meso- and bathypelagic species in particular appear widely distributed north of the Subtropical Front, even as far as Arctic regions. For instance, 66% of the 127 Antarctic deep-water calanoid copepods have been reported north of the Subtropical Front (Park and Ferrari 2009), 9 being considered bipolar. However, using a comprehensive molecular approach, Laakmann et al. (2012) found that the mesopelagic *Aetideopsis minor* represented two (Antarctic and Arctic) clades whereas no intraspecific genetic differences were found between populations of the bathypelagic *A. rostrata.*

The giant lysianassoid amphipod *Eurythenes gryllus* was long considered a cosmopolitan and eurybathic deep-sea species, but Havermans et al. (2013), using morphology and molecular methods, identified 12 species-level clades (d'Udekem d'Acoz and Havermans 2015, Havermans 2016). Bipolar pattern has been detected in some other pelagic and benthic taxonomic groups on morphological grounds (see Allcock and Griffiths 2014). More complete sampling, in particular for deep-sea species in tropical regions, and an approach combining molecular, morphological, and ecological data are needed to recognize the true distributions, detect potential cryptic species, and resolve this recurrent bipolar issue.

Antarctic pelagic species show a low rate of endemism (12% in copepods; 14% in euphausiids) in clear contrast to benthic species (Table 9.2). The biogeography of Antarctic benthic Crustacea is largely driven by the ACC and the front systems, as for pelagic crustaceans, but the timing of past continental connectivity, the geomorphology-associated physical and environmental factors, depth, sea temperature, and ice cover appear also to be key parameters (Koubbi et al. 2014).

Until recently, circumpolar distribution was a paradigm for most Antarctic benthic species (Hedgpeth 1969, 1970, Dell 1972, Arntz et al. 1997, Clarke and Johnston 2003). However, during the last decade, molecular phylogenetic and phylogeographic analyses revealed that this paradigm may be the exception rather than the rule: more and more so-called "circum-Antarctic" benthic species appeared to be complexes of cryptic species with restricted spatial and bathymetric distributions (Allcock and Strugnell 2012, Riesgo et al. 2015). Examples can be found in ostracods (Brandao et al. 2010), isopods (e.g., Leese and Held 2008), and amphipods (e.g., Havermans et al. 2010, Havermans 2012, Verheye et al. 2016).

On the other hand, a circumpolar distribution has been confirmed for some benthic species, e.g., the shrimps with planktotrophic larvae *Notocrangon antarcticus* and *Nematocarcinus lanceopes* (Raupach et al. 2010), and in some lysianassoid and epimeriid amphipods (Havermans 2012, 2014, Verheye et al. 2016). Circumpolar dispersal by rafting on the bull kelp *Durvillaea* has been observed in some subantarctic shallow-water crustaceans (Nikula et al. 2010).

The APF has long been recognized as the main biogeographic barrier for shelf and (upper) slope benthic species, clearly delimiting an Antarctic fauna pattern (De Broyer et al. 2014). However, in most crustacean groups, a significant fraction of the Antarctic species also occurs within the sub-antarctic region, supporting the concept of Southern Ocean *sensu lato*. This is the case for 20% (106 species) of benthic amphipods (De Broyer and Jażdżewska 2014), 25% of tanaidaceans (Błażewicz-Paszkowycz 2014), and 30% of pycnogonids (Soler-Membrives et al. 2014).

The Antarctic biogeographic region has usually been split in two—east and west—subregions (Ekman 1953, Hedgpeth 1969, De Broyer and Koubbi 2014). Relying on wider and improved recent data, Griffiths et al. (2009) suggested that the Antarctic region constitutes instead a "single functional unit," as initially proposed by Dell (1972), for Bivalvia, Bryozoa, Pycnogonida, Porifera, Ascidiacea, and Echinoidea. In contrast, this "single Antarctic province" does not hold true for Amphipoda (De Broyer and Jażdżewska 2014) with a high proportion of endemic species (24%) in the Antarctic Peninsula and Scotia Sea sub-region, interpreted as a discrete "West Antarctic" fauna distinguished from the continental Antarctic fauna. A clear distinction between the "Antarctic Peninsula and Scotia Sea Province" and the "Continental High Antarctic Province" (Clarke et al. 2007) was also observed in Hydrozoa (Peña Cantero 2004), Gastropoda (Linse et al. 2006), Actiniaria (Rodriguez et al. 2007), and less distinctly in Tanaidacea (Błażewicz-Paszkowycz 2014).

The isolating conditions of the SO, dating back to the Oligocene 33 Ma (Barker and Thomas 2004), have resulted in high species endemism of the shelf fauna (Dell 1972, De Broyer et al. 2014) (see Table 9.2). At the species level, Antarctic benthic crustaceans have no or few doubtful faunal links with the peripheral continental shelves of South Africa, Australia, or New Zealand but show strong affinities with southern South America due to long-lasting past and possibly ongoing connectivity (Arntz et al. 1999, 2005). Besides seafloor water temperature and geomorphological features, depth is a key parameter defining distribution patterns of benthic species. A clear distinction can be made between the shelf (approximately 0–1000 m) and the deep-sea faunas, but actual depth ranges of most species are unknown because of limited deep-sea sampling. Taking the well-documented amphipods as examples, some clear bathymetric clusters distinguish the shelf fauna (0–800 m), the upper slope (800–2,200 m), lower slope–upper abyssal (2,200–3,700 m), and abyssal fauna (3,700–4,500 m), keeping in mind that most records (89%) came from depths of less than 1,000 m (De Broyer and Jażdżewska 2014). On the other hand, many Antarctic benthic species show greater eurybathy than species in other oceans (Brey et al. 1996). This has been interpreted as an evolutionary adaptation or pre-adaptation to the oscillation of shelf ice during Cenozoic glacial-interglacial cycles. A fair number of Antarctic crustaceans show a wide bathymetric range of more than 1000 m (Table 9.5), assumed to be facilitated by the deep shelf habitats, the generation of cold, highly oxygenated Antarctic Bottom Water from shelf waters, and the mostly isothermal conditions of the water column. This view is challenged by the discovery of cryptic species along depth gradients, e.g., the isopods *Glyptonotus antarcticus* (Held and Wägele 2005) and *Betamorpha fusiformis* (Raupach et al. 2007), the ostracod complex *Macroscapha tensa* (Brandao et al. 2010), and the amphipod *Eurythenes gryllus* (Havermans et al. 2013).

Arctic

The contemporary Arctic fauna is characterized by low endemism (Table 9.2) and domination of species originating from the North Atlantic. Circumpolar taxa are numerous, especially within the sea ice and pelagic domain (e.g., *Apherusa glacialis, Gammarus wilkitzkii, Calanus glacialis*—key species of the Arctic food web) (Josefson et al. 2013). The biogeography of pelagic communities is impacted by the advection of Atlantic and Pacific waters bringing subarctic species of copepods and euphausiids into the Arctic basin (Buchholz et al. 2010, Kosobokova et al. 2011). Many abundant benthic taxa are widely distributed: *Ampelisca eschrichtii, Anonyx nugax, Arrhis phyllonyx, Byblis*

Table 9.5. Examples of Extreme Eurybathy in Polar Crustaceans

Species	Min.-max. Depth (m)	Depth Range (m)	Habitat
Antarctic			
Byblisoides juxtacornis (Amphipoda)	160–2315	2155	B
Liljeborgia homospora (Amphipoda)	1180–4392	3212	B
Orchomenella (Orchomenopsis) cavimanus (Amphipoda)	6–3683	3677	B
Pseudorchomene coatsi (Amphipoda)	0–2889	2889	B
Abyssorchomene scotianensis (Amphipoda)	385–3408	3023	BP
Parschisturella carinata (Amphipoda)	14–2081	2067	B
Necochea pardella (Amphipoda)	170–3725	3555	B
Alexandrella schellenbergi (Amphipoda)	340–2889	2549	B
Dodecasella elegans (Amphipoda)	68–2894	2826	B
Arctic			
Haliragesqvadridentatus (Amphipoda)	32–2681	2649	B
Andaniexis abyssi (Amphipoda)	174–2681	2507	B
Arrhis phyllonyx (Amphipoda)	30–2681	2651	B
Tarasovium cornutum (Cirripedia)	348–3941	3593	B
Munna acanthifera (Isopoda)	32–2681	2649	B
Eusergestes arcticus (Decapoda)	250–5030	4780	P
Hymenodora glacialis (Decapoda)	300–3900	3600	P
Caecognathia stygia (Isopoda)	180–2681	2501	B
Eurycope producta (Isopoda)	180–2681	2501	B
Parakanthophoreus longiremis (Tanaidacea)	50–2681	2631	B

gaimardi, Diastylis spp., and *Leucon nasica* (Piepenburg et al. 2011). The shelf seas adjoining the major gateways to the Arctic—the Barents and Kara Seas in the Atlantic section and the Chukchi Sea in the Pacific section—host the most diverse benthic communities. The share of Pacific and Atlantic species diminishes with increasing distance from the Bering Strait and Fram Strait, respectively (Josefson et al. 2013). Biogeographical zonation in the Russian Arctic is strongly influenced by freshwater discharge from the large Siberian Rivers. Cold brackish habitats of shallow estuaries established ~10,000 years ago after the final retreat of the ice sheet and harbor some Arctic relict species including the isopod *Saduria sibirica* (Josefson et al. 2013). The isolated populations of stenothermic cold-water species may also find refugia in the innermost basins of the glaciated fjords (Węsławski et al. 2011).

Polar Deep Seas

Despite recent progress, the Antarctic deep-sea fauna remains much less known than continental shelf faunas. The ANDEEP program (Brandt and Hilbig 2004, Brandt and Ebbe 2007) explored the deep-sea basins in the Weddell and Scotia seas and discovered significant numbers of unknown benthic crustacean species, in particular isopods and amphipods (De Broyer et al. 2004, Brandt et al. 2007b, 2012). In Munnidopsidae (Isopoda), 80% of collected species were unknown (Malyutina and Brandt 2007). While most taxa became less diverse with depth, the species richness of Isopoda is lowest on the shelf and upper slope (<400 species, mostly endemic) and greatest at bathyal and abyssal depths where ~600 mostly undescribed morphospecies have been recorded (Brandt et al. 2007b, 2009). In addition, this abyssal fauna was highly endemic, possibly simply reflecting undersampling but suggesting connectivity with the adjacent deep-sea basins in the Atlantic. This confirms that, in contrast to the shelf fauna, the Antarctic deep-sea benthos is not isolated from other deep-sea basins but connected via the cold isothermal environment and circulation of the Antarctic Bottom Water.

In the deep central Arctic Basin more than 300 benthic crustacean species occur (Table 9.2) with three tanaidacean species: *Parakanthophoreus gracilis, Pseudosphyrapus anomalus,* and *Pseudotanais affinis* being the most widely distributed. The deep Arctic fauna shows low endemism, high shares of subarctic species, and significant taxonomic overlap with the shelf fauna. Over 60% of macrobenthic species found below 500 m depth occur also on the shelf (Bluhm et al. 2011). The predominance of eurybathic species observed in the Atlantic section results from the glaciation history of this area, which favored survival of species able to escape to the great depths during the shelf glaciation and re-invade shallow waters when ice sheet retreated (Josefson et al. 2013). Brandt (1997) found only 10 exclusively deep-sea genera of Peracarida, while 60 genera had a wide bathymetric range in the deep Greenland Sea.

SEA ICE—A UNIQUE HABITAT

Sea-ice cover is a characteristic feature of both Polar regions but differs between the Arctic and the Antarctic. More than half of the sea ice in the Arctic is perennial ice (five to seven years), ranging from 2 to 4 m thick. In contrast, the majority of Antarctic ice is first-year ice less than 1 m thick that melts in the summer (Arndt and Swadling 2006). Both the extent and thickness of the Arctic perennial multi-year ice has been significantly decreasing over the past 30 years. At the same time, sea ice extent in the Southern Ocean has shown strong regional differences but overall is increasing (Parkinson 2014).

Sea ice provides a complex semisolid platform, which allows the development of diverse sympagic (ice-associated) communities unique to polar seas. The underside and interior network of channels and pores filled with brine constitute an "up-side down" benthic habitat colonized by numerous protozoan and metazoan grazers of pelagic and benthic origin that use ice as a breeding, nursery, and feeding ground and refuge area. Crustaceans represented by about 50 species in each area are prominent members of sympagic assemblages at both poles (Figure 9.4). In the Arctic, the fauna is dominated by four amphipod species inhabiting the ice–water interface: *Gammarus wilkitzkii, Apherusa glacialis, Onisimus nanseni,* and *O. glacialis* with abundances up to 500 individuals m^{-2} (Arndt and Pavlova 2005). In the upper layer of the water column beneath the ice, sibling species of predatory hyperiid amphipods, *Themisto libellula* and *T. abyssorum,* occur in swarms of up to 430 and 270 individuals m^{-2}, respectively. The ice interior in the Arctic is inhabited by small harpacticoid genera, *Harpacticus, Halectinosoma* and *Tisbe,* and cyclopoid copepods (*Cyclopina* spp.). Several species of calanoid copepods (including key Arctic species: *Calanus glacialis* and *C. hyperboreus*) perform diel vertical migrations to the ice subsurface where they graze on ice algae but do not enter the brine-channel system. Additionally, young stages of ice and pelagic copepods (eggs and nauplii) may accumulate in sea and ice in numbers of more than 100,000 m^{-2}. Other crustacean taxa (mysids, decapods. and euphausiids) do not contribute considerably to the sympagic communities in the Arctic (Arndt and Swadling 2006).

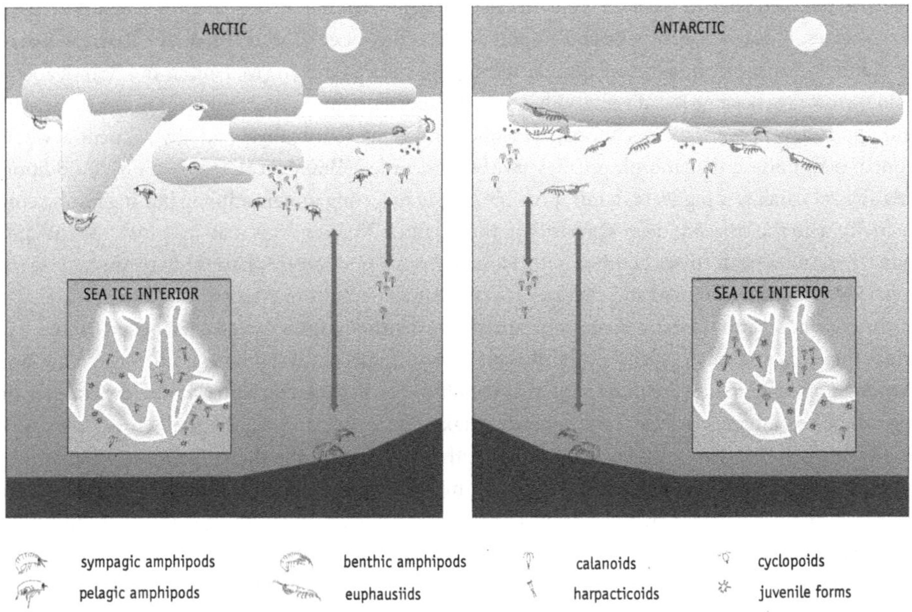

Fig. 9.4.
Sea-ice–associated crustaceans in the Arctic (left) and Antarctic (right). See color version of this figure in centerfold.

In contrast to the Arctic, euphausiids are important members of the Antarctic ice fauna. Pack ice has direct and indirect effect on krill distribution, life cycle, and population dynamics. *Euphausia crystallorophias,* the neritic "ice krill,"is found under the ice only over the continental shelf. The circumpolar distribution of Antarctic krill, *E. superba,* on the other hand, is limited to the area bound by the maximum sea ice extent, with the exception of populations around South Georgia (Siegel 2005, Atkinson et al. 2008; Figure 9.5). During winter, larval krill settles under Antarctic annual sea ice, preferably in complex under-ice habitats of eroded ice floes. Larval *E. superba* feeds on ice-associated biota, in particular microbial communities, a food source that is one to three orders of magnitude more abundant in winter than the food sources in the water column immediately below (Quetin and Ross 2009). The reproductive cycle, in particular ovarian development in spring and larval survival during the first winter, may be impacted by the timing and extent of sea ice in winter and/or spring, which mediates the food availability (Loeb et al. 1997, Quetin and Ross 2001). Stabilization of the water column by melting pack ice in late winter or early spring favorably affects planktonic diatom growth, the preferred food of krill, which in turn may enhance egg production during summer.

An important difference between the sympagic assemblages of the two Polar regions is the much higher abundance and biomass of copepods inhabiting the brine channels in the Antarctic. The three most frequently observed copepod species, calanoids *Stephos longipes* and *Paralabidocera antarctica,* and harpacticoid *Drescheriella glacialis* are circumantarctic, but with differing abundances. Cyclopoid copepods, which dominate the water column, remain loosely associated with the ice. Few amphipod species recorded from the Antarctic ice are thought to be strongly ice-associated, with *Eusirus* spp. being the most numerous (Krapp et al. 2008). Some benthic gammarids, however, (e.g., *Paramoera walkeri, Debroyerella fougneri, Gondogeneia antarctica, Pseudorchomene* cf. *plebs*), may periodically extensively use the ice–water interface as a feeding ground reaching densities close to 13,000 individuals m^{-2} (Swadling 2014).

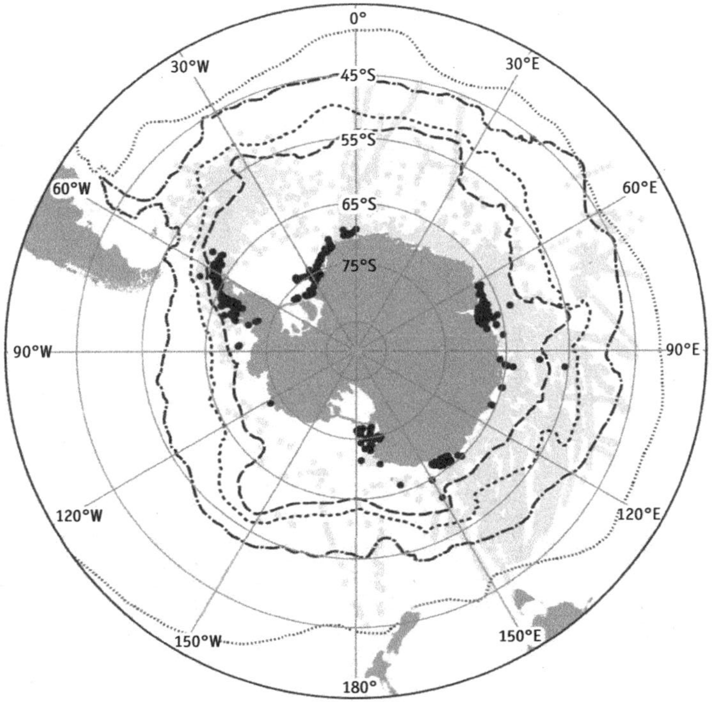

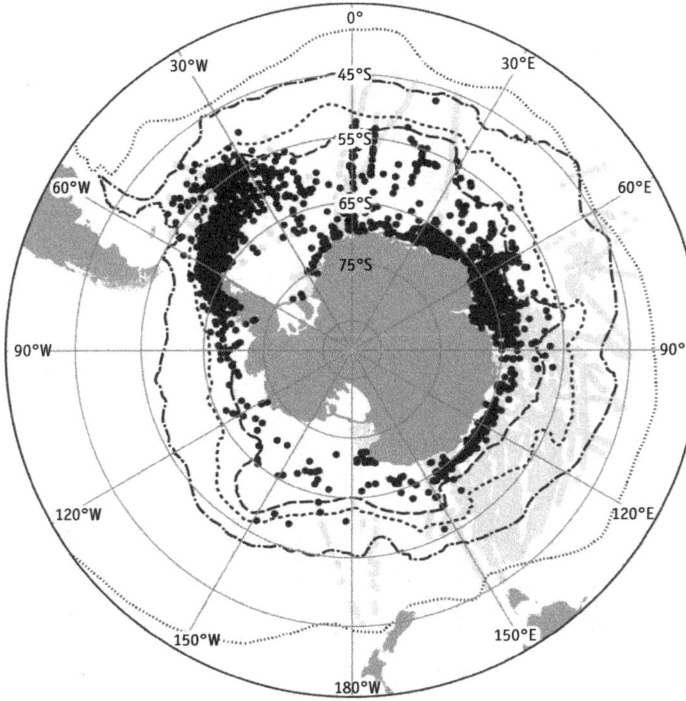

Fig. 9.5.
Comparative geographic distribution of *Euphausia crystallorophias* (left) and *Euphausia superba* (right). Dotted circumpolar lines: Polar Front (second from the south); Subtropical Front (northernmost line) From Cuzin-Roudy et al. 2014.

ON-GOING CHANGES

Loss of Sea Ice

Polar regions have experienced greater warming in recent decades than other oceans (ACIA 2004). Summer melting and ice export through the Fram Strait into the Greenland Sea strongly diminish ice cover in the Arctic each year, and it is predicted that the Arctic Ocean will be ice-free during the summer within 30 years (Berge at al. 2012). Rapid decrease of the extent and thickness of perennial ice is alarming, because due to its complicated underwater topography and good feeding conditions, it is a preferred habitat for amphipods, considered to be obligate residents of this habitat (Lønne and Gulliksen 1991). It has been estimated that the annual biomass loss of sympagic amphipods due to ice export through the Fram Strait could be up to 38×10^5 t WW (Arndt and Lønne 2002). Some released specimens, however, survive when advected to shallow coastal waters. Arndt and Pavlova (2005) identified continental shelves and coastal areas as temporary habitats for sympagic organisms during ice-free periods and as stepping stones for recolonization. *Gammarus wilkitzkii* may occur both in pelagial (Werner et al. 1999) and intertidal zones among macroalgae (Węsławski 1994). Berge et al. (2012) proposed that during ice melt *Apherusa glacialis*, the most numerous of ice-associated amphipods, may perform deep downward migrations, is transported by circum-Arctic currents, and ascends in areas more likely covered by ice.

Increasing Ultraviolet Radiation

Increased ultraviolet radiation (UVR) due to severe ozone loss has been recognized as an important ecological stressor for polar ecosystems (Dahms et al. 2011). Polar organisms are thought to be particularly vulnerable because they have adapted to substantially lower UV irradiance than that imposed by ozone depletion and sea ice thinning (Rautio et al. 2009, Dahms et al. 2011). Ozone depletion is less pronounced in the Arctic than in the Antarctic, where an ozone hole is observed every year (Manney et al. 2011). Biologically relevant doses of UV may penetrate the water and affect plankton and benthic taxa occurring down to 20–30 m depth (Agusti 2008). Exposure to UV radiation, especially UV-B radiation, is deleterious at molecular and cellular levels by damaging lipids, proteins, and DNA either by direct photochemical reaction or by indirect oxidative damage from reactive oxygen species (Llabres et al. 2013).

Pelagic crustaceans, particular early life stages present within the first few meters depth, are especially prone to physiological disorder and death from UVR. Browman et al. (2000) reported deformation, reduced hatching, and survival of first-stage nauplii of the subarctic copepod *Calanus finmarchicus*. Jarman et al. (1999) noted that Antarctic krill DNA may be particularly susceptible to damage from UV-B and that a 60-fold decline in abundance in the Southern Ocean observed between 1970 and 2003 attributed to reduction of sea-ice cover might have been partially caused by increased exposure to UV-B (Llabres et al. 2013). Organisms have evolved a range of behavioral, physiological, and biochemical adaptations that prevent or repair UVR-induced DNA damage. In crustaceans, UVR may be effectively blocked by the exoskeleton (Obermüller and Abele 2004, Van den Broecke et al. 2012). Some species avoid UVR by using unexposed habitats or by downward migrations during the day (Newman et al. 2003, Wold and Norrbin 2004, Dahms et al. 2011). Photoprotection (accumulation of light-filtering compounds) and antioxidant protection (quenching agents and antioxidant enzymes) are widespread (Obermüller et al. 2007, Rautio et al. 2009). Since most of the sun-screening compounds are acquired through an algal-rich diet, herbivorous species (e.g., *Gammarellus homari*, Obermüller et al. 2005; and Antarctic krill *Euphausia superba*, Newman et al. 2000) seem to be better protected than carnivorous ones (such as the lysianassoid amphipod *Anonyx nugax*, Obermüller et al. 2005). The last line of defense, when all above-mentioned protective measures are not sufficient, are DNA repair mechanisms that ensure the genomic integrity (Malloy et al. 1997).

Long-term effects of increasing UVR in polar systems are difficult to assess. Since different species differ in their sensitivity towards UVR shifts in species diversity and community composition may be a consequence. However, it has been suggested that possible consequences of indirect UVR effects via changes in trophic interactions and combined effects of UVR with other environmental stressors can have a more pronounced impact on polar ecosystems (Dahms et al. 2011).

Invasive Species

Ongoing warming is likely to increase the risk of non-indigenous species invasion to the Polar regions since it gradually reduces environmental barriers preventing the colonization by lower-latitude species and enables a rapid increase in human activity and marine traffic. Ballast water and hull biofouling are the two main anthropogenic vectors for marine species introduction, although plastic debris may also contribute to the spread of invasive species within Polar regions (Barnes 2002, Ware et al. 2016, Hughes and Ashton 2017). Invasive species are considered the important threat for native marine biodiversity in both Polar regions, especially in the Antarctic inhabited by highly endemic fauna (Barnes 2002).

Antarctic

A unique case of alien invasive species has been documented so far in Antarctic seas: two adults of the boreo-Arctic crab *Hyas araneus* were recorded off the Antarctic Peninsula (Tavares and De Melo 2004) with no subsequent records (Clarke et al. 2005). However, in the Southern Ocean warming context (Turner et al. 2009, 2013), subantarctic species expanding their range southward, emerging slope species, or organisms introduced by humans may be expected, with potentially harmful consequences for the native biota (Barnes et al. 2006, Aronson et al. 2007). Decapod crustaceans, lithodids in particular, relatively well represented at subantarctic latitudes in contrast to their rare Antarctic occurrences, have been considered potential invasive candidates (Thatje et al. 2005, Aronson et al. 2007, 2015, Smith et al. 2012, Griffiths et al. 2013, 2014). Among the 11 lithodid species so far recorded in Antarctic and subantarctic seas, none occurs on the cold Antarctic continental shelf. Their distribution was shown to be constrained by temperature, with no records from waters colder than 0°C (Hall and Thatje 2011). In contrast to brachyuran crabs, lithodid crabs of the Southern Ocean and adjacent waters present life-history adaptations (eurybathy, cold tolerance, lecithotrophic larvae) predisposing them to be "invaders" of the Antarctic continental shelf in cases where the shelf waters warm up (Aronson et al. 2015).

Arctic

As the Arctic fauna is relatively young, it has close relatives in the areas of its origin, the subarctic Atlantic and Pacific. Those species have a potential to expand their ranges with increasing sea temperature and stronger northward advection of North Atlantic waters along the west Spitsbergen or Pacific waters through the Bering Strait. The boreal hyperiid *Themisto compressa* has expanded northward by ~1000 km to Spitsbergen and is now able to reproduce and coexist in the Arctic with two sibling species, *T. abyssorum* and *T. libellula* (Kraft et al. 2013). Similarly, rapid northward movement of the North Atlantic copepods has been reported, with true Arctic species, such as *Calanus glacialis*, being replaced by temperate *Calanus finmarchicus* (Beaugrand et al. 2009). In addition to natural enhanced dispersal, greater shipping activity (e.g., opening of trans-Arctic shipping routes) and an extended shipping season are likely to increase the ship-mediated invasion risk in the region. Ware et al. (2016) detected 23 non-indigenous zooplankton species in the ballast waters from just eight ships visiting Svalbard in 2011. While current environmental conditions prevent their survival in this area, invasion threat will increase over the coming decades. Invasive benthic species have been discovered in the region at a rather slow pace. The littoral, boreal amphipod *Gammarus oceanicus* has been spreading along Svalbard for at least 20 years, sharing habitat with a local cold-water species, *G. setosus* (Węsławski et al. 2011). The decapod fauna of Isfjorden, Svalbard, has not changed during the last 150 years (Berge et al. 2009). On the other hand, over the last decades, two alien species of large crabs have established their populations in the Arctic. The king crab *Paralithodes camtschaticus* from the northern Pacific was introduced to the Murmansk area in the 1960s, spread in the southern Barents Sea, and then started to move south along the Norwegian coast and northeast of the Kola Peninsula (Falk-Petersen et al. 2011). The snow crab *Chionoecetes opilio*, native to the northwest Atlantic, the north Pacific, and the Japan Sea, was recorded for the first time the Barents Sea in 1996, presumably introduced through ballast water (Alsvåg et al. 2009). At present, it is expanding in the Barents Sea, and it was detected for the first time in Svalbard (Raudfjorden) in summer 2017 (University Centre in Svalbard data, www.unis.no). Both species are likely to have a significant impact both ecologically (predation and competition with local fauna) and economically (commercial fishery).

FUTURE DIRECTIONS

There are several possible avenues for future studies in polar Crustacea. On the one hand, additional sampling is needed to complete species inventories, particularly in hitherto poorly investigated deep seas, but also across wide geographic gradients since the quality and quantity of available information still varies broadly among both regions and taxa. It might be expected that an increased use of molecular methods would bring about identification of many unknown cryptic species. More research is necessary to better understand the role of crustaceans in the polar food web and ecosystem functioning, especially in the context of climate change. Here the question arises: What will be the ecological, physiological and genetic response of polar crustacean populations to the sea ice shrinking and water temperature increase? Sympagic species currently hold a crucial position in the food webs, hence resolving how they may react to changes in ice cover and duration is of fundamental importance to quantify the magnitude of future changes of polar ecosystems. Studies are also needed to assess possible consequences of the non-native species invasion from the boreal regions. Attention should be paid to the possibility of them becoming established residents and their ability to compete and replace native polar taxa, which may lead to the alterations in the food web.

CONCLUSIONS

Crustaceans are widespread and functionally important members of pelagic, benthic, and sea-ice communities of both Polar regions. The number of hitherto recorded species is slightly higher in the Antarctic (over 2,250) than in the Arctic (over 1,930). Species inventory, however, is still far from complete, and discoveries of new species, especially in poorly studied deep seas, may be expected. In comparison with the lower-latitude species, polar crustaceans typically exhibit slower growth and development, have extended life cycles, and attain larger final size. Despite many similarities, the two Polar regions are distinctly different ecosystems due to their different age, geographical and oceanographical settings, and degree of isolation. An old age and long history of isolation resulted in a high rate of endemism within the Antarctic benthic fauna, while the relatively young advective Arctic ecosystem contains few endemics and is dominated by widely distributed Arctic-boreal species. Abundant populations of pelagic microalgal grazers (large euphausiids in the Antarctic and much smaller copepods in the Arctic) play a crucial role in the polar ecosystems supporting upper trophic levels: fish, birds, and sea mammals. Sea-ice communities are similarly diversified but different at both poles. The most notable features of the Southern Ocean sympagic fauna are high abundance and biomass of calanoid copepods and euphausiids and the lack of any species that appear fully dependent on the ice. In the Arctic, tiny harpacticoids and cyclopoids are most numerous in the ice interior, while the ice underside is dominated by four species of amphipods that are thought to be obligate ice dwellers. Within the benthic domain, fewer species are observed in the littoral zone, while heterogeneous sublittoral habitats often host rich and diversified crustacean communities. In the Antarctic, amphipods and isopods are particularly numerous and speciose, while decapods and cirripeds are scare. In the Arctic, Amphipoda considerably outnumber other taxa in terms of species richness, followed by Cumacea and Isopoda. Changes inducted by ongoing warming, such as decline of sea-ice cover, increase of UV radiation, and advection of invasive species, may negatively affect biodiversity of polar crustaceans, but detailed studies are needed to assess the actual threat.

See also supplementary Figure 9.s1 (Antarctic species) and Figure 9.s2 (Arctic species), in the centerfold of the volume.

REFERENCES

ACIA. 2004. Impacts of a Warming Arctic: Arctic Climate Impact Assessment (ACIA) Overview Report. Cambridge University Press, London.

Agusti, S. 2008. The impact of increasing ultraviolet radiation on the polar seas. Pages 25–46 *in* C.M. Duarte, editor. Impacts of Global Warming on Polar Ecosystems. Fundación BBVA, Bilbao, Spain.

Allcock, A.L., and J.M. Strugnell. 2012. Southern Ocean diversity: new paradigms from molecular ecology. Trends in Ecology and Evolution 27:520–528.

Allcock, A.L., and H.J. Griffiths. 2014. Bipolarity. Pages 431–436 *in* C. De Broyer, P. Koubbi, H.J. Griffiths, B. Raymond, C. d'Udekem d'Acoz, A. van de Putte, B. Danis, B. David, S. Grant, J. Gutt, C. Held, G. Hosie, F. Huettmann, A. Post, and Y. Ropert-Coudert, editors. Biogeographic Atlas of the Southern Ocean. Scientific Committee on Antarctic Research, Cambridge, UK.

Alvsvåg, J., A. L. Agnalt, and K. E. Jørstad. 2009. Evidence for a permanent establishment of the snow crab (*Chionoecetes opilio*) in the Barents Sea. Biological Invasions 11:587–595.

Anderson, J.B., S.S. Shipp, A.L. Lowe, J. Smith Wellner, and A.B. Mosola. 2002. The Antarctic ice sheet during the last glacial maximum and its subsequent retreat history: a review. Quaternary Science Reviews 21:49–70.

Arnaud, P.M. 1977. Adaptations within the Antarctic marine benthic ecosystem. Pages 135–137 *in* G.A. Llano, editor. Adaptations within Antarctic Ecosystems. Proceedings of the Third SCAR Symposium on Antarctic Biology, Smithsonian Institution, Washington, DC.

Arndt, C.E., and O.J. Lønne. 2002. Transport of bioenergy by large scale Arctic ice drift. Pages 382–390 *in* V. Squire and P. Langhorne, editors. Ice in the Environment. Proceedings 16th IAHR International Symposium on Ice, Dunedin, New Zealand.

Arndt, C.E., and O. Pavlova. 2005. Origin and fate of ice fauna in the Fram Strait and Svalbard area. Marine Ecology Progress Series 301:55–66.

Arndt, C.E., and K.M. Swadling. 2006. Crustacea in Arctic and Antarctic sea ice: distribution, diet and life history strategies. Advances in Marine Biology 51:198–290.

Arntz, W.E., and C. Rios, editors. 1999. Magellan—Antarctic: ecosystems that drifted apart. Scientia Marina 63 (Suppl. 1):1–518.

Arntz, W.E., G. Lovrich, and S. Thatje, editors. 2005. The Magellan-Antarctic connection: Links and frontiers at high southern latitudes. Scientia Marina 69 (Suppl. 2):1–373.

Arntz, W.E., J. Gutt, and M. Klages. 1997. Antarctic marine biodiversity: an overview. Pages 3–14 *in* B. Battaglia, J. Valencia, and D.W.H. Walton, editors. Antarctic Communities: Species, Structure and Survival. Cambridge University Press, Cambridge, UK.

Aronson, R.B., M. Frederich, R. Price, and S. Thatje. 2015. Prospects for the return of shell-crushing crabs to Antarctica. Journal of Biogeography 42:1–7.

Aronson, R.B., S. Thatje, A. Clarke, L.S. Peck, D.B. Blake, C.D. Wilga, and B.A. Seibel. 2007. Climate change and invasibility of the Antarctic benthos. Annual Review of Ecology, Evolution, and Systematics 38:129–154.

Atkinson, A., V. Siegel, E. Pakhomov, and P. Rothery. 2004. Long-term decline in krill stock and increase in salps within the Southern Ocean. Nature 432:100–103.

Atkinson, A., V. Siegel, E.A. Pakhomov, P. Rothery, V. Loeb, R.M. Ross, L.B. Quetin, K. Schmidt, P. Fretwell, E.J. Murphy, G.A. Tarling, and A.H. Fleming. 2008. Oceanic circumpolar habitats of Antarctic krill. Marine Ecology Progress Series 362:1–23.

Bargelloni, L., L. Zane, N. Derome, G. Lecointre, and T. Patarnello. 2000. Molecular zoogeography of Antarctic euphausiids and notothenioids: from species phylogenies to intraspecific patterns of genetic variation. Antarctic Science 12:259–268.

Barker, P.F., and E. Thomas. 2004. Origin, signature and palaeoclimatic influence of the Antarctic Circumpolar Current. Earth Science Review 66:143–162.

Barnes, D. K. A. 2002. Invasions by marine life on plastic debris. Nature 416:808–809.

Barnes, D.K.A., D.A. Hodgson, P. Convey, C.S. Allen, and A. Clarke. 2006. Incursion and excursion of Antarctic biota: past, present and future. Global Ecology and Biogeography 15:121–142.

Beaugrand, G., C. Luczak, and M. Edwards. 2009. Rapid biogeographical plankton shifts in the North Atlantic Ocean. Global Change Biology 15:1790–1803.

Berge J., Ø. Varpe, M.A. Moline, A. Wold, P.E. Renaud, M. Daase, and S. Falk-Petersen. 2012. Retention of ice-associated amphipods: possible consequences for an ice-free Arctic Ocean. Biology Letters 8: 1012–1015.

Berge, J., P.E. Renaud, K. Eiane, B. Gulliksen, F.R. Cottier, Ø. Varpe, and T. Brattegard. 2009. Changes in decapod fauna of an Arctic fjord during the last 100 years (1908–2007). Polar Biology 32:953–961.

Błażewicz-Paszkowycz, M. 2014. Tanaidacea. Pages 173–180 *in* C. De Broyer, P. Koubbi, H.J. Griffiths, B. Raymond, C. d'Udekem d'Acoz, A. Van De Putte, B. Danis, B. David, S. Grant, J. Gutt, C. Held, G. Hosie, F. Huettmann, A. Post and Y. Ropert-Coudert, editors. Biogeographic Atlas of the Southern Ocean. Scientific Committee on Antarctic Research, Cambridge, UK.

Bluhm, B., W.G. Ambrose Jr, M. Bergmann, L.M. Clough, A.V. Gebruk, C. Hasemann, K. Iken, M. Klages, I.R. MacDonald, P.E. Renaud, I. Schewe, T. Soltwedel, and M. Włodarska-Kowalczuk. 2011. Diversity of the arctic deep-sea benthos. Marine Biodiversity 41:87–107.

Bocher, P., Y. Cherel, J.-P. Labat, P. Mayzaud, S. Razouls, and P. Jouventin. 2001. Amphipod-based food web: *Themisto gaudichaudii* caught in nets and by seabirds in Kerguelen waters, southern Indian Ocean. Marine Ecology Progress Series 223:261–276.

Brandao, S.N., J. Sauer, and I. Schön. 2010. Circumantarctic distribution in Southern Ocean benthos? A genetic test using the genus *Macroscapha* (Crustacea, Ostracoda) as a model. Molecular Phylogenetics and Evolution 55:1055–1069.

Brandt, A. 1992. Origin of Antarctic Isopoda (Crustacea, Malacostraca). Marine Biology 113:415–423.

Brandt, A. 1997. Biodiversity of peracarid crustaceans (Malacostraca) from the shelf down to the deep Arctic Ocean. Biodiversity and Conservation 6:1533–1556.

Brandt, A. 2000. Hypotheses on Southern Ocean peracarid evolution and radiation (Crustacea, Malacostraca). Antarctic Science 12:269–275.

Brandt, A., and B. Ebbe. 2007. ANDEEP III ANtarctic benthic DEEP-sea biodiversity: colonisation history and recent community patterns. Deep-Sea Research II 54:1645–1904.

Brandt, A., and B. Hilbig. 2004. ANDEEP (ANtarctic benthic DEEP-sea biodiversity: colonization history and recent community patterns)—a tribute to Howard L. Sanders. Deep-Sea Research II 51:1457–1919.

Brandt, A., W. Brökeland, M. Choudhury, S. Brix, S. Kaiser, and M. Malyutina. 2007a. Deep-sea isopod biodiversity, abundance and endemism in the Atlantic sector of the Southern Ocean—results from the ANDEEP I—III expeditions. Deep-Sea Research II 54:1760–1775.

Brandt, A., C. De Broyer, I. De Mesel, K.E. Ellingsen, A.J. Gooday, B. Hilbig, K. Linse, M.R.A. Thomson, and P.A. Tyler. 2007b. The biodiversity of the deep Southern Ocean benthos. Philosophical Transactions of the Royal Society of London Series B 362:39–66.

Brandt, A., C. De Broyer, I. De Mesel, K.E. Ellingsen, A.J. Gooday, B. Hilbig, K. Linse, M.R.A. Thomson, and P.A. Tyler. 2012. The biodiversity of the deep Southern Ocean benthos. Pages 291–334 *in* A.D. Rogers, N.M. Johnston, E.J. Murphy, and A. Clarke, editors. Antarctic Ecosystems: An Extreme Environment in a Changing World. Wiley-Blackwell, Chichester, UK.

Brandt, A., K. Linse, and M. Schüller. 2009. Bathymetric distribution patterns of Southern Ocean macrofaunal taxa: Bivalvia, Gastropoda, Isopoda and Polychaeta. Deep-Sea Research I 56:2013–2025.

Brey, T., C. Dahm, M. Gorny, M. Klages, M. Stiller, and W.E. Arntz. 1996. Do Antarctic benthic invertebrates show an extended level of eurybathy? Antarctic Science 8:3–6.

Britton, J.C., and B. Morton. 1994. Marine carrion and scavengers. Oceanography and marine biology: an annual review 32:369–434.

Browman, H. I., C. Alonso Rodriquez, F. Béland, J. J. Cullen, R. F., Davis, J. H. M. Kouwenberg, P. S. Kuhn, B. McArthur, J. A. Runge, J.-F. St.-Pierre, and R. D. Vetter. 2000. Impact of ultraviolet radiation on marine crustacean zooplankton and ichthyoplankton: a synthesis of results from the estuary and Gulf of St. Lawrence, Canada. Marine Ecology Progress Series 199:293–311.

Buchholz, F., C. Buchholz, and J.M. Węsławski. 2010. Ten years after: krill as indicator of changes in the macro-zooplankton communities of two Arctic fjords. Polar Biology 33:101–114.

Chapelle, G. 2001. Antarctic and Baikal amphipods. A key for understanding polar gigantism. Ph.D. dissertation, Université Catholique de Louvain, Louvain-la-Neuve, Belgium.

Chapelle, G., and L.S. Peck. 1999. Polar gigantism and oxygen. Nature 399:114–115.

Chapelle, G., and L.S. Peck. 2004. Amphipod crustacean size spectra: new insights in the relationship between size and oxygen. Oikos 106:167–175.

Clarke, A., and J.A. Crame. 1989. The origin of the Southern Ocean marine fauna. Pages 253–268 *in* J.A. Crame, editor. Origins and Evolution of the Antarctic Biota. The Geological Society, London.

Clarke, A., and J.A. Crame. 1992. The Southern Ocean benthic fauna and climate change: a historical perspective. Philosophical Transactions of the Royal Society of London, Series B 338:299–309.

Clarke, A., and N.M. Johnston. 2003. Antarctic marine benthic diversity. Oceanography and Marine Biology: An Annual Review 41:47–114.

Clarke, A., and L. S. Peck. 1991. The physiology of polar marine zooplankton. Polar Research 10:355–369.

Clarke, A., and P.A. Tyler. 2008. Adult Antarctic krill feeding at abyssal depths. Current Biology 18:282–285.

Clarke, A., D.K.A. Barnes, and D.A. Hodgson. 2005. How isolated is Antarctica? Trends in Ecology and Evolution 20:1–3.

Clarke, A., H.J. Griffiths, K. Linse, D.K.A. Barnes, and J.A. Crame. 2007. How well do we know the Antarctic marine fauna? A preliminary study of macroecological and biogeographical patterns in Southern Ocean gastropod and bivalve molluscs. Diversity and Distributions 13:620–632.

Conlan, K., E. Hendrycks, A. Aitken, B. Williams, S. Blasco, and E. Crawford. 2013. Macrofaunal biomass distribution on the Canadian Beaufort Shelf. Journal of Marine Systems 127:76–87.

Cuzin-Roudy, J., J.-O. Irisson, F. Penot, S. Kawaguchi, and C. Vallet, 2014. Southern Ocean euphausiids. Pages 309–320 *in* C. De Broyer, P. Koubbi, H.J. Griffiths, B. Raymond, C. d'Udekem d'Acoz, A. van de Putte, B. Danis, B. David, S. Grant, J. Gutt, C. Held, G. Hosie, F. Huettmann, A. Post, and Y. Ropert-Coudert, editors. Biogeographic Atlas of the Southern Ocean. Scientific Committee on Antarctic Research, Cambridge, UK.

Daase, M., S. Falk-Petersen, Ø. Varpe, G. Darnis, J. E. Søreide, A. Wold, E. Leu, J. Berge, B. Philippe, and L. Fortier. 2013. Timing of reproductive events in the marine copepod *Calanus glacialis*: a Pan-Arctic perspective. Canadian Journal of Fisheries and Aquatic Sciences 70:871–884.

Dahms, H.U., S. Dobretsov, and J. S. Lee. 2011. Effects of UV radiation on marine ectotherms in Polar regions. Comparative Biochemistry and Physiology, Part C 153:363–371.

Dayton, P.K., B.J. Mordida, and F. Bacon. 1994. Polar marine communities. American Zoologist 34:90–99.

De Broyer, C. 1977. Analysis of the gigantism and dwarfness of Antarctic and Subantarctic gammaridean Amphipoda. Pages 327–341 *in* G. A. Llano, editor. Adaptations Within Antarctic Ecosystems. Proceedings of the Third SCAR Symposium on Antarctic Biology, Smithsonian Institution, Washington, DC.

De Broyer, C., and B. Danis. 2011. How many species in the Southern Ocean? Towards a dynamic inventory of the Antarctic marine species. Deep-Sea Research II 58:5–17.

De Broyer, C., and A. Jażdżewska. 2014. Biogeographic patterns of Southern Ocean benthic amphipods. Pages 155--166 *in* C. De Broyer, P. Koubbi, H.J. Griffiths, B. Raymond, C. d'Udekem d'Acoz, A. van de Putte, B. Danis, B. David, S. Grant, J. Gutt, C. Held, G. Hosie, F. Huettmann, A. Post, and Y. Ropert-Coudert, editors. Biogeographic Atlas of the Southern Ocean. Scientific Committee on Antarctic Research, Cambridge, UK.

De Broyer, C., K. Jażdżewski, and P. Dauby. 2003. Biodiversity patterns in the Southern Ocean: lessons from Crustacea. Pages 201–214 *in* A.H.L. Huiskes, W.W.C. Gieskes, J. Rozema, R.M.L. Schorno, S.M. van der Vies, and W.J. Wolff, editors. Antarctic Biology in a Global Context. Proceedings of the SCAR Biology Symposium Amsterdam, Backhuys Publishers, Leiden, The Netherlands.

De Broyer, C., and P. Koubbi. 2014. The biogeography of the Southern Ocean. Pages 2–9 *in* C. De Broyer, P. Koubbi, H.J.Griffiths, B. Raymond, C. d'Udekem d'Acoz, A. Van De Putte, B. Danis, B. David, S. Grant, J. Gutt, C. Held, G. Hosie, F. Huettmann, A. Post and Y. Ropert-Coudert, editors. Biogeographic Atlas of the Southern Ocean. Scientific Committee on Antarctic Research, Cambridge, UK.

De Broyer, C., P. Koubbi, H.J. Griffiths, B. Raymond, C. d'Udekem d'Acoz, A. Van de Putte, B. Danis, B. David, S. Grant, J. Gutt, C. Held, G. Hosie, F. Huettmann, A. Post, and Y. Ropert-Coudert, editors. 2014. Biogeographic Atlas of the Southern Ocean. Scientific Committee on Antarctic Research, Cambridge, UK.

De Broyer, C., F. Nyssen, and P. Dauby. 2004. The crustacean scavenger guild in Antarctic shelf, bathyal and abyssal communities. Deep-Sea Research II 51:1733–1752.

De Broyer, C., Y. Scailteur, G. Chapelle, and M. Rauschert. 2001. Diversity of epibenthic habitats of gammaridean amphipods in the eastern Weddell Sea. Polar Biology 24:744–753.

Dell, R.K. 1972. Antarctic benthos. Advances in Marine Biology 10:1–216.

Dunton, K. 1992. Arctic biogeography: the paradox of the marine benthic fauna and flora. Trends in Ecology and Evolution 7:183–189.

Ekman, S. 1953. Zoogeography of the Sea. Sidgwick and Jackson, London.

Everson, I., editor. 2000. Krill. Biology, Ecology and Fisheries. Blackwell Science, Oxford, UK.

Falk-Petersen, J., P. E. Renaud, and N. Anisimova. 2011. Establishment and ecosystem effects of the alien invasive red king crab (*Paralithodes camtschaticus*) in the Barents Sea—a review. ICES Journal of Marine Science 68: 479–488.

Feldmann, R.M., D.M. Tshudy, and M.R.A. Thomson. 1993. Late Cretaceous and Paleocene decapod crustaceans from James Ross Basin, Antarctic Peninsula. Journal of Palaeontology 67:1–41.

Frederich, M., F.J. Sartoris, and H.-O. Portner. 2001. Distribution patterns of decapod crustaceans in polar areas: a result of magnesium regulation? Polar Biology 24:719–723.

Gorny, M. 1999. On the biogeography and ecology of the Southern Ocean decapod fauna. Scientia Marina 63 (Suppl. 1):367–382.

Graeve, M., G. Kattner, and D. Piepenburg. 1997. Lipids in Arctic benthos: does the fatty acid and alcohol composition reflect feeding and trophic interactions? Polar Biology 18:53–61.

Griffiths, H.J. 2010. Antarctic marine biodiversity—what do we know about the distribution of life in the Southern Ocean? PLoS ONE 5:e11683.

Griffiths, H.J., D.K.A. Barnes, and K. Linse. 2009. Towards a generalized biogeography of the Southern Ocean benthos. Journal of Biogeography 36:162–177.

Griffiths, H.J., R.J. Whittle, S.J. Roberts, M. Belchier, and K. Linse. 2013. Antarctic crabs: Invasion or endurance? PLoS ONE:e66981.

Griffiths, H.J., R.J. Whittle, S.J. Roberts, M. Belchier, K. Linse, and S. Thatje. 2014. Decapoda: crabs and lobsters. Pages 185–189 *in* C. De Broyer, P. Koubbi, H.J. Griffiths, B. Raymond, C. d'Udekem d'Acoz, A. van de Putte, B. Danis, B. David, S. Grant, J. Gutt, C. Held, G. Hosie, F. Huettmann, A. Post, and Y. Ropert-Coudert, editors. Biogeographic Atlas of the Southern Ocean. Scientific Committee on Antarctic Research, Cambridge, UK.

Gutt, J., V. Cummings, P. Dayton, E. Isla, A. Jentsch, and S. Schiaparelli. 2015. Antarctic animal forests: three-dimensional communities in Southern Ocean ecosystems. Pages 1–30 *in* S. Rossi, editor. Marine Animal Forests. Springer International, Basel, Switzerland.

Hall, S., and S. Thatje. 2011. Temperature-driven biogeography of the deep-sea family Lithodidae (Crustacea: Decapoda: Anomura) in the Southern Ocean. Polar Biology 34:363–370.

Havermans, C. 2012. DNA barcoding, phylogeography and phylogeny of the Lysianassoidea (Crustacea: Amphipoda) from the Southern Ocean and the world's deep seas. Ph.D thesis, Université Catholique de Louvain, Louvain-la-Neuve, Belgium.

Havermans, C. 2014. Phylogeographic patterns of the Lysianassoidea (Crustacea: Peracarida: Amphipoda). Pages 441–447 *in* C. De Broyer, P. Koubbi, H.J. Griffiths, B. Raymond, C. d'Udekem d'Acoz, A. van de Putte, B. Danis, B. David, S. Grant, J. Gutt, C. Held, G. Hosie, F. Huettmann, A. Post, and Y. Ropert-Coudert, editors. Biogeographic Atlas of the Southern Ocean. Scientific Committee on Antarctic Research, Cambridge, UK.

Havermans, C. 2016. Have we so far only seen the tip of the iceberg? Exploring species diversity and distribution of the giant amphipod *Eurythenes*. Biodiversity 17:12–25.

Havermans, C., Z.T. Nagy, G. Sonet, C. De Broyer, and P. Martin. 2010. Incongruence between molecular phylogeny and morphological classification in amphipod crustaceans: a case study of Antarctic lysianassoids. Molecular Phylogenetics and Evolution 55:202–209.

Havermans, C., G. Sonet, C. d'Udekem d'Acoz, Z.T. Nagy, P. Martin, S. Brix, T. Riehl, S. Agrawal, and C. Held. 2013. Genetic and morphological divergences in the cosmopolitan deep-sea amphipod *Eurythenes gryllus* reveal a diverse abyssal and a bipolar species. PLoS ONE 8: e74218.

Hedgpeth, J.W. 1969. Introduction to Antarctic zoogeography. Pages 1–29 *in* V.C. Bushnell and J.W. Hedgpeth, editors. Distribution of selected groups of marine invertebrates in waters south of 35°S latitude. Antarctic Map Folio Series, Folio 11. American Geographical Society, New York.

Hedgpeth, J.W. 1970. Marine biogeography of the Antarctic regions. Pages 97–104 *in* M.W. Holdgate, editor. Antarctic Ecology, Vol. 1. Academic Press, London.

Held, C. 2000. Phylogeny and biogeography of serolid isopods (Crustacea, Isopoda, Serolidae) and the use of ribosomal expansion segments in molecular systematics. Molecular Phylogenetics and Evolution 15:165–178.

Held, C., and J.W. Wägele. 2005. Cryptic speciation in the giant Antarctic isopod *Glyptonotus antarcticus* (Isopoda: Valvifera: Chaetiliidae). Scientia Marina 69:175–181.

Hirche, H.J. 1997 Life cycle of the copepod *Calanus hyperboreus* in the Greenland Sea. Marine Biology 128:607–618.

Huang, Y.M., M.O. Amsler, J.B. McClintock, C.D. Amsler, and B.J. Baker. 2007. Patterns of gammaridean amphipod abundance and species composition associated with dominant subtidal macroalgae from the western Antarctic Peninsula. Polar Biology 30:1417–1430.

Huenerlage, K., M. Graeve, C. Buchholz, and F. Buchholz. 2015. The other krill: overwintering physiology of adult *Thysanoessa inermis* (Euphausiacea) from the high-Arctic Kongsfjord. Aquatic Biology 23:225–235.

Hughes, K. A., and G. V. Ashton. 2017. Breaking the ice: the introduction of biofouling organisms to Antarctica on vessel hulls. Aquatic Conservation: Marine and Freshwater Ecosystems 27: 158–164.

Ingram, C.L., and R.R. Hessler. 1987. Population biology of the deep-sea amphipod *Eurythenes gryllus*: inferences from instar analyses. Deep Sea Research Part A 34:1889–1910.

Jarman, S., N. Elliott, S. Nicol, A. McMinn, and S. Newman. 1999. The base composition of the krill genome and its potential susceptibility to damage by UV-B. Antarctic Science 11:23–26.

Josefson, A.B., V. Mokievsky, M. Bergman, M.E. Blicher, B. Bluhm, S. Cochrane, N. V. Denisenko, C. Hasemann, L.L. Jorgensen, M. Klages, I. Schewe, M.K. Sejr, T. Soltwedel, and M. Wlodarska-Kowalczuk. 2013. Marine Invertebrates. Pages 277–309 *in* H. Melforte,editor. Arctic Biodiversity Assessment. Status and Trends in Arctic Biodiversity. Conservation of Arctic Flora and Fauna, Akureyri, Iceland.

Kosobokova, K.N, R.R. Hopcroft and H.-J. Hirche. 2011. Patterns of zooplankton diversity through the depths of the Arctic's central basins. Marine Biodiversity 41:29–50.

Koubbi, P., C. De Broyer, H.J. Griffiths, B. Raymond, C. d'Udekem d'Acoz, A. Van de Putte, B. Danis, B. David, S. Grant, J. Gutt, C. Held, G. Hosie, F. Huettmann, A. Post, Y. Ropert-Coudert, M. Stoddart, K.M. Swadling, and V. Wadley. 2014. Conclusions: present and future of Southern Ocean biogeography. Pages 469–476 *in* C. De Broyer, P. Koubbi, H.J. Griffiths, B. Raymond, C. d'Udekem d'Acoz, A. van de Putte, B. Danis, B. David, S. Grant, J. Gutt, C. Held, G. Hosie, F. Huettmann, A. Post, and

Y. Ropert-Coudert, editors. Biogeographic Atlas of the Southern Ocean. Scientific Committee on Antarctic Research, Cambridge, UK.

Kouwenberg, J.H.M., C. Razouls, and N. Desreumaux. 2014. Southern Ocean pelagic copepods. Pages 290–296 *in* C. De Broyer, P. Koubbi, H.J. Griffiths, B. Raymond, C. d'Udekem d'Acoz, A. van de Putte, B. Danis, B. David, S. Grant, J. Gutt, C. Held, G. Hosie, F. Huettmann, A. Post, and Y. Ropert-Coudert, editors. Biogeographic Atlas of the Southern Ocean. Scientific Committee on Antarctic Research, Cambridge, UK.

Kraft, A., E.-M. Nöthig, E. Bauerfeind, D. J. Wildish, G. W. Pohle, U. V. Bathmann, A. Beszczynska-Möller, and M. Klages. 2013. First evidence of reproductive success in a southern invader indicates possible community shifts among Arctic zooplankton. Marine Ecology Progress Series 493:291–296.

Krapp, R. H., J. Berge, H. Flores, B. Gulliksen, and I. Werner. 2008. Sympagic occurrence of eusirid and lysianassoid amphipods under Antarctic pack ice. Deep-Sea Research Part II 55:1015–1023.

Laakmann, S., H. Auel, and M. Kochzius. 2012. Evolution in the deep sea: biological traits, ecology and phylogenetics of pelagic copepods. Molecular Phylogenetics and Evolution 65:535–546.

Lawson, G.L., P.H. Wiebe, C.J. Ashjian, S.M. Gallager, C.S. Davis, and J.D. Warren. 1994. Acoustically-inferred zooplankton distribution in relation to hydrography west of the Antarctic Peninsula. Deep-Sea Research II 51:2041–2071.

Lee R.F., W. Hagen, and G. Kattner. 2006. Lipid storage in marine zooplankton. Marine Ecology Progress Series 307:273–306.

Leese, F., and C. Held. 2008. Identification and characterization of microsatellites from the Antarctic isopod *Ceratoserolis trilobitoides*: nuclear evidence for cryptic species. Conservation Genetics 9:1369–1372.

Legeżyńska J., M. Włodarska-Kowalczuk, M. Gluchowska, M. Ormańczyk, M. Kędra, and J.M. Węsławski. 2017. The malacostracan fauna of two Arctic fjords (west Spitsbergen): the diversity and distribution patterns of its pelagic and benthic components. Oceanologia 59:541–564.

Legeżyńska, J., M. Kędra, and W. Walkusz. 2012. When season does not matter: summer and winter trophic ecology of Arctic amphipods. Hydrobiology 684:189–214.

Linse, K., H.J. Griffiths, D.K.A. Barnes, and A. Clarke. 2006. Biodiversity and biogeography of Antarctic and subantarctic Mollusca. Deep-Sea Research II 53:985–1008.

Llabrés, M., S. Agustí, M. Fernández, A. Canepa, F. Maurin, F. Vidal, and C. M. Duarte. 2013. Impact of elevated UVB radiation on marine biota: a meta-analysis. Global Ecology and Biogeography 22:131–144.

Loeb, V., V. Siegel, O. Holm-Hansen, R. Hewitt, W. Fraser, W. Trivelpiece, and S. Trivelpiece. 1997. Effects of sea-ice extent and krill or salp dominance on the Antarctic food web. Nature 387:897–900.

Lønne, O.J., and B. Gulliksen. 1991. On the distribution of sympagic macro-fauna in the seasonally ice covered Barents Sea. Polar Biology 11:457–469.

Lörz, A.-N., and C. Held. 2004. A preliminary molecular and morphological phylogeny of the Antarctic Epimeriidae and Iphimediidae (Crustacea, Amphipoda). Molecular Phylogenetics and Evolution 31:4–15.

Malloy, K.D., M. A. Holman, D. Mitchell, and H. W. Detrich III. 1997. Solar UVB-induced DNA damage and photoenzymatic DNA repair in Antarctic zooplankton. Proceedings of the National Academy of Sciences 94:1258–1263.

Malyutina, M., and A. Brandt. 2007. Diversity and zoogeography of Antarctic deep-sea Munnopsidae (Crustacea, Isopoda, Asellota). Deep-Sea Research II 54:1790–1805.

Manney, G. L., M. L. Santee, M. Rex, N. J. Livesey, M. C. Pitts, P. Veefkind, E. R. Nash, I. Wohltmann, R. Lehmann, L. Froidevaux, L. R. Poole, M. R. Schoeber, D. P. Haffner, J. Davies, V. Dorokhov, H. Gernandt, B. Johnson, R. Kivi, E. Kyro, N. Larsen, P. F. Levelt, A. Makshtas, C. T. McElroy, H. Nakajima, M. C. Parrondo, D. W. Tarasick, P. von der Gathen, K. A.Walker, and N. S. Zinoviev. 2011. Unprecedented Arctic ozone loss in 2011. Nature 478:469–475.

Moran, A.L., and H.A. Woods. 2012. Why might there be giants? Towards an understanding of polar gigantism. Journal of Experimental Biology 215:1995–2002.

Newman, S. J., D. Ritz, and Nicol S. 2003. Behavioural reactions of Antarctic krill (*Euphausia superba* Dana) to ultraviolet and photosynthetically active radiation. Journal of Experimental Marine Biology and Ecology 297:203–217.

Newman, S. J., W. C. Dunlap, S. Nicol, and D. Ritz. 2000. Antarctic krill (*Euphausia superba*) acquire a UV-absorbing mycosporine-like amino acid from dietary algae. Journal of Experimental Marine Biology and Ecology 255:93–110.

Newman, W.A., and A. Ross. 1971. Antarctic Cirripedia. Antarctic Research Series 14. American Geophysical Union, Washington, DC.

Nicol, S., I. Forster, and J. Spence. 2000. Products derived from krill. Pages 262–283 *in* I. Everson, editor. Krill. Biology, Ecology and Fisheries. Blackwell Science, Oxford, UK.

Nikula, R., C.I. Fraser, H.G. Spencer, and J.M. Waters. 2010. Circumpolar dispersal by rafting in two subantarctic kelp-dwelling crustaceans. Marine Ecology Progress Series, 405:221–230.

Nygård, H., M. Vihtakari, and J. Berge. 2009. Life history of *Onisimus caricus* (Amphipoda: Lysianassoidea) in a high Arctic fjord. Aquatic Biology 5:63–74.

Obermüller, B., U. Karsten, and D. Abele. 2005. Response of oxidative stress parameters and sun screening compounds in Arctic amphipods during experimental exposure to maximal natural UVB radiation. Journal of Experimental Marine Biology and Ecology 323:100–117.

Obermüller, B., and D. Abele. 2004. Different UVB-tolerance in herbivorous versus carnivorous amphipods from Kongsfjorden. Berichte zur Polar- und Meeresforschung 492:222–230.

Obermüller, B., S.A. Morley, D.K.A. Barnes, and L.S. Peck. 2010. Seasonal physiology and ecology of Antarctic marine benthic predators and scavengers. Marine Ecology Progress Series 415:109–126.

Obermüller, B., S. Puntarulo, and D. Abele. 2007. UV-tolerance and instantaneous physiological stress responses of two Antarctic amphipod species *Gondogeneia antarctica* and *Djerboa furcipes* during exposure to UV radiation. Marine Environmental Research 64:267–85.

Orsi, A.H., T. Whitworth III, and W.D. Nowlin Jr. 1995. On the meridional extent and fronts of the Antarctic Circumpolar Current. Deep-Sea Research I 42:641–673.

Padovani, L.N., M. Delia Viñas, F. Sánchez, and H. Mianzan. 2012. Amphipod-supported food web: *Themisto gaudichaudii*, a key food resource for fishes in the southern Patagonian Shelf. Journal of Sea Research 67:85–90.

Park, T., and F.D. Ferrari. 2009. Species diversity and distributions of pelagic calanoid copepods from the Southern Ocean. Pages 143–179 *in* I. Krupnik, M.A. Lang, and S.E. Miller, editors. Smithsonian at the Poles: Contributions to International Polar Year Science. Smithsonian Institution Scholarly Press, Washington, DC.

Parkinson, C. L. 2014. Global sea ice coverage from satellite data: annual cycle and 35-yr trends. Journal of Climate 27:9377–9382.

Peck L.S., P. Convey, and D.K.A. Barnes. 2006. Environmental constraints on life histories in Antarctic ecosystems: tempos, timings and predictability. Biological Review 81:75–109.

Peck, L.S., and G. Chapelle. 1999. Amphipod gigantism dictated by oxygen availability?: A reply. Ecology Letters 2:401–403.

Peña, Cantero, A.L. 2004. How rich is the deep-sea Antarctic benthic hydroid fauna? Polar Biology 27:767–774.

Piepenburg, D., P. Archambault, W.G. Ambrose Jr., A.L. Blanchard, B.A. Bluhm, M.L. Carroll, K.E. Conlan, M. Cusson, H.M. Feder, J.M. Grebmeier, S.C. Jewett, M. Lévesque, V.V. Petryashev, M.K. Sejr, B.I. Sirenko, and M. Włodarska-Kowalczuk. 2011. Towards a pan-Arctic inventory of the species diversity of the macro- and megabenthic fauna of the Arctic shelf seas. Marine Biodiversity 41:51–70.

Poltermann, M. 2000. Growth, production and productivity of the Arctic sympagic amphipod *Gammarus wilkitzkii*. Marine Ecology Progress Series 193:109–116.

Post, A.L., A.J.S. Meijers, A.D. Fraser, K.M. Meiners, J. Ayers, N.L. Bindoff, H.J. Griffiths, A.P. Van de Putte, P.E. O'Brien, K.M. Swadling, and B. Raymond. 2014. Environmental setting. Pages 45–63 *in* C. De Broyer, P. Koubbi, H.J. Griffiths, B. Raymond, C. d'Udekem d'Acoz, et al., editors. Biogeographic Atlas of the Southern Ocean. Scientific Committee on Antarctic Research, Cambridge, UK.

Quetin, L.B., and R.M. Ross. 2001. Environmental variability and its impact on the reproductive cycle of Antarctic krill. American Zoologist 41:74–89.

Quetin, L.B., and R.M. Ross. 2009. Life under Antarctic pack ice: A krill perspective. Pages 285–298 *in* I. Krupnik, M.A. Lang, and S.E. Miller, editors. Smithsonian at the Poles: Contributions to International Polar Year Science. Smithsonian Institution Scholarly Press, Washington, DC.

Raupach, M.J., M. Malyutina, A. Brandt, and J.W. Wägele. 2007. Molecular data reveal a highly diverse species flock within the munnopsoid deep-sea isopod *Betamorpha fusiformis* (Barnard, 1920) (Crustacea: Isopoda: Asellota) in the Southern Ocean. Deep-Sea Research II 54:1820–1830.

Raupach, M.J., S. Thatje, J. Dambach, P. Rehm, B. Misof, and F. Leese. 2010. Genetic homogeneity and circum-Antarctic distribution of two benthic shrimp species of the Southern Ocean, *Chorismus antarcticus* and *Nematocarcinus lanceopes*. Marine Biology 157:1783–1797.

Rautio, M., S. Bonilla, and W.F. Vincent. 2009. UV photoprotectants in Arctic zooplankton. Aquatic Biology 7:93–105.

Riddle, M.J., M. Craven, P.M. Goldsworthy, and F. Carsey. 2007. A diverse benthic assemblage 100 km from open water under the Amery Ice Shelf, Antarctica. Paleoceanography 22:1–8.

Riesgo, A., S. Taboada, and C. Avila. 2015. Evolutionary patterns in Antarctic marine invertebrates: an update on molecular studies. Marine Genomics 23:1–13.

Rodríguez, E., P.J. López-González, and J.M. Gili. 2007. Biogeography of Antarctic sea anemones (Anthozoa, Actiniaria): What do they tell us about the origin of the Antarctic benthic fauna? Deep-Sea Research II 54:1876–1904.

Rogers, A.D. 2012. Evolution and biodiversity of Antarctic organisms: a molecular perspective. Pages 417–467 *in* A.D. Rogers, N.M Johnston., E.J. Murphy, and A. Clarke, editors. Antarctic Ecosystems: An Extreme Environment in a Changing World. Wiley-Blackwell, Chichester, UK.

Sainte-Marie, B. 1991. A review of the reproductive bionomics of aquatic gammaridean amphipods: variation of life history traits with latitude, depth, salinity and superfamily. Hydrobiologia 223:189–227.

Sainte-Marie, B., I. Bérubé I., S. Brillon, and F. Hazel. 2006. Observations on the growth of the sculptured shrimp *Sclerocrangon* Boreas (Decapoda: Caridea). Journal of Crustacean Biology 26:55–62.

San Vicente, C. 2010. Species diversity of Antarctic mysids (Crustacea: Lophogastrida and Mysida). Pages 1–80 *in* T.J. Mulder, editor. Antarctica: Global, Environmental and Economic Issues. Nova Science Publishers, New York.

Shea, J.R., and J.A. Percy. 1990. Salinity tolerance and osmoregulation of the Arctic marine amphipods *Onisimus litoralis* (Krøyer) and *Anonyx nugax* (Phipps). Polar Biology 10:275–281.

Sieg, J. 1986. Crustacea Tanaidacea of the Antarctic and the Subantarctic. 1. On material collected at Tierra del Fuego, Isla de los Estados, and the west coast of the Antarctic Peninsula. Pages 1–180 *in* L.S. Korniker, editor. Biology of the Antarctic Seas 18. Antarctic Research Series 45. American Geophysical Union, Washington, DC.

Sieg, J. 1992. On the origin and age of the Antarctic tanaidacean fauna. Pages 421–429 *in* V.A. Gallardo, Ferretti, O., and H.I. Moyano, editors. Oceanografía en Antártica. Centro Eula, Concepción, Chile.

Siegel, V. 2005. Distribution and population dynamics of *Euphausia superba*: summary of recent findings. Polar Biology 29:1–22.

Sirenko, B.I. 2001. List of species of free living invertebrates of Eurasian Arctic seas and adjacent deep waters. Explorations of the Fauna of the Seas 51:1–125.

Sirenko, B. I. 2009. Main differences in macrobenthos and bentic communities of the Arctic and Antarctic, as illustrated by comparison of the Laptev and Weddell Sea Faunas. Russian Journal of Marine Biology 35:445–453.

Sirenko, B.I., Clarke, R. Hopcroft, Huettmann, B. Bluhm, and R.Gradinger, editors. 2016. The Arctic Register of Marine Species (ARMS) compiled by the Arctic Ocean Diversity (ArcOD). http://www.marinespecies.org/arms.

Smetacek, V., P. Assmy, and J. Henjes. 2004. The role of grazing in structuring Southern Ocean pelagic ecosystems and biogeochemical cycles. Antarctic Science 16:541–558.

Smith, C.R., L.J. Grange, D.L. Naudts, B. Huber, and L. Guidi. 2012. A large population of king crabs in Palmer Deep on the west Antarctic Peninsula shelf and potential invasive impacts. Proceedings of the Royal Society of London B 279:1017–1026.

Smith, C.R., S. Mincks, and D.J. DeMaster. 2006. A synthesis of bentho-pelagic coupling on the Antarctic shelf: food banks, ecosystem inertia and global climate change. Deep-Sea Research II 53:875–894.

Soler-Membrives, A., T. Munilla, C.P. Arango, and H. Griffiths. 2014. Southern Ocean biogeographic patterns in Pycnogonida. Pages 138–141 *in* C. De Broyer, P. Koubbi, H.J. Griffiths, B. Raymond, C. d'Udekem d'Acoz, A. van de Putte, B. Danis, B. David, S. Grant, J. Gutt, C. Held, G. Hosie, F. Huettmann, A. Post, and Y. Ropert-Coudert, editors. Biogeographic Atlas of the Southern Ocean. Scientific Committee on Antarctic Research, Cambridge, UK.

Svavarsson, J. 1997. Diversity of isopods (Crustacea): new data from the Arctic and Atlantic Oceans. Biodiversity and Conservation 6:1571–1579.

Swadling, K.M. 2014. Sea-ice fauna. Pages 321–325 *in* C. De Broyer, P. Koubbi, H.J. Griffiths, B. Raymond, C. d'Udekem d'Acoz, A. van de Putte, B. Danis, B. David, S. Grant, J. Gutt, C. Held, G. Hosie, F. Huettmann, A. Post, and Y. Ropert-Coudert, editors. Biogeographic Atlas of the Southern Ocean. Scientific Committee on Antarctic Research, Cambridge, UK.

Tavares, M., and G.A. De Melo. 2004. Discovery of the first known benthic invasive species in the Southern Ocean: the North Atlantic spider crab *Hyas araneus* found in the Antarctic Peninsula. Antarctic Science 16:129–131.

Thatje, S., K. Anger, J.A. Calcagno, G.A. Lovrich, H.-O. Portner, and W.E. Arntz. 2005. Challenging the cold: crabs reconquer the Antarctic. Ecology 86:619–625.

Taylor, J., T. Krumpen, T. Soltwedel, J. Gutt, and M. Bergmann. 2017. Dynamic benthic megafaunal communities: assessing temporal variations in structure, composition and diversity at the Arctic deep-sea observatory HAUSGARTEN between 2004 and 2015. Deep–Sea Research Part I 122:81–94.

Thomson, M.R.A. 2004. Geological and paleoenvironmental history of the Scotia Sea region as a basis for biological interpretation. Deep-Sea Research II 51:1467–1487.

Thurston, M. 1972. The Crustacea Amphipoda of Signy Island, South Orkney Islands. British Antarctic Survey Scientific Reports 71:1–135.

Torres, J.J., J. Donnelly, T.L. Hopkins, T.M. Lancraft, A. Aarset, and D.G. Ainley, 1994. Proximate composition and overwintering strategies of Antarctic micronectonic Crustacea. Marine Ecology Progress Series 113:221–232.

Turner, J., R. Bindschadler, P. Convey, G. di Prisco, E. Fahrbach, J. Gutt, D. Hodgson, P. Mayewski, and C. Summerhayes, editors. 2009. Antarctic climate change and the environment—a contribution to the International Polar Year 2007–2008. Scientific Committee on Antarctic Research, Cambridge, UK.

Turner, J., N.E. Barrand, T.J. Bracegirdle, P. Convey, D.A. Hodgson, M. Jarvis, A. Jenkins, G. Marshall, M.P. Meredith, H. Roscoe, J. Shanklin, J. French, H. Goosse, M. Guglielmin, J. Gutt, S. Jacobs, M.C. Kennicutt II, V. Masson-Delmotte, P. Mayewski, F. Navarro, S. Robinson, T. Scambos, M. Sparrow, C. Summerhayes, K. Speer, A. Klepikov, 2013. Antarctic climate change and the environment: an update. Polar Record 50:237–259.

Udekem d'Acoz, C. d', and C. Havermans. 2015. Contribution to the systematics of the genus *Eurythenes* S.I. Smith in Scudder, 1882 (Crustacea: Amphipoda: Lysianassoidea: Eurytheneidae). Zootaxa 3971:1–80.

Udekem d'Acoz, C. d', and M.L. Verheye, 2017. *Epimeria* of the Southern Ocean with notes on their relatives (Crustacea, Amphipoda, Eusiroidea). European Journal of Taxonomy 359:1–553.

Vader, V., J. R. Johnsen, and J. Berge. 2005. Studies on the genus *Onisimus* Boeck, 1871 (Crustacea, Amphipoda, Lysianassoidea, Uristidae). Part I. The *brevicaudatus* and *sextonae* species groups. Organisms, Diversity & Evolution 5:161–164.

Van Den Broecke, L., K. Martens, V. Pieri, and I. Schön. 2012. Ostracod valves as efficient UV protection. Journal of Limnology 71:119–124.

Vassilenko, S.V., and V.V Petryashov, editors. 2009. Illustrated Keys to Free-Living Invertebrates of Eurasian Arctic Seas and Adjacent Deep Waters, Vol. I. University of Alaska Fairbanks.

Verheye, M.L., T. Backeljau, and C. d'Udekem d'Acoz. 2016. Looking beneath the tip of the iceberg: diversification of the genus *Epimeria* on the Antarctic shelf (Crustacea, Amphipoda). Polar Biology 39:925–945.

Voronkov, A., H. Hop, and B. Gulliksen. 2013. Diversity of hard-bottom fauna relative to environmental gradients in Kongsfjorden, Svalbard. Polar Research 32:11208.

Wägele, J.W. 1994. Notes on Antarctic and South American Serolidae (Crustacea, Isopoda) with remarks on the phylogenetic biogeography and description of new genera. Zoologische Jahrbücher Systematik 121:3–69.

Waller, C.L. 2013. Zonation in a cryptic Antarctic intertidal macrofaunal community. Antarctic Science 25:62–68.

Waller, C.L., M.R Worland, P. Convey, and D.K.A. Barnes. 2006. Ecophysiological strategies of Antarctic intertidal invertebrates faced with freezing stress. Polar Biology 29:1077–1083.

Ware, C., J. Berge, A. Jelmert, S. M. Olsen, L. Pellissier, M. Weisz, D. Kriticos, G. Semenov, S. Kwaśniewski, and I. G. Alsos. 2016. Biological introduction risks from shipping in a warming Arctic. Journal of Applied Ecology 53:340–349.

Watling, L.E., and M.H. Thurston. 1989. Antarctica as an evolutionary incubator: evidence from the cladistic biogeography of the amphipod family Iphimediidae. Pages 297–313 *in* J.A. Crame, editor. Origins and Evolution of the Antarctic Biota. The Geological Society, London.

Watts, J., and G.A. Tarling. 2012. Population dynamics and production of *Themisto gaudichaudii* (Amphipoda, Hyperiidae) at South Georgia, Antarctica. Deep-Sea Research II 59–60:117–129.

Werner I., H. Auel, C. Garrity, and W. Hagen. 1999. Pelagic occurrence of the sympagic amphipod *Gammarus wilkitzkii*—dead end or part of life-cycle? Polar Biology 22:56–60.

Węsławski, J.M., and J. Legeżyńska. 2002. Life cycles of some Arctic amphipods. Polish Polar Research 23:253–264.

Węsławski, J.M., J. Wiktor, M. Zajączkowski, and S. Swerpel. 1993. Intertidal zone of Svalbard, Macroorganism distribution and biomass. Polar Biology 13:73–79.

Węsławski, J.M. 1994. *Gammarus* (Crustacea, Amphipoda) from Svalbard and Franz-Josef Land—distribution and density. Sarsia 79:145–150.

Węsławski, J.M., A. Opanowski, J. Legeżyńska, B. Maciejewska, M. Włodarska-Kowalczuk, and M. Kędra. 2010. Hidden diversity in Arctic crustaceans. How many roles can a species play? Polish Polar Research 31:205–216.

Wesławski, J.M., M. Kendall, M. Włodarska-Kowalczuk, K. Iken, M. Kędra, J. Legeżyńska, and M.K. Sejr. 2011. Climate change effects on Arctic fjord and coastal macrobenthic diversity—observations and predictions. Marine Biodiversity 41:71–85.

Winkler, H. 1994. Characterisierung der Isopodenfauna (Crustacea, Malacostraca) des Scotia-Bogens aus biogeographischer Sicht: Ein multivariater Anzatz. Berichte zur Polar- und Meeresforschung 139:1–196.

Wittmann, A.C., C. Held, H.-O. Pörtner, and F.J. Sartoris. 2010. Ion regulatory capacity and the biogeography of Crustacea at high southern latitudes. Polar Biology 33:919–928.

Wold, A., and F. Norrbin. 2004. Vertical migration as a response to UVR stress in *Calanus finmarchicus* females and nauplii. Polar Research 23:27–34.

10

COLONIZATION OF COASTAL AND ESTUARINE ENVIRONMENTS

Joseph L. Staton, Brian A. Canada, Stephen A. Borgianini, and Karen M. Barkel

Abstract

Coastal and estuarine environments are some of the best-known and most well-studied ecosystems in the world in that these regions lie in close proximity to much of the world's human population. The crustaceans that inhabit these environments, both as adults and larvae, are adapted to the high productivity that characterizes such areas. We summarize their adaptations and behaviors and the physical characteristics of coastal zones and estuaries in shallower waters (<200 m). In an attempt to objectively review worldwide diversity and endemism within the Crustacea from coastal and estuarine environments, we have used open access global collection data and developed a novel application of an informatics principle (term frequency-inverse document frequency [*TF-IDF*]) to identify regions with unique faunal assemblages that typify some coastal, shallow waters to assess potential endemism (as assessed by our technique) across groups at differing taxonomic levels. Crustaceans, as a whole, show highest species richness and endemism in three clusters (using the *TF-IDF* assessment): the eastern temperate north Atlantic, the western temperate north Atlantic, and the western tropical south Pacific. Peracarid and decapod species dominate the collection data, making up 94% of all species analyzed. Peracarids dominate species richness across all temperate zones, yet their highest predicted coastal endemism appears in the eastern north Atlantic by our index. Our analyses using this new method focused on species from less than 200 m depth worldwide. Upcoming impacts of changing sea levels and increases in global temperature will likely have their greatest impact on the fauna of these zones.

Evolution and Biogeography. Edited by Gary C. B. Poore and Martin Thiel.

INTRODUCTION

Coastal and estuarine environments are considered ephemeral, in a geologic sense (Bianchi et al. 2013), and rapidly changing (Curray 1977, Kennett 1982) in that they change with eustatic sea level and extent of riverine input during recent times. Such recent alterations have led to major changes in the coastal physical environment over the last 20,000 years (Graham et al. 2003). The physical environment of a given coastal and estuarine region has a large impact on its ecological communities, including marine and estuarine invertebrate assemblages. As the changes in coastal zones can be rapid, adaptations are limited to broad estuarine/shelf life history patterns affecting dispersal and recolonization, especially after local extinctions. This pattern, which evolved millions of years ago, has been successful for a broad range of environmental conditions that have occurred in the most recent millennia and may override local adaptations through gene flow. Present-day biogeography for many crustaceans, therefore, is the result of maintenance and repositioning of previously adapted species via their individual dispersal and recruitment strategies that evolved prior to modern estuarine conditions. This pattern has proven resilient to local extinctions and has proven to be functionally sufficient in the context of classical rates of environmental change in the coastal zone.

Biodiversity within estuaries and nearshore shelf habitats are impacted by the interaction of multiple physical factors (physiography, geomorphology, estuarine circulation, currents, tidal prism, seasonality, sediment type, freshwater inflow, nutrient flux, etc.) and organisms' adaptations. Living and maintaining populations in these environments requires a unique set of life history strategies. All crustaceans living in these margins are adapted to these unique regions both as adults and as larvae and have developed morphological, physiological, and behavioral traits to survive. These adaptations make crustaceans successful in colonizing zones where temperature, salinity, and other physical characteristics vary spatially and temporally at small and large scales. When larval stages are present in crustacean species, they often differ radically from the adult, being distinct in form, physiology, and behavior. Such complex life history strategies can create sustainable dispersal and recruitment patterns that often overcome local environmental variation and displacement, thereby preventing potential extinction. Abbreviated or direct development in crustacean species is considered to be an advanced evolutionary strategy for estuarine retention (Anger 2001), but it is also associated with vulnerability to local extinction (Valentine and Jablonski 1986, Wares and Cunningham 2001), which can result in slow recolonization of habitats where species have been eliminated.

In the first part of this chapter, we provide an overview of the regionalized biodiversity of specialist and generalist crustaceans in this environment, highlighting adaptations to this zone. Biological traits such as thermal tolerance, osmoregulatory ability (Peterson and Ross 1991), and sediment preferences (Thorson 1950, Probert 1984) interact with the physical environment as well as latitudinal variation in total shelf area (Schopf et al. 1977) to determine the large-scale diversity patterns in some marine groups.

Our review concentrates on benthic crustaceans with a focus on the interaction of their adult and larval stages. This is not to say that other crustaceans (i.e., holoplankton) do not have significant impacts on ecological communities, but we have chosen for the sake of the habitat-based structure to review those species/groups that are associated with the benthic, demersal and shore environments. Previous articles summarizing biogeographic patterns of crustaceans have covered biogeographic diversity (e.g., Ōmori 1976, Brandt 1992, Frey 1986, Macpherson 1991, O'Hara and Poore 2000), but with a bias toward specific taxonomic groups within particular regional zones or restricted to freshwater/terrestrial environments.

In the second part of our review, we summarize coastal diversity of the Crustacea, using a novel analysis we developed on a global collection dataset. We believe this analysis enables an objective, high-level overview of the global patterns associated with the coastal regions along continental margins.

PHYSICAL FACTORS ASSOCIATED WITH DIFFERENT COASTAL ENVIRONMENTS

Physical factors affecting local diversity include temperature variation, substratum, sediment size and type, wave/coastal interaction, and the chemical nature of overlying and interstitial water (Dahl 1952, McLachlan 1990, McLachlan et al. 1993). Temperature and oxygen concentration also have a major role, but the former is correlated with latitude in part while seasonal and diel variation in the latter is often a complex combination of biotic and abiotic forces. Classically, the physiological tolerance ranges of a species, measured through viability and reproductive potential at differing temperatures, was considered predictive of its potential geographic range. Thus, global biogeographic patterns can be predicted from global patterns of the physical environment (Hutchins 1947). The global ranges of most species result to some level from an interaction between adult survival from season to season and the relative ability to restock sink populations or recolonize once occupied habitats based on reproductive constraints (e.g., reproductive mode/season, dispersal, and recruitment to the adult habitat).

Sandy Beaches

Some continental margins are lined with sandy shorelines that mark the transition between marine and terrestrial habitats. These beaches consist of unconsolidated particles of terrigenous sediment like sand, gravel, and cobbles, and carbonate biogenous marine material made up of shell and weathered coral, especially in more tropical zones. Waves, tides, and currents work to sort particles, thereby influencing the overall structure and conditions of these sedimentary environments (Bascom 1951, McLachlan 1996). Dramatic changes in sea level during the Pleistocene Age have contributed to the complexity of the coastal sandy beaches, as well as other coastal environments, due to multiple episodes of shoreline transgression and regression (Vail et al. 1977), and the regional temperature minima and maxima of differing regions associated with glaciation.

Multiple abiotic and biotic conditions continually shape sandy shorelines. Abiotic factors include grain size, degree of compaction, wave action, erosion and accretion, longshore transport, interstitial flushing rates, porosity, dissolved oxygen, organic content, and depth of light penetration. Biotic factors include bioturbation, burrowing, burying, infaunal gallery interconnectedness, microalgal productivity, and feeding behaviors of the organisms living in these environments. On dynamic (usually wave-exposed) sandy beaches, the effects of physical factors commonly override the biological effects of other sandy beach inhabitants.

The impact of these factors on sediment biota is tied to the duration of tidal inundation and to wave exposure. Ultimately, stress associated with occupying sandy shorelines has led to adaptations that allow organisms to overcome desiccation, wave action forces, and sand particle abrasion, all the while maintaining their position to avoid being swept away by tidal currents.

The adaptive value of small size is exhibited in crustacean meiofauna, notably harpacticoid copepods. Small size allows an individual to easily migrate through the interstices between sand particles, allowing them to feed and avoid predation while minimizing the variety of physical forces a larger animal would experience. Some larger crustaceans minimize the effects of physical stress by burrowing, forming semi-permanent burrow galleries (e.g., ghost shrimps, *Callichirus,* and other genera). This burrowing activity mitigates some of the physical stress to which an organism is subject, while providing elaborate pathways for oxygen-rich seawater to penetrate the subsurface gallery network. For example, ghost shrimps have modified pleopods that form flattened fan-like appendages which pump seawater through their galleries. This hydraulic pumping greatly increases the penetration of oxygen-rich water into their semi-permanent burrows and facilitates the removal of solid and liquid wastes, as well as supporting other infaunal commensals, such as pinnotherid crabs.

For crustaceans, an exoskeleton provides protection of soft tissues from sand particle scouring. It also provides the added strength and adaptive abilities necessary to dig into soft thixotropic sand as well as sorted and compact dilatant sand. However, in species that maintain relatively permanent burrow galleries as adults (e.g., callianassids), a thinner, clear exoskeleton has evolved in absence of UV threats and to enable greater flexibility to maneuver in their burrows.

Rapid burying and digging behavior employed by some crustaceans occupying sandy beaches is an alternative to burrow construction. Excavations can be shallow and non-permanent, e.g., by soldier crabs, *Mictyris* spp.; this activity allows organisms to move both laterally and perpendicularly to the changing tide, thereby maximizing foraging behaviors and minimizing physiochemical stresses while in the intertidal and subtidal zones (e.g., mole crabs, *Emerita* spp.). Burying behavior (Faulkes 2013) is used by many decapods as a tactic for ambush predation as well as effective camouflage for predation (in swimming crabs such as *Ovalipes* spp., *Arenaeus* spp.) as well as predator avoidance (e.g., ghost crabs, *Ocypode* spp.). Typically, these animals bury by utilizing a variety of uniquely modified posterior appendages that allow the majority of their body to be covered, but leaves exposed their anterior sensory structures, such as eyes and antennae, and their feeding apparatus, including claws and mouthparts.

Some sandy shore crustaceans such as cirolanid isopods combine burrowing behavior as tidal refugia while exhibiting circa-tidal pelagic foraging behaviors (see Fig. 10.1 for examples). *Eurydice pulchra* utilizes an endogenous circa-tidal swimming behavior to compensate for difference in neap tidal and spring tidal cycles (Reid and Naylor 1985). Wave action has been correlated with the distribution of some Amphipoda. Talitrid amphipods are often found in the supra-tidal zone associated with wrack and other organic debris that washes ashore with flood tides (Behbehani and Croker 1982, Marsden 1991). Certain species of haustoriid amphipods prefer low energy sandy beaches (Croker 1967, Ruppert and Fox 1988) while others (e.g. *Parahaustorius*) prefer the lower intertidal beaches where higher, direct wave energy requires them to have significant burrowing capabilities (Howard and Elders 1970).

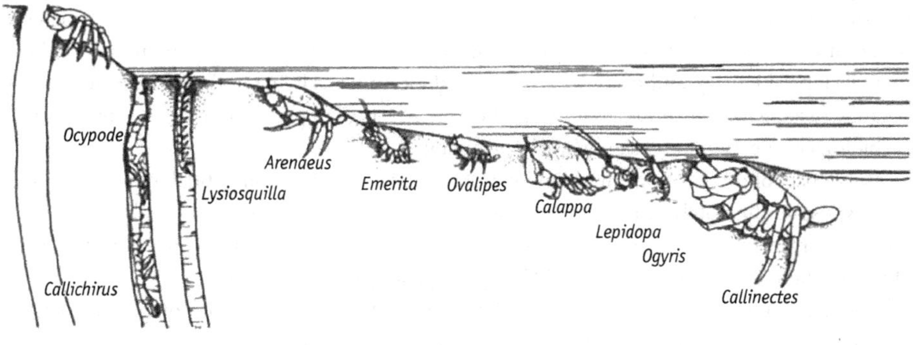

Fig. 10.1
Examples of digging and burrowing sandy beachfront crustacean species in their respective postures (from Pearse et al. 1942). As a rule, animals higher in the intertidal and exposed regions create burrows that are permanent to semi-permanent, whereas other species exhibit "burying" behaviors. With permission from John Wiley and Sons ©.

Intertidal Muddy Shores

Muddy intertidal zones can exist in the quiescent zones behind barrier islands, on the margins of erosional coasts, in delta-bar systems, or at the intertidal base of mangroves or salt marshes. They support numerous burrowing species and surface foragers as well as scavengers among the mangrove ecosystems (i.e., mangals). As most are associated with depositional rather than erosional coasts, the most extensive systems occur on passive continental margins, as a general rule. Mudflats occur where tidal currents are more dominant than wave action, which produces an inverse sediment gradient of finer sediments at higher elevations and coarser sediments at mid-tide levels (Little 2000).

Mudflats

Marine mudflats are formed when sand, clay, and suspended silt particles settle slowly in sheltered embayments. This gentle settling allows for the formation of mixtures of unconsolidated sediment and bound interstitial water that forms an intertidal mudflat, which is exposed to the air during ebb tide. The frequency and duration of inundation of tidal waters affect both sediment makeup of the mudflat and its degree of compaction.

Bioturbation by a wide variety of epibenthic and infaunal animals significantly alters the microtopography of the mudflat, increasing the number of microhabitats present, and helps to channel flood and ebb tidal waters encroaching onto and exiting the mudflat. The mudflat surface is highly productive (Elliott and Taylor 1989) due to the presence of photosynthetic benthic microalgae, cyanobacteria, ciliates, and euglenophytes (microphytobenthos). Large numbers of harpacticoid copepods and several marine ostracod species live at the sediment/water interface. When exposed at low tide, animals may experience temperatures exceeding 40°C at the surface and hypersaline conditions. Sporadic rains can cause salinity to drop to almost freshwater levels within seconds (Giere 2008).

The mud shrimps of the genus *Upogebia* construct extensive underwater galleries that can extend more than two meters deep, often creating an oxygenated microenvironment for other animals like attached phoronid worm tubes, small bivalves, pinnotherid crabs, and gobies (Kinoshita 2002). By pumping water through their galleries the shrimps enhance total organic carbon to nearly six times greater than that in the surrounding sediment (Papaspyrou et al. 2005). Likewise, callianassid and upogebid shrimps have also shown respiratory adaptation to low oxygen tensions (Thompson and Pritchard 1969, Felder 1979). Many hermit crabs roam the shallow, lower-water-energy zones, while other species (panopeid, grapsoid and xanthoid crabs) live in the muddy depressions and traverse the shallows at low tide, seeking refuge in nearby marshes or hunting those seeking refuge. Tubicolous amphipods (notably species of Corophioidea) burrow shallowly, allowing a combination of suspension- and deposit-feeding (See Little 2000, Faulkes 2013).

Fiddler crabs (species of *Uca*) utilize mudflats as foraging habitat (Salmon and Hyatt 1983, Christy and Salmon 1984). As deposit feeders, they utilize the spoon-like tip on their smaller claw to transfer surficial sediment to their mouthparts that sort sediment particles, removing benthic microalgae, cyanobacteria, protists, and organic matter adsorbed to the surface of the sediment particles. Many secondary and tertiary consumers eat fiddler crabs, making them a critical element in the detrital and grazing food chains (Teal 1962, Aveline 1980, Dye and Lasiak 1986, Koch and Wolf 2002). Intense foraging activity, coupled with their vast numbers, can result in turning over nearly the entire mudflat surface between successive high tides. Non-tidal fiddler species use burrows to avoid the peak heat of day, which reaches the terminal depths of the burrow in the late afternoon, where the trapped heat can help sustain nocturnal activities (Powers and Cole 1976). Australian fiddler crabs have been shown to avoid thermal extremes behaviorally, and the major chela of the males of *Uca mjoebergi* could function as a heat sink (Munguia et al. 2017). These crabs are also fossorial, and the bioturbation caused by their extensive burrowing activity allows surface water and oxygen to enter near-surface sediments (Bertness 1985). Similarly, dotillid crabs, like soldier crabs (*Mictyris* spp.) and sand bubbler crabs (*Scopimera* spp.), forage on surface sediments of tropical mudflats.

Mangroves

In tropical, subtropical, and some temperate regions, mangrove forests are the dominant transitional ecological community. Mangrove ecosystems are among the most productive on earth and exhibit relatively high biodiversity (Saenger et al. 1983, Farnsworth and Ellison 1996, Field et al. 1998, Komiyama et al. 2008). Part of the reason for this significant biodiversity is that mangroves are most often found in estuarine environments and demonstrate marked patterns of zonation along a salinity gradient. The growth habit of mangroves as well as their patterns of salinity-mediated zonation result in significant habitat heterogeneity. Of the invertebrate fauna within the mangrove ecosystem, decapod crustaceans are a large component (Aveline 1980). Diverse assemblages of crabs and shrimps play an important role in the food-web dynamics in the mangroves. Through a combination of herbivorous consumption and bioturbation, they not only form the foundation of the grazing food chain in the mangroves but also contribute to increased primary production by translocating nutrients, organic material, and sediment back to the surface of the mangrove forest.

Semiterrestrial and highly mobile swimming decapods dominate these systems, fiddler crabs (Family Ocypodidae) and sesarmid crabs (Family Sesarmidae) are plentiful, while gecarcinids, grapsids, ocypodids, and portunids are less abundant (Hartnoll et al. 2002). Many burrow in the muddy sediments trapped by the prop-root systems during high tide and emerge as active foragers on the mud surface when the tide is out (Macintosh 1984, Hartnoll 1988). By strategic burrow placement, mangrove burrowers have been shown to avoid thermal stress (Eshky et al. 1995). Behaviors interacting with the burrows can reduce seasonal risks of lethal high and low temperature. In Florida mangroves, fiddlers were able to reduce their temperature in peak heat as much as 10°C and raise temperature in colder conditions by as much as 5°C by entering their burrow (Smith and Miller 1973).

In the Caribbean, mangrove trunks, stems, and prop roots also provide attachment space and protection for epibentic crustaceans like barnacles (*Chthamalus* spp. and *Balanus* spp.), *Ligia* sp. (Isopoda), snapping shrimps (*Alpheus* spp.), and the mangrove tree crab (*Aratus pisonii*). In some Australian mangrove forests, mangrove–crab interactions are quite complex. Burrowing activity of semiterrestrial fossorial crabs such as *Heloecius* (Ocypodoidea) alters the microtopography of the *Avicennia* mangrove swamp surface. Composition of burrow tailings show that these crabs bring sandy sediments to the surface and modify the sediment surface makeup (Warren and Underwood 1986). Other crabs such as sesarmids feed directly on fallen mangrove leaves, enhancing the physical breakdown of the leaf fall and establishing a critical link between the detrital and grazing food chains (Lee 1998). Biological waste produced by these crabs fertilizes the marsh sediment, thus supporting organic recycling in the mangrove forest. Also isopods from the genera *Limnoria* and *Sphaeroma* burrow directly into mangrove prop roots, with one species (*S. terebrans*) possibly affecting the distributional patterns of *Rhizphora mucronata* (Svavarsson et al. 2002).

Saltmarsh Estuaries

Saltmarshes are a dominant estuarine community in temperate zones of the world, replaced closer to the tropics by coastal mangroves and limited in their poleward expansion by minimum winter sea surface temperatures. The current extent and level of development of local saltmarshes are based on slowed rates of sea-level rise, beginning 3,000 to 4,000 years ago, and these reduced rates created a balance whereby sedimentation rates within marshes could keep pace with the decreased rise in sea levels (Niering and Warren 1980).

Within their estuaries, saltmarshes are limited to regions that have periodic inundation by tidal estuarine waters, ranging from low salinity where the terrestrial plants are outcompeted by freshwater semi-aquatic plants to the margins of high-salinity soils that get limited saltwater inundation that can concentrate salts (100 psu soils or higher). Relative proximity to tidal flushing (i.e., main and lesser tidal creeks) controls the overall levels of soil anoxia (Howes et al. 1981, Mendelssohn et al. 1981), and salinity (Chapman 1974, Jefferies 1977).

Crustaceans exhibit patterns of osmoregulation directly related to their tolerance of salinity (Péqueux 1995). The limits of salinity tolerance often determine their distribution within the estuary and shelf habitats (Mantel and Farmer 1983). Fluctuations in salinity can prove extremely stressful. A large influx of fresh water can rapidly lower salinity and lower oxygen affinity (Mangum and Towle 1977).

Classically, the estuarine biodiversity has been viewed as a result of two biotas: one that has some level of freshwater tolerance (Odum 1988) and one that can tolerate some salinity intrusion into the fresh headwaters. The so-called horohalinicum (Kinne 1971) is the proposed maximum osmotic stress window in the range of 5–7 psu, which greatly limits overall biodiversity along the estuarine gradient pattern that was first noted by Remane (1934, see Remane and Schlieper 1971, Fig 10.2A) and is still cited as a driving force in estuarine diversity (e.g., Vuorinen et al. 2015) despite the challenge of this view over 20 years ago on the lack of physiological merit (Deaton and Greenberg 1986). However, these trends have been largely biased toward macrofaunal assemblages. In a more limited study, Attrill (2002, see Fig. 10.2B) noted that temporal variation in salinity within different regions of the estuary had a greater impact on local diversity patterns than predicted by the classic spatial gradient of mean salinities for an estuary. Indeed, adult crustaceans, as well as their larvae, are influenced by these patterns as different species release their planktonic larvae in different regions to comply with species-specific osmoregulatory ability or limitation of the species. This pattern and

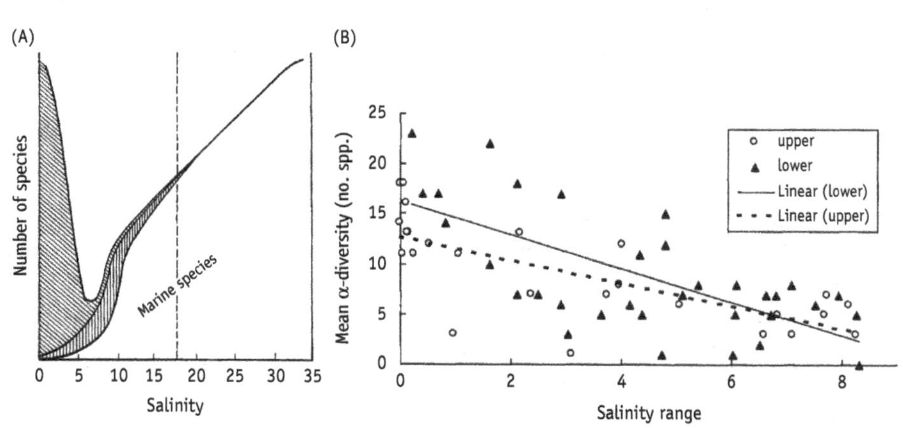

Fig. 10.2.
Early views of diversity in estuarine systems view most diversity and distribution as an overlap of physiological tolerant species based on salinity tolerance, with a physiological minimum diversity zone at 5–7% (A, after Remane 1934, in Attrill 2002), which has been termed the horohalinicum (Kinne 1971). Recent tests across diverse taxa suggest that salinity range is a more important structure of diversity and distribution of species within an estuary (B, Attrill 2002). With permission from John Wiley and Sons ©.

timing of larval release has implications for the effectiveness of dispersal and recruitment of estuarine decapod crustaceans, as discussed later, which can lead to larval retention within or dispersion outside to the continental shelf.

Subtidal Channels

The subtidal channel is the main estuarine exchange of the ocean with the associated river system that produces the estuarine system. As rivers widen and slow, waters intermix, leading to the characteristic range of hydrologic mixing patterns classically described by Pritchard (1952), ranging from well mixed and unstratified to the more highly stratified salt-wedge and fjord types. Slower-moving estuaries can supply new sediments to the inundated section of saltmarsh or to the smaller, superficial, surface drainages in the muddy high marsh and tidal creeks. Many species of crustaceans make their adult home here. For example, tube-dwelling and free-living species of amphipods, as well as species of panopeid and xanthid crabs, live in the muddy channels at low tide but venture onto the saltmarsh to hunt or hide at high tide. Large numbers of palaemonid shrimps can be dominant in these channels, often reaching hundreds to thousands of individuals per m^2 in the southeastern United States (Leight et al. 2005). Juvenile penaeid shrimps invade in the autumn of the year and grow out during periods of higher food availability of late summer/autumn (Beseres Pollack et al. 2008) before returning to the continental shelf as adults to mate and spawn eggs, leading to new naupliar larval pools the following spring (Bauer 2018). Juveniles of many species of portunid crabs and other small species move back and forth from the channels to the nearby marsh during each tidal cycle, both for hunting other prey and refuge from predation themselves. In many areas, subtidal populations of alpheid shrimps can be extensive, as evidenced by the near-constant sound pressure levels of their cumulative snapping recorded by subtidal recording units (Monczak et al. 2019).

Higher productivity in summer months can produce short to long persisting hypoxic conditions in estuaries (and other benthic zones offshore); portunid crabs (Das and Stickle 1994) and penaeid shrimps (Renaud 1986) have been shown to detect and actively avoid hypoxia or are absent from hypoxic areas in the field (Rabalais et al. 2001). Many crustacean species avoid these behaviorally by moving into shallows or higher intertidal marsh grass beds as a refuge from these conditions.

Intertidal Marsh

The tidally inundated region of saltmarshes are often characterized by higher borders of the main channel and connected tidal creeks, with low-lying marsh behind these extending toward the higher back marsh region (Fig. 10.3). The higher margins are even more exaggerated, as marsh grasses that have free tidal flushing on this perimeter grow taller and trap more sediment due to regular inundation of water with higher nutrients and oxygen levels and reduced levels of sulfides (Mendelssohn and Morris 2000). As one proceeds away from these channels, desiccation increases, oxygen levels drop, and salt content of soils increases. Along these margins, burrowing decapods create greater opportunities for water–nutrient exchange. Burrowing ocypodid and sesarmid species mine the root systems for food and detritus and secondarily increase soil oxygenation and flushing of reducing compounds (the result of bacterial decomposition) along the tidally flushed channels.

Toward the center of saltmarsh patches, tidal flushing is restricted to surface exposure and marsh plants experience increases in anoxia, salt content, and soil sulfides (Mendelssohn and Morris 2000) that produce stressed/stunted forms of the dominant marsh species (e.g., *Spartina alterniflora*). These areas transition to halophytes that are more salt tolerant, and in many places large, bare salty regions are bordered by succulents that represent the most resistant to salt concentrations

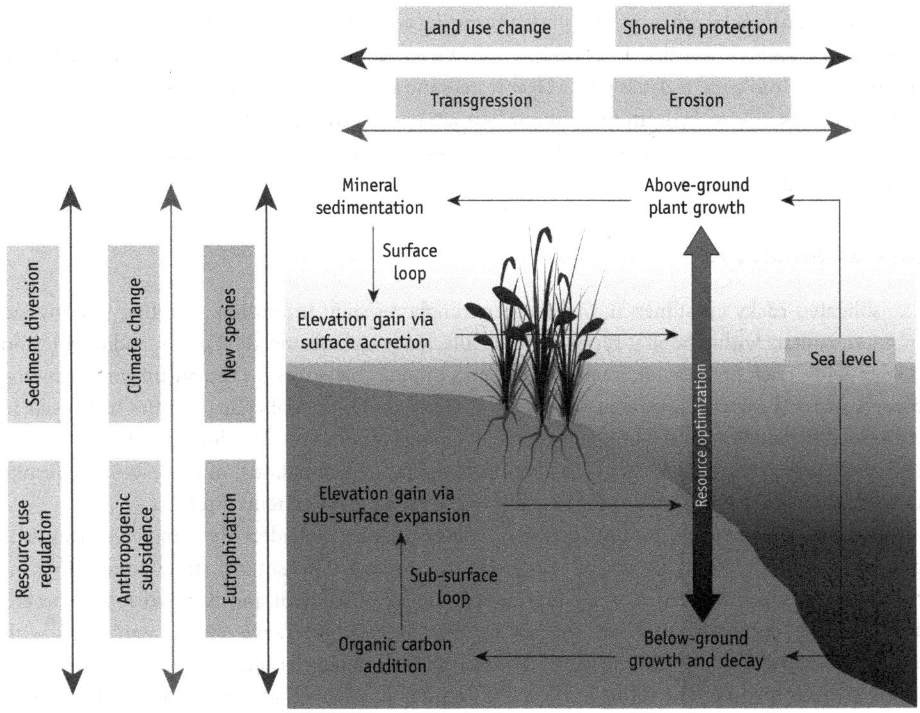

Fig. 10.3.
Structure of a saltmarsh estuary. The system can be altered by local factors (shown in green), climatic and geomorphologic factors (shown in blue), and the political/social/economic factors (shown in tan), which affects overall resilience and stability of the saltmarsh system at different scales (Kirwan and Megonigal 2013). With permission from Springer Nature ©. See color version of this figure in centerfold.

in the soil. Here, fiddler crabs (Ocypodidae) predominate and create large numbers of shallow burrows. Species of fiddler crabs and peracarids often help reprocess refractory carbon from marsh grass to increase the flow of carbon cycling throughout the marsh. During periods of tidal flooding, juvenile portunid crabs leave the deeper channels not only to hunt for molluscan prey (Boesch and Turner 1984) but also to take refuge in the stunted marsh grass, avoiding predation by larger fish and other crab species. In the saltmarsh of the Southwestern Atlantic, the varunid crab *Neohelice granulata* facilitates root growth in the local cordgrass *Spartina densiflora* (Daleo and Iribarne 2009).

Hard-Bottom Communities

Rocky shore communities have been described as hard islands in a sea of soft sediments (Young 2009). They occur inter- or subtidally as reefs on the continental shelf, colonized by algae, oysters, mussels, and encrusting worms. In temperate areas, their base can be basalt/granitic as a subtidal continuum with tectonic coasts, with either rocky intertidal zones or fjords. In lower temperate to tropical zones, they are often exposed regions of karst bedrock. The best studied of these rocky shores are granite and basalt, extending from shore to subtidal in the northwestern Atlantic, combinations of granite, sandstone, shale, etc., in regions like the Cape of Good Hope (McQuaid and Branch 1984) and a mix of ophiolitic and exotic terranes in other tectonic regions (e.g., California). In other temperate regions (e.g. western South America, southern Australia and the

Mediterranean), rocky outcrops and their encrusting communities have been studied extensively (e.g. Andrew 1999, Coll et al. 2010, Navarrete and Wieters 2000). Those in most tropical shallow waters transition into coral reef habitats in clear waters and sometimes into worm reef habitats in more erosional areas. For rocky bottoms without hermatypic invertebrates, most diversity is measured in encrusting forms of algae and sponges and (among crustaceans) for those species of encrusting barnacles inhabiting them.

Rocky Intertidal Zones

Consolidated rocky coastlines and bluffs are strongly resistant to erosion, resulting in a myriad of microhabitats within habitat types such as cliffs, ledges, platforms, tidal pools, and cobble and bolder fields. Each of these microhabitats presents unique challenges for the organisms occupying those spaces. These multiple habitat types have spatial and temporal characteristics that result in considerable biodiversity and unique patterns of biological zonation. The duration of tidal inundation, desiccation stress, intensity of wave action, temperature fluctuations, and salinity stress results in specialized ecological communities capable of tolerating these extreme stressors.

Stressors and their importance on rocky coastlines vary considerably along a vertical tidal gradient, allowing different sets of organisms to colonize different intertidal zones. This intertidal zonation may be driven by thermal physiology (Stillman and Somero 1996, Somero 2002) where certain crustacean species exhibit thermal tolerance limits associated with energetics-based limitations. These limitations establish the upper limit of species distribution in the rocky intertidal. The result is a suite of unique organisms that appear to form horizontal bands, resulting in vertical zonation along tidal gradients that extend from the supratidal spray zone through the intertidal zone until these biological communities are relieved of many of these physical stressors in the subtidal zone, where biotic stressors (competitors, predators) become more important. Rocky tide pools can serve as desiccation refugia in intertidal zones filled with seawater. They are ephemeral and allow organisms to avoid stagnation, desiccation, or temperature stress until renewed and flushed by the next high tide. Biological factors such as predation (i.e., Paine 1966, Menge 1976, 2000), attachment space competition (*sensu* Connell 1961), postsettlement selection (Menge 1991), and active and passive inhibition of settlement (sensu Roberts et al. 1991) can shape rocky shore communities.

Specialized adaptations allow organisms to survive in these stressful conditions. Animals can avoid desiccation and thermal stress by hiding under the damp blades of macroalgae. Others avoid these stresses and predation by occupying cracks and crevices in the heterogeneous three-dimensional rock matrix when not foraging. Sessile intertidal crustaceans such as barnacles only feed at high tide when covered by water and close up tightly to avoid desiccation when exposed. Significant wave action in the intertidal zone has the potential to dislodge, displace, and injure exposed organisms. Such organisms must either be firmly attached to the hard substratum, have the ability to hold on tightly, or reduce the size of their exposed surface area.

Some rocky shore isopods avoid the physical stresses of wave action by essentially occupying the supratidal zone (*Cirolana* spp., *Ligia* spp.). Hermit crabs are commonly found on intertidal ledges, platforms, and tide pools where standing water provides a relief from desiccation and potentially concentrates prey. Grapsid crabs are quite common in cracks, fissures, and burrows of the lower intertidal and rocky subtidal zone from tropical to boreal rocky shores. They exhibit the capability to hold tight to rocky substrata, and their low profile (i.e., the near-cosmopolitan tropical/subtropical Sally lightfoot crab *Grapsus grapsus*) minimizes the shearing effects of wave action. Many porcellanid and majid crabs occupy similar habitats to grapsids where they can be very abundant in tropical, subtropical, and temperate zones (e.g. Rivadeneira et al. 2010).

Bivalve and Worm Reefs and Other Biogenic Hard Bottoms

In temperate/boreal zones, the rocky intertidal habitat grades into similar subtidal rock formations, but tropical hard bottoms can be sporadic combinations of hardscape limestone, oyster reef, and even larger hermatypic worm reefs. In temperate regions extensive mussel beds also harbor a wide diversity of crustaceans (Buschbaum et al. 2009). In these complex habitats, crustaceans are crevice dwellers/burrowers (xanthoid crabs and rock lobsters) that use shell crevices/rock cracks as refugia (as with porcellanids) from predators or can even be intra-shell commensals (pinnotherid "pea crabs" in oysters and mussels). When these reefs are colonized by macroalgae, peracarid crustaceans (amphipods and isopods especially) can also be abundant in these biogenic reefs. At lower latitudes, these areas can be algal- or sponge-dominated communities with associated trophic and physical characteristics; in clear, oligotrophic waters, coral reefs dominate (see also Chapter 6).

COASTAL AND ESTUARINE HABITATS ARE LINKED BY LIFE HISTORY ADAPTATIONS

Many crustaceans have complex life history strategies where adults occupy discrete benthic habitats while larval life stages are planktonic and capable of maintaining metapopulation connectivity through a variety of dispersal and recruitment mechanisms (Palumbi 2003, Cowen and Sponaugle 2009). Because these larvae are planktonic, they can and often become spatially disconnected from their natal habitats. They are transported within, and potentially outside, the estuary by tidal currents. Larval behavior, coupled with physical transport mechanisms such as tidal currents, river flow, and rates of water exchange, can influence the in-estuary retention as well as offshore export of planktonic life stages (Morgan 2020). Estuaries possess unique physiographic and geomorphologic signatures, hence organisms found within estuaries exhibit their own distinctive dispersal patterns of retention (see Fig. 10.4) and export strategies (Bilton et al. 2002).

Estuarine larval retention strategies frequently result in species down-estuary ranges being limited to the oligohaline and limnetic environments within the estuary/river system (Bauer 2018) effectively isolating these populations and thereby reducing connectivity with other estuaries (Fig. 10.4). This resulting isolation may be due to phylogenetic or ontogenetic (i.e., delayed metamorphosis sensu Gebauer et al. 2003) constraints in adapting to fresh water by marine and estuarine species (Spooner 1947, Charmantier et al. 2001, Bilton et al. 2002).

Many estuarine species are able to control their position within the water column to exploit tidal currents to enhance transport or position maintenance, a behavior known as selective tidal stream transport (STST sensu Walker et al. 1978). STST is especially useful early in life history when minute larval stages are able to use their vertical position to achieve unidirectional horizontal movement (sensu Bilton et al. 2002; see Fig. 10.4).

To date, STST has only been examined in ocean and estuarine environments where the difference in duration and intensity of ebb and flood tidal flow is not significant. However, it may not be fully applicable to narrow, salt-wedge, river-dominated estuaries that have significant and variable net downstream flow. Here, periodically large river flow can dampen the intensity and duration of upriver tidal flow, especially in the narrower, upper river channels. Tidal currents in these narrow river channels are amplified by topographic funneling of these currents into the fluvial geometry of the rivers (Borgianini et al. 2012).

Indeed, larval retention and export strategies (Bilton et al. 2002; see Fig. 10.4 for larval retention) come into even greater long-term biogeographic patterns, given the impacts on the geographic range of saltmarshes due to climate change and the concomitant rise in sea level. Modern marshes are predicted at risk of dramatic losses over the next 100 years (Craft et al. 2009) through models

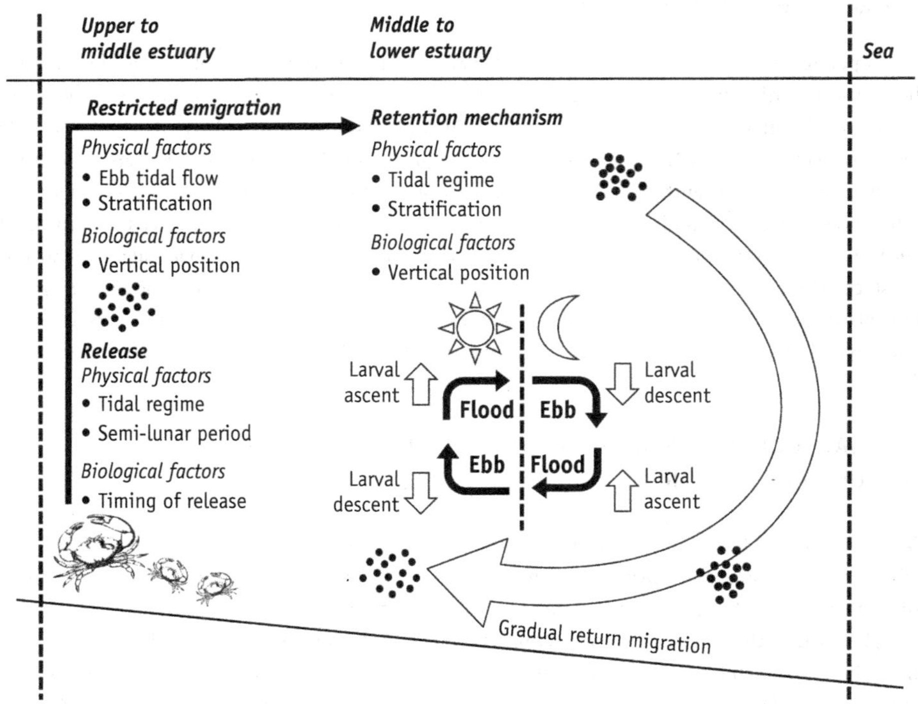

Fig. 10.4.
Model for the life cycle of estuarine organisms possessing a larval retention strategy to wholly remain within an estuary. Larvae (dots) ascend on flood tides and descend on ebb tides. By cycling vertically (as do holoplanktonic organisms) on day/night cycles within the estuary, they can actively feed and escape being consumed by visual predators. Shifts in post-larval behavior favors a return to higher in the estuary to recruit back to adult habitats of origin. Redrawn after Bilton et al. (2002). With permission from Elsevier ©.

estimating impacts of climate changes suggested by the Intergovernmental Panel on Climate Change (IPCC) Special Report on Emissions Scenarios (Meehl et al. 2007). Changes in these regions over time are likely to have a major impact on coastal crustacean species ranges and diversity over time.

These strategies are sharply contrasted with groups that have limited larval development or direct development, such as peracarids that completely lack larval stages. In a comparative study within coastal southern Africa, species with any level of planktonic development showed lower fluctuations in population sizes, higher recolonization rates, lower extinction rates, and less genetic structure than direct-developing species (Teske et al. 2007). Peracarids in coastal Chile increased in diversity with increasing latitude, while crustaceans with larval development decreased in diversity over the same gradient (Fernández et al. 2009). It has been theorized that direct developers will suffer worse in term of local extinctions with changing climate, a pattern that has been well demonstrated in mollusk faunas in the Gulf of Mexico (Hansen 1978).

DIVERSITY HOTSPOTS

Although there are variations (Rivadeneira et al. 2011, Pappalardo and Fernandez 2014), many coastal crustaceans follow the general trend of highest diversity near the equator, and decreasing species richness with increasing latitude (Pohle at al. 2011; see Chapter 15). However in crustaceans, latitudinal gradients are imperfect (see Chapter 15). Peak numbers of species just above or below the equator, such as 10° N and 20° S in *Uca* species (Levinton and Mackie 2013), suggest multiple factors are at work. In the Atlantic Ocean, species richness trends for coastal decapod and stomatopod crustaceans differed between the western and eastern basins, likely reflecting the influence of the Caribbean and Mediterranean Seas as regional diversity hotspots (Macpherson 2002).

Explanations for this pattern vary, with some suggesting that the tropics serve as a main origin of endemism and as a refuge during periods of changing climate (Renema et al. 2008). The tropics have been cited as being the greatest in diversity for all major groups, except the peracarids, which are highest in diversity in the cool temperate zone. This pattern has been noted in both northern (Watling 1979) and southern (Barnard 1991, Rivadeneira et al. 2011) hemispheres, and preference of peracarids for colder habitats may explain their increased penetration and diversification in deeper ocean habitats (Hessler and Wilson 1983).

Depth has also been examined as a factor structuring communities. Gray (1997, 2001, 2002) summarized studies of soft-sediment macroinvertebrates that support (and contrast) the paradigm that diversity increases with depth to about 2000 m, then declines. Variability in the size and population densities of meiofauna (both smaller in the deep sea) may enhance the influence of changing sediment characters on species richness patterns (Gray 2002). The effect of depth may vary with the environment; for example, on the rocky shorelines in the Gulf of Alaska, richness of decapod crustaceans was higher at a subtidal depth of 1 m than at 10 m below low tide (Pohle et al. 2011).

Coastal environments represent challenges (less stability and increased human impact) as well as benefits (increased food supply), factors that may decrease with increasing depth. For benthic decapods and stomatopods in the Atlantic Ocean, richness of coastal species varied more with changing latitude than does that of species that live on the shelf or abyssal plains (Macpherson 2002). In contrast, species richness patterns for pelagic euphausiids and decapods generally plateaued from about 50°N to 40°S (Macpherson 2002). For species with bathymetric ranges that bring them into contact with the oxygen minimum zone, species richness may be suppressed, a possible factor contributing to the inverse latitudinal gradient of peracarid species richness along the Chilean coast (Rivadeneira et al. 2011).

Species richness–area relationships are well documented for terrestrial (MacArthur and Wilson 1967) and marine organisms (Schopf 1977, Tittensor et al. 2010, Rivadeneira et al. 2011). The proportional pattern of regional continental shelf size is of a similar rank order to the diversity gradient in crustaceans reported by Abele (1982), which documented an increase of species richness in the tropics from Eastern Atlantic, to Eastern Pacific, to Western Atlantic, with the highest diversity on the Indo-West Pacific shelf areas. There are, however, crustacean groups that differ from this general trend. Species richness was not strongly correlated with shelf area in stomatopod (Macpherson 2002) and decapod crustaceans (O'Hara and Poore 2000, Macpherson 2002), including anomuran (Astorga et al. 2003) and brachyuran crabs (O'Hara and Poore 2000, Astorga et al. 2003, Levinton and Mackie 2013). For some non-crustacean species groups, adjacent marine regional faunal provinces have been estimated to differ by 20 to 40% of their invertebrate species diversity (Schopf 1979). This difference is influenced by current boundary interactions and geographic features interrupting water masses (capes and peninsulas) that often subdivide latitudinal clines in temperature minima and maxima.

Sea surface temperature has been shown to be a strong positive predictor of species richness for coastal marine species (Tittensor et al. 2010), a pattern that is also supported by several studies examining crustaceans specifically (Astorga et al. 2003, Levinton and Mackie 2013, Pappalardo and Fernández 2014; but see also Rivadeneira et al. 2011). But this prediction may be complicated by proximity to areas of high diversity such as in the Indo-West Pacific (McCallum et al. 2013). As originally summarized by Thorson (1950), mode of larval development varies with latitude for marine organisms. Richness of species with planktotrophic larvae generally increases at lower latitudes

(Astorga et al. 2003), while the opposite trend was observed in species with direct development or lecithotrophic larvae (Pappalardo and Fernandez 2014). Sea surface temperature, which is correlated with primary productivity for a region, predicts the availability of food at many trophic levels. This pattern suggests that primary productivity may be a driving force for increasing biodiversity. Indeed, primary productivity combined with organic pollution were the best predictors of family-based assemblages in rocky nearshore habitats (Pohle et al. 2011). In the Atlantic, sea surface temperature was significantly correlated with species richness of benthic decapod and stomatopod crustaceans, but its effect on pelagic crustaceans (euphausids and decapods) was not significant (Macpherson 2002).

The predictive power of sea surface temperature suggests that climate change could impact crustacean diversity trends. Stenothermal species, especially at the high-latitude ends of continents, may be more prone to extinctions (O'Hara and Poore 2000). Peracarids, which show an inverse relationship between species richness and sea surface temperature (Rivadeneira et al. 2011), could be adversely affected. Warming surface waters could increase the success of biological invasions and range extensions (Occhipinti-Ambrogi 2007) and may be most pronounced in species with wider tolerances to thermal and hypoxic stress (Lenz et al. 2011). Subtropical reefs have been suggested as a conservation target as they may serve as a refuge for more tropical species (Beger et al. 2014).

A METHOD TO ESTIMATE GLOBAL ENDEMISM IN CRUSTACEAN BIODIVERSITY

Methodological Approach

To attempt to characterize and summarize global diversity and biogeographic variation in marine crustaceans, a group comprising >1000 extant families in 42 orders and 6 classes (Ahyong et al. 2011), is a daunting if not impossible task in these brief pages. A summary for the Order Decapoda, alone, estimated 14,335 extant decapod species in 72 superfamilies (De Grave et al. 2009). For those familiar with the literature, concepts of endemism and inferences drawn about centers of origin for specific taxa are anecdotal, largely, in marine environments and are often discussed as a tangent to systematic work and often within a limited taxonomic group. With this in mind, we attempted to create an analysis of occurrence data from the Ocean Biogeographic Information System (OBIS) database (https://obis.org/) to construct an objective assessment across ocean basins. We extracted data for all crustacean collections recorded at ≤200 m depth (1,979,734 records) of 12,016 species (see Table 10.1), representing sampling frequencies ranging anywhere from only a few for many species to thousands of collections for others.

For each crustacean taxon analyzed, we parsed the sampling data into 219 geographic segments, dividing longitudinal distribution at 10° intervals and latitudinal distribution into six broad regions: Polar North (90° N to 66.5° N; *PolarN*), Temperate North (66.5° N to 23.5° N; *TempN*), Tropic North (23.5° N to 0°; *TropN*), Tropic South (0° to 23.5° S; *TropS*), Temperate South (23.5° S to 66.5° S; *TempS*), and Polar South (66.5° S to 90° S; *PolarS*). Within the Tropic North group, three zones (between 100° W and 70° W) are composed of isolated Atlantic and Pacific Ocean subregions, so these were partitioned into Atlantic and Pacific "subzones" to reflect their modern isolation by Central America.

Not all taxa have been equally sampled, as there is bias due to the relative ease of sampling, taxonomic identification, and commercial importance (fishery data). To compensate for sampling bias among taxa (e.g., 547,790 records were for *Homarus americanus* alone), we assigned the frequency data in each segment an integer value ranging from 1 to 4 based on the quartile rank of observations for that one taxon worldwide. For example, if the frequency of observations of a particular taxon in a given zone was greater than the 25th percentile and less than or equal to the 50th percentile (i.e., between the 1st and 2nd quartiles) of observations of that same taxon for the entire world, then we assign it a "normalized quartile rank" of 2. If no observations of that taxon occurred within a zone, then we assigned that zone a normalized quartile rank value of 0 for that taxon.

To estimate signals of potential endemism (for a given taxon in a given geographic zone or cluster of zones), each of the normalized values computed was converted to a "weight of importance," determined by computing its *TF-IDF* (term frequency-inverse document frequency) value. This method is commonly used in information retrieval and text mining to weigh the relative importance of a specific item or term to a given collection within a larger collection of terms (such as the frequency of a specific word within a given book as compared to its frequency across an entire library; Salton and McGill 1983, Salton and Buckley 1988). For a given item in a subset, the higher frequency of the item within a defined subset will increase its importance, while higher frequencies of the item throughout the entire corpus will reduce this importance. For example, a document-specific word like "endemism" would be ranked higher for a given document than common words like "the" or "that," and even within those documents where instances of the word "endemism" are found, such occurrences may vary in frequency. In short, the *TF-IDF* score can be used as a measure of the degree to which an item occurs specifically within a given division of a large database. For our purposes, we treat the species (or other taxon) as the "item" and the geographic zone or group of zones (a segment of the world) as the "division of the database."

We summarize the computation of the *TF-IDF* metric for a given taxon in a given zone as follows:

$TF_{taxon,zone}$ = normalized quartile rank (an integer ranging from 0 to 4, as discussed above)

$$IDF_{taxon} = log_{10}\frac{N}{n_{taxon}}$$

where N equals the total number of geographic zones in our analysis (i.e., 219) and n_{taxon} is equal to the total number of zones in which the taxon is observed (i.e., the number of zones for which the taxon has a nonzero *TF*). And finally, for each of the 219 geographic zones, we compute:

$$TF\text{-}IDF_{taxon,zone} = TF_{taxon,zone} \cdot IDF_{taxon}$$

In population genetics, this approach would be analogous to finding a greater number of unique alleles in separate subpopulations that would lead to a higher fixation index (F_{ST}; Wright 1951). In our case, the *TF-IDF* score represents the degree to which the worldwide observations of a species are limited restricted to a given, defined segment of the world. Further, because the *TF-IDF* score of a species increases when it is observed at higher frequencies within a specific segment, *TF-IDF* is not necessarily a measure of species rarity or scarcity. Consequently, we propose that higher *TF-IDF* values for species distributions serve as objective measures of species endemism to a given geographic segment, whereas lower *TF-IDF* values signal a greater degree of cosmopolitanism.

Our approach stands in contrast to previously proposed methods of estimating endemism such as Parsimony Analysis of Endemism (PAE) (Morrone 1994), Endemicity Analysis (EA) (Szumik and Goloboff 2004), and Biotic Element Analysis (BE) (Hausdorf and Hennig 2003). While a detailed comparison of these approaches is beyond the scope of the current work (for an example of

Table 10.1 Summary data of all *TwF-IDF* data at taxonomic level of Order (with portunids listed, as a family level example) to identify level of endemism for each level. For each taxonomic level we report numbers of species treated (from the OBIS database) overall, the global (across all zones) *TF-IDF* score, and the number of species in zones where the average *TF-IDF* value within a zone exceeds the global average, which we access as a locally endemic region. 219 zones (characterized in the text, *e.g,* 90°W is listed as 90W) are 10° in longitude wide and are divided into six latitudinal divisions of tropical (0°-23.5°) north/south (*TropN/TropS*), temperate (23.5°-66.5°) north/south (*TempN/TempS*), and polar (66.5°-90°) north/south (*PolarN/PolarS*) with three zones subdivided by the central Americas. For each taxonomic level, we rank the three highest individual zones by *TF-IDF*, and the three highest clusters of zones by Canberra distances (as described in the text). Complete data and an unabridged table with more clustering data are archived at ResearchGate (https://www.researchgate.net/publication/323398647). Overall rankings can be biased upwards by a low number of species are found in a single zone cluster, which tends to inflate the *TF-IDF* estimates of endemism. (**Note:* For a given taxonomic rank, the grand *TF-IDF* value is computed across all *spp.* and all zones.)

			Number (%) where Average *TF-IDF* ≥ Grand *TF-IDF*		**Ranking of single zone by *TF-IDF*$_{avg}$**					
Taxonomic Rank	**Total *Spp.***	**Grand *TF-IDF****	***Spp.***	**Zones**	**Highest**	**Second**	**Third**	**Notable clusters of contiguous/adjacent zones (w/cluster average of *TF-IDF*$_{avg}$)**		
SubP Crustacea	12016	3.33	4532(38%)	34(16%)	90W-80W, *TempN* (*TF-IDF*$_{avg}$ = 4.477, 1589 *spp.*)	10W-0E, *TempN* (*TF-IDF*$_{avg}$ = 4.429, 761 *spp.*)	0E-10E, *TempN* (*TF-IDF*$_{avg}$ = 4.337, 880 *spp.*)	10W-0E, *TempN*; 0E-10E, *TempN* (*TF-IDF*$_{avg}$ = 4.380, 1047 *spp.*)	80W-70W, *TempN*; 70W-60W, *TempN* (*TF-IDF*$_{avg}$ = 3.935, 1357 *spp.*)	140E-150E, *TropS*; 150E-160E, *TropS* (*TF-IDF*$_{avg}$ = 3.762, 694 *spp.*)
SubCl. Thecostraca	376	3.116	122(32%)	37(17%)	140E-150E, PolarS (*TF-IDF*$_{avg}$ = 8.158, 1 *spp.*)	150E-160E, *TropN* (*TF-IDF*$_{avg}$ = 8.158, 1 *spp.*)	30W-20W, *TempN* (*TF-IDF*$_{avg}$ = 8.158, 1 *spp.*)	20W-10W, PolarN; 30W-20W, *TempN* (*TF-IDF*$_{avg}$ = 4.599, 2 *spp.*)	130E-140E, *TempN*; 150E-160E, *TropN*;140E-150E, *TropN*; 130E-140E, PolarN; 140E-150E, *TempN* (*TF-IDF*$_{avg}$ = 4.538, 11 *spp.*)	130E-140E, *TempN*; 130E-140E, PolarN; 140E-150E, *TempN* (*TF-IDF*$_{avg}$ = 4.408, 9 *spp.*)
O. Kentrogonida	32	2.935	11(34%)	6(3%)	90W-80W, *TempN* (*TF-IDF*$_{avg}$ = 6.328, 8 *spp.*)	70W-60W, *TempN* (*TF-IDF*$_{avg}$ = 4.161, 2 *spp.*)	0E-10E, *TempN* (*TF-IDF*$_{avg}$ = 3.987, 5 *spp.*)	0E-10E, *TempN*; 10E-20E, *TempN* (*TF-IDF*$_{avg}$ = 3.390, 7 *spp.*)		

O. Sessilia	194	3.198	75(39%)	33(15%)	150E-160E, *TropN* (*TF-IDF*$_{avg}$ = 8.158, 1 *spp.*)	30W-20W, *TempN* (*TF-IDF*$_{avg}$ = 8.158, 1 *spp.*)	110W-100W, *TropN* (*TF-IDF*$_{avg}$ = 8.158, 1 *spp.*)	40W-30W, *TempN*; 30W-20W, *TempN* (*TF-IDF*$_{avg}$ = 5.099, 1 *spp.*)	150E-160E, *TropN*; 140E-150E, *TropN*; 140E-150E, *TempN*; 130E-140E, *TempN* (*TF-IDF*$_{avg}$ = 4.607, 7 *spp.*)	140E-150E, *TropN*; 140E-150E, *TempN*; 130E-140E, *TempN* (*TF-IDF*$_{avg}$ = 4.100, 6 *spp.*)
C. Malacostraca	11640	3.337	4428(38%)	34(16%)	90W-80W, *TempN* (*TF-IDF*$_{avg}$ = 4.474, 1517 *spp.*)	10W-0E, *TempN* (*TF-IDF*$_{avg}$ = 4.435, 729 *spp.*)	0E-10E, *TempN* (*TF-IDF*$_{avg}$ = 4.367, 849 *spp.*)	10W-0E, *TempN*; 0E-10E, *TempN* (*TF-IDF*$_{avg}$ = 4.398, 1007 *spp.*)	80W-70W, *TempN*; 70W-60W, *TempN* (*TF-IDF*$_{avg}$ = 3.962, 1311 *spp.*)	140E-150E, *TropS*; 150E-160E, *TropS* (*TF-IDF*$_{avg}$ = 3.775, 686 *spp.*)
SupO. Peracarida	5891	3.307	2186(37%)	31(14%)	140W-130W, *TropS* (*TF-IDF*$_{avg}$ = 6.566, 1 *spp.*)	110W-100W, *TempN* (*TF-IDF*$_{avg}$ = 5.249, 2 *spp.*)	10W-0E, *TempN* (*TF-IDF*$_{avg}$ = 4.386, 567 *spp.*)	10W-0E, *TempN*; 0E-10E, *TempN* (*TF-IDF*$_{avg}$ = 4.333, 747 *spp.*)	140W-130W, *TropS*; 150W-140W, *TempS* (*TF-IDF*$_{avg}$ = 4.278, 5 *spp.*)	80W-70W, *TempN*; 70W-60W, *TempN* (*TF-IDF*$_{avg}$ = 4.024, 824 *spp.*)
O. Amphipoda	2823	3.294	1183(42%)	25(11%)	140W-130W, *TropS* (*TF-IDF*$_{avg}$ = 6.566, 1 *spp.*)	110E-120E, *TropN* (*TF-IDF*$_{avg}$ = 4.864, 21 *spp.*)	10W-0E, *TempN* (*TF-IDF*$_{avg}$ = 4.614, 295 *spp.*)	10W-0E, *TempN*; 0E-10E, *TempN* (*TF-IDF*$_{avg}$ = 4.427, 400 *spp.*)	150W-140W, *TempS*; 140W-130W, *TropS* (*TF-IDF*$_{avg}$ = 4.278, 5 *spp.*)	80W-70W, *TempN*; 70W-60W, *TempN* (*TF-IDF*$_{avg}$ = 4.163, 475 *spp.*)
O. Cumacea	440	3.243	168(38%)	27(12%)	180W-170W, *TempS* (*TF-IDF*$_{avg}$ = 6.954, 1 *spp.*)	140E-150E, *TempS* (*TF-IDF*$_{avg}$ = 5.688, 3 *spp.*)	70E-80E, *TempS* (*TF-IDF*$_{avg}$ = 5.354, 12 *spp.*)	130W-120W, *TempN*; 120W-110W, *TempN* (*TF-IDF*$_{avg}$ = 4.432, 47 *spp.*)	60E-70E, *TempS*; 70E-80E, *TempS* (*TF-IDF*$_{avg}$ = 4.409, 13 *spp.*)	10W-0E, *TempN*; 0E-10E, *TempN* (*TF-IDF*$_{avg}$ = 4.385, 83 *spp.*)

(*Continued*)

Table 10.1. Continued

Taxonomic Rank	Total *Spp.*	Grand *TF-IDF**	Number (%) where Average *TF-IDF* ≥ Grand *TF-IDF*: *Spp.*	Zones	Ranking of single zone by $TF\text{-}IDF_{avg}$: Highest	Second	Third	Notable clusters of contiguous/adjacent zones (w/cluster average of $TF\text{-}IDF_{avg}$)		
O. Isopoda	1969	3.333	600(30%)	33(15%)	70E-80E, PolarS ($TF\text{-}IDF_{avg}$ = 5.960, 12 *spp.*)	150W-140W, PolarN ($TF\text{-}IDF_{avg}$ = 5.392, 2 *spp.*)	10E-20E, *TempS* ($TF\text{-}IDF_{avg}$ = 4.795, 157 *spp.*)	10W-0E, *TempN*; 0E-10E, *TempN* ($TF\text{-}IDF_{avg}$ = 4.261, 132 *spp.*)	70E-80E, *TempS*; 70E-80E, PolarS ($TF\text{-}IDF_{avg}$ = 4.240, 25 *spp.*)	10E-20E, *TempS*; 20E-30E, *TempS* ($TF\text{-}IDF_{avg}$ = 4.215, 196 *spp.*)
O. Lophogastrida	22	3.422	11(50%)	8(4%)	140W-130W, *TempN* ($TF\text{-}IDF_{avg}$ = 8.158, 1 *spp.*)	70W-60W, *TempS* ($TF\text{-}IDF_{avg}$ = 8.158, 1 *spp.*)	180W-170W, *TropS* ($TF\text{-}IDF_{avg}$ = 7.453, 2 *spp.*)	160E-170E, *TropS*; 170E-180E, *TempS* ($TF\text{-}IDF_{avg}$ = 5.197, 5 *spp.*)	70W-60W, *TempS*; 60W-50W, *TempS* ($TF\text{-}IDF_{avg}$ = 5.099, 1 *spp.*)	80W-70W, *TempN*; 90W-80W, *TempN*; 80W-70W, *TropN*, Atlantic ($TF\text{-}IDF_{avg}$ = 3.886, 2 *spp.*)
O. Mysida	402	3.358	150(37%)	28(13%)	110W-100W, *TempN* ($TF\text{-}IDF_{avg}$ = 8.158, 1 *spp.*)	90W-80W, *TempS* ($TF\text{-}IDF_{avg}$ = 6.299, 2 *spp.*)	100W-90W, *TempS* ($TF\text{-}IDF_{avg}$ = 5.392, 2 *spp.*)	40E-50E, *TropN*; 50E-60E, *TropS* ($TF\text{-}IDF_{avg}$ = 4.509, 3 *spp.*)	40E-50E, *TempS*; 50E-60E, *TropS*; 30E-40E, *TempS* ($TF\text{-}IDF_{avg}$ = 4.488, 5 *spp.*)	90W-80W, *TempN*; 80W-70W, *TempN* ($TF\text{-}IDF_{avg}$ = 4.346, 64 *spp.*)
O. Tanaidacea	235	3.372	80(34%)	23(11%)	50W-40W, *TropS* ($TF\text{-}IDF_{avg}$ = 5.647, 2 *spp.*)	160E-170E, *TempS* ($TF\text{-}IDF_{avg}$ = 5.141, 3 *spp.*)	120W-110W, *TempN* ($TF\text{-}IDF_{avg}$ = 5.009, 13 *spp.*)	130W-120W, *TempN*; 120W-110W, *TempN* ($TF\text{-}IDF_{avg}$ = 4.493, 18 *spp.*)	70W-60W, *TempS*; 60W-50W, *TempS* ($TF\text{-}IDF_{avg}$ = 4.121, 34 *spp.*)	10W-0E, *TempN*; 0E-10E, *TempN* ($TF\text{-}IDF_{avg}$ = 3.986, 33 *spp.*)

O. Decapoda	5422	3.361	2083(38%)	38(17%)	170E-180E, PolarS (*TF-IDF*$_{avg}$ = 5.196, 1 *spp.*)	160E-170E, PolarS (*TF-IDF*$_{avg}$ = 4.677, 2 *spp.*)	60E-70E, *TropS* (*TF-IDF*$_{avg}$ = 4.676, 2 *spp.*)	10W-0E, *TempN*; 0E-10E, *TempN* (*TF-IDF*$_{avg}$ = 4.592, 258 *spp.*)	140W-130W, *TempN*; 130W-120W, *TempN* (*TF-IDF*$_{avg}$ = 4.077, 201 *spp.*)	70E-80E, *TempS*; 60E-70E, *TropS* (*TF-IDF*$_{avg}$ = 3.897, 3 *spp.*)
F. Portunidae	207	3.116	83(40%)	27(12%)	130E-140E, *TempN* (*TF-IDF*$_{avg}$ = 5.249, 2 *spp.*)	110E-120E, *TropN* (*TF-IDF*$_{avg}$ = 5.099, 24 *spp.*)	20E-30E, *TempN* (*TF-IDF*$_{avg}$ = 4.613, 2 *spp.*)	130E-140E, *TropS*; 140E-150E, *TropS* (*TF-IDF*$_{avg}$ = 3.980, 39 *spp.*)	100E-110E, *TropN*; 110E-120E, *TropN* (*TF-IDF*$_{avg}$ = 3.946, 38 *spp.*)	160E-170E, *TropS*; 150E-160E, *TropS*; 140E-150E, *TropS* (*TF-IDF*$_{avg}$ = 3.832, 73 *spp.*)
O. Stomatopoda	318	3.448	143(45%)	24(11%)	80W-70W, *TempS* (*TF-IDF*$_{avg}$ = 8.158, 1 *spp.*)	10W-0E, *TempN* (*TF-IDF*$_{avg}$ = 6.954, 1 *spp.*)	20E-30E, *TempS* (*TF-IDF*$_{avg}$ = 6.686, 2 *spp.*)	10E-20E, *TempN*; 10W-0E, *TempN*; 0E-10E, *TempN* (*TF-IDF*$_{avg}$ = 5.483, 3 *spp.*)	10E-20E, *TempN*; 0E-10E, *TempN* (*TF-IDF*$_{avg}$ = 5.189, 3 *spp.*)	20E-30E, *TempS*; 10E-20E, *TempS* (*TF-IDF*$_{avg}$ = 4.811, 4 *spp.*)

such a comparative analysis, see Casagranda et al. 2012), what these approaches all have in common is that they are fundamentally based on the binary classification of a given taxon's presence or absence in a particular geographic area, without taking into account the frequency of observations of the taxon within such an area. Our approach, however, allows for the frequency of observations of a taxon in a given area to influence the estimated degree of endemism of the taxon to that area relative to other areas around the globe. We recognize the potential for sampling bias effects, and therefore our approach attempts to normalize the observation frequencies by assignment into quartile ranks. Our approach may also be contrasted with previously proposed metrics of taxon rarity (Grenyer et al. 2006) as well as range rarity and proportional range rarity (Selig et al. 2014), all of which are also fundamentally based on a binary presence-versus-absence classification, which is probably appropriate for analyses involving the presence of taxa in geographic zones of relatively fine resolution, such as the congruent icosahedral hexagons in Selig et al.'s (2014) analysis, in which each zone measured just under 2,600 square kilometers. We do not necessarily purport our analysis to be a replacement for or improvement upon any of these previously proposed methods. Rather, we suggest that it be considered as an alternative or complementary approach, particularly for situations in which the resolution of geographic zones is relatively coarse (potentially spanning multiple longitudinal and latitudinal degrees), as is the case in our present analysis. If anything, our approach ostensibly shares more in common with the Shannon-Wiener index of biodiversity (Keylock 2005), which does account for the frequency of individual observations of a given taxon relative to the count of all observations of all taxa in a given geographic area, but unlike our approach, the Shannon-Wiener index does not account for between-taxon frequency in a given geographical area and the frequency of that same taxon across the entire globe, which we believe is necessary for computing metrics of global biodiversity.

To create a summary of this endemism worldwide, we computed average *TF-IDF* signal values across all of the Crustacea by members within the orders and, separately, by members within families, for each of the 219 zones. For visualization purposes (Fig. 10.5), each zone was assigned a color value, the tint of which is proportional to the number of orders/families for a taxonomic group within a region. *TF-IDF* values were plotted as horizontal bars within each zone on a Mercator-projected world map generated by the *Basemap* toolkit, an enhancement of the Python-based *matplotlib* function library for plotting 2D graphs (see https://matplotlib.org/basemap/). For each bar plotted, the width represents the average *TF-IDF* signal for all order or family entities within that zone relative to the zone's width on the map (which itself represents the maximum zone-average *TF-IDF* value, as computed for all orders or families, respectively). Thus, Fig. 10.5 can be used to estimate patterns of likelihood of endemism for crustacean species to certain geographic zones, and it also may serve as an estimator of the richness of co-occurring taxa within each region. Several patterns are evident by observation; for example, there is a high diversity of taxa in the western Atlantic region, and except for northern Australia and Indonesian waters, the temperate zone biodiversity in any one longitudinal region is generally higher than for the tropical zones. However, while this procedure gives insight into the biogeographic diversity and relative endemism in each region, it cannot by itself directly estimate the connectivity or shared richness of neighboring regions. This is analogous to an Analysis of Variance that estimates differences among all means simultaneously without specifying which means contribute most to the significant *F*-value.

To identify adjacent geographic zones with similar *TF-IDF* signals across taxa within a rank, we applied a hierarchical clustering approach. For each taxonomic rank, we define m as the number of species in that rank that meet the previously discussed criteria for inclusion in the study, and n as the number of geographic zones in the analysis (which, in our case, is equal to 219). For the resulting $m \times n$ matrix of *TF-IDF* values for that rank, we generated a condensed matrix of Canberra distances between the vectors of *TF-IDF* values for each pair of geographic zones. Using a complete-linkage clustering algorithm, we produced a dendrogram that clusters geographic zones according to the similarity of their respective *TF-IDF* vectors.

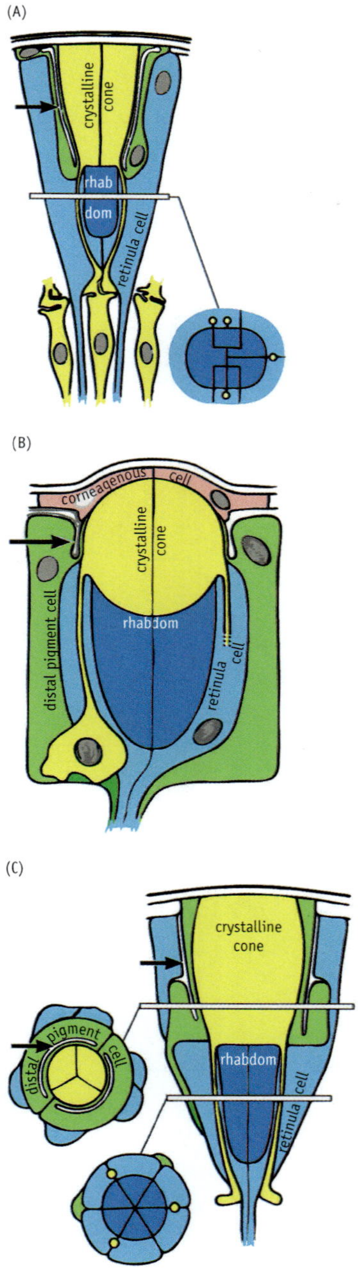

Fig. 1.4.
Ommatidia in three different maxillopodan species. (A) *Argulus foliaceus* (Branchiura). Overview with transverse section through the rhabdom. Retinula cell between two cone cell processes. (B) *Cypridina norvegica* (Ostracoda). Overview. (C) *Balanus crenatus* (Cirripedia). Overview with two transverse sections through the crystalline cone and the distal pigment cells and through the rhabdom. Three cone cell processes are present. Arrows (A, B, C) indicate extracellular space between the distal pigment cells in all three species. After Richter (2002).

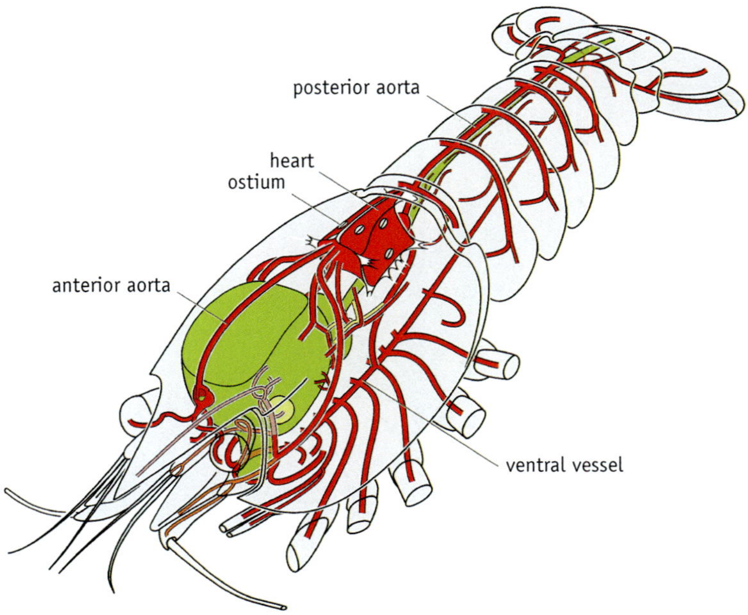

Fig. 1.5.
Tagmatization in malacostracan and branchiopod crustaceans.

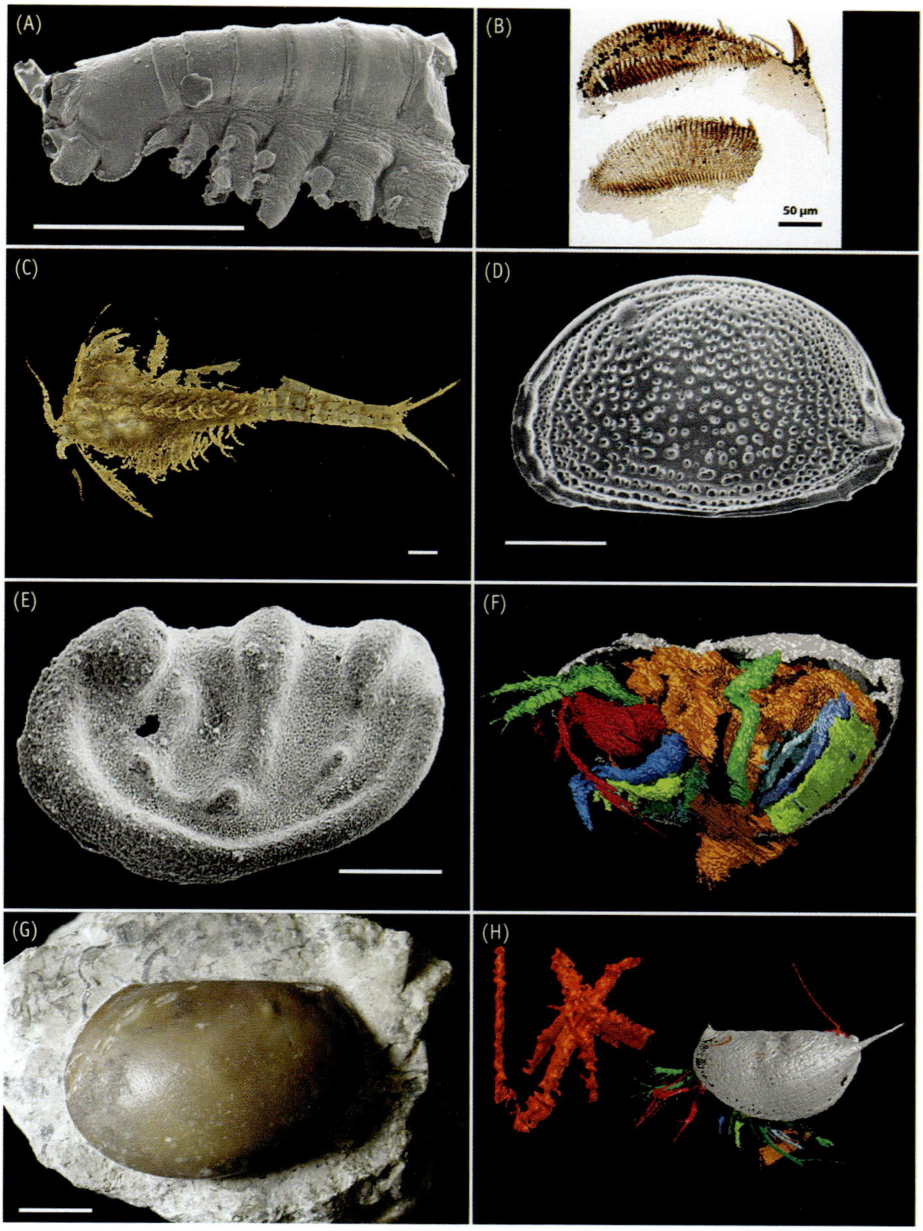

Fig. 2.2.
Cambrian fossils and Oligostracans. (A) *Yicaris dianensis* (Key Laboratory for Palaeobiology, Yunnan University (YKLP) 10840, holotype) from the Lower Cambrian of Yunnan, China. Scale bar is 300 μm. (B) Branchiopod-type mandibles (Geological Survey of Canada (GSC) 135369, 135370) from the Middle and Late Cambrian of Saskatchewan, Canada. Scale bar is 300 μm. (C) A virtual reconstruction of *Ercaicunia multinodosa* (YKLP 16201) from the Lower Cambrian of Yunnan, China. Scale bar is 1 mm. (D) Left valve of a female specimen of the podocopid ostracod *Mutilus speyeri* (British Museum of Natural History IO 4977) from Pliocene-Pleistocene aged sediments off the south coast of Turkey. Scale bar is 250 μm. (E) Left valve of a female specimen of the palaeocope ostracod *Homeokiesowia epicopa* (British Museum of Natural History OS 6695) from the Ordovician of Wales. Scale bar is 500 μm. (F) A virtual reconstruction of the fossil myodocope ostracod *Colymbosathon ecplecticos* from the Silurian of Herefordshire, England. (G) The myodocope-like leperditicopid *Leperditia baltica* (Yale Peabody Museum Invertebrate Paleontology (YPMIP) 7300) from the Late Silurian of Gotland, Sweden. Scale bar is 5 mm. (H) A virtual reconstruction of the fossil pentastomid *Invavita piratica*, shown isolated (left) and on its host Nyphatelina gravida (Ostracoda, Myodocopa) (right), both from the Middle Silurian (Wenlock) of Herefordshire, England. A, from Zhang et al. (2007), with permission from Springer Nature; B, from Harvey et al. (2012), with permission from PNAS; D, from Doruk (1973); E, from Siveter (1982); F, from Siveter et al. (2003); G, courtesy of the Yale Peabody Museum; H, from Siveter et al. (2015). Figures 2.2D–F, H courtesy of D. J. Siveter.

Fig. 2.4.

Multicrustaceans. (A) The phyllocarid *Echinocaris socialis* (Yale Peabody Museum Invertebrate Paleontology (YPMIP) 9674) from the Late Devonian of Pennsylvania. Scale bar is 1 cm. (B) The pygocephalomorph *Hoplita ginsburgi* (Collection of Fossil Invertebrates of the Facultad de Ciencias of Uruguay (FC-DPI) 4826) from the Early Permian of Uruguay. Scale bar is 5 mm. (C) The hoplocarid? *Sculda pennata/spinosa* (collection of Roger Frattigiani, Laichingen) from the late Jurassic Solnhofen Plattenkalk of Germany. Scale bar is 3 mm. (D) The decapod *Imocaris tuberculata* (San Diego Society of Natural History 25139) from the Middle Carboniferous of Arkansas. Scale bar is 5 mm. Photo courtesy of R. M. Feldmann. (E) The meiuran *Eocarcinus praecursor* (British Museum of Natural History In 18425) from the Early Jurassic of Great Britain. Scale bar is 1 cm. (F) An unidentified female ectinosomatid copepod (IHNFG—4899/Cop1) preserved in amber from the Miocene of Chiapas, Mexico. Scale bar is 200 μm. (G) A virtual reconstruction of the fossil cirripede cypris *Rhamphoverritor reduncus* from the Middle Silurian (Wenlock) of Herefordshire, England. (H) The putative turtle barnacle *Emersonius cybosyrinx* (American Museum of Natural History 28462) from the Eocene of Florida. Scale bar is 1 cm. A, from Beecher (1902), courtesy of the Yale Peabody Museum; B, from Piñeiro et al. (2012), courtesy of G. Piñeiro; C, from Haug et al. (2010a), courtesy of C. Haug; D, from Schram and Mapes (1984), courtesy of R. M. Feldmann; E, from Feldmann and Schweitzer (2010a), courtesy of R. M. Feldmann; F, from Huys et al. (2016), courtesy of F. J. Vega; G, from Briggs et al. (2005), courtesy of Yale Peabody Museum; H, from Ross and Newman (1967), courtesy of the American Museum of Natural History.

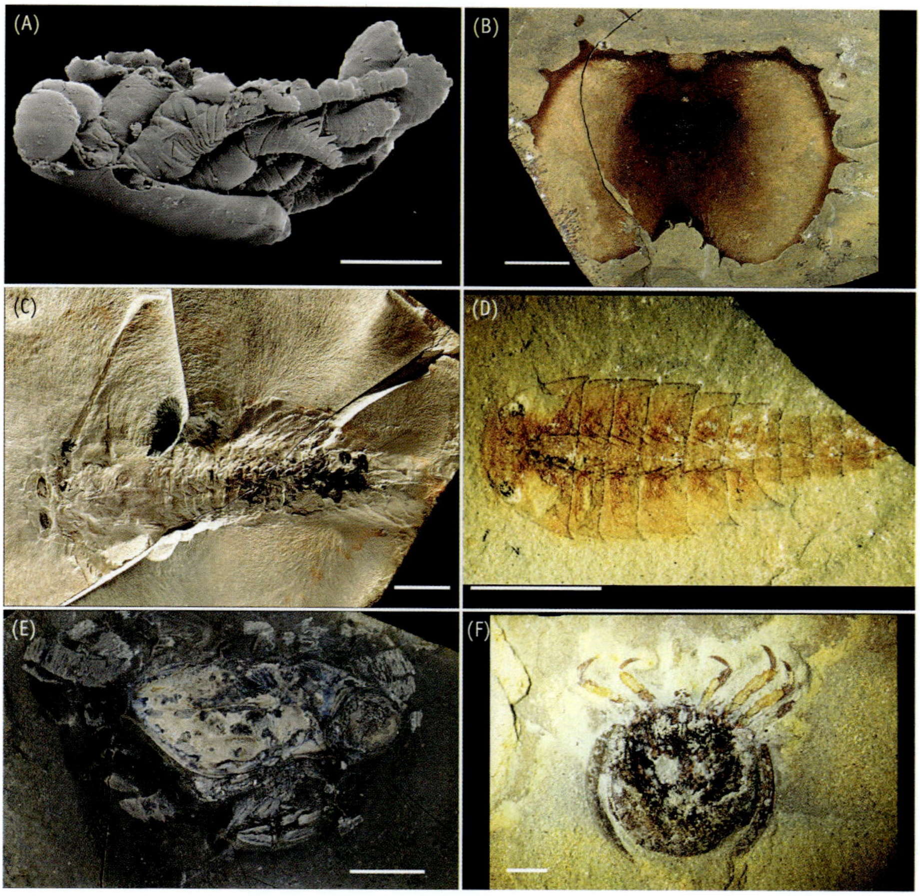

Fig. 2.5.
Allotriocarida and problematic groups. (A) The stem-group branchiopod *Rehbachiella kinnekullensis* from the Upper Cambrian of Sweden. Scale bar is 200 μm. Photo courtesy of D. Waloszek. (B) The kazacharthran branchiopod *Panacanthocaris ketmenica* (Nanjing Institute of Geology Paleontology and Stratigraphy K102, 84838) from the Late Triassic of Xinjiang, China. Scale bar is 5 mm. (C) Holotype of the stem-group remipede *Tesnusocaris goldichi* (United States National Museum 124173a) from the Late Carboniferous of Texas. Scale bar is 1 cm. (D) The enigmatic crustacean-like euthycarcinoid arthropod *Euthycarcinus kessleri* (Louis Grauvogel Collection, Staatliches Museum für Naturkunde in Stuttgart) from the middle Triassic of Bas-Rhin, France. Scale bar is 5 mm. Photo courtesy of L. Grauvogel-Stamm. (E) The enigmatic crustacean-like thylacocephalan arthropod *Dollocaris ingens* (Muséum National d'Histoire Naturelle, France 50939) from La Voulte sur Rhône, France. Scale bar is 1 cm. Photo courtesy of S. Charbonnier. (F) The enigmatic crustacean-like cyclid arthropod *Halicyne ornata* (Louis Grauvogel Collection, Staatliches Museum für Naturkunde in Stuttgart) from the middle Triassic of Bas-Rhin, France. Scale bar is 3 mm. A, from Walossek (1993), courtesy of D. Waloszek; B, from Chen and Zhou (1985), courtesy of L. Gang, photographed by TAH; C, from Neiber et al. (2011); D, from Gall and Grauvogel (1964) and Gall et al. (2006), courtesy of L. Grauvogel-Stamm; E, from Secretan (1985), courtesy of S. Charbonnier—Investissements d'avenir (ANR-11-INBS-0004-RECOLNAT); F, from Gall and Grauvogel (1967) and Gall et al. (2006), courtesy of L. Grauvogel-Stamm.

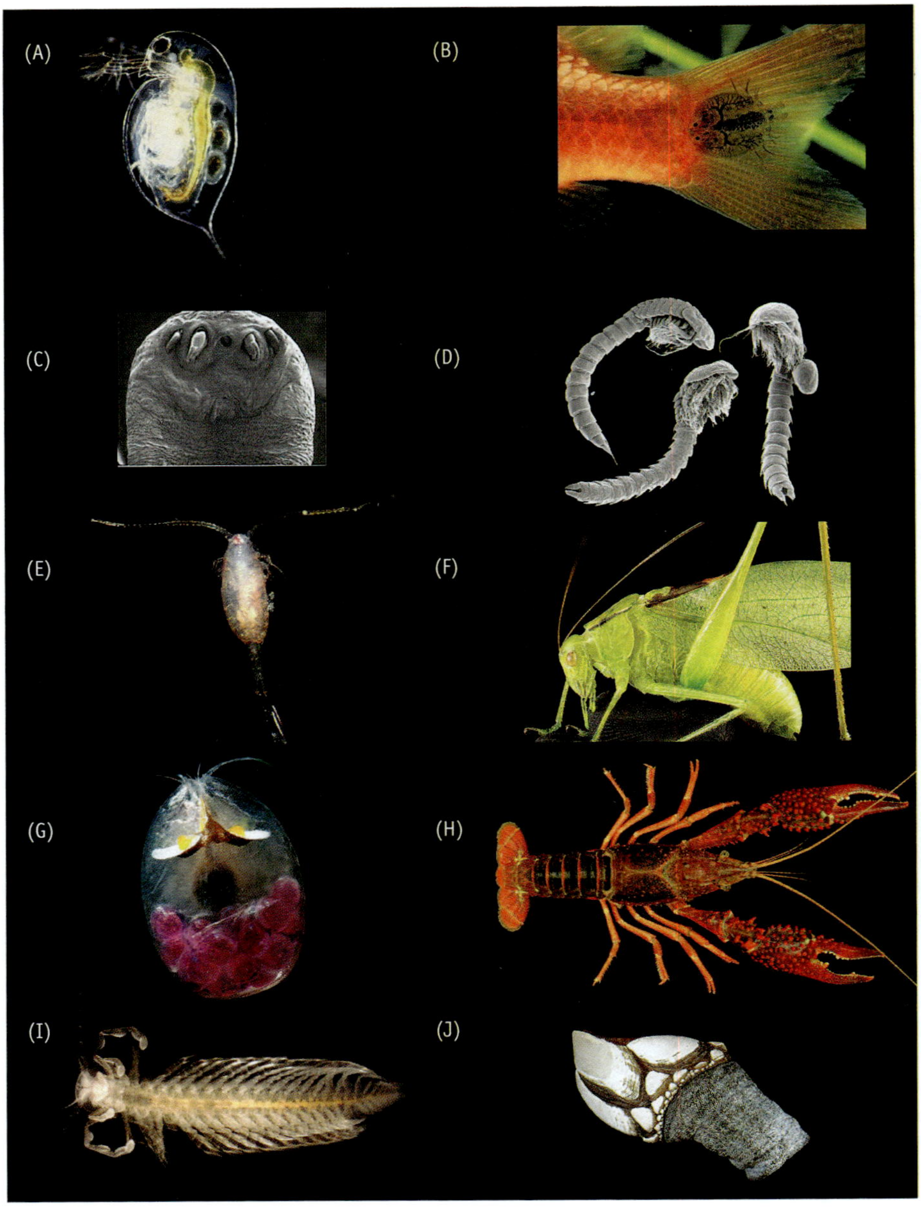

Fig. 4.1.
Examples demonstrating the morphological diversity across Pancrustacea. (A) *Daphnia pulex*, Branchiopoda (from Gewin 2005, under License CC-BY); (B) *Argulus foliaceus*, Branchiura (© G. Boxshall, under License CC-BY-NC-SA); (C) *Porocephalus crotali*, Pentastomida (from Abele et al. 1989, with permission from Oxford University Press); (D) *Lightiella monniotae*, Cephalocarida (© D. Waloszek and J. Olesen); (E) *Ridgewayia* sp., Copepoda (© T. Iliffe); (F) *Amblycorypha oblongifolia*, Hexapoda (from USGS Bee Inventory and Monitoring Lab); (G) *Gigantocypris* sp., Ostracoda (© D. Fenolio/ DEEPEND); (H) *Procambarus clarkii*, Malacostraca (© D. Felder); (I) *Cryptocorynetes* sp., Remipedia (© T. Iliffe); (J) *Pollicipes cornucopia*, Thecostraca (© H. Hillewaert, under License CC-BY-SA).

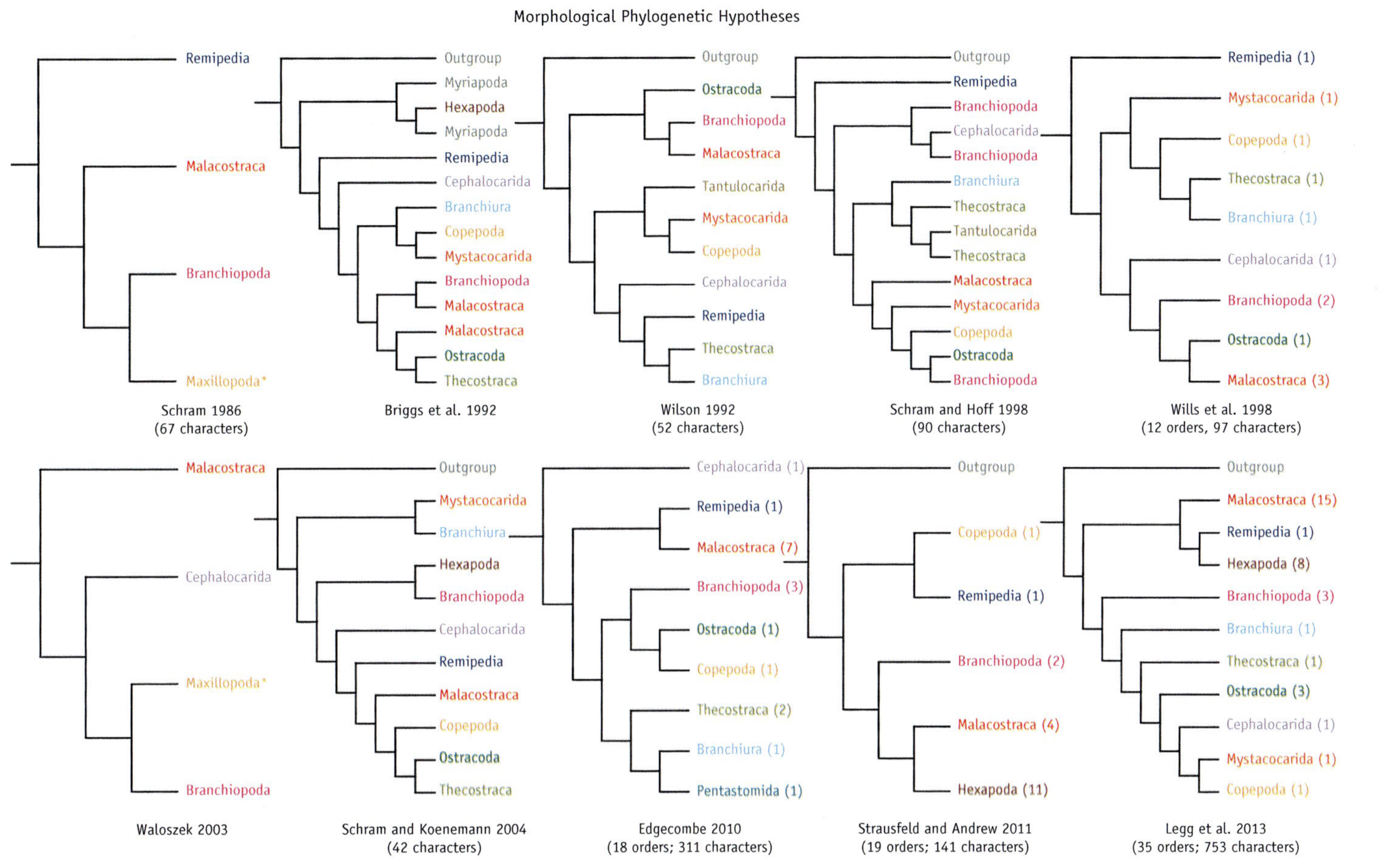

Fig. 4.2.
Alternative hypotheses of the Pancrustacean phylogeny based on morphological data. Numbers adjacent to taxon names in the tree represent how many orders for that taxon were included in the analyses (when available). Caption below citation indicates the total number of orders and characters used in the analyses. Fossil taxa excluded from phylogram for clarity. Class Maxillopoda, denoted by an asterisk, no longer constitutes a valid taxon. However, the source literature did not specify the maxillopodan taxa employed in said analyses.

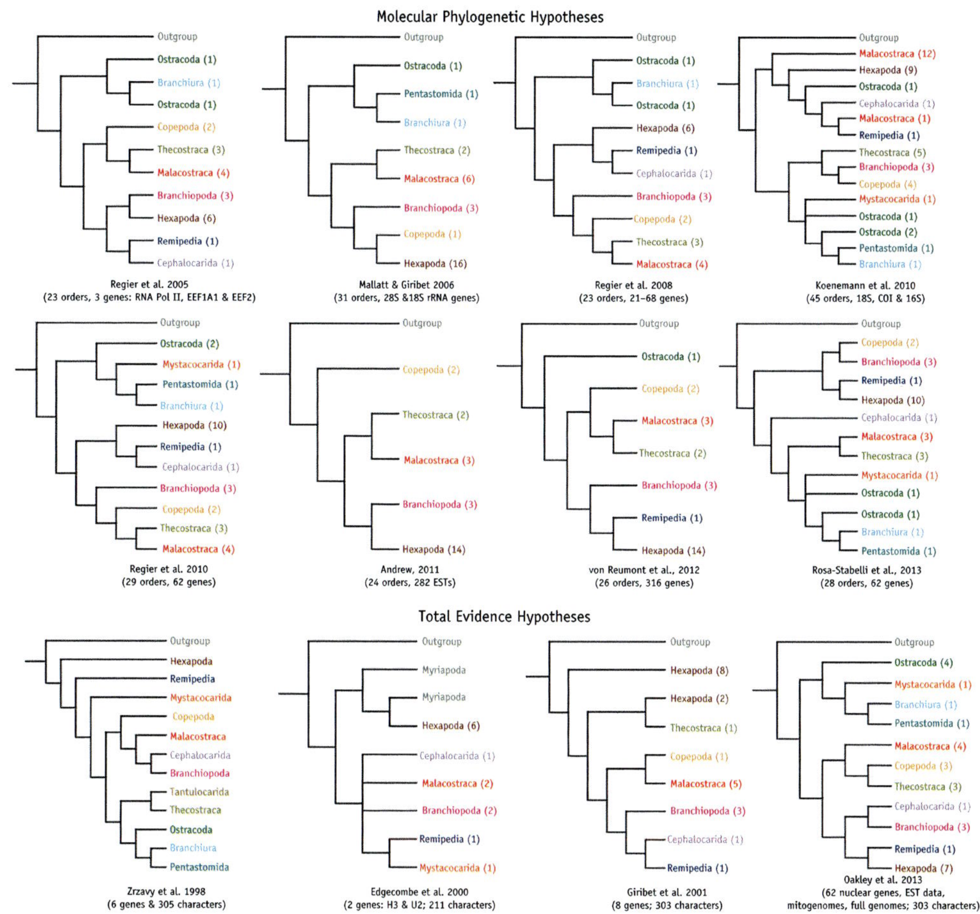

Fig. 4.3.

Alternative hypotheses of the Pancrustacean phylogeny based on genetic data (rows 1 and 2) and combined morphological and genetic data (row 3). Numbers adjacent to taxon names in the tree represent how many orders for that taxon were included in the analyses (when available). Caption below citation indicates the total number of orders, genes, and/or characters used in the analyses. Fossil taxa excluded from phylogram for clarity. Abbreviations: RNA Pol II, RNA polymerase II; EEF1A1, elongation factor alpha subunit 1; EEF2, elongation factor 2; 28S, 28S ribosomal mitochondrial gene; 18S, 18S ribosomal mitochondrial gene; COI, cytochrome oxidase I; 16S, 16S ribosomal mitochondrial gene; EST, expressed sequence tags; U2, small nuclear RNA.

Fig. 5.1.
Diversity and morphological disparity of crustaceans. Many crustaceans are parasites and show derived body shapes, adapted to their hosts. An extreme example is pentastomids (tongue worms), such as (A) *Raillietiella* sp., which lives in lung tracts of vertebrates. (B) Two pentastomids in vivo in the right lung wing of a house gekko. The branchiuran fishlouse *Argulus foliaceaus* (C) parasitizes fish; interestingly the sealouse *Caligus* also parasitizes the gills of salmonids, and its parasitic, blood-sucking feeding mode is convergent to branchiurans. Many taxa are extremely modified and miniaturized. These include (D) ostracods (*Heterocypris* sp.), (E) copepods (*Caligus rogeresseryi*), and (G) mystacocarids (*Derocheilocaris typicus*). (F) *Caprella causa* (male), a caprellid or skeleton shrimp, a rather derived group of small malacostracans that inhabit seagrass beds. (H, J) Examples of two relatively unknown malacostracans that contradict the typical "lobster like" image of the group. (H) *Nebalia antarctica*, a leptostracan species representing an older malacostracan lineage with seven pleonite segments. The (J, K) remipede *Xibalbanus tulumensis*, collected by the first author in a Mexican anchialine cave system (I). Abbreviations: Ant, antenna; Gnp, gnathopod; Max, maxilla; Mc, mouth cone; Mxp, maxilliped; Ps, preoral spine; Thp, thoracopod; TrA, trunk appendages. All images copyright B. von Reumont.

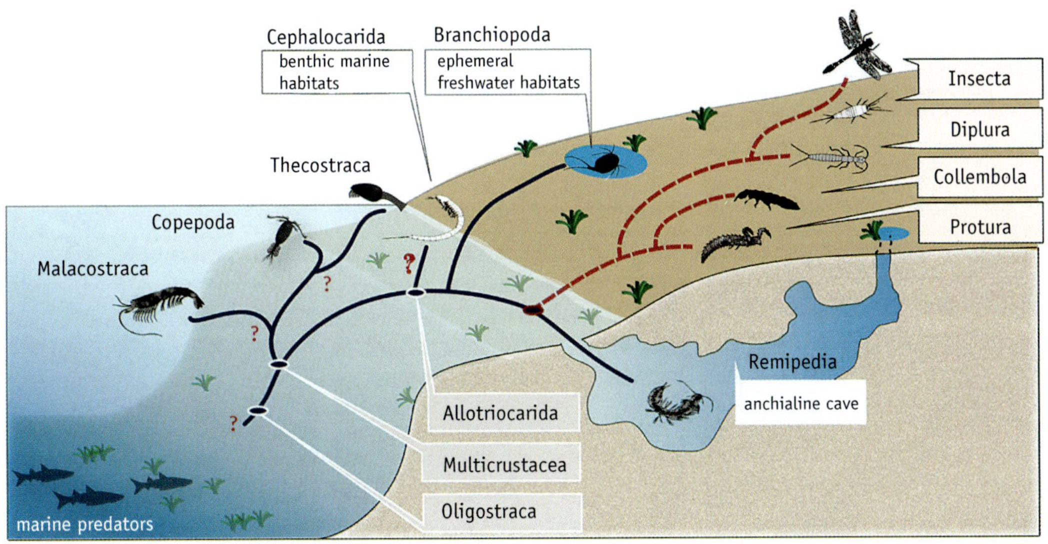

Fig. 5.7.
Evolutionary scenario depicting the transition from marine to terrestrial habitats by pancrustaceans. Shown is the most likely process supported by current data. Question marks indicate lineages that are still not robustly resolved. Hexapod lineages are illustrated by dotted branches, crustaceans by continuous lines. This figure is simplified (e.g., malacostracans such as hermit crabs also conquered land later). Modified from von Reumont et al. (2012), with permission from Oxford University Press.

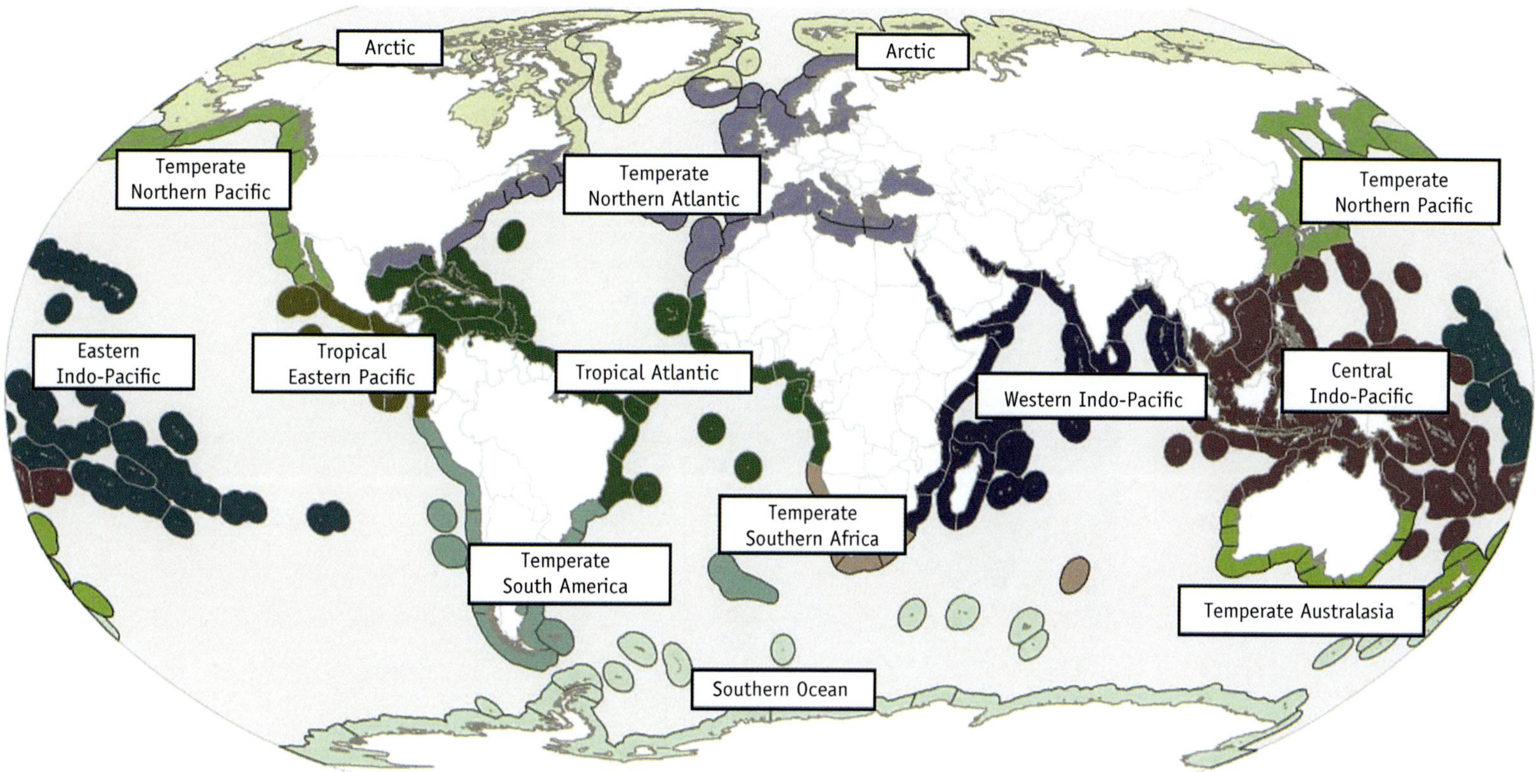

Fig. 6.1.
Biogeographic realms with ecoregion boundaries according to Spalding et al. 2007 (their Fig. 2a).

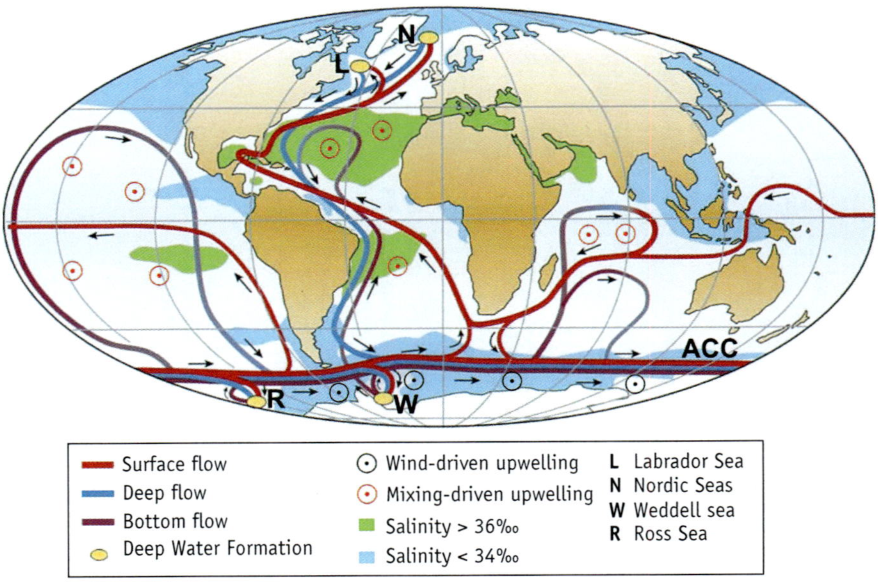

Fig. 6.2.
The global overturning circulation system. From Kuhlbrodt et al. (2007), with permission from the American Geophysical Union.

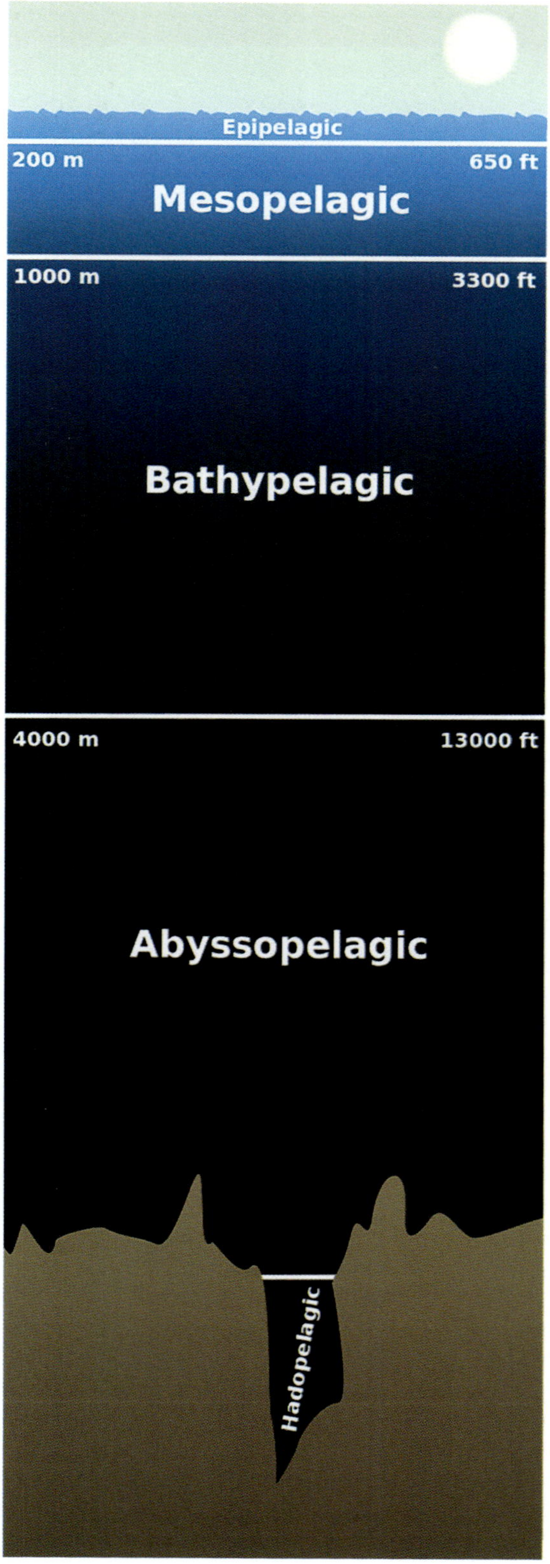

Fig. 6.3.
Depth zonation of the pelagos. Creative commons license—Wikipedia.

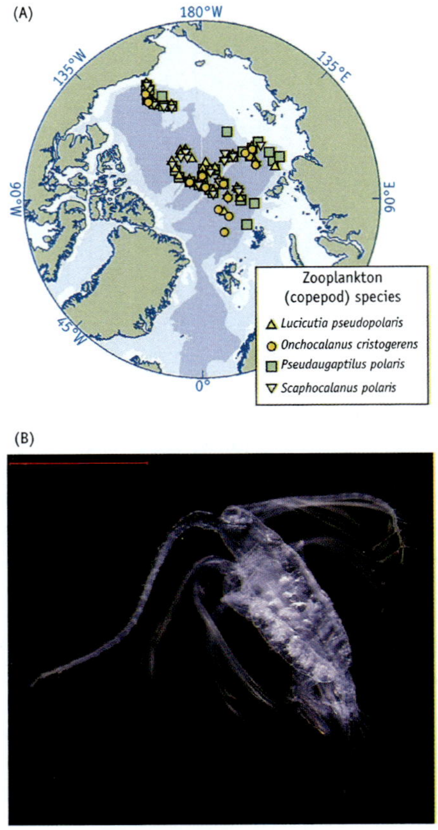

Fig. 6.5.
Distribution of Arctic pelagic crustaceans, including the endemic bathypelagic copepod *Pseudaugaptilus polaris* (photo: Russ Hopcroft). From Bluhm et al. (2015).

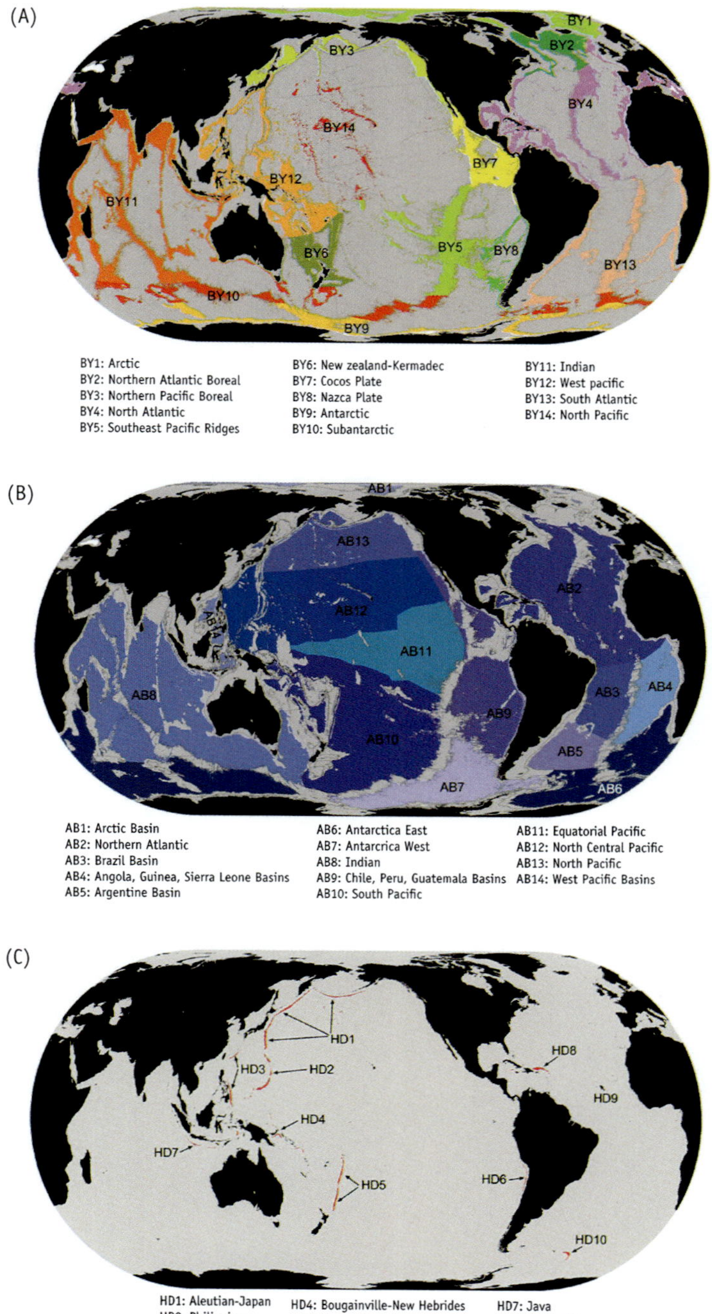

Fig. 7.1.
Biogeographic regions of the deep sea. (A) Lower bathyal provinces including continental slopes and ridges, as well as island slopes, and several thousand seamounts. (B) Abyssal provinces separated by mid-ocean ridges and other bathyal topography creating deep basins as well as by varying levels of POC flux to the seafloor. (C) Hadal provinces include all areas deep areas below 6,500 m. Modified after Watling et al. (2013), with permission from Elsevier.

Fig. 7.2.

Megafaunal Crustacea from the Peru Basin taken at >4,000 m depth in the DISCOL manganese nodule area. Many megafaunal Crustacea seem to rely on some form of hard substrate for a habitat, such as nodules or structures provided by living as well as dead sessile animals. (A) Arcturid isopods (ai), scalpelliform cirriped (sc), and mysid (m) associated on a gorgonian octocoral. Scale bar = 25 cm. (B) Polychelid decapod dug into the sediment. Scale bar = 10. cm (C) Caridean shrimp, Stylodactiloidea. Scale bar = 25 cm. (D) Parapagurid decapod (*Probeebei mirabilis*) with anemone on pleon. (E) Association on manganese nodule: coral stalk, isopods (i), lyssianassoid amphipod (la), ophiuroid (o), mysid (m), sponge (s), and munnopsid isopod (mi). Scale bar = 10 cm. (F) Gammaridean amphipods in a manganese nodule field. Scale bar = 10 cm. (G) *Munidopsis* sp. on manganese nodule. Scale bar = 10 cm. (H) scalpelliform barnacle. Scale bar = 10 cm. (I) *Paropsurus* sp. (Isopoda, Munnopsidae) feeding on salp. Scale bar = 10 cm. Photos courtesy of Alfred-Wegener-Institute/SONNE SO242-2, project JPIO/MIDAS; Ocean Floor Observation System (OFOS) imaging, used with permission under the CC4-BY-NC-SA license.

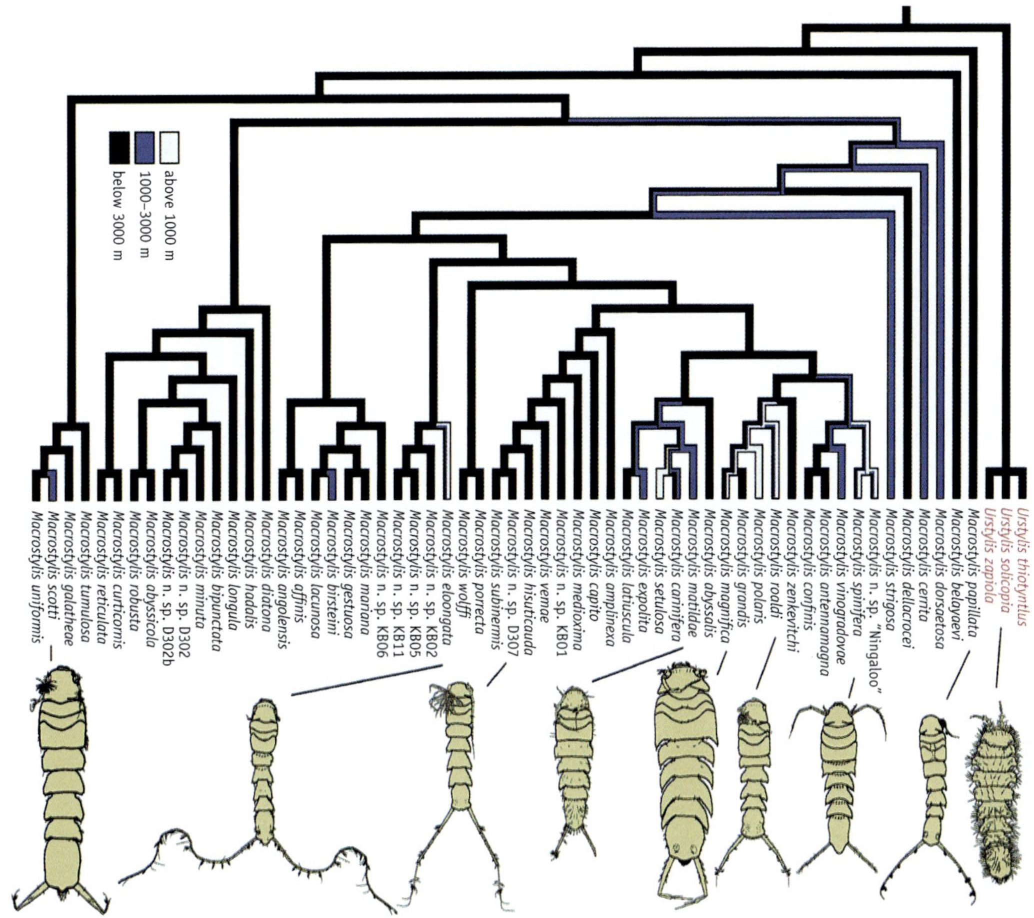

Fig. 7.3.

The abyss-shelf colonization history of Macrostylidae (Isopoda: Asellota). An ancestral state reconstruction was conducted based on a morphological phylogeny using depth zones "sublittoral" (0–1,000 m), "bathyal" (1,000–3,000 m), and "abyssal" + "hadal" (>3,000 m depth) as discrete characters. A parsimony reconstruction scheme as implemented in Mesquite was applied. White branches indicate sublittoral, grey branches indicate bathyal, black bars represent abyssal and hadal. Species (and predecessors) occurring across more than one depth zone have mixed coloration. An abyssal origin of Macrostylidae is most parsimonious. Within Macrostylidae, six emergence events into the bathyal became apparent, of which four led to shelf emergence. Potential reversals (submergence) have occurred, and one species (*Macrostylis magnifica*) has a eurybathic distribution across all three depth zones. Three species of Urstylidae formed the outgroup. Modified from Riehl et al. (2014), with permission from Oxford University Press.

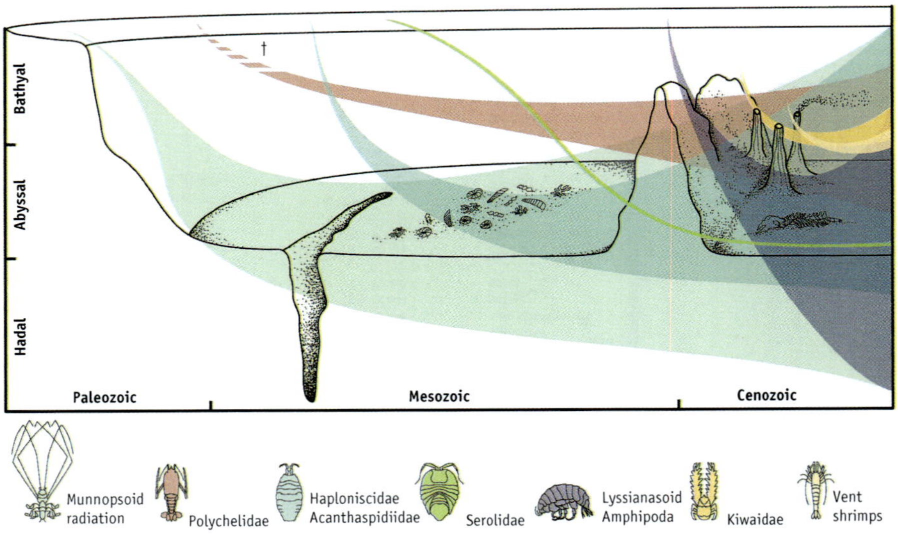

Fig. 7.6.
Overview of selected deep-sea crustacean lineages showing approximate times of origin, radiation, and depth distribution in a depth-time scheme from Paleozoic (541–252.2 Ma) until today. The time scale is vague on purpose, acknowledging the high uncertainties involved. Each lineage is represented by a colored beam. The height of each beam at the right margin indicates the approximate core depth distribution today. Phylogenetic links to potential shallow-water ancestors are indicated at the top left of each beam. Sea-floor structures are not correlated to the time line. The isopods of the munnopsoid radiation as well as the Haploniscidae/Acanthaspidiidae and Serolidae have colonized the abyss independently (Raupach et al. 2009). While the first originated before 250 Ma, the latter invaded the deep sea during the Mesozoic (252–66 Ma) (Lins et al. 2012). The first three groups are diverse on soft sediments, which may have been the key to their success. Subsequently, they colonized shallower depth zones and the hadal. The Serolidae has colonized abyssal depth without radiation at great depth. Polychelidae became extinct on continental shelves but inhabit intermediate depths today (Galil 2000). Among the lyssianassoid amphipods an end-Mesozoic–early-Cenozoic adaptive radiation resulted in a clade of specialized scavengers (Corrigan et al. 2014). Kiwaidae and other seep and vent decapods radiated in extreme habitats after the Paleocene (66–56 Ma) of the Cenozoic (66–0 Ma) (Roterman et al. 2013). With permission from Nele Heitland.

Fig. 7.10.
Common families of abyssal isopoda collected during the Vema-TRANSIT expedition in the North Atlantic. These families and genera are considered to be specialized deep-sea taxa because of their main distribution and diversity at great depths as well as the reduction of eyes. (A) Macrostylidae, *Macrostylis* aff. *magnifica*; (B) Munnopsidae, Eurycopinae, *Eurycope* sp.; (C) Munnopsidae, Betamorphinae, *Betamorpha* sp.; (D) Ischnomesidae, *Gracilimesus* sp.; (E) Mesosignidae, *Mesosignum* sp.; (F) Nannoniscidae, *Nannoniscus* sp.; (G) Desmosomatidae, *Chelator* sp.; (H) Dendrotionidae, *Acanthomunna* sp. All scale bars = 1 mm. From Devey et al. (2015), with permission from GEOMAR.

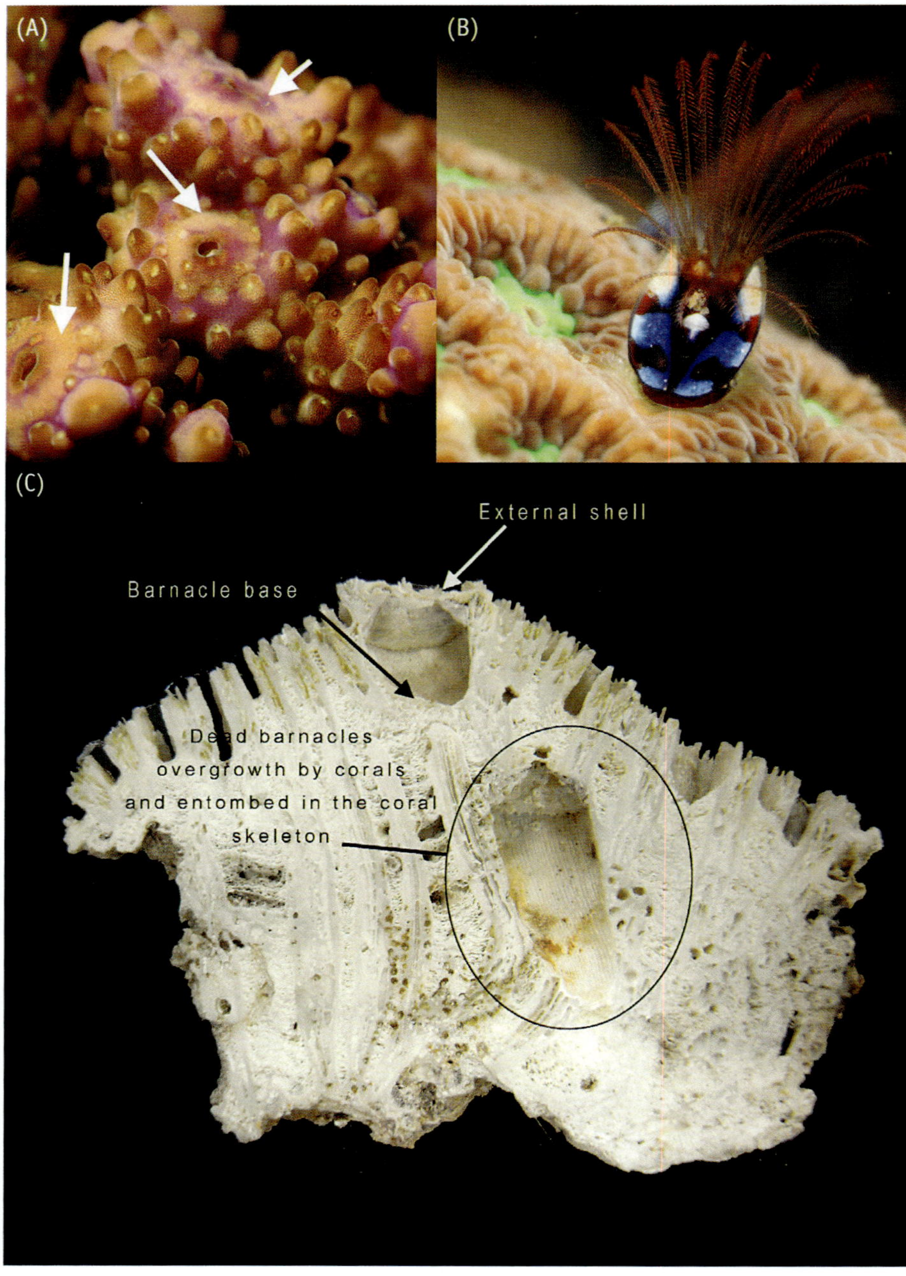

Fig. 8.1.
(A) The Prygomatid barnacles (Throacica) *Cantellius secundus* on *Acropora* corals (indicated by white arrows). (B) The Acrothoraican barnacle *Berndtia purpurea* bore in *Lepastrea* corals. (C) Cross section of a coral skeleton (*Alveopora japonica*) showing the barnacles *Nobia grandis*. On top, the live *Nobia* has its mantle cavity and base embedded in the coral skeleton and the shell is on the surface of corals. There is a dead *Nobia* entombed by coral. Once the barnacle is dead, the coral will gradually overgrow the surface and entomb the barnacle inside the skeleton.

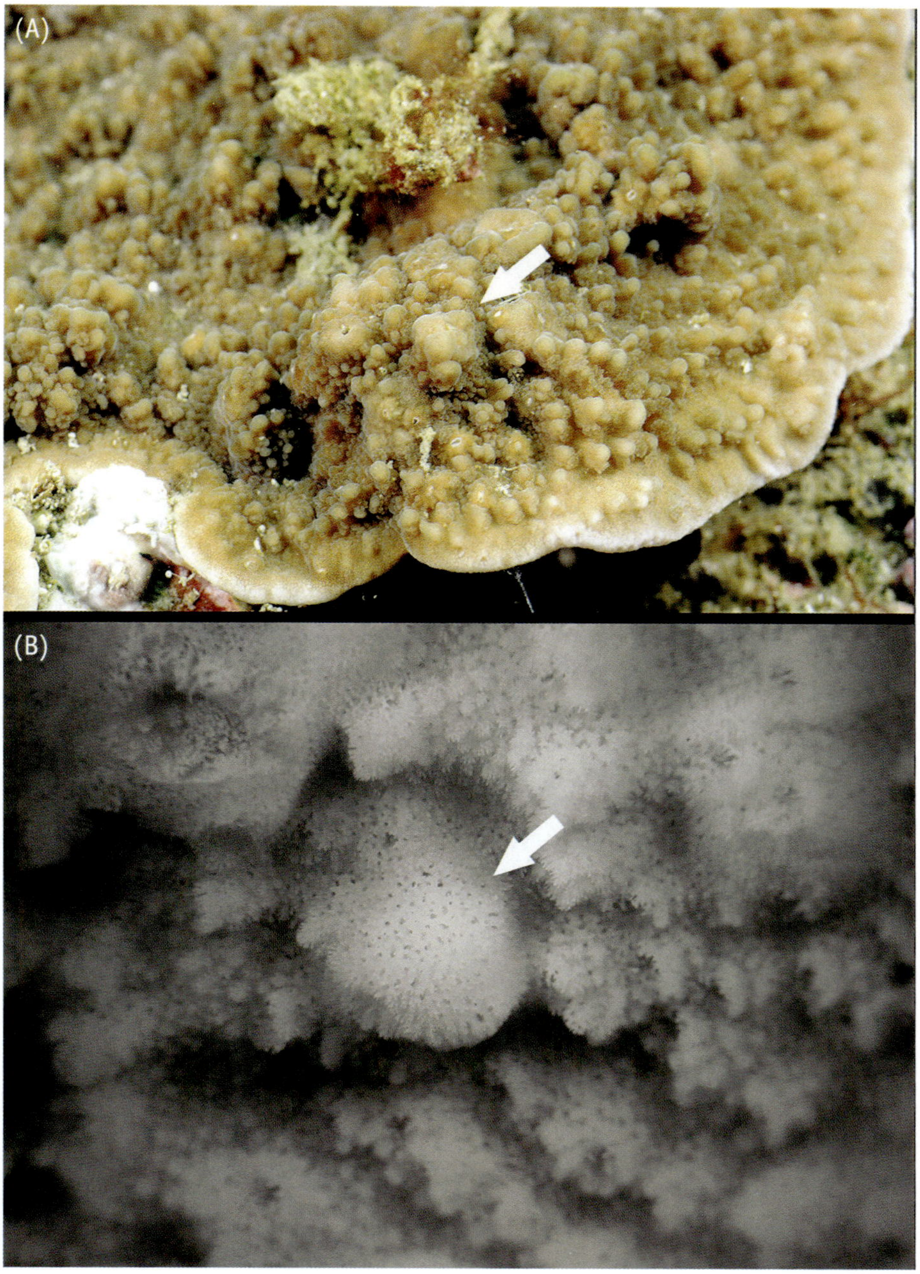

Fig. 8.8.
Photographs of *Montipora aequituberculata* with numbers of dome-shaped galls. (a) A colony showing several galls each inhabited by *Allopodion ryukyuensis* and *Hoplomolgus montiporae*. (b) A closer view of dome-shaped gall. White arrows indicate gall structure.

Fig. 8.9.
Live observations of reintroduced coral-associated copepod, *Xarifia obesa*, to its coral hosts. (a) An individual of female copepod carrying six eggs in dorsal view. (b) Copepod was seeking polyps after reintroduction to a coral colony. (c, d) Coral polyps started to extend their tentacles when in contact with copepod (probably antennule or antenna) and allowed copepod to enter its gastrovascular cavity. Scale bars = 0.5 mm. Arrows indicate the head of copepod.

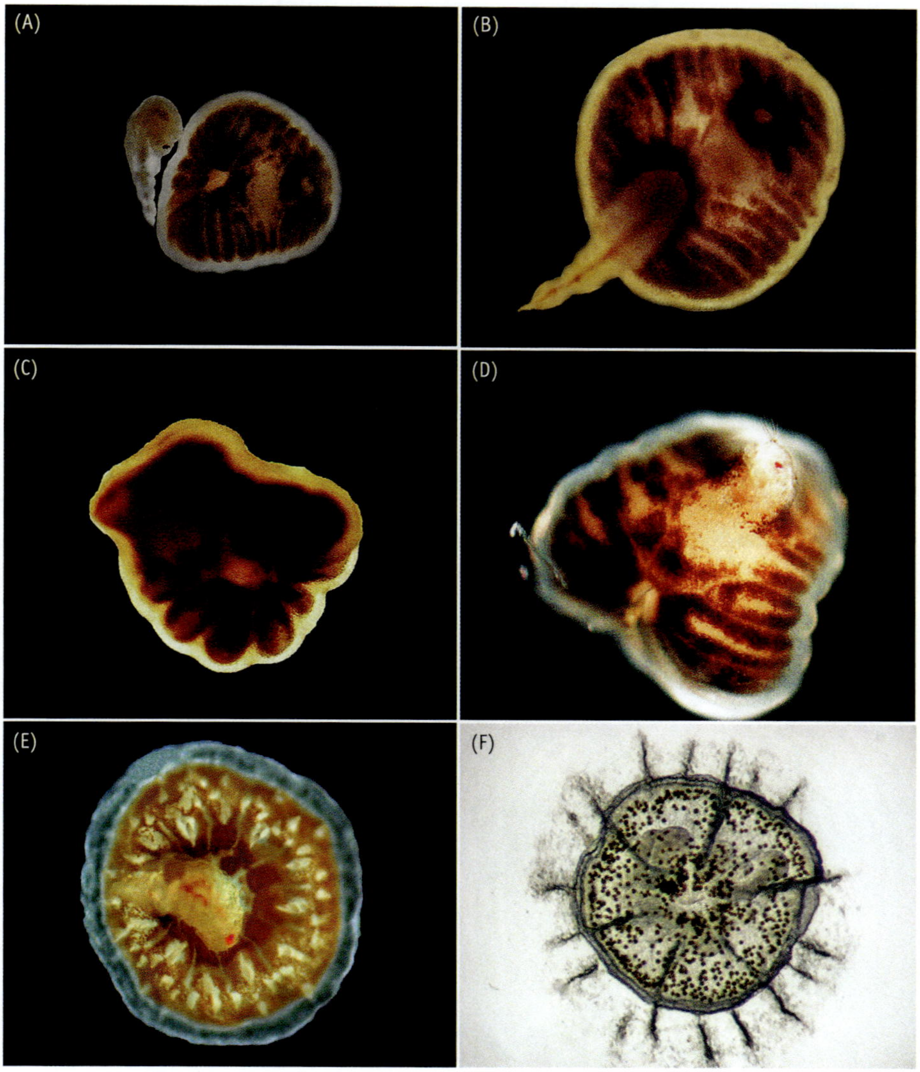

Fig. 8.12.
The reintroducing process of coral-associated copepods to bail-out polyps (a–d) and primary polyp (e, f). (a) Copepod was starting to look for polyps. (b) It tries to hide in the gastrovascular cavity of the polyp via an oral opening when approaching a coral polyp. (c) The infected bail-out polyp appeared an odd, outside shape with a swell in the lateral view. (d) Copepod was tearing the polyp suddenly and wanting to leave when we added a drop of 95% alcohol. (e) Copepod was starting to hide in the newly primary polyp. (f) Copepod infection might influence the development and skeleton forming of primary polyp of coral host.

Fig. 8.14.
Gall crabs in domicile. (A) *Pseudocryptochirus viridis* dwelling in a shallow crescent-shaped crevice. (B) *Dacryomaia edmonsonsi* in coral *Psammocora* sp. (C) Vertical cylindrical domicile and its occupant, *Cryptochirus* or *Lithoscaptus* sp. (D) Round domicile openings of the previous in *Favia* sp.

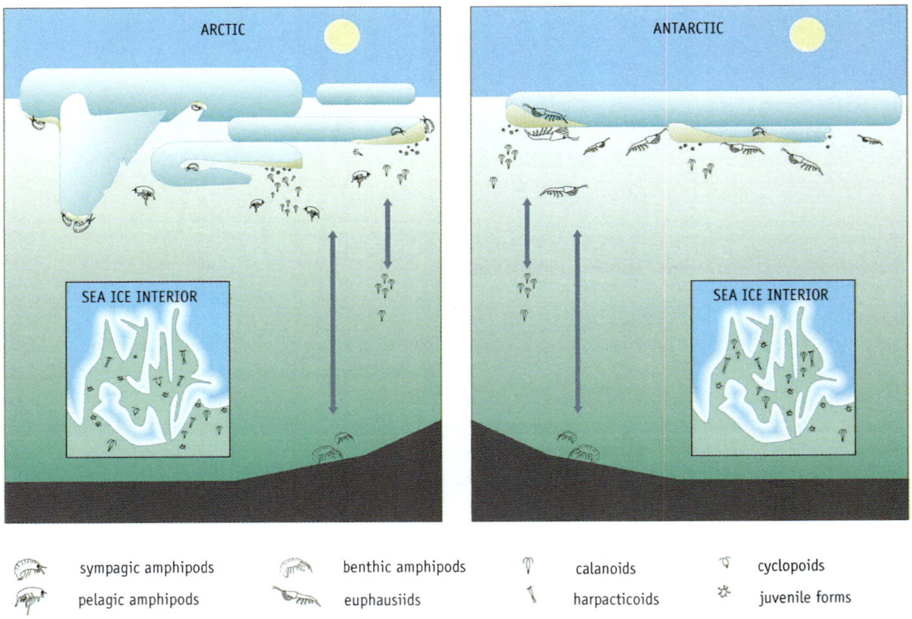

Fig. 9.4.
Sea-ice–associated crustaceans in the Arctic (left) and Antarctic (right).

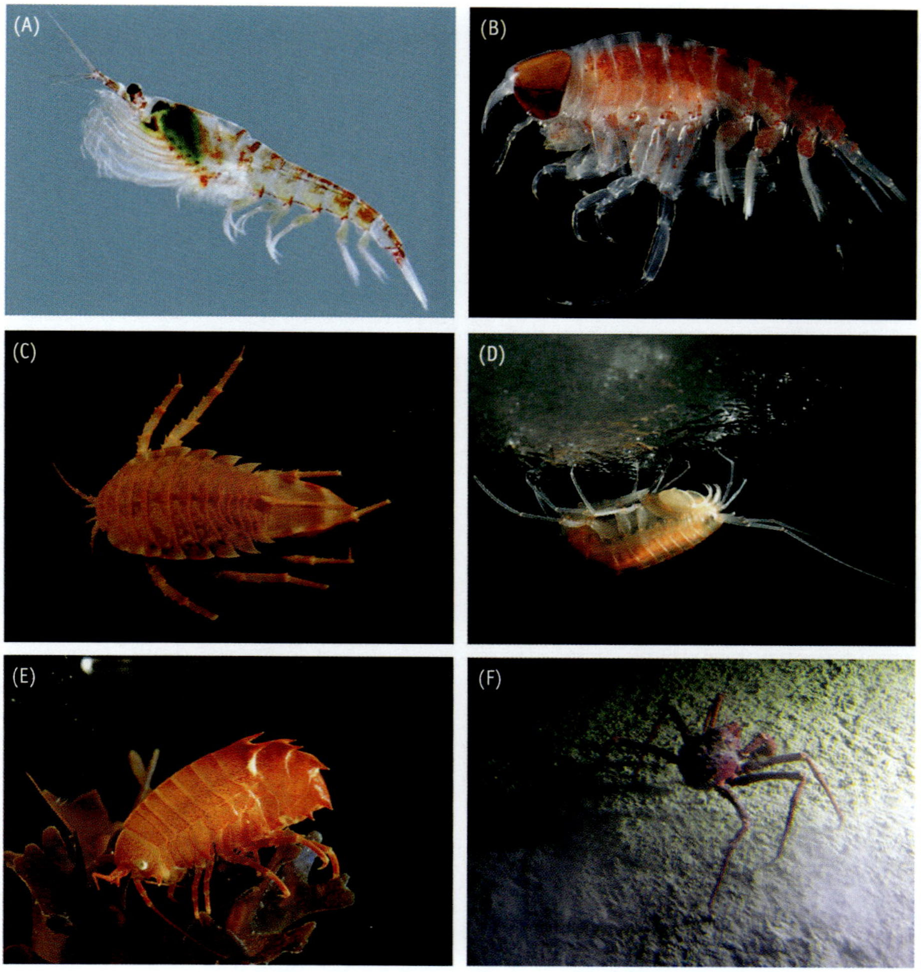

Fig. 9.s1.
Antarctic crustaceans: (A) Antarctic krill (*Euphausia superba*); photo courtesy of Volker Siegel; (B) pelagic hyperiid amphipod *Cyllopus lucasii*; photo courtesy of Martin Rauschert; (C) benthic giant isopod *Glyptonotus antarcticus*, 120 mm; photo courtesy of Cédric d'Udekem d'Acoz; (D) cryopleagic apmhipod *Eusirus microps*; photo courtesy of Ingo Arndt & Rupert Krapp; (E) benthic amphipod *Gnathiphimedia fuschi*; photo Claude de Broyer; (F) a potential invader of the Antarctic shelf, the king crab *Neolithodes yaldwini*, detected at 980 m in Palmer Deep (West Antarctic Peninsula) by the NHK submarine expedition 2017; photo Claude De Broyer, courtesy of NHK.

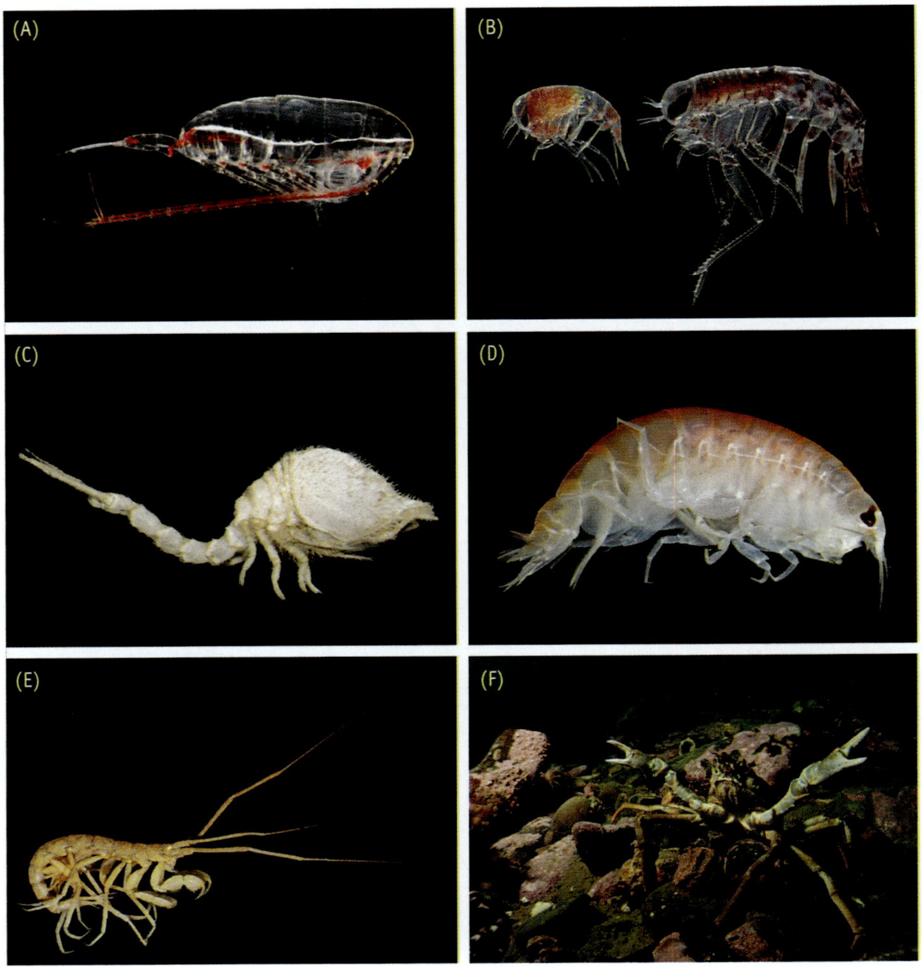

Fig. 9.s2.
Arctic crustaceans: (A) copepod *Calanus hyperboreus*; photo courtsey of Maria Włodarska-Kowalczuk, IO PAN; (B) hyperiid amphipods *Themisto abyssorum* (left) and *Themisto libellula* (right); photo courtsey of Maria Włodarska-Kowalczuk, IO PAN; (C) the largest world's cumacean, *Diastylis goodsiri*; photo courtsey of Kajetan Deja, IO PAN; (D) necrophagic amphipod *Anonyx nugax*; photo courtsey of Kajetan Deja, IO PAN; (E) benthic amphipod *Neohela monstrosa*; photo courtsey of Kajetan Deja, IO PAN; (F) crab *Hyas araneus*; photo courtsey of Piotr Bałazy, IO PAN.

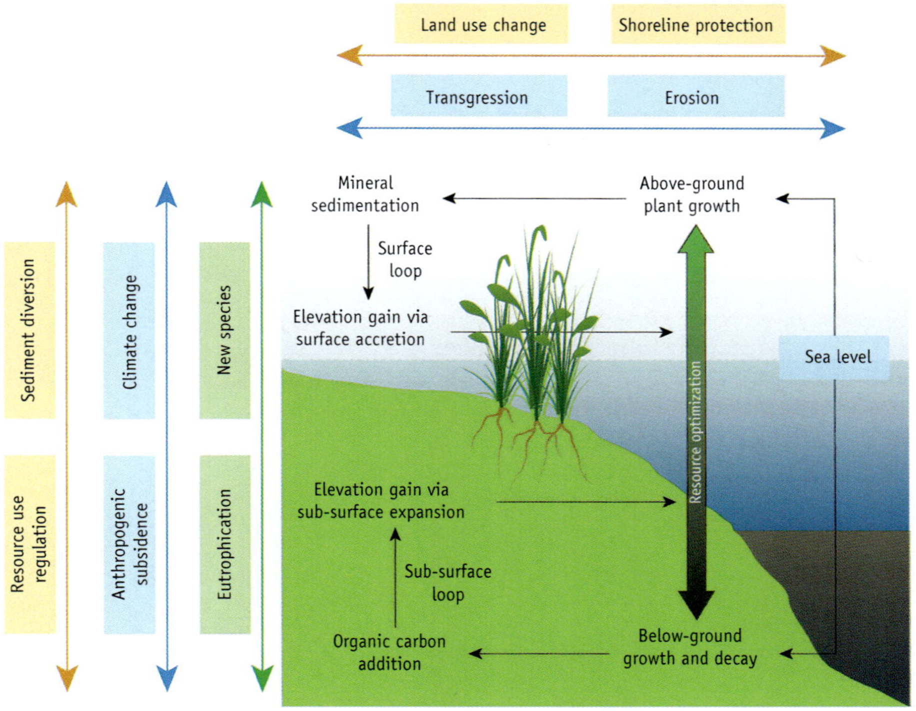

Figure 10.3.
Structure of a saltmarsh estuary. The system can be altered by local factors (shown in green), climatic and geomorphologic factors (shown in blue), and the political/social/economic factors (shown in tan), which affects overall resilience and stability of the saltmarsh system at different scales (Kirwan and Megonigal 2013). With permission from Springer Nature ©.

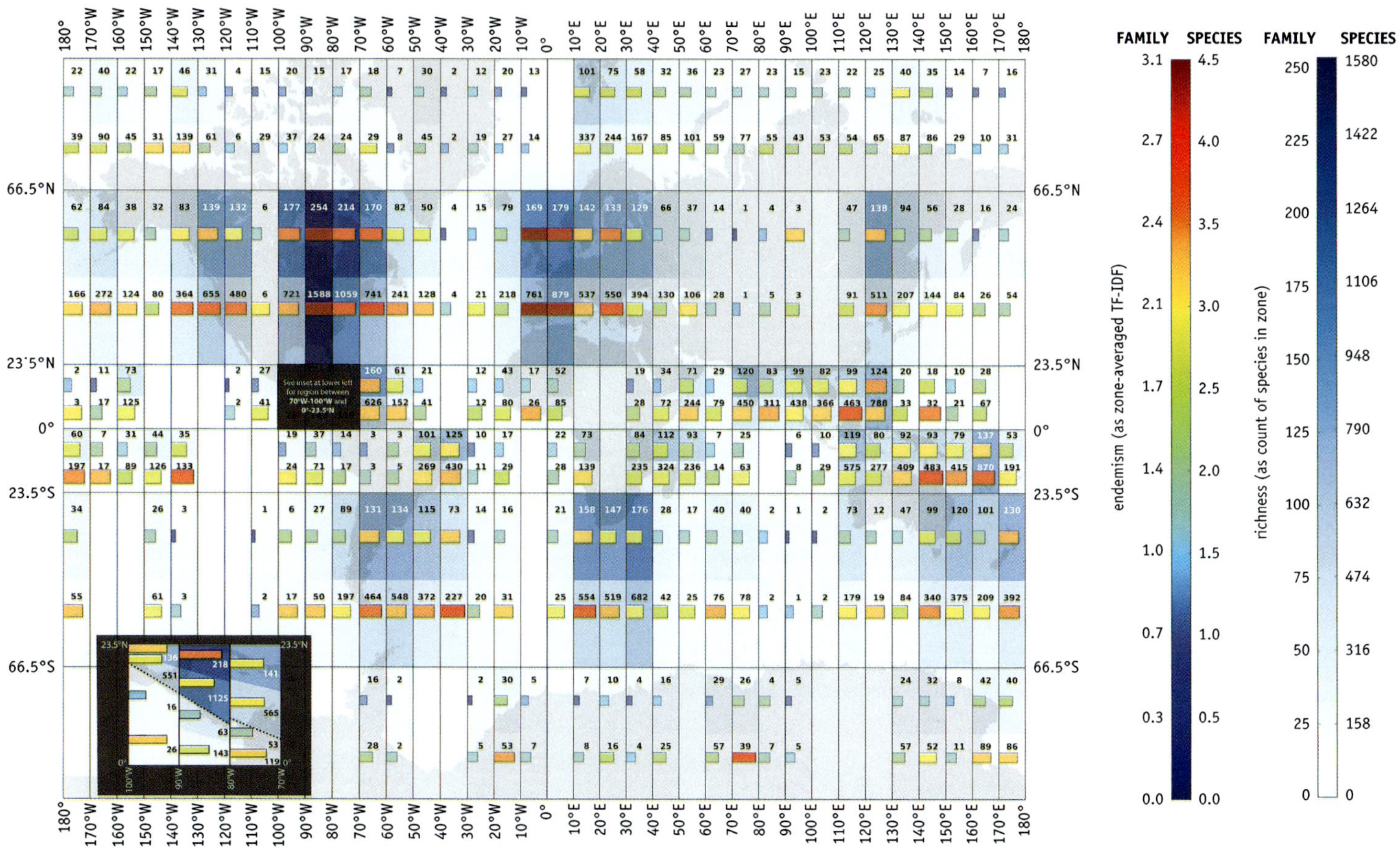

Fig. 10.5.
Plot of *TF–IDF* (term frequency–inverse document frequency) signal values for all crustacean taxa summarized by order level (upper zone and bar) and family level (lower zone and bar) diversity for each of 219 zones. Zones are delineated by 10° longitudinal windows for each group of tropical (0°–23.5° N/S for *TropN*/*TropS*), temperate (23.5°–66.5° N/S for *TempN*/*TempS*), or polar (66.5°–90° N/S for *PolarN*/*PolarS*) zones, for a total of 216 zones. Three of these zones overlap the Atlantic and Pacific oceans (as a result of bisection by the Isthmus of Panama) and were further divided into Atlantic and Pacific sub-zones. Each zone is tinted based on the numbers of orders or families present (with actual numbers printed), and the lengths of the horizontal bars within each zone represent average *TF–IDF* signals for crustaceans in that zone using the order or family diversity with the width of the zone (and color tint) scaled to the percentage of a maximum obtainable value (in this case *TF–IDF* = 8.158). The observed taxonomic diversity per zone is indicated by the tinted scale depicted on the right, with richness at the family levels (top) and species levels (bottom) explicitly listed for each zone.

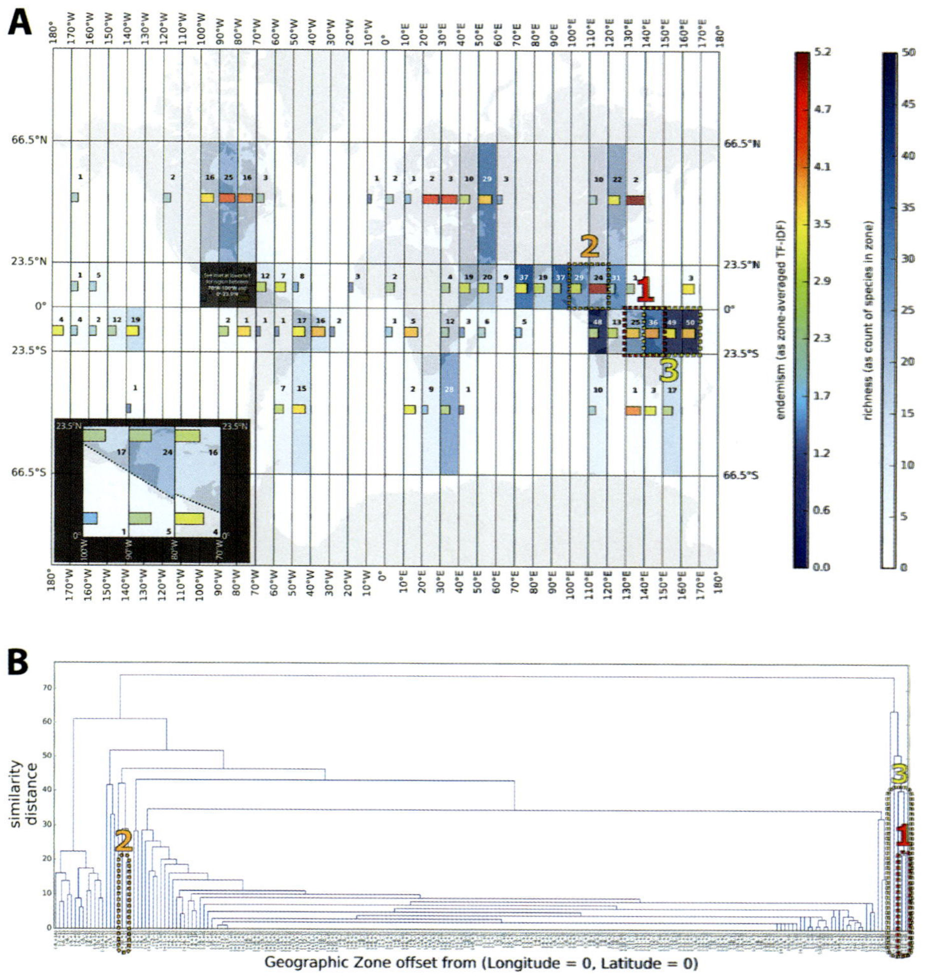

Fig. 10.6.

An objective, iterative method for assessing zone groups for unique endemic signatures via *TF–IDF* clustering. (A) Three strongest zone clusters by *TF–IDF* signal profiles for all species in the Portunidae, worldwide. For each zone we report summaries of the crustaceans from the OBIS database (200 m or less): the number of species (N) and the average *TF–IDF* value (plotted in each zone in A). The bar in each is the average *TF–IDF* value across species. Outlined zones are contiguous geographic groups identified by the algorithm applied to the zone clustering (part B). Similar scaling and tinting descriptions from Fig. 10.4 apply. (B) Zones clustered by species using Canberra distances (Lance and Williams 1967) and algorithms, as described in the text. The three zones rank highest in the region of the Arafura Sea, second in the South China Sea, and third also off the coast of northern Australia. Note that numeric labels (1, 2, and 3) are added to aid in visualizing cluster correspondence between (A) and (B).

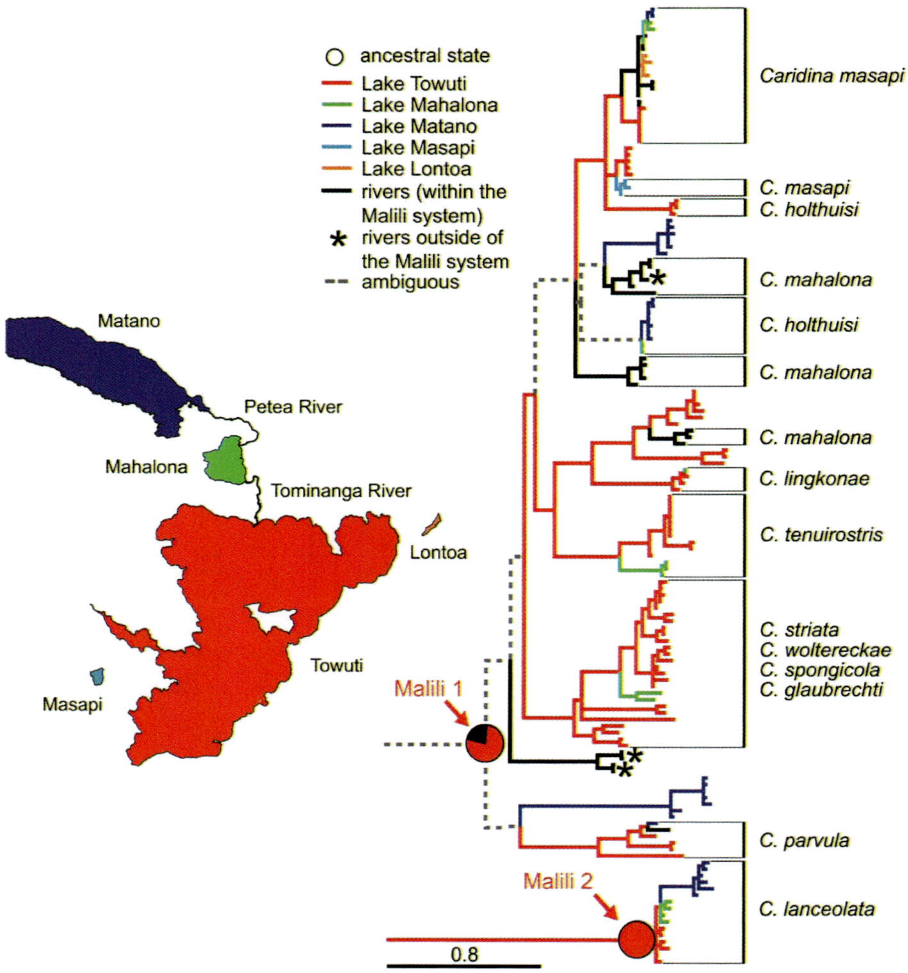

Fig. 11.4.
Adaptive radiation of the Malili lakes species flock of atyid freshwater shrimps of the genus *Caridina*. A molecular phylogeny (Bayesian Inference phylogram, mtDNA, 16S, and COI) revealed two independent clades within the Malili species flock. The majority of species are found within one large clade (Malili 1); the second clade consists of only a single species, *Caridina lanceolata* (Malili 2), suggesting two independent lake colonizations (for details, see von Rintelen et al. 2010). Within the species flock there is a clear geographic pattern (the Malili lakes consist of five interconnected lakes), i.e., several species show an allopatric distribution, either in a single (e.g., *C. tenuirostris*) or in more than one clade (e.g., *C. holthuisi*). Modified from von Rintelen et al. 2010, with permission from John Wiley and Sons.

Fig. 12.2.
Selected freshwater crustaceans: (A) notostracan branchiopod, *Triops* sp., temporary pool, northwestern New South Wales, Australia; (B) mixed branchiopods from temporary pool, northwestern New South Wales, Australia; (C) atyid shrimp, *Neocaridina* sp., from stream in southern China; (D) palaemonid shrimp (*Macrobrachium* sp.) with ectoparasitic isopod (*Tachaea* sp.), Pahang, peninsular Malaysia. (A, B, © M. Schwentner; C, © C. Huang; D, © M. Lo.)

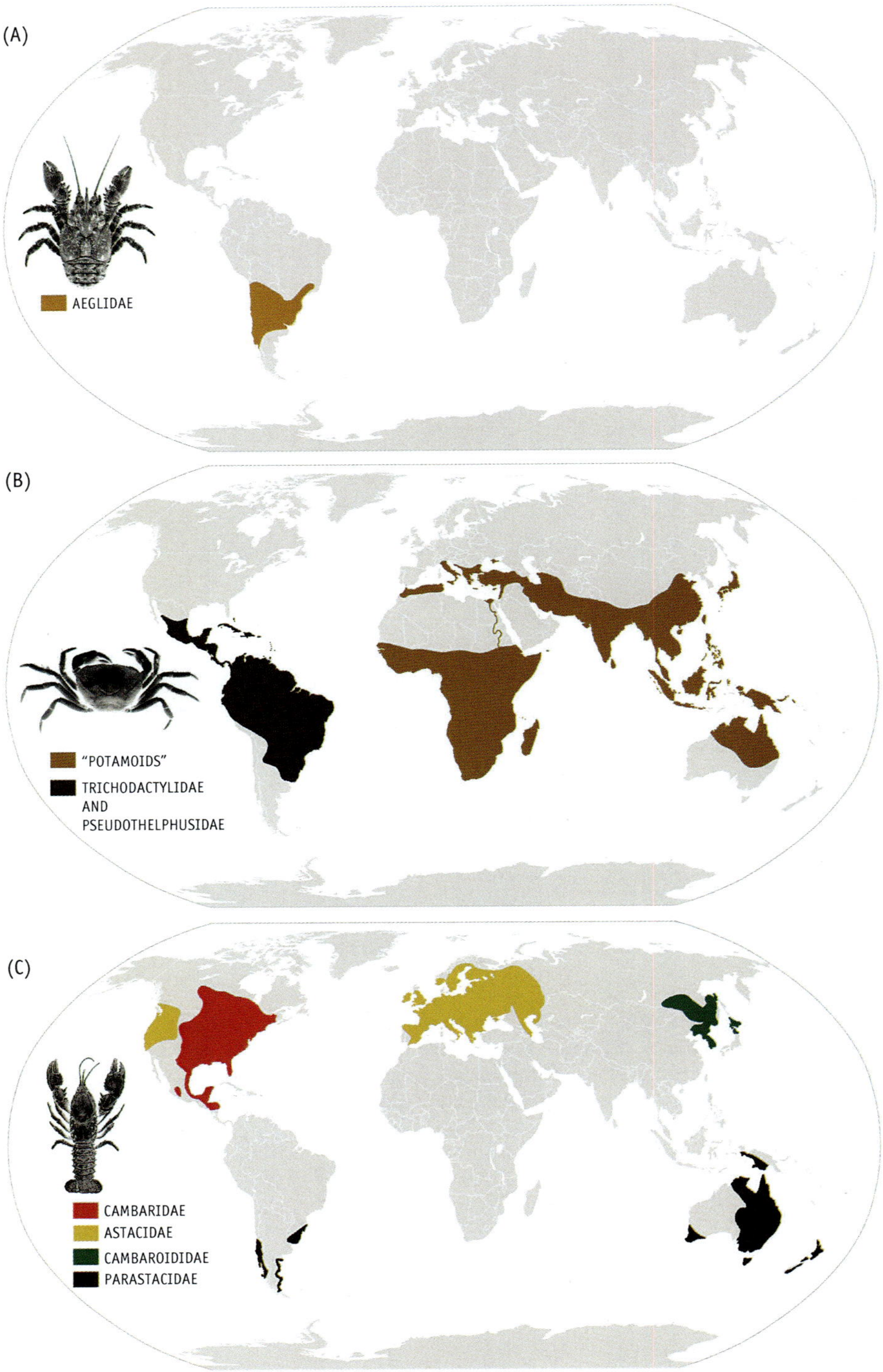

Fig. 12.3.
Global distribution of selected freshwater decapods: (A) Aeglidae; (B) primary freshwater crabs; (C) freshwater crayfish. Based on Ortmann (1902), Klaus et al. (2011), Bueno et al. (2016), Kawai and Crandall (2016).

Fig. 12.4.
Selected freshwater crustaceans: (A) *Aegla* sp., South America, aquarium trade; (B) potamid crab, *Yuebeipotamon calciatile*, hill stream in southern China; (C) parastacid crayfish, *Engaeus cymus*, near Wee Jasper, New South Wales, Australia; (D) parastacid crayfish, *Euastacus spinifer*, Wilson River, New South Wales, Australia. (A, © A. Anker; B, © C. Huang; C, © R. McCormack; D, © S. Ahyong.)

Fig. 12.5.
Selected freshwater crustaceans: (A) anaspididean syncarid, *Anaspides richardsoni*, Tarn Shelf, Mount Field, Tasmania; (B) gammarid amphipod, *Gammarus roseli*, Moravia, Czech Republic; (C) phreatoicidean isopod, Shannon Lagoon, Tasmania, Australia; (D) cave isopod, *Trogloniscus trilobatus*, Huangjiang County, Guangxi, China. (A, © S. Grove/ Tasmanian Museum and Art Gallery; B, © Michal Maňas (User:sneko1)—own work, CC BY 2.5; C, © S. Ahyong; D, © S. Huang and M. Tian, https://commons.wikimedia.org/w/index.php?curid=1869909.)

Fig. 13.1.
Croesus Cave, Tasmania, showing potential stygobiont cave habitats. Tree roots penetrating such systems and organic matter stranded on the stream bank during high water flow could provide habitat for troglobionts such as oniscidean isopods. From Boulton et al. (2014), with permission from John Wiley and Sons.

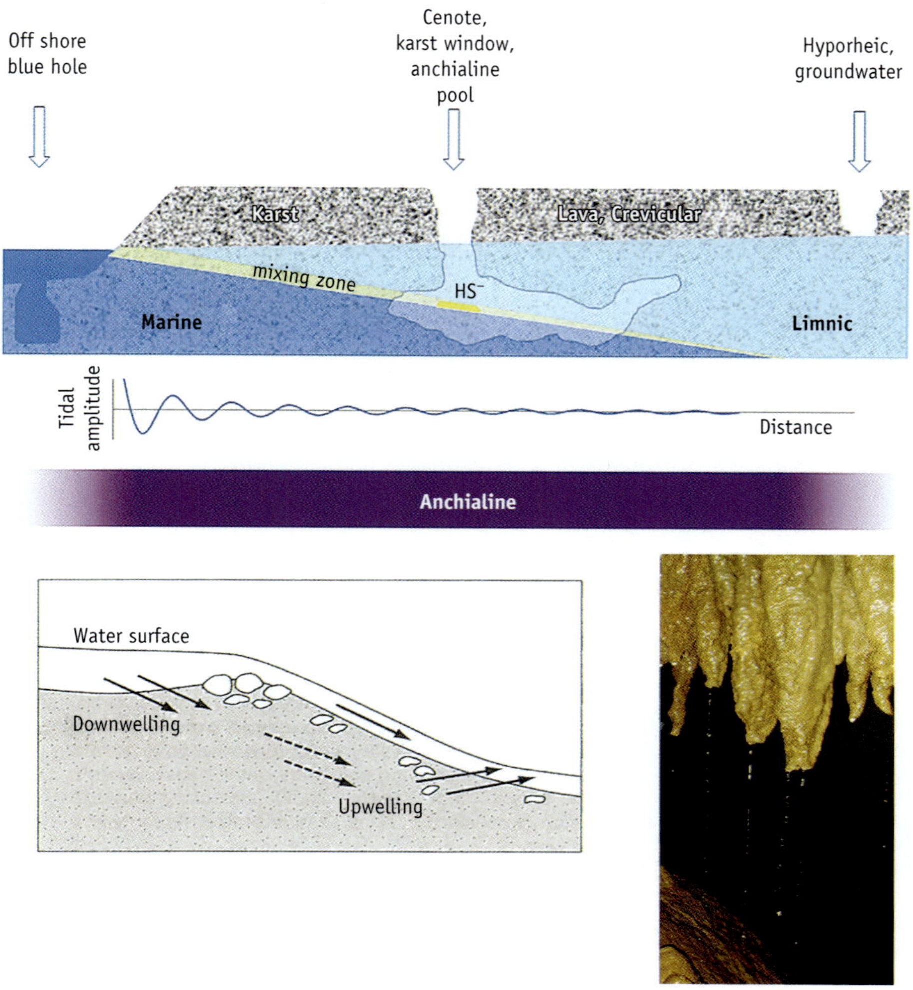

Fig. 13.2.
Subterranean habitats. (Top) Anchialine habitat. (Lower left) Hyporheic flowpaths at one section of a riffle (from Boulton et al. 2014). (Lower right) Water percolating from the epikarst across flowstone to form drips. This habitat is rich in copepods. Organ Cave, West Virginia. Top, modified from Bishop et al. (2015), with permission from Oxford University Press; bottom left, modified from Pipan and Culver (2012), with permission from Elsevier; bottom right, modified from Boulton et al. (2014), with permission from John Wiley and Sons.

Fig. 13.3.
Cave diver Richard "Harry" Harris seeking anchialine fauna in Whip Cave, Christmas Island, Indian Ocean. Photo courtesy of Bill Humphreys.

Fig. 13.6.
Representative crustaceans inhabiting anchialine waters. (A) *Parhippolyte sterreri* (Decapoda); (B) *Pseudoniphargus grandimanus* (Amphipoda); (C) *Bahalana caicosana* (Isopoda); (D) *Tulumella* sp. (Thermosbaenacea); (E) *Mictocaris halope* (Mictacea); (F) 1527 *Bermudamysis speluncola* (Mysida); (G) *Cumella abacoensis* (Cumacea); (H) *Ridgewayia* sp. (Calanoida); (I) *Spelaeoecia* sp. (Ostracoda); (J) *Cryptocorynetes* sp. (Remipedia). Photographs of anchialine crustaceans by Thomas M. Iliffe, from Pérez-Moreno et al. (2016).

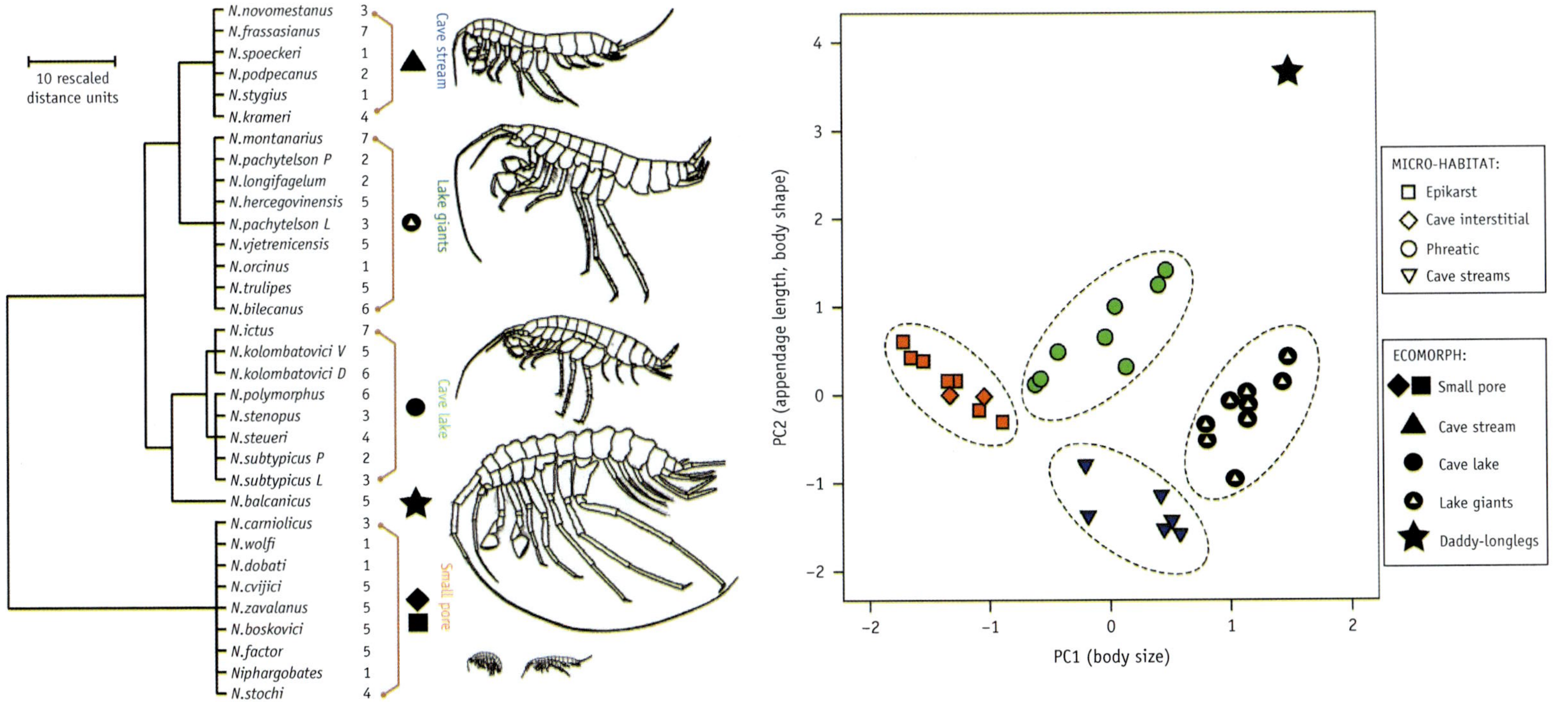

Fig. 13.9.
The relationship between morphology and habitat in *Niphargus*. (Left side) Phenogram of morphological similarity between 33 *Niphargus* species and populations with typical representatives of each morph illustrated approximately to scale. (Right side) First two principal components of the morphological traits of these species, explaining 97.5% of the total variation. Cave microhabitats (symbols) and proposed ecomorphs (shading) are only partially in agreement. From Trontelj et al. (2012), with permission from John Wiley and Sons.

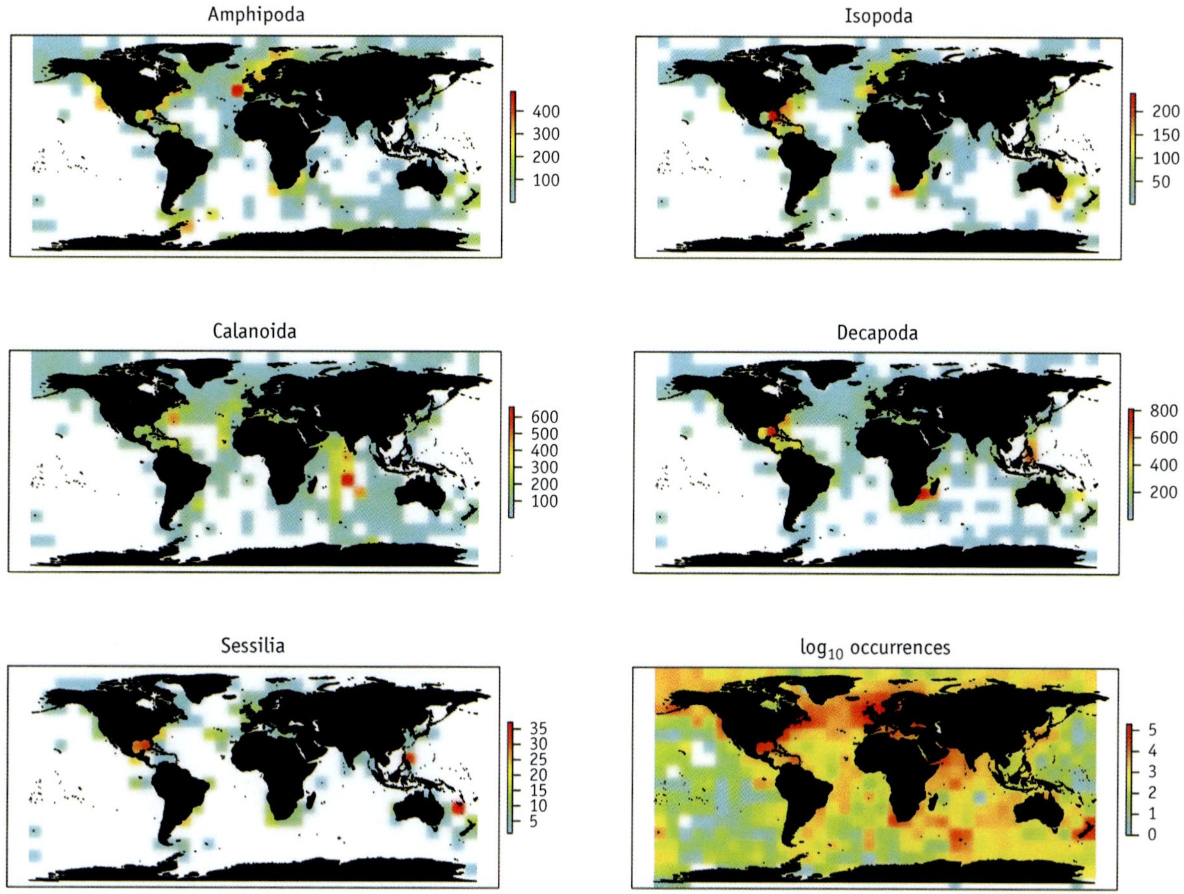

Fig. 15.5.
Global patterns of extrapolated species richness (in scale of colors) of Amphipoda, Isopoda, Calanoida, Decapoda, and Sessilia, in 10° resolution grid cells, based on OBIS database. Empty cells correspond to cells with less than 500 georeferenced occurrences (irrespective of the number of species). Also shown the total number of occurrences per cell (in $\log_{10}$ scale, summed across these five orders).

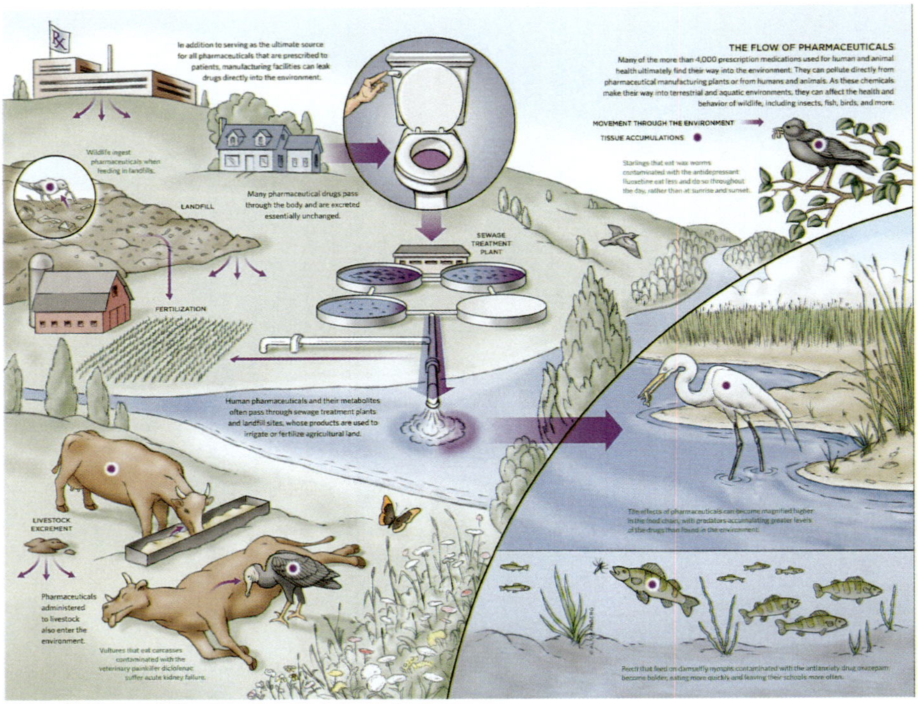

Fig. 18.3.
The flow of pharmaceuticals. Many of the more than 4,000 prescription medications used for human and animal health ultimately find their way into the environment. They can pollute directly from pharmaceutical manufacturing plants or from humans and animals. As these chemicals make their way into terrestrial and aquatic environments, they can affect the health and behavior of wildlife, including insects, fish, birds, and more. Photo courtesy of Al Granberg.

Fig. 18.5.
The aquatic pet: obligatory parthenogenetic marbled crayfish = Marmorkrebs (*Procambarus virginalis*). Photo courtesy of Piotr Gadawski.

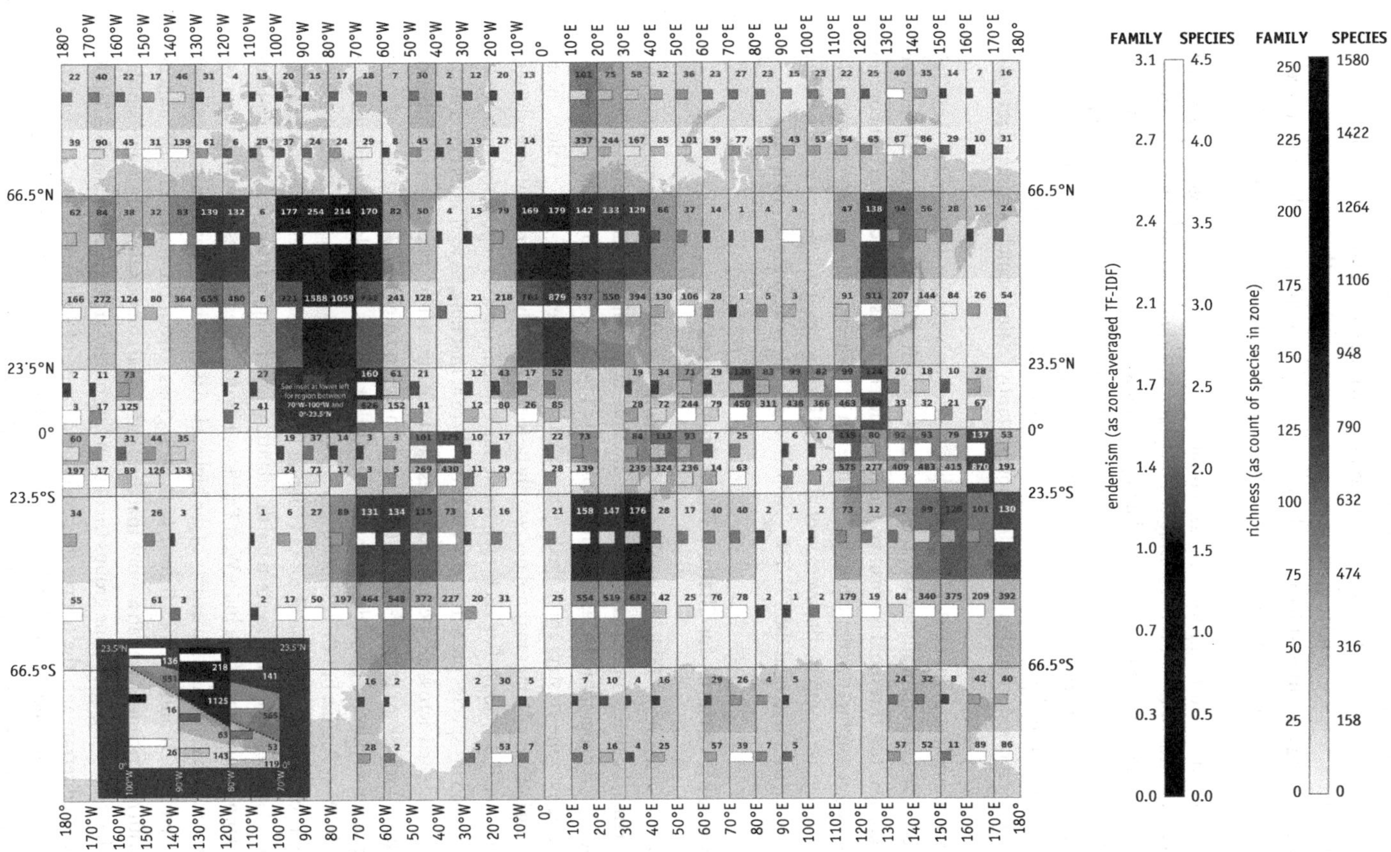

Fig. 10.5.
Plot of *TF–IDF* (term frequency–inverse document frequency) signal values for all crustacean taxa summarized by order level (upper zone and bar) and family level (lower zone and bar) diversity for each of 219 zones. Zones are delineated by 10° longitudinal windows for each group of tropical (0°–23.5° N/S for *TropN/TropS*), temperate (23.5°–66.5° N/S for *TempN/TempS*), or polar (66.5°– 90° N/S for *PolarN/PolarS*) zones, for a total of 216 zones. Three of these zones overlap the Atlantic and Pacific oceans (as a result of bisection by the Isthmus of Panama) and were further divided into Atlantic and Pacific sub-zones. Each zone is tinted based on the numbers of orders or families present (with actual numbers printed), and the lengths of the horizontal bars within each zone represent average *TF–IDF* signals for crustaceans in that zone using the order or family diversity with the width of the zone (and color tint) scaled to the percentage of a maximum obtainable value (in this case *TF–IDF* = 8.158). The observed taxonomic diversity per zone is indicated by the tinted scale depicted on the right, with richness at the family levels (top) and species levels (bottom) explicitly listed for each zone. See color version of this figure in centerfold.

We chose the Canberra similarity metric over others as it gives greater weight to differences between geographic zones when a given species is present in one zone (to any extent) while being completely absent in the other. If two geographic zones are adjacent, they should be more likely to contain the same species than two zones that are more distant, whether separated by open water or by a land mass (such as an isthmus). From our observations, many of the dendrogram subtrees generated using Canberra distances tend to group contiguous oceanic zones in the same body of water. However, the clustering algorithm still can result in singletons (i.e., a distinct single-zone subtree) as well as some subtrees whose constituent zones do exhibit similar *TF-IDF* values yet they are not directly adjacent to one another.

Following hierarchical clustering for each taxonomic rank, each of the 219 geographic zones is assigned to its own "leaf node" in the resulting dendrogram. We then "prune" the dendrogram's branches, when possible, into groups of related and geographically adjacent zones, thereby identifying a set of "zone clusters" for that taxonomic rank. Pruning is accomplished by computing an inconsistency coefficient (*IC*) for each non-leaf node (i.e., branch point) in the tree following the approach described in Zahn (1971). The *IC* represents the dissimilarity of the branches descending from a specific node; a higher *IC* implies more distantly related branches descendant from that node.

To aid in the identification of clusters of zones that are not only similar (in terms of their *TF-IDF* vectors) but also geographically adjacent, we set a threshold *IC* value of 1.0, based on iterative results across groups. This threshold is sensitive enough to detect at least three clusters that contain zones that form a geographically contiguous group. In each of the remaining clusters of geographically adjacent zones, we compute the average of all *TF-IDF* values over all species in each of the cluster's contiguous zones, termed the $TF\text{-}IDF_{avg}$, which we report by taxonomic grouping.

Results: Geographic Patterns for Shallow-Water Crustaceans

We summarize these results in Table 10.1 as the top three ranked individual geographic zones and the top three ranked clusters of adjacent zones for multiple taxonomic ranks in the subphylum Crustacea. Individual geographic zones are ranked by their $TF\text{-}IDF_{avg}$ values in descending order. Similarly, geographic zone clusters are ranked by their "cluster average" of $TF\text{-}IDF_{avg}$ values. Note that high-ranking individual zones may not necessarily be found in the highest-ranked zone clusters by this process (see Table 10.1). For each taxonomic rank listed, we also report the number of shallow-water (≤200 m) species present in the OBIS database at that rank, as well as its grand mean *TF-IDF* value (computed across all species and zones) and the number of species (and number of zones, respectively) whose average *TF-IDF* value exceeds this grand mean *TF-IDF* value.

Our mapping of worldwide data (Figure 10.5) suggests that there is a high level of species richness in the Indo-West Pacific, but this diversity spreads over a large tropical region encompassing approximately 13 of our defined tropical zones (ranging from 191 to 788 species for each zone). In comparison, the Gulf of Mexico and western north Atlantic (comprising four tropical and four temperate zones) far outweigh any of the Indo-West Pacific zones in species richness (ranging from 551 to 1,588 species for each zone) and have some of the highest levels of estimated endemism, as shown by our calculated *TF-IDF* index for each zone in these regions. Given the lower *TF-IDF* indices across zones with high richness, we can assume that their individual compositions are highly similar for many species. Following these zone groups (in descending order) are the eastern north Atlantic (including the Mediterranean), the temperate convergence zone of southern Africa, the eastern temperate Pacific, and the western south Atlantic. The estimates of endemism by *TF-IDF* are highest (as are the species richnesses in our study dataset) in the western temperate north Atlantic region (90° W–80° W, *TempN* as the highest-ranking single zone and 80° W–60° W, *TempN* as the second highest-ranking cluster), second highest in the eastern temperate north Atlantic zones (10° W–0° E,

TempN and 0° E–10° E, *TempN* as the second and third highest-ranking single zones, respectively, and together as the highest-ranking zone cluster), and third highest (140° E–160° E, *TropS* as the third highest-ranking zone cluster) in southwest tropical south Pacific (encompassing the Great Barrier Reef; see Table 10.1 for details).

The high diversity levels in western oceans is not surprising given the flow of geostrophic gyres and their possible effect on larval transport. The higher endemism of the eastern north Atlantic, the temperate convergence zone of southern Africa, the eastern temperate Pacific, and the western south Atlantic, seems to result largely from the high temperate zone peracarid diversity, which Abele (1982) noted previously. Of the total species we examined, peracarids comprise 49% of all species in the database, with the decapods the next largest portion at 45%. Yet of the highest three zones on potential endemism, for the western temperate north Atlantic, peracarids comprise 51% of the total species richness in that zone, and 15% of the richness of tropical Pacific near the Great Barrier Reef, while in the eastern temperate north Atlantic, they comprise nearly 72% of the species present. Even with the low level of species richness for peracarids, the highest cluster of zones of endemism are predicted by *TF-IDF* to be the eastern Atlantic (10° W–10° E, *TempN*, with 747 species) followed by the temperate western Atlantic (60° W–80° W, *TempN*, with 824 species, see Table 10.1). As with all studies using species diversity data, results can be biased dependent on the relative intensity of study by taxonomists for a region, which is analogous to the "monograph effect" described for paleontology (Newman and Palmer 2003, p. 5). Comparison of global richness and endemism most probably has a regional effect for regions that have received greater intensity of study in the world, which is certainly in play to explain the bias in domination of the north Atlantic in these collection data.

For a family-level example, we present results for the family Portunidae, which has a large number of shallow-water species (n = 207 species; Table 10.1) and samples within the OBIS database (Fig. 10.6). Here, we can see that there is one broadly distributed subcluster of zones that show less divergence (a level of cosmopolitanism) and three other distinct, divergent clusters of zones that are unique from one another, reflecting a relatively high level of possible endemism. In the case of portunid crabs, the top three zone clusters identified as having the highest average *TF-IDF* values were in the relatively well-documented tropical region north of Australia, including parts of the Arafura (1) and Coral Seas (3) and parts of the South China Sea (2) (see Fig. 10.6A). The high species richness in the Gulf of Mexico grouping may be, in part, also due to the intensive taxonomic attention that North American researchers have paid to it, although the Mediterranean can lay similar claim with no apparent effect on our measure for portunids. The endemic groups (tropical region north of Australia and the South China Sea) interestingly fall within the region that Abele (1982) identified as the most diverse for crustaceans, yet it does not lump all contiguous zones with higher numbers of species (see Fig. 10.6A). We hope that such discrimination between adjacent groups might lead to fruitful hypotheses for further investigation by specialists working in each region or taxonomic grouping.

We performed similar computations for multiple families and higher taxonomic groups, and their details are summarized in Table 10.1, where each row corresponds to a given taxonomic group. For each taxonomic group, we report the averages of *TF-IDF* for the species contained in each zone and characterize the individual zones (as well as clusters of zones) for the groups that exceed those average values, i.e., rank higher than average (or at least higher than an arbitrary threshold *TF-IDF* value of 3.0) and are therefore distinctively similar in taxonomic diversity for that group. We have observed several single-zone geographic segments with fewer species (25 or less) with relative collection abundance having produced high endemism estimates for our data, so these zones might be considered outliers that are either anomalous or may warrant further investigation. All original maps and corresponding clusters summarized in abridged fashion in Table 10.1 are available from ResearchGate (https://www.researchgate.net/publication/323398647).

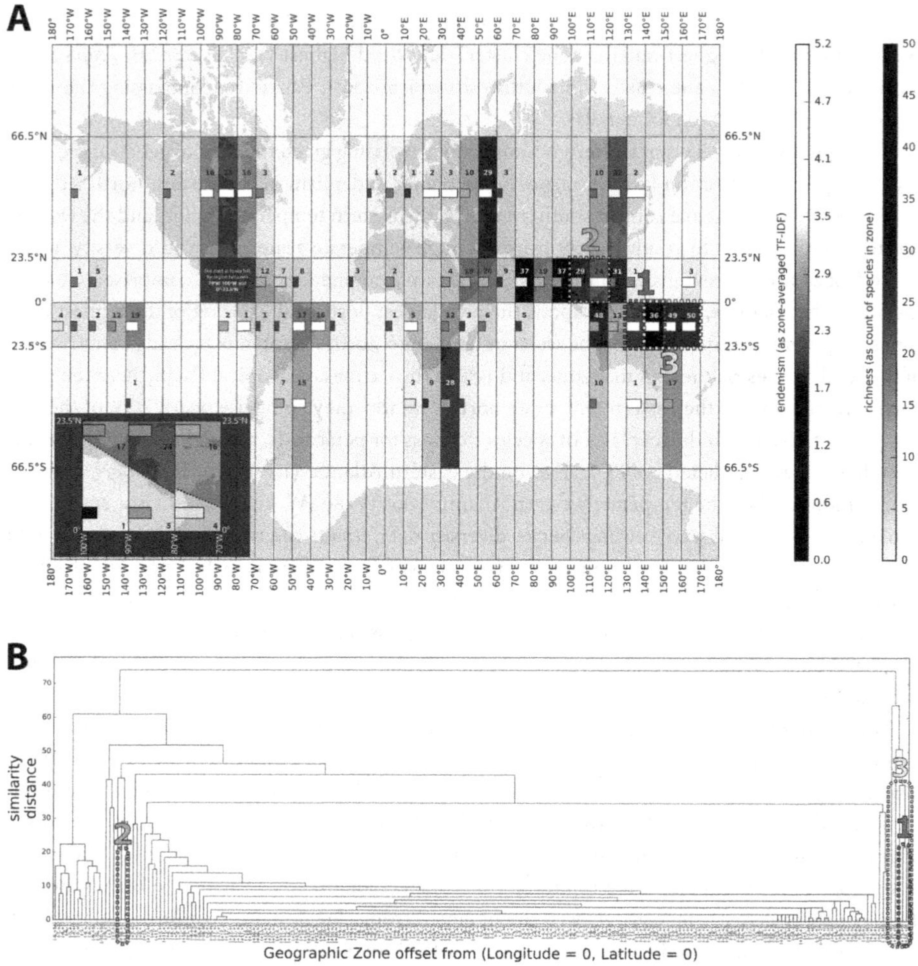

Fig. 10.6.
An objective, iterative method for assessing zone groups for unique endemic signatures via *TF–IDF* clustering. (A) Three strongest zone clusters by *TF–IDF* signal profiles for all species in the Portunidae, worldwide. For each zone we report summaries of the crustaceans from the OBIS database (200 m or less): the number of species (N) and the average *TF–IDF* value (plotted in each zone in A). The bar in each is the average *TF–IDF* value across species. Outlined zones are contiguous geographic groups identified by the algorithm applied to the zone clustering (part B),. Similar scaling and tinting descriptions from Fig. 10.4 apply. (B) Zones clustered by species using Canberra distances (Lance and Williams 1967) and algorithms, as described in the text. The three zones rank highest in the region of the Arafura Sea, second in the South China Sea, and third also off the coast of northern Australia. Note that numeric labels (1, 2, and 3) are added to aid in visualizing cluster correspondence between (A) and (B). See color version of this figure in centerfold.

ENDEMISM AND SPECIES RICHNESS RECONSIDERED

In his earlier review of biogeography and diversity, Abele (1982, p. 290) reported a decreasing "number of crustaceans [species richness] within the tropics" (in descending order): the Indo-West Pacific shelf, the western Atlantic, the eastern Pacific, and the eastern Atlantic, in general. He also noted the exception was that peracarids were most diverse in temperate zones rather than the tropics. With the availability of worldwide ocean collection data, we have attempted to reassess global diversity and endemism using these data. Since 1982, collection record data have increased more than three times (from ~232,000 records to over 729,000 for Crustacea; OBIS 2017).

The result of our analysis shifts the pattern of decreasing overall crustacean species richness and our estimates endemism across regions as follows (in descending order): western Atlantic, eastern Atlantic, followed by parts of the Indo-West Pacific. While Abele only ranked "number of crustaceans [species richness?] within the tropics," we note that our species richness data are loosely correlated with estimated level of endemism (0.63 Spearman's *r* for all non-zero cells), so a similar pattern exists for both. It is important to note, again, that many previous works are determining species richness to assess diversity or endemism (including rarity of occurrence), or both, but not across all crustaceans or not globally. Here the *TF-IDF* has no arbitrary minima, but instead uses context-dependent frequency records to assess locally important species in taxonomic classes. For the peracarids, as a whole, our method still identifies the greatest estimated endemism (as well as species richness) and in the world's temperate regions, when defined zones that possess limited numbers of species are discounted as outliers with low richness. The highest-species-richness zone for both amphipods and isopods occurred in the temperate western Atlantic (80° W–90° W; see online heatmap archives) with large pockets of isopod species richness in temperate Australia (140° E–150° E) and coastal temperate Africa (10° E–20° E). For amphipods, the second highest single zone for species richness is in the eastern temperate Atlantic (0° W–10° W), followed by temperate South America (50° W–60° W). Our endemism data, although correlated with our species richness data, however, might not be expected to completely concur with these patterns of richness or those of previous studies. Indeed, for one of the more extensive worldwide studies on isopods, Poore and Bruce (2012) found temperate Australasia to have the highest species richness, followed by Central Indo-West Pacific, Temperate Northern Pacific, and Tropical Atlantic in descending richness. Our richness data may not directly compare because our numbers are not summed over the same large areas as theirs (and they used a different database). Our estimate of endemism produces highest single clusters with few species in the polar zones (70° E–80° E *PolarS* and 150° W–140° W *PolarN*) but with few species that would appear to be regionalized and dominate the collection data. But it also highlights two highly endemic clusters in the Eastern Atlantic (10° W–10° E, *TempN*) and southern coastal Africa (10° E–30° E *TempS*). This region is off the Agulhas Bank, which is a convergence zone for fauna associated with the Benguela and Agulhas currents of the south Atlantic and Indian coasts, respectively. Given these low species counts for other identified zones, the *TF-IDF* index will tend to be inflated, so these outliers may or may not be an area of local endemism but may be of concern for research focused on isopods. Note that our dataset was also delimited to coastal and shallow-water crustaceans, so diversity of groups with extensive deep-water representatives are not included.

CONCLUSIONS

Crustacean diversity and the concept of relative endemism, or its complement of cosmopolitanism, are rarely the central focus of research in marine crustaceans. Most publications on the subject are presented as limited additions to or speculations on the relative range size, large or small, inherent in a specific taxonomic group under study or are lumped into much larger datasets of broad community diversity measures. For the first part of this review, we give an overview of important environmental factors that affect crustacean diversity in coastal environments and summarize the evolutionary responses to those factors that allow crustaceans to succeed under those conditions. Following this overview, we analyze shallow-water (<200 m) crustacean collection data, worldwide, using *TF-IDF* (an information retrieval-inspired approach) as an objective measure of endemism in crustaceans. This is accomplished by assessing localization of species for all crustaceans and subgroups of taxa worldwide, with respect to their longitudinal representation in the tropics, temperate, and polar regions. Our analysis locates the single zone with the highest species richness and estimated endemism for all of crustaceans to be in the western Atlantic in the Gulf of Mexico, with clusters of coastal crustaceans decreasing in endemism (from highest to lowest): eastern Atlantic, western Atlantic, followed by parts of the Indo-West Pacific. We also show that the general species richness and endemism of peracarids is still dominant in the temperate zones of the world, although diversity and predicted endemism for coastal species (<200 m) is unexpectedly high in the eastern and western north Atlantic, as compared with previous studies focused on a larger-scale, whole-ocean approach. We generated a new estimator of endemism (a *TF-IDF* index) for selected taxonomic groups displayed in Table 10.1, and display example heatmaps with species counts per zone. The entire expanded table of taxonomic database and their associated heatmaps are all freely available in an online archive. We hope that this objective assessment of relative endemism among taxonomic groups may serve as a springboard to an integrative biogeographic approach to crustacean diversity within and among taxa.

Crustaceans, as a group, evolved diverse developmental and behavioral aspects that impact their dispersal and recruitment and ultimately their biogeography since the first records of their occurrence. These patterns have been resilient for maintaining and even leading to some of the nearshore diversity that we see today. However, we should be keenly aware that the last 50 years have produced great changes in coastal zones through the effect of sea-level rise (Defeo et al. 2009), which alter local coastlines (and their estuaries) and will continue to impact associated estuaries and their wetlands. During the last 50 years, estimates of human-induced loss of coastal wetlands are 25–50% worldwide and regional assessments estimate a further 20–45% loss of saltmarsh in this century due to sea-level rise (Kirwan and Megonigal 2013). The coastal zone will be the most highly impacted regions of the oceans by climate change and rising sea level, with estuaries, coasts, and shallow continental shelf biotas among the most affected. To understand the ultimate impacts of climatic change to the coastal zone biota, future work should continue to establish baseline estimates of current conditions and develop new methods to help address these large-scale data.

REFERENCES

Abele, L.G., 1982. Biogeography. Pages 242–304 *in* L.G. Abele, editor. The Biology of Crustacea, Vol. 1, Systematics, the Fossil Record, and Biogeography. Academic Press, New York.

Ahyong, S.T., J.K. Lowry, M. Alonso, R.N. Bamber, G.A. Boxshall, P. Castro, S. Gerken, G. Karaman, J.W. Goy, D.S. Jones, K. Meland, D.C. Rogers, and J. Svavarsson. 2011. Subphylum Crustacea Brünnich, 1772. *In* Z.-Q. Zhang, editor. Animal Biodiversity: An Outline of Higher-Level Classification and Survey of Taxonomic Richness. Zootaxa 3148:164–191.

Andrew, N., editor. 1999. Under Southern Seas: The Ecology of Australia's Rocky Reefs. University of New South Wales Press, Sydney, Australia.

Anger, K. 2001. The biology of decapod crustacean larvae. Pages 263–318 *in* R. Vonk, editor. Crustacean Issues, Vol. 14. AA Balkema, Tokyo, Japan.

Astorga, A., M. Fernández, E.E. Boschi, and N. Lagos. 2003. Two oceans, two taxa and one mode of development: latitudinal diversity patterns of South American crabs and test for possible causal processes. Ecology Letters 6:420–427.

Attrill, M.J. 2002. A testable linear model for diversity trends in estuaries. Journal of Animal Ecology 71:262–269.

Aveline, L.C. 1980. Fauna dos manguezais brasileiros. Revista Brasileira de Geografia 42:786–821.

Barnard, J.L. 1991. Amphipodological agreement with Platnick. Journal of Natural History 25:1675–1676.

Bascom, W.N. 1951. The relationship between sand size and beach-face slope. Eos, Transactions American Geophysical Union 32:866–874.

Bauer, R.T. 2018. Life cycle and seasonal migrations. Pages 203–229 *in* G. Wellborn and M. Thiel, editors. The Natural History of The Crustacea, Vol. 5: Life Histories. Oxford University Press, New York.

Beger, M.B., P.L. Sommer, S. Harrison, D.A. Smith, and J.M. Pandolfi. 2014. Conserving potential coral reef refuges at high latitudes. Diversity and Distributions 20:245–257.

Behbehani, M.I., and R.A. Croker. 1982. Ecology of beach wrack in northern New England with special reference to *Orchestia platensis*. Estuarine, Coastal and Shelf Science 15:611–620.

Bertness, M. 1985. Fiddler crab regulation of *Spartina alterniflora* production on a New England salt marsh. Ecology 66:1042–1055.

Beseres Pollack, J., R. M. Mroch, and R. J. Feller. 2008. Juvenile white shrimp predation on macrobenthic and zooplanktonic prey. Journal of Shellfish Research 27:1247–1253.

Bianchi, T.S. 2013. Estuaries: where the river meets the sea. Nature Education Knowledge 4:1–6.

Bilton, D.T., J. Paula, and J.D.D. Bishop. 2002. Dispersal, genetic differentiation and speciation in estuarine organisms. Estuarine, Coastal and Shelf Science 55:937–952.

Boesch, D.F., and R.E. Turner. 1984. Dependence of fishery species on salt marshes: the role of food and refuge. Estuaries 7:460–468.

Borgianini, S.A., R. Styles, and R.J. Brodie. 2012. Simple model of megalopal transport in narrow river-dominated estuaries. Marine Ecology Progress Series 452:179–191.

Brandt, A. 1992. Origin of Antarctic Isopoda (Crustacea, Malacostraca). Marine Biology 113:415–423.

Buschbaum, C., S. Dittmann, J.-S. Hong, I.-S. Hwang, M. Strasser, M. Thiel, N. Valdivia, S.P. Yoon, and K. Reise. 2009. Mytilid mussels: global habitat engineers in coastal sediments. Helgoland Marine Research 63:47–58.

Casagranda, M.D., L. Taher, and C.A. Szumik. 2012. Endemicity analysis, parsimony and biotic elements: a formal comparison using hypothetical distributions. Cladistics 28:645–654.

Chapman, V.J. 1974. Salt Marshes and Salt Deserts of the World. Cramer, Lehre, Germany.

Charmantier, G., C. Haond, J. Lignot, and M. Charmantier-Daures. 2001. Ecophysiological adaptation to salinity throughout a life cycle: a review in homarid lobsters. Journal of Experimental Biology 204:967–977.

Christy, J.H., and M. Salmon. 1984. Ecology and evolution of mating systems of fiddler crabs (genus *Uca*). Biological Review 59:483–509.

Coll, M., C. Piroddi, J. Steenbeek, K. Kaschner, F. Ben Rais Lasram, J. Aguzzij, E. Ballesteros, C.N. Bianchi, J. Corbera, T. Dailianis, R. Danovaro, M. Estrada, C. Froglia, B.S. Galil, J.M. Gasol, R. Gertwagen, J. Gil, F. Guilhaumon, K. Kesner-Reyes, M.-S. Kitsos, A. Koukouras, N. Lampadariou, E. Laxamana, C.M. López-Fé de la Cuadra, H.K. Lotze, D. Martin, D. Mouillot, D. Oro, S. Raicevich, J. Rius-Barile, J.I. Saiz-Salinas, C. San Vicente, S. Somot, J. Templado, X. Turon, D. Vafidis, R. Villanueva, and E. Voultsiadou. 2010. The biodiversity of the Mediterranean Sea: estimates, patterns, and threats. PLoS ONE 5:e11842.

Connell, J.H. 1961. The influence of interspecific competition and other factors on the distribution of the barnacle *Chthamalus stellatus*. Ecology 42:710–723.

Cowen, R.K., and S. Sponaugle, 2009. Larval dispersal and marine population connectivity. Annual Review of Marine Science 1:443–466.

Craft, C., J. Clough, J. Ehman, S. Joye, R. Park, S. Pennings, H. Guo, and M. Machmuller. 2009. Forecasting the effects of accelerated sea-level rise on tidal marsh ecosystem services. Frontiers in Ecology and the Environment 7:73–78.

Croker, R.A. 1967. Niche diversity in five sympatric species of intertidal amphipods (Crustacea: Haustoriidae). Ecological Monographs 37:173–200.

Curray, J.R. 1977. Modes of emplacement of prospective hydrocarbon reservoir rocks of outer continental marine environments. Pages E1–E14 *in* Geology of Continental Margins Course Notes, Series 5, AAPG Continuing Education, Tulsa, Oklahoma.

Dahl, E. 1952. Some aspects of the ecology and zonation of the fauna on sandy beaches. Oikos 4:1–27.

Daleo, P., and O. Iribarne. 2009. The burrowing crab *Neohelice granulata* affects the root strategies of the cordgrass *Spartina densiflora* in SW Atlantic salt marshes. Journal of Experimental Marine Biology and Ecology 373:66–71.

Das, T., and W.B. Stickle, 1994. Detection and avoidance of hypoxic water by juvenile *Callinectes sapidus* and *C. similis*. Marine Biology 120:593–600.

Deaton, L.E., and M.J. Greenberg. 1986. There is no horohalinicum. Estuaries 9:20–30.

Defeo, O., A. McLachlan, D.S. Schoeman, T.A. Schlacher, J. Dugan, A. Jones, M. Lastra, and F. Scapini. 2009. Threats to sandy beach ecosystems: a review. Estuarine, Coastal and Shelf Science 81:1–12.

De Grave, S., N.D. Pentcheff, S.T. Ahyong, T.-Y. Chan, K.A. Crandall, P.C. Dworschak, D.L. Felder, R.M. Feldmann, C.H.J.M. Fransen, L.Y.D. Goulding, R. Lemaitre, M.E.Y. Low, J.W. Martin, P.K.L. Ng, C.E. Schweitzer, C.G.S. Tan, S.H.D. Tshudy, and R. Wetzer. 2009. A classification of living and fossil genera of decapod crustaceans. Raffles Bulletin of Zoology (Suppl. 21):1–109.

Dye, A.H., and T.A. Lasiak. 1986. Microbenthos, meiobenthos and fiddler crabs: trophic interactions in a tropical mangrove sediment. Marine Ecology Progress Series 32:259–267.

Elliott, M., and C.J.L. Taylor. 1989. The production ecology of the subtidal benthos of the Forth Estuary, Scotland. Scientia Marina (Barcelona) 53:531–541.

Eshky, A.A., R.J.A. Atkinson, and A.C. Taylor. 1995. Physiological ecology of crabs from Saudi Arabian mangrove. Marine Ecology Progress Series 126:83–95.

Farnsworth, E.J., and A.M. Ellison. 1996. Scale-dependent spatial and temporal variability in biogeography of mangrove root epibiont communities. Ecological Monographs 66:45–66.

Faulkes, Z. 2013. Morphological adaptations for digging and burrowing. Pages 276–295 *in* L. Watling and M. Thiel, editors. The Natural History of The Crustacea, Vol. 1: Functional Morphology and Diversity. Oxford University Press, New York.

Felder, D.L. 1979. Respiratory adaptations of the estuarine mud shrimp, *Callianassa jamaicense* (Schmitt, 1935) (Crustacea, Decapoda, Thalassinidea). The Biological Bulletin 157:125–137.

Fernández, M., A. Astorga, S.A. Navarrete, C. Valdovinos, and P.A. Marquet. 2009. Deconstructing latitudinal species richness patterns in the ocean: does larval development hold the clue? Ecology Letters 12:601–611.

Field, C.B., J.G. Osborn, L.L. Hoffman, J.F. Polsenberg, D.D. Ackerly, J.A. Berry, O. Björkman, A. Held, P.A. Matson, and H.A. Mooney. 1998. Mangrove biodiversity and ecosystem function. Global Ecology & Biogeography Letters 7:3–14.

Frey, D.G. 1986. The non-cosmopolitanism of chydorid Cladocera: implications for biogeography and evolution. Pages 237–256 *in* R.H. Gore, editor. Crustacean Biogeography. Balkema, Rotterdam, The Netherlands.

Gebauer, P.A., K.U. Paschke, and K. Anger. 2003. Delayed metamorphosis in decapod crustaceans: evidence and consequences. Revista Chilena de Historia Natural 76:169–75.

Giere, O. 2008. Meiobenthology: The Microscopic Motile Fauna of Aquatic Sediments. Springer Science & Business Media.

Graham, M.H., P.K. Dayton, and J.M. Erlandson. 2003. Ice ages and ecological transitions on temperate coasts. Trends in Ecology & Evolution 18:33–40.

Gray, J.S. 2002. Species richness of marine soft sediments. Marine Ecology Progress Series 244:285–297.

Gray, J.S. 2001. Marine diversity: the paradigms in patterns of species richness examined. Scientia Marina 65:41–56.

Gray, J.S., G.C.B. Poore, K.I. Ugland, R.S. Wilson, F. Olsgard, and O. Johannessen. 1997. Coastal and deep-sea benthic diversities compared. Marine Ecology Progress Series 159:97–103.

Grenyer, R., C.D.L. Orme, S.F. Jackson, G.H. Thomas, R.G. Davies, T.J. Davies, K.E. Jones, V.A. Olson, R.S. Ridgely, P.C. Rasmussen, and T.S. Ding. 2006. Global distribution and conservation of rare and threatened vertebrates. Nature 444:93–96.

Hansen, T.A. 1978. Larval dispersal and species longevity in Lower Tertiary gastropods. Science 199:885–887.

Hartnoll, R.G. 1988. The eco-ethology of mangroves. Pages 477–489 *in* G. Chelazzi and M. Vannini, editors. Behavioural Adaptations to Intertidal Life. NATO ASI Series, Series A, Vol. 151. Plenum Press, New York.

Hartnoll, R.G., S. Cannici, W.D. Emmerson, S. Fratini, A. Macia, Y. Mgaya, F. Porri, R.K. Ruwa, J.P. Shunula, M.W. Skov, and M. Vannini. 2002. Geographic trends in mangrove crab abundance in East Africa. Wetlands Ecology and Management 10:203–213.

Hausdorf, B., and C. Hennig. 2003. Biotic element analysis in biogeography. Systematic Biology 52:717–723.

Hessler, R.R., and G.D.F. Wilson. 1983. The origin and biogeography of malacostracan crustaceans in the deep sea. Pages 227–254 *in* R.W. Sims, J.H. Price, and P.E. Whalley, editors. Evolution, Time, and Space: The Emergence of the Biosphere. Academic Press, New York.

Howard, J.D., and C.A. Elders. 1970. Burrowing patterns of haustoriid amphipods from Sapelo Island, Georgia. *In* T.P. Crimes and J.C. Harper, editors. Trace Fossils. Geological Journal Special 3:243–262.

Howes, B.L., R.W. Howarth, J.M Teal, and I. Valiela. 1981. Oxidation-reduction potentials in a salt marsh: spatial patterns and interactions with primary production. Limnology and Oceanography 26:350–360.

Hutchins, L.W. 1947. The bases for temperature zonation in geographical distribution. Ecological Monographs 17:325–335.

Jefferies, R.L. 1977. Growth responses of coastal halophytes to inorganic nitrogen. Journal of Ecology 65:847–865.

Kennett, J.P. 1982. Marine Geology. Prentice-Hall, Englewood Cliffs, NJ.

Keylock, C. J. 2005. Simpson diversity and the Shannon–Wiener index as special cases of a generalized entropy. Oikos 109:203–207.

Kinne, O. 1971. Salinity: animals—invertebrates. Pages 821–996 *in* O. Kinne, editor. Marine Ecology, Vol. 1: Environmental Factors, Part 2. John Wiley, New York.

Kinoshita, K. 2002. Burrow structure of the mud shrimp *Upogebia major* (Decapoda: Thalassinidea: Upogebiidae). Journal of Crustacean Biology 22:474–480.

Kirwan, M.L., and J.P. Megonigal. 2013. Tidal wetland stability in the face of human impacts and sea-level rise. Nature 504:53–60.

Komiyama, A., J.E. Ong, and S. Poungparn. 2008. Allometry, biomass, and productivity of mangrove forests: A review. Aquatic Botany 89:128–137.

Koch, V., and M. Wolf. 2002. Energy budget and ecological role of mangrove epibenthos in the Caeté estuary, North Brazil. Marine Ecology Progress Series 228:119–130.

Lance, G.N., and W.T. Williams. 1967. Mixed-data classificatory programs I—Agglomerative systems. Australian Computer Journal 1:15–20.

Lee, S.Y. 1998. Ecological role of grapsid crabs in mangrove ecosystems: a review. Marine and Freshwater Research 49:335–343.

Leight, A.K., G.I. Scott, M.H. Fulton, and J.W. Daugomah. 2005. Long term monitoring of grass shrimp *Palaemonetes* spp. population metrics at sites with agricultural runoff influences. Integrative and Comparative Biology 45:143–150.

Lenz, M., B.A. da Gama, N.V. Gerner, J. Gobin, F. Gröner, Lenz, M., da Gama, B.A., Gerner, N.V., Gobin, J., Gröner, F., A. Harry, S.R. Jenkins, P. Kraufvelin, C. Mummelthei, J. Sareyka, and E.A. Xavier. 2011. Non-native marine invertebrates are more tolerant towards environmental stress than taxonomically related native species: results from a globally replicated study. Environmental Research 111:943–952.

Levinton, J., and J. Mackie. 2013. Latitudinal diversity relationships of fiddler crabs: biogeographic differences united by temperature. Global Ecology and Biogeography 22:1050–1059.

Little, C. 2000. The Biology of Soft Shores and Estuaries. Oxford University Press, New York.

MacArthur, R.H., and E.O. Wilson. 1967. The Theory of Island Biogeography. Princeton University Press, Princeton, NJ.

MacIntosh, D.J. 1984. Ecology and productivity of Malaysian mangrove crab populations (Decapoda: Brachyura). Pages 354–377 *in* E. Soepadmo, A.N. Rao, and D.J. Macintosh, editors. Proceedings of the Asian Symposium on Mangrove Environment—Research and Management. University of Malaya and UNESCO, Kuala Lumpur.

Macpherson, E. 1991. Biogeography and community structure of the decapod crustacean fauna off Namibia (Southeast Atlantic). Journal of Crustacean Biology 11:401–415.

Macpherson, E. 2002. Large-scale species-richness gradients in the Atlantic Ocean. Proceedings of the Royal Society B: Biological Sciences 269:1715–1720. http://doi.org/10.1098/rspb.2002.2091.

Mangum, C., and Towle, D. 1977. Physiological adaptation to unstable environments. American Scientist 65:7–75.

Mantel, L.H., and L.L. Farmer. 1983. Osmotic and ionic regulation. Internal Anatomy and Physiological Regulation 5:53–161.

Marsden, I.D. 1991. Kelp-sandhopper interactions on a sand beach in New Zealand. II. Population dynamics of *Talorchestia quoyana* (Milne-Edwards). Journal of Experimental Marine Biology and Ecology 152:75–90.

McCallum, A.W., G.C.B. Poore, A. Williams, F. Althaus, and T. O'Hara. 2013. Environmental predictors of decapod species richness and turnover along an extensive Australian continental margin (13–35°S). Marine Ecology 34: 298–312.

McLachlan, A. 1990. Dissipative beaches and macrofauna communities on exposed intertidal sands. Journal of Coastal Research 6:57–71.

McLachlan, A., E. Jaramillo, T.E. Donn, and F. Wessels. 1993. Sandy beach macrofauna communities and their control by the physical environment: a geographical comparison. Journal of Coastal Research, Special Issue No. 15. Beach and Surf Zone Morphodynamics:27–38.

McLachlan, A. 1996. Physical factors in benthic ecology: effects of changing sand particle size on beach fauna. Marine Ecology Progress Series 131:205–217.

McQuaid, C.D., and Branch, G.M. 1984. Influence of sea temperature, substratum and wave exposure on rocky intertidal communities: an analysis of faunal and floral biomass. Marine Ecology Progress Series 19:145–151.

Meehl, G.A., T.F. Stocker, W.D. Collins, P. Friedlingstein, T. Gaye, J.M. Gregory, A. Kitoh, R. Knutti, J.M. Murphy, A. Noda, S.C.B. Raper, I.G. Watterson, A.J. Weaver, and Z.C. Zhao. 2007. Global climate projections. Pages 747–846 *in* S. Solomon, D. Qin, M. Manning, Z. Chen, M. Marquis, K.B. Averyt, M. Tignor, and H.L. Miller, editors. IPCC, 2007: Climate Change 2007: The Physical Science Basis. Contribution of Working Group I to the Fourth Assessment Report of the Intergovernmental Panel on Climate Change. Cambridge University Press, Cambridge, UK.

Mendelssohn, I.A., K.L. McKee, and W.H. Patrick. 1981. Oxygen deficiency in *Spartina alterniflora* roots: metabolic adaptation to anoxia. Science 214:439–441.

Mendelssohn, I.A., and J.T. Morris. 2000. Eco-physiological controls on the productivity of *Spartina alterniflora* Loisel. Pages 59–80 *in* M.P. Weinstein and D.A. Kreeger, editors. Concepts and Controversies in Tidal Marsh Ecology. Kluwer, Dordrecht, The Netherlands.

Menge, B.A. 1976. Organization of the New England rocky intertidal community: role of predation, competition, and environmental heterogeneity. Ecological Monographs 46:355–393.

Menge, B.A. 1991. Relative importance of recruitment and other causes of variation in rocky intertidal community structure. Journal of Experimental Marine Biology and Ecology 146:69–100.

Menge, B. 2000. Bottom-up: top-down determination of rocky intertidal shorescape dynamics. Pages 62–81 *in* G.A. Polis, M.E. Power, and G. Huxel, editors. Food Webs at the Landscape Level. University of Chicago Press, Chicago.

Monczak, A., C. Mueller, M.E. Miller, Y. Ji, S. A. Borgianini, E. W. Montie. 2019. Sound patterns of snapping shrimp, fish, and dolphins in an estuarine soundscape of the southeastern USA. Marine Ecology Progress Series 609:49–68.

Morgan, S.G. 2020. Dispersal. In K. Anger, S. Harzsch and M. Thiel, editors. The Natural History of the Crustacea, Volume 7: Developmental Biology and Larval Ecology. Oxford University Press, New York.

Morrone, J.J. 1994. On the identification of areas of endemism. Systematic Biology 43:438–441.

Munguia, P., P. Backwell, and Z. Darnell. 2017. Thermal constraints on microhabitat selection and mating opportunities. Animal Behaviour 123:259–265.

Navarrete, S.A., and E.A. Wieters. 2000. Variation in barnacle recruitment over small scales: larval predation by adults and maintenance of community pattern. Journal of Experimental Marine Biology and Ecology 253:131–148.

Niering, W.A., and R.S. Warren. 1980. Vegetation patterns and processes in New England salt marshes. Bioscience 30:301–307.

Newman, M.E., and R.G. Palmer. 2003. Modeling Extinction. Santa Fe Institute Studies on the Sciences of Complexity. Oxford University Press, New York.

OBIS. 2017. Global biodiversity indices from the Ocean Biogeographic Information System. Intergovernmental Oceanographic Commission of UNESCO. http://www.iobis.org.

Occhipinti-Ambrogi, A. 2007. Global change and marine communities: alien species and climate change. Marine Pollution Bulletin 55:342–352.

Odum, W.E. 1988. Comparative ecology of tidal freshwater and salt marshes. Annual Review of Ecology and Systematics 19:14–176.

O'Hara, T.D., and G.C.B. Poore. 2000. Patterns of distribution for southern Australian marine echinoderms and decapods. Journal of Biogeography 27:1321–1335.

Ōmori, M., 1976. The systematics, biogeography, and fishery of epipelagic shrimps of the genus *Acetes* (Crustacea, Decapoda, Sergestidae). Ocean Research Institute, University of Tokyo, Tokyo, Japan.

Paine, R.T. 1966. Food web complexity and species diversity. The American Naturalist 100:65–75.

Palumbi, S.R., 2003. Population genetics, demographic connectivity, and the design of marine reserves. Ecological Applications 13:S146–S158.

Papaspyrou, S., T. Gregersen, R.P. Cox, M. Thessalou-Legaki, and E. Kristensen. 2005. Sediment properties and bacterial community in burrows of the ghost shrimp *Pestarella tyrrhena* (Decapoda: Thalassinidea). Aquatic Microbial Ecology 38:181–190.

Pappalardo, P., and M. Fernández. 2014. Mode of larval development as a key factor to explain contrasting effects of temperature on species richness across oceans. Global Ecology and Biogeography 23:12–23.

Pearse, A.S., H.J. Humm, and G.W. Wharton. 1942. Ecology of sand beaches at Beaufort, NC. Ecological Monographs 12:135–190.

Péqueux, A. 1995. Osmotic regulation in crustaceans. Journal of Crustacean Biology 15:1–60.

Peterson, M.S., and S.T. Ross. 1991. Dynamics of littoral fishes and decapods along a coastal river-estuarine gradient. Estuarine, Coastal and Shelf Science 33:467–483.

Pohle, G., K. Iken, K.R. Clarke, T. Trott, B. Konar, J.J. Cruz-Motta, M. Wong, L. Benedetti-Cecchi, A. Mead, P. Miloslavich, N. Mieszkowska, R. Milne, L. Tamburello, A. Knowlton, E. Kimani, and Y. Shirayama. 2011. Aspects of benthic decapod diversity and distribution from rocky nearshore habitat at geographically widely dispersed sites. PLoS ONE 6:e18606.

Poore, G.B.C. and N.L. Bruce. 2012. Global diversity of marine isopods (except *Asellota* and crustacean symbionts). PLoS One 7:e43529.

Powers, L.W., and J.F. Cole. 1976. Temperature variation in fiddler crab microhabitats. Journal of Experimental Marine Biology and Ecology 21:141–157.

Pritchard, D.W. 1952. Estuarine hydrography. Advances in Geophysics 1:243–280.

Probert, P.K. 1984. Disturbance, sediment stability, and trophic structure of soft-bottom communities. Journal of Marine Research 42:893–921.

Rabalais, N. N., D.E. Harper Jr., and R.E. Turner. 2001. Responses of nekton and demersal and benthic fauna to decreasing oxygen concentrations. Pages 115–128 *in* N.N. Rabalais and R.E. Turner, editors. Coastal Hypoxia: Consequences for Living Resources and Ecosystems. Coastal and Estuarine Studies 58. American Geophysical Union, Washington, DC.

Reid, D.G., and E. Naylor. 1985. Free-running, endogenous semilunar rhythmicity in a marine isopod. Journal of the Marine Biological Association of UK 65:85–91.

Remane, A. 1934. Die Brackwasserfauna. Verzeichnis der Veröffentlichungen Goldsteins 36:34–74.

Remane, A., and C. Schlieper. 1971. The Biology of Brackish Waters, 2nd ed. Wiley Interscience, New York.

Renema, W., D.R. Bellwood, J.C. Braga, K. Bromfield, R. Hall, K.G. Johnson, P. Lunt, C.P. Meyer, L.B. McMonagle, R.J. Morley, A. O'dea, J.A. Todd, F.P. Wesselingh, M.E.J. Wilson, and J.M Pandolfi. 2008. Hopping hotspots: global shifts in marine biodiversity. Science 321:654–657.

Renaud, M.L. 1986. Detecting and avoiding oxygen deficient sea water by brown shrimp, *Penaeus aztecus* (Ives), and white shrimp, *Penaeus setiferus* (Linnaeus). Journal of Experimental Marine Biology and Ecology 98:283–292.

Rivadeneira, M.M., P. Hernáez, J.A. Baeza, S. Boltaña, M. Cifuentes, C. Correa, A. Cuevas, E. del Valle, I.A. Hinojosa, N. Ulrich, N. Valdivia, N. Vásquez, A Zander, and M. Thiel, 2010. Testing the abundant-centre hypothesis using intertidal porcelain crabs along the Chilean coast: linking abundance and life history variation. Journal of Biogeography 37:486–408.

Rivadeneira, M.M., M. Thiel, E.R. González, and P.A. Haye. 2011. An inverse latitudinal gradient of diversity of peracarid crustaceans along the Pacific Coast of South America: out of the deep south. Global Ecology and Biogeography 20:437–448.

Roberts, D., D. Rittschof, E. Holm, and A.R. Schmidt. 1991. Factors influencing initial larval settlement: temporal, spatial and surface molecular components. Journal of Experimental Marine Biology and Ecology 150:203–221.

Ruppert, E.E., and R.S. Fox. 1988. Seashore Animals of the Southeast: A Guide to Common Shallow-Water Invertebrates of the Southeastern Atlantic Coast. University of South Carolina Press, Columbia.

Saenger, P., E.J. Hegerl, and J.D.S. Davie. 1983. Global status of mangrove ecosystems. Commission on Ecology Papers, no. 3. International Union for Conservation of Nature and Natural Resources, Gland, Switzerland. The Environmentalist 3:88.

Salmon, M., and G.W Hyatt. 1983. Spatial and temporal aspects of reproduction in North Carolina fiddler crabs (*Uca pugilator* Bosc). Journal of Experimental Marine Biology and Ecology 70:21–43.

Salton, G., and C. Buckley. 1988. Weighting approaches in automatic text retrieval. Information Processing and Management 24:513–523.

Salton, G., and M. McGill. 1983. *Introduction to Modern Information Retrieval.* McGraw-Hill, New York.

Schopf, T.J.M. 1979. The role of biogeographic provinces in regulating marine faunal diversity through geologic time. Pages 449–457 *in* J. Gray and A.J. Boucot, editors. Historical Biogeography, Plate Tectonics and the Changing Environment. Biology Colloquium, Oregon State University, Corvallis.

Schopf, T.J.M., J.B. Fisher, and C.A.F. Smith. 1977. Is the marine latitudinal diversity gradient merely another example of the species area curve? Pages 365–386 *in* B. Battaglia and J.A. Beardmore, editors. Marine Organisms: Genetics, Ecology and Evolution. Plenum Press, New York.

Selig, E.R., Turner, W.R., Troëng, S., Wallace, B.P., Halpern, B.S., Kaschner, K., Lascelles, B.G., Carpenter, K.E. and R. A. Mittermeier. 2014. Global priorities for marine biodiversity conservation. PLoS ONE 9:e82898.

Smith, W.K., and P.C. Miller. 1973. The thermal ecology of two south Florida fiddler crabs: *Uca rapax* Smith and *U. pugilator* Bosc. Physiological Zoology 46:186–207.

Somero, G.N. 2002. Thermal physiology and vertical zonation of intertidal animals: optima, limits, and costs of living. Integrative and Comparative Biology 42:780–789.

Spooner, G.M. 1947. The distribution of *Gammarus* species in estuaries. Part I. Journal of the Marine Biological Association of the United Kingdom 27:1–52.

Stillman, J.H., and G.N. Somero. 1996. Adaptation to temperature stress and aerial exposure in congeneric species of intertidal porcelain crabs (genus *Petrolisthes*): correlation of physiology, biochemistry and morphology with vertical distribution. Journal of Experimental Biology 199:1845–1855.

Svavarsson, J., M. K. Osore, and E. Ólafsson. 2002. Does the wood-borer *Sphaeroma terebrans* (Crustacea) shape the distribution of the mangrove *Rhizophora mucronata*? AMBIO: A Journal of the Human Environment 31:574–579.

Szumik, C., and P.A. Goloboff. 2004. Areas of endemism: an improved optimality criterion. Systematic Biology 53:968–977.

Teal, J.M. 1962. Energy flow in the salt marsh ecosystem of Georgia. Ecology 43:614–624.

Teske, P.R., I. Papadopoulos, G.I. Zardi, C.D. McQuaid, M.T. Edkins, C.L. Griffiths, and N.P. Barker. 2007. Implications of life history for genetic structure and migration rates of southern African coastal invertebrates: planktonic, abbreviated and direct development. Marine Biology 152:697–711.

Thompson, R.K., and A.W. Pritchard. 1969. Respiratory adaptations of two burrowing crustaceans, *Callianassa californiensis* and *Upogebia pugettensis* (Decapoda, Thalassinidea). Biological Bulletin 136:274–287.

Thorson, G. 1950. Reproductive and larval ecology of marine bottom invertebrates. Biological Reviews 25:1–45.

Tittensor, D.P., C. Mora, W. Jetz, H.K. Lotze, D. Ricard, E.V. Berghe, and B. Worm. 2010. Global patterns and predictors of marine biodiversity across taxa. Nature 466:1098–1101.

Vail, P.R., R.M. Mitchum Jr., and S. Thompson III. 1977. Seismic stratigraphy and global changes of sea level: Part 3. Relative changes of sea level from Coastal Onlap: section 2. Application of Seismic Reflection Configuration. AAPG Special Volumes 165:63–81.

Valentine, J.W., and D. Jablonski. 1986. Mass extinctions: sensitivity of marine larval types. Proceedings of the National Academy of Sciences 83:6912–6914.

Vuorinen, I., J. Hänninen, M. Rajasilta, P. Laine, J. Eklund, F. Montesino-Pouzols, F. Corona, K. Junker, H.M. Meier, and J.W. Dippner. 2015. Scenario simulations of future salinity and ecological consequences in the Baltic Sea and adjacent North Sea areas—implications for environmental monitoring. Ecological Indicators 50:196–205.

Walker, M.G., F.H. Jones, and G.P. Arnold. 1978. The movements of plaice (*Pleuronectes platessa* L.) tracked in the open sea. Journal du Conseil 38:58–86.

Wares, J.P., and C.W. Cunningham. 2001. Phylogeography and historical ecology of the North Atlantic Intertidal. Evolution 55:2455–2469.

Warren, J.H., and A.J. Underwood. 1986. Effects of burrowing crabs on the topography of mangrove swamps in New South Wales. Journal of Experimental Marine Biology and Ecology 102:223–235.

Watling, L. 1979. Marine flora and fauna of the Northeastern United States, Crustacea: Cumacea. National Oceanic and Atmospheric Administration Technical Report, National Marine Fisheries Circular 423:1–23.

Wright, S. 1951. The genetical structure of populations. Annals of Eugenics 1:323–334.

Young, C.M. 2009. Communities on deep-sea hard bottoms. Pages 39–60 *in* M. Wahl, editor. Marine Hard Bottom Communities. Springer, Berlin.

Zahn, C.T. 1971. Graph–theoretical methods for detecting and describing Gestalt clusters. IEEE Transactions on Computers C–20:68–86.

11

STANDING WATERS, ESPECIALLY ANCIENT LAKES

Kristina von Rintelen, Patricio De los Ríos, and Thomas von Rintelen

Abstract

Crustacea in standing waters are a diverse taxonomic assemblage with representatives in all available habitats from the benthic zone to the pelagial in larger water bodies. While most higher taxa are widespread and occasionally cosmopolitan, this is only partially true at the genus and species level. The crustacean fauna of geologically young lakes, or ponds, is characterized by widespread species that are not even necessarily restricted to lentic habitats. These species generally have good to excellent dispersal capabilities, especially those dwelling in ephemeral habitats. Small groups such as branchiopods and copepods dominate under these conditions among obligate still-water dwellers. In contrast, endemism and occasional striking adaptations are the hallmarks of crustacean species flocks, especially in the radiations of amphipods, decapods, and ostracods in the fewer than 10 ancient lakes worldwide. These radiations have arisen in situ through the diversification of unspecialized ancestors. All comparatively well-studied radiations for which molecular phylogenetic, taxonomic, and ecological data are available show particular adaptations of trophic morphology correlated to specific habitats. Prime examples are the species flocks of amphipods in Lake Baikal and of atyid shrimps in Lake Tanganyika and in two Indonesian lakes. These groups have most likely evolved through adaptive radiation. A major challenge for research on crustaceans in ancient lakes, and in standing waters generally outside Europe and North America, is the lack of basic data from species diversity to genetics for many, if not most, taxa. Getting a grip on species diversity, distributions, ecology, and, at a different level, genomics will be a research priority for coming decades.

Evolution and Biogeography. Edited by Gary C. B. Poore and Martin Thiel.

INTRODUCTION

Standing or lentic waters include all non-marine aquatic habitats such as lakes, lagoons, wetlands, and pools (ephemeral and permanent) with little or no horizontal water movement in contrast to lotic waters (rivers, streams, etc.). Although the physical and chemical characteristics of lentic habitats, particularly large (and often ancient) lakes, can be similar to those found in marine environments, there are marked differences in the factors regulating their biotas, including those of crustaceans (Elser et al. 2000, Seuront and Strutton 2004). This environmental heterogeneity in terms of mainly physical and chemical parameters is the main cause of differences in crustacean communities across both local (Bos et al. 1996, Soto and De los Ríos 2006) and continental scales (Gillooly and Dodson 2000).

The vast majority of standing water bodies (lakes) are not long-lived and are young geologically (<10,000 years), e.g., all Holocene lakes in the northern hemisphere. In contrast are the long-lived so-called ancient lakes, which have persisted for several hundred thousand to millions of years (Gorthner et al. 1994). This fundamental difference is reflected in the faunas of lentic habitats. Ephemeral and short-lived water bodies are characterized by low species diversity and generalist species with excellent active or passive dispersal capabilities (e.g., Schwentner et al. 2012). This sharply contrasts with the high diversity of overwhelmingly endemic specialist species in ancient lakes. Typically, species of groups dominating ephemeral and short-lived lentic habitats (e.g., branchiopods) have wide distribution ranges and often the capability to generate diapause eggs that facilitate their dispersal or allow them to stay dormant during adverse environmental conditions such as droughts or cold spells (Alekseev et al. 2007). These groups also commonly occur in ancient lakes, e.g., in the zooplankton, but lack the high degree of endemicity and the occasionally spectacular morphological adaptations that characterize the "true" ancient lake species. The latter have frequently evolved in situ into endemic flocks of highly specialized species. While the most famous examples of ancient lake species flocks are found among species flocks of cichlid fishes in the East African lakes (Barlow 2000), this is in a less spectacular way also true for crustaceans such as the amphipods of Lake Baikal (Martens 1997).

TAXONOMIC ACCOUNT

A general taxonomic synopsis of the diversity and origin of all non-marine groups of crustaceans is given in Chapter 7. All major taxa treated there are also found in standing waters. They range from predominantly microscopic groups of crustaceans (e.g., Branchiopoda, Copepoda, and Ostracoda) to conspicuous macroscopic groups within Malacostraca: amphipods (Fig. 11.1V), isopods (Fig. 11.1W), and decapods (Fig. 11.1.X–Y). The major groups are listed in Table 11.1. Species and adaptations specific to standing waters are addressed in detail below. Groups with a marginal representation in fresh water and standing waters in particular are the entirely parasitic fishlice (Branchiura) and the malacostracan crustacean groups Syncarida and Mysida, which are only briefly treated here (freshwater representatives of Spelaeogriphacea, Thermosbaenacea, Cumacea, and Tanaidacea inhabit brackish water, subterranean, or interstitial habitats and are not covered here).

Branchiura

Freshwater fishlice primarily live as ectoparasites on fishes but occasionally also on other hosts (Poly 2008) in both lotic and lentic habitats. Host specificity is low, and species such as the native

Table 11.1. Major Crustacean Taxa in Standing Waters*

Taxa		Remarks	Source(s)
Branchiopoda, Latreille, 1817	Anostraca, Sars, 1867	Elongated body and lack a carapace; constantly swimming on their backs in the water columns; approx. one quarter of all ~300 species on all continents are known only from the type localities	Brendonck et al. 2008
	Notostraca, Sars, 1867	Plasticity in external morphology and different modes of reproduction; wide geographic distribution; considered as "living fossils"	Brendonck et al. 2008
	Diplostraca, Gerstaecker, 1866: Laevicaudata, Spinicaudata, Cyclestherida + Cladocera	Laterally compressed bodies enclosed by a bivalve carapace; common worldwide but not very well studied	Brendonck et al. 2008, Forro et al. 2008
Copepoda, Milne-Edwards, 1840		Occur in almost all and are extremely abundant in freshwater habitats; more than 90% are endemic to a single zoogeographical region except Antarctica; major component of most planktonic, benthic, and groundwater communities	Boxshall and Defaye 2008
Ostracoda, Latreille, 1802 ("mussel shrimps")		Small and bivalved; very common in inland waters; 90% of all species occur in one zoogeographical region only; ancient lakes hold up to 25% of the total ostracod diversity	Martens et al. 2008
Malacostraca, Latreille, 1802	Amphipoda, Latreille, 1816	Approx. 20% of the total known amphipod diversity in fresh or inland waters worldwide; most abundant in cool and temperate environments and rare in the tropics; particularly diversified in subterranean environments, running waters and in temperate ancient lakes	Väinölä et al. 2008
	Isopoda, Latreille, 1817	Nearly 950 described species (~9% of all isopods) in continental waters; many freshwater inhabitants are narrow range endemics	Wilson 2008
	Decapoda, Latreille, 1802	Freshwater taxa: caridean shrimps (Atyidae and Palaemonidae); crayfish (Astacidae); crabs (Brachyura); Aeglidae, the only freshwater anomuran crab family in freshwater	Bond-Buckup et al. 2008, Crandall and Buhay 2008, De Grave et al. 2008, Yeo et al. 2008

[a] Compare Fig. 11.1.

palearctic *Argulus foliaceus* (Walker et al. 2011) can be extremely widespread. An invasive species, *Argulus japonicus*, has been spread globally from its native range in East Asia through the ornamental fish trade. Fishlice can severely affect fish stocks in lentic waters with high fish densities such as fish ponds (Poly 2008).

Mysida

Less than 10% of the small shrimp-like peracarid Mysida live in non-marine inland waters, the majority (a radiation of 24 species within the genus *Paramysis* of the family Mysidae) occurring in the Ponto-Caspian region (Porter et al. 2008). Species in standing waters are active predators of zooplankton. *Mysis* species mostly inhabit oligotrophic lakes of northern North America and Eurasia, while most species of *Neomysis* are lotic (Covich et al. 2009), even though the estuarine *Neomysis mercedes*, ranging from California to Alaska, can thrive in lakes and reservoirs as well (Haskell and Stanford 2006).

Syncarida

The majority of extant Syncarida (there are also two orders with fossil representatives) inhabit interstitial groundwaters worldwide. Only a few members of the order Anaspidacea live in open freshwater habitats (i.e., not in interstitial groundwater but in streams, ponds, lakes, or caves) and lentic species are only found in the endemic Tasmanian Anaspidesidae (Ahyong and Alonso-Zarazaga 2017), e.g., the two species of *Allanaspides* (moorland pools and a lake; Driessen et al. 2014), and at least three out of the seven species of *Anaspides* with one exclusively lentic species, *A. spinulae*, endemic to a single lake (Ahyong 2016).

Branchiopoda

Among the major groups of crustaceans (Fig. 11.1), branchiopods (Fig. 11.1A–H) are a monophyletic, morphologically very diverse group of ecologically important, largely freshwater organisms (Brendonck et al. 2008), including Anostraca ("fairy shrimps"), Notostraca ("tadpole shrimps"), and Diplostraca (see overview in Table 11.1). Diplostraca (Fig. 11.1B, E–H) comprise the paraphyletic assemblage of three lineages (Laevicaudata, Spinicaudata, Ciclestherida), formerly classified as Conchostraca (Brendonck et al. 2008) constituting the "clam shrimps," and the small Cladocera or "water fleas" (Fig. 11.1E–H; Forro et al. 2008). With the exception of some anostracans, most branchiopods are widely distributed and can be used for assessing the quality and function of temporary wetlands (Brendonck et al. 2008). Branchiopods are characteristic and usually dominating inhabitants of ephemeral pools (and other extreme habitats such as Antarctic lakes and saltwater ponds or lakes) and share adaptations such as short life cycles, the production of desiccation-resistant dormant eggs (cysts), and being tolerant to highly fluctuating water parameters (Benvenuto et al. 2015). These taxa also possess excellent (local) dispersal capabilities, either through wind dispersal of cysts, through occasional floods, or through animal vectors (see, e.g., Figuerola et al. 2005, Hulsman et al. 2007).

The water fleas are perhaps the showcase taxon of crustaceans in standing waters. This is best exemplified by *Daphnia* (Daphniidae), the largest and best known genus with about 75 described and an estimated 200 species worldwide (Forro et al. 2008). With representatives on all continents (Adamowicz et al. 2009), *Daphnia* plays an enormous ecological role in lake and pond ecosystems and has consequently become a model taxon for studies linking ecology and genomics (Miner et al. 2012). *Daphnia* possesses a wide range of anti-predator adaptations (shape, spines, vertical migration patterns) depending on the particular environment, and large pelagic species or morphs are lacking in waterbodies with fish, as they are preferentially preyed upon (Riessen and Young 2005).

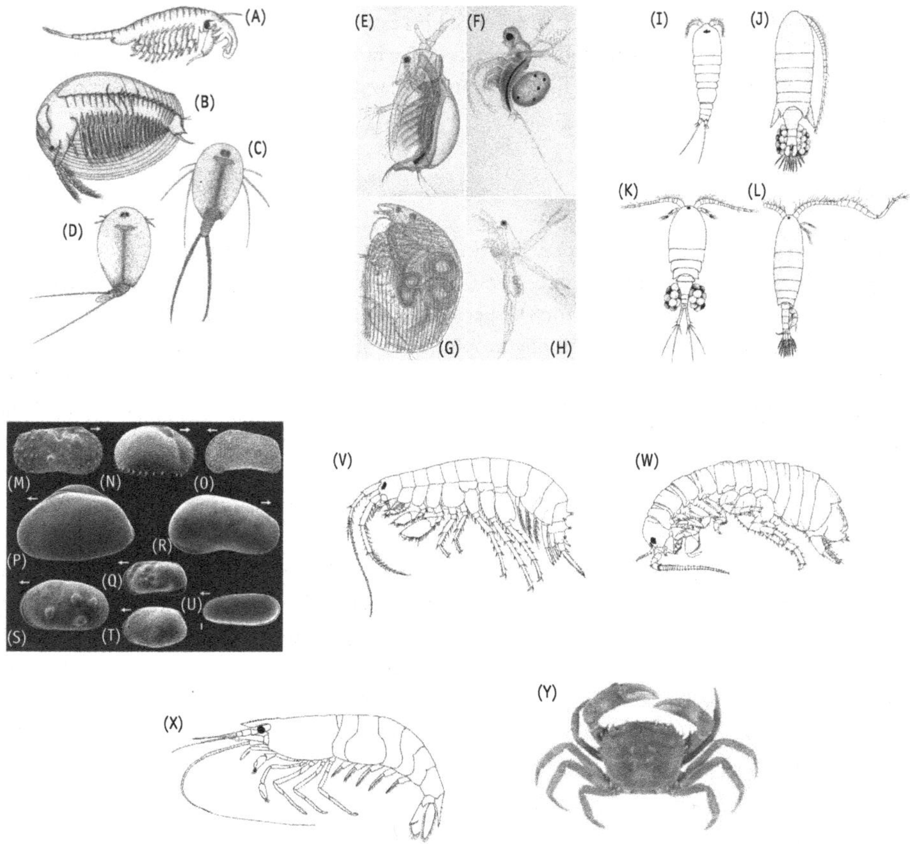

Fig. 11.1.
Overview of major groups of crustaceans occurring in standing waters (see Table 11.1) and with species flocks in ancient lakes (see Table 11.2). (A–D) Examples of larger Branchiopoda: (A) Anostraca (*Chriocephalus diaphanous*); (B) Diplostraca (*Limnadia lenticularis*); (C) Notostraca (*Triops cancriformis*); (D) Notostraca (*Lepidurus apus*). (E–H) Diplostraca: Cladocera "water fleas": (E) *Sida crystallina*; (F) *Bythotrephes longimanus*; (G) *Alona* sp.; (H) *Leptodora kindti*. (I–L) Examples of Copepoda: (I) Cyclopidae: *Graeteriella unisetigera*; (J) Centropagidae: *Boeckella triarticulata*; (K) Cyclopidae: *Cyclops abyssorum*; (L) Diaptomidae: *Eudiaptomus cracilis*. (M–U) Examples of main extant mon-marine groups of Ostracoda: (M) Ilyocyprididae: *Ilyocypris*; (N) Notodromadidae: *Centrocypris*; (O) Cyprididae: *Potamocypris*; (P) Cyrpididae: *Cyprinotus*; (Q) Limnocytheri dae: *Limnocythere*; (R) Candonidae: *Candona*; (S) Cytherideidae: *Cyprideis*; (T) Limnocytheridae: *Metacypris*; (U) Darwinulidae: *Darwinula*. (V) Example of Amphipoda: *Dikerogammarus oskari*. (W) Example of Isopoda: the epigean freshwater isopod *Eophreatoicus* sp. (X, Y) Examples of Decapoda: (X) the freshwater shrimps *Caridina weberi*, Atyidae; (Y) the freshwater crab *Plathythelphusa armata*, Platythelphusidae. With permissions from Springer: A–D, modified from Brendonck et al. 2008; E–H, modified from Forro et al. 2008; I–L, modified from Boxshall and Defaye 2008; M–U, modified from Martens et al. 2008; V, modified from Väinölä et al. 2008; W, modified from Wilson 2008; X, modified from De Grave et al. 2008; Y, modified from Marijnissen et al. 2008.

Copepoda

Copepods (Fig. 11.1I–L) are another important and highly diverse group in standing waters. With 5 orders and 25 families, they occur in almost all freshwater habitats worldwide (Boxshall and Defaye 2008). They are oviparous and, like branchiopods, produce dormant eggs that facilitate the successful colonization of new environments (Boxshall and Jaume 2000, Adamowicz et al. 2007, 2009).

Copepods can become problematic as introduced species, e.g., during the uncontrolled exchange of ballast water or as intermediate hosts for a variety of parasites that can affect human health or commercial freshwater aquaculture (Boxshall and Defaye 2008).

Ostracoda

Ostracoda (Fig. 11.1M–U), "mussel shrimps" or "seed shrimps," are small, bivalved crustaceans mainly inhabiting benthic environments in standing waters. Recent non-marine ostracods comprise about 200 genera with nearly 2,000 species, the majority (c. 75%) of which belong to the families Cyprididae and Candonidae (11 other families comprise c. 25%). They originated ~450–370 Ma, and their fossil record can be utilized for paleoecological studies, as their presence can be indicative of the physical and chemical properties of their habitats (Martens et al. 2008). This group can alternate sexual and parthenogenetic reproduction and generate isolated clones specialized for specific environmental conditions (Scheihing et al. 2011, Vandekerkhove et al. 2013). Ostracods form an important prey of littoral fishes (Guzman and Sielfield 2009, Riveros et al. 2012, Guerrero et al. 2015).

Amphipoda

Amphipods, like all peracarid crustaceans including the isopods (see below), undergo direct development with no independent larval stages. About 20% of all amphipod species live in inland waters, where they inhabit a wide spectrum of freshwater habitats and are most abundant in cool and temperate environments; most species are epibenthic, benthic, or (45% of all freshwater taxa) subterranean. Amphipods are usually associated with the littoral vegetation of lakes and rivers and the bottom matrix of small streams and springs. Some taxa are nektobenthic, whereas pelagic and parasitic or commensal species are rare in fresh waters. Their feeding habits are diverse; they can be herbivores, detritivores, carnivores, or omnivores. They often play a critical role in aquatic food webs, acting as conduits of nutrients and energy to higher trophic levels (Väinölä et al. 2008, Wellborn et al. 2015).

The best known lentic amphipods are the endemic and speciose species flocks in ancient lakes (see below). In contrast, there are few lacustrine species in younger lakes, and these are comparatively widespread and not endemic to single lakes. *Gammarus lacustris* has the widest native distribution range of any freshwater amphipod, ranging from northwestern North America across Europe to the Himalayas, but it has been suggested that in Asia at least, *G. lacustris* may represent a species complex rather than a single species (Katouzian et al. 2016). Recently, a new sister species endemic to and widespread in the European Alps, *G. alpinus*, has been described based on both molecular and morphological evidence (Fig. 11.2; Alther et al. 2016, but see Väinölä et al. 2017), underlining the necessity of more research on species diversity (not just in amphipods). There is only one other strictly lacustrine species (Hou and Sket 2016) among the 304 species of Gammaridae (Väinölä et al. 2008), while some species do occur in both lotic and lentic habitats. The same is true for *Hyalella* (Hyalellidae), the most speciose freshwater amphipod genus in the Americas with potentially more than 500 species (Väinölä et al. 2008). Among the plethora of putative undescribed species in the genus, there are several strictly lacustrine forms, in addition to the long-known endemic species flock in Lake Titicaca (see below).

Isopoda

Isopods are as diverse a group of crustaceans as are amphipods, but less than 10% (~950 species) of all isopods live in continental waters (Wilson 2008). Freshwater isopods are predominantly omnivorous or detritivorous and, if found in standing waters, mainly inhabit the littoral and pelagial zones of ponds and lakes (Wilson 2008, Wellborn et al. 2015), but never occur in temporary ponds

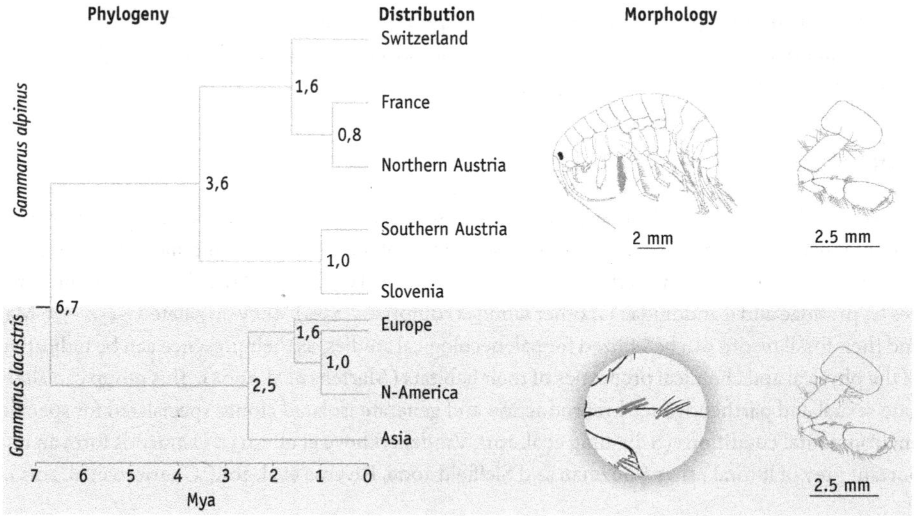

Fig. 11.2.
Distribution of the widespread amphipod species *Gammarus lacustris*. A combination of morphological and molecular data led to the detection of the new amphipod species *Gammarus alpinus* (but see Väinölä et al. 2017 for a taxonomic update), formerly reported as *G. lacustris*, but geographically clearly separated from each other. Modified from Figure 3 in http://www.eawag.ch/de/news-agenda/news-plattform/news/news/neu-entdeckt-und-schon-gefaehrdet/ (and according to Figure 2 of the original research article); drawings by Peter Penicka, Eawag. With permission from Eawag, Switzerland, and The Linnean Society of London, modified after Alther et al. 2016.

(Wetzer 2015). By far the largest strictly freshwater family, and at the same time the group with most species in lentic habitats, Asellidae (379 species), are the most diverse family in the Palearctic, where two-thirds of all species are found, but are also well represented in the Nearctic (Wilson 2008). Species of *Asellus* are commonly also found in lakes, but there are no strictly lacustrine species. In Australia, the Gondwanan Phreatoicidea are the most diverse isopod taxon (Wetzer 2015).

Decapoda

The best known Malacostraca are the Decapoda (Fig. 11.1.X–Y), including the infraorders Caridea, Astacidea, Brachyura, and Anomura. The main groups of freshwater caridean shrimps are the near-exclusively freshwater Atyidae (>360 species) and the Palaemonidae (subfamily Palaemoninae; almost 300 species in fresh water) (Moore 2006, De Grave et al. 2008). Atyidae are characterized by unique feeding chelipeds with complex brushes on the first and second pereopods, filtering out suspended matter or sweeping up microbial films. Species of both families commonly occur in standing water but, with the exception of endemic atyids in the ancient lakes of Sulawesi and Lake Tanganyika (see below), never exclusively so, and the vast majority of species are found only in lotic habitats.

Freshwater crayfishes (Astacidea) are a highly diverse group, and many species also inhabit ponds and lakes (Crandall and Buhay 2008, Covich 2015). They reproduce sexually, although hermaphroditic and even parthenogenetic species are known (Scholtz et al. 2003, Feria and Faulkes 2011). Freshwater crayfish are omnivorous and typically nocturnal. They are predators and can be extremely destructive when introduced to non-native habitats (Reynolds and Souty-Grosset 2012,

Kawai et al. 2016). Relatively few species are known to be strictly confined to standing waters, most notably the eight species of *Cherax* described from the Wissel Lakes in Indonesian New Guinea (Holthuis 1982), which would constitute a species flock if confirmed.

The freshwater crabs (Brachyura, Fig. 11.1Y) with more than 1,280 species occur in the tropics and subtropics in most parts of the world in a wide variety of aquatic and terrestrial habitats such as peat and freshwater swamps, stagnant ponds, and rice fields, and even in pools in tree holes and leaf axils (Moore 2006, Yeo et al. 2008). Freshwater crabs have direct development or short larval stages (Vogt 2013). They are primarily nocturnal, mostly omnivores, mainly feeding on vegetal matter but are also sometimes opportunistic carnivores feeding either on live prey such as fish and prawns or on dead animals. Crabs are often important food resources for other animals (Yeo et al. 2008). As all other decapods, most species of crabs found in standing waters are only opportunistic dwellers in lentic habitats and occur predominantly in rivers and streams. Notable exceptions are the ancient lake species flocks in Lake Tanganyika and the Sulawesi lakes (see below). The Aeglidae (Anomura; see above on Table 11.1) are abundant and widely distributed freshwater decapod "crabs" restricted to southern South America. The single genus *Aegla* inhabits lakes, streams, and caves (Bond-Buckup et al. 2008).

ANCIENT LAKES—SPECIES FLOCKS

Globally, about a dozen lakes stand out because of their age and biodiversity (Martens 1997). These so-called ancient lakes are considered to be at least 100,000 years old (Gorthner et al. 1994), but all existing ancient lakes are mostly more than 2 million years old (Cristescu et al. 2010). All are characterized by marked species endemism and species richness (Dumont 1998, Schön and Martens 2004, Cristescu et al. 2010, Vaillant et al. 2013). While the spectacular cichlid species flocks in the East African lakes are the most famous examples of ancient lake species flocks, ancient lakes in general also harbor largely endemic radiations of other fish taxa, mollusks, poriferans, diatoms, and crustaceans (Cristescu et al. 2010). The overview of crustaceans shown here (Table 11.2) is far from complete, especially for microscopic crustacean groups (often with a high number of cryptic or as yet undescribed species) and less well-studied lakes like Lake Inle in Myanmar, but illustrates the dominant species flocks in each lake: Amphipoda, Cladocera, Copepoda, Decapoda, Isopoda, and Ostracoda. In the following we provide a brief lake-wise account of the groups found in ancient lakes.

Ancient Lakes and Their Crustacean Species Flocks

Siberian Lake Baikal (Fig. 11.3) is the largest, deepest and oldest freshwater lake (Macdonald et al. 2005, Väinölä et al. 2008). With more than 350 species of endemic amphipods, 200 species of endemic ostracods (Martens et al. 2008), over 120 species of largely endemic copepods (Boxshall and Defaye 2008), and 6 endemic species of isopods (Hidding et al. 2003), it has by far the highest crustacean diversity of all ancient lakes (compare figures in Table 11.2). The amphipod species so far described belong to 76 genera and 11 families, 10 of which are autochthonous: Carinogammaridae, Crypturopodidae, Macrohectopodidae, Micruropodidae, Baikalogammaridae, Ommotagammaridae, Acanthogammaridae, Eulimnogammaridae, Pachyschesidae, and Pallaseidae. The amphipods of Lake Baikal are extremely diverse (e.g., morphologically, in depth preferences, etc.), and their taxonomic revision is still far from complete (Romanova et al. 2016). For example, members of the two major families, Acanthogammaridae and Micruropodidae, can be abyssal, planktonic, or (semi-)parasitic (Naumenko et al. 2017). Ostracods are similarly speciose in the lake with the two suborders Cytherocopina (genera

Table 11.2. Ancient Lakes with Major Species Flocks of Crustacea

Lake[a]	Taxon[b]	N species	Endemicity	Source(s)
Lake Baikal	Amphipoda	340	340 (100%)	Väinölä et al. 2008, Naumenko 2017
	Copepoda	120	>60 (>50%)	Boxshall and Defaye 2008
	Isopoda	6	6 (100%)	Hidding et al. 2003
	Ostracoda	200	200 (100%)	Martens et al. 2008
Caspian Sea	Amphipoda	95	40 (42%)	Väinölä et al. 2008
Lake Biwa	Cladocera	55	1 (1.8%)	Nishino 2012
	Copepoda	33	0 (0%)	Nishino 2012
	Ostracoda	40	16 (40%)	Nishino 2012
Lake Tanganyika	Copepoda	69	34 (49%)	Boxshall and Defaye 2008
	Decapoda: Atyidae	13	13 (100%)	Fryer 2006
	Decapoda: Platythelphusidae and Potamonautidae	10	10 (100%)	Marijnissen et al. 2006
	Ostracoda	200	200 (100%)	Martens and Behen 1994
Lake Malawi	Ostracoda	40	c. 24 (c. 60%)	Martens 2003
Lake Ohrid	Cladocera	31	1 (3%)	Albrecht and Wilke 2008
	Decapoda	2	0 (0%)	Albrecht and Wilke 2008
	Ostracoda	52	33 (63%)	Albrecht and Wilke 2008
	Copepoda	36	6 (17%)	Albrecht and Wilke 2008
	Isopoda	4	3 (75%)	Albrecht and Wilke 2008
	Amphipoda	10	9 (90%)	Albrecht and Wilke 2008
Malili lakes	Decapoda: Atyidae	15	15 (100%)	von Rintelen et al. 2012
	Decapoda: Geocarcinidae	5	5 (100%)	von Rintelen et al. 2012
Lake Poso	Decapoda: Atyidae	6	6 (100%)	von Rintelen et al. 2012
	Decapoda: Geocarcinidae	3	3 (100%)	von Rintelen et al. 2012
Lake Titicaca	Amphipoda	17	16 (94%)	Gonzalez and Coleman 2002, Coleman and Gonzalez 2006, De Los Ríos-Escalante et al. 2012
	Ostracoda	50	50 (100%)	Martens and Behen 1994

[a] *Compare Fig. 11.2.*
[b] *Compare Fig. 11.1.*

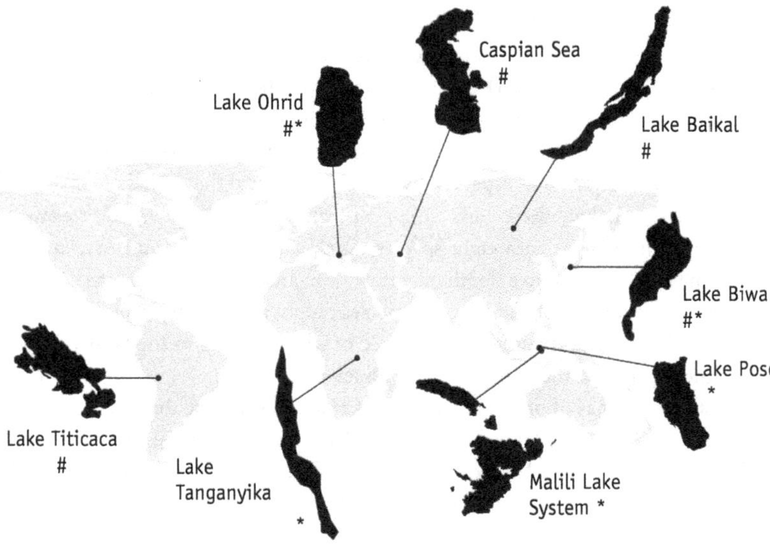

Fig. 11.3.
Ancient lakes of the world with major species flocks and radiations of Crustacea (except for Lake Malawi with a species flock of ostracods only). Macroscopic groups (without Isopoda): *, Decapoda (shrimps and crabs); #, Amphipoda (not to scale; for details on species flocks, see Table 11.1).

Limnocythere and *Cytherissa*) and Cypridocopina (family Candonidae with the genera *Candona, Pseudocandona,* and *Baicalocandona*) (Karanovic and Sitnikova 2017). A recent molecular study on the genus *Cytherissa* (Schön et al. 2017) suggested a high number of cryptic, yet undescribed species. Species of this genus occur from the littoral down to 500 m depth (Schön et al. 2017). The copepods are represented by five orders: Calanoida, Cyclopoida, Siphonostomatoida, Poecilostomatoida, and Harpacticoida. Only the cyclopoid family Cyclopidae and the harpacticoid family Canthocamptidae have large species flocks in the lake (Boxshall and Evstigneeva 1994). In Lake Baikal cyclopoid copepods inhabit pelagic areas from the littoral to the abyss (Mayor et al. 2010). The endemic isopods of the family Asellidae comprise three endemic genera with six species, all of which occur on rocky substrata with different depth preferences ranging from shallow water (3–10 m for *Baicalasellus angarensis*) to a depth of >100 m (*Mesoasellus dybowskii*) (see Hidding et al. 2003 for more details).

The Caspian Sea is the world's largest saline lake. As the major group of crustaceans there, amphipods comprise 95 species from several lineages, 40 of them endemic, including a large endemic species flock of the family Pontogammaridae (Väinölä et al. 2008). Further crustaceans occurring in the Caspian Sea are non-marine cumaceans, cladocerans, copepods, and mysids (see review in Mordukhai-Boltovskoi 1964 for details).

Lake Biwa is Japan's largest freshwater lake. It harbors a rich crustacean fauna, including 55 species of cladocerans (one endemic), 33 species of copepods (none endemic), 40 species of ostracods (16 endemic), and 16 species of malacostracans (three endemic). These are either planktonic (cladocerans and copepods), benthic (cladocerans,ostracods, malacostracans), or parasitic (some malacostracans) (Nishino 2012). The largest group are ostracods that dwell in the shallow depths of the lake; only three species seem to go deeper than 50 m (Karanovic 2015).

Lake Tanganyika harbors large species flocks of endemic ostracods and decapods (crabs and shrimps). Another major group is the 69 species of copepods (essentially from three suborders,

Calanoida, Cyclopoida, and Harpacticoida; Cirhuza and Plisnier 2016), about half endemic to the lake (Boxshall and Defaye 2008), including a rare account of a single parasitic freshwater copepod species (Boxshall and Strong 2006). Martens (1994) reported 64 described species of ostracods, 60 endemic; the total estimated number of species is around 200 species with 7 (out of 16) endemic genera in 5 families (Candonidae and Cypridopsidae dominate with 41 species) from the lake. If the estimate proves correct, ostracod species diversity would equal that of Lake Baikal.

In Lake Tanganyika, there are three endemic genera of atyid freshwater shrimps (Decapoda): *Limnocaridina* with eight species, *Caridella* with three, and *Atyella* with two. Atyid shrimps seem to occur down to a depth of at least 50 m and possibly deeper as they make up about 20% of the stomach content of endemic fish species caught at 120 m depth (Fryer 2006). Freshwater crabs (Decapoda) include 10 endemic species, 9 of which belong to the family Platythelphusidae and one to the family Potamonautidae (Marijnissen et al. 2006).

The crustacean faunas of other African rift lakes are less diverse than that of Lake Tanganyika. Martens (2003) mentioned a large species flock of 40 lacustrine species of Ostracoda (subfamilies Cypridopsinae, Cyprinotinae, Limnocytherinae, Timiriaseviinae) with about 60% endemism from Lake Malawi/Nyassa. Martens (1994) estimated 50 species in 20 genera.

The crustacean fauna of Lake Ohrid in the Balkans seems to have been better surveyed than that of other ancient lakes. According to the review in Albrecht and Wilke (2008), it comprises 31 species of cladocerans with only one endemic species, a rich ostracod fauna with 52 species in the Candoninae and Cytheroidea of which 63% are endemic, 36 species of copepods (families Ameiridae, Canthocamptidae, Cyclopidae, Diaptomidae, Ectinosomatidae; Petkovski and Karanovic 1997) of which 6 species (17%) are endemic, only 2 non-endemic decapods (a widespread European crayfish: *Astacus astacus*, and a freshwater crab: *Potamon fluviatilis*), 4 species of isopods with 3 endemic species belonging to the genus *Proasellus* (Asellidae), and 10 species of amphipods with 9 endemic species from the genus *Gammarus* (Gammaridae; Albrecht and Wilke 2008, Wysocka et al. 2013).

The ancient lakes of the Indonesian island Sulawesi, the Malili lakes and Lake Poso, harbor endemic species flocks of decapod crustaceans (see review in von Rintelen et al. 2012). In the Malili lakes, 15 endemic species of the freshwater shrimp genus *Caridina* (Atyidae) and endemic freshwater crabs of the family Gecarcinucidae with the genera *Parathelphusa* (2 species), *Nautilothelphusa* (one species), and *Syntripsa* (2 species) are found. Lake Poso harbors six endemic species of the freshwater shrimp genus *Caridina* and endemic crab species (Gecarcinidae), one species each of the genera *Migmathelphusa, Parathelphusa,* and *Sundathelphusa.*

Both groups of decapods are abundant throughout the benthos of all Sulawesian lakes (von Rintelen et al. 2012), some shrimp species are pelagic or occur in different depths, either in shallow-water regions (0–3 m) or deeper (for details, see revision by von Rintelen and Cai 2009). There are a few species of isopods in the Malili lakes as well (collected by KvR and TvR, unpublished data). Vaillant et al. (2013) identified one species of copepod in the Malili lakes (genus *Eodiaptomus*), possibly occurring at up to 100 m depths, and another species (genus *Neodiaptomus*) in Lake Poso.

Peru's Lake Titicaca's crustacean fauna comprises 16 endemic and one non-endemic amphipod species of the genus *Hyalella* (Hyalellidae; Gonzalez and Coleman 2002, Coleman and Gonzalez 2006, De Los Ríos-Escalante et al. 2012) and 3 endemic plus one non-endemic ostracod species in 4 genera of Limnocytheridae (with estimates of up to 50 species in 15 genera) (Martens 1994). The *Hyalella* species described and revised by Gonzalez and Coleman (2002) and Coleman and Gonzalez (2006) show various depth preferences, e.g., occurring epibenthic at approximately 0.5–10 m or epigean, normally in more than 10 m depth. The Lake Titicaca *Hyalella* species flock is not monophyletic, though, as the lake was colonized at least five times independently and intralacustrine radiation is confined to two largely endemic clades (Adamowicz et al. 2018).

Diversity Patterns and Adaptive Radiation of Ancient Lake Crustaceans

Interestingly, the most diverse macroscopic crustacean taxa in any of these lakes (Fig. 11.3) are either decapods (shrimps and crabs) in the tropical ancient lakes (in East Africa and Southeast Asia) or amphipods in the temperate (Europe and Asia) or high-altitude tropical regions (Lake Titicaca in South America). Only Lake Biwa in Japan (Nishino 2012) and Lake Ohrid in the central Balkans harbor both groups (Table 11.2). In these cases, however, there is always one dominant, endemic group of amphipods and only one (Biwa) or two non-endemic species (Ohrid) of decapods. Amphipods are known to be more abundant in cool and temperate environments, including temperate ancient lakes (Väinölä et al. 2008). A hypothesis (climatic factors, ecological or other competition) for this reciprocal absence of either amphipods or crabs and shrimps in most ancient lakes remains to be tested. Another group of macroscopic crustaceans, freshwater isopods, are known to be narrow-range endemics (Wilson 2008) such as in Lake Ohrid (Table 11.2), but not many records have been published from other lakes.

For microscopic groups (e.g., Ostracoda, Copepoda, Cladocera), there is no such clear pattern (see Table 11.2, Fig. 11.1). Only cladocerans, abundant in Lake Baikal and the Caspian Sea, for instance (Forro et al. 2008), seem to be largely absent in the tropics. Similarly, the degree of endemism does not seem to be as high as in the large-sized taxa (Table 11.2). However, these groups tend to be much less well known and studied than decapods and amphipods (with the possible exception of ostracods).

In terms of overall lacustrine species diversity and endemism, amphipods stand out. Their radiation in Lake Baikal with 340 described species is the largest crustacean species flock in an ancient lake, or perhaps anywhere, followed by ostracods in Lake Baikal and Lake Tanganyika (Table 11.2). Adaptive radiations have occurred frequently in ancient lakes elsewhere. Studies on the evolution of ancient lake organisms have continuously resulted in important insights into general patterns of speciation and radiation (e.g., Streelman and Danley 2003, Kocher 2004), including new insights from modern, integrative methods (e.g., Glor 2010, Naumenko et al. 2017). Like classic examples of adaptive radiations in fishes and mollusks, ancient lake species flocks of crustaceans are characterized by morphological and ecological adaptations: comprehensively studied examples of endemic adaptive radiations are decapods (shrimps and crabs) from the ancient lakes of Sulawesi (Malili lakes, Lake Poso; von Rintelen et al. 2012), decapods (mainly crabs, partly shrimps) from Lake Tanganyika (Marijnissen et al. 2008), and amphipods from Lake Baikal (Macdonald et al. 2005, Naumenko et al. 2017).

Adaptive radiation in general is considered a response to natural selection and ecological opportunities involving diversification of species and associated adaptations (Glor 2010). Schluter (2000) defined four criteria of adaptive radiation: common ancestry and rapid speciation as two phylogenetic criteria, and phenotype-environment correlation and trait utility as ecological criteria. Streelman and Danley (2003) suggested that adaptive radiation occurs in three general stages (at least in vertebrates): (1) divergences in habitat, (2) divergence in trophic morphology (stages 1–2: ecological selection is suggested as driving force), and (3) diversification in communication (sexual selection suggested as a driving force). These stages of adaptive radiation can also be detected in the crustacean lake radiations from Lake Baikal, the Sulawesian lakes and Lake Tanganyika.

As a first parallel, all species flocks show an endemicity of 100% (Table 11.2), suggesting that they have radiated within their respective lake. This does not necessarily mean that each lake has been colonized only once by each group. The Lake Titicaca amphipods (*Hyalella*) are the product of five lake colonizations (Adamowicz et al. 2018). In Lake Baikal, Naumenko et al. (2017) suggest two independent invasions of gammaridean amphipods into the lake leading to two endemic radiations: the more recent, more speciose, and more (morphologically and ecologically) diverse group including members of the family Acanthogammaridae and the older, less diverse group including members of the family Micruropidae. Similar examples are the species flocks of atyid shrimps in the Malili lakes and crabs from both Sulawesian lakes. These originated from independent lake colonizations instead of a single colonization event with subsequent radiation (see review in von Rintelen et al. 2012). Molecular data revealed that atyid shrimps of the genus *Caridina* colonized the Malili lakes twice independently (von Rintelen et al. 2010, 2012). Molecular data of gecarcinucid crabs have revealed at least two independent lake colonizations of each Sulawesian lake, i.e., the species flocks in each lake are not monophyletic (Schubart and Ng 2008, von Rintelen et al. 2012).

Only the shrimp species flock in Lake Poso and the species flock of crabs in Lake Tanganyika colonized their respective lake only once: molecular studies revealed a common ancestry (monophyletic clade) of all shrimp species of the genus *Caridina* in Lake Poso (von Rintelen et al. 2012) and of the crab species of the genus *Platythelphusa* in Lake Tanganyika (Marijnissen et al. 2008). No molecular phylogeny is available yet for shrimps from Lake Tanganyika to test the number of lake invasions, but similar to Lake Baikal's amphipods, different endemic genera (Atyidae: *Limnocaridina, Caridella,* and *Atyella*) occur in the lake (Fryer 2006), and thus, more than one lake invasion is rather likely.

Among the non-monophyletic species flocks, not all lineages radiated into several species. For example, in the Malili shrimps (Atyidae: *Caridina*), one colonization event was without subsequent radiation (with a single lacustrine species, *C. lanceolata*), while the other gave rise to a radiation of 14 species that seem to be strongly influenced by the lakes' geography (von Rintelen et al. 2010; see Fig. 11.4 for details). Similarly, in Lake Titicaca only two *Hyalella* lineages out of five radiated (Adamowicz et al. 2018). However, in those lineages with common ancestry and rapid speciation into several species within the respective lake, Schluter's phylogenetic criteria seem to apply. This is basically the case in all examples given here with the exception of Lake Tanganyika shrimps, for which a molecular phylogeny is still pending.

In all lake radiations discussed here, authors have suggested the colonization of an unspecialized ancestor that radiated into several ecologically specialized species. This led to the evolution of two extreme examples in atyid shrimps: *Caridina spongicola*, an exclusively sponge-dweller from the Malili lakes (von Rintelen et al. 2007), and *Limnocaridina iridinae*, a shrimp from Lake Tanganyika associated with a bivalve (Roth-Woltereck 1958). This leads to the question whether Schluter's ecological criteria also apply.

All these lake radiations match the first and second stages of an adaptive radiation as suggested in Streelman and Danley (2003): they all show divergence of habitat preferences and divergence in trophic and other morphological characteristics (Fig. 11.5), including the less-well-studied example of the Lake Tanganyika shrimps (Fig. 11.5A). In other words, they meet at least one of Schluter's ecological criteria as well, the phenotype–environment correlation. This kind of correlation has been well studied for the substratum preferences and subsequent adaptation of feeding appendages of the Sulawesian lake decapod radiations of shrimps (Atyidae; Fig. 11.5C) and crabs (Gecarcinidae) (von Rintelen et al. 2012) and the trophic morphology, habitat specificity, diet composition, and stable isotope analyses for the Lake Tanganyika crabs (Platythelphusidae) (Fig. 11.5D; Marijnissen et al. 2008).

Phenotype–environment correlation has been assumed for other ancient lake crustaceans as well. In the extremely diverse amphipods of Lake Baikal, several morphological features (body

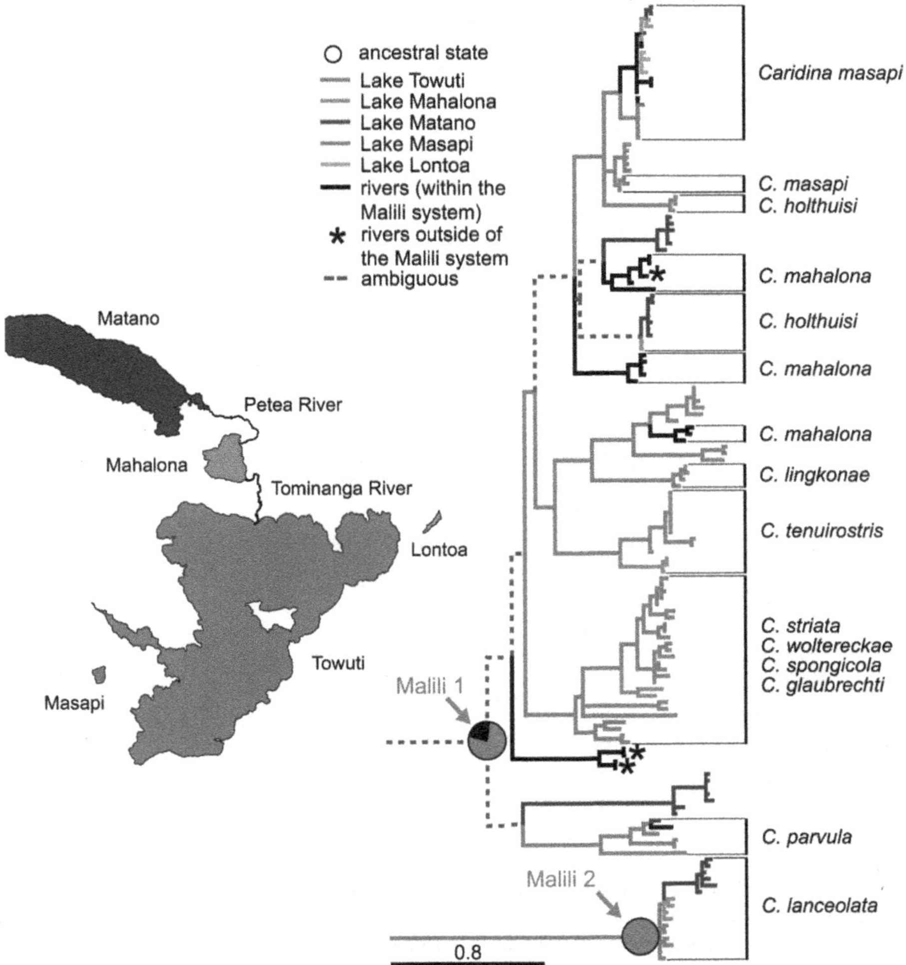

Fig. 11.4.
Adaptive radiation of the Malili lakes species flock of atyid freshwater shrimps of the genus *Caridina*. A molecular phylogeny (Bayesian Inference phylogram, mtDNA, 16S, and COI) revealed two independent clades within the Malili species flock. The majority of species are found within one large clade (Malili 1); the second clade consists of only a single species, *Caridina lanceolata* (Malili 2), suggesting two independent lake colonizations (for details, see von Rintelen et al. 2010). Within the species flock there is a clear geographic pattern (the Malili lakes consist of five interconnected lakes), i.e., several species show an allopatric distribution, either in a single (e.g., *C. tenuirostris*) or in more than one clade (e.g., *C. holthuisi*). Modified from von Rintelen et al. 2010, with permission from John Wiley and Sons. See color version of this figure in the centerfold.

armor and reduction in appendages and sensory organs) evolved in several lineages in parallel (Fig. 11.5B; Naumenko et al. 2017). In Naumenko et al. 2017, the authors argued, however, that it is difficult to explain the extant species diversity by niche availability alone, as numerous species still occupy similar niches; they suggested the application of genome- and transcriptome-wide approaches to study adaptive radiations. In a similar vein, the lack of evidence for niche differentiation in some groups, such as Baikalian ostracods (Schön et al. 2000), has been used to entertain notions of non-adaptive radiation in ancient lakes (Schön and Martens 2004).

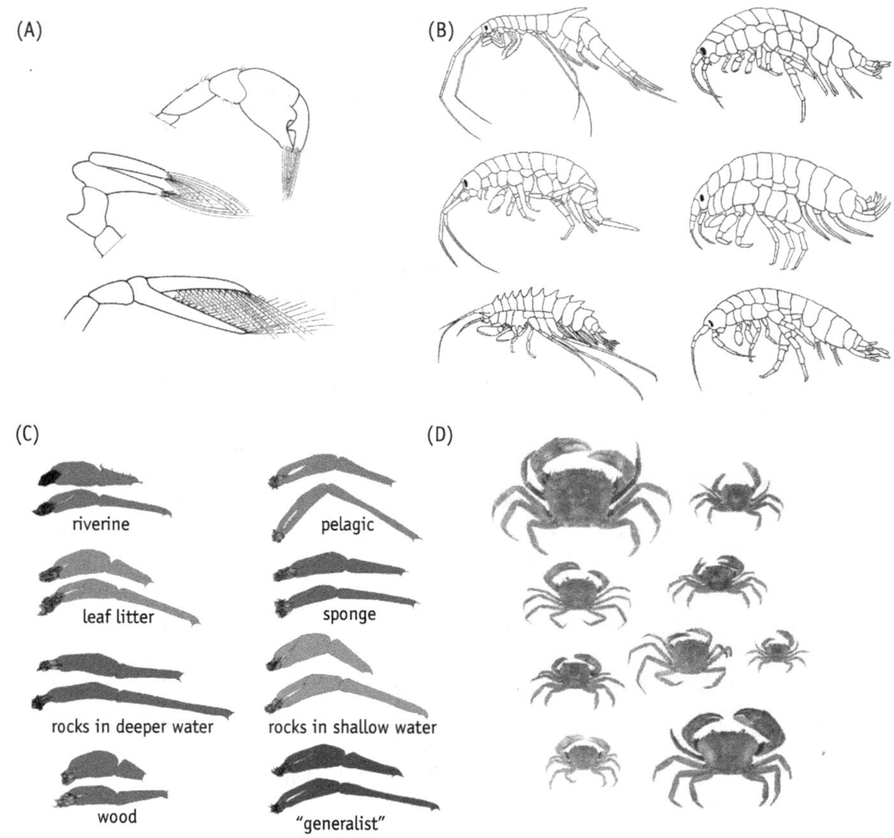

Fig. 11.5.
Examples of major adaptive radiations of Crustacea in ancient lakes. (A) The chelipeds (feeding appendages) of atyid shrimps in Lake Tanganyika (from top to bottom: examples of the genera *Caridella, Atyella*, and *Limnocaridina*). (B) Representatives of Lake Baikal's amphipod fauna with generalized and specialized morphologies. (C) The morphologically adapted chelipeds (feeding appendages) of atyid shrimps of the genus *Caridina* in the Malili lakes ("generalist" = unspecialized chelipeds of riverine and the lacustrine species *Caridina lanceolata*). (D) Similar representatives of the Lake Tanganyika crab radiation. A, modified from Fryer 2006, with permission from Springer; B, modified from Macdonald et al. 2005, with permission from Elsevier; C, modified from von Rintelen et al. 2010, with permission from John Wiley and Sons; D, modified from Marijnissen et al. 2008, with permission from Springer.

Trait utility, i.e., "evidence that morphological and physiological traits of species are indeed useful," proved difficult to test (Schluter 2000, p. 11). No study cited here has tested it so far, although in most cases three of four criteria clearly apply and are specifically mentioned as such (e.g., in Marijnissen et al. 2008 for Lake Tanganyika crabs (Platythelphusidae)). In addition, at least the Sulawesian atyid shrimp radiations (*Caridina*) seem to have reached the third stage of adaptive radiation according to Streelman and Danley (2003): they exhibit unusual and species-specific coloration patterns (von Rintelen et al. 2012) that suggests sexual selection as a driving force, although this color-driven diversification needs critical testing. Ecological speciation is highly likely in all cases (ecological selective pressure). Marijnissen et al. (2006) discussed allopatric factors and sympatric speciation mechanism as other potential drivers for the Tanganyika crab species flock of the genus *Platythelphusa* but also concluded that this remains speculative.

In summary, a few general patterns hold true for all ancient lake crustacean species flocks: (1) several lakes were colonized by potential unspecialized crustacean ancestors and subsequently radiated into several specialized species; (2) in each group and lake, at least one lineage meets three of four criteria for a truly adaptive radiation; (3) all radiations seem to have reached at least two stages of adaptive radiations, and only in at least one case (Sulawesian lake shrimps) color-driven sexual selection seems likely; and (4) generally, potential driving forces of these adaptive radiations in Crustacea are mainly ecological and sexual selection. Further, allopatric and sympatric speciation mechanisms have been hypothesized.

OTHER LAKES

In stark contrast, relatively short-lived lakes and ponds lack the highly endemic and specialized fauna of ancient lakes (see also "Introduction"). Interestingly, a significant proportion of species in these water bodies also occurs in lotic habitats and in some groups (certainly decapods, but presumably also amphipods, isopods, and potentially also some smaller-sized taxa) there seem to be few to no species that live exclusively in standing water outside of ancient lakes. This will be rather difficult to quantify given the lack of respective reliable data on ecology and distribution for most groups, though (see "Future Directions"). If confirmed, this would not be overly surprising given the young age and/or temporary nature (in the case of ponds) of these water bodies. Species specialized for ephemeral water bodies present an extreme case of adaptation to this particular type of standing water (see also above, systematic account of Branchiopoda). This centers on their capability to produce dormant eggs that serve the double purpose of generating a high potential for the colonization of new habitats and for ensuring survival over periods of desiccation once an ephemeral habitat has been colonized (Ripley and Simovich 2009). Adaptations to less extreme lentic habitats and particularly lacustrine settings include those to a pelagic lifestyle. In *Daphnia*, for example, planktonic species or forms are small, have a thin transparent cuticle and small eyes and spines, all characters that will help these species to remain floating. These characters, however, might equally well be interpreted as anti-predator adaptations (Fryer 1991).

Species of microscopic crustaceans that are predominantly found in lakes and ponds (Cladocera, Copepoda, Ostracoda) can show significant habitat specificity and adaptations. In temperate Lake Stechlin (Germany), e.g., benthic species of these taxa can be divided into nine habitat groups plus some generalists (Flößner et al. 1985). A similar pattern of niche separation in Cladocera were recovered by Adamczuk (2014), and competitive exclusion was found for eight pairs or triplets of closely related species of Cladocera in the Stechlin study (Flößner et al. 1985). Among larger crustaceans, niche specificity has been investigated in North American *Hyalella* (Amphipoda), where cryptic species in the *H. azteca* species complex co-exist with grazing periphyton on submerged macrophytes in several lakes. While evidence for species-specific predation risk and spatial distribution was found in a study on three species (Wellborn and Cothran 2007), a stable isotope-based analysis did not support niche partitioning for food resources (Dionne et al. 2017). Despite the similarities to patterns observed in ancient lakes in these and a few other studies, the geographic scale is fundamentally different here, as species are endemic to catchments at best and usually occur at a wider geographic scale and, depending on the taxon, occasionally also in rivers.

Given the higher propensity for dispersal, the rate of endemism is generally surprisingly high at all geographic levels in crustaceans of standing waters (excluding ancient lakes) at the species level. Even in taxa with dormant dispersal stages and consequently high passive dispersal potential such as cladocerans, only a handful of species are truly widespread (across continents/zoogeographic regions) or even cosmopolitan (not considering human-mediated dispersal), and most are at least endemic to zoogeographic regions or geographic subdivisions of these (see, e.g., study by Adamwicz et al. 2009 on

Daphnia). When it comes to taxa lacking such dispersal stages, species ranges tend to become smaller with the abbreviation of development and habitat specialization (e.g., groundwater, caves) and local or regional endemism may extend to the genus level, as exemplified by atyid shrimps (von Rintelen et al. 2012) or phreatoicidean isopods (Wilson and Johnson 1999). At the family level, endemism even at the level of zoogeographic regions is less frequent (but applies, e.g., to 24 of 53 amphipod families with freshwater representatives; Väinölä et al. 2008), presumably because many groups are old and tectonic vicariance plays a role and some taxa have lineages with ocean-crossing dispersal stages such as the atyid or palaemonid shrimps. However, generalizations are difficult at this stage as fully resolved molecular phylogenies are still lacking for many taxa, and there are no universal criteria for assigning ranks such as genus and family to a group of related species, making comparisons at higher levels less useful than those at the species level. In addition, all this holds true for crustaceans in lentic and lotic habitats, with the exception of the very few groups such as *Daphnia* entirely confined to standing water.

FUTURE DIRECTIONS

Oddly enough after centuries of taxonomic research, the single most fundamental global research priority, not just for crustaceans in standing waters, is to strive toward a complete species inventory. Obviously, this must include fundamental morphological and ecological data for all species. All recent studies on macroscopic crustaceans in ancient lakes (see above and Table 11.2) have led to a significant increase in the number of known species, even though many of these flocks have been known for more than 100 years and some have been cited as textbook examples of intralacustrine radiations. For the most diverse groups such as amphipods in Lake Baikal and ostracods in Lake Baikal and Lake Tanganyika, species numbers are still underestimates. In this and other groups with a good fossil record in sediments, this taxonomic shortcoming also limits their potential for assessing long-term ecological changes, climate change, etc. (Martens 1994). In ancient lakes, some doubt has been voiced about the adaptive origin of species flocks. Such controversy and other evolutionary hypotheses can obviously only meaningfully be addressed or tested if sufficient data are available, and the most fundamental data in this context are taxonomic data. Beyond ancient lakes, this also applies to studying fundamental questions of ecology, such as what enables species coexistence, as exemplified by the study of Wellborn and Cothran (2007) into niche diversity of recently discovered cryptic species of North American lake amphipods.

Knowing the real diversity of crustaceans in standing waters is generally crucial for meaningful comparisons of species diversity between regions, habitats, and taxa. This even applies to a core point underlying this chapter: the number of crustaceans occurring in standing waters, which is almost impossible to quantify for many taxa at present. As always, the overwhelming diversity of the tropics constitutes the biggest challenge. Apart from the specific situation in ancient lakes, the European, North American, and Australian faunas are comparatively well investigated across all taxa, while large parts of Asia, South America, and Africa are generally poorly known. Larger taxa such as decapods are probably proportionally better known, but even this is underresearched. In addition, even in well-studied regions and groups, fundamental and immensely useful genetic data such as DNA barcodes (Hebert et al. 2003) are still widely lacking. Given the poor state of taxonomic research on many freshwater crustaceans, which is to a significant part a result of having too few people doing taxonomy, DNA barcodes can help to overcome this impediment and enable non-specialists to reliably identify species. This will become increasingly crucial given the need to deal with shifting species ranges in the wake of climate change but also because of more intentional and accidental species introductions by humans. Invasive species are increasingly becoming an issue, and decapod crustaceans have an infamous history in this respect (Strayer 2010), but less conspicuous taxa like cladocerans and copepods have also been introduced to and have affected lake ecosystems such as in North America (Cordell 2012, Strecker 2012).

This basic task of taxonomy will be complemented (and aided) by the application of modern sequencing techniques (next generation sequencing, NGS). The ongoing genomic revolution through the rapid spread and development of NGS has the potential to allow deep insights into the genetic basis of adaptations, not just in the context of ancient lakes, which harbor some of the most striking examples of morphological evolution in limnic crustaceans. Ichthyology is leading the way in the aquatic realm with landmark studies on the cichlid radiations in African lakes, which have also shown the potential of such work to challenge the interpretation of long-standing textbook systems (Martin et al. 2015). Genomic techniques can not only "dig deep" but are actually also increasingly employed to help in the task of inventorying diversity in the first place through meta-barcoding approaches or environmental barcoding. The scope for the application of genomics to taking crustacean studies to the next level is immense, and so far this potential has been barely touched, even for model groups such as *Daphnia* (Miner et al. 2012).

SUMMARY AND CONCLUSIONS

Crustacea are important elements of standing waters and can make up much of the biomass (e.g., in the pelagial of northern hemisphere lakes) and/or species diversity (e.g., in some ancient lakes). While the ecological role of, e.g., branchiopods or copepods in young lakes or ephemeral habitats is undisputed, hard data are still lacking for the sometimes striking crustacean species flocks in ancient lakes, whose evolution has nevertheless received a lot of interest and contributed to a better understanding of adaptive radiations in general. A lot of the potential of crustaceans to inform ecology and evolutionary biology is yet unrealized due to a lack of basic taxonomic and ecological data, and future research efforts should aim at overcoming this hindrance for a better understanding of global diversity patterns and biogeography.

ACKNOWLEDGMENTS

The contribution of PDlR was supported by project MECESUP UCT 0804. The authors KvR and TvR were supported by BMBF (project INDOBIOSYS MfN Berlin, 16GW0111K) with regard to the crustacean biodiversity of ancient lakes in Indonesia.

REFERENCES

Adamczuk, M. 2014. Niche separation by littoral-benthic Chydoridae (Cladocera, Crustacea) in a deep lake—potential drivers of their distribution and role in littoral-pelagic coupling. Journal of Limnology 73:490–501.

Adamowicz, S.J., M.C. Marinone, S.M. Marque, J.W. Martin, J. D.C. Allen, M.N. Pyle, P.R. De los Ríos-Escalante, C.N. Sobel, C. Ibañez, J. Pinto, and J.D.S. Witt. 2018. The *Hyalella* (Crustacea: Amphipoda) species cloud of the ancient Lake Titicaca originated from multiple colonizations. Molecular Phylogenetics and Evolution 125:232–242.

Adamowicz, S.J., S. Menu-Marque, P.D.N. Hebert, and A. Purvis. 2007. Molecular systematics and patterns of morphological evolution in the Centropagidae (Copepoda: Calanoida) of Argentina. Biological Journal of the Linnaean Society 90:279–292.

Adamowicz, S.J., A. Petruzek, J.K. Colbourne, P.D.N. Hebert, and J.D.S. Witt. 2009. The scale of divergence: a phylogenetic appraisal of intercontinental allopatric speciation in a passively dispersed freshwater zooplankton genus. Molecular Phylogenetics and Evolution 50:423–436.

Albrecht, C., and T. Wilke. 2008. Ancient Lake Ohrid: biodiversity and evolution. Hydrobiologia 615:103–140.

Ahyong, S.T. 2016. The Tasmanian mountain shrimps, *Anaspides* Thomson, 1894 (Crustacea, Syncarida, Anaspididae). Records of the Australian Museum 68:313–364.

Ahyong, S.T., and M.A. Alonso-Zarazaga. 2017. Anaspidesidae, a new family for syncarid crustaceans formerly placed in Anaspididae Thomson, 1893. Records of the Australian Museum 69:257–258.

Alekseev, V.R., O. Ravera, and B.T. De Stasio. 2007. Introduction to diapause. Pages 3–10 *in* V.R. Alekseev, B. De Stasio, and J.J. Gilbert, editors. Diapause in Aquatic Invertebrates: Theory and Human Use. Springer, Dordrecht, The Netherlands.

Alther, R., C. Fišer, and F. Altermatt. 2016. Description of a widely distributed but overlooked amphipod species in the European Alps. Zoological Journal of the Linnean Society 179:751–766.

Barlow, G.W. 2000. The Cichlid Fishes: Nature's Grand Experiment in Evolution. Perseus, Cambridge, MA.

Benvenuto, C., B. Knott, and S.C. Weeks. 2015. Crustaceans of extreme environments. Pages 379–417 *in* M. Thiel and L. Watling, editors. Lifestyles and Feeding Biology. The Natural History of Crustacea, Vol. 2. Oxford University Press, Oxford.

Bond-Buckup, G., C.G. Jara, M. Pérez-Losada, L. Buckup, and K.A. Crandall. 2008. Global diversity of crabs (Aeglidae: Anomura: Decapoda) in freshwater. Hydrobiologia 595:267–273.

Bos, D.G., B.F. Cumming, C.F. Watters, and J.P. Smol. 1996. The relationship between zooplankton conductivity and lake-water ionic composition in 111 lakes from the interior of pleateau of British Columbia, Canada. International Journal of Salt Lake Research 5:1–15.

Boxshall, G.A., and D. Defaye. 2008. Global diversity of copepods (Crustacea: Copepoda) in freshwater. Hydrobiologia 595:195–207.

Boxshall, G.A., and T.D. Evstigneeva. 1994. The evolution of species flocks of copepods in Lake Baikal: a preliminary analysis. Archiv für Hydrobiologie, Beihefte, Ergebnisse der Limnologie 44:235–245.

Boxshall, G.A., and D. Jaume. 2000. Making waves: the repeated colonization of freshwater by copepod crustaceans. Advances in Ecological Research 31:61–79.

Boxshall, G.A., and E.E. Strong. 2006. An extraordinary shift in life habit within a genus of cyclopid copepods in Lake Tanganyika. Zoological Journal of the Linnean Society 146:275–285.

Brendonck, L., D.C. Rogers, J. Olesen, S. Weeks, and W.R. Hoeh. 2008. Global diversity of large branchiopods (Crustacea: Branchiopoda) in freshwater. Hydrobiologia 595:167–176.

Cirhuza, D.M., and P.D. Plisnier. 2016. Composition and seasonal variations in abundance of copepod (Crustacea) populations from the northern part of Lake Tanganyika. Aquatic Ecosystem Health & Management 19:401–410.

Coleman, C.O., and E.R. Gonzalez. 2006. New hyalellids (Crustacea, Amphipoda, Hyalellidae) from Lake Titicaca. Organisms Diversity and Evolution 6, Electronic Supplement 10:1–28.

Cordell, J.R. 2012. Invasive freshwater copepods of North America. Pages 161–172 *in* R.A. Francis, editor. A Handbook of Global Freshwater Invasive Species, Vol. 1: Introduced Freshwater Organisms. Earthscan, New York.

Covich, A.P. 2015. Freshwater crustaceans: adaptations to complex inland habitats and species interactions. Pages 337–378 *in* M. Thiel and L. Watling, editors. Natural History of Crustacea, Vol. 2: Life Styles and Feeding Biology of the Crustacea. Oxford University Press, New York.

Covich, A.P., J.H. Thorp, and D.C. Rogers. 2009. Introduction to the Subphylum Crustacea. Pages 695–723 *in* J.H. Thorp and A.P. Covich, editors. Ecology and Classification of North American Freshwater Invertebrates, 3rd ed. Academic Press, London.

Crandall, K.A., and J.E. Buhay. 2008. Global diversity of crayfish (Astacidae, Cambaridae and Parastacidae-Decapoda) in freshwater. Hydrobiologia 595:295–301.

Cristescu, M.E., S.J. Adamowicz, J.J. Vaillant, and D.G. Haffner. 2010. Ancient lakes revisited: from the ecology to the genetics of speciation. Molecular Ecology 19:4837–4851.

De Grave, S., Y. Cai, and A. Anker. 2008. Global diversity of shrimps (Crustacea: Decapoda: Caridea) in freshwater. Hydrobiologia 595:287–293.

De Los Ríos-Escalante, P., J.J. Morrone, and R. Rivera. 2012. Distributional patterns of South American species of *Hyalella* (Amphipoda: Hyalellidae). Gayana 76:153–161.

Dionne, K., F. Dufresne, and C. Nozais. 2017. Overlapping trophic niches among co-occurring amphipods from a cryptic species complex. Freshwater Biology 62:1052–1062.

Driessen, M.M., S.A. Mallick, S. Thurstans, B. Smith, and R. Brereton. 2014. Distribution and conservation status of two endemic Tasmanian crustaceans, *Allanaspides hickmani* and *Allanaspides helonomus* (Syncarida: Aanaspididae). Papers and Proceedings of the Royal Society of Tasmania 148:1–9.

Dumont, H.J. 1998. The Caspian lake: history, biota, structure and function. Limnology and Oceanography 43:44–52.

Elser, J., W.F. Fagan, R.F. Denno, D.R. Dobberfuhl, A. Folarin, A. Huberty, S. Interlandi, S.S. Kilham, E. MacCauley, K.L. Schulz, E.H. Slemann, and R.W. Sterner. 2000. Nutritional constraint in terrestrial and freshwater food webs. Nature 408:578–580.

Feria, T.P., and Z. Faulkes. 2011. Forecasting the distribution of Marmorkrebs, a parthenogenetic crayfish with high invasive potential, in Madagascar, Europe, and North America. Biological Invasions 6:55–67.

Figuerola, J., A.J. Green, and T.C. Michot. 2005. Invertebrate eggs can fly: evidence of waterfowl-mediated gene flow in aquatic invertebrates. American Naturalist 165:274–280.

Flößner, D. P. Kasprzak, G. Mothes, D. Ronneberger, and W. Schönborn. 1985. The invertebrate communities. Pages 213–259 *in* S.J. Casper, editor. Lake Stechlin. A Temperate Oligotrophic Lake. Monographiae Biologicae, Vol. 58. Springer, Dordrecht, The Netherlands.

Forro, L., N.M. Korovchinsky, A.A. Kotov, and A. Petrusek. 2008. Global diversity of cladocerans (Cladocera; Crustacea) in freshwater. Hydrobiologia 595:177–184.

Fryer, G. 1991. Functional morphology and the adaptive radiation of the Daphniidae (Branchiopoda:Anomopoda). Philosophical Transactions: Biological Sciences 331:1–99.

Fryer, G. 2006. Evolution in ancient lakes: radiation of Tanganyikan atyid prawns and speciation of pelagic fishes in Lake Malawi. Hydrobiologia 568:131–142.

Gillooly, J.F., and S.I. Dodson. 2000. Latitudinal patterns in the size distribution and seasonal dynamics of new world, freshwater cladocerans. Limnology and Oceanography 45:22–30.

Glor, R.E. 2010. Phylogenetic insights on adaptive radiation. Annual Review of Ecology, Evolution, and Systematics 41:251–270.

Gonzalez, E.R., and C.O. Coleman. 2002. *Hyalella armata* (Crustacea, Amphipoda, Hyalellidae) and the description of a related new species from Lake Titicaca. Organisms Diversity and Evolution 2, Electronic Supplement 6:1–19.

Gorthner, A., K. Martens, B. Goddeeris, and G. Coulter. 1994. What is an ancient lake? Pages 97–100 *in* K. Martens, B. Goddeeris, and G. Coulter, editors. Speciation in Ancient Lakes. E. Schweizerbart'sche Verlagsbuchhandlung, Stuttgart, Germany.

Guerrero, C.J., E. Poulin, M.A. Méndez, and I. Vila. 2015. Caracterización trófica de *Orestias* (Teleostei: Cyprinodontidae) en el Parque Nacional Lauca. Gayana 79:18–25.

Guzman, J.A., and W. Sielfield. 2009. Diet of *Orestias agassii* (Cuvier and Valenciennes, 1846) (Teleostei: Cyprinodontidae) of "El Huasco" salt lake, high Andes of northern Chile. Gayana 73:28–32.

Haskell, C.A., and J.A. Stanford. 2006. Ecology of an estuarine mysid shrimp in the Columbia River (USA). River Research and Applications 22:739–753.

Hebert, P. D., S. Ratnasingham, and J.R. DeWaard. 2003. Barcoding animal life: cytochrome c oxidase subunit 1 divergences among closely related species. Proceedings of the Royal Society of London B 270 (Suppl.):S96–S99.

Hidding, B., E. Michel, A.V. Natyaganova, and D.Y. Sherbakov. 2003. Molecular evidence reveals a polyphyletic origin and chromosomal speciation of Lake Baikal`s endemic asellid isopods. Molecular Ecology 12:1509–1514.

Hou, Z., and B. Sket. 2016. A review of Gammaridae (Crustacea: Amphipoda): the family extent, its evolutionary history, and taxonomic redefinition of genera. Zoological Journal of the Linnean Society 176:323–348.

Holthuis, L.B. 1982. Freshwater Crustacea Decapoda of New Guinea. Pages 603– 619 *in* J.L. Gressitt, editor. Biogeography and Ecology of New Guinea. Monographiae Biologicae, Springer, Dordrecht, The Netherlands.

Hulsmans, A., K. Moreau, L. De Meester, B.J. Riddoch, and L. Brendonck. 2007. Direct and indirect measures of dispersal in the fairy shrimp *Branchipodopsis wolfi* indicate a small-scale isolation-by-distance pattern. Limnology and Oceanography 52:676–684.

Karanovic, I. 2015. Barcoding of ancient lake ostracods (Crustacea) reveals cryptic speciation with extremely low distances. PLoS ONE 10:e0121133.

Karanovic, I., and T.Y. Sitnikova. 2017. Phylogenetic position and age of Lake Baikal candonids (Crustacea, Ostracoda) inferred from multigene sequence analyzes and molecular dating. Ecology and Evolution 2017:7091–7103.

Katouzian, A.R., A. Sari, J.N. Macher, M. Weiss, A. Saboori, F. Leese, and A.M. Weigand. 2016. Drastic underestimation of amphipod biodiversity in the endangered Irano-Anatolian and Caucasus biodiversity hotspots. Scientific Reports 6:22507.

Kawai, T., Z. Faulkes, and G. Scholtz. 2016. Freshwater Crayfish. A Global Overview. CRC Press, Boca Raton, FL.

Kocher, T.D. 2004. Adaptive evolution and explosive speciation: the cichlid fish model. Nature Review Genetics 5:288–298.

Macdonald, K.S., Yampolsky, L., and J.E. Duffy. 2005. Molecular and morphological evolution of the amphipod radiation in Lake Baikal. Molecular Phylogenetics and Evolution 35:323–343.

Marijnissen, S.A.E., E. Michel, S.R. Daniels, D. Erpenbeck, S.B.J. Menken, and F.R. Schram. 2006. Molecular evidence for recent divergence of Lake Tanganyika endemic crabs (Decapoda: Platythelphusidae). Molecular Phylogenetics and Evolution 40:628–634.

Marijnissen, S.A.E., E. Michel, M. Kamermans, K. Olaya-Bosh, M. Kars, D.F. Cleary, E.E. van Loon, P.G. Rachello Dolmen, and S.B.J. Menken. 2008. Ecological correlates of species differences in the Lake Tanganyika crab radiation. Hydrobiologia 615:81–94.

Martens, K. 1994. Ostracod speciation in ancient lakes: a review. Archiv für Hydrobiologie–BeiheftErgebnisse der Limnologie 44:203–222.

Martens, K. 1997. Speciation in ancient lakes. TREE 12:177–182.

Martens, K. 2003. On the evolution of *Gomphocythere* (Crustacea, Ostracoda) in Lake Nyassa/Malawi (East Africa), with the description of 5 new species. Hydrobiologia 497:121–144.

Martens, K., and F. Behen. 1994. A checklist of the recent non-marine ostracods (Crustacea, Ostracoda) from the inland waters of South America and adjacent islands. Travaux scientifiques du Musée national d'histoire naturelle Luxembourg 22:1–81.

Martens, K., I. Schön, C. Meisch, and D.J. Horne. 2008. Global diversity of ostracods (Ostracoda, Crustacea) in freshwater. Hydrobiologia 595:185–193.

Martin, C. H., J.S. Cutler, J.P. Friel, C. Dening Touokong, G. Coop, and P.C. Wainwright. 2015. Complex histories of repeated gene flow in Cameroon crater lake cichlids cast doubt on one of the clearest examples of sympatric speciation. Evolution 69:1406–1422.

Mayor, T.Y., N.G. Shevelea, L.V. Sukhanova, O.A. Timoshkin, and S.V. Kiril`chik. 2010. Molecular-phylogenetic analysis of cyclopoids (Copepoda: Cyclopoida) from Lake Baikal and its water catchment basin. Russian Journal of Genetika 46:1556–1564.

Miner, B.E., L.D. Meester, M.E. Pfrender, W. Lampert, and N.G. Hairston Jr. 2012. Linking genes to communities and ecosystems: *Daphnia* as an ecogenomic model. Proceedings of the Royal Society B 279:1873–1882.

Moore, J. 2006. An Introduction to the Invertebrates. Cambridge University Press, Cambridge, UK.

Mordukhai-Boltovskoi, D. 1964. Caspian fauna beyond the Caspian Sea. Internationale Revue der gesamten Hydrobiologie und Hydrogeograhie 49:139–176.

Naumenko, S.A., M.D. Logacheva, N.V. Popova, A.V. Klepikova, A.A. Penin, G.A. Bazykin, A.E. Etingova, N.S. Mugue, A.S. Kondrashov, and L.Y. Yampolsky. 2017. Transcriptome-based phylogeny of endemic Lake Baikal amphipod species flock: fast speciation accompanied by frequent episodes of positive selection. Molecular Ecology 26:536–553.

Nishino, M. 2012. Crustacea of Lake Biwa. Pages 89–102 *in* H. Kawanabe, M. Nishino, and M. Maehata, editors. Lake Biwa: Interactions Between Nature and People. Springer, Dordrecht, The Netherlands.

Petkovski, T.K., and T. Karanovic. 1997. Two new copepod species (Crustacea: Copepoda) from the Ohrid Lake. Annales de Limnologie 33:245–253.

Poly, W.J. 2008. Global diversity of fishlice (Crustacea: Brachiura: Argulidae) in freshwater. Hydrobiologia 595:209–212.

Porter, M.L., K. Meland, and W. Price. 2008. Global diversity of mysids (Crustacea-Mysida) in freshwater. Hydrobiologia 595:213–218.

Reynolds, J., and C. Souty-Grosset. 2012. Management of Freshwater Biodiversity. Crayfish as Bioindicators. Cambridge University Press, Cambridge, UK.

Riessen, H.P., and J.D. Young. 2005. Daphnia defense strategies in fishless lakes and ponds: one size does not fit all. Journal of Plankton Research 27:531–544.

Ripley, B.J., and M.A. Simovich. 2009. Species richness on islands in time: variation in ephemeral pond crustacean communities in relation to habitat duration and size. Hydrobiologia 617:181–196.

Riveros, J., I. Vila, and M.A. Méndez. 2012. Trophic niche of *Orestias agassii* (Cuvier and Valenciennes, 1846) in the streams system of Salar de Huasco (20 degrees 05' S, 68 degrees 15' W). Gayana 76:79–81.

Romanova, E.V., V.V. Aleoshin, R.M. Kamaltynov, K.V. Mikhailov, M.D. Logacheva, E.A. Sirotinina, A.Y. Gornov, A.S.Anikin, and D.Y. Sherbakov. 2016. Evolution of mitochondrial genomes in Baikalian amphipods. BMC Genomics 17(Suppl. 14):1016.

Roth-Woltereck, E. 1958. *Limnocaridina iridinae*, n. sp. Eine interessante Garnele aus dem Tangangyika-See (Decapoda, Atyidae). Zoologischer Anzeiger 161:188–192.

Schwentner, M., B.V.Timms, and S. Richter. 2012. Flying with the birds? Recent large-area dispersal of four Australian *Limnadopsis* species (Crustacea: Branchiopoda: Spinicaudata). Ecology and Evolution 2:1605–1626.

Seuront, L., and P.G. Strutton. 2004. Handbook of Scaling Methods in Aquatic Ecology. Measurements, Analysis, Simulation. CRC Press, Boca Raton, FL.

Scheihing, R., L. Cardenas, P. Labarca, and R.F. Nespolo. 2011. Spatial and temporal analysis of the genetic diversity and population structure of a freshwater ostracod from the high Andean plateau. Evolutionary Ecology Research 13:91–104.

Schluter, D. 2000. The Ecology of Adaptive Radiation. Oxford University Press, Oxford.

Scholtz, G., A. Braband, L. Tolley, A. Reimann, B. Mittmann, C. Lukhaup, F. Steuerwald, and G. Vogt. 2003. Parthenogenesis in an outsider crayfish. Nature 421:806.

Schön, I., and K. Martens. 2004. Adaptive, pre-adaptive and non-adaptive component of radiation in ancient lakes: a review. Organism Diversity and Evolution 4:137–156.

Schön, I., E. Verheyen, and K. Martens. 2000. Speciation in ancient lake ostracods: comparative analysis of Baikalian *Cytherissa* and Tanganyikan *Cyprideis*. Verhandlungen der Internationalen Vereinigung für Limnologie 27:2674–2681.

Schön, I., V. Pieri, D.Y. Sherbakov, and K. Martens. 2017. Cryptic diversity and speciation in endemic *Cytherissa* (Ostracoda, Crustacea) from Lake Baikal. Hydrobiologia 800:61–79.

Schubart, C.D., and P.K.L. Ng. 2008. A new molluscivore crab of Lake Poso confirms multiple colonization of ancient lakes in Sulawesi by freshwater crabs (Decapoda: Brachyura). Zoological Journal of the Linnean Society 154:211–221.

Soto, D., and P. De los Ríos. 2006. Influence of trophic status and conductivity on zooplankton composition in lakes and ponds of Torres del Paine National Park (Chile). Biologia Bratislava 61:541–546.

Strayer, D.L. 2010. Alien species in fresh waters: ecological effects, interactions with other stressors, and prospects for the future. Freshwater Biology 55:152–174.

Strecker, A.L. 2012. An overview of invasive freshwater cladocerans: *Bythostrephes longimanus* Leydig as a case study. Pages 149–156 *in* R.A. Francis, editor. A Handbook of Global Freshwater Invasive Species, Vol. 1: Introduced Freshwater Organisms. Earthscan, New York.

Streelman, J.T., and P.D. Danley. 2003. The stages of vertebrate evolutionary radiation. TREE 18:126–131.

Vaillant, J.J., D.G. Bock, G.D. Haffner, and M.E. Cristescu. 2013. Speciation patterns and processes in the zooplankton of the ancient lakes of Sulawesi island, Indonesia. Ecology and Evolution 3:3083–3094.

Väinölä, R., J.K. Vainio, S. Usjak, R. Alther, C. 'Fiišer, F. Altermatt, and J.D.S. Witt. 2017. Global phylogeography of Gammarus lacustris G.O. Sars, 1864, the most widespread freshwater amphipod spcies. Biodiversity Journal 8:561.

Vandekerkhove, J., K. Martens, G. Rossetti, F. Mesquita-Joanes, and T. Namiotko. 2013. Extreme tolerance to environmental stress of sexual and parthenogenetic resting eggs of Eucypris virens (Crustacea, Ostracoda). Freshwater Biology 58:237–247.

von Rintelen, K., and Y. Cai. 2009. Radiation of endemic species flocks in ancient lakes: systematic revision of the freshwater shrimp *Caridina* H. Milne Edwards, 1837 (Crustacea: Decapoda: Atyidae) from the ancient lakes of Sulawesi, Indonesia, with the description of eight new species. The Raffles Bulletin of Zoology 57:343–452.

von Rintelen, K., M. Glaubrecht, C.D. Schubart, A. Wessel, and T. von Rintelen. 2010. Adaptive radiation and ecological diversification of Sulawesi's ancient lake shrimps. Evolution 64:3287–3299.

von Rintelen, T., K. von Rintelen, M. Glaubrecht, C.D. Schubart, and F. Herder. 2012. Aquatic biodiversity hotspots in Wallacea: the species flocks in the ancient lakes of Sulawesi, Indonesia. Pages 290–315 *in* D.J. Gower, K.G. Johnson, J.E. Richardson, B.R. Rosen, L. Rüber, and S.T. Williams, editors. Biotic Evolution and Environmental Change in Southeast Asia. Cambridge University Press, Cambridge, UK.

von Rintelen, K., T. von Rintelen, M. Meixner, C. Lüter, Y. Cai, and M. Glaubrecht. 2007. Freshwater shrimp-sponge association from an ancient lake. Biology Letters 3:262–264.

Walker, P.D., I.J. Russon, R. Duijf, G. van der Velde, and S.E.Wendelaar Bonga. 2011. The off-host survival and viability of a native and non-native fish louse (*Argulus*, Crustacea: Branchiura). Current Zoology 57:828–835.

Wellborn, G.A., and R.D. Cothran. 2007. Niche diversity in crustacean cryptic species: complementarity in spatial distribution and predation risk. Oecologia 154:175–183.

Wellborn, G.A., J.D.S. Witt, and R.D. Cothran. 2015. Class Malacostraca, Superoders Peracarida and Syncarida. Pages 871–796 *in* J.H. Thorp, editor. Thorp and Covish's Freshwater Invertebrates, 4th ed. Academic Press, San Diego, CA.

Wetzer, R. 2015. Collecting and preserving marine and freshwater Isopoda (Crustacea: Peracarida). Biodiversity Data Journal 3:e4912.

Wilson, G.D.F. 2008. Global diversity of Isopod crustaceans (Amphipoda; Crustacea) in freshwater. Hydrobiologia 595:231–240.

Wilson, G.D.F., and R.T. Johnson. 1999. Ancient endemism among freshwater isopods (Crustacea, Phreatoicidea). Pages 264–268 *in* W. Ponder and D. Lunney, editors. The Other 99%. The Conservation and Biodiversity of Invertebrates. Royal Zoological Society of New South Wales, Mosman, Australia.

Wysocka, A., M. Grabowski, L. Sworobowicz, A. Burzynski, A. Kilikowska, G. Kostoski, and J. Sell. 2013. A tale of time and depth: instralacustrine radiation in endemic Gammarus species flock from the ancient Lake Ohrid. Zoological Journal of the Linnean Society 167:345–359.

Yeo, D.C.J., P.K.L. Ng, N. Cumberlidge, C. Magalhaes, S.R. Daniels, and M.R. Campos. 2008. Global diversity of crabs (Crustacea: Decapoda: Brachyura) in freshwater. Hydrobiologia 595:275–286.

12

COLONIZATION, ADAPTATION, RADIATION, AND DIVERSITY IN FRESH WATER

Shane T. Ahyong and Chao Huang

Abstract

About 12,000 of the 67,000 described species of crustaceans occur in fresh water. Crustaceans have colonized almost every type of freshwater environment in most parts of all continents. A common theme in marine-to-freshwater transitions is not only acquisition of osmoregulatory capabilities to cope with hyposalinity, but also optimizing reproductive strategies to cope with ecological and environmental variability. A key reproductive adaptation for fresh water is direct rather than extended planktonic development. Some groups, such as peracarids, were preadapted, already having direct development, whereas others, such as decapods, had to acquire it. Other crustaceans, such as branchiopods, are adapted not only to hyposalinity (and hypersalinity) but also to surviving in transient waters. Crustaceans have been colonizing fresh waters since the Middle Cambrian to Early Ordovician and have independently adapted to life in inland waters many times throughout geological history. The pattern and timing of invasions has shaped present-day distributions. Contemporary distributions and diversity of crustaceans in surface waters are surveyed in the context of their paleohistory. Different groups of crustaceans have very different current distributions that reflect the differential influence of different patterns of colonization, geological history, ecology, and the constraints (or benefits) of their evolutionary heritage.

INTRODUCTION

Crustaceans (excluding hexapods) are predominantly marine, but many groups have colonized fresh water. Today, of the some 67,000 described species of crustacean, about 12,000 occur in fresh water (Balian et al. 2008, Ahyong et al. 2011). This colonization has involved almost every

Evolution and Biogeography. Edited by Gary C. B. Poore and Martin Thiel.

type of freshwater environment, from temporary pools to permanent flowing waters and hot springs, in most parts of all continents. The distribution of freshwater crustaceans in terms of taxonomic and geographic spread, however, is not uniform in space or in time. Some groups flourish in habitats where others do not, and different groups have colonized fresh waters at different times in geological history. The present chapter examines the contemporary worldwide distribution of freshwater crustaceans in epigean (surface) waters and explores why they occur where they do.

CHALLENGES TO ENTERING FRESH WATER

Living in inland waters poses several significant challenges over living in the sea. The chief distinctions between marine and inland waters are in salinity and relative stability of environmental parameters. Crustaceans evolved in the ocean as osmoconformers, so marine species have little difficulty in maintaining their internal ion concentrations close to that of the surrounding seawater. Any transition to inland waters requires overcoming physiological challenges posed by the very low ionic concentrations of fresh water (hyposaline) (or sometimes higher than seawater salinities, termed hypersaline). Therefore, the ability to maintain stable internal ionic concentrations (osmoregulation) is essential to surviving in inland waters (Covich 2015, Lignot and Charmantier 2015). Freshwater crustaceans osmoregulate via physical barriers, such as a tough cuticle and impermeable external membranes, as well as active physiological mechanisms, such as ion pumps on the gills and epipods to retain salts, and active excretory organs to pump water out. Branchiopods, ostracods, copepods, and some malacostracans (syncarids, isopods, and tanaidaceans) have the excretory ducts on the maxillae, the maxillary glands. In decapods, amphipods, and mysidaceans, the excretory ducts are on the antenna, the antennal glands (Charmantier et al. 2009).

Marine habitats typically have relatively stable temperatures and water chemistry, particularly dissolved oxygen and pH, compared to inland waters. Even with the seasonal and cyclical variation, of annual, lunar, and daily tidal cycles, marine habitats generally have lesser extremes in environmental variables than in inland waters. Marine waters have significant thermal stability owing to the effectively "infinite" water mass. The considerably smaller water volume of inland waters, especially surface waters, is subject to much greater daily and annual variation in temperature and dissolved oxygen concentration, as well as being subject to the effects of drying or drought. Crustaceans in freshwater habitats, much like estuarine species, typically have acquired a higher tolerance to thermal stress and anoxia than their marine counterparts.

Other significant adaptations in many freshwater crustaceans are in the radical changes to their life history (Vogt 2013). This is especially pronounced in groups, such as copepods, ostracods, and branchiopods, which can produce resting eggs and enter diapause, or in decapods, which have replaced extended planktonic larval development and large clutch sizes with abbreviated or direct development with fewer but larger eggs and brooding of juveniles. A long planktonic larval phase is maladaptive in inland waters with unidirectional, often turbulent flow and insufficient long-term productivity to support developing larvae. As such, reduction or elimination of long planktonic larval phases is highly beneficial in inland waters. Jalihal et al. (1993) termed this tendency toward reduction or even elimination of the larval phase in adaptation to a limnic lifestyle as "freshwaterization."

Clearly, the environmental challenges posed by living in fresh water have entailed significant adaptations for survival that affect physiology, reproduction, and somatic morphology. Different crustacean groups have independently colonized fresh water over geological time and have handled that process in similar and different ways.

EPIGEAN HABITATS

Crustaceans occur in almost all inland waters, from high-altitude alpine areas to lowlands at or near sea level, in both standing (lentic) and flowing (lotic) waters. Lentic waters include ponds, swamps, bogs, fens, and lakes, as well as hypersaline lakes. The distinctions between lakes and ponds are subjective, but for present purposes, the most important distinction is that lakes are typically permanent (even if the water level fluctuates), whereas ponds are smaller and often temporary, requiring inhabitants to have life histories that accommodate the absence of water. Therefore, the length of time water is present (hydroperiod) substantially influences the types of crustaceans found. The temporary nature of ponds, especially in arid areas, requires special adaptations in their inhabitants. Swamps and bogs, sometimes associated with ponds and lakes, are dominated by aquatic vegetation and a high accumulation of organic content. Bogs develop in low-lying areas, usually at high latitudes, and receive water mostly from precipitation. Slow anaerobic decomposition in bogs forms peat, releases humic acid, and acidifies the water. As a result, species diversity in bogs may be relatively low, though the resident species may be highly specialized. In these low-pH habitats, Cladocera and copepods are the prevalent crustaceans and other groups are rare, particularly strongly calcified groups such as decapods (Batzer et al. 2016). Swamps are characterized by emergent aquatic vegetation ranging from grass-like forms to shrubs and trees. The standing water level varies seasonally, but swamps are not typically strongly acidified as in bogs and support higher species diversity. The stillness of lentic waters means that crustaceans living there can be structurally less robust than lotic forms but may require greater physiological tolerances. As with many aquatic inland species, precise distributions of species are also significantly affected by geochemistry that can drive fine, landscape-scale bioregionalization patterns (Rogers 2014). A common feature of moderately deep lentic habitats, owing to the lack of water movement, is stratification in temperature and oxygen concentration, both being highest at the water surface. Here, seasonal effects can be particularly pronounced; at high latitudes, for instance, the surface may be frozen in the winter and warm in the summer. Crustaceans living in such habitats must be able to navigate both extremes through adaptations, be they behavioral or physiological.

Lotic waters include creeks, streams, and rivers, some of which may be subterranean as in waters associated with karsts and other types of caves. Movement of lotic waters usually minimizes or even eliminates stratification so temperature and dissolved oxygen concentration is more constant at any given point in a stream or river. Like lakes and ponds, the differences between rivers, creeks, and streams are subjective but essentially relate to longevity and size. Creeks and streams starting at high elevations may expand in size to become rivers at lower elevations, ultimately draining into the sea. Environmental conditions differ considerably along the length of such water courses, from flow volume and velocity to substratum type and sediment load, along with water temperature, salinity, and biota. Aside from the physical impact of water movement on its inhabitants requiring more robust morphology than in lotic forms, crustaceans in streams and rivers frequently have special adaptations to function and reproduce under flow conditions.

MODES OF COLONIZATION

Crustaceans have colonized inland waters via two primary pathways, either separately or in combination: active migration and passive isolation (Notenboom 1991, Anger 2016). Active migration refers to incursions from marine to inland waters via intermediate habitats that have intermediate environmental conditions, such as brackish estuarine waters. Passive isolation occurs when marine populations become landlocked and isolated from the marine system, becoming gradually exposed to increasingly hyposaline conditions via terrestrial freshwater inputs.

Active Migration

The primary route of active migration has been hypothesized as direct upstream migration from the sea, through estuaries, and ultimately into the upper reaches of rivers, streams, and inland lakes, following gradual development of osmoregulatory capability and life history adaptations. Epigean tanaidaceans, cumaceans, and mysidaceans, already having direct development, are thought to have taken such a route (Jaume and Boxshall 2008, Porter et al. 2008). Among decapods, owing to the more extensive adaptations required for larval development for "freshwaterization," Anger (2016) considered ecologically transitional habitats such as brackish mangrove swamps, salt marshes, seasonally inundated wetlands, and anchialine caves to be more likely entrance portals to fresh water than direct invasion through estuaries into rivers. Other proposed routes of active migration are from the sea to land (such as shoreline habitats) to fresh water, or from estuary to fresh water to land. These latter two routes are exemplified in "terrestrialized" shore crabs such as Ocypodidae, Gecarcinidae, Sesarmidae, and Coenobitidae; all, however, remain reproductively tied to the sea. Although such routes of active migration have been demonstrated, they are not considered major pathways for radiation into limnic habitats (Schubart et al. 1998, Anger 2016).

Passive Isolation

As a result of glacial and interglacial cycles, tectonic subsidence, or other abiotic processes, sea level rises can result in marine transgressions and inundation of continental lowlands. When sea levels fall, marine populations can become landlocked and isolated (Notenboom 1991, Hoorn et al. 1995, Väinölä et al. 2008, Yang et al. 2013). These landlocked habitats could be in the form of coastal lakes, saltmarshes, or something similar. Over time, epigean or subterranean freshwater inputs create increasingly hyposaline conditions leading to gradual adaptive changes in the fauna (Notenboom 1991) and ultimately wider colonization of inland fresh waters. A similar process may also occur in subterranean habitats, with coastal marine interstitial faunas initially stranded inland following marine regressions and then acquiring adaptations to increasingly hyposaline conditions brought by groundwater. Calanoid copepods (Suárez-Morales 2003) and a number of stygobiotic hadziid amphipods (Notenboom 1991, Väinölä et al. 2008) and mysids (Porter et al. 2008) may have followed this route. The alternating marine transgressions and regressions over evolutionary time have been effective isolating mechanisms, leading to extensive speciation, as documented, for instance, in the Neotropics and East Asia (Valentí 2011, Yang et al. 2013).

INLAND CRUSTACEAN GROUPS

Most major crustacean groups occur in inland epigean waters in most biogeographic regions (Fig. 12.1): Branchiopoda, Ostracoda and Copepoda, Branchiura, and Malacostraca. Two euryhaline species of marine barnacle (*Amphibalanus amphitrite* and *A. improvisus*: Cirripedia) have also become established in fresh water via estuarine pathways and are not further treated here (Suárez-Morales 2015). Malacostraca are widely represented in inland waters and include the orders Decapoda, Syncarida, and most peracaridan orders (Isopoda, Amphipoda, Cumacea, Mysida, Spelaeogriphacea, Tanaidacea, and Thermobaenacea).

Branchiopoda

The class Branchiopoda, with orders Anostraca (fairy shrimp), Notostraca (tadpole shrimps), Laevicaudata (smooth clam shrimps), and Diplostraca (water fleas, spiny and tropical clam shrimps), can justifiably be considered (colloquially) as the "original" freshwater crustaceans. Despite their marine

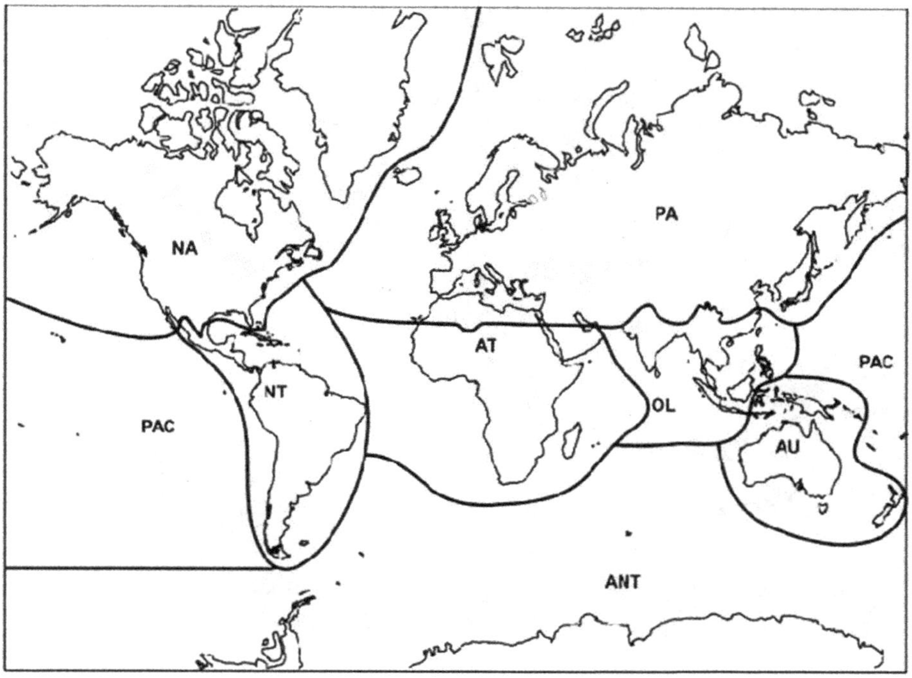

Fig. 12.1.
Major world biogeographic regions. Based on Balian et al. (2008, Fig. 1).

origins, branchiopods may have been the earliest crustaceans to colonize inland waters. Molecular divergence analyses estimate Middle Cambrian to Early Ordovician (478–512 Ma) origins of the freshwater branchiopod biota (Sun et al. 2016). Ostracods, although pre-dating Branchiopoda, appear not to have colonized fresh water until the Devonian, around 400–370 Ma (Martens et al. 2008). Either way, branchiopods have almost exclusively occupied fresh water for virtually their entire existence; the few oceanic genera are secondarily marine (Vernberg and Vernberg 1983). Branchiopods are usually informally divided into "small branchiopods" (containing Cladocera, a diverse group of about 1,000 species within Diplostraca), and "large branchiopods" (containing the remainder; about 600 described species).

Small branchiopods, Cladocera, almost all occur in continental fresh or saline waters; fewer than a dozen species occur in brackish or marine waters. They inhabit the benthic and pelagic parts of lakes, permanent and temporary standing waters on loose substratum or rock; they can form a significant component of the plankton of inland waters worldwide. They are virtually cosmopolitan at a continental scale, including Antarctica. Notwithstanding the need for further taxonomic scrutiny, especially of putatively widespread or cosmopolitan species, generic diversity is comparable across most biogeographic regions, but species level diversity is highest in the Palaearctic followed by the Nearctic and Neotropics. Notably, the Australasian region has high generic and species endemicity, compared to low Afrotropical generic and species endemicity (Forró et al. 2008). Broadly speaking, however, Cladocera exhibit a bipolar disjunct distribution, containing both wide-ranging and narrow-range endemic species or complexes, but with species richness concentrated in the warm temperate to subtropics between about 25° and 50° latitude of both hemispheres. This pattern has been attributed to the effects of climatic changes in the Tertiary leading to extinctions at tropical and boreal latitudes (Korovchinsky 2006).

Large branchiopods (Fig. 12.2A, B) primarily live in temporary standing waters such as playas, vernal pools, rock pools, alkali lakes, and saline lakes, especially in arid or semi-arid areas.

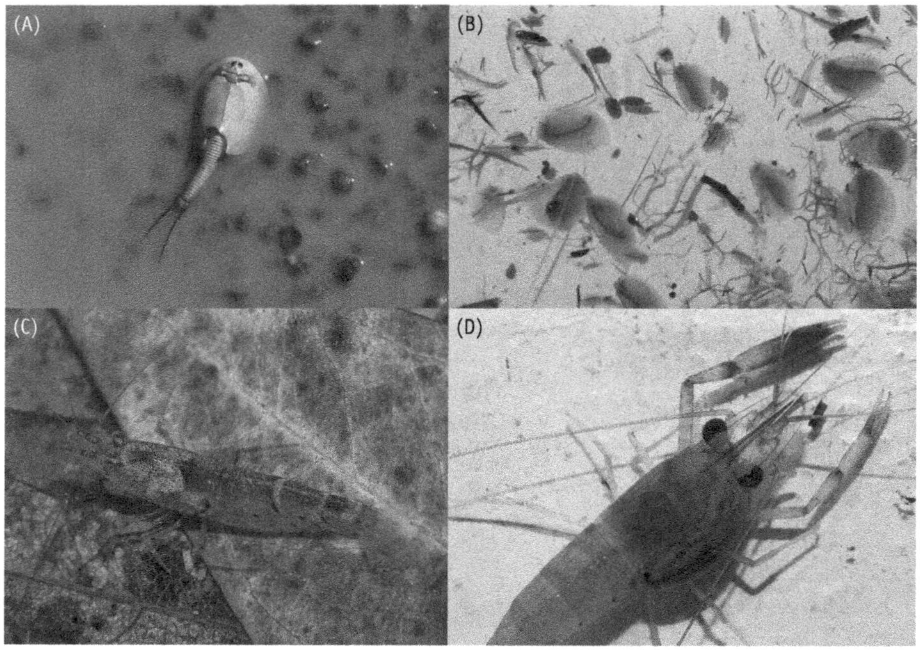

Fig. 12.2.
Selected freshwater crustaceans: (A) notostracan branchiopod, *Triops* sp., temporary pool, northwestern New South Wales, Australia; (B) mixed branchiopods from temporary pool, northwestern New South Wales, Australia; (C) atyid shrimp, *Caridina* sp., from stream in southern China. D, palaemonid shrimp (*Macrobrachium* sp.) with ectoparasitic isopod (*Tachaea* sp.), Pahang, peninsular Malaysia. (A, B, © M. Schwentner; C, © C. Huang; D, © M. Lo.) See color version of this figure in centerfold.

Such habitats are typically seasonal and may be dry for many months to years. Like cladocerans, large branchiopods occur worldwide, including the Antarctic Peninsula. Anostracans occur on all continents and major island systems except New Zealand, favoring mid- to high latitudes, with very few tropical species. Similarly, spinicaudatans favor temperate regions, and cyclestheridans occur in the tropics and subtropics. Notostracans and laevicaudatans are cosmopolitan (except Antarctica) (Cáceres and Rogers 2015). A bipolar distribution is also evident in large branchiopods, related to habitat type. Unlike cladocerans, maximum overall species richness and abundance of large branchiopods occurs at mid-latitudes in deserts and steppes where temporary surface water is most common. At a high taxonomic level, most groups are widely distributed, consistent with the antiquity of the clades, and the highest species richness is in the Palaearctic. However, some low-level taxa have more confined distributions, such as the putatively Laurasian genera *Cyzicus* and *Eocyzicus* occurring in Eurasia, Africa, and North America (Brendonck et al. 2008).

Adaptations variously present among branchiopods for surviving in temporary waters include short generation times with rapid growth, wide dietary range, high tolerance to salinity characteristic of inland pools and salt lakes, multiple modes of reproduction including sexual and asexual modes depending on specific environmental conditions, and drought or frost-resistant resting eggs ("cysts"). These diapausing eggs are usually highly resistant to desiccation, freezing, and anoxia, may remain viable for decades or possibly centuries, and are readily dispersed by wind or other physical processes. Moreover, the "egg banks" in the substratum only hatch differentially, extending the potential longevity of any given season's reproductive output (see Chapter 4 in Volume 5). A general adaptation to low oxygen concentration, which may be especially pronounced in shallow, warm, stagnant waters, is the facility for gas exchange across a significant proportion of the body surface, rather than only the gills (Cáceres and Rogers 2015).

Ostracoda

The 7,600 Recent species of Ostracoda fall into two major groups, Myodocopida and Podocopida, and the monotypic Paleocopida. Only Podocopida have colonized fresh water, with some 2,000 described species worldwide. They occur in all types of surface and subterranean waters including lakes, ponds, saline pools, and groundwater and special habitats such as phytotelmata and crayfish burrows. Some have even become semi-terrestrial.

Marine podocopidans from three main lineages (Darwinuloidea, Cytheridoidea, and Cypridoidea) have invaded fresh and other inland waters independently and at different times. Darwinuloidea, the oldest and smallest group in fresh water today, with about 30 species in a single family, entered non-marine habitats in the Devonian (400–370 Ma); the primarily marine Cytheridoidea entered non-marine waters at the end of the Permian (~250 Ma), with further incursions in the Neogene or Quaternary; and the Cypridoidea, containing the majority of non-marine species, probably entered fresh water by the Mid- to Late Jurassic (~175–150 Ma). Altogether, at least 9 to 12 independent incursions into non-marine habitats have occurred since the Paleozoic (Martens et al. 2008). Today, ostracods occur across all continents and in all aquatic habitats, both permanent and temporary.

The highest species and generic richness, concentrated within two cypridoid families (Cyprididae and Candonidae), is in the Palaearctic region, followed by the Afrotropical, Nearctic, and Neotropical regions. Average overall regional endemicity is high, around 60% and 90% at genus and species level, respectively, meaning very few species but many genera have intercontinental distributions. Notably, although the Australian region has lower absolute generic and species richness than other continents, it has the highest proportional endemicity (alongside the Afrotropical region) (Karanovic 2012). Within these numbers, however, some habitats stand out. Just as large branchiopod diversity peaks in transient water habitats (steppes and deserts), ancient lakes hold extensive ostracod species flocks; some 200 species of ostracod are present in Lake Baikal and 100 in Lake Tanganyika, for instance. Indeed, 20–25% of non-marine ostracod diversity is held by ancient lakes (Martens et al. 2008; see Chapter 11). The vernal rock-pool fauna of southwest Africa also exhibits high generic endemism, as will likely be confirmed for similar faunas in Australia when they are adequately studied. Notable, but as yet underexplored, are faunas of subterranean habitats in Australia, especially those presumably derived from strandings following the substantial marine regressions during the Mesozoic (Wilson 2008). Studies to date of calcretes on northwestern Australia have revealed numerous narrow range endemics of all types of invertebrates, including ostracods (Karanovic 2007, Humphreys 2008).

Non-marine ostracods present a range of adaptations to harsh environments and transient water supply. Chief among these is a range of reproductive strategies, which, much like the branchiopods, include both parthenogenetic and sexual reproduction (Innes and Dufresne 2020; see Chapter 9 in Volume 6), short generation times, resting eggs resistant to desiccation, and extended brood care. Other adaptations include high tolerance to changing salinity characteristic of temporary inland pools and wide dietary range allowing exploitation of varying food resources.

Copepoda

Around 3,000 of the over 14,000 species of copepod occur in fresh water, where they can be the most abundant invertebrates in both the benthos and plankton. They occur in almost every freshwater habitat from ancient lakes to thermal springs and saline lakes and, because of their typically tiny size, can exploit microhabitats unavailable to larger forms, such as surface films on wet mosses or damp leaf litter (Suárez-Morales 2015).

Copepods have marine origins dating to at least the Mid- to Late Cambrian. Copepods from five of the nine recognized orders have colonized fresh water (the free-living Calanoida, Cyclopoida, Harpacticoida, and Gelyelloida and the parasitic Siphonostomatoida). Within these orders, at least 22 independent colonization events have occurred (Boxshall and Jaume 2000). Most of the families with freshwater members also include marine and estuarine brackish water species, suggesting the likely pathway to colonization of fresh water. Most families are cosmopolitan or nearly so, though some distributional trends are evident.

Calanoids are primarily planktonic and usually found in open-water (limnetic) regions of lakes and large ponds. Diaptomidae, with about 440 freshwater species, are the dominant calanoid family in inland waters of North and South America, Eurasia, and Africa, i.e., all major biogeographic regions except the Australasian and Antarctic regions. The diaptomids initially colonized continental waters of Laurasia following the breakup of Gondwana. They are then believed to have colonized the African Plate after the separation of Gondwana and Laurasia in two successive colonizations in which the second largely displaced the first, starting in the north and diversifying southwards (Boxshall and Jaume 2000). Freshwater colonization of South America is thought to have occurred more recently, from North America following the formation of the Isthmus of Panama in the Pliocene (3 Ma). In the Australasian and Antarctic regions (primarily the Subantarctic Islands), the Centropagidae are the dominant calanoid group in fresh water; they are absent from the Afrotropics and believed to have colonized Gondwanan fresh waters following the separation of Africa (~95 Ma) (Boxshall and Jaume 2000, Boxshall and Defaye 2008).

The freshwater Cyclopoida are dominated by Cyclopidae with about 800 species, of which the majority of families and genera are widespread, if not cosmopolitan; they are thought to have colonized fresh waters in Pangaean time (late Paleozoic). Cyclopoids are primarily associated with substrata, especially the margins of standing water, and are seldom planktonic. Species of some genera, such as *Paracyclops, Ectocyclops*, and *Tropocyclops*, are semi-terrestrial.

The harpacticoids are the most diverse freshwater copepods with more than 1,120 freshwater species. Like cyclopoids, they are typically benthic, especially in bottom sediments and detritus, but may be associated with all types of substrata, including floating vegetation. The dominant families, the epigean Canthocamptidae and mainly hypogean Parastenocarididae and Ameiridae, account for about 90% of the freshwater species and are cosmopolitan at the genus level except for the apparent absence of parastenocaridids from New Zealand and Antarctica. Canthocamptids and parastenocaridids invaded fresh waters of Pangea, whereas ameirids appear to have invaded fresh water starting in Laurasian time (early Mesozoic). The fish parasitic Siphonostomatoida (from high Palaearctic and Nearctic latitudes) and Gelyelloida (from groundwater in northwestern Europe) probably entered fresh water in postglacial Pleistocene time.

At a continental scale, the Palaearctic region, with more than 1,200 freshwater species, holds the highest number of species and genera, with species richness similar to the combined totals from the Nearctic, Neotropical, and Afrotropical regions. The diversity of the Oriental and Australasian regions is comparatively low, with around 380 and 200 species, respectively. The substantial aridity of Australia may account for the low species numbers, though exploration of subterranean faunas, especially in Western Australia, will probably markedly increase taxonomic estimates (Karanovic 2004, Karanovic and Cooper 2012). To each of these regions, however, Cyclopidae contribute about 30% of species richness in varying combinations of genera (Suárez-Morales 2015). Regional species endemism is very high; more than 90% of the freshwater species occur in a single region. The Neotropical, Afrotropical, Palaearctic, and Australasian regions all have species endemicity exceeding 80%, being highest in the Neotropics. Not surprisingly, the Antarctic and Pacific Islands regions have both the lowest species richness and endemicity.

Freshwater copepods are generally morphologically similar to their marine counterparts, though interstitial hypogean forms often have varying degrees of reduction that assist mobility in narrow spaces. Semi-terrestrial forms have additional spinules and setae on the hind body and legs, which probably resist adhesion of detritus and assist mobility, presumably assisting individuals to climb into phytotelms (small pools of water held by some plants, such as tree holes and water trapped in the axils of bromeliads) (Dole-Olivier et al. 2000).

Physiological adaptations to unfavorable environmental conditions, such as fluctuating water levels, anoxia, and thermal extremes, include the ability to regulate their metabolic rate, tolerate low oxygen concentrations, and also enter diapause at egg or late developmental stages. Some planktonic species avoid anoxic stratification in the water column by undertaking vertical migrations (Williamson and Magnien 1982). Diapausing eggs, resistant to desiccation and temperature extremes, are formed by taxa occurring in arid or semi-arid regions where a short hydroperiod and inundation followed by desiccation is the norm. The ability to form drought-resistant eggs also allows their dispersal by the wind (Cáceres and Soluk 2002). Aside from parasitic forms, most freshwater copepods are generally herbivorous or omnivorous but may also be facultative carnivores (calanoids and cyclopoids) or detritivores (harpacticoids), allowing exploitation of a wide range of food sources (Suárez-Morales 2015).

Branchiura

As obligate ectoparasites, usually of fish, the ~160 species and 4 genera of branchiurans occur worldwide (except for Antarctica). It is not known how or when branchiurans entered fresh water, but the group is of considerable age given that its sister group, the endoparasitic Pentastomida, have a Cambrian-Ordovician fossil record (Walossek and Müller 1994). The majority of species belong to the cosmopolitan genus *Argulus*, with the highest diversity in the Afrotropical and Neotropical regions (Poly 2008, Neethling and Avenant-Oldewage 2016). The genus *Dolops*, being present in Africa, Australia (Tasmania), and South America, was speculated to be Gondwanan, but affinities among species of this genus are not congruent with the sequence of the Gondwanan breakup, suggesting dispersal as a better explanation of their distributions than vicariance (Møller 2009).

Morphological adaptations of branchiurans pertain to their specialized ectoparasitic habit: strong swimming ability and anterior appendages modified to form suckers and/or hooks, enabling them to rapidly find and attach to a host. Branchiurans are subject to the environment of their hosts, and some species can tolerate rapid salinity shifts.

Decapoda

Four major groups of decapods represented in fresh water—crabs (Brachyura), aeglids (Anomura), shrimps (Caridea), and crayfish (Astacidea)—are covered here. One species of hermit crab (Anomura) occurs in freshwater springs, *Clibanarius fonticola* from Vanuatu, but nothing is known about its development (McLaughlin and Murray 1990).

Caridea

The caridean shrimps have colonized almost all freshwater habitats ranging from creeks and lakes to caves and thermal springs. About 22% of the 3,500 caridean species occurs in fresh water. Freshwater carideans occur in all biogeographical regions except Antarctica but favor the tropics and subtropics. Favored habitats are permanent lotic waters of virtually any size. The center of diversity is in Southeast Asia; the highest species richness, by at least a factor of three, is in the Oriental Region with more than 350 species, followed by the Neotropical, Afrotropical, and Australasian regions.

Despite its large physical size, the Palaeartic freshwater caridean fauna is comparatively depauperate, with about 50 species (De Grave et al. 2008).

Atyidae (Fig. 12.2C) and Palaemonidae (Fig. 12.2D) are the dominant caridean families in fresh water, together accounting for 98% of species and 93% of genera. The majority of species and genera are restricted to a single biogeographic region. Several genera, such as the palaemonid *Macrobrachium* and the atyids *Caridina, Neocaridina,* and *Paratya* are variously represented in the Oriental, Afrotropical, Australasian, and Pacific regions. Of these, *Caridina* (>280 spp.) and *Macrobrachium* (>240 spp.) are numerically the most important genera.

At least two major independent invasions of fresh water have occurred, but the timing of invasions is not well established. Fossil Atyidae are known from Cretaceous freshwater facies, and origins estimated by molecular divergence range from the Permian to Early Cretaceous (von Rintelen et al. 2012). Palaemonids have probably invaded fresh water on several independent occasions, and the dominant freshwater palaemonid genus, *Macrobrachium*, is inferred to have originated in the Late Oligocene or Early Miocene (Murphy and Austin 2005, Anger 2013). Very minor invasions of fresh water have also occurred in the families Alpheidae (four species) and Xiphocarididae (two species).

Like other decapods, the major adaptations to fresh water in caridean shrimps are in reproductive strategies (but also in many cases, euryhalinity). Specific morphological adaptations (compared to marine forms) are few, although a number of atyids, e.g., *Australatya* and *Atya*, have robust, muscular walking legs that enable individuals to grip the stream substratum under high-flow conditions. Unlike primary freshwater crabs and crayfish, few freshwater shrimps have direct development, instead being abbreviated or sometimes even extended. Moreover, although many species, especially atyids, can complete their life cycle in fresh water, many others are amphidromous and therefore euryhaline, especially among *Macrobrachium*, with breeding and grow-out occurring in estuaries or the sea. Consequently, reproductive strategies among freshwater shrimps are much less uniform than in other freshwater decapods, even within genera, with many seemingly transitional species. For instance, the Oriental atyid genus *Neocaridina* includes species with both direct and indirect development, and the cosmopolitan *Macrobrachium* includes species in which adults live in fresh water but variously have direct, indirect, and even extended development (Anger 2013, Cumberlidge et al. 2015, Irving et al. 2017). As such, freshwater shrimps can be considered to be less "freshwaterized" than other inland decapods. Not surprisingly, freshwater shrimps require permanent water, and those that are amphidromous require a migration route to the sea. Long-distance dispersal of euryhaline shrimps is much more likely than for landlocked freshwater species, accounting for the many species and genera with transoceanic ranges, albeit usually still within biogeographic regions (Page et al. 2005).

Aeglid Crabs

The aeglid crabs or freshwater squat lobsters of the family Aeglidae are a small group of Neotropical decapods (one genus, 84 species) restricted to rivers, streams, lakes, and caves of southern South America: Chile, Brazil, Bolivia, Uruguay, and Argentina (Fig. 12.3A, Fig. 12.4A). Freshwater aeglids are thought to have diverged from their marine ancestors in the Late Cretaceous (about 75 Ma), and colonized fresh water from the western side of South America, probably Chile during the Tertiary, i.e., the "Pacific origin hypothesis" (Ortmann 1902, Pérez-Losada et al. 2004). Ancestral marine aeglids are thought to have gradually acquired freshwater adaptations through passive isolation of landlocked populations in the west (Bueno et al. 2016). Western aeglids significantly radiated around 40–45 Ma, dispersing eastwards with further radiations in central and eastern South America 23–35 Ma, entering all of the major southern river systems except the most southern Patagonian drainages (Bond-Buckup et al. 2008). Today, aeglids are distributed over a wide altitudinal (near sea level to

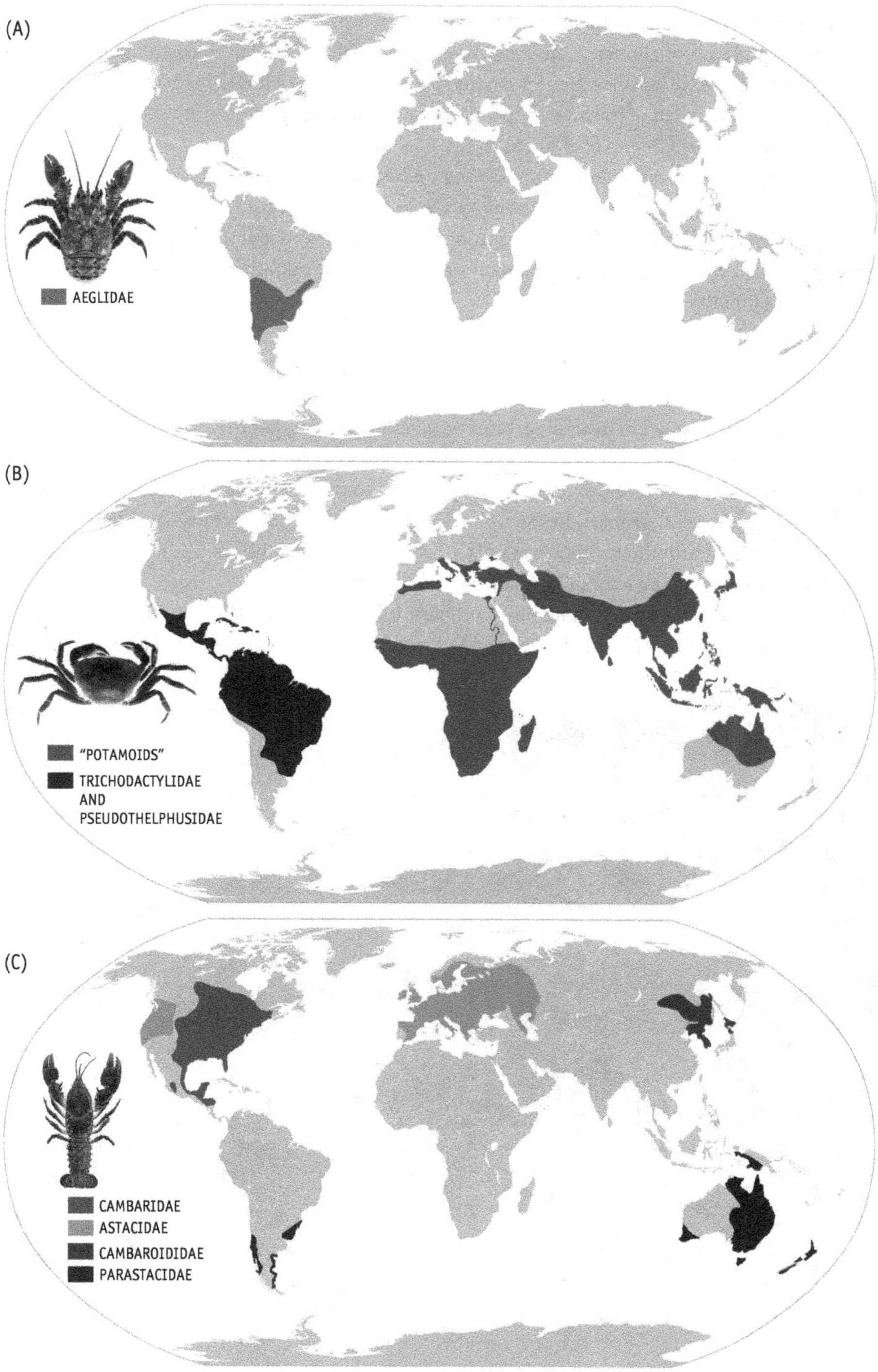

Fig. 12.3.
Global distribution of selected freshwater decapods: (A) Aeglidae; B) primary freshwater crabs; (C) freshwater crayfish. Based on Ortmann (1902), Klaus et al. (2011), Bueno et al. (2016), Kawai and Crandall (2016). See color version of this figure in centerfold.

3,613 m) and latitudinal range spanning subtropical to temperate zones (20–50° S). Species, however, show high levels of allopatric endemicity and are usually restricted to catchments within hydrographic basins. Also, species are not evenly distributed, but richness increases from north to south with more than half of the known species occurring in southern or southeastern Brazil (with about 90% endemism) (Bueno et al. 2016).

Aeglids typically live in flowing waters, often on bedrock substratum, and their compact, dorsoventrally flattened body may assist mobility around and under rocks in high-flow conditions. Although reproductive output is strongly contingent on climate and temperature, aeglids hatch their eggs during the dry season when flow conditions are mildest, optimizing recruitment conditions for juveniles (Bueno et al. 2016).

Brachyuran Crabs

The Brachyura (true crabs) in fresh waters can be divided into lineages that are wholly independent of the sea (primary freshwater crabs) and primarily marine lineages with euryhaline members that enter fresh water, but, with few exceptions, must return to the sea to breed (secondary freshwater crabs). We focus here on primary freshwater crabs. Primary freshwater crabs have direct development with maternal brooding, which not only greatly increases survival in fresh water but severely restricts dispersal, resulting in high levels of endemicity (Ng 1988).

With over 1,300 species in five families, primary freshwater crabs (Fig. 12.4B) comprise close to 20% of all brachyurans (Yeo et al. 2008) and occur circumtropically in the Palaearctic, Oriental,

Fig. 12.4.
Selected freshwater crustaceans: (A) *Aegla* sp., South America, aquarium trade; (B) potamid crab, *Yuebeipotamon calciatile*, hill stream in southern China; (C) parastacid crayfish, *Engaeus cymus*, near Wee Jasper, New South Wales, Australia; (D) parastacid crayfish, *Euastacus spinifer*, Wilson River, New South Wales, Australia. (A, © A. Anker; B, © C. Huang; C, © R. McCormack; D, © S. Ahyong.) See color version of this figure in centerfold.

Australasian, Afrotropical, and Neotropical regions (Fig. 12.3B). The freshwater crab fossil record dates from the Late Oligocene (less than 30 Ma), but molecular divergence estimates place the origin of freshwater crabs at around 135 Ma. The circumtropical distribution of freshwater crabs has been speculated to be possibly Gondwanan. Phylogenetic relationships of the group, however, do not correspond to the sequential fragmentation of Gondwana, making post-Gondwanan dispersal a better explanation of contemporary global distributions in fresh water (Klaus et al. 2011, Tsang et al. 2014). In addition, some primary freshwater crabs tolerate marine salinities for extended periods, so transoceanic dispersal remains a possible component in their historical expansion (Esser and Cumberlidge 2011, Klaus et al. 2013). As such, elucidating the continental scale colonization patterns of primary freshwater crabs requires fine-scale examination of regional distributions over time. Brachyurans have independently colonized fresh water at least twice: Trichodactylidae in the New World, in the common ancestor of the Old World families Potamidae, Gecarcinucidae, and Potamonautidae—the "potamoids", and possibly Pseudothelphusidae in the New World, should they prove only distantly related to the "potamoids" (Tsang et al. 2014).

Potamidae is the largest family with around 550 known species and two subfamilies, essentially split east–west: Potaminae (Palaearctic) and Potamiscinae (eastern Palaearctic and Oriental) (Yeo and Ng 2003). The differentiation of the two subfamilies is believed to be the result of vicariance from the collision of the Indian tectonic plate with the Asian continent around 40–50 Ma, which subsequently caused the rapid uplift of the Himalayan region. Potaminae extend over a vast but narrow longitudinal range from Morocco to Myanmar, with its range in the north constrained by cold climate and in the south and west by a combination of aridity and possible competition from gecarcinucoid crabs on the Indian subcontinent (Shih et al. 2009). The more easterly Potamiscinae extend from northern China to Southeast Asia, including the Sunda Shelf islands of Borneo, Sumatra, and Java as far east as Wallace's Line. Anomalously, the three species on the Afrotropical island of Socotra are potamiscines (Shih et al. 2009) and may have dispersed via the eastern Arabian Peninsula in earlier times (Apel and Brandis 2000).

Gecarcinucidae (with almost 400 species) occur throughout Southeast Asia, India, and northeastern Australia, encompassing the Oriental, Palaearctic, and Australasian regions. Australian gecarcinids probably dispersed from New Guinea across the Torres Strait during periods of low sea level during the Pleistocene.

Potamonautidae (about 150 species) dominate the Afrotropics but are also present at the southern extremes of the western Palaearctic; they are estimated to have entered inland Africa during the upper Cretaceous (78 Ma; Daniels et al. 2006), with subsequent radiation throughout most of sub-Saharan Africa.

The Neotropical Pseudothelphusidae (~270 species) and Trichodactylidae (~50 species), both thought to have originated in the Cretaceous, occur only in the warmer parts of Central and South America (Cumberlidge 2016). Pseudothelphusids typically live at altitudes above 300 m with two centers of radiation, Colombia and southern Mexico (Villalobos and Alvarez 2010, Magalhães et al. 2016). This pattern is believed to have been initiated by tectonic activity during the Late Mesozoic and Cenozoic, which divided the ancestral stock leading to subfamilial differentiation. The trichodactylids are generally lowland species with a more southerly distribution than pseudothelphusids, extending to near Buenos Aires, Argentina (Rodríguez 1986, Magalhães et al. 2016).

Primary freshwater crabs occur from near sea level to an altitude of almost 3,000 m (Dai 1999). Most are strongly aquatic and live in burrows in or near hillstreams, rivers, lakes, and ponds. Nevertheless, many species are amphibious, foraging on land, especially after heavy rain, enabling better retention of water in their branchial cavities. Further, some highly terrestrialized species have vascularized linings of the branchial chambers that increase aerial gas exchange (Cumberlidge 1991). Perhaps the most highly adapted amphibious freshwater crabs are some of the Australian species of

Austrothelphusa, especially *A. transversa*, which ventilate their branchial chambers by lateral oscillation of the thoracic walls, thereby actively breathing air (Greenaway et al. 1983). *Austrothelphusa transversa* ranges into the semi-arid zone and survives drought (sometimes for several years) by estivating in deep burrows sealed with clay (Waltham 2016).

Arboreal species, such as *Neotiwaripotamon whiteheadi, Globonautes macropus*, and *Kani maranjandu*, dwell in tree holes and also rely on collected rainwater. Troglobitic species usually have reduced eyes and more elongated legs than epigean forms, and others such as *Socotra* and *Tiwaripotamon*, live in karsts where no permanent surface water can be observed; these crabs are assumed to rely on rainwater collecting in crevices and holes in the eroded calcareous substratum (Cumberlidge and Wranik 2000). Despite being able to live on land, young freshwater crabs require an aquatic environment for molting, and it seems that even adults must be at least in contact with water to successfully molt (P. K. L. Ng, personal communication). Extended brood care is common; juveniles of different broods of the African *Potamonautes raybouldi* have been documented together with the female parent in the same water-filled tree hole. The female cleans out the tree holes and introduces empty mollusc shells that buffer against low pH levels (Bayliss 2002, Cumberlidge and Vannini 2004). In the New World, similar maternal brood care is practiced by the sesarmid crab, *Metopaulias depressus*, which lives in the water filled axils of Jamaican bromeliads (Diesel and Schuh 1993).

Most freshwater crabs are omnivorous scavengers and opportunistic predators, but some have very specialized feeding habits, such as *Sinopotamon cochlearidigitum*, with spoon-shaped chela tips that are presumably used for scraping algae from rock surfaces, and the molluscivorous *Syntripsa flavichela*, which possesses molariform teeth on the chelae for cracking snail shells (Dai 1999, Chia and Ng 2006).

Astacidea

Freshwater crayfish (Fig. 12.4C, D) are perhaps the most familiar of all freshwater decapods given their ubiquity, often large size, and culinary popularity. Morphologically, freshwater crayfish greatly resemble their marine relatives, nephropoid lobsters, though their chief adaptation to life in continental waters is direct development. Unlike the nephropoid lobsters, with multiple planktonic larval stages, freshwater crayfish hatchlings are fully developed as benthic juveniles. They occur almost worldwide with 669 species in 38 genera known, divided among 2 superfamilies and 4 families (Crandall and De Grave 2017). Body fossils of freshwater crayfish are known from the Mesozoic, but trace fossils and molecular divergence estimates indicate that they originated in the Triassic of Pangea. The origins of the two major freshwater crayfish lineages (Astacoidea and Parastacoidea) and their broad modern distributions are deeply rooted in the separation of Laurasia and Gondwana (183 Ma). Today, the highest diversity is present in the Nearctic and Australasian regions, containing the centers of diversity of the Astacoidea and Parastacoidea, respectively (Fig. 12.3C).

The Laurasian Astacoidea, comprising three families (Astacidea, Cambaridae, and Cambaroididae), are restricted to the Northern Hemisphere. Cambaroidids (one genus) are sister to the remaining astacoids and occur only in East Asia. Astacids (four genera) occur in Europe and North America (Pacific Northwest), diverging from the North American cambarids 90 Ma (Breinholt et al. 2009). The geographical center of astacoid diversity, and primary domain of Cambaridae, is in the southern Appalachian Mountains, southeastern United States. Cambarids further extend northward into southern Canada and southward through Mexico. Compared to the parastacids, much of the taxonomic diversity in the Cambaridae is of relatively recent origin, seemingly driven by the isolating effects of pre- and post-Pleistocene changes in river drainages (Crandall and Buhay 2008). As a result, significantly more taxonomic ambiguity is evident in the cambarid genera than among the much older, better-defined parastacid genera.

The Gondwanan Parastacoidea, with one family, Parastacidae, all occur in the Southern Hemisphere: Australia, Papua New Guinea, New Zealand, Madagascar, and South America. The interrelationships of the eastern Gondwanan freshwater crayfish (i.e., Australia, New Guinea, New Zealand, and Madagascar) have not been robustly resolved, but current studies indicate an east–west pattern of Gondwanan divergence, with radiation of the eastern Gondwanan crayfish occurring around 152 Ma (Toon et al. 2010). Whereas Madagascar, New Zealand, and Papua New Guinea include one genus each, and South America (Chile, Brazil, Uruguay) 3 genera, 10 genera and more than 138 species occur in Australia (Ahyong 2014).

Like the extant anaspidacean syncarids, parastacids predominantly occur in cool temperate streams and rivers but also in tropical or subtropical areas at elevations where water temperatures remain low. Thus, the modern distribution of Australian freshwater crayfish is largely relictual and reflects the increasing aridity of Australia since the Mid-Miocene (Martin 2006). The center of generic and species diversity is in the moist southeast of the continent, with some in the southwestern corner, and some in the wet tropics of the northeast. Crayfish are absent from the arid and semi-arid zone of central Australia, apart from *Cherax destructor*, which ranges across much of the eastern half of the continent following major inland river drainages. Most other parastacid species have limited, catchment-related ranges in coastal drainages. The Australian parastacids include the world's largest (*Astacopsis gouldi*) and near-smallest (*Tenuibranchiurus glypticus*) species of freshwater crayfish (Ahyong 2014).

Despite diverging in the Late Triassic (183 Ma), the astacoids and parastacoids parallel each other in habitat preferences and associated morphological adaptations. Freshwater crayfish segregate ecologically into largely aquatic and largely terrestrial forms. The largely aquatic forms live in creeks, streams, swamps, and rivers. They may forage on land, especially at night, but most time is spent in the water, whether out in the open or in the burrow; these are often termed "moderate" or "tertiary" burrowers. Morphologically, these crayfish best reflect the marine nephropoid phenotype in the elongated, muscular pleon and large chelipeds with palms held horizontally; examples include *Astacus* (Astacidae) and *Cherax* (Parastacidae). These crayfish are highly mobile and, when faced with unfavorable environmental conditions, such as poor water quality or low water levels, variously respond by burrowing more deeply, migrating up or downstream, or even leaving the water in search of other water bodies (for example, in the case of anoxia).

The semi-terrestrial crayfish, often termed "primary burrowers," are highly specialized for extensive burrowing. They burrow in marginal areas of streams, seepages, swamps, or other areas with either minimal, intermittent, or no surface water, though water is retained in their burrows. Particular burrowing adaptations include a compact, short, and slender pleon, chelipeds held directly forward with vertically opening dactyls, a shortened rostrum, small eyes, and body surface ornamentation (Ahyong 2014). Given that surface water may be absent from the local environs for extended parts of the year, primary burrowers can tolerate oxygen deficit, and some of the most "terrestrialized" species have a high, vaulted carapace with enlarged gill chambers and a much reduced pleon. Examples include *Fallicambarus* (Cambaridae) and *Engaeus* (Parastacidae); they seldom leave their burrows, and their presence is often indicated at the surface only by a mud mound or chimney at the burrow entrance. A related adaptation to variable water supply among burrowing crayfish is extended parental care, in which broods, often of more than one generation, may remain in the burrow with the parents (Richardson 2007, Palaoro et al. 2016).

The global distributions of freshwater crayfish can be largely attributed to vicariance, moderated by historical climatic and landscape changes (Koizumi et al. 2012, Lodge et al. 2012). The absence of native crayfish in India, Southeast Asia, and Africa, however, is conspicuous (Fig. 12.3C). These same areas are rich in primary freshwater crabs (Fig. 12.3B), and given that both groups are ecologically comparable, Ortmann (1902) hypothesized replacement of the crayfish by the more recently diverged freshwater crabs (Rodríguez 1986, Pretzmann 1987, Barbaresi and Gherardi 1997). Taking their lead from Ortmann (1902), several authors have widened consideration to the general potential of primary freshwater crabs as biotic barriers to crayfish. The lack of overlap between crayfish and primary freshwater crabs in Mexico, Costa Rica, Guatemala, and the Mediterranean region has been attributed to competitive exclusion (Holt 1964, Rodriguez 1986, Pretzmann 1987). Holt (1964) interpreted the presence of branchiobdellid annelids (typical epibionts on freshwater crayfish) on Costa Rican pseudothelphusid crabs as evidence of host-switching from cambarine crayfish, presumed to have been extirpated by the crabs during the Pleistocene. Behavioral experiments have shown the Mediterranean *Potamon fluviatile* to dominate and exclude the native crayfishes, *Austropotamobius*, from shared resources (Barbaresi and Gherardi 1997, Gherardi and Cioni 2004). According to Barbaresi and Gherardi (1997), both crab and crayfish formerly occupied the same range, with current regional distribution patterns of the crayfish affected more by biotic (competition) than environmental conditions. Such results corroborate the possibility of competitive exclusion of freshwater crayfish by primary freshwater crabs in India and Africa, though at present it remains to be demonstrated that freshwater crayfishes occurred in these regions in the past.

Syncarida

The syncarids comprise of two disparate groups, Anaspidacea and Bathynellacea. Although they were much more diverse in the Paleozoic, syncarids today have a minor presence in surface waters, with their main diversity in groundwater interstices. Bathynellaceans all occur interstitially, whereas anaspidaceans include both interstitial and epigean forms. Bathynellaceans lack a fossil record, but their global distribution in continental fresh water (and secondarily in anchialine habitats; Schminke 2014) indicates ancient derivation: Bathynellidae occur primarily in the Northern Hemisphere, Leptobathynellidae primarily in the Southern Hemisphere (especially South America), and Parabathynellidae are widespread in both hemispheres (Camacho and Valdecasas 2008). Given that bathynellaceans are hypogean, they are not further discussed here (see Chapter 13).

Anaspidaceans originated in the Carboniferous, had an extensive Late Paleozoic (260 Ma) Laurentian distribution in fresh or brackish water, and had been restricted to Gondwana since the Mesozoic, where they occupied vegetated freshwater habitats (Schram 1984). Today anaspidaceans are a small group of 5 families and 31 described species with a relictual distribution on Gondwanan terrains: southeastern Australia, New Zealand, and South America (Chile and Argentina). Like the Gondwanan parastacid crayfish, southeastern Australia is the center of diversity of Anaspidacea, with four of five families present—Anaspidesidae, Koonungidae, Psammaspididae, and Stygocarididae. Stygocaridids occur in Australia, New Zealand, and South America, suggesting an origin prior to the fragmentation of Gondwana (160–180 Ma), but Anaspidesidae, Koonungidae, and Psammaspididae are uniquely Australian and apparently evolved in situ; likewise, the fifth family, Patagonaspididae, is uniquely South American (Coineau and Camacho 2013).

Most Recent anaspidaceans are subterranean or occur in shallow hypogean sediments; epigean forms occur only in southeastern Australia (Tasmania and Victoria). The limited distribution of modern anaspidaceans, either in special habitats such as caves or groundwater or in cool temperate epigean habitats, is probably the result of predation pressure over evolutionary time and habitat contraction as a result of the aridification of Australia since the Tertiary. Low water temperature is critical to modern anaspidaceans, probably as a holdover from their Gondwanan past. Only Anaspidesidae and Koonungidae have epigean representatives. Anaspidesidae, endemic to Tasmania, occur in high altitude surface waters (or occasionally at low altitudes in caves), and usually only in habitats free of predatory fish. *Allanaspides* lives in pools and crayfish burrows in saturated button grass moors, as well as lakes. The largest anaspidaceans, *Anaspides* (to almost 60 mm body length; Fig. 12.5A), occur in most types of freshwater habitat (small creeks, runnels, tarns, lakes) but almost exclusively in the absence of native galaxiid fishes (Galaxiidae) or introduced

Fig. 12.5.
Selected freshwater crustaceans; (A) anaspididean syncarid, *Anaspides richardsoni*, Tarn Shelf, Mount Field, Tasmania; (B) gammarid amphipod, *Gammarus roseli*, Moravia, Czech Republic; (C) phreatoicidean isopod, Shannon Lagoon, Tasmania, Australia; (D) cave isopod, *Trogloniscus trilobatus*, Huanjiang County, Guangxi, China. (A, © S. Grove/ Tasmanian Museum and Art Gallery; B, © Michal Maňas (User:sneko1)—own work, CC BY 2.5; C, © S. Ahyong; D, © S. Huang and M. Tian https://commons.wikimedia.org/w/index.php?curid=1869909.) See color version of this figure in centerfold.

trout (Salmonidae) (Ahyong 2016). In rare cases where anaspidesids co-exist with fish, as in the lacustrine genus *Paranaspides* or *Anaspides spinulae* from Lake St. Clair (Tasmania), significant refugia are present in the form of dense, extensive vegetation and exfoliating rock sheets (O'Brien 1990, Ahyong et al. 2017). Koonungidae primarily occur in groundwater, especially in the hyporheic zone, though some species of *Koonunga* occur in small muddy creeks (without fish) among vegetation and detritus, but with strong connection to groundwater. Overall, anaspidaceans are ill-adapted to survive fast-moving predators, lacking a sustained caridoid escape reaction or even the well-calcified body armor of many other crustaceans. Syncarids may have escaped fish predation and environmental warming on mainland Australia, New Zealand, and South America by moving underground, as appears to be the case with Koonungidae in southeastern mainland Australia (Leijs et al. 2015). Similarly, Williams (1965) suggested that in Tasmania, *Anaspides* entered caves as a defensive response to climate warming in the Tertiary. Recent molecular divergence estimates, however, suggest that subterranean colonization by *Anaspides* occurred later, in the Pleistocene, probably as an expansion into new cave habitats created by the action of voluminous interglacial meltwaters on existing karst formations (Richter et al. 2018).

Amphipoda

The approximately 2,000 species of freshwater amphipods make up around 20% of all known amphipods, occurring in most epigean and hypogean habitats with more or less permanent water including lakes, streams, marshes, springs, groundwater, and karst. Epigean forms favor lotic waters, however, not tolerating low levels of dissolved oxygen. Most freshwater amphipods are epibenthic detritivores or scavengers and, where abundant, often play a major role in processing and breakdown of detritus and organic matter (Väinölä et al. 2008).

Little is known of the early diversification of amphipods and the subsequent colonization of fresh water. The fossil record is currently poor, none older than the Eocene. That amphipods invaded fresh water on multiple independent occasions, however, is undisputed. Global distribution patterns of amphipods suggest the group to be at least 250 million years old with colonization of fresh water in Pangaean time (Bousfield 1983). Bogidielloids, gammaroids, and crangonyctoids are believed to be among the earliest groups to enter fresh water (Lowry and Myers 2013). Väinölä et al. (2008) identified three broad distributional patterns of freshwater amphipods: (1) ancient continental distributions, reflecting former distributions on Pangaea, or later on Laurasia and Gondwana; (2) marine relicts, as exemplified by the Tethyan distribution of hadzioids, which appear to have colonized continental waters through cycles of marine transgressions and regressions; and (3) recent incursions into fresh water through estuaries or lagoons, with little diversification, as in corophioids and eusiroids.

Today, freshwater amphipods are most abundant in temperate waters but are rare or absent in the tropics. Whether this reflects physiological constraints or ecological factors, such as fish predation or competition from other invertebrates, is not known (Glazier 2009). The highest diversity is in the temperate Palaearctic Region (with 70% of all freshwater amphipod species), followed by the Nearctic (13%); here, the gammaroids and crangonyctoids are dominant (Väinölä et al. 2008). In the Southern Hemisphere, the Australasian and Neotropical regions hold 6% and 7% of the amphipod freshwater fauna, respectively. Notably, the Afrotropical region holds about half the diversity of the Australasian Region, and the Oriental region, less than half as much again.

The approximately 800 freshwater species of the superfamily Gammaroidea account for the bulk of the Palaeartic epigean fauna (Väinölä et al. 2008; Fig. 12.5B). In the Neotropics, the epigean amphipods are talitroids, Hyalellidae. The Crangonyctoidea, a mostly subterranean group, have a widespread distribution suggesting ancient continental ancestry in Pangaea, although family differentiation correlates regionally. Crangonyctidae and Niphargidae are the major freshwater crangonyctoid families in the Nearctic and Palaearctic, respectively. Crangonyctoids also dominate the Australasian fauna, but in the Gondwanan families Neoniphargidae (also South America) and Paramelitidae (also South Africa). The hypogean Bogidielloidea are distributed worldwide and are thought to have colonized fresh water in great antiquity.

The Hadzioidea and Corophioidea have also independently invaded fresh waters since the Mesozoic, sometimes as a result of marine transgressions and regressions (Bradbury and Eberhard 2000). Indeed, many of the groundwater amphipod groups are thought to have entered these habitats via such processes, though in subterranean substrata (Notenboom 1991). The presence of numerous hadzioids in anchialine caves may indicate that freshwater invasion is an ongoing process (Holsinger 1994).

Epigean freshwater amphipods are morphologically similar to their nearest marine relatives, usually more closely reflecting phylogenetic proximity than ecology. Hypogean forms, however, show typical adaptations as in eye reduction and, in the case of interstitial forms, adopt a slender habitus and appendages and enhanced sensory structures (Notenboom 1991). Abyssal forms in ancient lakes, such as in Lake Baikal, show structural modifications similar to deep-sea marine forms with pronounced spination and elongated pereopods (Takhteev 2000).

Cumacea

Like tanaidaceans, cumaceans have only a marginal presence in non-marine waters. Most of the 21 non-marine species of cumaceans occur in the Ponto-Caspian region, principally the Caspian

Sea, belonging to 8 genera of Pseudocumatidae. One species each from the Nannastacidae occurs in low salinity or fresh water in the northeastern United States and another in Argentina (Jaume and Boxshall 2008). The Caspian cumaceans are believed to be derived from estuarine forms that occurred in rivers that discharged into the proto-Mediterranean Sea prior to the closure of the brackish Sarmatian Basin lake, which covered the Ponto-Caspian region during the Miocene (14.5–8.3 Ma) (Dumont 2000, Jaume and Boxshall 2008). Presumably, their estuarine osmoregulatory adaptations paved the way for life in less saline waters.

Isopoda

The isopods are the most speciose of the peracarids (about 10,500 species) in all marine through to most terrestrial habitats worldwide. Some 950 described species live in continental waters (Wilson 2008). They occur widely in epigean lentic (e.g., Phreatoicidea; Fig. 12.5C) and lotic habitats (e.g., Asellidae) as well as caves, groundwater interstices, and even hot springs. Not surprisingly, isopods are among the most morphologically diverse groups of crustaceans. Phreatoicideans, although usually epigean, are typically benthic or infaunal, living among plant roots and organic detritus; and asellids are the dominant freshwater isopods in most epigean habitats in the Neartic and Neotropics; other groups, such as epicarideans and cymothoids are obligate parasites; cirolanids and sphaeromatids are free-living scavengers but in fresh water are largely hypogean.

Molecular divergence estimates suggest an Ordovician origin for the Isopoda and colonization of the deep sea as far back as the Early Permian (~290 Ma) (Lins et al. 2012). Entry into fresh water has occurred independently on multiple occasions in Isopoda. Based on their presumptive ages and the degree of adaptation to fresh water, Wilson (2008) divided the freshwater isopods into "primary" and "secondary" freshwater groups. "Primary freshwater" groups contain exclusively freshwater higher-level taxa. These include the entire order Phreatoicidea (Fig. 12.5C) and several asellotan families, principally Asellidae, Stenasellidae, and Protojaniridae, and are thought to have colonized fresh waters early in their phylogenetic history. The fossil record indicates that phreatoicideans are known to have colonized fresh water by at least the Triassic (Wilson and Edgecombe 2003), and today all known species occur in fresh water. The freshwater asellotan families do not have a fossil record, but their modern distributions provide clues about their evolutionary timing. Like phreatoicideans, Protojaniridae have a Gondwanan distribution, occurring in Sri Lanka, South Africa, Australia, and Chile (Wilson 2008). The largely Northern Hemisphere distribution of Asellidae and Stenasellidae is Laurasian, suggesting pre-Mesozoic origins; this is consistent with molecular divergence estimates indicating a Triassic origin for the common ancestor of the two families (Lins et al. 2012). The "primary freshwater" group may also include the Australian hypogean Tainisopidae whose higher-level taxonomic position is unresolved (Wilson 2008). Many questions remain unanswered about freshwater isopod distributions in the deep past; these require more paleontological data but also more detailed and better resolved phylogenies (Wilson 2008).

"Secondary freshwater" isopods are smaller freshwater-adapted groups within major marine lineages. They occur across the continents and include microcerberids and various "flabelliferan" isopods (chiefly cirolanids and sphaeromatids), many of which have penetrated fresh water via estuaries and are now largely or fully freshwater, including free-living scavengers (e.g., some Cirolanidae) and specialist parasites such as *Probopyrus* (Bopyridae, infecting decapod shrimps) and various Cymothoidae (infecting fish). Species of the corallanid genus *Tachaea* are ectoparasites on caridean shrimps (Fig. 12.2D). The Oniscidea are primarily terrestrial forms, though fewer than 10 species and genera, probably secondarily aquatic, occur in fresh water worldwide (Wilson 2008; Fig. 12.5D).

The differing ages of the radiations of freshwater isopods are reflected in their modern distributions. The phreatoicideans are a Gondwanan group, now occurring in southern Africa,

India, New Zealand, and Australia. The fossil record indicates that Paleozoic and Mesozoic phreatoicideans occupied vegetated freshwater lacustrine environments, much like their modern descendants in lakes, bogs, swamps, and occasionally karsts, in cool-temperate rainforest areas today. In Australia, as with other Gondwanan groups, the increasing aridity of the continent owing to the northward movement of the Australian plate since the Mesozoic has largely restricted the range of phreatoicideans to the wetter parts of the cool-temperate southeast. Although phreatoicideans already live in moist or wet environments, in areas of more transient surface water, such as bogs and swamps, they escape desiccation by burrowing.

Mysida

Of the approximately 1,100 species of mysids, about 70 species live in inland waters and, like their marine relatives, are epibenthic or pelagic. Unlike isopods and amphipods, mysids are relatively minor components of fresh waters worldwide, in terms of diversity, though they can be ecologically significant where abundant. Mysids have independently colonized inland waters on multiple occasions. Porter et al. (2008) identified four main patterns: (1) Tethyan subterranean relicts, as stygofauna, in India, the Mediterranean, and the Caribbean area, as a result of passive isolation after strandings following marine regressions in the Miocene resulting in endemic species in widespread genera; (2) Pleistocene glacial relicts, probably also by passive isolation following independent strandings on multiple continents dating to 3–7 Ma; (3) autochthonous Ponto-Caspian endemics, which, like the cumaceans from the same area, evolved in the inland seas of the Ponto-Caspian basin that have become increasingly hyposaline over the last 20 million years; and (4) euryhaline estuarine fauna, representing recent or incipient incursions into fresh water through estuaries. Today, the highest diversity is in the Palaeartic Region (centered in the Ponto-Caspian Basin) and the Neotropics (comprising subterranean Tethyan relicts in the Caribbean).

Tanaidacea

Tanaidaceans are an almost exclusively marine group. Although they date to at least the Lower Carboniferous, tanaidaceans have not successfully colonized non-marine waters to the same extent as other peracarid groups. Fewer than 20 species occur in brackish estuarine waters and only five species have been recorded from fresh water or landlocked brackish water worldwide. The three species of *Pseudohalmyrapseudes,* all occurring in relatively close proximity in the southwest Pacific (Sulawesi, northern Australia, and the Bismarck Archipelago), have been recorded from a karst stream, spring, and brackish lake, respectively (Bamber 2008). A deep-occurring species has been reported from Lake Baikal, questionably identified as *Parakanthophoreus longiremus* (a deep-sea trench species) (Jaume and Boxshall 2008). *Sinelobus stanfordi* has putatively extensive circumtropical and temperate distribution that includes New Zealand, Florida, Argentina, and Brazil in marine through freshwater habitats. Records of *S. stanfordi* worldwide may refer to more than one species, but the possibility of translocations by human or aquatic birds, as is possibly the case with the freshwater population in Lake Rotorua, New Zealand, cannot be excluded (Chapman et al. 2011).

Spelaeogriphacea

Extant spelaeogriphaceans have a relictual, largely Gondwanan distribution (Australia, Brazil, South Africa), occurring in ancient cratons that have not been affected by marine transgressions since the Early Cretaceous, around 140 Ma. This suggests that spelaeogriphaceans had colonized continental waters prior to the beginning of the breakup of Gondwana and that modern distributions are the result of vicariance. Colonization of cavernicoulous and groundwater habitats is hypothesized to have taken place via an intermediate lacustrine step at some time between the Carboniferous and Late Jurassic (Shen et al. 1998, Poore 2015). Suffice it to say, physiological and morphological adaptation to fresh water in spelaeogriphaceans commenced no later than the Mesozoic. Modern spelaeogriphaceans are colorless, blind, and minute, in keeping with their dark subterranean habitat; marked morphological differences from known fossils are not otherwise evident.

Thermosbaenacea

The modern distribution of thermosbaenaceans (3 families, 5 genera, at least 19 species), taking in the Caribbean and Mediterranean seas, northwestern Australia, and Southeast Asia, displays a relictual Tethyan pattern, suggesting a former wide range in the Tethys Sea during the Cretaceous. Thermosbaenaceans are believed to have colonized groundwater as a result of stranding following marine regressions during the Mesozoic (Jaume 2008). Today, thermosbaenaceans are restricted to groundwater and anchialine caves, and their distributions are usually attributed to plate tectonics (Poore and Humphreys 1992). Thermosbaenacea currently lacks a fossil record, but molecular divergence estimates suggest, at least for the genus *Halosbaena*, that although deep generic divergences are ancient, the species are too young for their distributions to be explained by Tethyan vicariance alone (Page et al. 2018). The single known species of the Thermosbaenidae is notable for its high thermal tolerance, occurring in a thermo-mineral spring in an area of Tunisia that, until the Late Pliocene, was inundated by the sea (Wagner 1994). As such, Thermosbaenidae might be comparatively young.

FUTURE DIRECTIONS

Understanding the basis for contemporary distributions of freshwater crustaceans requires detailed knowledge not only of life histories and current ecology but also of phylogenetic history and paleodistributions. Unfortunately, the current fossil record for most freshwater crustaceans, especially in the deep past, is sparse and likely to remain a relatively small proportion of available freshwater fossil sites (relative to marine facies) (Ng et al. 1995). Little detailed paleohistorical information is available for most groups of freshwater crustaceans, and future research efforts should include extensive phylogenetic emphasis on identifying the nearest marine relatives of freshwater lineages and timing and rates of diversification among inland lineages. These data will also form important underpinnings for process models explaining modern distributions, such as a recent proposal for anostracan branchiopods (Rogers 2015). At a more fundamental level, the modern freshwater crustacean fauna remains highly underdocumented, especially in interstitial and subterranean habitats, so significantly more taxonomic effort is required in order to properly apprehend global freshwater biodiversity. This is especially important given current rates of habitat modification and worldwide translocations of species that frequently become invasive and can have radical effects on native species.

SUMMARY AND CONCLUSIONS

Although proportionally few species occur in inland versus marine waters, crustaceans are ubiquitous globally. A common theme in marine to freshwater transitions is not only acquisition of osmoregulatory capabilities but also optimizing reproductive strategies to cope with ecological and environmental variability, "freshwaterization." A key reproductive adaptation for fresh water is direct rather than extended planktonic development. Some groups, such as peracarids, were already "blessed" with direct development and, in this respect, were preadapted for freshwater living, whereas others, such as decapods, had to acquire it. Crustaceans have been colonizing fresh waters since the Middle Cambrian to Early Ordovician and have independently adapted, or become "freshwaterized," many times.

The branchiopods have the distinction of being the first crustaceans to invade fresh water, in which they now almost exclusively occur; those now in the sea are secondarily marine. Like the copepods and ostracods, freshwater branchiopods occur in all major biogeographic regions including Antarctica. More so than any other group, branchiopods can be considered the crustaceans best adapted to surviving in transient surface waters (both fresh and sometimes hypersaline) through very short generation times and resting eggs resistant to heat and drought. In many ways, rather than being adapted for fresh water per se, they are perhaps better thought of as adapted for life in transient waters. Small branchiopod (Cladocera) diversity peaks in the anti-equatorial tropics and subtropics, whereas large branchiopods are most diverse at higher latitudes on deserts and steppes in temporary surface pools. Branchiopods first invaded the fresh waters of Pangea in the Paleozoic and with the subsequent fragmentation of the supercontinent were carried worldwide. Similar processes also underlie the largely cosmopolitan freshwater distributions of ostracods and copepods, which first colonized inland waters in Devonian through Carboniferous time, followed by subsequent independent invasions from within. Today, both groups are most diverse in the Palearctic region, with around 90% species-level endemism, though with proportionally the highest levels in Australia. The tiny size of copepods and ostracods has allowed them to exploit almost every habitat with available water, including damp forest litter; their wide range of reproductive strategies includes parthenogenesis, diapause, and production of resting eggs when faced with unfavorable environmental conditions.

The malacostracans probably first invaded fresh water in the Carboniferous through the syncarids. Although much more diverse in the Paleocene, syncarids are now largely confined to groundwater worldwide. Epigean syncarids are now rare and relictual, occurring exclusively in southeastern Australia. Consistent with their former Gondwanan distribution, anaspidaceans require cool temperate waters, and their current distributions reflect contraction of habitat and perhaps the influence of predatory fish over evolutionary time.

Peracarid malacostracans, already having direct development, were preadapted to life in inland waters. The most successful groups in fresh water are Amphipoda and Isopoda, which independently invaded fresh water multiple times, probably initially in the Triassic. Both groups now occur in all types of epigean waters and hypogean habitats worldwide and include successful terrestrial lineages. Other peracaridans have a much more limited presence in fresh water: tanaidaceans, cumaceans, and mysidans with a number of epigean representatives that probably colonized fresh waters via active penetration of estuaries, largely during the Cenozoic.

The remaining malacostracans, Decapoda, have more constrained large-scale freshwater distributions, reflecting more recent inland invasions. The majority of freshwater decapods are epigean, occurring in lotic and, less commonly, lentic waters. Although many are troglobitic, they are excluded from interstitial groundwater by their comparatively large body size. Key adaptations independently derived from among freshwater decapod orders are abbreviated or direct modes of development. The disjunct worldwide distribution of freshwater crayfish reflects ancient vicariance, being deeply split phylogenetically and biogeographically between the Northern (Laurasian) and Southern Hemisphere (Gondwanan) faunas. Freshwater crayfish favor temperate waters, with centers of distribution in North America and southeastern Australia. The freshwater crabs and shrimps are of more recent origin than the crayfish, having invaded fresh water on multiple occasions since the Cretaceous; they are primarily tropical animals. Freshwater shrimps and primary freshwater crabs are most diverse in the Oriental, Afrotropical, and Neotropical regions. Notably, freshwater crayfish are entirely absent from Africa, India, and Southeast Asia and may have been competitively excluded by freshwater crabs. The unique freshwater squat lobsters, Aeglidae, diverged from their marine ancestors about 75 Ma and now occur exclusively to southern South America.

Although the timing, mode of entry, and details of freshwater occupation differ widely between groups, several common solutions have been independently evolved to overcome the challenges of moving from the relative stability of marine environments to the much more variable and transient inland waters. Chief among these adaptations are reproductive strategies that reduce larval phasing and shorten generation time, with development abbreviated or direct, or sometimes even punctuated through diapause and the production of resting eggs; and physiological and behavioral adaptations to cope with hyposalinity, thermal extremes, and oxygen deficit. Each of these adaptations has been acquired over evolutionary time under different environmental circumstances. Most of the major crustacean lineages first colonized fresh waters during the Paleozoic, with multiple independent invasions of fresh water ever since. Thus, different groups of freshwater crustaceans have very different current distributions that do not follow any single pattern. Rather, they reflect the differential influence of different patterns of colonization, geological history, ecology, and the constraints (or benefits) of their evolutionary heritage.

REFERENCES

Ahyong, S.T. 2014. Diversity and distribution of Australian freshwater crayfish with a check-list of the world Parastacidae and key to the genera (Decapoda: Astacidea: Parastacoidea). Pages 245–271 *in* D.C.J. Yeo, N. Cumberlidge, and S. Klaus, editors. Advances in Freshwater Decapod Crustacean Systematics and Biology: Crustaceana Monographs 19. Brill, Leiden, The Netherlands.

Ahyong, S.T. 2016. The Tasmanian Mountain Shrimp, *Anaspides* Thomson, 1894 (Crustacea: Syncarida: Anaspididae). Records of the Australian Museum 68:313–364.

Ahyong, S.T., J.K. Lowry, M. Alonso, R.N. Bamber, G.A. Boxshall, P. Castro, S. Gerken, G.S. Karaman, J.W. Goy, D.S. Jones, K. Meland, D.C. Rogers, and J. Svavarsson. 2011. Subphylum Crustacea Brünnich, 1772. Zootaxa 3148:165–191.

Ahyong, S.T., M. Schwentner, and S. Richter. 2017. The Tasmanian Lake Shrimps, *Paranaspides* Smith, 1908 (Crustacea, Syncarida, Anaspidesidae). Records of the Australian Museum 69:259–275.

Alvarez, F., and J.L. Villalobos. 2016. Freshwater Decapod Diversity and Conservation in Mexico. Pages 237–266 *in* T. Kawai and N. Cumberlidge, editors. A Global Overview of the Conservation of Freshwater Decapod Crustaceans. Springer International, Cham, Switzerland.

Anger, K. 2013. Neotropical *Macrobrachium* (Caridea: Palaemonidae): on the biology, origin, and radiation of freshwater-invading shrimp. Journal of Crustacean Biology 33:151–183.

Anger, K. 2016. Adaptation to life in fresh water by decapod Crustaceans: evolutionary challenges in the early life-history stages. Pages 127–168 *in* T. Kawai and N. Cumberlidge, editors. A Global Overview of the Conservation of Freshwater Decapod Crustaceans. Springer International, Cham, Switzerland.

Apel, M., and D. Brandis. 2000. A new species of freshwater crab (Crustacea: Brachyura: Potamidae) from Socotra Island and description of *Socotrapotamon* n. gen. Fauna Arabia 18:133–144.

Balian, E. V., H. Segers, C. Lévêque, and K. Martens. 2008. The Freshwater Animal Diversity Assessment: an overview of the results. Hydrobiologia 595:627–637.

Bamber, R.N. 2008. A new species of the freshwater tanaidacean genus *Pseudohalmyrapseudes* (Crustacea: Tanaidacea: Parapseudidae) from Sulawesi. Records of the Western Australian Museum 24:421–428.

Barbaresi, S., and F. Gherardi. 1997. Italian freshwater decapods: exclusion between the crayfish *Austropotamobius pallipes* (Faxon) and the crab *Potamon fluviatile* (Herbst). Bulletin française de la Pêche et de la Piscicolture 347:731–747.

Batzer, D., H. Wu, T. Wheeler, and S. Eggert. 2016. Peatland invertebrates. Pages 219–250 *in* D. Batzer and D. Boix, editors. Invertebrates in Freshwater Wetlands. Springer International, Cham, Switzerland.

Bayliss, J. 2002. The East Usambara tree—hole crab (Brachyura: Potamoidea: Potamonautidae)—a striking example of crustacean adaptation in closed canopy forest, Tanzania. African Journal of Ecology 40:26–34.

Bond-Buckup, G., C.G. Jara, M. Perez-Losada, L. Buckup, and K.A. Crandall. 2008. Global diversity of crabs (Aeglidae: Anomura: Decapoda) in freshwater. Hydrobiologia 595:267–273.

Bousfield, E. L. 1983. An updated phyletic classification and palaeohistory of the Amphipoda. Pages 257–277 *in* F. R. Schram, editor. Crustacean Phylogeny: Crustacean Issues 1. A.A. Balkema, Rotterdam, The Netherlands.

Boxshall, G. A., and D. Defaye. 2008. Global diversity of copepods (Crustacea: Copepoda) in freshwater. Hydrobiologia 595:195–207.

Boxshall, G.A., and D. Jaume. 2000. Making waves: the repeated colonization of fresh water by copepod crustaceans. Advances in Ecological Research 31:61–79.

Bradbury, J.H., and S. Eberhard. 2000. A new stygobiont melitid amphipod from the Nullarbor Plain. Records of the Western Australian Museum 20:39–50.

Breinholt, J., M. Pérez-Losada, and K.A. Crandall. 2009. The timing of the diversification of the freshwater crayfishes Pages 343–355 *in* J. W. Martin, K. A. Crandall, and D. L. Felder, editors. Crustacean Issues Vol. 18: Advances in Decapod Phylogenetics. CRC Press, Boca Raton, FL.

Brendonck, L., D.C. Rogers, J. Olesen, S. Weeks, and W.R. Hoeh. 2008. Global diversity of large branchiopods (Crustacea: Branchiopoda) in freshwater. Hydrobiologia 595:167–176.

Bueno, S. L. S., R. M. Shimizu, and J. C. B. Moraes. 2016. A remarkable anomuran: the taxon *Aegla* Leach, 1820. Taxonomic remarks, distribution, biology, diversity and conservation. Pages 23–64 *in* T. Kawai and N. Cumberlidge, editors. A Global Overview of the Conservation of Freshwater Decapod Crustaceans. Springer International, Cham, Switzerland.

Cáceres, C.E., and D.C. Rogers. 2015. Class Branchiopoda. Pages 687–708 *in* J. H. Thorp and D. C. Roger, editors. Thorp and Covich's Freshwater Invertebrates, Vol. 1: Ecology and General Biology. Academic Press, San Diego, CA.

Cáceres, C.E., and D.A. Soluk. 2002. Blowing in the wind: a field test of overland dispersal and colonization by aquatic invertebrates. Oecologia 131:402–408.

Camacho, A.I., and A.G. Valdecasas. 2008. Global diversity of syncarids (Syncarida; Crustacea) in freshwater. Hydrobiologia 595:257–266.

Chapman, M. A., M. H. Lewis, and M. J. Winterbourn. 2011. Guide to the Freshwater Crustacea of New Zealand. New Zealand Freshwater Sciences Society, Christchurch, New Zealand.

Charmantier, G., M. Charmantier-Daures, and D. Towle. 2009. Osmotic and ionic regulation in aquatic arthropods. Pages 165–230 *in* D. H. Evans, editor. Cells and Animals. CRC Press, Boca Raton, FL.

Chia, O.K.S., and P.K.L. Ng. 2006. The freshwater crabs of Sulawesi, with description of two new genera and four new species (Crustacea: Decapoda: Brachyura: Parathelphusidae). Raffles Bulletin of Zoology 52:381–428.

Coineau, N., and A. I. Camacho. 2013. Superorder Syncarida Packard, 1885. Pages 357–449 *in* F. R. Schram and J. C. Von Vaupel Klein, editors. Treatise on Zoology—Anatomy, Taxonomy, Biology. The Crustacea. Brill, Leiden, The Netherlands.

Crandall, K.A., and J.E. Buhay. 2008. Global diversity of crayfish (Astacidae, Cambaridae, and Parastacidae––Decapoda) in freshwater. Hydrobiologia 595:295–301.

Crandall, K.A., and S. De Grave. 2017. An updated classification of the freshwater crayfishes (Decapoda: Astacidea) of the world, with a complete species list. Journal of Crustacean Biology 37:615–653.

Covich, A. P. 2015. Freshwater crustaceans: adaptations to complex inland habitats and species interactions. Pages 337–378 *in* M. Thiel and L. Watling, editors. The Natural History of the Crustacea, Vol. 2: Lifestyles and Feeding Biology. Oxford University Press, New York.

Cumberlidge, N. 1991. The respiratory system of *Globonautes macropus* (Rathbun, 1898), a terrestrial freshwater crab from Liberia (Gecarcinucoidea, Gecarcinucidae). Crustaceana 61:69–80.

Cumberlidge, N. 2016. Global diversity and conservation of freshwater crabs (Crustacea: Decapoda: Brachyura). Pages 1–22 *in* T. Kawai and N. Cumberlidge, editors. A Global Overview of the Conservation of Freshwater Decapod Crustaceans. Springer International, Cham, Switzerland.

Cumberlidge, N., and M. Vannini. 2004. Ecology and taxonomy of a tree-living freshwater crab (Brachyura: Potamoidea: Potamonautidae) from Kenya and Tanzania, East Africa. Journal of Natural History 38:681–693.

Cumberlidge, N., and W. Wranik. 2000. A new genus and species of freshwater crab (Potamoidea, Potamidae) from Socotra Island, Yemen. Journal of Natural History 36:51–64.

Cumberlidge, N., H.H. Hobbs, Jr., and D.M. Lodge. 2015. Class Malacostraca, Order Decapoda. Pages 797–847 *in* J. H. Thorp and D. C. Roger, editors. Thorp and Covich's Freshwater Invertebrates—Vol. 1. Ecology and General Biology. Academic Press, San Diego, CA.

Dai, A.Y. 1999. Fauna Sinica: Arthropoda Crustacea Malacostraca Decapoda Parathelphusidae Potamidae. Science Press, Beijing, China.

Daniels, S. R., N. Cumberlidge, M. Pérez-Losada, S. A. E. Marijnissen, and K. A. Crandall. 2006. Evolution of Afrotropical freshwater crab lineages obscured by morphological convergence. Molecular Phylogenetics and Evolution 40:227–235.

De Grave, S., Y. Cai, and A. Anker. 2008. Global diversity of shrimps (Crustacea: Decapoda: Caridea) in freshwater. Hydrobiologia 595:287–293.

Dole-Olivier, M. J., D. M. P. Galassi, P. Marmonier, and M. Creuzé Des Châtelliers. 2000. The biology and ecology of lotic microcrustaceans. Freshwater Biology 44:63–91.

Dumont, H.J. 2000. Endemism in the Ponto-Caspian fauna, with special emphasis on the Onychopoda (Crustacea). Advances in Ecological Research 31:181–196.

Diesel, R., and M. Schuh. 1993. Maternal care in the Bromeliad Crab *Metopaulias depressus* (Decapoda): maintaining oxygen, pH and calcium levels optimal for the larvae. Behavioral Ecology and Sociobiology 32:11–15.

Esser, L.J., and N. Cumberlidge. 2011. Evidence that salt water may not be a barrier to the dispersal of Asian freshwater crabs (Decapoda: Brachyura: Gecarcinucidae and Potamidae). Raffles Bulletin of Zoology 59:259–268.

Forró, L., N. M. Korovchinsky, A. A. Kotov, and A. Petrusek. 2008. Global diversity of cladocerans (Cladocera; Crustacea) in freshwater. Hydrobiologia 585:177–184.

Gherardi, F., and A. Cioni. 2004. Agonism and interference competition in freshwater decapods. Behaviour 141:1297–1324.

Glazier, D. S. 2009. Amphipoda. Pages 89–115 *in* G. F. Likens, editor. Encyclopedia of Inland Waters. Academic Press, San Diego, CA.

Greenaway, P., H.H. Taylor, and J. Bonaventura. 1983. Aerial gas exchange in Australian freshwater/land crabs of the genus *Holthuisana*. Journal of Experimental Biology 103:237–251.

Holsinger, J.R., 1994. Pattern and process in the biogeography of subterranean amphipods. Hydrobiologia 287:131–145.

Holt, P.C. 1964. A new branchiobdellid (Annelida) from Costa Rica. Tulane Studies in Zoology 12:1–4.

Hoorn, C., J. Guerrero, G.A. Sarmiento, and M.A. Lorente. 1995. Andean tectonics as a cause for changing drainage patterns in Miocene northern South America. Geology 23:237–240.

Humphreys, W. F. 2008. Rising from down under: developments in subterranean biodiversity in Australia from a groundwater perspective. Invertebrate Systematics 22:85–101.

Innes. D., and F. Dufresne. 2020. Parthenogenesis. In R. Cothran and M. Thiel, editors. The Natural History of the Crustacea: Reproductive Biology. Oxford University Press, New York.

Irving, R.A., T.P. Dawson, and D. Wowor. 2017. An amphidromic prawn, *Macrobrachium latimanus* (von Martens, 1868) (Decapoda: Palaemonidae), discovered on Pitcairn, a remote island in the southeastern Pacific Ocean. Journal of Crustacean Biology 37:503–506.

Jalihal, D. R., K.N. Sankolli, and S. Shenoy. 1993. Evolution of larval developmental patterns and the process of freshwaterization in the prawn genus *Macrobrachium* Bate, 1868 (Decapoda, Palaemonidae). Crustaceana 65:365–376.

Jaume, D. 2008. Global diversity of spelaeogriphaceans & thermosbaenaceans (Crustacea; Spelaeogriphacea & Thermosbaenacea) in freshwater. Hydrobiologia 595:219–224.

Jaume, D., and G.A. Boxshall. 2008. Global diversity of cumaceans and tanaidaceans (Crustacea: Cumacea and Tanaidacea) in freshwater. Hydrobiologia 595:225–230.

Karanovic, I. 2007. Candoninae Ostracodes from the Pilbara Region in Western Australia. Crustaceana Monographs 7:1–432.

Karanovic, I. 2012. Recent Freshwater Ostracods of the World. Crustacea, Ostracoda, Podocopida. Springer, Heidelberg, Germany.

Karanovic, T., 2004. Subterranean Copepoda from Arid Western Australia. Brill, Leiden, The Netherlands.

Karanovic, T., and S. J. B. Cooper. 2012. Explosive radiation of the genus *Schizopera* on a small subterranean island in Western Australia (Copepoda: Harpacticoida): unravelling the cases of cryptic speciation, size differentiation and multiple invasions. Invertebrate Systematics 26:115–192.

Kawai, T., and K. Crandall. 2016. Global diversity and conservation of freshwater crayfish (Crustacea: Decapoda: Astacoidea). Pages 65–114 *in* T. Kawai and N. Cumberlidge, editors. A Global Overview of the Conservation of Freshwater Decapod Crustaceans. Springer International, Cham, Switzerland.

Klaus, S., D.C.J. Yeo, and S.T. Ahyong. 2011. Freshwater crab origins—laying Gondwana to rest. Zoologischer Anzeiger 250:449–456.

Klaus, S., S. Selvandran, J.W. Goh, D. Wowor, D. Brandis, P. Koller, C.D. Schubart, B. Streit, R. Meier, P.K.L. Ng., and D.C.J. Yeo. 2013. Out of Borneo: Neogene diversification of Sundaic freshwater crabs (Crustacea: Brachyura: Gecacinucidae: *Parathelphusa*). Journal of Biogeography 40:63–74.

Koizumi, I., N. Usio, T. Kawai, N. Azuma, and R. Masuda. 2012. Loss of genetic diversity means loss of geological information: the endangered japanese crayfish exhibits remarkable historical footprints. PLoS ONE 7:e33986.

Korovchinsky, N.M. 2006. The Cladocera (Crustacea: Branchiopoda) as a relict group. Zoological Journal of the Linnean Society 147:109–124.

Leijs, R., T. Bradford, J. G. Mitchell, W. F. Humphreys, S. J. B. Cooper, P. Goonan, and R. A. King. 2015. The evolution of epigean and stygobitic species of *Koonunga* Sayce, 1907 (Syncarida: Anaspidacea) in southern Australia, with the description of three new Species. PLoS ONE 10:e0134673.

Lignot, J.H., and G. Charmantier 2015. Osmoregulation and excretion. Pages 249–284 *in* E. Chang and M. Thiel, editors. The Natural History of the Crustacea, Vol. 4: Physiology. Oxford University Press, New York.

Lins, L. S. F., S. Y. W. Ho, G. D. F. Wilson, and N. Lo. 2012. Evidence for Permo-Triassic colonization of the deep sea by isopods. Biology Letters 8:979–982.

Lodge, D.M., A. Deines, F. Gherardi, D.C.J. Yeo, T. Arcella, A.K. Baldridge, M.A. Barnes, W.L. Chadderton, J.L. Feder, C.A. Gantz, G.W. Howard, C.L. Jerde, B.W. Peters, J.A. Peters, L.W. Sargent, C.R. Turner, M.E. Wittmann, and Y. Zeng. 2012. Global introductions of crayfishes: evaluating the impact of species invasions on ecosystem services. Annual Review of Ecology, Evolution, and Systematics 43:449–472.

Lowry, J.K., and A. A. Myers. 2013. A phylogeny and classification of the Senticaudata subord. nov. (Crustacea: Amphipoda). Zootaxa 3610:1–80.

Magalhães, C., M. R. Campos, P. A. Collins, and F. L. Mantelatto. 2016. Diversity, distribution and conservation of freshwater crabs and shrimps in South America. Pages 303–322 *in* T. Kawai and N. Cumberlidge, editors. A Global Overview of the Conservation of Freshwater Decapod Crustaceans. Springer International, Cham, Switzerland.

Martens, K., I. Schön, C. Meisch, and D.J. Horne. 2008. Global diversity of ostracods (Ostracoda, Crustacea) in freshwater. Hydrobiologia 595:185–193.

Martin, H.A. 2006. Cenozoic climatic change and the development of the arid vegetation in Australia. Journal of Arid Environments 66:533–563.

McLaughlin, P. A., and T. Murray. 1990. *Clibanarius fonticola*, new species (Anomura: Paguridea: Diogenidae), from a fresh-water pool on Espiritu Santo, Vanuatu. Journal of Crustacean Biology 10:695–702.

Møller, O. S. 2009. Branchiura (Crustacea)—survey of historical literature and taxonomy. Arthropod Systematics and Phylogeny 67:41–55.

Murphy, N.P., and C.M. Austin. 2005. Phylogenetic relationships of the globally distributed freshwater prawn genus *Macrobrachium* (Crustacea: Decapoda: Palaemonidae): biogeography, taxonomy and the convergent evolution of abbreviated larval development. Zoologica Scripta 34:187–197.

Neethling, L. A. M., and A. Avenant-Oldewage. 2016. Branchiura— a compendium of the geographical distribution and a summary of their biology. Crustaceana 89:1243–1446.

Ng, P.K.L. 1988. The freshwater crabs of Peninsular Malaysia and Singapore. National University of Singapore. Shinglee Press, Singapore.

Ng, P.K.L., Z. Števčić, and G. Pretzmann. 1995. A revision of the family Deckeniidae, Ortmann 1897 (Crustacea, Decapoda, Brachyura, Potamoidea) from the Seychelles, Indian Ocean. Journal of Natural History 29:581–600.

Notenboom, J. 1991. Marine regressions and the evolution of groundwater dwelling amphipods (Crustacea). Journal of Biogeography 18:437–454.

O'Brien, D. P. 1990. The conservation status of the mountain shrimp (*Anaspides tasmaniae* and *Anaspides spinulae*): a report on its distribution, ecology and taxonomy, including recommendations for management. Department of Parks, Wildlife and Heritage, Tasmania.

Ortmann, A.E. 1902. The geographical distribution of freshwater decapods and its bearing upon ancient geography. Proceedings of the American Philosophical Society 41:267–400.

Page T.J., A.M. Baker, B.D. Cook, and J.M. Hughes. 2005. Historical transoceanic dispersal of a freshwater shrimp: the colonization of the South Pacific by the genus *Paratya* (Atyidae). Journal of Biogeography 32:581–593.

Page, T.J., J.M. Hughes, K.M. Real, M.I. Stevens, R.A. King, and W.F. Humphreys. 2018. Allegory of a cave crustacean: systematic and biogeographic reality of *Halosbaena* (Peracarida: Thermosbaenacea) sought with molecular data at multiple scales. Marine Biodiversity 48:1185–1202.

Palaoro, A.V., E. Del Valle, and M. Thiel. 2016. Life history patterns are correlated with predictable fluctuations in highly seasonal environments of semi-terrestrial burrowing crayfish. Hydrobiologia 767:51–63.

Pérez-Losada, M., G. Bond-Buckup, C.G. Jara, and K.A. Crandall. 2004. Molecular systematics and biogeography of the southern South American freshwater "crabs" *Aegla* (Decapoda: Anomura: Aeglidae) using multiple heuristic tree search approaches. Systematic Biology 53:767–780.

Poly, W.J. 2008. Global diversity of fishlice (Crustacea: Branchiura: Argulidae) in freshwater. Hydrobiologia 595:209–212.

Poore, G.C.B. 2015. Orders Bochusacea, Mictacea and Spelaeogriphacea. Pages 77–92 *in* J. C. Von Vaupel Klein, M. Charmantier-Daures, and F. R. Schram, editors. Treatise on Zoology—Anatomy, Taxonomy, Biology. The Crustacea. Brill, Leiden, The Netherlands.

Poore, G.C.B., and W.F. Humphreys .1992. First record of Thermosbaenacea (Crustacea) from the Southern Hemisphere: a new species from a cave in tropical Western Australia. Invertebrate Taxonomy 6:719–725.

Porter, M.L., K. Meland, and W. Price. 2008. Global diversity of mysids (Crustacea-Mysida) in freshwater. Hydrobiologia 595:213–218.

Pretzmann, G. 1987. A contribution to a historic analysis of Mediterranean freshwater decapods chorology. Investigación Pesquera 51:17–25.

Richter, S., M. Schwentner, C.S. Wirkner, and S.T. Ahyong. 2018. Phylogeny and diversity of Tasmanian mountain shrimps (Crustacea, Anaspidesidae)—independent immigration from surface to subterranean habitats. Zoologica Scripta 47:84–105.

Richardson, A.M. 2007. Behavioral ecology of semiterrestrial crayfish. Pages 319–338 *in* J.E. Duffy and M. Thiel, editors. Evolutionary Ecology of Social and Sexual Systems: Crustaceans as Model Organisms. Oxford University Press, New York.

Rodríguez, G. 1986. Centers of radiation of freshwater crabs in the Neotropics. Pages 51–67 *in* R.H. Gore and K.L. Heck, editors. Biogeography of the Crustacea, Crustacean Issues 3. A. A. Balkema, Rotterdam, The Netherlands.

Rogers, D.C. 2014. Anostracan (Crustacea: Branchiopoda) zoogeography II: relating distribution to geochemical substrate properties in the USA. Zootaxa 3856:1–49.

Rogers, D.C. 2015. A conceptual model for anostracan biogeography. Journal of Crustacean Biology 35:686–699.

Schminke, H.K. 2014. Freshwater origin of Bathynellacea (Malacostraca). Crustaceana 87:1225–1242.

Schram, F.R. 1984. Fossil Syncarida. Transactions of the San Diego Society of Natural History 20:189–246.

Schubart, C. D., R. Diesel, and S. B. Hedges. 1998. Rapid evolution to terrestrial life in Jamaican crabs. Nature 393:363–365.

Shen, Y.B., R. S. Taylor, and F. R. Schram. 1998. A new spelaeogriphacean (Crustacea: Peracarida) from the Upper Jurassic of China. Contributions to Zoology 68:19–35.

Shih, H.T., D. C. J. Yeo, and P. K. L. Ng. 2009. The collision of the Indian plate with Asia: molecular evidence for its impact on the phylogeny of freshwater crabs (Brachyura: Potamidae). Journal of Biogeography 36:703–719.

Suárez-Morales, E. 2003. Historical biogeography and distribution of the freshwater calanoid copepods (Crustacea: Copepoda) of the Yucatan Peninsula, Mexico. Journal of Biogeography 30:1851–1859.

Suárez-Morales, E. 2015. Class Maxillopoda. Pages 709–755 *in* J. H. Thorp and D. C. Roger, editors. Thorp and Covich's Freshwater Invertebrates, Vol. 1: Ecology and General Biology. Academic Press, San Diego, CA.

Sun, X.Y., X. Xia, and Q. Yang. 2016. Dating the origin of the major lineages of Branchiopoda. Palaeoworld 25:303–317.

Takhteev, V.V. 2000. Trends in the evolution of Baikal amphipods and evolutionary parallels with some marine malacostracan faunas. Advances in Ecological Research 31:197–220.

Toon, A., M. Pérez-Losada, C.E. Schweitzer, R.M. Feldmann, M. Carlson, and K.A. Crandall. 2010. Gondwanan radiation of the Southern Hemisphere crayfishes (Decapoda: Parastacidae): evidence from fossils and molecules. Journal of Biogeography 37:2275–2290.

Tsang, L.M., C.D. Schubart, S.T. Ahyong, J.C.Y. Lai, E.Y.C. Au, T.Y. Chan, P.K.L. Ng, and K.H. Chu. 2014. Evolutionary history of true crabs (Crustacea: Decapoda: Brachyura) and the origins of freshwater crabs. Molecular Biology and Evolution 31:1173–1187.

Väinölä, R., J.D.S. Witt, M. Grabowski, J.H. Bradbury, K. Jazdzewski, and B. Sket. 2008. Global diversity of amphipods (Amphipoda; Crustacea) in freshwater. Hydrobiologia 595:241–255.

Valentí, R. 2011. Neotropical biodiversity: timing and potential drivers. Trends in Ecology and Evolution 26:508–513.

Varpe, Ø., and M.J. Ejsmond, 2018. Semelparity and iteroparity. Pages 97–124 in G.A. Wellborn and M. Thiel, editors. The natural history of the Crustacea: life histories. Oxford University Press, New York.

Vernberg, W.B., and F.J. Vernberg. 1983. Freshwater adaptations. Pages 109–177 *in* W.B. Vernberg and F.J. Vernberg, editors. The Biology of Crustacea, Vol. 8: Environmental adaptations. Academic Press, New York.

Villalobos, J. L., and F. Alvarez. 2010. Phylogenetic analysis of the Mexican freshwater crabs of the tribe Pseudothelphusini (Decapoda: Brachyura: Pseudothelphusidae). Zoological Journal of the Linnean Society 160:457–481.

Vogt, G. 2013. Abbreviation of larval development and extension of brood care as key features of the evolution of freshwater Decapoda. Biological Reviews 88:81–116.

von Rintelen, K., T.J. Page, X. Cai, K. Roe, B. Stelbrink, B.R. Kuhadja, T.M. Iliffe, J. Hughes, and T. von Rintelen. 2012. Drawn to the dark side: a molecular phylogeny of freshwater shrimps (Crustacea: Decapoda: Caridea: Atyidae) reveals frequent cave invasions and challenges current taxonomic hypotheses. Molecular Phylogenetics and Evolution 63:82–96.

Wagner, H.P. 1994. A monographic review of the Thermosbaenacea (Crustacea: Peracarida). Zoologische Verhandelingen 291:1–338.

Walossek, D., and K.J. Müller. 1994. Pentastomid parasites from the Lower Paleozoic of Sweden. Transactions of the Royal Society of Edinburgh: Earth Sciences 85:1–37.

Waltham, N.J. 2016. Unravelling life history of the inland freshwater crab *Austrothelphusa transversa* in seasonal tropical river catchments. Australian Zoologist 38:217–222.

Williams, W.D. 1965. Ecological notes on Tasmanian Syncarida (Crustacea: Malacostraca), with a description of a new species of *Anaspides*. Internationale Revue der Gesamten Hydrobiologie 50:95–126.

Williamson, C. E., and R. E. Magnien. 1982. Diel vertical migration in *Mesocyclops edax*: implications for predation rate estimates. Journal of Plankton Research 4:329–339.

Wilson, G.D.F. 2008. Global diversity of Isopod crustaceans (Crustacea; Isopoda) in freshwater. Hydrobiologia 595:231–240.

Wilson, G.D F., and G.D. Edgecombe. 2003. The Triassic isopod *Protamphisopus wianamattensis* (Chilton) and comparison with extant taxa (Crustacea, Phreatoicidea). Journal of Paleontology 77:454–470.

Yang, L., Z. Hou, and S. Li. 2013. Marine incursion into East Asia: a forgotten driving force of biodiversity. Proceedings of the Royal Society of London B 280:20122892.

Yeo, D.C.J., and P.K.L. Ng. 2003. Recognition of two subfamilies in the Potamidae Ortmann, 1896 (Brachyura, Potamidae) with a note on the genus Potamon Savigny, 1816. Crustaceana 76:1219–1235.

Yeo, D.C.J., P.K.L. Ng, N. Cumberlidge, C. Magalhães, S.R. Daniels, and M.R. Campos. 2008. Global diversity of crabs (Crustacea: Decapoda: Brachyura) in freshwater. Hydrobiologia 595:275–286.

13

SUBTERRANEAN AND ANCHIALINE WATERS

Renée E. Bishop, William Humphreys, and Damià Jaume

Abstract

Crustaceans have successfully colonized the subterranean habitat, and many have become obligate inhabitants, occurring virtually everywhere there are interconnected voids underground. With the exception of most oniscidean isopods and a few talitrid amphipods, subterranean crustaceans inhabit water (stygobionts), where they dominate the stygofauna both in biomass and diversity of species. Four major taxonomic groups predominate: amphipods, isopods, copepods, and ostracods. Although most higher crustacean taxa have representatives in both epigean and subterranean habitats, some groups such as remipedes, thermosbaenaceans, spelaeogriphaceans, bathynellaceans, mystacocarids, and gelyelloid copepods are known only from the subterranean environment. Subterranean habitats vary physically and range from organically rich shallow habitats around seeps to cave systems more than a kilometer deep. Water quality, which can range from fresh to marine to hypersaline, static to flowing, and oxic to suboxic, impacts species distribution. Dispersal patterns in subterranean crustaceans are also diverse. Freshwater stygobiont crustaceans have narrow endemic ranges, and their dispersal is limited by saltwater. The distribution of several freshwater taxa might reflect the movement of tectonic plates. The extraordinarily diverse anchialine fauna, initially distributed along the Tethyan coast, was likely spread by vicariance due to movement of tectonic plates. Originating from epigean ancestors, many stygobionts have a marine origin. While the existence of preadaptations does not necessarily guarantee successful colonization of the subterranean habitat, a suite of characteristics is frequently observed in subterranean crustaceans, with most being weakly chitinized, lacking or with reduced eyes and pigments, and enhanced non-optic sense organs. Metabolic rates tend to be lower than in epigean crustaceans. Limited evidence indicates subterranean crustaceans are longer lived with lower reproductive potential. These adaptations make subterranean crustacean populations particularly vulnerable to anthropogenic impacts. The morphological, physiological, and life history adaptations to a subterranean existence are most likely common responses to the physical

Evolution and Biogeography. Edited by Gary C. B. Poore and Martin Thiel.

environment of each subterranean ecosystem. Extensive biodiversity and phylogeography studies are still required, and there is a pressing need to comprehend the functional role of stygofauna in subterranean waters.

INTRODUCTION

Some members of most higher crustacean taxa have adapted to living in subterranean habitats to the extent that they have become obligate inhabitants of the subterranean realm. They are called stygobionts if they inhabit the water-filled voids and troglobionts if they inhabit the air-filled voids. Crustaceans predominate over other taxa in subterranean waters in terms of both taxonomic diversity and biomass, the latter often an order of magnitude greater than for non-crustaceans (Sket 2012). Various crustacean taxa have different propensities to adapt to subterranean environments. Three classes have no recorded stygobiont species (Cephalocarida, Branchiura, and Cirripedia), two have only stygobiont species (Remipedia and the marine interstitial Mystacocarida), and five, including the three most diverse classes, contain both stygobiont and epigean members (Branchiopoda, Malacostraca, Copepoda, Tantulocarida, and Ostracoda). Within each class, some taxonomic groups are especially adept at invading and establishing lineages within subterranean aquatic habitats, but a few higher taxa are known only as stygobionts (e.g., Spelaeogriphacea, Thermosbaenacea, Bathynellacea, and gelyelloid copepods).

Subterranean crustaceans live in a wide variety of conditions both in terms of habitat matrix and water quality. They occupy inland and coastal caves, including anchialine systems, shallow and deep phreatic networks, and infiltration zones, which include interstitial and hyporheic substrata and epikarst. They inhabit water ranging from fresh (flowing or static) to hypersaline. Because subterranean crustaceans comprise many higher taxa and occupy diverse habitats on land and in water, it is beyond our scope to comprehensively review all subterranean crustaceans, and only representative examples are given. Much is already known and has been reviewed in a series of general texts in English: Botosananu (1986), Juberthie and Decu (1994, 2001), Wilkens et al. (2000), Iliffe and Kornicker (2009), White and Culver (2012), and two dedicated journals (*Subterranean Biology* and the *International Journal of Speleology*).

DESCRIPTION OF SUBTERRANEAN HABITATS

Geology and Hydrology

Subterranean crustaceans are best known from carbonate karst systems because the extensive caves within the limestone make these habitats most accessible (Fig. 13.1 depicts a karst stygobiontic habitat). However, subterranean crustaceans are found not only in large cave systems but almost anywhere there are interconnected subsurface voids, ranging in geology from lava tubes through scree, gravels, and finer alluvial deposits formed from a wide variety of minerals. These habitats range from caves and artesian aquifers 1–2 km deep, through fractured rock and pisolites, to shallow subterranean habitats (SSHs), such as groundwater calcretes, closely connected with the surface (Culver and Pipan 2011).

Most subterranean crustaceans inhabit aquatic systems. The only crustaceans to have emigrated from terrestrial habitats and diversified in non-aquatic subterranean systems are oniscidean isopods and a few talitrid amphipods. A few species from other groups may be found in caves, but they lack overt cave adaptations (e.g., the robber crab *Birgus latro* and several

Fig. 13.1.
Croesus Cave, Tasmania, showing potential stygobiont cave habitats. Tree roots penetrating such systems and organic matter stranded on the stream bank during high water flow could provide habitat for troglobionts such as oniscidean isopods. From Boulton et al. (2014), with permission from John Wiley and Sons. See a color version of this figure in the centerfold.

terrestrial brachyurans). Terrestrial cave habitats experience much lower temperature variation than the external climate and typically have saturated or near-saturated relative humidity. The damping effect increases with distance from cave entrances and with cave depth. Thus, caves may provide refuges for species from unfavorable surface conditions, such as aridity, that have caused the local, regional, or global extinction of surface ancestors of troglobionts. Troglobionts depend on organic matter transported into caves by agents such as water or root growth, or exceptionally on bacterial chemoautotrophy known from caves (e.g., Frassassi, Italy; Movile, Romania), aquifers (Edwards Aquifer, Texas), and anchialine systems (Yucatan, Mexico).

Subterranean habitats are found all around the world and encompass a wide variety of form, ranging from the water film on cave walls to immense complex systems such as massive aquifers sustained by chemoautrophy.

The cave hygropetric, which comprises the water film flowing on cave walls, supports species that specialize in harvesting food from this resource (Sket 2004). Crustaceans (mainly amphipods such as *Typhlogammarus mrazeki*) found in this habitat, however, while well adapted to moving in the water film, are neither restricted to nor specialists of this habitat.

The hypotelminorheic habitat (Culver et al. 2006) comprises permanent superficial seepages rich in organic matter (wet leaf patches) above clay layers that support fully stygobiont species including crangonyctid and gammarid amphipods in the United States and Europe and niphargid and pseudoniphargid amphipods in Europe.

Epikarst, the zone of rainfall infiltration, is located at the top of the aerated zone and consists of heavily weathered carbonate or gypsum bedrock directly beneath the soil. In some cases, when soil is absent, the epikarst is exposed at the surface. The depth of the epikarst region is typically 3–10 m but depends on the geomorphology and lithology of the region (Williams 2008). For example, at high latitudes Pleistocene glaciers have removed the soil and reduced the epikarst to less than a meter thick. In the Dinaric region of Slovenia, Pipan and Culver (2007) identified an exceptionally rich diversity of endemic copepods in the epikarst.

Stream flow may form subterranean habitats as well. The hyporheic is the space below the stream flow where there is short-term interaction between the stream and groundwater (in the stream bank; it is known as the parafluvial, which has long-term interactions with the stream flow). Defined by the hydrology, these short hyporheic flow paths, on a centimeter-to-meter scale, may form from rough features of the stream bed while long hyporheic flow paths (a meter or more) may create upwelling groundwater (such as below riffles and bends) allowing stygobiont crustaceans to rise toward the stream water. Down-welling stream water (such as above riffles and bends), being relatively nutrient-rich, allows non-stygobiont crustaceans to more deeply penetrate into the stream bed. The hyporheic comprises a zone of transition from a mixed insect and crustacean-dominated system in the stream to a crustacean-dominated groundwater (well depicted in Table 2 in Stanford et al. 1994). The most thoroughly studied areas are the Lobau B wetland in Vienna (Danielopol 1984), the Rhone River (Marmonier and Creuze des Chateliers 1992), the Flathead River, Montana (Stanford et al. 1994), and the Hunter River, New South Wales, Australia (Hancock 2006).

Anchialine environments (Figure 13.2) consist of bodies of haline waters, usually with a restricted exposure to open air, that are found in volcanic or limestone bedrock with subterranean connections to the sea and showing noticeable marine as well as terrestrial influences (Stock et al. 1986). As such, they serve as transitional habitats on a continuum between fresh groundwater and marine environments. The description of these ecosystems has evolved over decades. Riedl (1966) originally introduced these systems as Randhöhlen (marginal caves), defined mainly by the position and origin of the cave. Holthuis (1973) applied the term anchialine based on the morphology of the systems having no surface connection with the sea but containing brackish or salt water and fluctuating with the tide. Stock et al. (1986) refined the definition to include subterranean connections and restricted surface interactions. To further emphasize the extent of subterranean influences, Bishop et al. (2015) realigned the term "anchialine" to support the concept that these systems should be examined on a continuum as a "subterranean estuary," as first characterized by Moore (1999) and used in another context as "groundwater estuaries" as a direct parallel with anchialine systems (Humphreys et al. 2009). Envisaged as subterranean estuaries (Bishop et al. 2015; Figs. 13.2 and 13.3), they are commonly found in tropical oceanic islands in the Caribbean, Hawaii, Bermuda, Sinai Peninsula, Mediterranean coast, South Pacific Ryukyus and Phillipines, and Indian Ocean (Australia: Christmas and Barrow Islands and Cape Range).

Water Chemistry

The water quality in subterranean ecosystems varies from freshwater to hypersaline but may vary within the same system from fresh to fully marine water as in the anchialine system of Túnel de La Atlántida in Lanzarote. Similarly, in arid zones, water salinity may vary by marked vertical stratification over several meters from nearly fresh to hypersaline, as found in some species-rich groundwater calcrete communities such as those alongside the playa Lake Way, Wiluna, Western Australia (Watts and Humphreys 2001). As a result, the water chemistry of each type of system can be quite complex.

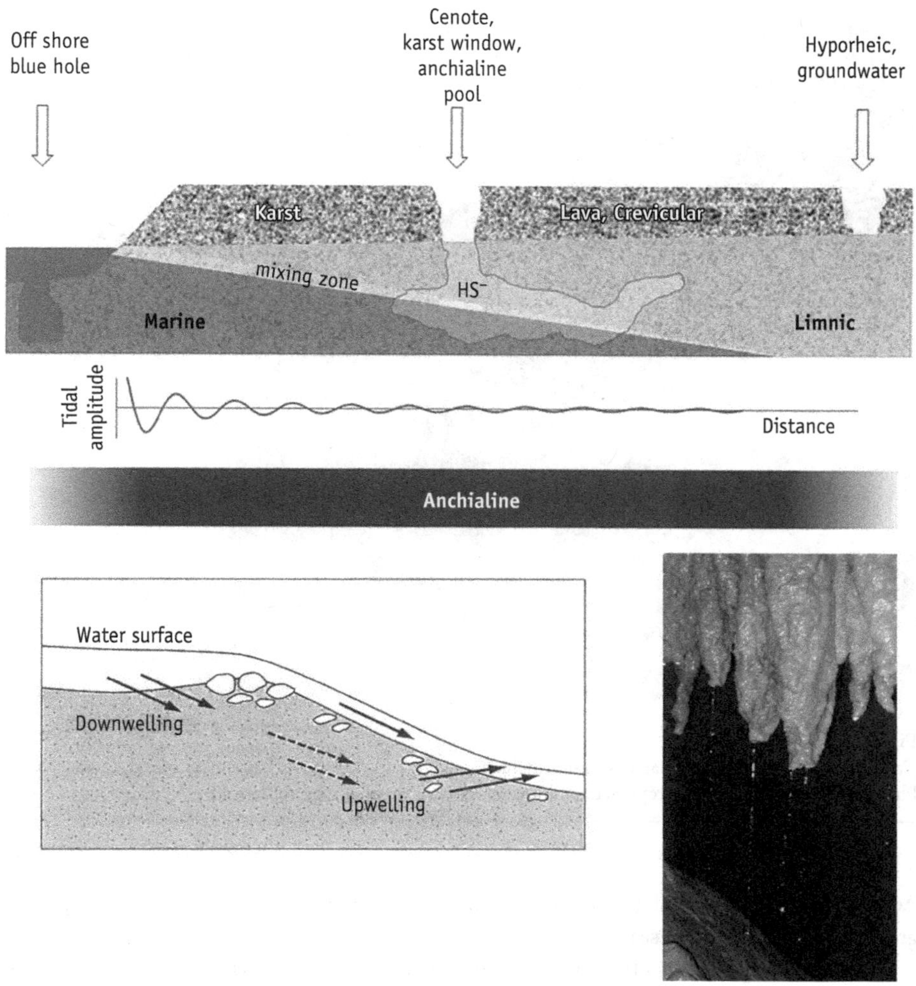

Fig. 13.2.
Subterranean habitats. (Top) Anchialine habitat. (Lower left) Hyporheic flowpaths at one section of a riffle (from Boulton et al. 2014). (Lower right) Water percolating from the epikarst across flowstone to form drips. This habitat is rich in copepods. Organ Cave, West Virginia. Top, modified from Bishop et al. (2015), with permission from Oxford University Press; bottom left, modified from Pipan and Culver (201), with permission from Elsevier; bottom right, modified from Boulton et al. (2014), with permission from John Wiley and Sons. See a color version of this figure in the centerfold.

Organic carbon is typically sparse in subterranean systems with the exception of shallow surface habitats SSHs in which there are sometimes high concentrations of dissolved and particulate organic carbon (DOC and POC). Categorized as ultra- to oligotrophic (Alcocer et al. 1999), the DOC values in both tropical anchialine (cenotes) and temperate US coastal groundwater systems vary dynamically with the season (Alcocer et al. 1999, Goñi and Gardner 2003).

The features of a highly stratified water column and potentially slow exchange rate characteristic of subterranean, particularly anchialine, ecosystems can increase concentrations of heavy metals, such as mercury, cobalt, and copper, toxic to crustaceans. In a study of anchialine systems along the

Fig. 13.3.
Cave diver Richard "Harry" Harris seeking anchialine fauna in Whip Cave, Christmas Island, Indian Ocean. Photo courtesy of Bill Humphreys. See a color version of this figure in the centerfold.

Adriatic Coast, Kwokal et al. (2014) found total mercury levels (THg) three orders of magnitude greater than in the adjacent sea.

Dissolved oxygen concentration within subterranean ecosystems is generally lower than in surface waters and declines with depth and along the groundwater flowpath. Calcrete aquifers are often anoxic at depth, but in anchialine systems dissolved oxygen levels can reach suboxic to anoxic levels at the halocline and then, unusually, increase with depth owing to oxygen being transported within the underlying marine water.

It is difficult to isolate the impact of a single chemical feature on the distribution of subterranean crustaceans. For example, in the Pilbara region of northwestern Australia, the composition of diverse groundwater ostracod communities was impacted by pH and carbonate saturation of the host waters, and ostracods were rarely present below pH 6, in reducing environments, or where total nitrogen concentrations exceeded 10 mg l^{-1}. Where ostracods were common, surface species were found in the more saline waters (Reeves et al. 2007). Some higher taxa in anchialine systems, such as remipedes and thaumatocyprid ostracods, are restricted to suboxic waters (typically less than 1 mg l^{-1} oxygen) within fully marine waters, whereas others, such as the atyid shrimp genus *Stygiocaris*, are euryhaline. Bathynellacea are the epitome of a stygobiont order, being present globally only in continental groundwaters. Parabathynellids are normally minute (<2 mm) but the exceptionally large (3.6 mm) *Brevisomabathynella uramurdahensis* and *B. clayi* occur in groundwater of salinity >30 PSU, close to marine salinity, adjacent to playas in arid Western Australia. They occur together with species of Oniscidea (*Haloniscus*), ameirid harpacticoids (*Parapseudoleptomesochra*), and cyclopid copepods (*Mesocyclops* and *Halicyclops*; Fig. 13.4).

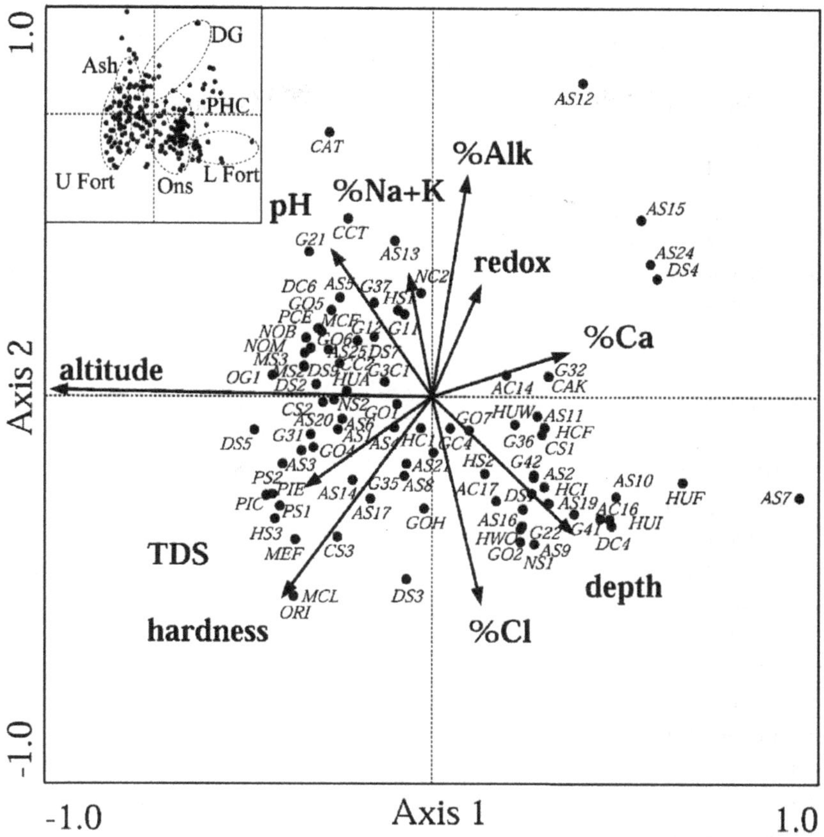

Fig. 13.4.
Influence of water chemistry on the composition of groundwater ostracod communities in the Pilbara district of Western Australia. Canonical correspondence analysis of 110 species and environmental variables among 445 sites. The small inset plot refers to the ordination of samples among five drainage systems. From Reeves et al. (2007), with permission from Springer.

DIVERSITY, ENDEMICITY, AND BIOGEOGRAPHY OF SUBTERRANEAN CRUSTACEANS

Crustacea are as ubiquitous in subterranean waters as insects are on land. The dominance of crustaceans in subterranean waters can be gleaned from a worldwide compilation of all subterranean taxa (Botosaneanu 1986). Of 5,204 taxa of stygofauna (including the marine interstitial) encompassing all invertebrate phyla, 2,861 (55%) comprise crustaceans, of which 43% are copepods, 24% amphipods, and 21% isopods. More recently, Hobbs (2012, Table 1, summarized here as Table 13.1) enumerated >4,570 described species of crustaceans inhabiting subterranean waters, indicating the large increase in knowledge in the intervening quarter century reducing the proportion of both copepods and amphipods to 24% and 17%, respectively. The significance of subterranean crustaceans is such that stygobiontic species represent ~40% of the European continental crustacean fauna (Danielopol et al. 2000).

Freshwater stygobionts were, until recently, thought to have extensive linear ranges exceeding 2,000 km, but recent molecular studies of diverse European taxa (*Niphargus virei,*

Table 13.1. Distribution of 3,748 Stygobiont Species Among Crustacean Orders

Order or Subclass	No. Stygobiont Species
Order Diplostraca	5
Class Remipedia	29
Subclass Tantulocarida	1
Subclass Mystacocarida	13
Order Platycopioida	2
Order Calanoida	39
Order Misophrioida	22
Order Cyclopoida	231
Order Gelyelloida	2
Order Harpacticoida	787
Order Myodocopida	8
Order Halocyprida	29
Order Platycopida	1
Order Podocopida	1,201
Order Leptostraca	1
Order Bathynellacea	329
Order Anaspidacea	21
Order Spelaeogriphacea	4
Order Thermosbaenacea	37
Order Mysida	45
Order Mictacea	1
Order Bochusacea	3
Order Amphipoda	784
Order Isopoda	973
Order Tanaidacea	0
Order Cumacea	0
Order Decapoda	222

Source: Summarized and modified from Hobbs 2012, with permission from Elsevier.

N. rhenorhodanensis, Troglocaris anophthalmus, T. hercegovinensis, Spelaeocaris pretneri) have shown that these ranges include numerous cryptic lineages and actual ranges rarely exceeded 200 km in length, with half of the cryptic lineages being recorded from one site only (Trontelj et al. 2009). This concurs with studies from arid zone groundwaters in Australia (summarized by Humphreys 2008, 2012, 2017).

Freshwater stygobionts must be generally thought of as narrow-range endemics and, as their dispersal is limited by seas surrounding continents, they are likely to be exceptionally good biogeographic indicators. Some higher taxa of freshwater crustaceans are considered sufficiently old to reflect the movement of tectonic plates, leaving clear trails of related taxa on different plates. For example, of the four known species of the freshwater order Spelaeogriphacea, one is restricted to the top of Table Mountain, South Africa, a location at ~750 m altitude that has not been near sea level since before the fragmentation of Gondwana. The remaining species occur in western Brazil and two in the Fortescue valley, northwestern Australia (Poore 2015). Similarly, stygobiont hypsimetopodine phreatoicidean isopods, including *Nichollsia*, occur in eastern India and, as *Nichollsia* is nested phylogenetically within the Australian Hypsimetopodinae, Wilson and Keable (2001) contended that this stygobiont clade has a minimum age of 130 million years, coinciding with the separation of Greater India from Australia. Further evidence of the Gondwanan connection is provided by other phreatoicidean isopods that occur presently in India, South Africa, New Zealand, and Australia, although fossils indicate a former cosmopolitan distribution. The genus *Crenisopus* in northern Australia is phylogenetically basal to most families in the Phreatoicidea, suggesting divergence after they entered fresh water but prior to the fragmentation of East Gondwana during the Mesozoic era, and so provides a link between African and Australasian lineages of phreatoicideans (Wilson and Keable 1999; Fig. 13.5). Vicariance evidenced from distributional trails within lineages is undermined by increasing evidence of transcontinental dispersal within unlikely lineages (de Queiroz 2016). None of the examples above has yet been tested using time-calibrated molecular phylogenetics.

Anchialine ecosystems support an extraordinary diversity of specialized crustaceans largely endemic to this environment and which mostly represent biogeographic or phylogenetic relicts, or both (Iliffe 1992). They have a highly vicariant distribution across the full Tethyan track (Stock 1993) as far apart as northwestern Australia, the Canary Islands, and Caribbean islands plus Yucatan, most on the North American Plate. Over this 18,000 km range, the generic composition of this assemblage is highly predictable (Wagner 1994) and is characteristically composed of atyid

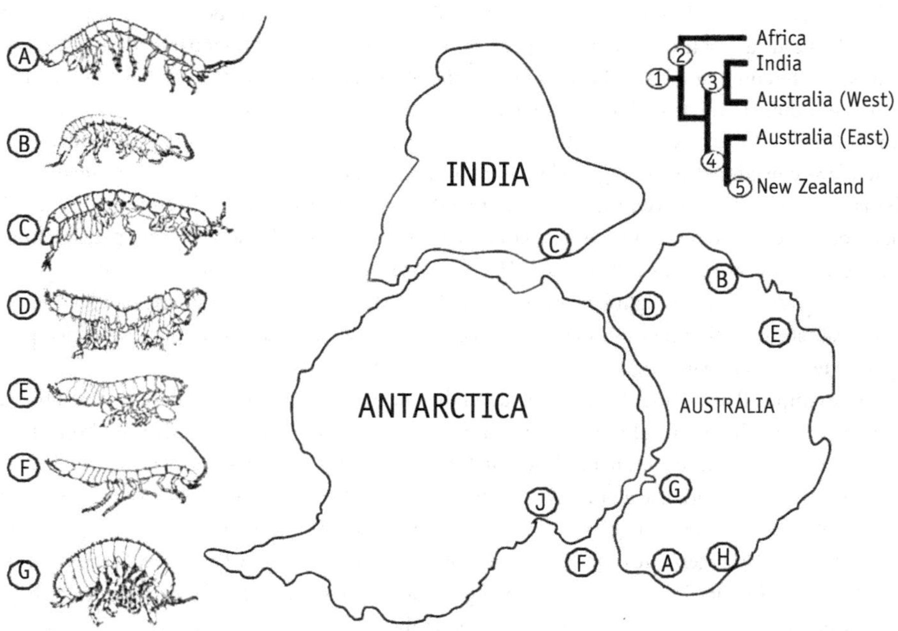

Fig. 13.5.
Distribution of selected members of the Phreatoicidea in East Gondwana. (Bottom right) Triassic distribution of modern continents in East Gondwana, letters as for the modern distribution. (Top right) Area cladogram representing vicariant events in phreatoicidean phylogeny occurring during the breakup of East Gondwana; Antarctic presumed basal to both Australian branches. (A) *Phreatoicoides gracilis*; (B) *Pilbaraphreatoicus platyarthricus*; (C) *Nicollsia kashiensis*; (D) *Hyperoedesipus plumosus*; (E) *Crenisopus acinifer*; (F) *Phreatoicus typicus*; (G) *Gariwerdeus ingletonesis*; (H) *Crenoicus* sp.; (J) Jurassic fossil isopod *Protamphisopus* sp. Modified from Wilson (2008), with permission from CSIRO Publishing.

shrimps, thermosbaenaceans, hadziid amphipods, cirolanid isopods, remipedes, thaumatocypridid ostracods, and an array of copepods such as epacteriscid, pseudocyclopiid, and ridgewayiid calanoids, halicyclopine cyclopoids, speleophriid misophrioids, and superornatiremid harpacticoids. Stock (1993) suggested that the present distribution may have arisen from vicariance by plate tectonics. This explanation would encompass (1) the constraints of high taxonomic predictability, (2) the great barriers to dispersal between congeneric species across ocean basins, and (3) their stygobiont habits, and provides a distribution that would fit within the full extent of the Tethys in the late Mesozoic and early Tertiary (fully expounded in Boxshall et al. 2014). There is as yet no convincing test of this hypothesis in which the timings of vicariance and plate movement are concordant (see also amphi-Atlantic discussion below); indeed, the evidence points to a multi-staged process with the inferred ages of divergence between species of *Halosbaena* (Thermosbaenacea) not old enough to be associated with the spread of the Tethys (Page et al. 2016).

Representing subterranean crustaceans with global distribution are the Bathynellacea, typically minute (<1 mm) stygobiont syncarid crustaceans. They are considered to have established in the continental groundwater habitat across Pangaea (Schminke 2011) and to have dispersed by means of plate tectonics, having congeneric species distributed across continents. They are absent only from oceanic islands of the Pacific and elsewhere. Most species are endemic to small areas, known only from their type locality and immediate surroundings (Schminke 2011). Detailed sampling and molecular research in the Iberian Peninsula (Camacho et al. 2012) and Australia (Abrams et al. 2012) have revealed a plethora of short-range endemic species. The morphologically well-characterized Gondwanan genus *Atopobathynella* (Parabathynellidae) is found in Australasia, India, Madagascar, and South America, and similar distributions have been observed in several stygobiont crustacean taxa, including copepods (Ranga Reddy 2011). Other parabathynellid genera also have intercontinental distributions (e.g., *Notobathynella* and *Chilibathynella*), while the distribution of others in the tribe Iberobathynellini is amphi-Atlantic and related to the opening of the North Atlantic Ocean (Camacho et al. 2000), whereas *Hexabathynella* occurs worldwide. Other genera are considered to be regionally endemic (e.g., *Octobathynella,* Australia; *Iberobathynella,* Iberian Peninsula). However, these intercontinental associations have not been tested by molecular clock evidence of vicariance events concordant with geological evidence from plate tectonics (unlike some terrestrial lineages; Rix and Harvey 2012). Nowhere is this more obvious than in the oft-cited and seemingly straightforward amphi-Atlantic vicariance of stygobiontic lineages. Despite large-scale studies using genomic methods, congruence between genetic divergence and vicariance by plate tectonics remains vigorously debated.

For example, Bauzà-Ribot (2012) concluded that vicariance by plate tectonics explained the genetic divergence between subterranean metacrangonyctid amphipods currently found on islands separated by the Atlantic Ocean. Phillips et al. (2013), when attempting to verify these conclusions, obtained results indicating a later trans-Atlantic dispersal date or else the divergence would require a series of later invasions of fresh water from a common marine ancestor. A most thorough study, using mitochondrial genomes, was conducted on atyid cave shrimps (*Typhlatya, Stygiocaris,* and *Typhlopatsa*) (Jurado-Rivera et al. 2017). Despite these genera having a highly disjunct distribution throughout the Tethyan track, their results imply a mixture of both vicariance (the Atlantic opening) and dispersal, since the Australia/Madagascar shrimp divergence postdates Gondwanan breakup.

By contrast, the predictable anchialine crustacean assemblage is present at intervals throughout the Tethyan realm indicating that they dispersed initially along Tethyan shores, and subsequent vicariance was the result of plate tectonic movement. Although the remipede-type anchialine assemblage closely fits this model, recent discoveries of some of its members, such as *Halosbaena* (Thermosbaenacea) and *Humphreysella* (Thaumatocyprididae) on isolated seamount islands (Humphreys and Danielopol 2006), and the inferred age of divergence from molecular phylogenetics (Page et al. 2016) indicate that at least some elements are capable of oceanic dispersal (Humphreys 2014).

DOMINANT CRUSTACEAN GROUPS

Amphipoda

At least 30 families of gammaridean amphipods inhabit subterranean waters (Holsinger 1993), some having almost worldwide distributions (e.g., Bogidiellidae, Hadziidae), while others are markedly regional (Allocrangonyctidae, United States; Crangonyctidae, Holarctic, mainly North America; Niphargidae, Europe and Asia; Calliopiidae, New Zealand; Phreatogammaridae, New Zealand, South America, and New Caledonia; Perthiidae, Neoniphargidae, and Paramelitidae, Australia) or restricted to anchialine systems (Liljeborgiidae, Lysianassoidea, Pardaliscidae). Together, they encompass at least 778 species (Holsinger 1993, Hobbs 2012).

Isopoda

At least 17 families of isopods from the 8 suborders (Phreatoicidea, Microcerberidea, Tainisopidea, Sphaeromatidea, Cymothoida, Asellota, Calabozoidea, and Oniscidea) inhabit subterranean waters and include at least 948 species (Hobbs 2012). Many higher taxa have ancient biogeographic patterns found only in Laurasia (Asellidae, Stenasellidae) or Gondwana (Phreatoicidea, Protojaniridae, and the janirid *Heterias*). The subterranean representatives of the predominantly marine superfamily Anthuroidea (Cymothoida) and Microcerberidea are mostly marine with a broad distribution and largely interstitial lifestyle but are also represented in both freshwater and anchialine habitats (Wägele 1983, Poore and Humphreys 2013). Cirolanidae (Cymothoida) are an extremely diverse cosmopolitan marine family, but their subterranean representatives have a Tethyan distribution and are very diverse in the Mediterranean region, with internal divergences as far back as 160 mya (Baratti et al. 2010), with species in the Maghreb and especially in the High Atlas Mountains reflecting the uplift of this mountain range (Baratti et al. 2004). *Haptolana* occurs in settings covering the full extent of the Tethys seaway (Bruce and Humphreys 1993), including continental inland waters. The family Sphaeromatidae (Sphaeromatidea) is primarily marine but appears in subterranean fresh waters of Laurasia and in New Zealand. Nine of the many families of Asellota contain subterranean species, seven of these in the superfamily Aselloidea. They are largely Laurasian with the exception of the Protojaniridae and *Heterias* (Janiridae) in Gondwanan fragments. The only two known species of Calabozoidea are endemic to South America. The Oniscidea (~950 subterranean species in 17 families; Hobbs 2012) are fully adapted to terrestrial environments, many as troglobionts, including members of the family Paraplatyarthridae of Australia and South America. They are the most successful group of crustaceans living on land, with very few species, such as *Haloniscus* or *Typhlotricholigioides*, adapted to aquatic habitats, living in either groundwaters or hypersaline lakes (Schmidt 2008).

Copepoda

Copepods are, with 1,077 species recorded thus far, the major component of groundwater fauna (Hobbs 2012). Six of the eight orders of copepods contain subterranean species, namely Platycopioida, Misophrioida, Calanoida, Cyclopoida, Harpacticoida, and Gelyelloida, with gelyelloids being exclusively stygobiont. Stygobiont platycopioids and misophrioids are known only from anchialine caves and seem unable to cross the osmotic barrier enabling the colonization of fresh inland groundwaters. The only two species of Gelyelloida known live in fresh groundwaters in the European Alps. The Calanoida are represented by a few members of the exclusively limnic family Diaptomidae in fresh subterranean waters, but stygobiont representatives of at least seven families (Arietellidae, Boholinidae, Epacteriscidae, Fosshageniidae, Pseudocyclopiidae, Ridgewayiidae, and Stephidae) have been reported from anchialine caves. The Cyclopoida and Harpacticoida represent the bulk of species of copepods in groundwater habitats with at least 897 species recorded in fresh water, mostly belonging to the harpacticoid families Canthocamptidae, Parastenocarididae, and Ameiridae and the cyclopoid family Cyclopidae. Anchialine caves harbor remarkable representatives of the cyclopoid family Cyclopinidae as well as the few known members of the family Speleoithonidae, whereas most members of the harpacticoid families Superornatiremidae and Novocriniidae, or the single species of Rotundiclipeidae, inhabit anchialine environments (Galassi 2001).

Ostracoda

All four extant subclasses of Ostracoda contain subterranean species, namely Halocyprida, Myodocopida, Platycopida, and Podocopida, the latter including the "ancient asexuals" of the family Darwinulidae (Smith et al. 2006); analysis of sexual dimorphism of the valves indicated that members of Darwinulidae had been asexual since the late Triassic yet subsequent work by Smith et al. (2006) demonstrated the occurrence of rare male specimens. Together the ostracods encompass at least 976 species (Hobbs 2012), occurring in all types of subterranean waters found in interstitial, aquifer, cave, and anchialine systems.

Anchialine systems support a high diversity of ostracods, especially in the Bermuda-Yucatan axis (Kornicker et al. 2007). That region alone possesses at least 15 species of Halocyprida of the families Thaumatocyprididae, Deeveyidae, and Polycopidae, the former occupying the full Tethyan track to Australia (Humphreys and Danielopol 2006). In addition, there are at least 13 species of Myodocopida belonging to the families Cypridinidae, Philomedidae, Rutidermatidae, Sarsiellidae, and Cylindroleberididae (Fig. 13.6).

The degree of endemicity of even microcrustacean stygobionts is illustrated from the work of I. Karanovic (Ostracoda: Podocopida) and T. Karanovic (Copepoda) (see references cited in Humphreys 2008) on the fauna of the contiguous Pilbara and Yilgarn regions of the Western Shield of Australia, an area continuously emergent from the sea since at least the early Palaeozoic (>450 Ma). Both higher taxa show minimal overlap at the species and genus levels between these contiguous regions. Among the candonid ostracods, there is an extraordinary radiation in the Pilbara with 2 of the 3 tribes and 12 of the 13 genera endemic, and a clear Gondwanan distribution, with the Pilbara having as much generic overlap with South America as with the contiguous Yilgarn.

Decapoda

This diverse order includes the most conspicuous invertebrates in subterranean waters, and members are represented in temperate and tropical waters, especially in karst and volcanic terrains. All the stygobionts are members of the suborder Pleocyemata, including the Stenopodidea, Caridea, Astacidea, Anomura, and Brachyura. Caridean shrimps are well represented in anchialine caves (Procarididae, Agostocarididae, Hippolytidae, Alpheidae, Maxillocarididae) together with Atyidae and Palaemonidae, which are also well represented in continental subterranean waters. The Procarididae and Agostocarididae are known only from subterranean habitats.

Thirteen families of brachyuran crabs contain stygobionts comprising more than 67 species from 33 genera. They are especially species-rich in Central and South America (20 species of

Fig. 13.6.
Representative crustaceans inhabiting anchialine waters. (A) *Parhippolyte sterreri* (Decapoda); (B) *Pseudoniphargus grandimanus* (Amphipoda); (C) *Bahalana caicosana* (Isopoda); (D) *Tulumella* sp. (Thermosbaenacea); (E) *Mictocaris halope* (Mictacea); (F) *Bermudamysis speluncola* (Mysida); (G) *Cumella abacoensis* (Cumacea); (H) *Ridgewayia* sp. (Calanoida); (I) *Spelaeoecia sp.* (Ostracoda); (J) *Cryptocorynetes* sp. (Remipedia). Photographs of anchialine crustaceans by Thomas M. Iliffe, from Pérez-Moreno et al. (2016). See a color version of this figure in the centerfold.

Pseudothelphusidae) and Indonesia and the Philippine islands (10 species of Parathelphusidae) (Hobbs 2012).

Subterranean Astacidea occur only in North America and Cuba, where c. 32 stygobiont species have been reported thus far.

Regarding stenopodideans and anomurans, their presence in subterranean waters is almost anecdotal. The only stenopodidean shrimp known thus far to be exclusively stygobiont is *Macromaxillocaris bahamensis*, the only representative of its own family Macromaxillocarididae, from anchialine caves in the Bahamas. Anomurans are represented by the galatheid *Munidopis polymorpha* from lava tubes in the Canary Islands and by four *Aegla* species (Aeglidae) in South America.

ORIGIN OF AQUATIC SUBTERRANEAN CRUSTACEANS

General Adaptations and Preadaptations

Assuming that stygobiont crustaceans derived from epigean ancestors, the question arises: What impelled their colonization of subterranean waters? Since there are remarkable differences in geology, climate, and diversity of subterranean biotopes and taxa among the various regions of the world harboring subterranean crustaceans, it is unlikely that a single explanation applies for their origin. We know that almost everywhere, crustacean species may adapt in ways that allow them to exist simultaneously in suitable surface and subterranean waters. Species that fit this profile include epigean cyclopoid and harpacticoid copepods in wells or at the hyporheic medium (Gourbault and Lescher-Moutoué 1979, Rouch 1991), freshwater epigean gammaridean amphipods (Culver et al. 1995), or asellid isopods (Sket 1994) in caves, as well as marine littoral cyclopoid and harpacticoid copepods in the anchialine environment (Jaume and Boxshall 1996). Preadaptations to subterranean life include omnivory vs. food specialization, prevalence of chemical and tactile communication in reproductive behavior, photo-negativity, locomotion in or at the substratum or swimming in close contact to it, ability to cope with low oxygen concentrations, and euryhalinity (in the case of marine taxa).

Given the right set of circumstances, animals with such attributes may evolve to become stygobiontic. On the contrary, stygobionts derived from specialized ancestors such as filter feeders, grazers, highly specialized predators, parasites, commensals of other invertebrates, sand burrowers, or with nestling, tube-dwelling, or pelagic mating reproductive strategies, are exiguous. For example, Notenboom (1991) pointed out that the majority of stygobiont amphipods belong to 10 families distributed over only 5 of the 19 gammaridean superfamilies; the remaining 14 superfamilies that include highly specialized taxa hardly have any subterranean representation despite their broad distribution in shallow marine environments.

In any event, possession of a set of preadaptations is not sufficient to guarantee successful colonization of subterranean waters, and generalizations about which taxa are prone to the development of stygobitic characteristics are not possible. Thus, entire higher taxa such as cumaceans and tanaidaceans that include many apparently hypogean-preadapted species (depigmented, anophthalmous) have no or hardly any stygobiont representatives. The same holds at lower taxonomic levels; for instance, anthuroid isopods are mostly marine, and only 5 of 66 described genera include stygobionts (Wägele 1989, Poore and Humphreys 2013).

Terrestrial Origin

A few members of the terrestrial isopod suborder Oniscidea are fully aquatic in subterranean waters. These are the monotypic *Typhlotricholigioides* and *Mexiconiscus*, both from Mexico, *Cantabroniscus* from northern Spain, *Thailandoniscus* (Thailand) and *Utopioniscus* (Sardinia), as well as the three known species of *Macedonethes* (Balkans and Spain) and numerous species of *Haloniscus* from New Caledonia and Australia (Taiti and Humphreys 2001, Cooper et al. 2008). The occurrence of these taxa in subterranean waters must be considered a secondary adaptation since all display remnants of a water-conducting system, consisting of belts of scale-rows along the pereopods and antennae, proving that the ancestors of these taxa lived at least temporarily out of water (Tabacaru 1999).

Freshwater Origin

In general, colonization of inland groundwaters by limnic epigean crustaceans is purported to proceed via two main portals: karst systems and associated transitional habitats such as springs and hyporheic habitats linked to water courses. Stygobiont representatives of typical freshwater crustacean groups such as koonungid syncaridans, aselloid and phreatoicidoid isopods, crangonyctid amphipods, cyclopoid copepods of the cyclopid subfamilies Cyclopinae and Eucyclopinae, diaptomid calanoid copepods, anomopod branchiopods, and decapod groups such as cambarid crayfishes and potamoidean brachyurans are presumed to derive directly from freshwater epigean taxa since they have epigean relatives and are absent from the sea. In addition, entire high-level taxa restricted to fresh inland groundwaters and known nowhere else, such as the order Spelaeogriphacea (Peracarida), the Gelyelloidea (Copepoda), and the suborders Bathynellacea (Syncarida) and Calabozoida (Isopoda), are presumed to have an old freshwater origin since they neither have marine relatives nor appear in regions invaded by the sea in geologically recent times. They are not euryhaline either, except in the case of Bathynellacea, in areas of secondary salinity (Humphreys 2008). The same holds for families such as the Parastenocarididae (Copepoda Harpacticoida), Protojaniridae (Isopoda), and Stygocarididae, Psammanaspididae, and Patagonaspididae (Syncarida).

A few cases are known of stygobionts that appear to have colonized marine stygohabitats directly from fresh inland groundwaters. Wägele (1983, 1990) considered the Microcerberidea, a suborder of highly modified elongate micro-isopods adapted to live in the interstices of unconsolidated sediments, to have diversified primarily in freshwater stygohabitats to secondarily invade marine coastal groundwaters. Seemingly, Vonk and Schram (2003) proposed that the primary diversification of the ingolfiellidean amphipods took place in fresh subterranean waters, from where they ultimately colonized marine littoral subterranean environments and even the deep sea.

A few limnostygobiont crustaceans of marine origin are presumably derived from ancestors that first colonized surface fresh waters (probably proceeding upstream from coastal springs), then went through a short epigean phase, and finally entered subterranean waters. Such an active colonization scenario has been proposed by Sket (1970) to explain the occurrence of a diverse assemblage of stygobiont sphaeromatid isopods (i.e., *Caecosphaeroma, Monolistra*) in southern Europe and could also hold for the three species of another genus in the same family, *Bilistra*, in New Zealand (Sket and Bruce 2004).

Marine Origin

A substantial fraction of stygobiont crustacean species derive directly from marine ancestors. This origin is suggested by their taxonomic affinities, comparison of their geographic distribution with palaeogeographic data, and their tolerance to elevated salinities. Well-known examples are thermosbaenaceans (Fig. 13.7), atlantasellid and microparasellid isopods, hadziid amphipods, and stygobiont representatives of groups such as eriopisid amphipods and palaemonid and atyid shrimps. Direct colonization of inland subterranean waters by these marine taxa may have proceeded via submarine coastal karstic springs, the interstitial medium associated with unconsolidated sediments developed in the marine littoral and sublittoral, and anchialine caves (Stock 1980, Notenboom 1991, von Reumont et al. 2012: Fig. 13.5). In each example, the process first requires the development of populations adapted to interstitial or crevicular habitats and, above all, to euryhaline conditions. These preadapted populations will then have colonized coastal and subsequently inland, subterranean waters through active migration or passive isolation, or a combination of both. For example, Danielopol and Bonaduce (1989) proposed that the presence of representatives of the typically marine euryhaline cytherid ostracod subfamilies Pseudolimnocytherinae and Psammocytherinae in continental groundwaters of southern Europe is the result of an active colonization process through submarine karstic springs. According to these authors, the mixohaline conditions prevalent during the submarine phases of discharge shifted into freshwater conditions after a marine regression, when the springs remained disconnected from the sea, enabling the active inland migration of the founder population of marine euryhaline ancestors.

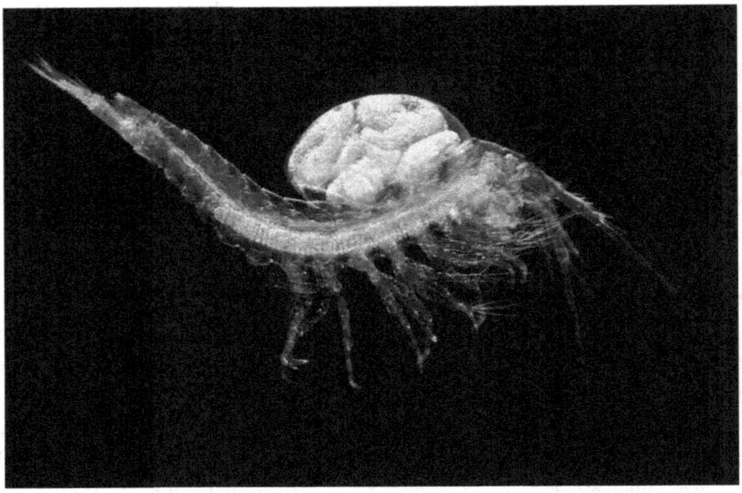

Fig. 13.7.
Tethysbaena scabra, an anchialine thermosbaenacean from the Balearic Islands (West Mediterranean). Thermosbaenaceans are exclusively subterranean and are characterized by the display of a dorsal brood pouch in gravid females; this structure contrasts remarkably with the condition exhibited in other peracarid groups, where females carry the embryos in a ventral marsupium. The subterranean environment has acted as a refuge for ancient lineages completely vanished from the epigean medium. See a color version of this figure in the centerfold.

Many groups of stygobiont crustaceans of marine origin show distributional patterns concordant with former (Tertiary and sometimes Late Cretaceous) land–sea configurations. Remarkable examples are the highly diversified stygobiont amphipod families Pseudoniphargidae and Metacrangonyctidae or thermosbaenaceans of the genus *Tethysbaena*, all of them showing the same amphi-Atlantic distribution over areas previously covered by ancient epicontinental seas (Notenboom 1988, Wagner 1994, Bauzà-Ribot et al. 2012). Moreover, these groups exhibit high levels of endemicity and low frequencies of sympatry, pointing to a low vagility and dispersal ability. If these organisms had actively colonized the subterranean waters, it is hard to explain why they have become so static in their new environment. It may be that in such groups, marine regressions have induced the passive settlement of marine populations into inland waters (Holsinger 1991, Notenboom 1991). Ancestors of these freshwater forms may have populated the interstitial littoral environment, which has a dynamic salinity or oxygen content, and thus could be preadapted to the colonization of inland subterranean waters. Marine regression vicariant events may have led to the isolation of small peripheral populations of these preadapted species in coastal stygo-habitats, where they ultimately speciated. This may have taken place simultaneously in several parts of the distribution range of the same ancestor and in a recurrent way, giving rise to swarms of closely related new species sharing the same direct ancestor (Notenboom 1988, Holsinger 1991).

"Thalasso-stygobionts"

Some crustacean groups, including all members of the Class Remipedia, the entire peracarid order Mictacea, tulumellid thermosbaenaceans, and procaridid and maxillocaridid shrimps, never penetrated into inland fresh waters or into the upper, low-salinity waters of anchialine caves, remaining always within fully marine water. This is also apparent for stygobiont representatives of another peracarid order, Bochusacea, and of the copepod orders Misophrioida and Platycopioida, all but one representative of the ostracod order Halocyprida, and the cave-dwelling representatives of stephid, pseudocyclopid, and pseudocyclopiid calanoid copepods. All presumably share physiological limitations to their ability to cross the osmotic barrier represented by the fresh–marine water interface and are typical inhabitants of oceanic islands or of portions of mainland submitted to subsidence. Stock (1986) hypothesized that these animals could be derived from or even be part of a hidden marine fauna adapted to an interstitial or crevicular life, where the stagnant, oligoxic waters and the darkness prevailing in these habitats emulate the environmental conditions found in anchialine caves. Indeed, some of them (for example, the calanoids *Exumella, Ridgewayia, Metacalanus,* and *Paramisophria*) still maintain congeners in the shallow marine interstitial medium, especially in tropical seas or in deep-sea suprabenthic layers. A representative of the genus *Exumella* closely related to, if not conspecific with, *Exumella mediterranea*, a species known only from anchialine caves of Sardinia and the Balearics (Mediterranean), has been reported to occur in the community hovering around the halocline developed in a hypersaline, oligoxic basin of the Aegean Sea at 3,500 m depth (Alexander et al. 2009). This finding raises the possibility that the peculiar crustacean community of oligoxic waters of anchialine caves was present also in similar extreme environments of the deep ocean.

ADAPTATIONS TO A SUBTERRANEAN EXISTENCE

Many crustacean stygobionts demonstrate morphological, physiological, and life history adaptations to the subterranean environment (Franz et al. 1994, Culver et al. 1995). The nature of the adaptations is habitat dependent, with animals inhabiting open water of caves being in general large and their body appendages elongate, while those found in interstitial voids are small with elongate, vermiform bodies.

Morphological Adaptations

Stygobiont crustaceans tend to have a convergent morphology, including having a reduction or loss of eyes and pigments where the body is characteristically translucent, being weakly chitinized and having enhanced non-optic sense organs including elongated appendages (Fig. 13.8; Culver et al. 1995). Those species inhabiting interstitial spaces are often elongate, even vermiform, which facilitates their movement between particles. Figure 13.9 illustrates the correlation between *Niphargus* spp. appendage length, body shape, and size with microhabitat.

The suite of morphological traits possessed by cave organisms is the accepted paradigm of troglomorphy (by extension stygomorphy); however, Pipan and Culver (2012), in a review on patterns of troglomorphy, challenged this with two counterarguments: troglomorphic features can be found in ecosystems that are not considered caves or cave-like, and not all organisms in caves, even those in permanently reproducing populations, are troglomorphic. This means that neither the lack of food nor the absence of external environmental cues is the selective force for troglomorphic characteristics. Troglomorphic organisms can be found in the previously described shallow subterranean habitats (SSHs) that are close to the surface (<10 m), and these habitats are aphotic yet abundant in organic material due to their intimate connection with the surface. The rich energy supply may enable the abundant fauna in SSHs and allow the coexistence of, for example, three genera of similar-sized amphipods and three sister species of stygobiont dytiscid beetles in the Sturt Meadows groundwater calcrete aquifer, Australia (Guzik et al. 2009, King et al. 2012; Fig. 13.10).

The evolution of troglomorphic characters, first raised by Darwin (1859), remains the most contentious issue in subterranean biology. At issue are which characters are really related to subterranean life, the process by which they arise, and the attributes guiding the evolution of troglomorphic

Fig. 13.8.
The thaumatocypridid ostracod *Humphreysella baltanasi* (Kornicker 2009), an anchialine species showing characteristic adaptation to subterranean life such as lack of pigment and translucence. Photo courtesy of Ross Anderson. See a color version of this figure in the centerfold.

characteristics—additionally, what drives the evolutionary patterns of parallelism, convergence, and divergence subsumed within the concept of troglomorphism and how best to equate with general evolutionary terminology. Pipan and Culver (2012) made a strong case that the absence of light, and that alone, is the common factor. Research in subterranean biology still has a major focus on whether the reduction or loss of eyes, for example, is the result of directional selection, with eyes being disadvantageous in the dark, or is the result of neutral mutation resulting from the lack of selection in the now-useless character, vision, in the total darkness of subterranean systems.

The presence of venom has evolved repeatedly in the arthropods, appearing in three of the four main groups (Chelicerata, Myriapoda, and Hexapoda) but until von Reumont et al. (2014) reported venom glands in remipedes, no venomous crustaceans had been described. The presence of venom and a controlled delivery system in the Remipedia indicates that venom has in fact evolved in crustaceans. *Speleonectes tulumensis* possesses the ability to inject venom via the hypodermic claw of the maxillule, which is connected to a large gland (Schram and Lewis 1989, van der Ham and Felgenhauer 2007). The gland injects a venom which is a unique mix of mostly known venoms including proteinases, chitinases, and an assumed paralytic neurotoxin.

Physiological Adaptations

Physiological adaptations to a subterranean existence include the loss of circadian periodicity and reduced metabolic rate (Langecker 2000). It is widely recognized that a low metabolic rate is one of the adaptations to the oligotrophic subterranean milieu, and this adaptation is found in both troglobionts and stygobionts. Numerous studies indicate that stygobionts have lower metabolic rates than their epigean counterparts (citations in Bishop et al. 2004, Simčič et al. 2005). The point of contention is the cause of the metabolic depression.

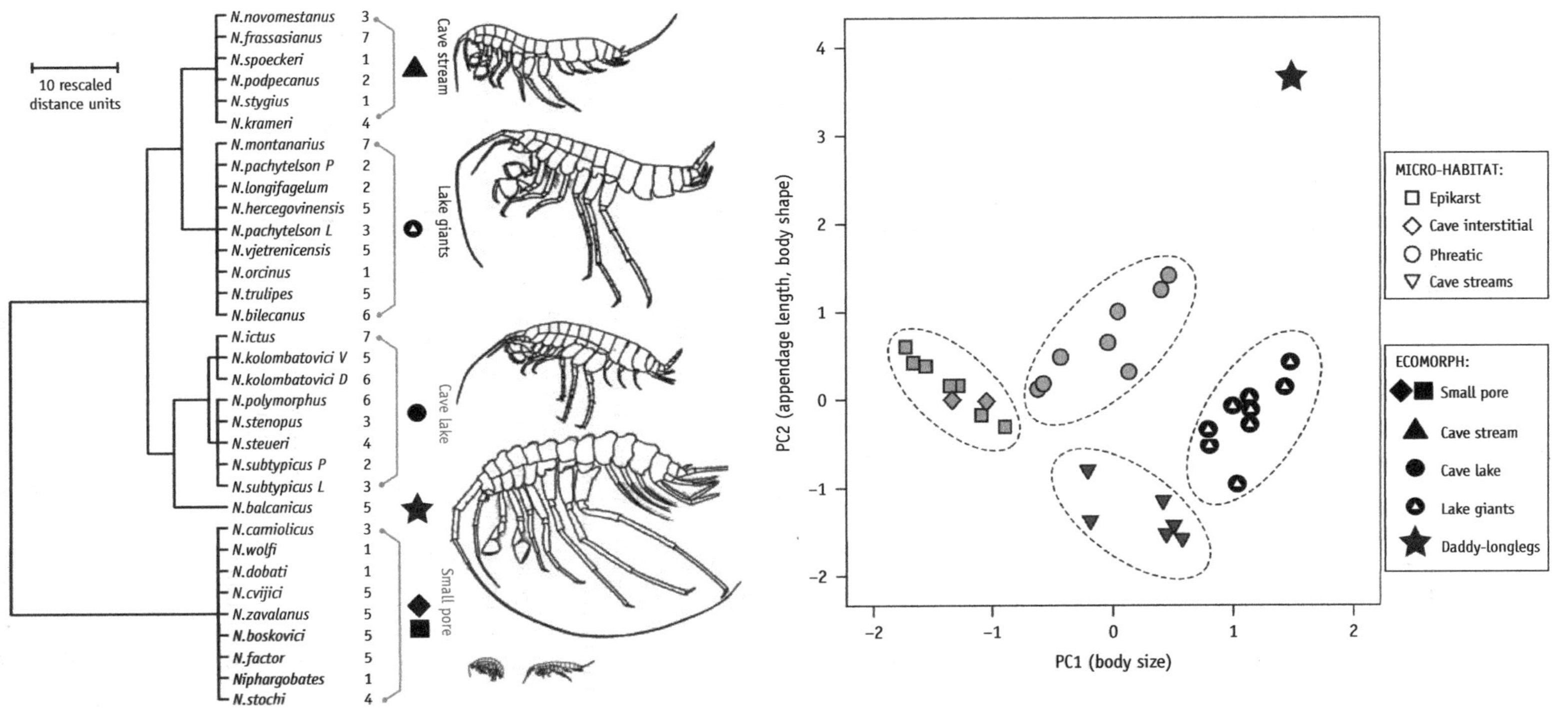

Fig. 13.9.
The relationship between morphology and habitat in *Niphargus*. (Left side) Phenogram of morphological similarity between 33 *Niphargus* species and populations with typical representatives of each morph illustrated approximately to scale. (Right side) First two principal components of the morphological traits of these species, explaining 97.5% of the total variation. Cave microhabitats (symbols) and proposed ecomorphs (shading) are only partially in agreement. From Trontelj et al. (2012), with permission from John Wiley and Sons. See a color version of this figure in the centerfold.

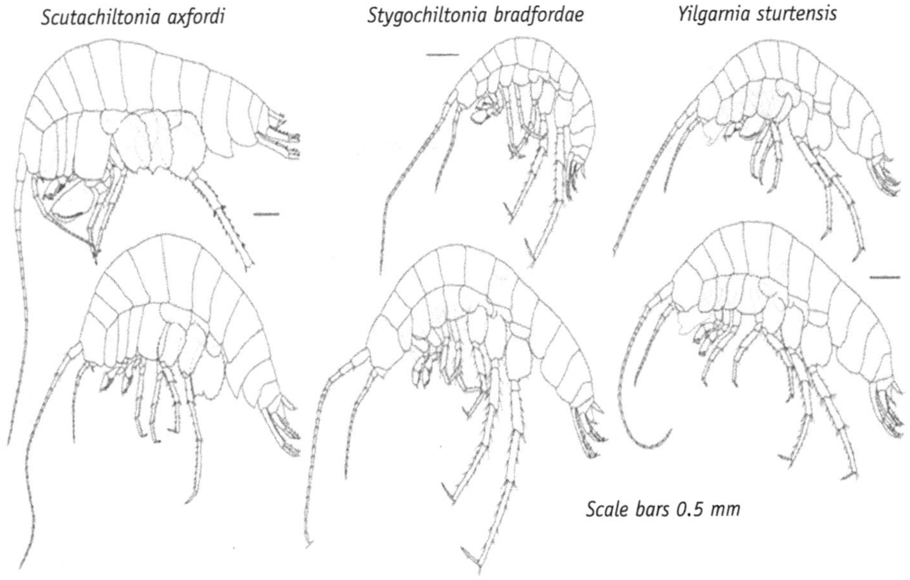

Fig. 13.10.
Sympatric species of chiltoniid amphipods from Sturt Meadows calcrete aquifer, a shallow subterranean habitat. Males upper; females lower. From King et al. (2012), with permission from Oxford University Press.

Species inhabiting habitats with low productivity are expected to have a more generalist diet to meet their energy needs. However, in an oligotrophic groundwater habitat, *Proasellus* showed a high degree of trophic specialization, feeding preferentially on biofilm, a finding that challenges the prediction that in environments with low productivity, species would adopt generalist feeding strategies (Francois et al. 2016).

Another hypothesis proposes that the general reduction in stygobiont metabolism is a response to a low-oxygen environment. Crustaceans conserve their energy in response to environmental anaerobiosis by depressing ATP demand and the reduction of other processes, such as digestion and motility, resulting in significant physiological impacts upon stygobionts. Yet cave organisms have been shown to demonstrate reduced metabolism in well-oxygenated cave systems (Bishop and Iliffe 2009).

The assumption that subterranean organisms have reduced metabolic demands as an evolutionary response to a lack of food or reduced environmental oxygen levels now seems simplistic, and the cause may very well be a dynamic result of environmental factors and a stress response to the aphotic environment. Pipan and Culver (2012) proposed that adaptation to subterranean habitats is a dynamic response to varying selective forces beginning with the absence of light, the strength of environmental stressors, the amount and flux of carbon, and finally the intensity of competition within the subterranean system.

Life History Adaptations

Subterranean crustaceans have fewer offspring and are longer-lived than surface relatives, attributes long considered adaptations to the low-energy aphotic environment but now questionable and they typically lack resting stages (Coineau 2000). These attributes should contribute to the stability of populations but render them slow to recover from perturbations.

For example, among bathynellaceans, the millimeter-long *Antrobathynella stammeri* lays a single egg, and embryonic and postembryonic development takes nine months, while the life span of *Iberobathynella* exceeds 2.5 years (Coineau 2001). In contrast, Venarsky et al. (2012) reported that although longevity of the stygobiont crayfish *Orconectes australis* could exceed 22 years, most did not, and females reached maturity and first reproduction at ages comparable to those of epigean crayfish species.

The features that characterize the life history of subterranean organisms may be a result of stressors of the subterranean habitat, including darkness, a lack of environmental cues (Howarth 1993), and episodic flushing (Matsuda 1982). Light levels have been shown to impact the amount of antennal gland hormones regulating development in talitrid amphipods. Matsuda (1982) proposed that hormonal shifts resulting from stress reduced development, increased mutation rates, and gave rise to neoteny in subterranean crustaceans.

CONSERVATION OF STYGOBIONT CRUSTACEANS

Stygobiont crustaceans are typically short-range endemic species, often exceptionally so, being limited to single caves or small aquifers. Hence species, distributions may be within the boundary of a single moderately sized resource project, such as large dams and mining, especially open cast. They are threatened by the removal of habitat (removal of the geological matrix within which they live or by making subterranean water into surface water by removing the overburden) (see Figures 3 and 4 in Humphreys 2009 and 2011, respectively). The lowering of groundwater levels may leave stygobionts stranded, and an increase in level may place the favorable water depth in an unsuitable rock stratum. Above all, most attention has been paid to changes in the groundwater quality whereby pollutants of many types may threaten stygobionts. Excessive groundwater extraction in continental waters or in near-coastal waters may result in salinity exceeding the physiological tolerance of stygobionts, especially in inland waters (Humphreys 2009). Prolonged withdrawal of groundwater or reduced rainfall, will result in a declining water table that may strand stygofauna in dewatered habitat or drop the water below the substrate matrix inhabitable by stygofauna. Fracking may mix waters from different aquifers of differing water quality unsuitable to sustain the stygofauna or contaminate groundwater with a diverse suite of additives added to facilitate the fracking.

FUTURE DIRECTIONS

A strategic direction for the conservation of stygobionts, among which crustaceans predominate, is to establish the details of the ecological role in maintaining groundwater quality and providing ecosystem services, a daunting task. The issue has been addressed in numerous papers (among them Boulton 2001, Humphreys 2006, Boulton et al. 2008). A study of stygobiontic fauna, including *Paraleptamphopus* amphipods, in sewage-infiltrated groundwaters in New Zealand indicated that pollution may have serious repercussions for stygofauna community structure with potentially irreversible consequences (Hartland et al. 2011).

Comprehensive fieldwork is required to assess the composition of crustacean fauna in the subterranean waters of most of the world. In Europe, where the subterranean environment has been well studied, a comprehensive six-nation study of stygobionts found that 75% of the 379 species were recorded for the first time in the region and that 42% of the species added in the survey were new to science (Deharveng et al. 2009). Groundwater fauna in the United States is scarcely explored outside karst areas, in marked contrast to Australia, where by far the greatest stygobiont and troglobiont diversity is known from non-karst areas. South America, Africa, and India remain largely unexplored in this regard. The lack of scientific representation between the continents impedes another major research direction, molecular phylogenetic verification of the hypothesized congeneric condition of members of many stygobiont lineages of inland groundwater distributed over different continents. In addition to contributing to the resolution of phylogenies, such studies will contribute to our knowledge of the age of the lineages and their history, whether this has been driven by dispersion through the oceans or through vicariance associated with plate tectonics.

Of utmost importance is to address the vulnerability of the subterranean environment to anthropogenic impacts. There is a need to determine and develop improved ways of evaluating these fragile habitats non-destructively and on a finer scale to further examine processes without disturbing the environment. The extent to which stygobitic organisms are vulnerable to anthropogenic-induced environmental changes has not been critically assessed; however, given the known life history traits, endemic nature of populations, and specificity of the stygobitic environment, the conclusion that these unique habitats require urgent protection cannot be ignored.

SUMMARY AND CONCLUSIONS

The physical features of subterranean habitats vary substantially, ranging from shallow seeps to deep cave systems. Concomitantly, subterranean crustaceans, found virtually everywhere there is a subterranean void, are diverse in habitat range, origin, and adaptations to the subterranean realm.

Crustaceans are ubiquitous in the subterranean habitats, comprising 55% of the stygofauna, yet within Crustacea, the tendency toward successful colonization of the subterranean realm varies widely between the higher taxonomic groups. Of the families of isopods with subterranean representatives, many have ancient biogeographic patterns and are found associated with land masses derived from Laurasia or Gondwana. Ostracoda are prevalent in subterranean waters, occupying a range of stygobitic habitats.

There is also significant diversity in the habitat range of subterranean organisms. Freshwater stygobionts, limited in their dispersal, tend to be narrow-range endemics with the evolution of several taxa being old enough to reflect global plate tectonics. Many groups, such as the amphipod families Pseudoniphargidae and Metacrangonyctidae and the thermosbaenaceans *Tethysbaena,* have distributional patterns that correspond to Tertiary and Late Cretaceous land–sea arrangements. The distribution pattern of the diverse anchialine fauna initially followed the coasts of the Tethys Sea to be affected subsequently by vicariance and plate tectonics.

Originating from epigean ancestors, many species of stygobiont crustaceans can theoretically exist in the epigean and subterranean habitat simultaneously. The presence of preadaptations alone does not guarantee success in colonizing the subterranean environment. Based on geographic distribution, taxonomic affinities, and salt tolerance, most stygobionts have a marine origin. Thalasso-stygobionts are restricted to marine waters and could have been derived from marine fauna adapted to life in oligoxic waters with similar environmental conditions to those of anchialine caves.

Adaptations to a subterranean existence encompass morphological, physiological, and life history characteristics; yet in some instances, what are thought to be troglomorphic characteristics can be observed in epigean species as well. Paradoxically, not all stygobiotic species demonstrate troglomorphism. One primary adaptation to the stygobiotic environment is the lower metabolism observed in subterranean organisms than in epigean counterparts, but like all other areas in subterranean research, there are exceptions. Reduced metabolic rates are most likely a complex response to the environmental parameters of each habitat and not a result of a single characteristic such as reduced food availability.

It is clear that continued interdisciplinary research is necessary to better understand the distribution, origin, and faunistic composition of the world's subterranean habitats and the adaptations of subterranean organisms. Additionally, collaborative research is needed to raise awareness of the fragility of subterranean ecosystems.

REFERENCES

Abrams, K.M., M.T. Guzik, S.J. Cooper, W.F. Humphreys, R.A. King, J.L. Cho, and A.D. Austin. 2012. What lies beneath: molecular phylogenetics and ancestral state reconstruction of the ancient subterranean Australian Parabathynellidae (Syncarida, Crustacea). Molecular Phylogenetics and Evolution 64:130–144.

Alcocer, J., A. Lugo, M. del Rosario Sánchez, E. Escobar, and M. Sánchez. 1999. Bacterioplankton from cenotes and anchialine caves of Quintana Roo, Yucatan Peninsula, Mexico. Journal of Tropical Biology and Conservation 47:19–25.

Alexander, E., A. Stock, H.W. Breiner, A. Behnke, J. Bunge, M.M. Yakimov, and T. Stoeck. 2009. Microbial eukaryotes in the hypersaline anoxic L'Atalante deep-sea basin. Environmental Microbiology 11:360–381.

Baratti, M., M. Yacoubi Khebiza, and G. Messana. 2004. Microevolutionary processes in the stygobitic genus *Typhlocirolana* (Isopoda Flabellifera Cirolanidae) as inferred by partial 12S and 16S rDNA sequences. Journal of Zoological Systematics and Evolutionary Research 42:27–32.

Baratti, M., M. Filippelli, F. Nardi, and G. Messana. 2010. Molecular phylogenetic relationships among some stygobitic cirolanid species (Crustacea, Isopoda). Contributions to Zoology 79:57–67.

Bauzà-Ribot, M.M., C. Juan, F. Nardi, P. Oromí, J. Pons, and D. Jaume. 2012. Mitogenomic phylogenetic analysis supports continental-scale vicariance in subterranean thalassoid crustaceans. Current Biology 22:1–6.

Bishop, R.E., and T.M. Iliffe. 2009. Metabolic rates of stygobiontic invertebrates from the Túnel de la Atlántida, Lanzarote. Marine Biodiversity 39:189–194.

Bishop, R.E., B. Kakuk, and J.J. Torres. 2004. Life in the hypoxic and anoxic zones: metabolism and proximate composition of Caribbean troglobitic crustaceans with observations on the water chemistry of two anchialine caves. Journal of Crustacean Biology 24:379–392.

Bishop, R.E., W.F. Humphreys, N. Cukrov, V. Žic, G.A. Boxshall, M. Cukrov, T. Iliffe, F. Krsinic, W.S. Moore, J. Pohlman, and B. Sket. 2015. "Anchialine" redefined as a subterranean estuary in a crevicular or cavernous geological setting. Journal of Crustacean Biology 35:511–514.

Botosaneanu, L., editor. 1986. Stygofauna Mundi: A Faunistic, Distributional, and Ecological Synthesis of the World Fauna Inhabiting Subterranean Waters (including the Marine Interstitial). E.J. Brill, Leiden, The Netherlands.

Boulton, A.J. 2001. Twixt two worlds: taxonomic and function biodiversity at the surface water/groundwater interface. Records of the Western Australian Museum, Supplement No. 64:1–13.

Boulton, A.J., G.D. Fenwick, P.J. Hancock, and M.S. Harvey. 2008. Biodiversity, functional roles and ecosystem services of groundwater invertebrates. Invertebrate Systematics 22:117–126.

Boulton, A.J., M.A. Brock, B.J. Robson, D.S. Ryder, J.M. Chambers, and J.A. Davis. 2014. Australian Freshwater Ecology: Processes and Management, 2nd ed. John Wiley, Oxford, UK.

Boxshall, G.A., S. Zylinski, D. Jaume, T.M. Iliffe, and E. Suárez-Morales. 2014. A new genus of speleophriid copepod (Copepoda: Misophrioida) from a cenote in the Yucatan, Mexico with a phylogenetic analysis at the species level. Zootaxa 3821:321–336.

Bruce, N. L., and W.F. Humphreys. 1993. *Haptolana pholeta*, sp. nov., the first subterranean flabelliferan isopod crustacean (Cirolanidae) from Australia. Invertebrate Systematics 7:875–884.

Camacho, A.I., E. Serban, and N. Guil. 2000. Phylogenetical review and biogeographic remarks on the interstitial and subterranean freshwater iberobathynells (Crustacea, Syncarida, Parabathynellidae). Journal of Natural History 34:563–585.

Camacho, A.I., B.A. Dorda, and I. Rey. 2012. Undisclosed taxonomic diversity of Bathynellacea (Malacostraca: Syncarida) in the Iberian Peninsula revealed by molecular data. Journal of Crustacean Biology 32:816–826.

Coineau, N. 2000. Adaptations to interstitial groundwater life. Pages 189–210 *in* D. Wilkens, C. Culver, and W.F. Humphreys, editors. Ecosystems of the World, Vol. 30: Subterranean Ecosystems. Elsevier, Amsterdam.

Coineau, N. 2001. Syncarida. Pages 863–876 *in* C. Juberthie and V. Decu, editors. Encyclopaedia Biospeologica. Tome 1. Societé Internationale de Biospéologie, Moulis, France.

Cooper, S.J., K.M. Saint, S. Taiti, A.D. Austin, and W.F. Humphreys. 2008. Subterranean archipelago: mitochondrial DNA phylogeography of stygobitic isopods (Oniscidea: Haloniscus) from the Yilgarn region of Western Australia. Invertebrate Systematics 22:195–203.

Culver, D.C., and T. Pipan. 2011. Redefining the extent of the aquatic subterranean biotope—shallow subterranean habitats. Ecohydrology 4:721–730.

Culver, D.C., T.C. Kane, and W. Fong. 1995. Adaptation and Natural Selection in Caves. The Evolution of *Gammarus minus*. Harvard University Press, Cambridge, MA.

Culver, D., T. Pipan, and S. Gottstein. 2006. Hypotelminorheic: a unique freshwater habitat. Subterranean Biology 4:1–7.

Danielopol, D.L. 1984. Ecological investigations on the alluvial sediments of the Danube in the Vienna area—a phreatobiological project. Verhandlungen der Internationalen Vereinigung für Limnologie 22:1755–1761.

Danielopol, D.L., and G. Bonaduce. 1989. The colonization of subsurface habitats by Loxoconchidae Sars and Psammocytheridae Klie (Ostracoda). Pages 437–458 *in* R. Whatley and C. Maybury, editors. Ostracoda and Global Events. Chapman and Hill, London.

Danielopol, D.L., P. Pospisil, and R. Rouch. 2000. Biodiversity in groundwater: a large-scale view. Trends in Ecology and Evolution 15:223–224.

Darwin, C. 1859. On the Origin of Species by Means of Natural Selection, or the Preservation of Favoured Races in the Struggle of Life. John Murray, London.

Deharveng, L., F. Stoch, J. Gibert, A. Bedos, D. Glassi, and M. Zagmajster. 2009. Groundwater biodiversity in Europe. Freshwater Biology 54:709–726.

de Queiroz, A. 2016. Jurassic primates, immobile ducks and other oddities: a reply to Heads' review of The Monkey's Voyage. Australian Systematic Botany 29:403–423.

Francois, C.M., F. Mermillod-Blondin, F. Malard, C. Fourel, C.J. Lécuyer, L. Douady, and L. Simon. 2016. Trophic ecology of groundwater species reveals specialization in a low-productivity environment. Functional Ecology 30:262–273.

Franz, R., J. Bauer, and T. Morris. 1994. Review of biologically significant caves and their faunas in Florida and South Georgia. Brimleyana 20:1–109.

Galassi, D.M. 2001. Groundwater copepods: diversity patterns over ecological and evolutionary scales. Hydrobiologia 453/454:227–253.

Goñi, M.A., and I.R. Gardner. 2003. Seasonal dynamics in dissolved organic carbon concentrations in a coastal water-table aquifer at the forest-marsh interface. Aquatic Geochemistry 9:209–232.

Gourbault, N., and F. Lescher-Moutoué. 1979. Faune des eaux souterraines de Majorque. Endins 5/6:43–54.

Guzik, M.T., S.J.B Cooper, W.F. Humphreys, and A.D. Austin. 2009. Fine-scale comparative phylogeography of a sympatric sister species triplet of subterranean diving beetles from a single calcrete aquifer in Western Australia. Molecular Ecology 18:3683–3698.

Hancock, P.J. 2006. The response of hyporheic invertebrate communities to a large flood in the Hunter River, New South Wales. Hydrobiologia 568:255–262.

Hartland, A., G.D. Fenwick, and S.J. Bury. 2011. Tracing sewage-derived organic matter into a shallow groundwater food web using stable isotope and fluorescence signatures. Marine and Freshwater Research 62:119–129.

Hobbs, H.H. 2012. Crustacea. Pages 177–194 *in* W.B. White, and D.C. Culver, editors. Encyclopedia of Caves. Academic Press, San Diego, CA.
Holsinger, J. R. 1993. Biodiversity of subterranean amphipod crustaceans. Global patterns and zoogeographic implications. Journal of Natural History 27:821–835.
Holsinger, J. R. 1991. What can vicariance biogeographic models tell us about the distributional history of subterranean amphipods? Hydrobiologia 223:43–45.
Holthuis, L.B. 1973. Caridean shrimps found in land-locked saltwater pools at four Indo-West Pacific localities (Sinai Peninsula, Funafuti Atoll, Maui and Hawaii Islands): with the description of one new genus and four new species. Zoologische Verhandelingen 128:1–55.
Howarth, F.G. 1993. High-stress subterranean habitats and evolutionary change in cave-inhabiting arthropods. American Naturalist 142:S65–S77.
Humphreys, W.F. 2006. Aquifers: the ultimate groundwater dependent ecosystems. Australian Journal of Botany 54:115–132.
Humphreys, W.F. 2008. Rising from down under: developments in subterranean biodiversity in Australia from a groundwater fauna perspective. Invertebrate Systematics 22:85–101.
Humphreys, W.F. 2009. Hydrogeology and groundwater ecology: does each inform the other? Hydrogeology Journal 17:5–21.
Humphreys, W.F. 2011. Management of groundwater species in Karst environments. Pages 283–318 *in* P. van Beynen, editor. Karst Management. Springer Verlag, Berlin.
Humphreys, W.F. 2012. Diversity patterns in Australia. Pages 203–219 *in* W.B. White and D.C. Culver, editors. Encyclopedia of Caves. Academic Press, San Diego, CA.
Humphreys, W.F. 2014. Subterranean fauna of Christmas Island: habitats and salient features. Raffles Bulletin of Zoology Supplement 30:29–44.
Humphreys, W.F. 2017. Australasian subterranean biogeography. Pages 269–293 *in* M.C. Ebach, editor. Handbook of Australasian Biogeography. CRC Press, Boca Raton, FL.
Humphreys, W.F., and D.L. Danielopol. 2006. *Danielopolina* (Ostracoda, Thaumatocyprididae) on Christmas Island, Indian Ocean, a seamount island. Crustaceana 78:1339–1352.
Humphreys, W.F., C.H.S. Watts, S.J.B. Cooper, and R. Leijs. 2009. Groundwater estuaries of salt lakes: buried pools of endemic biodiversity on the western plateau, Australia. Hydrobiologia 626:79–95.
Iliffe, T.M. 1992. Anchialine cave biology. Pages 614–636 *in* A.I. Camacho, editor. The Natural History of Biospeleology. Monografias Museo Nacional de Ciencias Naturales, Madrid.
Iliffe, T.M., and L.S. Kornicker. 2009. Worldwide diving discoveries of living fossil animals from the depths of anchialine and marine caves. Smithsonian Contributions to Marine Sciences 38:269–280.
Jaume, D., and G.A. Boxshall. 1996. Rare cyclopoid copepods (Crustacea) from Mediterranean littoral caves. Bulletin of The Natural History Museum, London (Zoology Series) 62:83–99.
Juberthie, C., and V. Decu, editors. 1994. Encyclopaedia Biospeologica. Tome 1. Societé Internationale de Biospéologie, Moulis, France.
Juberthie, C., and V. Decu, editors. 2001. Encyclopaedia Biospeologica. Tome III. Societé Internationale de Biospéologie, Moulis, France.
Jurado-Rivera, J.A., J. Pons, F. Alvarez, A. Botello, W.F. Humphreys, T.J. Page, T.M. Iliffe, E. Willassen, K. Meland, C Juan, and D. Jaume. 2017. Phylogenetic evidence that both ancient vicariance and dispersal have contributed to the biogeographic patterns of anchialine cave shrimps. Scientific Reports 7:2852.
King, R.A., T. Bradford, A.D. Austin, W.F. Humphreys, and S.J.B. Cooper. 2012. Divergent molecular lineages and not-so-cryptic species: the first descriptions of stygobitic chiltoniid amphipods (Talitroidea: Chiltoniidae) from Western Australia. Journal of Crustacean Biology 32:465–488.
Kornicker, L.S., T.M. Iliffe, and E. Harrison-Nelson. 2007. Ostracoda (Myodocopa) from anchialine caves and ocean blue holes. Zootaxa 1565:1–151.
Kwokal, Ž., N. Cukrov, and V. Cuculić. 2014. Natural causes of changes in marine environment: mercury speciation and distribution in anchialine caves. Estuarine, Coastal and Shelf Science 151:10–20.
Langecker, T.G. 2000. The effects of continuous darkness on cave ecology and cavernicolous evolution. Pages 135–158 *in* H. Wilkens, D.C. Culver, and W.F. Humphreys, editors. Ecosystems of the World, Vol. 30: Subterranean Ecosystems. Elsevier, Amsterdam.

Marmonier, P., and M. Creuze des Chateliers. 1992. Biogeography of the benthic and interstitial living ostracods (Crustacea) of the Rhone River (France). Journal of Biogeography 19:693–704.

Matsuda, R. 1982. The evolutionary process in talitrid amphipods and salamanders in changing environments, with a discussion of "genetic assimilation" and some other evolutionary concepts. Canadian Journal of Zoology 60:733–749.

Moore, W.S. 1999. The subterranean estuary: a reaction zone of ground water and sea water. Marine Chemistry 65:111–125.

Notenboom, J. 1988. Phylogenetic relationships and biogeography of the groundwater-dwelling amphipod genus *Pseudoniphargus* (Crustacea), with emphasis on the Iberian species. Bijdragen tot de Dierkunde 58:159–204.

Notenboom, J. 1991. Marine regressions and the evolution of groundwater dwelling Amphipoda. Journal of Biogeography 18:437–454.

Page, T.J., J.M. Hughes, K.M. Real, M.I. Stevens, R.A. King, and W.F. Humphreys. 2016. Allegory of the cave crustacean: systematic and biogeographic reality of *Halosbaena* (Peracarida: Thermosbaenacea) sought with molecular data at multiple scales. Marine Biodiversity 2016:1–18.

Pérez-Moreno, J.L, T.M. Iliffe, and H.D. Bracken-Grissom. 2016. Life in the underworld: Anchialine cave biology in the era of speleogenomics. International Journal of Speleology 45:149–170.

Phillips, M.J., T.J. Page, M. de Bruyn, J.A. Huey, W.F. Humphreys, J.M. Hughes, and J.M. Waters. 2013. The linking of plate tectonics and evolutionary divergence. Current Biology 23:R603–R605.

Pipan, T., and D.C. Culver. 2007. Regional species richness in an obligate subterranean dwelling fauna–epikarst copepods. Journal of Biogeography 34:854–861.

Pipan, T., and D.C. Culver. 2012. Convergence and divergence in the subterranean realm: a reassessment. Biological Journal of the Linnean Society 107:1–14.

Poore, G.C.B. 2015. Orders Bochusacea, Mictacea and Spelaeogriphacea. Pages 77–92 *in* J.C. von Vaupel Klein, M. Charmantier-Duares, and F.R. Schram, editors. Treatise on Zoology—Anatomy, Taxonomy, Biology. Revised and updated, as well as extended from the Traité de Zoologie 5. Brill, Leiden, The Netherlands.

Poore, G.C.B., and W.F. Humphreys. 2013. *Bunderanthura bundera* gen. et sp. nov. from Western Australia, first anchialine Leptanthuridae (Isopoda) from the Southern Hemisphere. Records of the Western Australian Museum 28:21–29.

Ranga Reddy, Y. 2011. Gondwanan heritage in groundwater crustaceans of peninsular India. Current Science 101:156–158.

Reeves, J. M., P. De Deckker, and S.A. Halse. 2007. Groundwater ostracods from the arid Pilbara region of northwestern Australia: distribution and water chemistry. Hydrobiologia 585:99–118.

Riedl, R. 1966. Biologie der Meereshöhlen. Paul Parey, Hamburg, Germany.

Rix, M.G., and M.S. Harvey. 2012. Phylogeny and historical biogeography of ancient assassin spiders (Araneae: Archaeidae) in the Australian mesic zone: evidence for Miocene speciation within Tertiary refugia. Molecular Phylogenetics and Evolution 62:375–396.

Rouch, R. 1991. Structure du peuplement des harpacticides dans le milieu hyporhéique d'un ruisseau des Pyrénées. Annales de Limnologie 27:227–241.

Schmidt, C. 2008. Phylogeny of the terrestrial Isopoda (Oniscidea): a review. Arthropod Systematics and Phylogeny 66:191–226.

Schminke, H.K. 2011. Arthropoda: Crustacea: Malacostraca: Bathynellacea: Parabathynellidae. Page 244 *in* Invertebrate Fauna of the World, Vol. 21: Number 1. National Institute of Biological Resources. Ministry of Environment, Incheon, Republic of Korea.

Schram, F.R., and C.A. Lewis. 1989. Functional morphology of feeding in the Nectiopoda. Crustacean Issues 6:115–122.

Simčič, T., S. Lukančič, and A. Brancelj. 2005. Comparative study of electron transport system activity and oxygen consumption of amphipods from caves and surface habitats. Freshwater Biology 50:494–501.

Sket, B. 1970. Über Struktur und Herkunft der unterirdischen Fauna Jugoslawiens. Bioloski vestnik 18:69–78.

Sket, B. 1994. Distribution of *Asellus aquaticus* (Crustacea: Isopoda: Asellidae) and its hypogean populations at different geographic scales, with a note on *Proasellus istrianus*. Hydrobiologia 287:39–47.

Sket, B. 2004. The cave hygropetric—a little known habitat and its inhabitants. Archiv für Hydrobiologie 160:413–425.

Sket, B. 2012. Anchihaline (anchialine) Caves and Fauna. Pages 17–24 *in* W.B. White and D.C. Culver, editors. Encyclopedia of Caves. Academic Press, San Diego, CA.

Sket, B., and N.L. Bruce. 2004. Sphaeromatids (Isopoda, Sphaeromatidae) from New Zealand fresh and hypogean waters, with description of *Bilistra* n. gen. and three new species. Crustaceana 76:1347–1370.

Smith, R.J., T. Kamiya and D.J. Horne. 2006. Living males of the "ancient asexual" Darwinulidae (Ostracoda: Crustacea). Proceedings of the Royal Society B: Biological Sciences 273:1569–1578.

Stanford, J.A., J.V. Ward, and B.K. Ellis. 1994. Ecology of the alluvial aquifers of the Flathead River, Montana. Pages 367–389 *in* J. Gibert, D.L. Danielopol, and J.A. Stanford, editors. Groundwater Ecology. Academic Press, London.

Stock, J.H. 1980. Regression model evolution as exemplified by the genus *Pseudoniphargus* (Amphipoda). Bijdragen tot de Dierkunde 50:105–144.

Stock, J.H. 1986. Two new amphipod crustaceans of the genus *Bahadzia* from "blue holes" in the Bahamas and some remarks on the origin of the insular stygofaunas of the Atlantic. Journal of Natural History 20:921–933.

Stock, J.H. 1993. Some remarkable distribution patterns in stygobiont Amphipoda. Journal of Natural History 27:807–819.

Stock, J.H., T.M. Iliffe, and D. Williams. 1986. The concept "anchialine" reconsidered. Stygologia 2:90–92.

Tabacaru, I. 1999. L'adaptation à la vie aquatique d'un remarquable trichoniscide cavernicole, *Cantabroniscus primitivus* Vandel, et le problem de la monophylie des isopodes terrestres. Travaux de l'Institut de Spéologie "Emile Racovitza" 37/38:115–131.

Taiti, S., and W.F. Humphreys. 2001. New aquatic Oniscidea (Crustacea: Isopoda) from groundwater calcretes of Western Australia. Records of the Western Australian Museum, Supplement No. 64:133–151.

Trontelj, P., C.J. Douady, C. Fišer, J. Gibert, P. Gorički, T. Lefébure, B. Sket, and V. Zakšek. 2009. A molecular test for cryptic diversity in ground water: how large are the ranges of macro-stygobionts? Freshwater Biology 54:727–744.

Trontelj, P., A. Blejec, and C. Fišer. 2012. Ecomorphological convergence of cave communities. Evolution 66:3852–3865.

van der Ham, J.L., and B.E. Felgenhauer. 2007. The functional morphology of the putative injecting apparatus of *Speleonectes tanumekes* (Remipedia). Journal of Crustacean Biology 27:1–9.

Venarsky, M.P., A.D. Huryn, and J.P. Benstead. 2012. Reexamining extreme longevity of the cave crayfish *Orconectes australis* using new mark-recapture data: a lesson on the limitations of iterative size at age models. Freshwater Biology 57:1471–1481.

Vonk, R., and F.R. Schram. 2003. Ingolfiellidea (Crustacea, Malacostraca, Amphipoda): a phylogenetic and biogeographic analysis. Contributions to Zoology 72:39–72.

von Reumont, B.M., R.A. Jenner, M.A. Wills, E. Dell'Ampio, G. Pass, I. Ebersberger, B. Meyer, S. Koenemann, T.M. Iliffe, A. Stamatakis, O. Niehuis, K. Meusemann, and B. Misof. 2012. Pancrustacean phylogeny in the light of new phylogenomic data: support for Remipedia as the possible sister group of Hexapoda. Molecular Biology and Evolution 29:1031–1045.

von Reumont, B.M., A. Blanke, S. Richter, F. Alvarez, C. Bleidorn, and R.A. Jenner. 2014. The first venomous crustacean revealed by transcriptomics and functional morphology: remipede venom glands express a unique toxin cocktail dominated by enzymes and a neurotoxin. Molecular Biology and Evolution 31:48–58.

Wägele, J.W. 1983. On the origin of the Microcerberidae (Crustacea: Isopoda). Zeitschrift fur Zoologische Systematik und Evolutionsforsch 21:249–262.

Wägele, J.W. 1989. Evolution and phylogenetisches System der Isopoda. Stand der Forschung und neue Erkenntnisse. Zoologica 140:1–262.

Wägele, J.W. 1990. Aspects of the evolution and biogeography of stygobiontic Isopoda (Crustacea: Peracarida). Bijdragen tot de Dierkunde 60:145–150.

Wagner, H.P. 1994. A monographic review of the Thermosbaenacea (Crustacea: Peracarida). Zoologische Verhandelingen 291:1–338.

Watts, C.H.S., and W.F. Humphreys. 2001. A new genus and six new species of Dytiscidae (Coleoptera) from underground waters in the Yilgarn palaeodrainage system of Western Australia. Records of the South Australian Museum 34:99–114.

Williams, P.W. 2008. The role of the epikarst in karst and cave hydrogeology: a review. International Journal of Speleology 37:1–10.

Wilkens, H., D.C. Culver, and W.F. Humphreys, editors. 2000. Ecosystems of the World, Vol. 30, Subterranean Ecosystems. Elsevier, Amsterdam.

Wilson, G.D. 2008. Gondwanan groundwater: subterranean connections of Australian phreatoicidean isopods (Crustacea) to India and New Zealand. Invertebrate Systematics 22:301–310.

Wilson, G.D., and S.J. Keable. 1999. A new genus of phreatoicidean isopod (Crustacea) from the north Kimberley region, Western Australia. Zoological Journal of the Linnean Society 126:51–79.

Wilson, G.D., and S.J. Keable. 2001. Systematics of the Phreatoicidea. Crustacean Issues 13:175–194.

White, W.B., and D.C. Culver. 2012. Encyclopedia of Caves. Academic Press, San Diego, CA.

14

TERRESTRIAL ENVIRONMENTS

Spyros Sfenthourakis, Alan A. Myers, Stefano Taiti, and James K. Lowry

Abstract

Among crustaceans, only Amphipoda, Isopoda, and Decapoda have invaded truly terrestrial environments, but only two groups show full adaptations to live on land: the family Talitridae among the Amphipoda and the suborder Oniscidea among the Isopoda. The Talitridae occur primarily in forest leaf litter, but a number of other habitats, including caves, are recorded. Talitrids are important ecological contributors to the litter fauna, often occurring in high densities. Their adaptations to a terrestrial way of life include the retention of the mitten-shaped second gnathopods, a neotenic condition among males; the first article of antenna 2 greatly enlarged and fixed to the side of the head; and enlarged gills and pleopods often reduced, sometimes to vestigial stumps. Talitrids have a skewed world distribution being at their most diverse in New Zealand, Tasmania, and Japan/Taiwan. They occur in the Caribbean and Central America but are absent from South and North America except as introduced taxa. Their distribution is largely a result of tectonic activity during the past 150 million years and of extinctions during the Tertiary due to increasing aridity of the climate. The Oniscidea (terrestrial isopods) are the only crustaceans that have managed to adapt to almost all habitat types on land and have become the most species-rich suborder of Isopoda. Although monophyly of the Oniscidea is generally accepted, current taxonomy, based almost entirely on morphological characters, needs extensive revision. Terrestrial isopods present a number of unique adaptations to life on land, some of which result from what can be considered as pre-adaptations of ancestral marine isopods, such as egg development in a marsupium, being dorso-ventrally oblate and having a pleopodal respiration. Other crucial adaptations of Oniscidea include the water-conducting system, the structure of their cuticle, and the "covered" type of pleopodal lungs, all of which are responses to the acute problem of desiccation. They are also among the most speciose taxa in caves, some species have even returned to an aquatic life, and a few species have evolved

Evolution and Biogeography. Edited by Gary C. B. Poore and Martin Thiel.

social behavior. Oniscidea are increasingly being used in biogeographical, phylogeographical, ecological, and evolutionary research and can become model organisms for a broad range of biological studies.

INTRODUCTION

Among crustaceans, three major groups have invaded terrestrial environments: the peracarid suborders, Amphipoda and Isopoda, and Decapoda (Richardson and Araujo 2015). A small number of Ostracoda and Copepoda might also be regarded as terrestrial, as they live in the film of water formed under fallen leaves and branches, among mosses or other very humid sites in dense forests. Nevertheless, given that these forms have not developed a full set of adaptations to the terrestrial environment, they can be considered semi-terrestrial. Among all these crustacean groups, only the family Talitridae among the Amphipoda and the suborder Oniscidea among the Isopoda show full adaptations to live on land with no need of an aquatic phase during their entire life. These two groups are treated in this chapter.

Amphipoda

The family Talitridae includes the only truly terrestrial amphipods, commonly called land-hoppers. There are currently about 358 recognized talitrid species in about 118 genera. Three-quarters of these, about 117 species in 90 genera, are terrestrial. Although the majority of land-hoppers live under forest litter, from sea level to 1,314 m in the Dominican Republic (*Caribitroides genaroi*) (Ortiz and Lalana 2009) and to 1,600 m in the cloud forests in Costa Rica (*Cerrorchestia hyloraina)* (Lindeman 1991a), some species have invaded other habitats. *Makawe hurleyi* lives in tussock grassland from sea level to 2,000 m altitude on South Island, New Zealand (Duncan 1994). A Tasmanian species, *Albidiator albidus*, burrows up to 20 cm in the soil (Friend 1987). Moeed and Meads (1983) reported talitrids as common in certain New Zealand trees up to 1.5 m above ground. Friend and Richardson (1986) reported talitrids from the rain forest canopy in north Queensland. Richardson (1991) reported *Laniporchestia lanipo* living in moss 2 m up the tree *Metrosideros collina*, and *Hawaiorchestia gagnei* living in the axils of *Freycinetia arborea* in mountainous terrain on Oahu, Hawaii. Although rare, talitrids are known to inhabit caves in Western Australia (Lowry and Springthorpe 2015b), off the coast of Japan (White et al. 2013), Hawaii (Bousfield and Howarth 1976), and the Canary Islands in the northeastern Atlantic (Stock and Martin 1988). Finally, talitrids have been collected under snow at 1,850 m altitude on Mt. Kosciusko in Australia (Friend and Richardson 1986).

Isopoda

Terrestrial isopods (suborder Oniscidea) are the only lower taxonomic group, i.e., below class or rather phylum level, that have living representatives of all evolutionary grades leading from marine to fully terrestrial life. As such, they constitute an important model group for comparative studies in a vast variety of subjects, such as evolution, ecology, and ecophysiology. The origin of Oniscidea has been estimated sometime in the Carboniferous (Broly et al. 2013). Sfenthourakis and Taiti (2015) reported 3,710 species in 527 genera and 37 families as of April 2014.

Among Oniscidea, there is a wide variation in evolutionary lines and forms, as shown also by a strong correlation between species richness and generic richness within families combined with the absence of correlation between species richness in genera and generic richness in families

(Sfenthourakis and Taiti 2015). This pattern emerges from the occurrence of many monotypic or small genera in large families and very species-rich genera in small families (e.g., Platyarthridae: 7 genera with 122 species, Trachelipodidae: 6 genera with 109 species, Ligiidae: 6 genera with 95 species). Nevertheless, Oniscidea classification might be radically revised in the near future under the light of molecular phylogenetic analyses.

According to the most widely accepted taxonomy today (see Fig. 14.1), Crinocheta are the richest in biodiversity (c. 80% of species) including groups with more advanced adaptations to almost all terrestrial ecosystems. Species richness is highly asymmetric, with most speciose clades generally being also those assumed to be the most derived, but not without exceptions. The largest family, Armadillidae, with some 580 species in 80 genera, belongs to a derived clade, but Trichoniscidae, belonging to the less derived Synocheta, is also among the richest, with some 500 species and 87 genera. In addition, Philosciidae, a relatively basal clade within Crinocheta, has around 500 species in 107 genera, although it likely is not monophyletic. Schmidt (2008), based on morphological evidence, proposed a slightly different classification, which is trying to reflect more accurately the assumed phylogenetic relationships within Oniscidea.

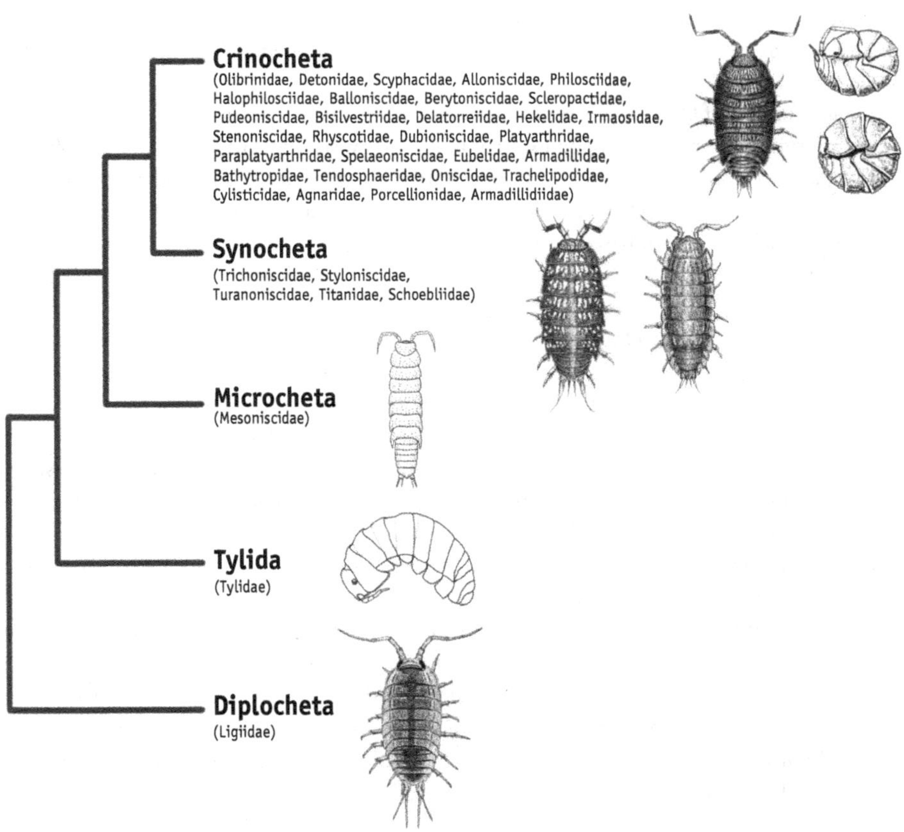

Fig. 14.1.
Phylogenetic relationships among the main oniscidean lineages. From Erhard 1998, with permission from Taylor & Francis ©.

PHYLOGENY AND PHYLOGEOGRAPHY

Up to now, we lack a comprehensive phylogeny or adequate phylogeographic analyses of amphipods. Suffice it to say that talitrid amphipods shared a common ancestor with the intertidal talitroid family Hyalidae (Bousfield 1984). This section focuses on isopods, for which some preliminary work has been published.

Isopoda

The relationship of Oniscidea to other suborders of Isopoda has been ambiguous, with early authors proposing Asellota, "Flabellifera," and Valvifera, or other combinations of taxa as their sister group (Schmidt 2008). A cladistic approach by Brusca and Wilson (1991) suggested Calabozoidea as the sister group of Oniscidea, while Tabacaru and Danielopol (1996) proposed Valvifera. A comprehensive phylogenetic analysis of Isopoda based on molecular evidence is lacking, but Wetzer (2002) used mtDNA markers that might not be appropriate for such a deep phylogeny, and Dreyer and Wägele (2002) used one nuclear DNA marker. The latter authors proposed the monophyletic clade, Scutocoxifera, that includes Oniscidea with Valvifera, Sphaeromatidea, Cymothoida, and Anthuridea (no longer recognized as a suborder; Brandt and Poore 2003). This arrangement was supported by certain morphological characters, with Oniscidea being the basal clade within this group. More robust analyses, based on more markers, are needed to fully resolve relationships among isopod suborders.

The monophyly of Oniscidea is generally supported by many morphological characters (for a general overview of terrestrial isopod external morphology, see Fig. 14.2). Before the application of principles of phylogenetic systematics, some authors supported an independent origin of various terrestrial isopod groups from different marine taxa (e.g., Vandel 1965). More recently, Michel-Salzat and Bouchon (2000), based on mtDNA markers but a similarity-based tree construction method, suggested that *Ligia* and *Tylos* are closer to Valvifera and Sphaeromatidea than to the rest of Oniscidea.

The synapomorphies of Oniscidea most relevant to terrestrial life (based on the review by Schmidt 2008) are: (1) presence of a water-conducting system, formed by rows of scales on the ventral side of coxal plates, (2) antennula with three or fewer articles, (3) first pereopod not subchelate, (4) first pleopod sexually dimorphic, and (5) tergites with scale-setae. The most comprehensive discussion of phylogenetic relationships (Schmidt 2008) reviewed previous analyses and presented a detailed account of the character states that define taxa. As expected, several questions remained open, and the monophyly of several taxa was unresolved. Nevertheless, the monophyly of many clades is well supported by several synapomorphies, and it remains to be seen whether they will be corroborated also by molecular evidence.

A well-established large clade within Oniscidea is the Crinocheta, originally based on a few key characters of the male reproductive apparatus, but supported by many other morphological structures. In addition, all authors agree on the monophyly of Tylidae and Mesoniscidae, despite ambiguity regarding their position (especially the former's) among Oniscidea (see below). The monophyly of Ligiidae is questionable. Similarly, the status of Synocheta is uncertain, although most included taxa do seem to form a monophyletic unit. Synapomorphies of Crinocheta include the muscular system, the structure of pleopods, the genital papilla (Erhard 1997), and the anatomy of the stomach (Tabacaru and Giurginca 2003).

The few terrestrial isopod genera or species groups studied from a phylogenetic perspective have revealed interesting patterns of speciation and evolutionary diversification. Enzymes were used to evaluate divergence among populations of the troglobiotic species *Schizidium perplexum*

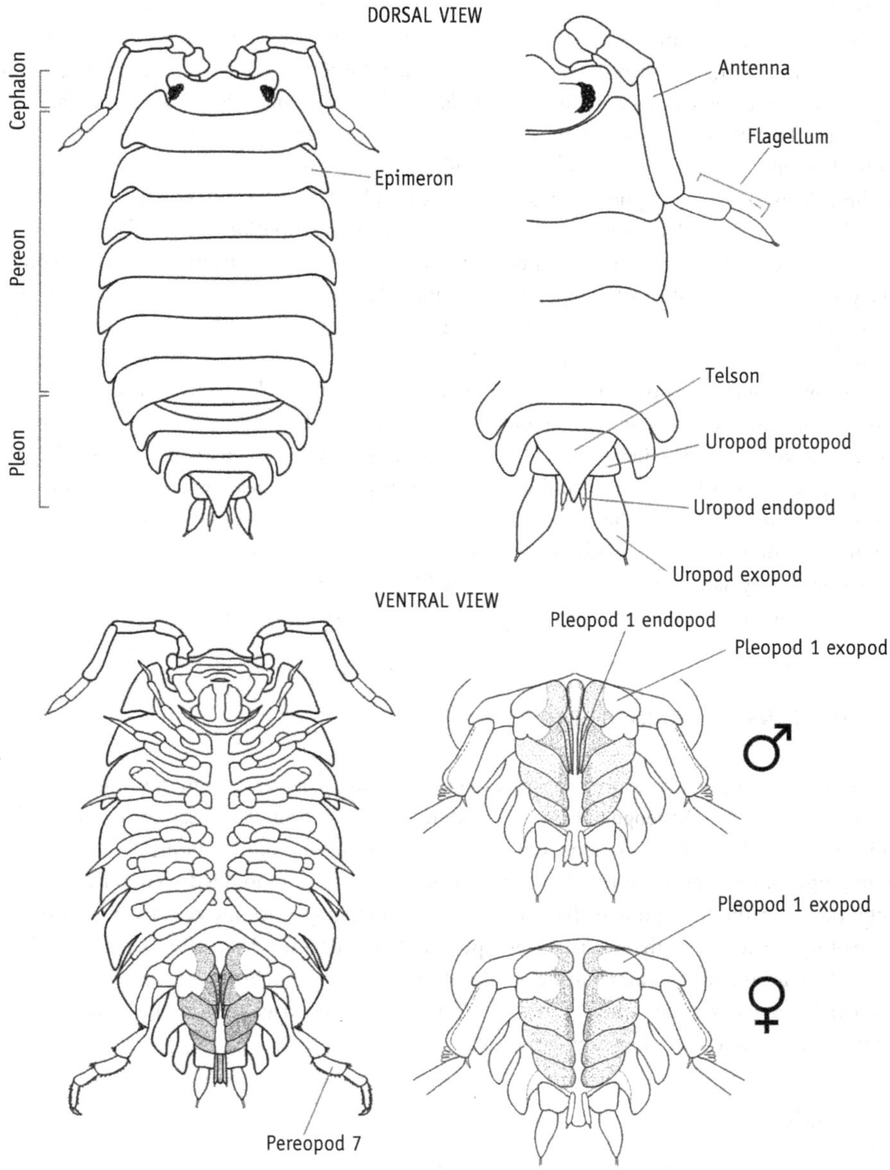

Fig. 14.2.
External features of a terrestrial isopod. Modified from Hopkin 1991, using the method by Montesanto (2015). With permission from Field Studies Council ©.

living in caves in Crete (Paragamian et al. 1987). Proteins were also used in investigations of evolutionary dynamics among *Oritoniscus paganus* populations and in the "*Trichoniscus pusillus*" complex (Cobolli Sbordoni et al. 1997) providing evidence for large genetic variation between epigean and cave-dwelling populations. Allozymes were also used to investigate patterns of gene flow between epigean and cave-dwelling populations of *Androniscus dentiger* (Gentile and Sbordoni 1998), finding support for an increased role of genetic drift in differentiation within this species.

Rivera et al. (2002) used mtDNA markers to address relationships among Hawaiian cavernicolous isopods and found no support for a vicariant mode of speciation but support for an adaptive shift from a marine littoral to a terrestrial subterranean habitat in *Littorophiloscia*. Charfi-Cheikhrouha (2003) explored genetic diversity in *Armadillidium pelagicum* using 16S rDNA, while Taiti et al. (2003) used mtDNA markers and morphological evidence to show that terrestrial forms of *Ligia* evolved independently from littoral ones on different Hawaiian islands. Molecular phylogenetic methods have also been applied on *Ligia* species to explore the effects of environmental factors (Eberl et al. 2013) and dispersal (Santamaria et al. 2013) on distribution patterns. Molecular phylogenetic studies in several oniscidean taxa found very high genetic divergence among conspecific populations even at short geographical distances (e.g., Klossa-Kilia et al. 2006 for *Ligidium*; Parmakelis et al. 2008 and Kamilari et al. 2014 for *Trachelipus*; Lee et al. 2014 for *"Spherillo" grossus*). Poulakakis and Sfenthourakis (2008) showed that *Orthometopon* species of the Aegean region have evolutionary patterns congruent with known major paleogeographic events but not with the current taxonomy and morphological evaluation of characters. Similarly, Hurtado et al. (2014) showed that diversification of the supralittoral genus *Tylos* is mostly explained by paleogeography and not by dispersal, and its overwater dispersal abilities are not as well developed as often assumed for halophilic species. On the other hand, Niikura et al. (2015) suggested that the distribution of the major clades within *Tylos granuliferus* can be explained by ocean currents, even though they date their divergence in the late Miocene/early Pliocene. Dimitriou et al. (2018) found evidence against the monophyly of Porcellionidae and the large genus *Porcellio*.

ADAPTATIONS TO TERRESTRIAL LIFE

Water is important for thermoregulation, excretion, osmoregulation, communication, and respiration. Thus, a major challenge of terrestrialization is the evolution of adaptations that offer alternative paths to these vital functions. Respiratory adaptations are more complex in isopods than in amphipods. In isopods, respiration is performed primarily by modified pleopods. Terrestrial amphipods breathe through gills that have become larger, often with increased surface area than those of aquatic forms. Both terrestrial amphipods and isopods resist desiccation generally by living in moist environments and maximizing uptake of water from the environment. In isopods, water is taken up by capillary action, while in amphipods, posterior appendages move water to the region of the mouthparts and anus where uptake occurs.

Amphipoda

The most spectacular adaptation to the terrestrial environment is the retention of the mitten-shaped second gnathopod, a neotenic condition among males in 32 of the 90 terrestrial genera (35%). This adaptation does not occur in supralittoral talitrids except for the European species *Talitrus saltator*. The neotenic condition (Fig. 14.3D) is geographically widespread and is found among taxa in Australia, New Zealand, the Philippines, Japan, Hawaii, the western Pacific, and the Caribbean and Canary Islands in the Atlantic. It appears that neotenic forms have evolved independently over a number of times. This phenomenon is associated with a change in sexual systems where mate guarding is probably not as important in the terrestrial environment as it is in the aquatic (see also Zimmer 2001). The energetic demands of carrying large gnathopods can be compensated, and the neotenic condition becomes more appropriate. As far as is known, actual mating among terrestrial neotenic talitrids has not been observed. In the neotenic supralittoral talitrid *Talitrus saltator*, the female is held by the male using stout second antennae (Williamson 1951)

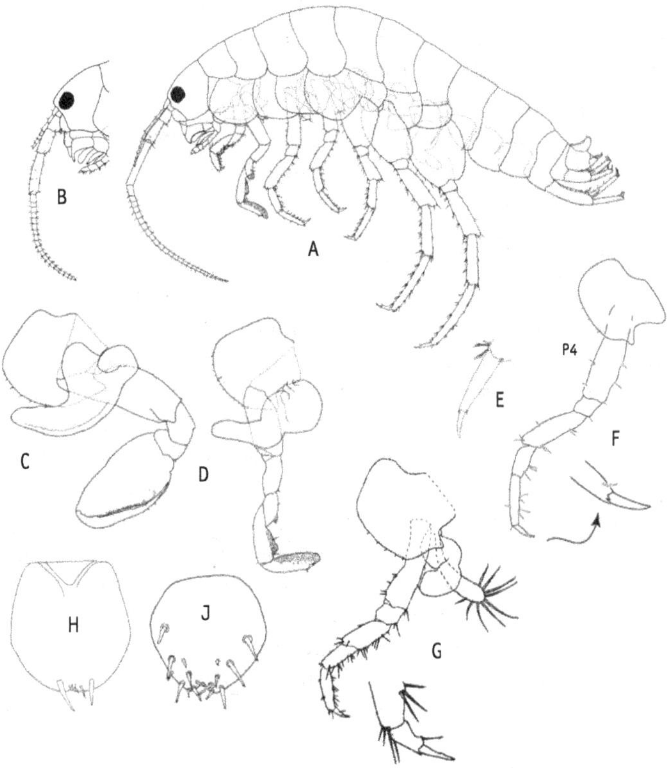

Fig. 14.3.
Neotenic characters in terrestrial Amphipoda. (A) *Austrotroides pectinalis*; (B) head of *Keratroides rex* showing short antenna 1; (C) enlarged male gnathopod 2 of *Agilestia hyperocha*; (D) neotenic male gnathopod 2 of *Fleuriella crenata*; (E) pereopod 4 with non "pinched" dactylus of *Pyatakovestia boninensis*; (F) cuspidactylate pereopod 4 of *Minamitalitrus zoltani*; (G) pereopod 4 with pinched dactylus of *Talitroides topitotum*; (H) weakly setiferous telson of *Richardsoniella maritima*; (J) heavily setiferous telson of *Platorchestia pickeringi*. A, C, D, modified from Friend (1982); B, H, modified from Friend (1987); E, modified from Morino and Miyamoto (2015); F, White et al. (2013); G, modified from Morino (2013); J, modified from Richardson (1991).

with the pereopods carrying out the function of the well-developed second gnathopod of other marine and supralittoral talitrids. But *Talitrus* is unique in being the only neotenic talitrid in having a stout second antenna.

In all talitrids (supralittoral or terrestrial), the first article of antenna 2 is greatly enlarged and fixed to the side of the head, and the small second article with a small gland cone partially fused to it acts as an excretory organ. Even though talitrids respire, their gills must remain damp. According to Lowry and Springthorpe (2015a) the exosmotic damp film around the gills is of fresh water, so it presents an osmoregulatory problem similar to that of freshwater amphipods (Lowry and Myers 2013). Apparently, it is solved in the same way as freshwater amphipods, by the enlargement of the "osmoregulatory organ" of antenna 2.

Gills are larger in terrestrial talitrids than they are in supralittoral talitrids. Supralittoral talitrids such as *Orchestia gammarellus* are just as efficient at respiring in seawater as they are in air. *Cryptorchestia garbinii* has respiration rates in air that are twice as efficient as they are in water (Ivlev and Suschenya 1961). *Arcitalitrus dorrieni* has lost the ability to sustain prolonged oxygen uptake in water, and this may be due to the reduction of the pleopods, which would impair oxygen uptake in water but probably not in air. Duncan (1969) found that terrestrial neotenic forms such as *Snaresorchestia patersoni* and *Makawe hurleyi* drown with prolonged immersion in water even though both have pleopods 1–2 well developed.

Nearly all supralittoral taxa have well-developed pleopods. In terrestrial species they are often reduced, sometimes to vestigial stumps. Species that retain well developed pleopods may beat them to move air through the ventral groove and over the exosmotic water covering the gills (Hurley 1968). In terrestrial talitrids, the gills become enlarged, often with lobes and folds to increase their surface area (Dahl 1977, Lincoln 1979). The reduction in gnathopod and pleopod size has facilitated an increase in gill size (Hurley 1968, Moore and Taylor 1984). However, the restrictions inherent in the amphipod respiratory system may be the main reason why talitrids, unlike oniscidean isopods, have not been fully successful at invading the driest terrestrial environments.

Supralittoral talitrids are considered more desiccation-tolerant than terrestrial species (Spicer et al. 1987). Moore and Richardson (1992) found that the landhopper *Mysticotalitrus tasmaniae* controls the amount of water in the ventral channel through uptake via special positioning of the body and the urosome, which allows dabbing the tips of uropods 1 and 2 on a wet substratum. In the case of cuspidate landhoppers, the ends of the third uropods and telson are also used for this function. Water moves from the abdomen to the thoracic region seemingly through movement of the pleopods or by the distal lobe of the pereopod 6 gill. Water is taken directly into the gut by "anal drinking." Water also enters through the mouth and into the esophagus.

The maintenance of a pool of water in the ventral channel is important because the channel is the location of the coxal gills and the brood pouch. The gills of talitrids are a major site of water loss and ion exchange (Moore and Francis 1985), and both these properties of the gills require a regular supply of water.

Talitrids are important ecological contributors to the litter fauna in wetter forests wherever they occur. In Irish coniferous forests, the introduced *Arcitalitrus dorrieni* probably consumes more leaf litter annually than all the other macrofaunal species in this coniferous forest combined (O'Hanlon and Bolger 1999). For many talitrid populations, food may not be limiting (Friend and Richardson 1986). On the other hand, in *Makawe hurleyi* and *Snaresorchestia patersoni* in New Zealand, populations decline in autumn due to reduction in litter thickness (Duncan 1969). Duncan (1969) hypothesized that in the late summer and autumn, amphipods eat far more food than is being produced, resulting in a population crash after the standing crop of litter has been utilized. Habitat space may also be a limiting factor. The densities of land-hoppers are exceptional for litter-dwelling animals of comparable size, exceeding 10,000 individuals m^{-2} in Tasmanian forests (Friend and Richardson 1977) where several species coexist, but densities of single species populations can reach more than 6,800 individuals m^{-2} for *Keratroides angulosus* in Tasmanian forests (Friend 1980). Introduced species such as *A. dorrieni* in the Isles of Scilly may also reach high densities of 3,265 individuals m^{-2} (Richardson 1980). Factors that hold population levels below food supply may also be climatic. Desiccation and osmotic stress caused by submergence in rainwater have been recorded as mortality factors (Duncan 1969). Lazo-Wasem (1984) recorded an upper thermal tolerance of 32.6°C in *A. sylvaticus* and suggested that temperature and humidity acting together affect the survival of this species, which must inhabit leaf litter with saturated (100% relative humidity) conditions. Hurley (1959, 1968) suggested that behavioral mechanisms account for the success enjoyed by the Amphipoda on land. Predation does not appear to be a strong regulator, although talitrids can be significant dietary items for a number of vertebrate predators (Searle 1927, Campbell and Gray 1942, Ashton 1975, Friend 1975, Hall 1980, Smith and Calver 1984). Where species coexist, they may partition the habitat vertically (Dickman 1986, Duncan 1969).

The diet of landhoppers is generally unspecialized and consists principally of decaying plant material (Devitt 1981, Morton and Richardson 1984), although observations in the laboratory suggest that they will consume carcasses of their own kind. Talitrids prefer leaves that have undergone some microbial conditioning over fresh ones, but it appears to be the chemical and physical effects of fungi and bacteria, rather than their actual presence, which attracts them (Morton and Richardson 1984).

Whether coprophagy is important is uncertain. One species, *K. angulosus*, has strong spines on the lateral ampullae of the proventriculus, which may serve to remove the peritrophic membrane from ingested fecal pellets (Weeks 1992).

Talitrids do not ingest a high proportion of the litter fall in forests: 25% of the total annual leaf litter production is consumed by *A. sylvaticus* in subtropical forest in New South Wales (Clark 1954), 20% by *K. vulgaris* and *K. angulosus* in eastern Tasmania (Friend 1980), but only 2.9% by *Mysticotalitrus tasmaniae* in southern Tasmania (Friend 1975). These rates correlate approximately with the densities of the animals.

In New Zealand, *Makawe hurleyi* and *Snaresorchestia patersoni* invade the litter of the long grass habitat. Duncan (1969) studied competition between these two species and considered that in situations of intense interspecific competition, *M. hurleyi* would be a poorer competitor than *S. patersoni* because reproduction in *M. hurleyi* is less efficient than that of *S. patersoni* since it carries a larger number of eggs, has a higher mortality rate, and has a relatively longer immature stage.

Isopoda

Terrestrial isopods have adopted a wide range of behavioral adaptations to live on land, avoiding water loss, predation, and competition. The other major problem the Oniscidea had to face on land is respiration in an aerial environment.

In the process of terrestrialization from marine ancestors, oniscideans went through many morphological, physiological and behavioral adaptations that reduced their water dependence. The first colonizers of land were most likely amphibious *Ligia*-like forms. *Ligia* species, most of which occur along rocky sea shores, exemplify the sea-to-land transition (Carefoot and Taylor 1995).

In terrestrial isopods a complex water-conducting system overcomes the water dependence problem (Hoese 1981, Hornung 2011). In the lower Oniscidea, Ligiidae to Synocheta, an open system uptakes water by capillarity from an external source using longitudinal grooves and rows of scales on the caudal and rostral surface of the sixth and seventh pereopods (Fig. 14.4A). Water is then forwarded to the other part of the body, especially to the pleopods for respiration. In the higher Oniscidea (Crinocheta), water with ammonia excreted by the maxillary nephridia is forwarded caudally through grooves with scales present on the body's ventral surface along the base of pereon epimera and pereopod insertions (Fig. 14.4B,C) (Hoese 1981). Through this route, ammonium evaporates. Species with this system can also uptake water from external sources through the mouth or uropods (Drobne and Fajgelj 1993).

Respiration in terrestrial isopods is performed mostly by pleopods. In primitive species, oxygen uptake occurs through the thin ventral surface of the exopods (Schmidt and Wägele 2001). This type of respiratory structure is present in Ligiidae, Synocheta, and in several Crinocheta—all living in very humid habitats, such as in the littoral zone, the forest litter layer, deep in the soil (endogean), or in caves. In forms more adapted to terrestrial environments (inside Crinocheta), respiration occurs in specialized structures situated on the pleopodal exopods. The evolution of the respiratory apparatus is correlated with the habitat in which the animals live along a gradient of humidity, with more complex respiratory organs in species living in drier environments. Three types of respiratory structures can be distinguished: dorsal respiratory fields, uncovered lungs, and covered lungs (Ferrara et al. 1994, 1997, Taiti et al. 1998, Paoli et al. 2002).

In the dorsal respiratory fields on the pleopodal exopods, *Oniscus*-type lungs (Hoese 1982b), the cuticle of the respiratory surface is much thinner than that of the rest of the exopod, smooth

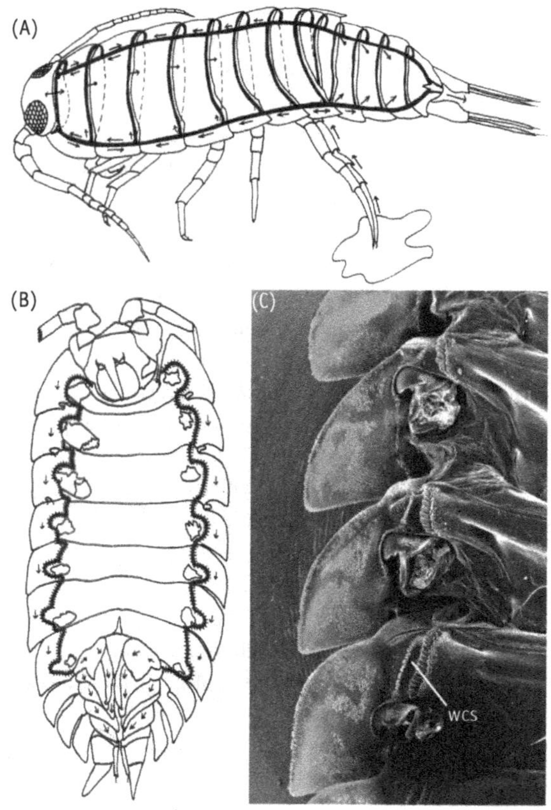

Fig. 14.4.
Water conducting system (WCS) in Oniscidea: (A) *Ligia*-type WCS; (B) *Porcellio*-type WCS; (C) detail of WCS on ventral side, showing longitudinal grooves and scales at base of epimera (*Adinda malaccensis*). A, modified from Hoese (1982a); B, modified from Hoese (1981), with permission from Springer; C, SEM photo courtesy of S. Taiti.

or weakly wrinkled. This kind is present in Crinocheta living in very humid habitats such as the supralittoral or in the forest litter (e.g., some Philosciidae, Scleropactidae, and Oniscidae).

In uncovered lungs (Fig. 14.5A,B) the respiratory field occupies an extensive part of the exopod dorsal wall and presents a strongly pleated surface with thinner cuticle (*Trachelipus*-type of Hoese 1982b). This type is present, for instance, in Trachelipodidae, Cylisticidae, and some genera of Eubelidae and Armadillidae.

Covered lungs are present in species inhabiting drier habitats. They are characterized by an internal region with a respiratory function, represented by cuticular invaginations forming respiratory trees and communicating with the exterior by means of spiracles. The lungs are located within the pleopodal exopods, with the spiracles usually on the dorsal surface close to the lateral margin. The external surface around and inside the spiracle is modified as the so-called perispiracular area consisting of verruca-like or honeycombed cells (Fig. 14.5D,F), which keep the humidity at the level required for respiration. Two types of covered lungs can be identified on the basis of the number of spiracles and respiratory trees inside the exopod: polyspiracular (Fig. 14.5C,D) and monospiracular (Fig. 14.5E,F). The monospiracular lung is by far the most common type. A particular type of monospiracular lung is present in the genus *Periscyphis* inhabiting extremely dry

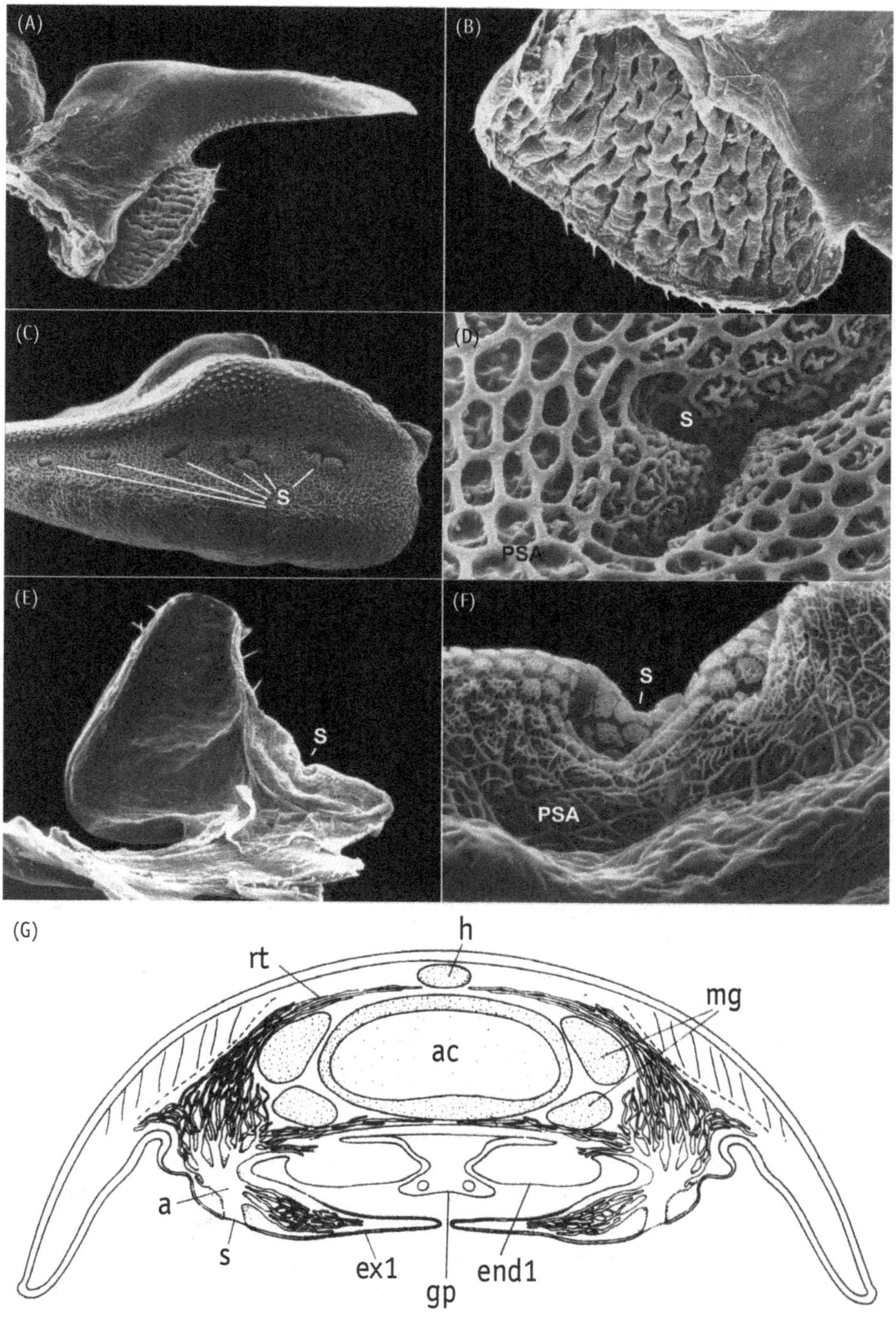

Fig. 14.5.
Pleopodal lungs in Oniscidea. Uncovered lung in *Atracheodillo marmorivagus*: (A) male pleopod 2 exopod; (B) detail of respiratory surface of male pleopod 1 exopod (polyspiracular covered lung in *Somaloniscus taramassoi*); (C) pleopod 4 exopod showing spiracles; (D) detail of a spiracle and perispiracular area (monospiracular covered lung in *Cubaris murina*); (D) male pleopod 1 exopod showing single spiracle; (F) detail of spiracle and perispiracular area, *Periscyphis* sp.; (G) cross section of pleon through male pleopod 1. a, atrium; ac, alimentary canal; end 1, pleopod 1 endopod; ex1, pleopod 1 exopod; gp, genital papilla; h, heart; mg, midgut gland; psa, perispiracular area; rt, respiratory tubules; s, spiracle. Modified from Ferrara et al. (1997). Reproduced with permission from Elsevier ©.

habitats (deserts and semi-deserts), in which the respiratory trees penetrate the body cavity in the pleon and seventh pereonite (Ferrara et al. 1997) (Fig. 14.5G). The fact that all kinds of lungs (uncovered and covered) can be found within different families, sometimes within a single genus (e.g., *Aphiloscia*, Philosciidae), shows that these structures evolved several times independently through convergence.

As in all Peracarida, the brood care of fertilized eggs and young is in a ventral marsupium formed by five pairs of overlapping lamellar oostegites at the base of the anterior pereopods (Fig. 14.6A) (Hoese and Jansen 1989, Hornung 2011). The amphibian type of marsupium is typical of the lower Oniscidea living in very humid environments (e.g., in species of *Ligia*) and is similar to that of aquatic species, i.e., open anteriorly and posteriorly, with water supplied by the open water-conducting system (Fig. 14.6B). The terrestrial type of marsupium is found in higher Oniscidea, the Crinocheta, occurring in more arid environments. It is completely closed and is not connected to the water-conducting system (Fig. 14.6C). The internal fluid of the marsupium is provided by finger-like extensions of the ventral sternites, the cotyledons, and contains the nutritive components and oxygen for embryo development (Fig. 14.6D, E). The number and position of the cotyledons vary according to family and seem to be correlated with the aridity of the habitat (Hoese and Jansen 1989), with the highest number present in species of Armadillidae (Lewis 1991). Terrestrial isopods also differ from aquatic isopods in having waxy and cement-like layers in addition to the usual exoskeleton layers which are considered adaptations to living in air (Hornung 2011). Nevertheless, the eco-morphological type (see below) does not seem to affect the structure of cotyledons (Csonka et al. 2015).

The dorsal surface of Oniscidea shows a large variety of sensory and non-sensory structures, e.g., scale-setae, microscales, pits, pores, ridges, tubercles, or spines. Scale-setae are unique to oniscideans and are probably linked to terrestrial adaptation. In many of the higher Crinocheta, special setae called *noduli laterales* are present on pereonites, sometimes also on the cephalon and pleonites, and are presumably tactile organs. The various cuticular structures such as scales, microridges, and protuberances (ridges, tubercles, and spines) present in many Oniscidea have been considered either as anti-adhesive structures facilitating their movement in the soil or as anti-predator mechanisms (Schmalfuss 1975).

Most terrestrial isopods living in epigean environments are nocturnal and, in temperate regions, reduce their activity during cold winter and hot summer months. To avoid desiccation and predation during the time of reduced activity, terrestrial isopods use dark, moist, cool shelters as in leaf litter or under tree bark and stones. In general, males are more active during the mating season and gravid females reduce activity and shelter during the incubation period. Aggregation in woodlice is common, an adaptive behavior to survive in harsh condition to avoid desiccation (Hornung 2011), but has been described in few widespread species (e.g., *Oniscus asellus, Porcellionides pruinosus*, and *Porcellio scaber*). Aggregation is influenced by thigmotaxis, negative phototaxis, and interattraction among conspecifics (Devigne et al. 2011). An aggregation pheromone seems also to be involved. It is secreted by the hindgut and is excreted with the fecal pellets (Takeda 1984).

Some species of terrestrial isopods can survive in dry deserts, subdeserts, and steppes, like the 20 species of *Hemilepistus* distributed from northern Africa to central Asia (Taiti and Schotte 2016). One species, *H. reaumurii,* occurring in subdesertic areas from Algeria to Syria, is large, up to 23 mm long, and has adapted to live in xeric habitats by adopting a social behavior (Linsenmair 2007). *Hemilepistus reaumurii* is a monogamous species which reproduces only once during its life. Adults dig deep burrows in spring where the couple lives and forages during the morning and evening, staying in the burrow during the hottest time of the day. They take care of their own brood for up to five months and are very aggressive toward juveniles not belonging to their own family (Linsenmaier 1994).

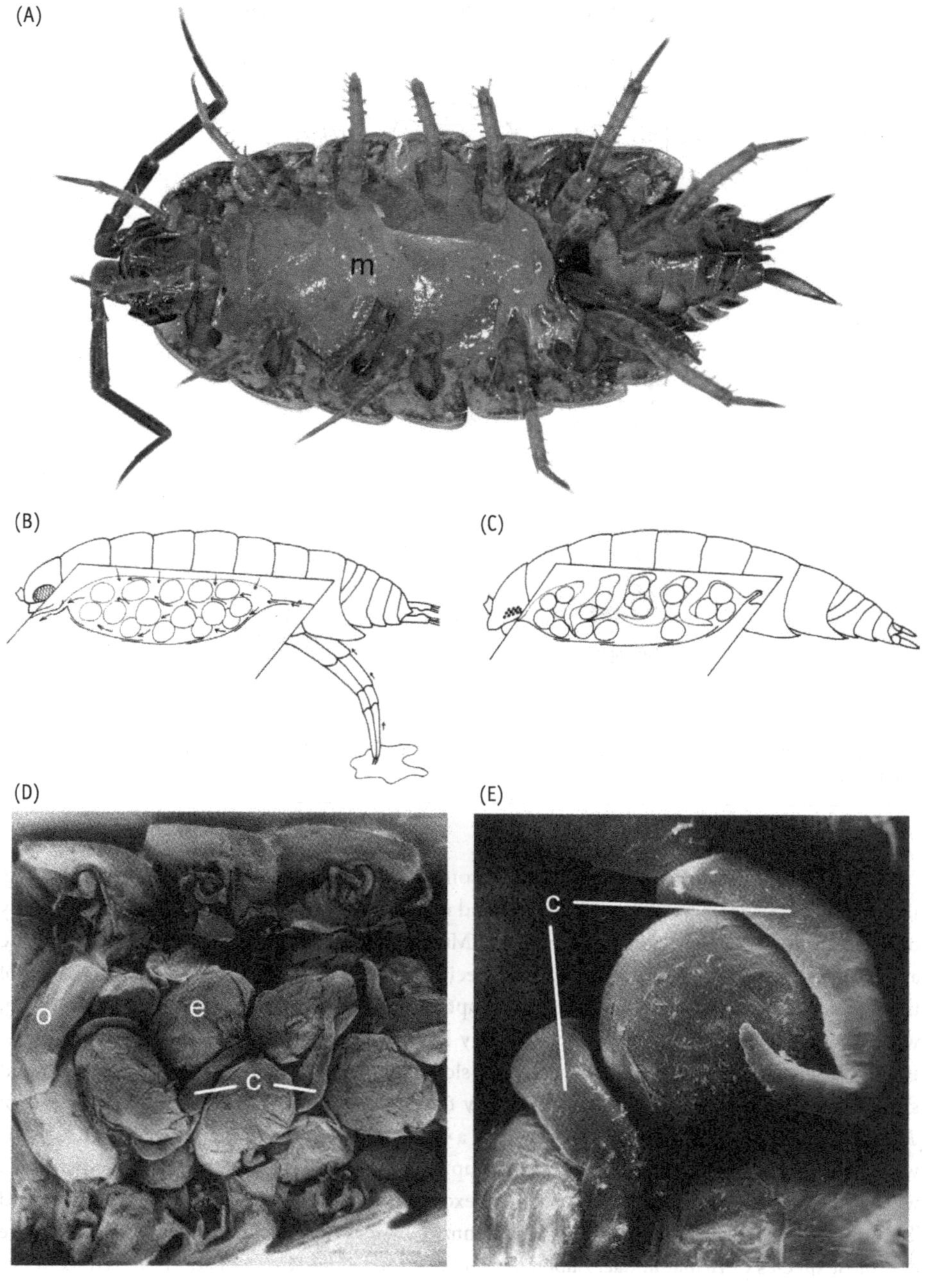

Fig. 14.6.
Marsupium in Oniscidea: (A) ventral view of a female specimen of *Orthometopon dalmatinum* showing marsupium; (B) amphibian type of marsupium; (C) terrestrial type of marsupium; (D, E) ventral view of marsupium in *Philoscia muscorum* showing embryos and cotyledons. c, cotyledons; e, embryos; m, marsupium; o, oostegite. A, photo by S. Taiti; B–E, modified from Hoese and Janssen (1989).

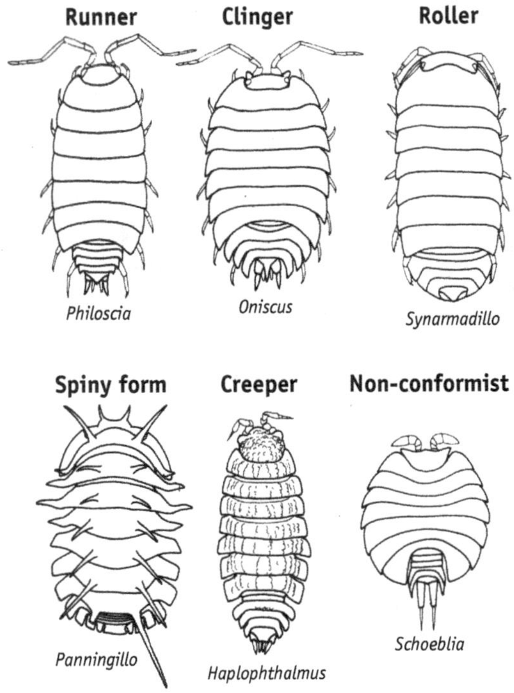

Fig. 14.7.
Eco-morphological categories in Oniscidea. From Schmalfuss (1984). Reproduced with permission of the Zoological Society of London ©.

Within Oniscidea, five main groups of ecomorphs are recognized (Schmalfuss 1984; Fig. 14.7): runners, clingers, rollers, spiny forms, and creepers. Forms that cannot be assigned to any of these groups are labeled "non-conformists." Most of these ecomorphs have analogs in marine and freshwater isopod suborders, but their specific role and detailed morphological features of the terrestrial forms are showing important adaptations to life on land. Runners are fast-moving, with smooth dorsal body surface and relatively long pereopods (e.g., most species of Ligiidae, Philosciidae, and Trichoniscinae). Clingers are slow-moving, with broad and flat body, short and strong pereopods with which they cling tightly to the substratum if threatened (e.g., species of *Trachelipus, Nagurus*, and *Porcellio*). Rollers have a very convex body (semi-circular in cross section) with often smooth dorsum and are able to roll up into a more or less perfect ball when disturbed, with antennae that remain outside the body (exoantennal conglobation), as in Cylisticidae and Tendosphaeridae, or inside the ball (endoantennal conglobation), as in Tylidae, Armadillidiidae and most species of Armadillidae, Eubelidae, and Scleropactidae. Conglobation can reduce water loss by about 35%. Spiny forms are conglobating species occurring in tropical forests; they bear conspicuous tergal spines and belong to Eubelidae and Armadillidae. All the above-mentioned categories are epigean species. Creepers are endogean species inhabiting the soil, crevices in rocky biotopes, and caves, and have a small-sized (<5 mm), very convex and elongated body, tergites with longitudinal ribs and short appendages (e.g., Bathytropidae, some Platyarthridae, Stenoniscidae, and Haplophthalminae); some creepers have also a conglobating ability, like species of Buddelundiellinae (Trichoniscidae). The so called non-conformists include all the terrestrial isopods which do not fit into the preceding categories (about 10% of all the species), such as commensals in termite and/or ant nests, or forms that can be classified as diggers (e.g., species of

Leptotrichus which dig into loose sand or *Hemilepistus* which dig permanent burrows in desert and semi-desert areas). Even though diggers could be a behaviorally distinct group, they do not exhibit some well-defined common morphological features, so they cannot be identified as a separate eco-morphological category.

A further category, aquatic species, especially important as it signifies a secondary return to water, could be added. Among the Oniscidea there are several species that live permanently or temporarily in water, e.g., some stygobiotic species of Trichoniscidae, Styloniscidae, Olibrinidae, and Philosciidae (Taiti and Humphreys 2001, Taiti and Ferrara 2004, Taiti and Xue 2012, Taiti et al. 2018). One species, *Haloniscus searlei,* occurs in the salt lakes of southern Australia. It is almost certain that all these aquatic species of Oniscidea are not primitive taxa but are specialized forms derived from terrestrial ancestors, secondarily returned and morphologically adapted to the aquatic environment (Taiti and Xue 2012, Taiti et al. 2018).

It is not surprising that a large number of Oniscidea have adapted to live in subterranean habitats. More than 300 species are strictly cave dwellers (troglobionts), many others commonly (troglophiles) or occasionally (trogloxenes) occur in caves (see Chapter 13). Many species also live in other endogean habitats, i.e., deep in the soil, or in ants' and termites' nests. Most species are exclusive to single subterranean environments, but a few species have been recorded both from caves and endogean habitats. Troglobiotic terrestrial isopods have been recorded from 16 families. The family with the largest number of troglobiotic forms is Trichoniscidae in the northern hemisphere, followed by Styloniscidae in the southern hemisphere and Philosciidae in the tropics. The Cylisticidae include both troglobionts (e.g., *Troglocylisticus cyrnensis* from Corsica) and endogean forms (most species of the *nasutus* group in the genus *Cylisticus*). All other families contribute only in a very small percentage (in general, less than 2% each) to the total number of troglobiotic Oniscidea (Taiti 2004).

The adaptive traits to cavernicolous life are typical of most troglobiotic arthropods, i.e., reduction or absence of eyes and body pigment, long and thin appendages, especially antennae and pereopods, reduced metabolism and low fertility. The most common eco-morphological categories in subterranean habitats are the runners in caves (e.g., Trichoniscinae, Styloniscidae, and Philosciidae) and the creepers in the soil (Haplophthalminae). Rollers are also common cave dwellers (Buddelundiellinae, Cylisticidae, Armadillidiidae, Scleropactidae, and Armadillidae).

BIOGEOGRAPHY

Terrestrial talitrids are most diverse in the southern hemisphere, where their uneven distribution is probably attributable to historical levels of aridity. The low diversity in the northern hemisphere is probably due to bolide impact at the KT boundary. Isopods, by contrast, have a more even global diversity, although there is some evidence of species-richness patterns that do not follow classical latitudinal gradients of species richness.

Amphipoda

While generic diversity of terrestrial Amphipoda is high, species diversity is low. The distribution of genera is skewed (Fig. 14.8). Most genera occur in Australia (mainly Tasmania) and New Zealand, but they are also well represented in Japan and Taiwan. Three genera occur in Southeast Asia, three occur in Europe and the northeast Atlantic, one in Central America and the Caribbean, one in South Africa, and none in North or South America. Southeast Asia, in particular, is poorly studied, and more genera should be expected. Five genera are cavernicolous, all in the northern hemisphere except *Australorchestia tantabiddyensis* from Western Australia; most have been found on islands.

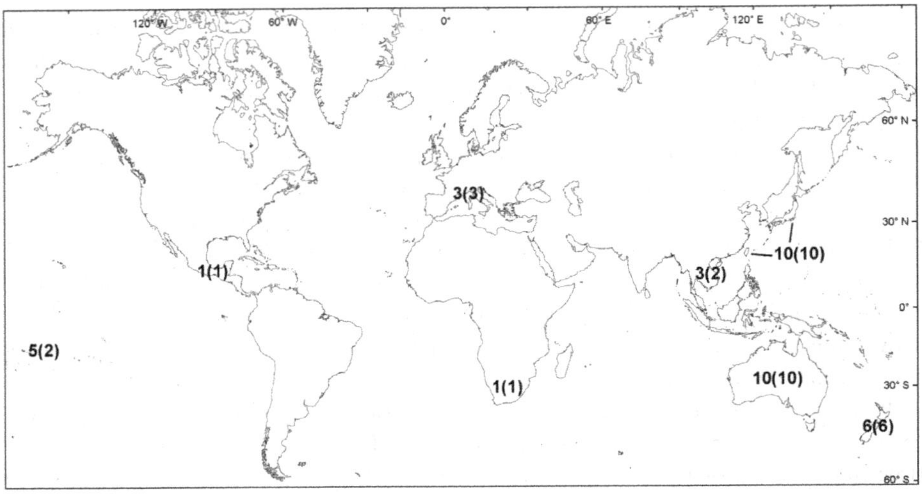

Fig. 14.8.
World distribution of terrestrial talitrid amphipod generic level diversity: numbers of genera (endemic genera in parenthesis).

Terrestrial talitrids are either simplidactylate (with simple dactyls on pereopods 3–7) or cuspidactylate (with cusps on the anterior margins of the dactyls of pereopods 3–7) (Bousfield 1984). Supralittoral talitrids are all cuspidactylate except for marshhoppers that can be either. Simplidactylate talitrids appear to be ancestral as they share a number of plesiomorphic character states: simplidactylate pereopods 3–7, pereopod 4 with simple dactylus, elongated first antennae, and weakly setiferous telson. These character states are also shared with their sister taxon, the marine Hyalidae. Simplidactylate talitrids are almost entirely terrestrial and are mainly limited to fragments of old Gondwana.

Cuspidactylate talitrids appear to be derived as they share apomorphic character states: cuspidactylate pereopods 3–7, pereopod 4 with thickened (pinched) dactylus, shortened first antenna, and strongly setiferous telson (Fig. 14.3J). Cuspidactylate talitrids are widespread, occur in both hemispheres, and are both terrestrial and supralittoral. Cuspidactylate talitrids also occur in New Zealand, which suggests that they were present in Gondwana. This is strengthened by the fact that the supralittoral cuspidactylate *Bellorchestia* from Australia and New Zealand is most closely related to a supralittoral cuspidactylate talitrid *Capeorchestia* from South Africa. In Tasmania, the terrestrial talitrid *Mysticotalitrus* (simplidactylate) appears to be phylogenetically close to the terrestrial genus *Talitriator* (simplidactylate) from South Africa (Friend 1987), again suggesting that these two taxa are derived from an ancestor that lived in Gondwanan forests at least 150 mya.

Thus, there is the belief that both simplidactylate and cuspidactylate talitrids were present in Gondwana as suggested by their prevalence in both hemispheres (although uncommon in the northern hemisphere), with their origins possibly stretching back to Pangaea. Simplidactylate talitrids (ancestral) evolved before cuspidactylate talitrids (derived) but may have been limited to eastern Gondwana, where simplidactylate terrestrial forms dominated forest litter. Cuspidactylate talitrids were able to outcompete simplidactylate taitrids in many parts of the world during the late Cretaceous and early Tertiary but were unable to entirely replace them in old Gondwana fragments.

The sparsity of terrestrial talitrids in Europe and their near absence from North and South America may be due to the impact of the bolide at Chicxulub at the KT Boundary. They would have been decimated by the hours-long, global pulse of intense IR flux that followed. This IR flux caused by reentering ejecta was stressful enough to kill all individual non-marine macroscopic organisms except those protected within soils, underground, under rocks, in water, or in dense aquatic vegetation (Robertson et al. 2004). Organisms living in unprotected leaf litter would have been annihilated (Robertson et al. 2004). Distant parts of the globe were probably spared some of the worst of the effects, in particular, fragments of Gondwana that were far from the impact site (Robertson et al. 2004).

Colonization of terrestrial environments would have continued through the Tertiary, and some extant terrestrial talitrids may have originated in the Tertiary. One such taxon is almost certainly *Floresorchestia*, which has colonized leaf litter on several oceanic islands. In most cases, evolution in this genus has not progressed to generic levels (there are unpublished exceptions) suggesting that these species are Tertiary (probably post-Miocene) in origin.

The higher diversity in Tasmania/New Zealand than in mainland Australia and in South Africa may be the result of the more recent arid environment found in these latter regions. In the early Tertiary, rainforest abounded in mainland Australia, but through the Miocene and later Pliocene, rainfall decreased and the continent became more arid (Martin 1998) with the consequent extinction of those terrestrial amphipods incapable of adapting. The dry coastal conditions would have also inhibited the evolution of new land colonizers. Moreover, in South Africa, during the Tertiary, the progressive aridity of the African continent resulted in the spread of savannahs and could have exterminated most of the surviving elements of the then-temperate Tertiary forests (Coetzee 1980).

Europe has only one terrestrial talitrid genus, *Cryptorchestia*, which appears to be derived from an *Orchestia*-like ancestor (Lowry and Fanini 2013). Until the Eocene–early Oligocene, the climate of Laurasia was humid, tropical, or subtropical and presumably suitable for the evolution of terrestrial amphipods. According to Bousfield (1984), derived, cuspidactylate talitrids evolved on the shores of the Tethys. As the Tethys Sea closed with tectonic movements, mountains rose and progressively cut off the northward passage of the southwest monsoon rains from the Tethys, and as a consequence, Laurasia became affected by ever-increasing aridity. Extinction has probably played a very significant role in establishing the extant distribution of terrestrial amphipods worldwide.

With the exception of the species-rich genus *Floresorchestia*, the high generic diversity of terrestrial amphipods indicates high antiquity and low vagility leading to high endemicity. Each terrestrial genus has evolved from a palustral (Bousfield 1968) or supralittoral (Wildish 1982) talitrid that has made its way into leaf litter adjacent to the marsh or shore habitat and then progressed inland, undergoing a long period of in situ evolution. There is virtually no dispersal between regions by terrestrial amphipods. Lindeman (1991b) has suggested rafting in some instances, but the rafting and then establishment of a population in a new region of terrestrial leaf litter is hard to conceive. In effect, the distribution of most terrestrial amphipods is tied to the distribution of their palustral or supralittoral progenitors.

Even within regions, different terrestrial species may have different palustral or supralittoral ancestors. In the Caribbean and Central America, for example, terrestrial talitrids are known in the genera *Caribitroides, Cariborchestia,* and *Cerrorchestia.* According to Lindeman (1990), *Caribitroides* shares a common ancestor with the palustral *Chelorchestia,* while according to Smith (1998), *Cariborchestia* is not closely related to *Caribitroides* and most likely derived from a sandhopper represented now in the region by *Tethorchestia.* For us, *Caribitroides* is more closely related to *Macarorchestia* from the European coast, which might indicate a common ancestor before the splitting of the Atlantic Ocean. Lindeman (1991a) considered the Costa Rica–Panama genus *Cerrorchestia* to be a relatively recent offshoot of the palustral *Chelorchestia* group.

In Japan, the cave-dwelling *Minamitalitrus* superficially resembles *Spelaeorchestia* from Hawaiian lava tubes, but according to White et al. (2013), *Minamitalitrus* is probably independently derived from an extant or extinct terrestrial taxon on Minami–Daitōjima, the similarities being due to convergent evolution as a result of their common subterranean existence.

In summary, terrestrial talitrids are considered to be of Gondwanan or even Pangaean origin and evolved independently from many different palustral or supralittoral talitrids. They are thought to have been widespread in Cretaceous wet forests around the world, but were virtually exterminated from North America, South America, and Western Europe by the effects of the bolide impact at the KT boundary. Increasing aridity during the Tertiary in Europe and much of Africa and Australia drove many more taxa to extinction, except in extreme southern regions such as Tasmania and New Zealand that remained sufficiently moist. During the Tertiary a number of terrestrial talitrids evolved from palustral and/or beach talitrids, in particular *Floresorchestia* species on Pacific islands.

Six species of terrestrial talitrid are considered to be anthropogenic introductions (*Talitroides topitotum, T. alluaudi, Brevitalitrus hortulanus, Arcitalitrus dorrieni, A. sylvaticus*, and *Hermaniator cylindripes*). They have been accidentally transported by humans within the tropics and from the southern temperate to the northern temperate zones.

Talitroides species are the most widely distributed, occurring in Europe (Vader 1972), North America (Shoemaker 1936), the East Atlantic (Andersson 1963, Dahl 1977), Australia (Friend 1980), the Pacific to Hawaii (Stephensen 1935, Bousfield and Howarth 1976), Madagascar and the Indian Ocean Islands (Ruffo 1958), and South America (de Castro 1972). In the warm temperate zone and the tropics, these invasive species (Bousfield and Howarth 1976, Biernbaum 1980) colonize indigenous habitats readily, and there is evidence that they compete successfully with native land-hoppers. In the Hawaiian Islands, they have colonized lava-tube caves and are threatening the restricted native species *Spelaeorchestia koloana* (Bousfield and Howarth 1976). On the Koolau Mountains of Oahu, *Talitroides topitotum*, and to a lesser extent *T. alluaudi*, dominate the fauna of the lower slopes and appear with the native species even at the highest altitudes. The site of origin of *Talitroides* species is unknown.

Brevitalitrus hortulanus was originally described from specimens collected at Kew Gardens in England (Calman 1912). It has since been reported from Vanuatu (Stephensen 1943, Bousfield 1971). Other species in the genus are known from the Pacific Islands (Stephensen 1943, Bousfield 1971). It appears to be a Pacific genus introduced to England.

Arcitalitrus species are native to southeastern Australia, but *A. dorrieni* is now widespread in gardens in Ireland and Great Britain (Rawlinson 1937, Ingle 1958) and *A. sylvaticus* has been introduced to California (Bousfield and Carlton 1967). According to Griffiths (1999), *Talitriator cylindripes* has been introduced from South Africa into Ascension and to the St. Helena Islands in the mid-Atlantic Ocean.

Why are these alien species such effective colonizers? In Mexico, *Talitroides topitotum* breeds year-round (ovigerous females having been found in all months of the year) and has a sex ratio biased toward females (Alvares et al. 2000). Both these characteristics are advantageous for effective colonization. In Hong Kong, *T. topitotum* has an average life span of 10 months and has three to four generations per year. It produces up to two broods during its life time (Lam and Ma 1989).

Arcitalitrus dorrieni is univoltine in Ireland with a summer–autumn peak in reproductive activity. It has a biased sex ratio in favor of females at reproduction but is biased toward males at recruitment (O'Hanlon and Bolger 1997a, b). *Arcitalitrus dorrieni* is less tolerant of low humidity than sandhoppers and, in addition, appears to be less tolerant to freezing than their sandhopper relatives, not surviving below 0°C (Cowling et al. 2003). On the other hand, it is quite tolerant of raised salinity and in Ireland, at least, occurs adjacent to the sea (A.M., personal observation). It seems likely that their high reproductive output due to multiple broods per year and preponderance of females in the population is at least in part responsible for their success, provided that temperatures remain above 0°C.

Isopoda

Terrestrial isopods have a worldwide distribution, being absent only from the polar regions and very high elevations (>4,800 m asl), and can be found in almost all ecosystems, from littoral habitats to deserts. Even though they are considered organisms with restricted dispersal ability, mainly due to their strong dependence on high levels of relative humidity in their micro-habitat, they have colonized all known isolated oceanic islands. Several species (e.g., *Porcellionides pruinosus, Porcellio laevis, P. scaber, Armadillidium vulgare, Trichorhina tomentosa, Nagurus cristatus, Cubaris murina*, etc.) are cosmopolitan, following anthropogenic introductions, or are pantropical, for reasons that are unclear.

The world catalog of terrestrial isopods by Schmalfuss (2003) gives coarse-grained data on species' distribution, but these data offer some insights into general biogeographical patterns. According to Sfenthourakis et al. (2007), there is strong evidence for insular tropical areas and regions with Mediterranean-type ecosystems being diversity hotspots for Oniscidea, resulting in mid-latitude peaks in species richness, rather than the generally accepted equatorial peak. Hotspots of diversity with large numbers of endemics are represented by the Mediterranean area (Iberian Peninsula, Italy, Balkan Peninsula, Greece), South Africa, Australia, and Brazil (Fig. 14.9). Furthermore, isopods do not seem to conform to Rapoport's Rule, suggesting a reduced role of biotic interactions in shaping their distribution patterns. This mid-latitudinal richness peak is corroborated by the global distribution of Armadillidiidae and Porcellionidae, which comprise ~16% of all Oniscidea species (Sfenthourakis and Taiti 2015). Both families are particularly species-rich at mid-latitudes of the northern hemisphere. In contrast, the three richest families (Armadillidae, Philosciidae, and Trichoniscidae), comprising around half of the known species, are distributed over a wider latitudinal range but do not much change the global pattern. A latitudinal gradient of decreasing species richness can be detected at a smaller scale within Europe where there is an almost linear trend of decreasing species richness from the Mediterranean to northern Europe (Hornung and Sólymos 2007). These authors also found a latitudinal cline in endemism, with endemics dominating the south (Iberian, Italian, and Balkan peninsulas), possibly

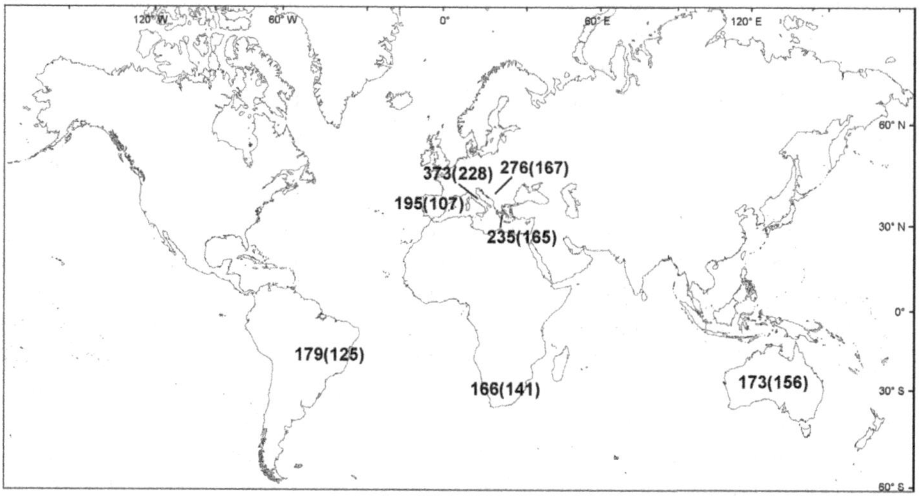

Fig. 14.9
Hotspots of diversity for terrestrial isopods: numbers of species (endemic species in parenthesis). From Schmalfuss (2003), updated to include records until December 2017.

due to dispersal barriers by high central European mountains and/or effects of glacial refugia. The gradient found within Europe agrees with the global pattern, as maximum richness occurs in Mediterranean countries lying in its south. A plausible explanation for the unusual latitudinal diversity pattern of Oniscidea could be high levels of habitat heterogeneity in Mediterranean countries, where favorable humid habitats are sparsely distributed and isolated by a dry "matrix," offering greater chances for allopatric speciation. A test of this hypothesis would be the richness patterns within the tropics, as it would predict lower species richness in contiguous tropical habitats than in insular ones.

Studies of elevation gradients in Oniscidea species richness are few. Schmalfuss and Ferrara (1982) found isopod diversity sharply declining above the tree line on Mt. Cameroon and studies on Greek mountains (see references in Sfenthourakis et al. 2012) generally showed a decrease of diversity with regard to elevation. Nevertheless, one species usually dominates communities at the highest elevations above the tree line, its identity differing between mountains. This effect might be partially due to ecological release (e.g., absence of lizards and/or invertebrate predators that cannot tolerate very cold, alpine environments). Strictly alpine species seem to be rare among isopods, but most species found on mountain tops occupy broad elevation ranges.

Karagkouni et al. (2016) explored latitudinal trends in body size and could only find weak support for the aridity resistance hypothesis (larger size in arid regions), whereas the relationship of size and temperature varied among families. Size of conglobating species was found to be affected by soil pH, while that of non-conglobating species conformed better to Bergmann's Rule. Overall, body size is constrained mostly by phylogeny and not environmental predictors. Furthermore, no trend conforming to the "island rule" (smaller species become larger and larger species become smaller on islands) could be found in Oniscidea (Karagkouni et al. 2017).

Biogeographical studies at a local scale are more common, but few have tested general patterns. Sutton and Harding (1989), in a detailed account of British isopod distribution, concluded that the most crucial factor in determining species richness is the availability of suitable habitats, while Purse et al. (2012) also showed habitat use to be the main driver of isopod distribution patterns. Hornung et al. (2008) found species richness in Hungary to decrease from undisturbed wet habitats to disturbed dry ones, as well as high compositional turnover among geographical regions and habitat characteristics. The importance of habitat diversity for isopod richness has been identified by several authors. A study by Sfenthourakis and Triantis (2009) showed that the occurrence of keystone habitats, e.g., riparian, is crucial for isopod diversity. Sfenthourakis et al. (2004) found that most species occurring at low abundance are local endemics, but there are also local endemics exhibiting high abundance. Phylogeographic analyses have revealed a strong effect of palaeogeographic history on current distribution (Poulakakis and Sfenthourakis 2008, Kamilari et al. 2014, Lee et al. 2014) but also effects of environmental factors such as seawater temperature for amphibious species (Eberl et al. 2013) or even evolutionary and dispersal events that may lead to unexpected distribution patterns (Santamaria et al. 2013). Such studies are expected to become more numerous in the future, given the rapid advance of molecular techniques.

At the community level, Oniscidea seem to exhibit low equitability due to the dominance of one or a few widespread and/or synanthropic species, such as *Atlantoscia floridana* in the Brazilian Itapuã State Park (Almerão et al. 2006) or *Armadillidium vulgare* in several European habitats (Hornung et al. 2007, Tajovský et al. 2012). Nevertheless, we still lack an explanation for this skewness.

Isopod communities also exhibit significant nestedness (species-poor communities hosting a subset of species present in richer communities; Sfenthourakis et al. 2004, Triantis et al. 2008, Baini et al. 2014) when species-rich communities are included in the data set, while species-poor communities are more heterogeneous, following the spatial distribution of habitats. Oniscidean communities seem to be assembled according to patterns of local habitat heterogeneity. Sfenthourakis et al. (2006), Pitta et al. (2012), and Baini et al. (2014) did not find any evidence for a role of interspecific competition in community structure, even among congeneric species. Some evidence for biotic interactions with other taxa shaping isopod communities, though, has been provided by Antonović et al. (2012), who found higher diversity near bog edges in Croatia attributed to both edge effects and predator pressure by ants of the genus *Myrmica* and lycosid spiders. Sfenthourakis and Triantis (2009) also showed interplay between the relative contribution of generalist and specialist species and keystone habitats in insular Mediterranean communities. Such communities comprise a small core group of generalist species, which are present even on very small islets, while a range of specialist species depend on the erratic occurrence of freshwater-related habitats, which may even explain the presence of the "Small Island Effect."

Tajovský et al. (2012) studied a heterogeneous sample of 13 forest fragments in the Czech Republic and found the highest density and epigeic activity in the smallest fragments despite the fact that larger fragments contained a wider range of habitats. Similarly, Hornung et al. (2007) did not find any evidence for a decrease in diversity due to habitat disturbance, and Vilisics et al. (2007a,b) determined that only a few species showed significant association with habitat features. On the contrary, Messina et al. (2014) found a strong association of isopod species with specific plant communities and soil types at a coastal area in Sicily. A plausible explanation of these contradictory findings may be that saprophagous isopods exploit heterogeneity at very fine scales, so that disturbance and fragmentation may not affect their diversity, as long as keystone habitats are not eliminated.

Troglobiotic oniscideans are important biogeographically, since, for example, in temperate regions they include relict forms of ancient faunas. The highest number of troglobiotic Oniscidea is found in zones on the border of Quaternary glaciations, i.e., southern United States, Mexico, southern Europe, Asia Minor, and Caucasus. Climatic changes occurring in the late Tertiary and Quaternary played a fundamental role in the adaptation of terrestrial isopods to life in caves, according to the Climatic Relict Hypothesis. Manicastri and Argano (1989) found the highest diversity of troglobiotic species in circum-Mediterranean countries and at similar latitudes in the New World, all regions near the southern borders of Pleistocene glacial ice cover (Jass and Klausmeier 2006). Troglobiotic species, though, are constrained by the availability of an appropriate geological setting, and so their distribution patterns may simply reflect the distribution of caves. The structure of karstic systems also affects endemism (Manicastri and Argano 1989), so that, for example, troglobionts in Greece are mostly endemic to one local cave system whereas in the northern Balkans, where caves form more extensive systems, local endemism is generally lower (with the striking exception of Romania). Recently, many troglobiotic terrestrial isopods have also been discovered in tropical and subtropical areas (Brazil, southern China, Hawaii) (Taiti and Howarth 1997, Taiti and Xue 2012, Campos-Filho et al. 2014) that were considered to be particularly unfavorable for speciation in caves due to the limited climatic changes in the past. Most of these tropical troglobionts still have extant epigean relatives and seem to have resulted from parapatric speciation, according to the Adaptive Shift Hypothesis (Rivera et al. 2002).

FUTURE DIRECTIONS

A deeper understanding of adaptation to the terrestrial environment requires detailed and robust phylogenies of modern taxa. The rapid advancement in DNA analysis greatly facilitates this effort but the main problem comes from the severe lack of specialists who could provide the necessary material for analysis. Furthermore, there are gaps in our knowledge on the diversity and distribution of both Talitridae and Oniscidea, especially in the tropics, but also in central and eastern Asia, several oceanic islands, and, importantly, from caves and other subterranean systems. Subterranean species often exhibit relictual distributions that can shed light on paleogeographic events and ancestral clades. There are currently few biologists available to make these studies. Thus, emphasis should be given to training taxonomists and promoting collecting expeditions in underexploited regions of the world. In addition, whole-genome sequencing and mapping of amphipod and isopod species representing main clades would facilitate phylogenetic reconstructions using NGS techniques but would also provide important input to comparative studies in the interplay among evolution, ecology, and development ("evo-devo," "eco-devo").

Another promising line of research is the comparative study of physiology, developmental biology, and neurobiology of modern species, representing the whole range of the transition from marine to arid terrestrial environments. Such studies could offer insight into evolutionary processes, including the origins of novel adaptations, especially for Oniscidea, which is the taxon that can provide such evidence at the scale of relatively closely related genera and species.

Even though extensive descriptive work on the reproductive systems and other aspects of terrestrial crustacean biology is available, we still lack a deep understanding of the actual range of reproductive behavior and population genetics. For example, for the Oniscidea, we still need more intensive research on the role and evolution of cotyledons (and their analogy to mammalian placenta), on the role of intracellular parasites in sex determination (see Dunn et al., 2020), and on mating systems and behavior of these remarkable creatures. All these future directions, together with further research on the ecological role of terrestrial crustaceans, will help us understand the unique evolutionary pathways that Crustacea followed in order to conquer land.

CONCLUSIONS

Among crustaceans, only the family Talitridae in Amphipoda and the suborder Oniscidea in Isopoda show full adaptations to live on land without the need of an aquatic phase during their life cycle.

Terrestrial talitrids are considered to have been established in Pangaea and have evolved from many different palustral or supralittoral talitrids independently. To date, 117 species in 90 genera of land-hoppers are recognized, occurring from sea level to >2,000 m under forest litter, in tussock grasslands, and in very wet areas of up to 1.5 m in trees. Six species are known to have become invasive species. They have been transported by humans within the tropics and usually from the southern temperate to the northern temperate zones. Talitrids are important ecological contributors to the litter fauna in wetter forests of southeastern Australia, New Zealand, and Japan. The diet of land-hoppers is generally unspecialized and appears to consist principally of decaying plant material. The most spectacular adaptation to the terrestrial environment is the retention of the mitten-shaped gnathopod 2, a neotenic condition among males. This condition is geographically widespread and is found among taxa in Australia, New Zealand, the Philippines, Japan, Hawaii, the western Pacific, and the Caribbean and Canary Islands in the Atlantic. It appears that neotenic forms have evolved independently a number of times. In response to living in air, talitrids have developed a greatly enlarged second article on the peduncle of antenna 2, similar to that of freshwater gammarid amphipods.

Terrestrial isopods have been, by far, the most successful group to invade land, leading to more than 3,700 described species in almost all terrestrial habitats with large morphological and physiological diversity. Oniscidea are the only known low-rank taxon that includes living representatives of most transition stages from marine to fully terrestrial life, such as amphibious, littoral, supralittoral, riparian, mesophilous, even desert-living species, as well as a large number of cave-dwelling and secondarily aquatic ones. Along this evolutionary process, the most advanced terrestrial subgroups developed amazing adaptations mainly in response to water loss and respiration in an aerial environment. These include the cotyledons (an egg-feeding apparatus in the female marsupium), a variety of pleopodal lung forms, a complex water-conducting system, and the ability to roll up into a ball that has evolved independently in several Oniscidea lineages. In addition, Oniscidea evolved a variety of behavioral adaptations, ranging from aggregation, humid micro-habitat selection, and a nocturnal lifestyle to the complex burrow-digging and semi-social behavior of desert-living species. Their latitudinal richness and elevation patterns at various scales are of great interest, with diversity hotspots at mid-latitudes and complicated habitat type–species richness interactions. As saprophagous animals, they exploit heterogeneity at fine scales depending strongly on keystone habitats. Overall, Oniscidea provide an excellent model for comparative studies in many fields, from physiology to behavior and biogeography, but it is of paramount importance to fill in the gaps in knowledge of their diversity and to apply modern phylogenetic and phylogenonics methods to resolve evolutionary relationships.

ACKNOWLEDGMENTS

We would like to thank the editors for inviting us to contribute to this volume and for their constructive comments on previous drafts of the text. We would also like to thank Dr. Giuseppe Montesanto, Pisa, for help with the oniscidean figures, and Miles Abadilla for adapting Figure 14.8.

REFERENCES

Almerão, M.P., D.S.Jr.M. Mendoça, A.F. Quadros, E. Pedó, L.G.R. Silva, and P.B. Araujo. 2006. Terrestrial isopod diversity in the subtropical Neotropics: Itapuã State Park, southern Brazil. Iheringia, Série Zoologia 96:473–477.

Alvares, F., I. Winfield, and S. Chazaro. 2000. Population study of the landhopper *Talitroides topitotum* (Crustacea: Amphipoda: Talitridae) in central Mexico. Journal of Natural History 34:1619–1624.

Andersson, A. 1963. On a collection of Amphipoda of the family Talitridae from the Canary Islands. Archiv für Zoologie und Zootomie, Series 2 15:211–18.

Antonović, I., A. Brigić, Z. Sedlar, J. Bedek, and R. Šoštarić. 2012. Terrestrial isopod community as indicator of succession in a peat bog. Zookeys 176:171–178.

Ashton, D.H. 1975. Studies of litter in *Eucalyptus regnans* forests. Australian Journal of Botany 23:413–433.

Baini, F., M.A. Bologna, M. Pitzalis, S. Taiti, A. Vigna Taglianti, and M. Zapparoli. 2014. Assessing patterns of co-occurrence and nestedness of arthropod assemblages in an artificial–natural Mediterranean forest mosaic (Isopoda Oniscidea, Coleoptera Carabidae). Rendiconti Lincei-Scienze Fisiche e Naturali 25:339–350.

Biernbaum, C.K. 1980. Occurrence of the tramp terrestrial amphipods *Talitroides alluaudi* and *T. topitotum* in S. Carolina. Brimleyana 3:107–111.

Bousfield, E.L. 1968. Terrestrial adaptations in Crustacea. American Zoologist 8:393–398.

Bousfield, E. L. 1971. Amphipoda of the Bismarck Archipelago and adjacent Indo-Pacific islands (Crustacea). Steenstrupia 1:255–93.

Bousfield, E.L. 1984. Recent advances in the systematics and biogeography of landhoppers (Amphipoda: Talitridae) of the Indo-Pacific region. Bishop Museum special publication 72:171–210.

Bousfield, E.L., and J. Carlton. 1967. New records of Talitridae (Crustacea: Amphipoda) from the central California coast. Bulletin of the Southern California Academy of Science 66:277–283.

Bousfield, E.L., and F.G. Howarth. 1976. The cavernicolous fauna of Hawaiian lava tubes. 8. Terrestrial Amphipoda (Talitridae), including a new genus and species with notes on its biology. Pacific Insects 17:144–154.

Brandt, A., and G.C.B. Poore. 2003. Higher classification of the flabelliferan and related Isopoda based on a reappraisal of relationships. Invertebrate Systematics 17:893–923.

Broly, P., P. Deville, and S. Maillet. 2013. The origin of terrestrial isopods (Crustacea: Isopoda: Oniscidea). Evolutionary Ecology 27:461–476.

Brusca, R., and G. Wilson. 1991. A phylogenetic analysis of the Isopoda with some classificatory recommendations. Memoirs of the Queensland Museum 31:143–204.

Calman, W.T. 1912. On a terrestrial amphipod from Kew Gardens. Annals and Magazine of Natural History, Series 8 10:132–137.

Campbell, A.G., and A. Gray. 1942. Lyrebirds of Sherbrooke. Emu 42:10–11.

Campos-Filho, I.S., P.B. Araujo, M.E. Bichuette, E. Trajano, and S.Taiti. 2014. Terrestrial isopods (Crustacea: Isopoda: Oniscidae) from Brazilian caves. Zoological Journal of the Linnean Society 172:360–425.

Carefoot, T., and B. Taylor 1995. *Ligia*: A prototypal terrestrial isopod. Crustacean Issues 9:47–60.

Charfi-Cheikhrouha, F. 2003. Genetic diversity in the mitochondrial 16S rDNA among five populations of *Armadillidium pelagicum* (Isopoda, Oniscidea). Crustaceana Monographs 2:365–380.

Clark, D.P. 1954. The ecology of the soil fauna in a rainforest with special reference to the Amphipod, *Talitrus sylvaticus* (Haswell). Ph.D. thesis. University of Sydney, Sydney, Australia.

Cobolli Sbordoni, M., V. Ketmaier, E. de Matthaeis, and S. Taiti. 1997. Biochemical systematics and evolutionary relationships in the *Trichoniscus pusillus* complex (Crustacea, Isopoda, Oniscidea). Heredity 79:463–472.

Coetzee, J.A. 1980. Tertiary environmental changes along the south-western African coast. Palaeontologia Africana 23:197–203.

Cowling, J.E., J.I. Spicer, J.M. Weeks, and K.J. Gaston. 2003. Environmental tolerances of an invasive terrestrial amphipod *Arcitalitrus dorrieni* (Hunt) in Britain. Comparative Biochemistry and Physiology, Part A 136:735–747.

Csonka, D., K. Halasy, and E. Hornung. 2015. Histological studies on the marsupium of two terrestrial isopods (Crustacea, Isopoda, Oniscidea). ZooKeys 515:81–92.

Dahl, E. 1977. The amphipod functional model and its bearing upon systematics and phylogeny. Zoologica Scripta 6:221–228.

de Castro, A. L. 1972. *Talitrus* (*Talitroides*) *pacificus* Hurley, anfipodo terrestre introduzido em Sao Paulo. Arquivos do Instituto Biologico, São Paulo 39:201–203.

Devigne, C., P. Broly, and J.L. Deneubourg. 2011. Individual preferences and social interactions determine the aggregation of woodlice. PLoS ONE 6:e17389.

Devitt, D.M. 1981. Aspects of the ecology of the euterrestrial amphipods of Mt. Wellington. Honors thesis, Department of Zoology, University of Tasmania, Hobart.

Dickman, C.R. 1986. An experimental study of competition between the dasyurid marsupials, *Antechinus stuartii* and *A. swainsonii*. Ecological Monographs 56:221–241.

Dimitriou, A.C., S. Taiti, H. Schmalfuss, and S. Sfenthourakis. 2018. A molecular phylogeny of Porcellionidae (Isopoda, Oniscidea) reveals inconsistencies with present taxonomy. ZooKeys 801:163–176.

Dreyer, H., and J. Wägele. 2002. The Scutocoxifera tax. nov. and the information content of nuclear SSU RDNA sequences for reconstruction of isopod phylogeny (Peracarida: Isopoda). Journal of Crustacean Biology 22:217–234.

Drobne, D., and A. Fajgelj. 1993. Use of Tc-99m-pertechnetate to follow liquid water uptake by *Porcellio scaber*. Journal of Experimental Biology 178:275–279.

Duncan, K.W. 1969. The ecology of two species of terrestrial Amphipoda (Crustacea: Family Talitridae) living in waste grassland. Pedobiologia 9:323–341.

Duncan, K.W. 1994. Terrestrial Talitridae (Crustacea: Amphipoda). Fauna of New Zealand 31:1–128.

Dunn, A.M., T. Rigaud, and A.T. Ford. 2020. Environmental influences on crustacean sex determination and reproduction: environmental sex determination, parasitism and pollution. Pages 396–430 *in* R. Cothran and M. Thiel, editors. The Natural History of the Crustacea, Vol. 6: Reproductive Biology. Oxford University Press, New York.

Eberl, R., M. Mateos, R.K. Grosberg, C.A. Santamaria, and L.A. Hurtado. 2013. Phylogeography of the supralittoral isopod *Ligia occidentalis* around the Point Conception marine biogeographical boundary. Journal of Biogeography 40:2361–2372.

Erhard, F. 1997. Das pleonale Skelet-Muskel-System von *Titanethes albus* (Synocheta) und weiterer Taxa der Oniscidea (Isopoda), mit Schlußfolgerungen zur Phylogenie der Landasseln. Stuttgarter Beiträge zur Naturkunde, Serie A 550:1–70.

Erhard, F. 1998. Phylogenetic relationships within the Oniscidea (Crustacea, Isopoda). Israel Journal of Zoology 44:303–309.
Ferrara, F., P. Paoli, and S. Taiti. 1994. Philosciids with pleopodal lungs? The case of the genus *Aphiloscia* Budde-Lund, 1908 (Crustacea: Isopoda: Oniscidea), with a description of six new species. Journal of Natural History 28:1231–1264.
Ferrara, F., P. Paoli, and S. Taiti. 1997. An original respiratory structure in the xeric genus *Periscyphis* Gerstaecker, 1873 (Crustacea: Oniscidea). Zoologischer Anzeiger 235:147–156.
Friend, J.A. 1975. A study of the energy flow through a natural population of euterrestrial talitrid amphipods. Honors thesis. Department of Zoology, University of Tasmania, Hobart.
Friend, J.A. 1980. The taxonomy, zoogeography and aspects of the ecology of the terrestrial amphipods (Amphipoda, Talitridae) of Tasmania. PhD thesis. University of Tasmania, Hobart.
Friend, J.A. 1982. New terrestrial amphipods (Amphipoda: Talitridae) from Australian forests. Australian Journal of Zoology 30:461–491.
Friend, J.A. 1987. Terrestrial amphipods (Amphipoda: Talitridae) of Tasmania: Systematics and zoogeography. Records of the Australian Museum Supplement 7:1–85.
Friend, J.A., and A.M.M. Richardson. 1977. A preliminary study of niche partition in two Tasmanian terrestrial amphipod species. Ecological Bulletin 25:24–35.
Friend, J.A., and A.M.M. Richardson. 1986. Biology of terrestrial Amphipods. Annual Review of Entomology 31:25–48.
Gentile, G., and V. Sbordoni. 1998. Indirect methods to estimate gene flow in cave and surface populations of *Androniscus dentiger* (Isopoda, Oniscidea). Evolution 52:432–442.
Griffiths, C.L. 1999. The terrestrial amphipods (Crustacea: Amphipoda) of South Africa. Annals of the South African Museum 105:345–362.
Hall, S. 1980. The diets of two coexisting species of *Antechinus* (Marsupialia: Dasyuridae). Australian Wildlife Research 7:365–372.
Hoese, B. 1981. Morphologie und Funktion des Wasserleitungssystems der terrestrischen Isopoden (Crustacea, Isopoda, Oniscoidea). Zoomorphology 98:135–167.
Hoese, B. 1982a. Der *Ligia*-Typ des Wasserleitungssystems bei terrestrischen Isopoden und seine Entwicklung in der Familie Ligiidae (Crustacea, Isopoda, Oniscoidea). Zoologische Jahrbücher, Abteilung für Anatomie und Ontogenie der Tiere 108:225–261.
Hoese, B. 1982b. Morphologie und Evolution der Lungen bei den terrestrischen Isopoden (Crustacea, Isopoda, Oniscoidea). Zoologische Jahrbücher, Abteilung für Anatomie und Ontogenie der Tiere 107:396–422.
Hoese, B., and H. Jansen. 1989. Morphological and physiological studies on the marsupium in terrestrial isopods. Monitore zoologico italiano, Nuova Serie, Monografia 4:153–173.
Hopkin, S. 1991. A key to the woodlice of Britain and Ireland. Field Studies 7:599–650.
Hornung, E. 2011. Evolutionary adaptation of oniscidean isopods to terrestrial life: Structure, physiology and behavior. Terrestrial Arthropod Reviews 4:95–130.
Hornung, E., and P. Sólymos. 2007. Macroecology and biogeography of terrestrial isopods in Europe: a preliminary report. Page 24 *in* Abstract Book of the 7th International Symposium on the Biology of Terrestrial Isopods, Tunisia.
Hornung, E., B. Tóthmérész, T. Magura, and F. Vilisics. 2007. Changes of isopod assemblages along an urban-suburban-rural gradient in Hungary. European Journal of Soil Biology 43:158–165.
Hornung, E., F. Vilisics, and P. Sólymos. 2008. Low alpha and high beta diversity in terrestrial isopod assemblages in the Transdanubian region of Hungary. Pages 1–13 *in* M. Zimmer, C. Charfi-Cheikrouha, and S. Taiti, editors. Proceedings of the International Symposium of Terrestrial Isopod Biology—ISTIB-7, Shaker Verlag, Aachen, Germany.
Hurley, D.E. 1959. Notes on the ecology and environmental adaptations of the terrestrial Amphipoda. Pacific Science 13:107–129.
Hurley, D.E. 1968. Transition from water to land in amphipod crustaceans. American Zoologist 8:327–353.
Hurtado, L.A., E.J. Lee, M. Mateos, and S. Taiti. 2014. Global diversification at the harsh sea-land interface: Mitochondrial Phylogeny of the Supralittoral Isopod genus *Tylos* (Tylidae, Oniscidea). PLoS ONE 9:e94081.

Ingle, R.W. 1958. A new British record of the amphipod. *Talitrus* (*Talitroides sylvaticus* (Haswell). Journal of Natural History 1:591–592.

Ivlev, V.S., and L.M. Suschenya. 1961. Intensity of aquatic and atmospheric respiration in some marine Crustacea. Zoologicheskii Zhurnal 40:1345–1353.

Jass, J.P., and B.R. Klausmeier. 2006. Transborder associations of terrestrial isopods (Crustacea, Isopoda, Oniscidea) of Mexico and the United States. Western North American Naturalist 66:132–134.

Kamilari, M., E. Klossa-Kilia, G. Kilias, and S. Sfenthourakis. 2014. Old Aegean palaeoevents driving the diversification of an endemic isopod species (Oniscidea, Trachelipodidae). Zoologica Scripta 43:379–392.

Karagkouni, M., S. Sfenthourakis, A. Feldman, and S. Meiri. 2016. Biogeography of body size in terrestrial isopods (Crustacea: Oniscidea). Journal of Zoological Systematics and Evolutionary Research 54:182–188.

Karagkouni, M., S. Sfenthourakis, and S. Meiri. 2017. The island rule is not valid in terrestrial isopods (Crustacea: Oniscidea). Journal of Zoology 301:11–16.

Klossa-Kilia, E., G. Kilias, G. Tryfonopoulos, K. Koukou, S. Sfenthourakis, and A. Parmakelis. 2006. Molecular phylogeny of Greek populations of the genus *Ligidium* (Isopoda, Oniscidea) using three mtDNA gene segments. Zoologica Scripta 35:459–472.

Lam, P.K.S., and H.H.T Ma. 1989. Some observations on the life cycle and population dynamics of *Talitroides topitotum* (Burt) (Amphipoda: Talitridae) in Hong Kong. Journal of Natural History 23:1087–1092.

Lazo-Wasem, E.A. 1984. Physiological and behavioural ecology of the terrestrial amphipod *Arcitalitrus sylvaticus* (Haswell, 1880). Journal of Crustacean Biology 4:343–355.

Lee, T.R.C., S.Y.W. Ho, G.D. Wilson, and N. Lo. 2014. Phylogeography and diversity of the terrestrial isopod *Spherillo grossus* (Oniscidea: Armadillidae) on the Australian East Coast. Zoological Journal of the Linnean Society 170:297–309.

Lewis, F. 1991. The relationship between broodpouch colyledons, aridity and advancement. Pages 81–88 *in* P. Juchault and J. Mocquard, editors. The Biology of Terrestrial Isopods III: Proceedings of the Third International Symposium on the Biology of Terrestrial Isopods, Poitiers, France.

Lincoln, R.J. 1979. British Marine Amphipoda: Gammaridea (No. 818). British Museum (Natural History), London.

Lindeman, D. 1990. New terrestrial amphipods (Crustacea: Amphipoda; Talitridae) from Mexico and Central America. Canadian Journal of Zoology 68:2323–2337.

Lindeman, D. 1991a. Natural History of the terrestrial amphipod *Cerrorchestia hyloraina* Lindeman (Crustacea: Amphipoda: Talitridae) in a Costa Rican cloud forest. Journal of Natural History 25:623–638.

Lindeman, D. 1991b. Phylogeny and zoogeography of the New World terrestrial amphipods (landhoppers) (Crustacea: Amphipoda: Talitridae). Canadian Journal of Zoology 69:1104–1116.

Linsenmair, K.E. 1994. Reproductive behaviour and chemical communication in the monogamous desert isopod *Hemilepistus reaumuri* and some other subsocial isopods. Ethology Ecology & Evolution 3:3–4.

Linsenmair, K.E. 2007. Sociobiology of terrestrial isopods. Pages 339–364 *in* J.E. Duffy and M. Thiel, editors. Evolutionary Ecology of Social and Sexual Systems: Crustaceans as Model Organisms. Oxford University Press, New York.

Lowry, J.K., and L. Fanini. 2013. Substrate dependent talitrid amphipods from fragmented beaches on the north coast of Crete (Crustacea, Amphipoda, Talitridae), including a redefinition of the genus *Orchestia* and descriptions of *Orchestia xylino* sp. nov. and *Cryptorchestia* gen. nov. Zootaxa 3709:201–229.

Lowry, J.K., and A.A. Myers. 2013. A phylogeny and classification of the Senticaudata subord. nov. (Crustacea: Amphipoda). Zootaxa 3610:1–80.

Lowry, J.K., and A.A. Myers. 2019. New genera of Talitridae in the revised Superfamily Talitroidea Bulycheva 1957 (Crustacea, Amphipoda, Senticaudata). Zootaxa 4553(1):1–100.

Lowry, J.K., and R.T. Springthorpe. 2015a. The tropical talitrid genus *Floresorchestia* (Amphipoda: Talitridae). Zootaxa 3935:1–68.

Lowry, J.K., and R.T. Springthorpe. 2015b. Coastal Talitridae (Amphipoda: Talitroidea) from north-western Australia to Darwin with a revision of the genus *Cochinorchestia* Lowry & Peart, 2010. Zootaxa 3985:151–202.

Manicastri, C., and R. Argano. 1989. An analytical synopsis of the troglobitic terrestrial isopods. Monitore Zoologico Italiano, Nuova Serie, Monografia 4:63–73.

Martin, H.A. 1998. Tertiary climatic evolution and the development of aridity in Australia. Proceedings of the Linnean Society of New South Wales 119:115–136.

Messina, G., G. Montesanto, E. Pezzino, S. Sciandrello, D. Caruso, and B.M. Lombardo. 2014. Plant communities preferences of terrestrial crustaceans (Isopoda: Oniscidea) in a protected coastal area of southeastern Sicily (Italy). Biologia 69:354–362.

Michel-Salzat, A., and D. Bouchon. 2000. Phylogenetic analysis of mitochondrial LSUrRNA in oniscids. Comptes Rendus de l'Académie des Sciences, Serie III—Sciences de la Vie 323:827–837.

Moeed, A., and M.J. Meads. 1983. Invertebrate fauna of four tree species in Orongorongo Valley, New Zealand, as revealed by trunk traps. New Zealand Journal of Ecology 6:39–53.

Montesanto, G. 2015. A fast GNU method to draw accurate scientific illustrations for taxonomy. ZooKeys 515:191–206.

Moore, P.G., and C.H. Francis. 1985. On the water relations and osmoregulation of the beach-hopper *Orchestia gammarellus* (Pallas) (Crustacea: Amphipoda). Journal of Experimental Marine Biology and Ecology 94:131–150.

Moore, M.L., and A.M.M. Richardson. 1992. Water uptake and loss via the urosome in terrestrial talitrid amphipods (Crustacea: Amphipoda). Journal of Natural History 26:67–77.

Moore, P.G., and A.A. Taylor. 1984. Gill area relationships in an ecological series of gammaridean amphipods (Crustacea). Journal of Experimental Marine Biology and Ecology 74:179–186.

Morino, H. 2013. New records of the land-hopper, *Talitroides topitotum* (Burt, 1934) (Crustacea, Amphipoda, Taltridae), from subtropical East Asia. Bulletin of the National Museum of Nature and Science Series A, Zoology 39:193–201.

Morino, H., and H. Miyamoto. 2015. Redefinition of *Paciforchestia* Bousfield, 1982 and description of *Pyatakovestia* gen. nov. (Crustacea, Amphipoda, Talitridae). Bulletin of the National Museum of Natural Sciences Series A 41:105–121.

Morton, H.P., and A.M.M. Richardson. 1984. Observations on the feeding preferences of two species of Tasmanian terrestrial amphipods (Crustacea: Amphipoda: Talitridae). Proceedings of the Royal Society of Tasmania 18:79–84.

Niikura, M., M. Honda, and K. Yahata. 2015. Phylogeography of semiterrestrial isopod, *Tylos granuliferus*, on east Asian coasts. Zoological Science 32:105–113.

O'Hanlon, R.P., and T. Bolger. 1997a. Size at maturity and sex ratio of *Arcitalitrus dorrieni* (Hunt, 1925) (Amphipoda, Talitridae) at two sites in County Galway, Ireland. Crustaceana 70:676–693.

O'Hanlon, R.P., and T. Bolger. 1997b. Aspects of the life history and reproductive biology of the introduced terrestrial amphipod *Arcitalitrus dorrieni* (Hunt) at two sites in Co. Galway, Ireland. Journal of Natural History 31:1175–1202.

O'Hanlon, R.P., and T. Bolger. 1999. The importance of *Arcitalitrus dorrieni* (Hunt) (Crustacea: Amphipoda: Talitridae) in coniferous litter breakdown. Applied Soil Ecology 11:29–33.

Ortiz, M., and R. Lalana. 2009. A new terrestrial amphipod (Crustacea, Amphipoda, Talitridae) from the Dominican Republic. Solenodon 8:25–32.

Paoli, P., F. Ferrara, and S. Taiti. 2002. Morphology and evolution of the respiratory apparatus in the family Eubelidae (Crustacea, Isopoda, Oniscidea). Journal of Morphology 253:272–289.

Paragamian, K., V. Galanopoulos, A. Legakis, and M. Spanakis. 1987. Ecological, electrophoretic and electron microscope investigations on *Schizidium perplexum* (Isopoda) from Crete. Biologia Gallo-Hellenica 13:165–168.

Parmakelis, A., E. Klossa-Kilia, G. Kilias, K.A. Triantis, and S. Sfenthourakis. 2008. Increased molecular divergence of two endemic *Trachelipus* (Isopoda, Oniscidea) species from Greece reveals patterns not congruent with current taxonomy. Biological Journal of the Linnean Society 95:361–370.

Pitta, E., S. Giokas, and S. Sfenthourakis. 2012. Significant pairwise co-occurrence patterns are not the rule in the majority of biotic communities. Diversity 4:179–193.

Poulakakis, N., and S. Sfenthourakis. 2008. Molecular phylogeny and phylogeography of the Greek populations of the genus *Orthometopon* (Isopoda, Oniscidea) based on mitochondrial DNA sequences. Zoological Journal of the Linnean Society 152:707–715.

Purse, B.V., S.J. Gregory, P. Harding, and H.E. Roy. 2012. Habitat use governs distribution patterns of saprophagous (litter-transforming) macroarthropods—a case study of British woodlice (Isopoda: Oniscidea). European Journal of Entomology 109:543–552.

Rawlinson, R. 1937. The occurrence of the amphipod *Talitrus dorrieni* Hunt in Co. Galway, Ireland. Annals and Magazine of Natural History 20:589–592.

Richardson, A.M.M. 1980. Notes on the occurrence of *Talitrus dorrieni* Hunt (Crustacea: Amphipoda: Talitridae) in south west England. Journal of Natural History 14:751–757.

Richardson, A.M.M. 1991. Two new species of landhoppers (Crustacea: Talitridae) from O'ahu, Hawaiian Islands, with redescription of *Platorchestia pickeringi* and key to landhoppers of O'ahu. Bishop Museum Occasional Papers 31:185–201.

Richardson, A., and P.B. Araujo 2015. Lifestyles of terrestrial crustaceans. Pages 299–336 *in* M. Thiel and L. Watling, editors. The Natural History of History f the Crustacea, Vol. 2: Lifestyles and Feeding Biology. Oxford University Press, New York.

Rivera, M., F. Howarth, S. Taiti, and G. Roderick. 2002. Evolution in Hawaiian cave-adapted isopods (Oniscidea: Philosciidae): vicariant speciation or adaptive shifts? Molecular Phylogenetics and Evolution 25:1–9.

Robertson, D.S., M.C. McKenna, O.B. Toon, S. Hope, and J.A. Lillegraven, 2004. Survival in the first hours of the Cenozoic. Geological Society of America Bulletin 116:760–768.

Ruffo, S. 1958. Amphipodes terrestres et des eaux continentales de Madagascar, des Comores et de la Reunion (Etudes sur les Crustaces Amphipodes). Mémoires de l'Institut Scientifique de Madagascar, Série A 12:35–66.

Santamaria, C.A., M. Mateos, S. Taiti, T.J. Dewitt, and K.A. Hurtado. 2013. A complex evolutionary history in a remote archipelago: phylogeography and morphometrics of the Hawaiian endemic *Ligia* isopods. PLoS ONE 8:e85199.

Schmalfuss, H. 1975. Morphologie, Funktion und Evolution der Tergithöcker bei Landisopoden (Oniscoidea, Isopoda, Crustacea). Zeitschrift für Morphologie der Tiere 80:287–316.

Schmalfuss, H. 1984. Eco-morphological strategies in terrestrial isopods. Symposia of the Zoological Society of London 53:49–63.

Schmalfuss, H. 2003. World catalog of terrestrial isopods (Isopoda: Oniscidea). Stuttgarter Beiträge zur Naturkunde, Serie A 654:1–341.

Schmalfuss, H., and F. Ferrara. 1982. Observations on the distribution and ecology of terrestrial isopods (Oniscoidea) in South-west Cameroon. Monitore Zoologico Italiano, Nuova Serie, Supplemento 17:243–265.

Schmidt, C. 2008. Phylogeny of the terrestrial Isopoda (Oniscidea): a review. Arthropod Systematics & Phylogeny 66:191–226.

Schmidt, C., and J. Wägele. 2001. Morphology and evolution of respiratory structures in the pleopod exopodites of terrestrial Isopoda (Crustacea, Isopoda, Oniscidea). Acta Zoologica 82:315–330.

Searle, J. 1927. Food of the lyre-bird. Victorian Naturalist 44:116.

Sfenthourakis, S., and S. Taiti. 2015. Patterns of taxonomic diversity among terrestrial isopods. ZooKeys 515:13–25.

Sfenthourakis, S., and K.A. Triantis. 2009. Habitat diversity, ecological requirements of species, and the Small Island Effect. Diversity and Distributions 15:131–140.

Sfenthourakis, S., S. Giokas, and E. Tzanatos. 2004. From sampling stations to archipelagos: investigating aspects of the assemblage of insular biota. Global Ecology and Biogeography 13:23–35.

Sfenthourakis, S., E. Tzanatos, and S. Giokas. 2006. Species co-occurrence: the case of congeneric species and a causal approach to patterns of species association. Global Ecology and Biogeography 15:39–49.

Sfenthourakis, S., C. Kassara, M. Mylonas, and H. Schmalfuss. 2007. An analysis of the global distribution of known Oniscidea. Page 44 *in* Abstracts Volume of the 7th International Symposium on the Biology of Terrestrial Isopods, Tunis, Tunisia.

Sfenthourakis, S., D. Skouras, and Y. Anastasiou. 2012. A comparison of terrestrial isopod communities among different habitat types on Mt. Chelmos (Peloponnisos, Greece). Journal of Biological Research 18:198–204.

Shoemaker, C.R. 1936. The occurrence of the terrestrial amphipods, *Talitrus alluaudi* and *Talitrus sylvaticus*, in the United States. Journal of the Washington Academy of Sciences 26:60–64

Smith, D.G. 1998. A new genus and species of terrestrial amphipod from Isla Mona in the Caribbean Sea (Crustacea, Amphipoda, Talitridae). Spixiana 21:179–186.

Smith, G.T., and M.C. Calver. 1984. The diet of the nestling Noisy Scrub-bird *Atrichornis clamosus*. Australian Wildlife Research 11:553–558.

Spicer, J.I., P.G. Moore, and A.C. Taylor. 1987. The physiological ecology of land invasion by the Talitridae (Crustacea: Amphipoda). Proceedings of the Royal Society B 232:95–124.

Stephensen, K. 1935. Terrestrial Talitridae from the Marquesas. Marquesan Insects 3. Pacific Entomological Survey Publications 8:19–34.

Stephensen, K. 1943. The terrestrial amphipod *Talitroides hortulanus* (Calman 1912) found in Samoa and in the New Hebrides. Entomologiske meddelelser 23:295–702.

Stock, J.H., and J.M. Martin. 1988. A new cave hopper (Amphipoda: Talitridae) from lava tubes in La Palma, Canary Islands. Journal of Natural History 22:1121–1133.

Sutton, S., and P. Harding. 1989. Interpretation of the distribution of terrestrial isopods in the British Isles. Monitore Zoologico Italiano, Nuova Serie, Monografia 4:43–61.

Tabacaru, I., and D. Danielopol. 1996. Phylogénie des isopodes terrestres. Comptes Rendus de l'Academie des Sciences, Serie III—Sciences de la Vie 319:71–80.

Tabacaru, I., and A. Giurginca. 2003. The stomach in Oniscidea: structure and evolution. Crustaceana Monographs 2:69–83.

Taiti, S. 2004. Crustacea: Isopoda: Oniscidea (woodlice). Pages 265–267 *in* J. Gunn, editor. Encyclopedia of Caves and Karst Science. Routledge, New York.

Taiti, S., and F. Ferrara. 2004. The terrestrial Isopoda (Crustacea: Oniscidea) of the Socotra Archipelago. Fauna of Arabia 20:211–325.

Taiti, S., and F. Howarth. 1997. Terrestrial isopods (Crustacea, Oniscidea) from Hawaiian caves. Mémoires de Biospéologie 24:97–118.

Taiti, S., and W. Humphreys. 2001. New aquatic Oniscidea (Crustacea; Isopoda) from groundwater calcretes of Western Australia. Records of the Western Australian Museum, Supplement 64:133–151.

Taiti, S., and M. Schotte. 2016. *Hemilepistus* Budde-Lund, 1879. *In* C.B. Boyko, N.L. Bruce, K.L. Merrin, Y. Ota, G.C.B. Poore, S. Taiti, M. Schotte, and G.D.F. Wilson, editors. World Marine, Freshwater and Terrestrial Isopod Crustaceans Database. http://www.marinespecies.org/aphia.php?p=taxdetails&id=248811.

Taiti, S., and Z. Xue. 2012. The cavernicolous genus *Trogloniscus* nomen novum, with descriptions of four new species from southern China (Crustacea, Oniscidea, Styloniscidae). Tropical Zoology 25:183–209.

Taiti, S., P. Paoli, and F. Ferrara. 1998. Morphology, biogeography, and ecology of the family Armadillidae (Crustacea, Oniscidea). Israel Journal of Zoology 44:291–301.

Taiti, S., M. Arnedo, S. Lew, and G. Roderick. 2003. Evolution of terrestriality in Hawaiian species of the genus *Ligia* (Isopoda, Oniscidea). Crustaceana Monographs 2:85–102.

Tajovský, K., J. Hošek, J. Hofmeister, and J. Wytwer. 2012. Assemblages of terrestrial isopods (Isopoda, Oniscidea) in a fragmented forest landscape in central Europe. Zookeys 176:189–198.

Taiti, S., R. Argano, P. Marcia, F. Scarpa, D. Sanna, and M. Casu. 2018. The genus *Alpioniscus* Racovitza, 1908 in Sardinia: taxonomy and natural history (Isopoda, Oniscidea, Trichoniscidae). ZooKeys 801:229–263.

Takeda, N. 1984. The aggregation phenomenon in terrestrial isopods. Symposia of the Zoological Society of London 53:381–404.

Triantis, K., S. Sfenthourakis, and M. Mylonas. 2008. Biodiversity patterns of terrestrial isopods from two island groups in the Aegean Sea (Greece): Species-area relationship, small island effect, and nestedness. Ecoscience 15:169–181.

Vader, W. 1972. Terrestrial Amphipoda collected in greenhouses in the Netherlands. Zoologica Bijdragen 13:32–36.

Vandel, A. 1965. Sur l'existence d'oniscoïdes très primitifs menant une vie aquatique et sur le polyphylétisme des isopodes terrestres. Annales de Spéléologie 20:489–518.

Vilisics, F., Z. Elek, G.L. Lövei, and E. Hornung. 2007a. Composition of terrestrial isopod assemblages along an urbanisation gradient in Denmark. Pedobiologia 51:45–53.

Vilisics, F., P. Sólymos, and E. Hornung. 2007b. A preliminary study on habitat features and associated terrestrial isopod species. Pages 195–199 *in* K. Tajovsky, J. Schlaghamersky, and L.V. Piz, editors. Contributions to Soil Zoology in Central Europe II, Ceske Budejovice. Czech Republic.

Weeks, J.M. 1992. The role of coprophagy in the maintenance of body copper and zinc concentrations in talitrid amphipods (Crustacea, Amphipoda, Talitridae). Comparative Biochemistry and Physiology 101A:313–318.

Wetzer, R. 2002. Mitochondrial genes and isopod phylogeny (Crustacea: Isopoda). Journal of Crustacean Biology 22:1–14.

White, K., J.K. Lowry, and H. Morino. 2013. A new cave-dwelling talitrid genus and species from Japan (Crustacea: Amphipoda: Talitridae). Zootaxa 3682:240–248.

Wildish, D. J. 1982. Talitroidea (Crustacea, Amphipoda) and the driftwood ecological niche. Canadian Journal of Zoology 60:3071–74.

Williamson, D.I. 1951. On the mating and breeding of some semi-terrestrial amphipods. Report of the Dove Marine Laboratory 12:49–62.

Zimmer, M. 2001. Why do male terrestrial isopods (Isopoda: Oniscidea) not guard females? Animal Behaviour 62:815–821.

15

LATITUDINAL GRADIENT OF DIVERSITY OF MARINE CRUSTACEANS: TOWARDS A SYNTHESIS

Marcelo M. Rivadeneira and Gary C. B. Poore

Abstract

The latitudinal diversity gradient (LDG) is a phenomenon acknowledged for over two centuries. The LDG of marine crustaceans has been studied often but without reaching consensus on its ultimate causative processes. We have undertaken a new synthesis to assess the generality of the LDG and evaluated how potential sampling and other biases, spatial scale, geographic regions, taxonomic aggregation, and differences between clades affect patterns. A meta-analysis of 186 datasets, encompassing 20 studies and 7 crustacean orders, revealed a strong effect size of the species richness-latitude correlation, supporting the existence of a "canonical" LDG. The effect size was sensitive to spatial scale, with studies conducted over shorter latitudinal ranges tending to show a weaker LDG. Correcting for sampling biases in the number of occurrences, taxonomic completeness and spatial heterogeneity did not affect the strength of the LDG, nor did the degree of taxonomic aggregation; effect sizes were similar at family and ordinal levels. However, between orders effect sizes varied strongly, with peracarid orders (Amphipoda, Cumacea, Isopoda) showing a weaker or inverse LDG compared with non-peracarid orders (Calanoida, Euphausiacea, Decapoda, Sessilia). Additional analyses based on a global dataset of >2 million occurrences of >13,000 species revealed patterns undetected by the meta-analysis, including: (1) the existence of a marked bi-modal LDG, with peaks of diversity in subtropical areas (Calanoidea, Decapoda, Sessilia) and in temperate areas (Amphipoda, Isopoda), (2) interhemispheric asymmetry, variable across groups and depths, and (3) ocean basin differences in the shape of the LDG, dependent on taxonomic clade. Both ecological and evolutionary processes play a part. The fossil record of Decapoda showed that its global canonical LDG can be explained by median and range of the age of genera, i.e., hotspots of diversity harbor both younger and older genera and contain a high proportion of genera originating during the Paleogene. In addition, the effect size was negatively related to family age, the LDG being stronger in older families of early Cenozoic and Mesozoic origin. Modes of larval development also played a significant

Evolution and Biogeography. Edited by Gary C. B. Poore and Martin Thiel.

part, taxa without planktonic larvae having weaker or inverse LDG compared with taxa with pelagic larvae. Because clades with direct development tend to show smaller bathymetric and latitudinal ranges than those with pelagic larvae, differences in diversification rates may be implied. Overall, our evidence suggested that the ultimate causes of the LDG are deeply tied to geographic differences in macro-evolutionary rates, i.e., greater rates of species origin and lower rates of extinction in the tropics than in higher latitudes combined with a strong tropical niche conservatism.

INTRODUCTION

The latitudinal diversity gradient (LDG; usually expressed as the numbers of species) remains one of the most prominent patterns in macroecology and biogeography, with hundreds of studies evaluating the LDG pattern in a variety of taxa and ecosystems at multiple spatiotemporal scales (Pianka 1966, Rohde 1999, Willig et al. 2003, Wiens and Graham 2005; see Chapter 10). Several previous meta-analyses confirmed the pervasiveness of the canonical pattern of species diversity, which is characterized by increasing diversity from poles toward the tropics (Willig et al. 2003, Hillebrand 2004a, 2004b).

In recent decades the LDG phenomenon has received increasing attention in marine taxa (Clarke 1992, Crame and Clarke 1997, Gray 2001, Jablonski et al. 2006, Rex and Etter 2010). However, the increase of species richness of the epifauna from the Arctic toward the tropics had already been noticed 60 years ago in the pioneer work of Gunnar Thorson (1957). His was indeed the first study showing the existence of a negative relationship between species richness and latitude in two crustacean groups (Brachyura and Amphipoda). However, Thorson's research also showed the lack of a clear LDG in Cumacea and other infaunal taxa. Fisher (1960) was the first to show the existence of latitudinal increase in the richness of calanoid copepods from the Arctic toward the tropical Pacific. While many subsequent studies have supported this "canonical" pattern of increasing diversity toward the tropics (e.g., Boschi 2000, Dworschak 2000, Macpherson 2002, Wood-Walker et al. 2002, Dworschak 2005, Rombouts et al. 2009, Levinton and Mackie 2013, Pappalardo and Fernández 2014), many other studies have failed to support this trend, showing either an inverse, flat, or even non-linear LDG (Pohle et al. 2011, Rivadeneira et al. 2011, Azovsky et al. 2012, Doti et al. 2014, Pappalardo and Fernández 2014, Yasuhara et al. 2017; see Chapter 10). The variety of shapes of LDG is by no means peculiar to Crustacea, being also reported in other marine and terrestrial taxa (Willig et al. 2003, Hillebrand 2004a, 2004b, Kinlock et al. 2018). A meta-analysis carried out by Hillebrand (2004b) confirmed the prevalence of a canonical LDG in marine organisms in which species richness decreases toward higher latitudes, although its strength was dependent on the scale of analysis, habitat, and characteristics of the organism studied. His study included 45 datasets for Crustacea based on 15 published studies. These studies were carried out mostly at ordinal level, were aimed at Decapoda, Isopoda, and Copepoda, and were mostly restricted to the Atlantic.

The existence or otherwise of a LDG is complicated by other global phenomena: (1) the possibility that the LDG may not be unimodal with a single peak in the tropics (Rutherford et al. 1999, Worm et al. 2005, Chaudhary et al. 2016), (2) the potential for the LDG to differ between the Northern and Southern Hemispheres (Rex et al. 1993, Rex and Etter 2010), (3) the promotion of high diversity in environmentally stable areas such as cold deep-sea regions (Rex and Etter 2010, Valentine and Jablonski 2015), and (4) the probability that the gradient in the Indo-West Pacific (IWP) is complicated by the biodiversity hotspot effect (Clarke and Crame 1997, Bellwood and Meyer 2009, Gaither and Rocha 2013, Brown 2014). Examples of these phenomena can be found among Crustacea (e.g., Rex et al. 1993), but a formal evaluation of its extent is still missing.

The underlying causes of the LDG have been subject of intense debate over the decades (Fischer 1960, Rohde 1992, Willig et al. 2003, Brown 2014). While causes have been sought for other marine taxa (Roy et al. 1998, Roy et al. 2000, Jablonski et al. 2006, Tittensor et al. 2010), the few works on marine crustaceans have focused mainly on the role of sea surface temperature (Macpherson 2002, Astorga et al. 2003, Rombouts et al. 2009). A growing body of literature acknowledges the importance of evolutionary processes shaping the LDG in marine organisms (Jablonski et al. 2006, Mittelbach et al. 2007), including some on crustaceans (e.g., in Ostracoda, Yasuhara et al. 2009, 2017). However, for most crustacean groups the role of evolutionary forces on LDG has been indirectly tested (Rivadeneira et al. 2011) or can be suggested by marked clade-effects associated with larval dispersal (Fernández et al. 2009, Pappalardo and Fernández 2014).

In this chapter we attempt to achieve a synthesis of patterns and processes associated with the LDG in marine crustaceans. First, we carried out a new meta-analysis based on an updated and much expanded dataset and evaluated how factors shape the strength of the LDG. Second, we used a large database of georeferenced species occurrences of four crustacean clades to evaluate four aspects of interest: (1) the uni-modality of the LDG, (2) the existence of interhemisphere asymmetry in the LDG, (3) differences between shallow and deep-sea regions, and (4) differences in the LDG across ocean basins, particularly between the IWP and the rest. Finally, we reviewed possible processes that may shape the LDG, exploring the role of ecological and evolutionary factors controlling richness patterns at a global scale in Decapoda.

A META-ANALYSIS APPROACH

We built a database of studies addressing the strength of the LDG, i.e., the Pearson product-moment correlation between absolute latitude and species richness. The final database included 20 studies and 186 datasets. We used as a starting point the information in 12 studies and 29 datasets compiled by Hillebrand (2004b) and added more recent studies and new analyses of global datasets. We included only studies for which we were able to obtain richness and latitude estimations directly from tables and based at least on three spatial points. No digitalization of points in figures was attempted. The new analyses were based on georeferenced records provided by the Ocean Biogeographic Information System (OBIS [www.iobis.org]; OBIS 2015, 2016) for five representative crustacean orders: Calanoida, Decapoda, Amphipoda, Isopoda, and Sessilia. The taxonomy of each taxon relied on the World Record of Marine Species (WoRMS Editorial Board, 2018). Only taxa identified to species level were included. More than 2,200,000 records and 13,500 species were analyzed. All records were assigned to 10° latitudinal bins (considering absolute latitudes). For each bin, we obtained face value (total counts) and rarefied (i.e., corrected for sampling effort) species richness. The correlation between species richness and absolute latitude was calculated for each order and families with more than 5,000 records and 15 species.

For each dataset we obtained information of several types of covariable (Table 15.1). The spatial scale of analysis (local/regional) indicates the scale at which richness information was obtained. Field surveys based on local collections aimed at particular habitats are considered of local scale. Contrarily, regional-scale studies tend to integrate multiple types of habitats. The hemisphere effect refers to whether the study included a single (north/south) or both hemispheres of the planet. The longitude indicates whether the study included a single (Pacific/Atlantic/Indian) or multiple oceans ("worldwide" or "near-worldwide").

For the new datasets built based on the OBIS database, different types of potential sampling biases were considered. The total number of occurrences is a proxy of overall sampling intensity. Spatial bias of occurrences indicates whether face value or corrected (rarefied) richness estimations are used. The completeness of the species richness was estimated as the ratio between observed and expected richness based on Chao 1 non-parametric index (Chao and Lee 1992). Green's dispersion coefficient measures the differences in sampling effort (occurrences) among latitudinal bins.

Table 15.1. Analysis of Heterogeneity for the Strength (r_z) of the LDG

Variable	Type	Levels (categorical) or Range of Variability (Continuous)	Q	df	I^2 (%)	p-value
Spatial scale	Categorical	Local/Regional	812.0854	174	79	**0.002**
Latitudinal extent	Continuous	—	779.7893	174	78	**<0.0001**
Hemisphere	Categorical	North/South/Both	813.6735	175	78	**<0.0001**
Longitude	Categorical	Pacific/Atlantic/Indian/ Worldwide	813.6735	175	78	**<0.0001**
Occurrences*	Continuous	—	152.4215	48	69	0.2318
Spatial bias of occurrences*	Categorical	Face value/rarefied	151.4427	48	68	0.1221
Species completeness*	Continuous	—	153.6606	48	69	0.5946
Green's coefficient*	Continuous	—	152.976	48	69	0.3423
Taxonomic aggregation	Categorical	Family/Superfamily/ Infraorder/Order/ Superorder/Subclass	813.2469	174	79	0.4728
Orders	Categorical	Calanoida/Decapoda/ Amphipoda/Isopoda/ Sessilia	678.7843	160	76	**<0.0001**
Mode of larval dispersal	Categorical	Pelagic/Direct	813.6735	175	78	**<0.0001**

*Only for datasets obtained from the OBIS database.

Effect size, a quantitative measure of the magnitude of a certain phenomenon, was calculated for each dataset by transforming correlation values to z-normalized values:

$$r_z = 0.5 * ln\left(\frac{1+r}{1-r}\right)$$

and with variance

$$\sigma^2_{r_z} = \frac{1}{n-3}$$

where r is the observed correlation coefficient and n the number of observations. We analyzed whether effect sizes varied according to the different covariables, using Cochrane's Q and I^2 heterogeneity index. We carried out only univariate comparisons; complex additive and multiplicative models were not attempted. All analyses were carried out using the libraries MAc (Del Re and Hoyt 2012) and Metaphor (Viechtbauer 2010) in R (R Core Team 2016). The entire datasets are available in the supplementary online information.

GENERALITY OF THE LDG IN THE META-ANALYSIS

The overall effect size of the meta-analysis was negative and significant (r_z = – 0.87, Q = 813.7, d.f. = 175, p < 0.0001; Fig. 15.1). The general negative trend between diversity and latitude was observed, although ~40% of the datasets did not show significant relationships. However, negative relationships prevailed in both significant (95%) and non-significant (80%) datasets.

These results confirm the pervasiveness of the LDG in marine crustaceans. Indeed, Hillebrand (2004b) reached similar results based on 29 datasets. The effect size was similar to those estimated for other marine datasets (r_z = –0.93 to –0.71; Hillebrand 2004b).

Spatial Scale and Extent

The spatial scale and extent of analysis had a significant impact on the effect size (Fig. 15.2, Table 15.1). Studies conducted at regional scale showed a stronger LDG than studies carried out at local scales (Q = 813.7, d.f. = 175, p < 0.0001). The overall relationship between r_z and spatial extent was significant (Q = 779.8, d.f. = 174, p < 0.0001, Fig. 15.3a), implying that studies carried out over broader latitudinal ranges tended to show stronger LDG.

Previous studies had pointed out the importance of spatial scale as a key factor explaining differences in the strength of LDG, where local scale studies consistently show a weaker LDG than more wide-ranging ones (Hillebrand 2004a, 2004b). Scale differences in the LDG could be related to prevalence of regional (i.e., historical/evolutionary) versus local (i.e., ecological) factors (Ricklefs 1987, Cornell and Lawton 1992, Rivadeneira et al. 2002, Witman et al. 2004). At local scales, the arrival of species to local assemblages from the regional pool may be restricted due to

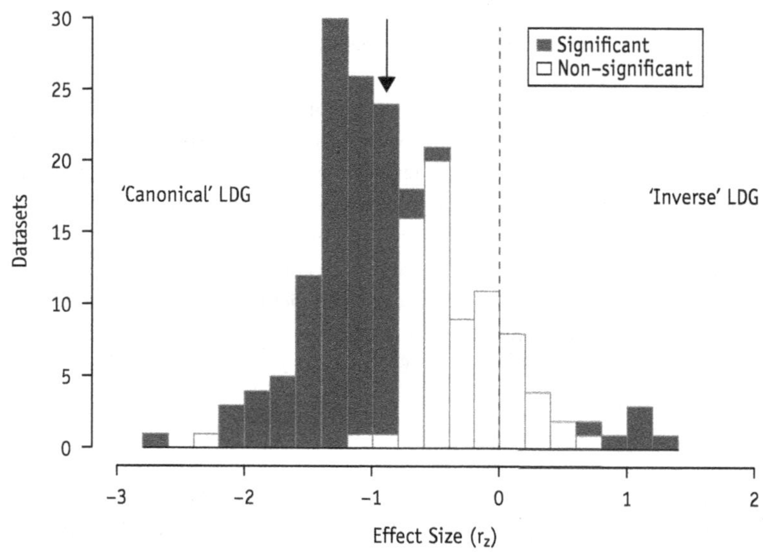

Fig. 15.1.
Frequency distributions of the effect sizes (Z-corrected correlation between richness and absolute latitude) of the LDG calculated for 186 datasets analyzed in the meta-analysis. Negative r_z values indicate a negative relationship between richness and absolute latitude ("canonical" LDG). The arrow shows the median value.

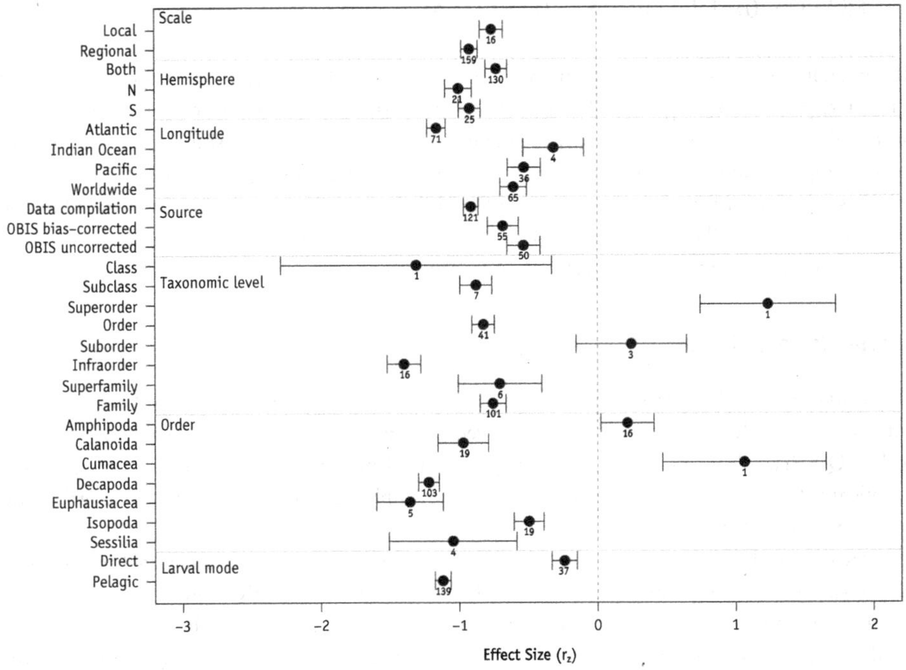

Fig. 15.2.
Effect size (r_z) of the LDG accordingly to the spatial scale of analyses, hemisphere, longitude, source of data, taxonomic level of aggregation, taxonomic order analyzed, and type of larval development. Bars indicate 95% confidence intervals.

species interactions, creating "saturated" assemblages. Thus, species saturation may lead to a rather flat or weak LDG at local scales.

The spatial extent of analysis also has an important impact on the strength of the LDG in marine taxa (Hillebrand 2004b). This could be explained by the pattern of spatial autocorrelation of species richness: at closer distances, species richness tends to be similar, and these differences become significant only at larger distances. For instance, Astorga et al. (2003) found that the characteristic scale of spatial autocorrelation of species richness of Brachyura and Anomura along Pacific and Atlantic coasts of South America occurs at c. 1,500–2,500 km (i.e., 14–23° latitude). This implies that studies covering a few degrees of latitude (e.g., <20°) may yield weaker or non-significant LDG. On the contrary, even studies carried out using local estimates of diversity may exhibit a strong LDG when carried out over large latitudinal gradients (Rex et al. 2000, Gage et al. 2004).

Hemisphere and Ocean Basin Effects

The geographic scope of datasets also influenced the strength of the LDG (Fig 15.2, Table 15.1). Studies restricted to particular hemispheres had stronger effect sizes than studies carried out over both hemispheres ($Q = 813.7$, d.f. = 175, $p < 0.0001$). In addition, r_z varied across different ocean basins ($Q = 813.7$, d.f. = 175, $p < 0.0001$;Fig. 15.2), with the Indian Ocean showing the weakest LDG and the Atlantic the strongest.

Differences across hemispheres and ocean basins may be due in part to the reduced geographic extent implied by regional studies (i.e., Indian Ocean). However, they may also imply biogeographic

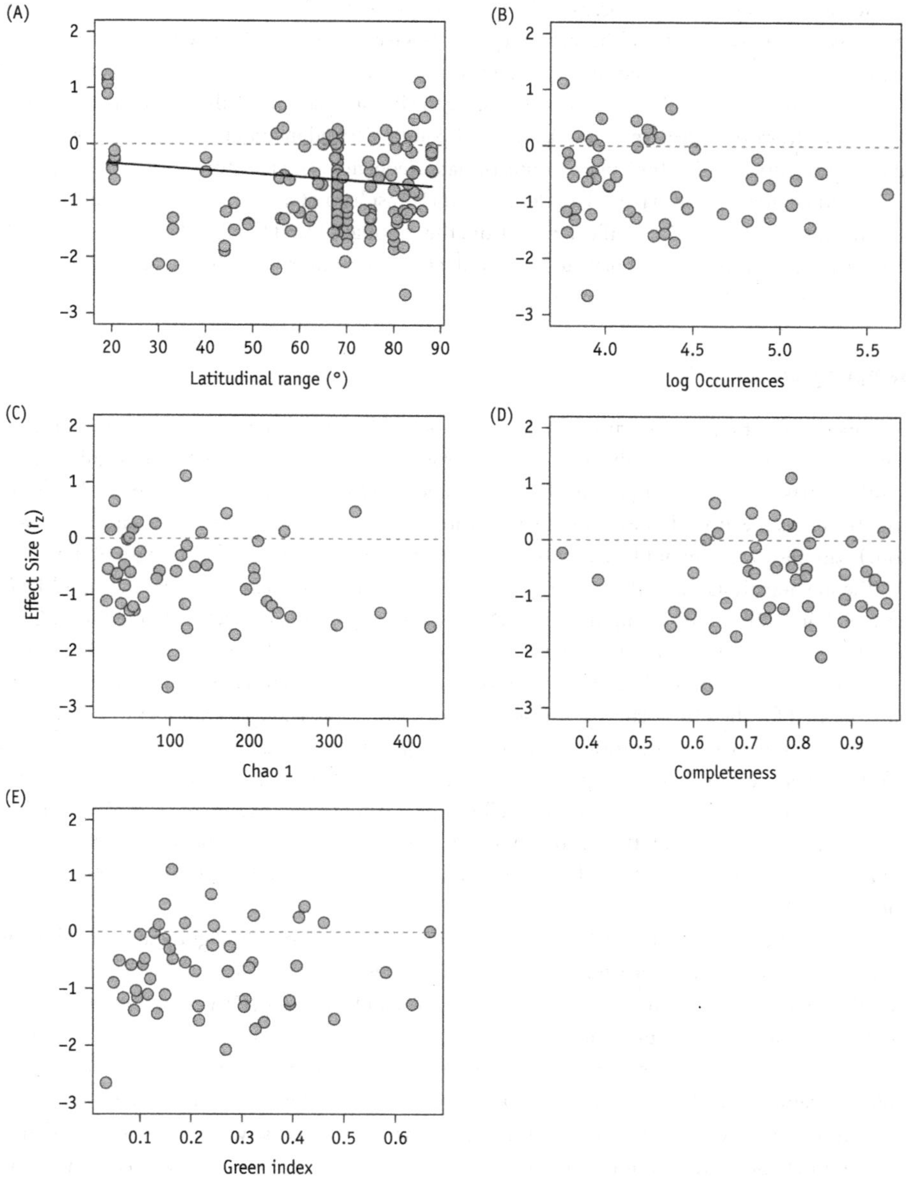

Fig. 15.3.
Effect size (r_z) of the LDG accordingly to: (a) the absolute latitudinal range analyzed, (b) number of occurrences (log-transformed), (c) total extrapolated richness of the group analyzed, (d) degree of completeness of species inventories, and (e) Green index of spatial heterogeneity. Analyses in b–e are based on information extracted from the OBIS database.

imprints produced by historical processes. The greater distances between coastlines in the Southern Hemisphere than in the Northern Hemisphere imply higher isolation of the shallow marine biotas between regions, notably across the Pacific, Atlantic, and Southern oceans. However, the onset of the Antarctic Circumpolar Current c. 30 Ma (Scher et al. 2015) may have allowed a strong connectivity among temperate areas of the Southern Hemisphere. Also, particular historical events may have shaped LDG patterns in certain regions. Gage et al. (2004) discussed how LDG of deep-sea Cumacea may be obscured by historical processes (i.e., Quaternary extinctions/dispersal) in the North Atlantic. Several marine taxa inhabiting the Pacific coast of South America exhibit strong departures from the negative (canonical) LDG pattern (Rivadeneira et al. 2015) attributed to a major biotic turnover that took place during the late Neogene–early Pleistocene there (Rivadeneira and Marquet 2007, Villafaña and Rivadeneira 2014). Yasuhara et al. (2017) showed that the LDG of ostracods in the Western Pacific departs from the canonical trend that had been observed in the Atlantic, likely due to different plate tectonic and temporal geomorphological dynamics promoting different diversification trends across regions.

Sampling Biases

Biological collections are intrinsically biased geographically and temporally. For instance, the number of collections in North America and Europe is higher by some orders of magnitude than in other parts of the world. In theory, because tropical areas are represented mostly by developing countries (i.e., the gross domestic product per capita is much lower between 10° N and 20° S, Sachs 2001), sampling effort might be considerably lower, potentially reducing the strength of the LDG.

The meta-analysis showed that the potential effect of sampling bias on the strength of LDG is limited. For instance, the strength of the LDG did not vary for the datasets extracted from OBIS, whether or not species richness was rarefied (Fig. 15.2). Using the OBIS information yielded trends similar to those from other datasets extracted directly from the literature. Other proxies of sampling bias assessed for the OBIS datasets showed similar trends (Fig. 15.3). The number of occurrences, completeness of species inventories, and coefficient of dispersion did not affect the strength of the effect size (Fig 15.3b–e, Table 15.1; $p > 0.05$ in all cases). While these results may reflect the fact that our analyses were focused on diverse and well-sampled families (i.e., >5,000 records of occurrences and >15 species per family), they also highlight the usefulness and robustness of OBIS and other data compilation initiatives at a global scale to address key biogeographic questions (but see below and Fig. 15.4).

Previous studies have also shown that the shape of the LDG is only slightly impacted by sampling bias. For instance, increasing taxonomic knowledge did not result in a drastic effect on the overall negative LDG of Thalassinidea (now Gebiidea and Axiidea) (Dworschak 2000, 2005), although it revealed that an apparent drop in species richness of these infraorders in the tropical area between 10° N and 5° S detected in a previous analysis (Dworschak 2000) was an artefact of insufficient sampling and taxonomic splitting (Dworschak 2005). Similarly, Rivadeneira et al. (2011) showed that the inverse LDG in peracarids along the Chilean coast was largely independent of the temporal completeness of the species inventories, i.e., the addition of new species described after 1950 did not affect the inverse shape of the LDG. Thus, while species inventories of marine crustaceans seem far from complete (Dworschak 2000, 2005, Poore and Bruce 2012, Poore et al. 2015), it could be argued that the overall LDG pattern may not be drastically changed by future taxonomic effort.

However, a source of bias in the OBIS data is the exclusion of undetermined species. For some taxa these may be many, and geographically biased to the Southern Hemisphere. Four comprehensive surveys in Australia have reported a high proportion of undescribed species. These studies, unlike some others, involved competent taxonomists and have made a special point of sorting samples to "morphospecies" and attempting to identify these species from the literature. They first sorted and attempted to identify 6,870 individuals of Isopoda from the southeastern continental slope. Ninety percent of the 359 morphospecies recognized were at that time undescribed (Poore et al. 1994). The second study, on the continental shelf of southern Australia, yielded 60,258 individuals

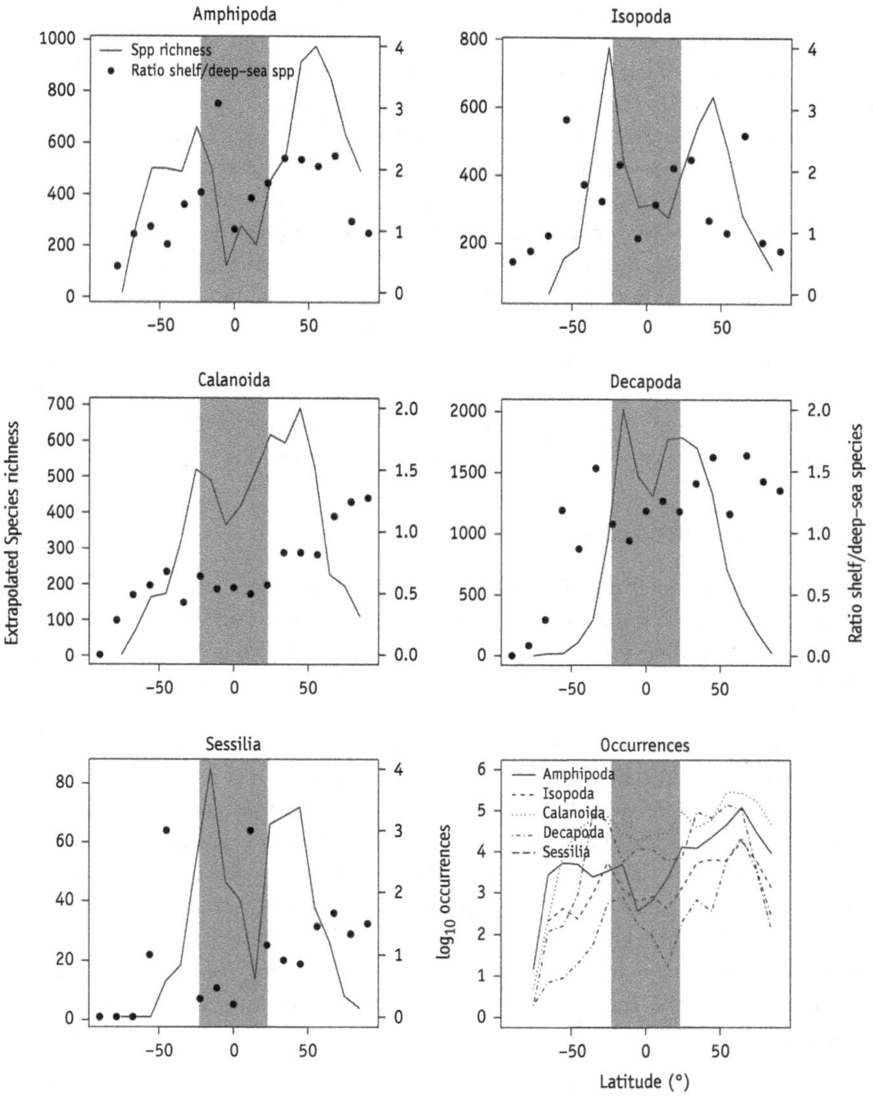

Fig. 15.4.
LDG of extrapolated species richness (Chao 1 non-parametric index) of Amphipoda (n = 3,276 species), Isopoda (n = 2,593 species), Calanoida (n = 1,295 species), Decapoda (n = 4,840), and Sessilia (n = 204 species) based on the OBIS database. Also shown is the ratio of species richness in the shelf (<200 m) and deep-sea areas (≥200 m). Values higher than 1 for the ratio indicate the prevalence of shelf-inhabiting species, whereas values <1 indicate that deep-sea species prevail. The gray area highlights the tropics (±23°).

from 104 samples, equivalent to an area of 10.4 m². Of the 803 species of molluscs, polychaetes, and crustaceans collected, 51% (409 species) were undescribed (Coleman et al. 1997). Even in taxonomically better known taxa, a significant fraction of the fauna remains undescribed, notably in a survey of western Australian decapods 30% of the 524 species found were undescribed (Poore et al. 2008). Finally, a survey along the entire continental slope of western Australia identified species in some taxonomic groups of small macrofaunal crustaceans found in 135 grab samples (Poore et al.

2014). These authors differentiated and attempted to identify 126 morphospecies of Isopoda (92% undescribed), 36 morphospecies of Decapoda (61% undescribed), and 60 species of myodocopid ostracods (75% undescribed). The same survey found 127 species of Amphipoda, 284 species of Tanaidacea, and 86 species of Cumacea, none of which were described species.

Similar fractions of undescribed species, around 85–90% of the total fauna, have been summarized for peracarids by Poore et al. (2015), from the Southern Ocean and South Atlantic (Brandt et al. 2004, Brandt et al. 2007a, 2007b, 2007c, Ellingsen et al. 2007, Malyutina and Brandt 2007) and a lower fraction (46% of isopod species) in the Gulf of Mexico (Wilson 2008). With one exception, these studies are from Australia and the Southern Ocean (comparable studies in the Northern Hemisphere are few; Poore et al. 2014), and it is difficult to avoid the conclusion that the data available for our analysis, especially on the number of species of small crustaceans in the Southern Hemisphere, have limited our capability to discover the true pattern. We suggest that inclusion of these data may modify the LDG pattern, at least for particular regions. For instance, any perceived depression of species number toward the South Pole may be compromised by incorporation of the large number of undescribed species in the Southern Hemisphere.

Taxonomic Aggregation and Trends among Clades

The meta-analysis showed that LDG is relatively unaffected by the level of taxonomic aggregation ($Q = 813.2$, d.f. = 174, $p = 0.47$;Fig. 15.2), i.e., the inverse LDG is concluded for studies carried out for particular families or whole suborders. However, particular orders tend to show contrasting LDG trends ($Q = 678.8$, d.f. = 160, $p < 0.0001$;Fig. 15.2); indeed, Amphipoda and Cumacea both exhibited an inverse LDG.

The lack of conformity to the canonical LDG in Peracarida is partially supported by previous studies. Thorson (1957) already noticed a week LDG for Cumacea, yet he reported a strong negative LDG for Amphipoda. Moreover, deep-sea studies (500–4,000 m depth) carried out at local scales have reported strong negative LDG for Isopoda and Cumacea (Rex et al. 2000, Gage et al. 2004). Pelagic amphipods of the family Hyperiidae exhibited markedly negative LDG across a wide oceanic transect along the Atlantic (Burridge et al. 2017). So, it is possible that regional-scale studies pooling multiple kinds of habitats may mask true LDG, at least in these taxa.

Studies carried out along the Pacific and Atlantic coasts of South America have stressed the existence of an inverse LDG, i.e., decreasing species richness toward lower latitudes, in some clades of Peracarida (Fernández et al. 2009, Rivadeneira et al. 2011, Doti et al. 2014), a trend that could be attributed to the restricted geographic area and/or particular evolutionary dynamics in the region. Rivadeneira et al. (2011) linked the inverse LDG reported for peracarids along the southeastern Pacific to the evolutionary success of the group around the Southern Ocean (Knox and Lowry 1977, Watling and Thurston 1989, Brandt 1999, Clarke et al. 2004).

GLOBAL ANALYSIS OF THE LDG

The Shape of the LDG

Our meta-analysis was based on the assumption that the functional relationship between taxonomic richness and latitude is linear, which is not warranted in all cases. This may be particularly important because in several groups the peak of richness seems slightly distant from the equator, as observed in other groups (Powell et al. 2012, Beaugrand et al. 2013), creating a non-linear hump-shaped LDG that could not be properly described by linear functions (Fig. 15.4). The use of absolute latitudes to estimate the strength of the LDG in the meta-analysis assumes that its shape is symmetrical between hemispheres, which again may be an oversimplification. For instance, there are several cases of interhemispherical asymmetries in the LDG (Valdovinos et al. 2003, Chaudhary et al. 2016), including marine isopods (Espinosa-Pérez and Hendrickx 2006).

The latitudinal environmental gradients, notably temperature and energy flux, that dominate terrestrial and shallow marine ecosystems are less evident in the uniformly cold deep sea. Yet it has been here too that LDGs have been debated (Hessler and Sanders 1967, Rex and Etter 2010, Valentine and Jablonski 2015). For instance, deep-sea asellote isopods have greater richness in the South Atlantic than in the North Atlantic (Wilson 1998). Brandt et al. (2005) showed that while in the Northern Hemisphere (where the Arctic Ocean has very few species) the LDG was reversed, in the Southern Ocean more species were found at higher latitudes (between 58°14′ S and 65°20′ S).

In order to tackle these problems, we used a more robust approach, analyzing geographic patterns of variation in species richness based on georeferenced occurrences of species from the OBIS database. We evaluated the LDG using 10° bins and using the Chao 1 extrapolation index to account/ameliorate for possible (and likely) bias introduced by differences in sampling effort across space.

Our new results showed that the shape of the LDG in crustaceans is: (1) not unimodal, (2) not symmetrical between hemispheres, and (3) dependent on bathymetry. First, species richness showed two maximum peaks in all five clades analyzed (Fig. 15.4). Yet the number of georeferenced occurrences was not uniform across latitudes for each clade; the bimodality in species richness cannot be solely attributed to latitudinal differences in sampling effort. The position of these two peaks differed between taxa but was located within or close to the tropics in the case of Calanoida, Decapoda, and Sessilia, hence explaining the general conformance to the canonical LDG pattern detected in the meta-analysis. However, both peaks of diversity for Amphipoda and Isopoda were placed well outside the tropics, around 30–50° N and S. Quite similar results were found by Chaudhary et al. (2016) for 65,000 species, including 250 amphipods. Second, the LDG was not symmetrical between hemispheres, as the position of these peaks differed between taxa. Third, the LDG seemed to be related to bathymetric patterns (i.e., the ratio of shelf/deep-sea species) at least in some orders. For instance, the shelf/deep-sea ratio followed a hump-shaped pattern in Amphipoda and Isopoda, characterized by the prevalence of shelf species in the tropics and deep-sea species at higher latitudes. However, the shelf/deep-sea species ratio showed a remarkable latitudinal south to north gradient in Calanoida, Decapoda, and Sessilia, i.e., the Southern Ocean is strongly dominated by deep-sea species, whereas shelf species prevailed in the temperate Northern Hemisphere.

In general, these results agree with previous studies that have raised questions about the generality of the unimodal LDG paradigm, particularly in peracarids. While decapods, brachyurans in particular, are significantly more diverse at low latitudes (near, rather than at the equator) than at high latitudes, as exemplified by data from both coasts of the Americas (Boschi 2000), the same is not true for all crustacean taxa. Isopods, for example, diverge from this pattern. Asellote isopods, a taxon abundant and diverse throughout the deep sea, are more diverse in the mid-South Atlantic (Argentine Basin) than in the North Atlantic, where their diversity falls away from the equator (Rex et al. 1993, Wilson 1998). Asellotes at shelf depths are more diverse south of 47° S in Argentina than they are north of this latitude (Doti et al. 2014). Historical factors may be responsible for the deep-sea anomaly, but the predominance of gravelly sediment promoting high diversity in the southern Patagonian shelf as well as the tectonic history of the southern tip of South America were hypothesized as being more significant in the Argentine study. Non-asellote diversity decreases from north to south across the entire deep Atlantic Ocean (Wilson 1998), again defying the expected trend. Further evidence for high isopod diversity in the Southern Hemisphere was provided by Poore et al. (1994), who noted that their collection of 359 species from south-eastern Australia was more diverse than had been recorded at similar latitudes in the North Atlantic, a finding repeated on the west coast of Australia (an additonal 126 species; Poore et al. 2014; and 354 species of Asellota; Doti et al. 2014). Some of the richest collections of asellote isopods have come from the Southern Ocean, 674 species reported by Brandt et al. (2007a). Ellingsen et al. (2007) concluded that their diversity was not related to latitude within the range 58–74° S.

Basin and Longitudinal Effects

The worldwide analysis of species richness of five major Crustacean clades based on OBIS occurrences pooled in 10° grid cells revealed that LDG is not consistent across different basins and longitudinal areas of the global ocean (Fig. 15.5). However, it should be noted that vast areas of the ocean lack, particularly in the open ocean, a proper representation in terms of number of occurrences in the OBIS database, which could likely mask LDG patterns. The extent of some hotspots of richness could also be underrepresented.

Any latitudinal gradient in the IWP can be interpreted as part of the effect of the biodiversity hotspot, or coral triangle, centered on the Malay Archipelago from which biodiversity decreases in all directions (Hoeksema 2007, Bellwood and Meyer 2009, Gaither and Rocha 2013). The "bullseye" pattern results in typical LDGs, in shallow water at least, away from a center north of the equator in the Western Pacific but has little impact in the Central or Eastern Pacific. Few studies of crustaceans in this area address diversity gradients specifically. Latitudinal species richness of deep-sea squat lobsters along the Western and Eastern Pacific has been shown to increase from higher latitudes toward the equator with significant high anomalies at 20° S (New Caledonia) and at 30° N in the Western Pacific (Macpherson et al. 2010). The Central Indo-Pacific is home to more species of isopods than any other region, their distribution decidedly asymmetrical latitudinally with 1.35 times as many species in the temperate Southern Hemisphere than in the temperate North Atlantic and in the Northern Pacific, and almost four times as many Antarctic as Arctic species (Poore and Bruce 2012). A recent study has shown that Ostracoda in the West Pacific also do not follow the canonical LDG, reaching a maximum richness at mid-latitudes, contrary to what has been observed in the Atlantic (Yasuhara et al. 2017).

UNDERLYING PROCESSES

Causes of the LDG pattern have been debated intensely for decades (Rohde 1999), with a recent focus on the creation of integrative and explanatory conceptual frameworks (Wiens and Graham 2005, Jablonski et al. 2006, Mittelbach et al. 2007, Brown 2014). The plethora of possible explanations (Pianka 1966, Rohde 1999, Willig et al. 2003) can be broadly framed into two major categories: evolutionary (or historical) and ecological (Mittelbach et al. 2007, Brown 2014). On the one hand, evolutionary explanations pose that geographic variation in diversification rates (i.e., origin, extinction, and dispersal) may shape the LGR. Ecological factors, on the other hand, assume that the origin and maintenance of the LGR is due to contemporary processes (i.e., carrying capacity). Of course, ecological and evolutionary hypotheses are not mutually exclusive (Belmaker and Jetz 2015).

To date, studies analyzing causes of the LDG in crustaceans are few. Most have considered only ecological explanations, based on correlations between species richness and proxies of environmental energy. For instance, several studies have reported that the bulk of variance in species richness of several clades of Decapoda could be explained by present-day sea surface temperature (Macpherson 2002, Astorga et al. 2003, Rombouts et al. 2009, Levinton and Mackie 2013). Local species richness of deep-sea Ostracoda also depends strongly on sea temperature over the last 400 ka, and the LDG vanishes during glaciation periods (Yasuhara et al. 2009). In a study of species richness of decapods over a wide latitudinal range in western Australia (c. 13–35° S), contradictory

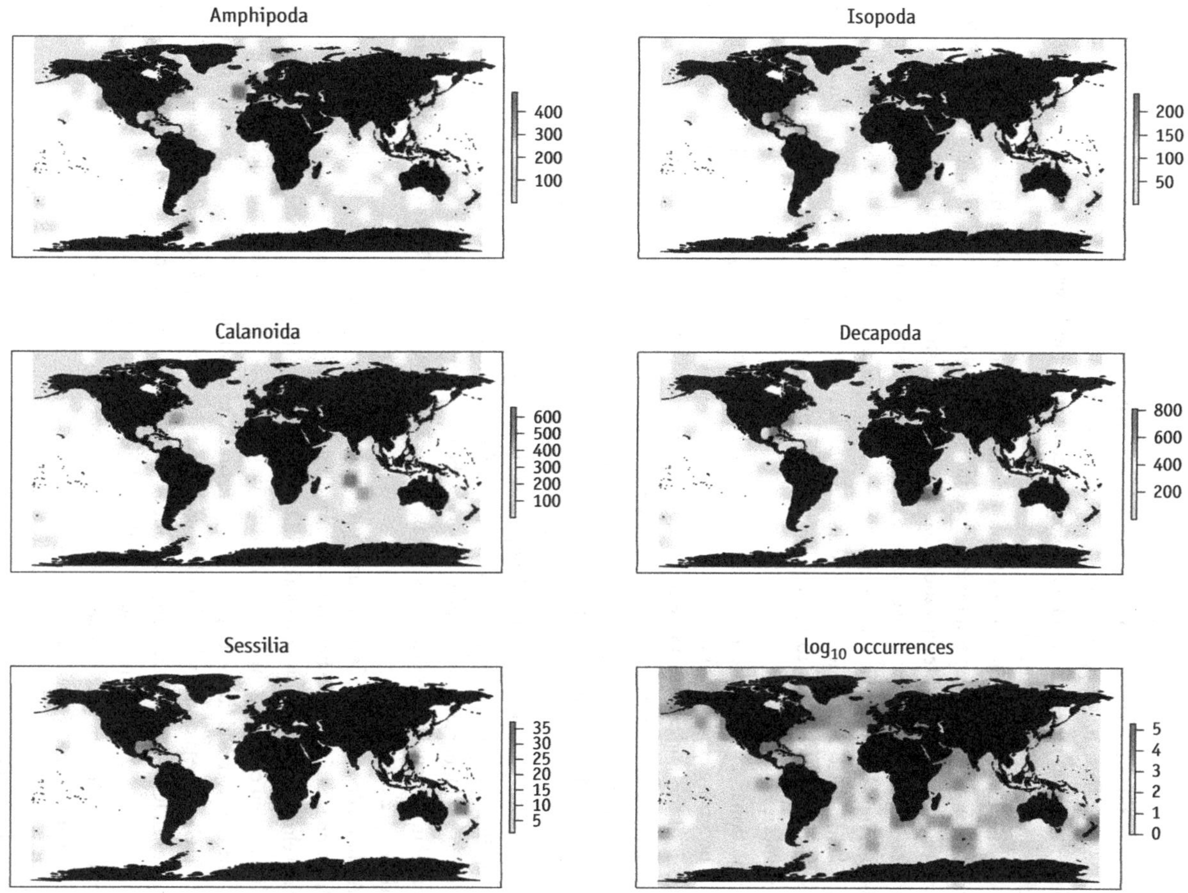

Fig. 15.5.
Global patterns of extrapolated species richness (in scale of colors) of Amphipoda, Isopoda, Calanoida, Decapoda, and Sessilia, in 10° resolution grid cells, based on OBIS database. Empty cells correspond to cells with less than 500 georeferenced occurrences (irrespective of the number of species). Also shown the total number of occurrences per cell (in $\log_{10}$ scale, summed across these five orders). See color version of this figure in centerfold.

results were found. McCallum et al. (2013) found that temperature was not associated with latitude, and the latter proved to be the best predictor of sample species richness along the western Australian margin. They concluded that proximity to the center of the highly diverse IWP region may be important, especially considering that almost 40% of species in their study had a wide IWP distribution. In a parallel survey of peracarids and polychaetes in the same area, McCallum et al. (2014) showed that spatial variability in productivity correlated with the patterns of benthic species richness. Productivity did not vary linearly with latitude but was determined by local hydrodynamic factors. The greatest species richness was recorded at the shallower and more southern sites. Despite the strong support for the "species-energy hypothesis" concluded by some studies, its underlying causes may depend on physiological, ecological, biogeographic, and even evolutionary mechanisms (Evans et al. 2005).

A connection between niche breadth and latitude has been long hypothesized, i.e., species niche is reduced or more packaged toward lower latitudes (McArthur 1972). The latitude–niche breadth hypothesis (Vazquez and Stevens 2004) has also been suggested as a potential explanation for the existence of a canonical LDG, based on the increasing proportion of small species toward tropical areas in several marine crustacean clades (Steele 1988, Papacostas and Freestone 2016). However, a recent study showed no general latitudinal trend in the trophic niche breadth of brachyuran crabs (Papacostas and Freestone 2016); niche breadth is positively related to latitude but only in temperate species.

Evolutionary/historical explanations for the LDG in crustaceans propose that diversification events occurring at scales of hundred thousands to millions of years have created the gradient. A robust test of these ideas demand either a good fossil record and/or nearly complete phylogenies for large clades, which are clearly unavailable for most crustaceans. However, these ideas could be put indirectly to a test based on the taxonomic structure of the biotas (e.g., species-genus ratios; Krug et al 2008). For instance, Rivadeneira et al. (2011) used taxonomic distinctness, a measure that integrates species richness with taxonomic structure, as a coarse proxy of diversification. These authors found that the inverse LDG of peracarids along the southeastern Pacific is linked to an increase in the taxonomic distinctness toward higher latitudes (i.e., there are proportionally more higher taxa at higher latitudes), likely reflecting the fact that the Southern Ocean could be an incubator for many crustaceans groups (Knox and Lowry 1977, Watling and Thurston 1989, Brandt 1999, Clarke et al. 2004).

Additionally, the role of evolutionary processes is indirectly suggested by the sharp differences in the strength of the LDG between groups with different reproductive strategies; groups with direct development tend to show a weaker LDG than groups with pelagic larvae (Fig. 15.2). Previous studies carried out along the Pacific and Atlantic coasts of South America provided an example of this sort of pattern: clades with pelagic larvae (Brachyura and Anomura) have a negative (canonical) LDG, whereas clades with brooding and direct development (Peracarida) showed a positive (inverse) LDG (Astorga et al. 2003, Fernández et al. 2009, Rivadeneira et al. 2011, Pappalardo and Fernández 2014). The ultimate causes of these differential patterns between groups with different reproductive strategies have been linked to clade-dependent physiological responses to environmental temperature (Fernández et al. 2009).

Differences in dispersal capabilities associated to clades with different larval types could be translated into different biogeographic patterns and evolutionary rates (Jablonski and Lutz 1983). We indirectly tested this idea by comparing the geographic range sizes (measured as latitudinal and bathymetric ranges) of species with direct development (5,869 species of Amphipoda and Isopoda) versus species with pelagic larvae (6,339 species of Calanoida, Decapoda, and Sessilia) based on data extracted from OBIS. Results show that both latitudinal and bathymetric ranges are wider in species with pelagic larvae than in species with direct development (Mann-Whitney test, $p < 0.001$ in both cases; Fig. 15.6). Interestingly, both groups exhibit contrasting latitudinal trends of

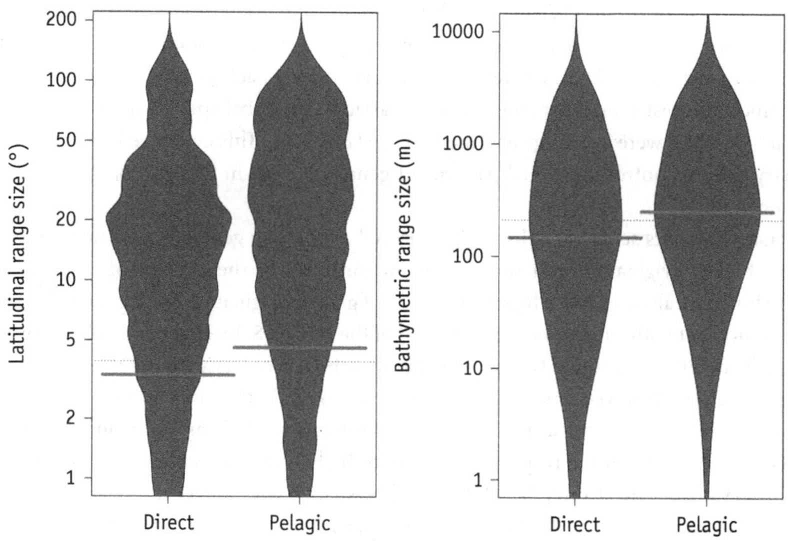

Fig. 15.6.
Density distributions of the latitudinal and bathymetric range of crustacean species with pelagic larvae (combining Calanoida, Decapoda, and Sessilia) and with direct development (combining Amphipoda and Isopoda), based on information collected from OBIS. Latitudinal range was estimated as the difference between the northernmost and southernmost latitudinal occurrences (n = 5,869 species with direct development, n = 6,339 species with pelagic larvae). Bathymetric range was estimated as the difference between the deepest and shallowest distribution (n = 1,042 species with direct development, n = 1,765 species with pelagic larvae). In order to improve clarity of the figures, values lower than 0.5° and 0.5 m, respectively, are not shown. Dotted horizontal lines represent the overall median values, whereas thick solid lines indicate group medians.

bathymetric distribution: shelf species prevail in the tropics and deep-sea forms at higher latitudes in the case of direct developers (Fig. 15.4), whereas in clades with pelagic larvae, shelf species dominate toward the Northern Hemisphere, and deep-sea forms prevail around the Southern Ocean. Because species with smaller geographic ranges tend to exhibit higher extinction and origination rates (Finnegan et al. 2015, Orzechowski et al. 2015), clades without a pelagic larval stage might have higher diversification rates than clades with pelagic larvae. A direct test of this idea is still missing given the limited fossilization potential of peracarids.

Despite most crustacean groups having a poor fossil record (Sepkoski 2000), groups such as Ostracoda do and have been used to evaluate LDG dynamics in the deep-time (Yasuhara et al. 2009, 2017). These authors demonstrated that the LDG pattern is strongly variable across scales of thousands to millions of years, i.e., species richness has not always peaked in the tropics (Yasuhara et al. 2009, 2017). Decapoda is another group whose fossil record has grown during the last decades (Schweitzer et al. 2010), but it has still not been used to investigate the role of evolutionary processes shaping the LDG.

Here we use a random forest analysis to evaluate the role of both ecological and evolutionary factors shaping global patterns of genus richness of Decapoda, in 10° resolution grid cells (n = 68). The analysis is based on a subset of 195 modern genera having a fossil representative. Ecological factors include mean sea surface temperature, sea-surface temperature range, mean primary productivity, coastline length, and mean depth (e.g., Tittensor et al. 2010). Evolutionary factors are median genus age and age range, extracted from the Paleobiology Database (https://paleobiodb.org/). The random forest approach is a powerful machine-learning classification and regression method (Liaw and Wiener 2002) that circumvents assumptions of traditional OLS and GLM regressions; it is being increasingly used in ecological and evolutionary analyses (Cutler et al. 2007, Fenberg et al. 2015). Analyses were carried out using the package rfPermute (Archer 2015) in R. The random forest model explained 63% of variation of global species richness, and the only significant variables were age range and median age (Fig. 15.7). This evidence favors the role of evolutionary factors controlling global patterns of genus richness in this group, at least at a coarse resolution (10°).

Diversity hotspots seem to harbor both very old and young genera, suggesting the combined effects of higher origination and reduced extinction rates in the control of richness patterns. Diversity hotspots also contain a high proportion of genera originating during the Paleogene (43–66 Ma) coincident with a major radiation period of the group (Schweitzer and Feldmann 2005). In addition, the poleward latitudinal endpoint of the distribution of Decapoda species shows a marked taxonomic conservatism (i.e., species per genus tends to retain the same latitudinal limit, Moran's $I = 0.36$, $p < 0.001$), suggesting a "tropical niche conservatism" (Wiens and Donoghue 2004) that prevents taxa occurring in the tropics to disperse to higher latitudes unless their niche evolves. If niche conservatism were strong, the LDG would be shaped over longer timescales, i.e., older taxa should tend to have a stronger LDG than younger taxa (Crame 2000), given more time for niche evolution within an old clade (Wiens and Donoghue 2004). We conducted a preliminary test of this idea, comparing the LDG size of 26 families (effect sizes calculated based on a global scale based on OBIS data) with its first appearance in the fossil record. Results show that older families (i.e., originating during the Mesozoic and early Paleogene) tend to exhibit a clear canonical LDG, whereas younger families (i.e., originating <50 Ma) show an inverse LDG (Fig. 15.8). Hence, the niche evolves and tropical clades are able to disperse toward high latitudes only once enough time has passed.

FUTURE DIRECTIONS

The results of our analyses may guide future studies, providing testable hypotheses regarding patterns and processes shaping the LDG in marine crustaceans. Studies should be carried out over large geographic scales, keeping in mind that the latitude is only one of the spatial axes of variation of a more complex phenomenon occurring in four dimensions: latitude, longitude, depth, and time. The availability of inventories of modern and fossil species occurrences opens this possibility, yet further taxonomic work is much needed, especially in particular regions of the global ocean, including wide areas of the Pacific and the Southern Ocean. Future LDG assessments should be part of more general studies addressing longitudinal and bathymetric gradients of diversity (particularly in slope and deep-sea areas) and its variability at multiple scales of time. For instance, our analyses suggest that the shape of the LDG seems to vary at different depths for particular clades. Future studies should also consider ecological and life-history traits of species, which can be used to evaluate trends of functional diversity (e.g., Berke et al. 2014). For instance, the proportion of carnivorous copepods seems to follow a clear latitudinal gradient, being much higher in the tropics than in high latitudes (Wood-Walker 2002). The mean body size of Decapoda also shows a marked latitudinal trend characterized by an increase in the proportion of small species toward lower latitudes (Steele 1988). On the other hand, evolutionary factors are key to understanding spatial patterns of species diversity (Jablonski et al. 2006, Valentine et al. 2008), and so further efforts should be also devoted to putting these ideas to a test. Despite most crustacean groups having a poor fossil record, some such as Decapoda, Ostracoda, and Sessilia have suitable records (Sepkoski 2000) for future exploration (e.g., Yasuhara et al. 2009, 2017). In addition, in the absence of fossil records, phylogenies based on molecular markers available for many clades (e.g., Davies et al. 2016) open the possibility of testing

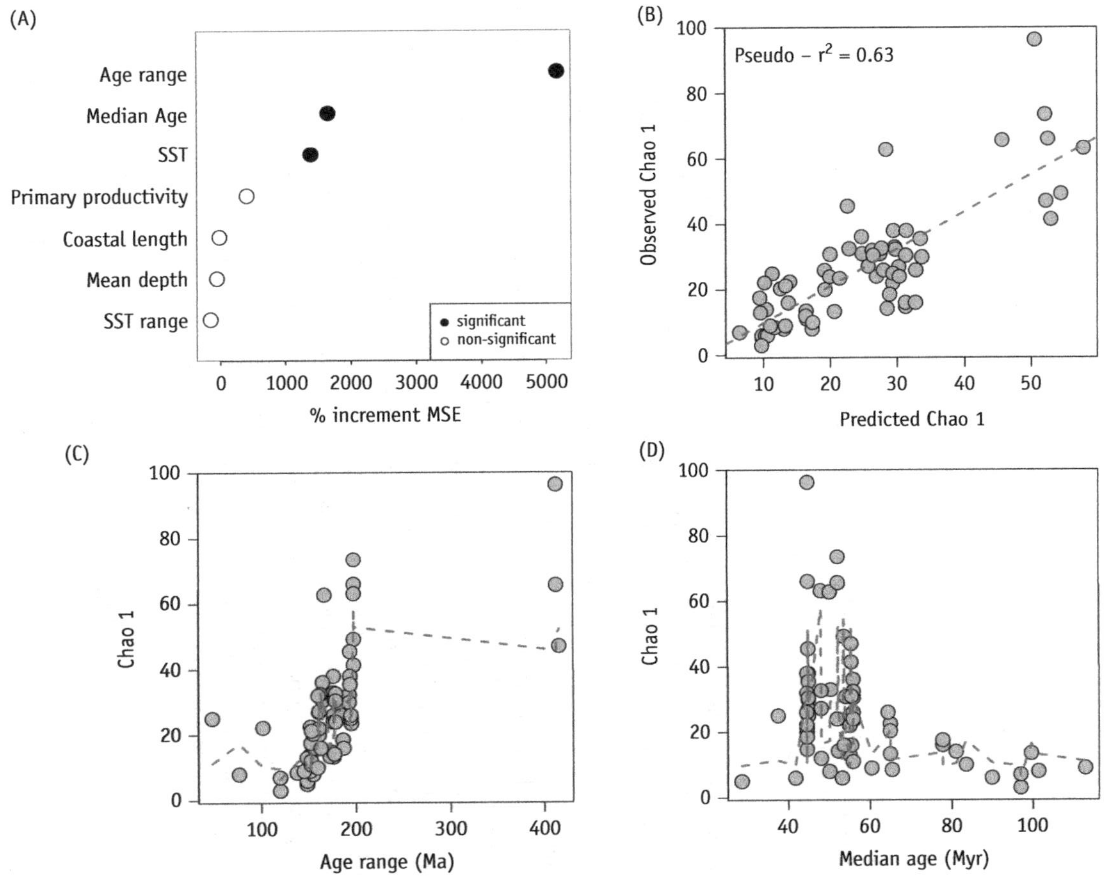

Fig. 15.7.
Ecological and evolutionary factors determining global patterns of extrapolated genus richness (Chao 1) of Decapoda in 10° grid cells worldwide (n = 68, only cells >50 occurrences): (a) variable importance (mean decreased accuracy) of the random forest analysis; (b) observed versus predicted genus richness; (c) observed genus richness versus genus age range; (d) observed genus richness versus median genus age.

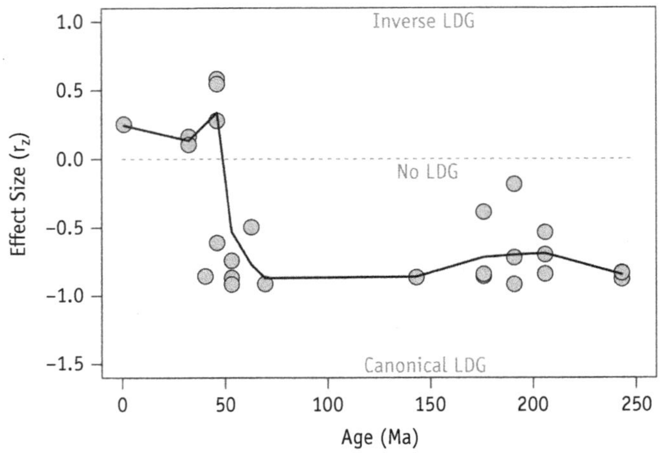

Fig. 15.8.
Relationship between the effect size (r_z) and the family age. The black line corresponds to a locally weighted scatterplot smoothing.

many evolutionary ideas based on robust statistical methods. This also seems crucial because the LDG seems to be impervious to the level of taxonomic aggregation (i.e., family/order/class). For instance, phylogenetic analyses could be used to evaluate the evolution of biogeographic states of species, i.e., tropical/extratropical origin of the group and the importance of biogeographic dispersal (Jansson et al. 2013). Species distribution models based on georeferenced species occurrences and oceanographic information could be combined with phylogenetic information to infer patterns of environmental niche conservatism (e.g., Verbruggen et al. 2009). Recent studies combine plate tectonic dynamics with phylogenetic methods to unveil the role of macroevolutionary processes creating geographic patterns of diversity (Leprieur et al. 2016). Finally, the search for the ultimate causes of the LDG should not exclude the importance of ecological processes maintaining diversity, which demands data gathered at a local scale over large geographic areas. Despite growing evidence demonstrating the importance of evolutionary and ecological factors (Schoch et al. 2006, Tittensor et al. 2010) shaping the LDG in marine organisms, a proper test of the combined effects of these factors, choosing the proper spatial scales of analyses, remains largely incomplete for most taxa (Belmaker and Jetz 2015). This generality is also true for marine crustaceans. Future studies may also assess the impact of oceanic rafting (e.g., hitching rides on seaweeds) on clades such as peracarids with direct development (Thiel and Haye 2006), and hence on LDG.

SUMMARY AND CONCLUSIONS

We have attempted a synthesis of patterns and processes related to the LDG in marine crustaceans, drawing our conclusions from the most common and well-studied groups. Importantly, although many crustacean groups remain largely unstudied (Appeltans et al. 2012) and many areas of the global ocean remain poorly studied (Poore et al. 2014; Fig. 15.5), our results strongly suggest that our general findings should be relatively robust to improvements in global and regional species inventories.

Special attention should be paid to the scale and geographic extent of analyses; the existence of a negative (canonical) LDG should be the working hypothesis only in studies carried out at a regional scale, preferably over large geographic areas. On the contrary, studies using local-scale estimators of species richness (i.e., α-diversity) will likely produce weak or no LDG, especially when surveys cover small geographic areas. However, and despite the value and usefulness of this new meta-analysis approach in achieving synthesis, the simple correlation between species richness and absolute latitude does not adequately describe key features of the LDG in crustaceans. Indeed, our new analyses in several groups confirm that:

1. The LDG is bimodal with peaks inside or outside the tropics.
2. Sharp interhemispheric asymmetries exist. The Northern Hemisphere peak is farther away from the equator than the Southern Hemisphere peak.
3. Patterns may depend upon bathymetry and also seem related to the type of larval development. The Southern Ocean is dominated by deep-sea forms,
4. The LDG is not uniform across different oceans and basins, creating a spatial mosaic of richness patterns. The well-documented IWP hotspot of species richness, not particularly evident in our datasets, reminds us that the study of LDG phenomenon needs to consider additional axes of spatial (and temporal) variation.

The ultimate processes driving the LDG seem intimately related to macroevolutionary processes, via geographic differences in evolutionary rates and species-selection processes. Indirect evidence for this is suggested by several pieces of evidence:

1. The strong differences in the shape of the LDG between clades with and without pelagic larvae (but see Rex et al. 2000).
2. Global patterns of variation of genus-level richness of Decapoda better explained by the median genus age and genus range, suggesting the legacy of evolutionary events occurred during the Paleogene on present day LDG.
3. Tropical niche conservatism, which prevent younger clades (e.g., families) from reaching high latitudes. Hence, the canonical LDG is observed only in families older than c. 50 Ma.

ACKNOWLEDGMENTS

The first author is deeply grateful for the invitation of the editors to collaborate in this chapter. Martin Thiel provided very thoughtful and constructive comments that forced us to sharpen our ideas and approaches. The research of MMR was funded by FONDECYT grants 1140841, 1150664, 1151094, and 1181300.

REFERENCES

Appeltans, W., S.T. Ahyong, G. Anderson, M.V Angel, Tom Artois, N. Bailly, R.N. Bamber, A. Barber, I. Bartsch, A. Berta, M. Błażewicz-Paszkowycz, P. Bock, G.A. Boxshall, C.B. Boyko, S.N. Brandão, R.A. Bray, N.L. Bruce, S.D. Cairns, T.-Y. Chan, L. Cheng, A.G. Collins, T. Cribb, M. Curini-Galletti, F. Dahdouh-Guebas, P.J.F. Davie, M.N. Dawson, O. De Clerck, W. Decock, S. De Grave, N.J. de Voogd, D.P. Domning, C.C. Emig, C. Erséus, W.N. Eschmeyer, K. Fauchald, D.G. Fautin, S.W. Feist, C.H.J.M. Fransen, H. Furuya, O. Garcia-Alvarez, S. Gerken, D. Gibson, A. Gittenberger, S. Gofas, L. Gómez-Daglio, Dennis P. Gordon, M.D. Guiry, F. Hernandez, B.W. Hoeksema, R.R. Hopcroft, D. Jaume, P.M. Kirk, N. Koedam, S. Koenemann, J.B. Kolb, R.M. Kristensen, A. Kroh, G. Lambert, D.B. Lazarus, R. Lemaitre, M. Longshaw, J. Lowry, E. Macpherson, L.P. Madin, C.L. Mah, G. Mapstone, P.A. McLaughlin, J. Mees, K. Meland, C.G. Messing, C.E. Mills, T.N. Molodtsova, R.D. Mooi, B. Neuhaus, P.K.L. Ng, C. Nielsen, J.L. Norenburg, D.M. Opresko, M. Osawa, G. Paulay, W. Perrin, J.F. Pilger, G.C.B.

Poore, P.J.A. Pugh, G.B. Read, J.D. Reimer, M. Rius, R.M. Rocha, J.I. Saiz-Salinas, V. Scarabino, B. Schierwater, A. Schmidt-Rhaesa, K.E. Schnabel, M. Schotte, P. Schuchert, E. Schwabe, H. Segers, C. Self-Sullivan, N. Shenkar, V. Siegel, W. Sterrer, S. Stöhr, B.J. Swalla, M.L. Tasker, E.V. Thuesen, T. Timm, M.A. Todaro, X. Turon, S. Tyler, P. Uetz, J. van der Land, B. Vanhoorne, L.P. van Ofwegen, R.W.M. van Soest, J. Vanaverbeke, G.K. Walker-Smith, T.C. Walter, A. Warren, G.C. Williams, S.P. Wilson, and M.J. Costello. 2012. The magnitude of global marine species diversity. Current Biology 22:2189–2202.

Archer, E. 2015. rfPermute: Estimate Permutation p-Values for Random Forest Importance Metrics. R package version 1.6.2. http://CRAN.R-project.org/package=rfPermute.

Astorga, A., M. Fernandez, E.E. Boschi, and N. Lagos. 2003. Two oceans, two taxa and one mode of development: latitudinal diversity patterns of South American crabs and test for possible causal processes. Ecology Letters 6:420–427.

Azovsky, A.I., L.A. Garlitska, and E.S. Chertoprud. 2012. Broad-scale patterns in local diversity of marine benthic harpacticoid copepods (Crustacea). Marine Ecology Progress Series 460:63–77.

Beaugrand, G., I. Rombouts, and R.R. Kirby. 2013. Towards an understanding of the pattern of biodiversity in the oceans. Global Ecology and Biogeography 22:440–449.

Bellwood, D.R., and C.P. Meyer. 2009. Searching for heat in a marine biodiversity hotspot. Journal of Biogeography 36:569–576.

Belmaker, J., and W. Jetz. 2015. Relative roles of ecological and energetic constraints, diversification rates and region history on global species richness gradients. Ecology Letters 18:563–571.

Berke, S. K., D. Jablonski, A. Z. Krug, and J. W. Valentine. 2014. Origination and immigration drive latitudinal gradients in marine functional diversity. PLoS ONE 9:e101494.

Boschi, E.E. 2000. Biodiversity of marine decapod brachyurans of the Americas. Journal of Crustacean Biology 20 (Special Number 2):337–342.

Brandt, A.1999. On the origin and evolution of Antarctic Peracarida (Crustacea, Malacostraca). Scientia Marina 63: 261–274.

Brandt, A., W. Brökeland, S. Brix, and M. Malyutina. 2004. Diversity of Southern Ocean deep-sea Isopoda (Crustacea, Malacostraca)—a comparison with shelf data. Deep Sea Research Part II: Topical Studies in Oceanography 51:1753–1768.

Brandt, A., K.E. Ellingsen, S. Brix, W. Brökeland, and M. Malyutina. 2005. Southern Ocean deep-sea isopod species richness (Crustacea, Malacostraca): influences of depth, latitude and longitude. Polar Biology 28:284–289.

Brandt, A., S. Brix, W. Brökeland, M. Choudhury, S. Kaiser, and M. Malyutina. 2007a. Deep-sea isopod biodiversity, abundance, and endemism in the Atlantic sector of the Southern Ocean--Results from the ANDEEP I-III expeditions. Deep Sea Research Part II: Topical Studies in Oceanography 54:1760–1775.

Brandt, A., C. De Broyer, I. De Mesel, K.E. Ellingsen, A.J. Gooday, B. Hilbig, K. Linse, M.R.A. Thomson, and P.A. Tyler. 2007b. The biodiversity of the deep Southern Ocean benthos. Philosophical Transactions of the Royal Society B: Biological Sciences 362:39–66.

Brandt, A., A. Gooday, S.N. Brandao, S. Brix, W. Brokeland, T. Cedhagen, M. Choudhury, N. Cornelius, B. Danis, I. De Mesel, R.J. Diaz, D.C. Gillan, B. Ebbe, J.A. Howe, D. Janussen, S. Kaiser, K. Linse, M. Malyutina, J. Pawlowski, M.J. Raupach, and A. Vanreusel. 2007c. First insights into the biodiversity and biogeography of the Southern Ocean deep sea. Nature 447:307–311.

Brown, J.H. 2014. Why are there so many species in the tropics? Journal of Biogeography 41:8–22.

Burridge, A.K., M. Tump, R. Vonk, E. Goetze, and K.T. Peijnenburg. 2017. Diversity and distribution of hyperiid amphipods along a latitudinal transect in the Atlantic Ocean. Progress in Oceanography 158:224–235

Chao, A., and S.M. Lee. 1992. Estimating the number of classes via sample coverage. Journal of the American Statistical Association 87:210–217.

Chaudhary, C., H. Saeedi, and M.J. Costello. 2016. Bimodality of latitudinal gradients in marine species richness. Trends in Ecology and Evolution 31:670–676.

Clarke, A. 1992. Is there a latitudinal diversity cline in the sea? Trends in Ecology and Evolution 7:286–287.

Clarke, A., and J.A. Crame. 1997. Diversity, latitude and time: patterns in the shallow sea. Pages 122–147 *in* R.F.G. Ormond, J.D. Gage, and M.V. Angel, editors. Marine Biodiversity: Patterns and Processes. Cambridge Univerity Press, Cambridge, UK.

Clarke, A., R.B. Aronson, J.A. Crame, J.-M. Gili, and D.B. Blake. 2004. Evolution and diversity of the benthic fauna of the Southern Ocean continental shelf. Antarctic Science 16:559–568.

Coleman, N., A.S.H. Gason, and G.C.B. Poore. 1997. High species richness in the shallow marine waters of south-east Australia. Marine Ecology Progress Series 154:17–26.

Cornell, H.V., and J.H. Lawton. 1992. Species interactions, local and regional processes, and limits to the richness of ecological communities: a theoretical perspective. Journal of Animal Ecology 61:1–12.

Crame, J.A. 2000. Evolution of taxonomic diversity gradients in the marine realm: evidence from the composition of recent bivalve faunas. Paleobiology 26:188–214.

Crame, J.A., and A. Clarke. 1997. The historical component of marine taxonomic diversity gradients. Pages 258–273 *in* R.F.G. Ormond, J.D. Gage, and M.V. Angel, editors. Marine Biodiversity: Patterns and Processes. Cambridge Univerity Press, Cambridge, UK.

Cutler, D.R., T.C. Edwards, K.H. Beard, A. Cutler, and K.T. Hess. 2007. Random forests for classification in ecology. Ecology 88:2783–2792.

Davis, K. E., J. Hill, T. I. Astrop, and M. A. Wills. 2016. Global cooling as a driver of diversification in a major marine clade. Nature Communications 7:13003.

Del Re, A., and W. Hoyt. 2012. MAc: Meta-analysis with correlations. R package version 1.1. http://CRAN. R-project. org/packageMAc.

Doti, B.L., D. Roccatagliata, and J.L. Gappa. 2014. An inverse latitudinal biodiversity pattern in asellote isopods (Crustacea, Peracarida) from the Southwest Atlantic between 35° and 56°S. Marine Biodiversity 44:115–125.

Dworschak, P.C. 2000. Global diversity in the Thalassinidea (Decapoda). Journal of Crustacean Biology 20:238–245.

Dworschak, P.C. 2005. Global diversity in the Thalassinidea (Decapoda): an update (1998–2004). Nauplius 13:57–63.

Ellingsen, K., A. Brandt, B. Ebbe, and K. Linse. 2007. Diversity and species distribution of polychaetes, isopods and bivalves in the Atlantic sector of the deep Southern Ocean. Polar Biology 30:1265–1273.

Espinosa-Pérez, M.d.C., and M.E. Hendrickx. 2006. A comparative analysis of biodiversity and distribution of shallow-water marine isopods (Crustacea: Isopoda) from polar and temperate waters in the East Pacific. Belgian Journal of Zoology 136:219.

Evans, K.L., P.H. Warren, and K.J. Gaston. 2005. Species-energy relationships at the macroecological scale: a review of the mechanisms. Biological Reviews 80:1–25.

Fenberg, P.B., B.A. Menge, P.T. Raimondi, and M.M. Rivadeneira. 2015. Biogeographic structure of the northeastern Pacific rocky intertidal: the role of upwelling and dispersal to drive patterns. Ecography 38:83–95.

Fernández, M., A. Astorga, S.A. Navarrete, C. Valdovinos, and P.A. Marquet. 2009. Deconstructing latitudinal species richness patterns in the ocean: Does larval development hold the clue? Ecology Letters 12:601–611.

Finnegan, S., S.C. Anderson, P.G. Harnik, C. Simpson, D.P. Tittensor, J.E. Byrnes, Z.V. Finkel, D.R. Lindberg, L.H. Liow, and R. Lockwood. 2015. Paleontological baselines for evaluating extinction risk in the modern oceans. Science 348:567–570.

Fischer, A.G. 1960. Latitudinal variations in organic diversity. Evolution 14:64–81.

Gage, J., P. Lambshead, J.D. Bishop, C. Stuart, and N.S. Jones. 2004. Large-scale biodiversity pattern of Cumacea (Peracarida: Crustacea) in the deep Atlantic. Marine Ecology Progress Series 277:181–196.

Gaither, M.R., and L.A. Rocha. 2013. Origins of species richness in the Indo-Malay-Philippine biodiversity hotspot: evidence for the centre of overlap hypothesis. Journal of Biogeography 40:1638–1648.

Gray, J.S. 2001. Marine diversity: the paradigms in patterns of species richness examined. Scientia Marina 65:41–56.

Hessler, R.R., and H.L. Sanders. 1967. Faunal diversity in the deep-sea. Deep-Sea Research 14:65–78.

Hillebrand, H. 2004a. On the generality of the latitudinal diversity gradient. American Naturalist 163:192–211.

Hillebrand, H. 2004b. Strength, slope and variability of marine latitudinal gradients. Marine Ecology Progress Series 273:251–267.

Hoeksema, B.W. 2007. Delineation of the Indo-Malayan centre of maximum marine biodiversity: the coral triangle. Pages 117–178 *in* W. Renema, editor. Biogeography, Time, and Place: Distributions, Barriers, and Islands. Springer, Dordrecht, The Netherlands.

Jablonski, D., and R. Lutz. 1983. Larval ecology of marine benthic invertebrates: Paleobiological implications. Biological Reviews 58:21–89.

Jablonski, D., K. Roy, and J.W. Valentine. 2006. Out of the tropics: evolutionary dynamics of the latitudinal diversity gradient. Science 314:102–106.

Jansson, R., G. Rodríguez-Castañeda, and L.E. Harding. 2013. What can multiple phylogenies say about the latitudinal diversity gradient? A new look at the tropical conservatism, out of the tropics, and diversification rate hypotheses. Evolution 67:1741–1755.

Kinlock, N. L., L. Prowant, E. M. Herstoff, C. M. Foley, M. Akin-Fajiye, N. Bender, M. Umarani, H. Y. Ryu, B. Şen, and J. Gurevitch. 2018. Explaining global variation in the latitudinal diversity gradient: Meta-analysis confirms known patterns and uncovers new ones. Global Ecology and Biogeography 27:125–141.

Knox, G.A., and J.K. Lowry. 1977. A comparison between the benthos of the Southern Ocean and the North Polar Ocean with special reference to the Amphipoda and the Polychaeta. Pages 423–462 *in* M.J. Dunbar, editor. Polar Oceans. Arctic Institute of North America, Calgary, Canada.

Krug, A. Z., D. Jablonski, and J. W. Valentine. 2008. Species-genus ratios reflect a global history of diversification and range expansion in marine bivalves. Proceedings of the Royal Society B: Biological Sciences 275:1117–1123.

Leprieur, F., P. Descombes, T. Gaboriau, P. F. Cowman, V. Parravicini, M. Kulbicki, C. J. Melián, C. N. de Santana, C. Heine, D. Mouillot, D. R. Bellwood, and L. Pellissier. 2016. Plate tectonics drive tropical reef biodiversity dynamics. Nature Communications 7:11461.

Levinton, J., and J. Mackie. 2013. Latitudinal diversity relationships of fiddler crabs: biogeographic differences united by temperature. Global Ecology and Biogeography 22:1050–1059.

Liaw, A., and M. Wiener. 2002. Classification and regression by randomForest. Rnews 2:18–22.

Macpherson, E. 2002. Large–scale species–richness gradients in the Atlantic Ocean. Proceedings of the Royal Society of London B: Biological Sciences 269:1715–1720.

Macpherson, E., B. Richer de Forges, K.E. Schnabel, S. Samadi, M.-C. Boisselier, and A. Garcia-Rubies. 2010. Biogeography of the deep-sea galatheid squat lobsters of the Pacific Ocean. Deep-Sea Research I 57:228–238.

Malyutina, M., and A. Brandt. 2007. Diversity and zoogeography of Antarctic deep-sea Munnopsidae (Crustacea, Isopoda, Asellota). Deep Sea Research II 54:1790–1805.

McArthur, R.H.1972. Geographical Ecology: Patterns in the Distribution of Species. Harper & Row, New York.

McCallum, A.W., G.C.B. Poore, A. Williams, F. Althaus, and T. O'Hara. 2013. Environmental predictors of decapod species richness and turnover along an extensive Australian continental margin (13–35°S). Marine Ecology 34:298–312.

McCallum, A.W., S. Woolley, M. Błażewicz-Paszkowycz, J. Browne, S. Gerken, R. Kloser, G.C.B. Poore, D. Staples, A. Syme, J. Taylor, G. Walker-Smith, A. Williams, and R.S. Wilson. 2014. Productivity enhances benthic species richness along an oligotrophic Indian Ocean continental margin. Global Ecology and Biogeography 24:462–471.

Mittelbach, G.G., D.W. Schemske, H.V. Cornell, A.P. Allen, J.M. Brown, M.B. Bush, S.P. Harrison, A.H. Hurlbert, N. Knowlton, and H.A. Lessios. 2007. Evolution and the latitudinal diversity gradient: speciation, extinction and biogeography. Ecology Letters 10:315–331.

Orzechowski, E.A., R. Lockwood, J.E. Byrnes, S.C. Anderson, S. Finnegan, Z.V. Finkel, P.G. Harnik, D.R. Lindberg, L.H. Liow, and H.K. Lotze. 2015. Marine extinction risk shaped by trait–environment interactions over 500 million years. Global Change Biology 21:3595–3607.

Papacostas, K.J., and A.L. Freestone. 2016. Latitudinal gradient in niche breadth of brachyuran crabs. Global Ecology and Biogeography 25:207–217.

Pappalardo, P., and M. Fernández. 2014. Mode of larval development as a key factor to explain contrasting effects of temperature on species richness across oceans. Global Ecology and Biogeography 23:12–23.

Pianka, E.R. 1966. Latitudinal gradients in species diversity: a review of concepts. American Naturalist 100:33–46.

Pohle, G., K. Iken, K.R. Clarke, T. Trott, B. Konar, J.J. Cruz-Motta, M. Wong, L. Benedetti-Cecchi, A. Mead, and P. Miloslavich. 2011. Aspects of benthic decapod diversity and distribution from rocky nearshore habitat at geographically widely dispersed sites. PloS ONE 6:e18606.

Poore, G.C.B., and N.L. Bruce. 2012. Global diversity of marine isopods (except Asellota and crustacean symbionts). PLoS ONE 7:e43529.

Poore, G.C.B., L. Avery, M. Błażewicz-Paszkowycz, J. Browne, N.L. Bruce, S. Gerken, C. Glasby, E. Greaves, A.W. McCallum, and D. Staples. 2015. Invertebrate diversity of the unexplored marine western margin of Australia: taxonomy and implications for global biodiversity. Marine Biodiversity 45:271–286.

Poore, G.C.B., J. Just, and B.F. Cohen. 1994. Composition and diversity of Crustacea Isopoda of the southeastern Australian continental slope. Deep-Sea Research Part I: Oceanographic Research Papers 41:677–693.

Poore, G.C.B., A.W. McCallum, and J. Taylor. 2008. Decapod Crustacea of the continental margin of southwestern and central Western Australia: preliminary identifications of 524 species from FRV *Southern Surveyor* voyage SS10-2005. Museum Victoria Science Reports 11:1–106.

Poore, G.C.B., L. Avery, M. Błażewicz-Paszkowycz, J. Browne, N.L. Bruce, S. Gerken, C. Glasby, E. Greaves, A.W. McCallum, D. Staples, A. Syme, J. Taylor, G. Walker-Smith, M. Warne, C. Watson, A. Williams, R.S. Wilson, and S. Woolley. 2014. Invertebrate diversity of the unexplored marine western margin of Australia: taxonomy and implications for global biodiversity. Marine Biodiversity 45:271–286.

Powell, M.G., V.P. Beresford, and B.A. Colaianne. 2012. The latitudinal position of peak marine diversity in living and fossil biotas. Journal of Biogeography 39:1687–1694.

R Core Team. 2016. R: A language and environment for statistical computing. In R Foundation for Statistical Computing, Vienna, Austria. http://www.R-project.org/.

Rex, M.A., and R.J. Etter. 2010. Deep-Sea Biodiversity. Pattern and Scale. Harvard University Press, Cambridge, MA.

Rex, M.A., C.T. Stuart, and G. Coyne. 2000. Latitudinal gradients of species richness in the deep-sea benthos of the North Atlantic. Proceedings of the National Academy of Sciences of the United States of America 97:4082–4085.

Rex, M.A., C.T. Stuart, R.R. Hessler, J.A. Allen, H.L. Sanders, and G.D.F. Wilson. 1993. Global-scale latitudinal patterns of species diversity in the deep-sea benthos. Nature 365:636–639.

Ricklefs, R.E. 1987. Community diversity: relative roles of local and regional processes. Science 235:167–171.

Rivadeneira, M.M., and P.A. Marquet. 2007. Selective extinction of late Neogene bivalves on the temperate Pacific coast of South America. Paleobiology 33:455–468.

Rivadeneira, M.M., M. Fernández, and S.A. Navarrete. 2002. Latitudinal trends of species diversity in rocky intertidal herbivore assemblages: spatial scale and the relationship between local and regional species richness. Marine Ecology Progress Series 245:123–131.

Rivadeneira, M.M., M. Thiel, E.R. Gonzalez, and P.A. Haye. 2011. An inverse latitudinal gradient of diversity of peracarid crustaceans along the Pacific Coast of South America: out of the deep south. Global Ecology and Biogeography 20:437–448.

Rivadeneira, M.M., A.H. Alballay, J.A. Villafaña, P.T. Raimondi, C.A. Blanchette, and P.B. Fenberg. 2015. Geographic patterns of diversification and the latitudinal gradient of richness of rocky intertidal gastropods: the "into the tropical museum" hypothesis. Global Ecology and Biogeography 24:1149–1158.

Rohde, K. 1992. Latitudinal gradients in species diversity: the search for the primary cause. Oikos 65:514–527.

Rohde, K. 1999. Latitudinal gradients in species diversity and Rapoport's rule revisited: a review of recent work and what can parasites teach us about the causes of the gradients? Ecography 22:593–613.

Rombouts, I., G. Beaugrand, F. Ibañez, S. Gasparini, S. Chiba, and L. Legendre. 2009. Global latitudinal variations in marine copepod diversity and environmental factors. Proceedings of the Royal Society of London B: Biological Sciences 276:3053–3062.

Roy, K., D. Jablonski, and J.W. Valentine. 2000. Dissecting latitudinal diversity gradients: functional groups and clades of marine bivalves. Proceedings of the Royal Society of London Series B: Biological Sciences 267:293–299.

Roy, K., D. Jablonski, J.W. Valentine, and G. Rosenberg. 1998. Marine latitudinal diversity gradients: Tests of causal hypotheses. Proceedings of the National Academy of Sciences of the United States of America 95:3699–3702.

Rutherford, S., S. D'Hondt, and W. Prell. 1999. Environmental controls on the geographic distribution of zooplankton diversity. Nature 400:749–753.

Sachs, J. 2001. Tropical underdevelopment. National Bureau of Economic Research Working Paper 8119.

Scher, H.D., J.M Whittaker., S.E. Williams, J.C. Latimer, W.E. Kordesch, and M.L. Delaney, 2015. Onset of Antarctic Circumpolar Current 30 million years ago as Tasmanian Gateway aligned with westerlies. Nature 523:580–583.

Schoch, G.C., B.A. Menge, G. Allison, M. Kavanaugh, S.A. Thompson, and S.A. Wood. 2006. Fifteen degrees of separation: Latitudinal gradients of rocky intertidal biota along the California Current. Limnology and Oceanography 51:2564–2585.

Schweitzer, C.E., and R.M. Feldmann. 2005. Decapod crustaceans, the K/P event, and Palaeocene recovery. Pages 17–53 *in* S. Koenemann and R.A. Jenner, editors. Crustacea and Arthropod Relationships. CRC Press, Boca Raton, FL.

Schweitzer, C.E., R.M. Feldmann, A. Garassino, H. Karasawa, and G. Schweigert. 2010. Systematic list of fossil decapod crustacean species. Crustaceana Monographs 10:1–222.

Sepkoski, J.J. 2000. Crustacean biodiversity through the marine fossil record. Contributions to Zoology 69:213–222.

Steele, D. 1988. Latitudinal variations in body size and species diversity in marine decapod crustaceans of the continental shelf. Internationale Revue der gesamten Hydrobiologie und Hydrographie 73:235–246.

Thiel, M., and P.A. Haye. 2006. The ecology of rafting in the marine environment. III. Biogeographical and evolutionary consequences. Oceanography and Marine Biology: An Annual Review 44:323–429.

Thorson, G. 1957. Bottom communities (sublittoral or shallow shelf). In: J.W. Hedgpeth (ed.) Treatise on Marine Ecology and Palaeoecology. *Memoir of the Geological Society of America* 67V1:461–534.

Tittensor, D.P., C. Mora, W. Jetz, H.K. Lotze, D. Ricard, E.V. Berghe, and B. Worm. 2010. Global patterns and predictors of marine biodiversity across taxa. Nature 466:1098–1101.

Valdovinos, C., S.A. Navarrete, and P.A. Marquet. 2003. Mollusk species diversity in the Southeastern Pacific: Why are there more species towards the pole? Ecography 26:139–144.

Valentine, J.W., and D. Jablonski. 2015. A twofold role for global energy gradients in marine biodiversity trends. Journal of Biogeography 42:997–1005.

Valentine, J.W., D. Jablonski, A.Z. Krug, and K. Roy. 2008. Incumbency, diversity, and latitudinal gradients. Paleobiology 34:169–178.

Vazquez, D. P., and R. D. Stevens. 2004. The latitudinal gradient in niche breadth: concepts and evidence. American Naturalist 164:E1–E19.

Verbruggen, H., L. Tyberghein, K. Pauly, C. Vlaeminck, K. V. Nieuwenhuyze, W. H. Kooistra, F. Leliaert, and O. D. Clerck. 2009. Macroecology meets macroevolution: evolutionary niche dynamics in the seaweed *Halimeda*. Global Ecology and Biogeography 18:393–405.

Viechtbauer, W. 2010. Conducting meta-analyses in R with the metafor package. Journal of Statistical Software 36:1–48.

Villafaña, J.A., and M.M. Rivadeneira. 2014. The rise and fall of the diversity of marine vertebrates during the Neogene on the temperate Pacific coast of South America. Paleobiology 40:659–674.

Watling, L., and M.H. Thurston. 1989. Antarctica as an evolutionary incubator: evidence from the cladistic biogeography of the amphipod family Iphimediidae. Geological Society, London, Special Publications 47:297–313.

Wiens, J.J., and M.J. Donoghue. 2004. Historical biogeography, ecology and species richness. Trends in Ecology and Evolution 19:639–644.

Wiens, J.J., and C.H. Graham. 2005. Niche conservatism: integrating evolution, ecology, and conservation biology. Annual Review of Ecology Evolution and Systematics 36:519–539.

Willig, M.R., D.M. Kaufman, and R.D. Stevens. 2003. Latitudinal gradients of biodiversity: Pattern, process, scale, and synthesis. Annual Review of Ecology Evolution and Systematics 34:273–309.

Wilson, G.D.F. 1998. Historical influences on deep-sea isopod diversity in the Atlantic Ocean. Deep-Sea Research 45:279–301.

Wilson, G.D.F. 2008. Local and regional species diversity of benthic Isopoda (Crustacea) in the deep Gulf of Mexico. Deep Sea Research Part II: Topical Studies in Oceanography 55:2634–2649.

Witman, J.D., R.J. Etter, and F. Smith. 2004. The relationship between regional and local species diversity in marine benthic communities: a global perspective. Proceedings of the National Academy of Sciences of the United States of America 101:15664–15669.

Wood-Walker, R.S., P. Ward, and A. Clarke. 2002. Large-scale patterns in diversity and community structure of surface water copepods from the Atlantic Ocean. Marine Ecology Progress Series 236:189–203.

Worm, B., M. Sandow, A. Oschlies, H.K. Lotze, and R.A. Myers. 2005. Global patterns of predator diversity in the open oceans. Science 309:1365–1369.

WoRMS Editorial Board. 2018. World Register of Marine Species. http://www.marinespecies.org.

Yasuhara, M., G. Hunt, T.M. Cronin, and H. Okahashi. 2009. Temporal latitudinal-gradient dynamics and tropical instability of deep-sea species diversity. Proceedings of the National Academy of Sciences 106:21717–21720.

Yasuhara, M., H. Iwatani, G. Hunt, H. Okahashi, T. Kase, H. Hayashi, T. Irizuki, Y.M. Aguilar, A.G.S. Fernando, and W. Renema. 2017. Cenozoic dynamics of shallow-marine biodiversity in the Western Pacific. Journal of Biogeography 44: 567–578.

16

INTERTIDAL TO ABYSS: CRUSTACEANS AND DEPTH

Anna McCallum and Torben Riehl

Abstract

Crustaceans occur from the shelf to hadal depths, but the immense environmental change that occurs along this depth gradient results in significant faunal change. One well-established pattern is the dramatic decline in biomass with depth, a result of an exponential decline in food availability. Average body size becomes smaller, despite observations of deep-sea gigantism in some crustaceans. Crustacean species tend to occupy a limited depth range, resulting in high faunal turnover. The depths of the greatest faunal turnover vary widely throughout the oceans, and there do not appear to be distinct bathymetric "zones" at ocean-wide scales. Molecular research at the species level confirms that small bathymetric changes are often more significant at promoting population differentiation than geographic distance. Observation of crustaceans in the laboratory demonstrates that the interaction between pressure and temperature is likely to act together in limiting the bathymetric range of many species. Debate continues around species richness and diversity gradients, and it remains unclear whether there are more crustacean species on the shelf compared to bathyal depths. Diversity patterns vary between taxa. Decapods are species rich on the shelf and upper slope and less so in the abyss. Isopods show high bathyal diversity, although this pattern varies between regions. For other crustaceans, it is difficult to make generalizations on diversity gradients as there are fewer studies, and results vary depending on geographic region and the method used to estimate diversity and richness. In cumaceans, amphipods. and harpacticoids, species richness is often highest on the shelf, while maximum species diversity occurs in deeper water. Food availability and temperature are good correlates for depth-diversity gradients.

INTRODUCTION

To appreciate the full diversity of marine crustaceans, it is necessary to look beyond the sun-lit coastal shelf and into the deeper waters of the oceans. It is on the continental slopes and seamounts

Evolution and Biogeography. Edited by Gary C. B. Poore and Martin Thiel.

and down into the abyss that the diversity of marine habitats and associated crustacean communities can really be grasped. From the shelf to the abyss, changes in depth are accompanied by immense environmental change. Water pressure increases steadily, light disappears, and temperature declines dramatically to less than 3°C in the abyss (Levin et al. 2001, Menot et al. 2010). Crustaceans have managed to adapt to all of these environmental extremes and can be found in the deepest and darkest corners of the oceans (see Chapter 7). It is no surprise then that the depth gradient from the shelf to the abyss is associated with patterns of major faunal change. The size of crustaceans, their abundance, community composition, species richness (number of species), and diversity (the relative abundance of species) are all strongly correlated with depth.

The deep sea is remote and costly to survey. After more than a century of deep-sea collecting, few datasets can be drawn on to make broad generalizations on faunal patterns from the shelf to the abyss (Rex and Etter 2010, O'Hara et al. 2016) and even fewer that report on crustaceans alone (Poore et al. 2013). Much of our understanding of bathymetric patterns is based on data from the northern Atlantic (Rex and Etter 2010), and fewer surveys have examined depth gradients in the Indian or the Western Pacific Ocean. Historically, the shelf and the upper slope have been much better sampled than the abyss (O'Hara et al. 2016). This unevenness in sampling efforts has implications for measuring gradients of biodiversity when using data compiled across studies (O'Hara et al. 2016). For instance, compiling known records of crustaceans into depth zones produces data that are skewed toward shallow waters where collection effort is high. Hence, this chapter focuses on studies that have attempted to evenly survey the crustacean fauna from shallow to deep waters along a transect or within a single region.

This chapter describes the major macro-ecological patterns that occur along bathymetric gradients for benthic and demersal crustaceans, with an emphasis on the bathyal habitats where environmental change is greatest. The chapter outlines the three major patterns associated with depth: (1) a change in abundance and body size with depth; (2) gradients of species richness and diversity; and (3) species distributions and faunal turnover. We then discuss the environmental variables most likely to play a role in the patterns seen for Crustacea. Crustacean evolutionary patterns associated with the deep sea are discussed in Chapter 7.

SEAFLOOR BATHYMETRY

Beyond the coastline, the adjacent seafloor is known as the "continental shelf" (Fig. 16.1). The shelf extends out from the continent to a variable extent and remains relatively shallow. At about 200 m depth, there is a steep increase in depth, which marks the "shelf break" and the beginning of the "continental slope." The continental slope possesses some of the steepest gradients of depth and the strongest environmental gradients. Sediment from the continent above cascades down the slope and accumulates as a pile of sediment at the base of the slope as the "continental rise." Continental shelf, shelf break, slope, and continental rise together are referred to as "continental margin." At approximately 3,500 m, the seafloor levels out into the expansive abyssal plain (Watling et al. 2013).

Bathymetric complexity occurs on seamounts and on mid-ocean ridges. Seamounts are typically formed from extinct volcanoes and rise several kilometers above the ocean floor; many have even steeper gradients of depth than continental margins. The relatively flat abyss is interrupted by trenches formed by subduction, where the oceanic crust buckles and deepens. Many trenches occur in areas of the Pacific Ocean which are tectonically active, including the Mariana Trench, which has the deepest seafloor recorded in the Challenger Deep at 10,890 m below sea level (Jamieson et al. 2010). The terms used above are geomorphic, and biologists (Gage and Tyler 1992, Rogers 2015) typically define the ecological "zones" associated with these features as (1) coastal, sublittoral

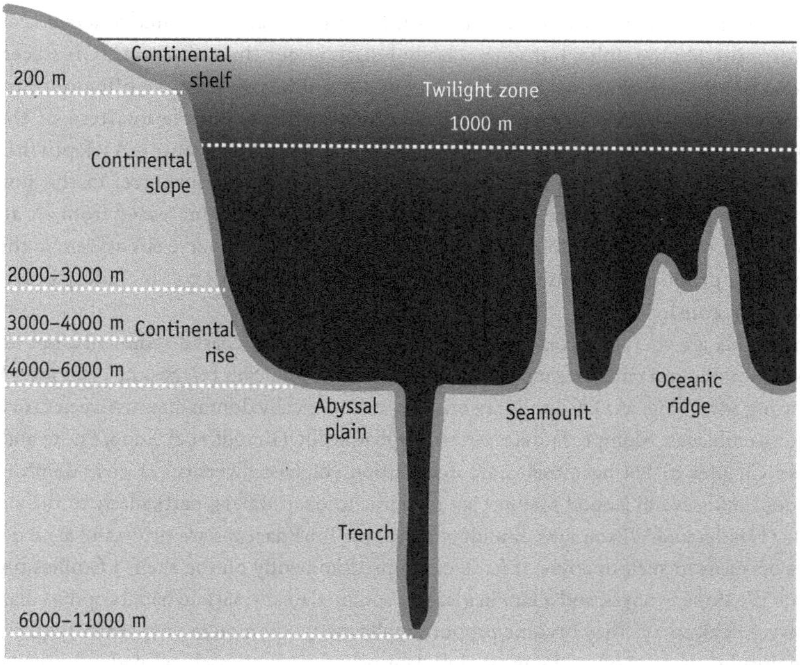

Fig. 16.1
Schematic illustration of seafloor bathymetry and geomorphic features.

or subtidal, 0–200 m, (2) bathyal, 200–2,000 m, (3) abyssal, 3,500–6,500 m, and (4) hadal, >6,500 m (Watling et al. 2013).

RELATIVE ABUNDANCE OF CRUSTACEANS ACROSS DEPTH

Although crustaceans occur from shelf to abyss, the dominant taxonomic groups change with depth. Large, bottom-walking decapods, such as brachyuran crabs and lobsters, are largely limited to shallow, near-shore environments in the temperate zone and to the tropics, where they play key ecological roles as top predators (Boudreau and Worm 2012). Stomatopod crustaceans are also largely limited to shallow tropical waters. Of the 483 known species, only 22 occur below 500 m depth (as listed on the World Register of Deep-Sea Species). As depth increases, these groups become rare. In contrast, larger swimming shrimps and prawns become the dominant megafauna with depth (Rosa et al. 2012). Anomurans have diversified in the deep sea. Squat lobsters are the most frequently encountered anomurans at bathyal and abyssal depths, with more than 800 species worldwide (Schnabel et al. 2011). A number of other anomurans occur down to abyssal depths, including king crabs (Lithodidae) and parapagurid hermit crabs (Wilson and Ahyong 2015). Decapods were long considered to be absent from the hadal zone, but a few species have been reported, the first being a spider crab *Teratomaia* collected from the Kermadec Trench by the Galathea expedition in the 1950s (Wilson and Ahyong 2015), and more recently, prawns of the dendrobranchiate genera *Benthesicymus* and *Plesiopenaeus* (as *Cerataspis)* have been observed down to 7,703 m (Jamieson et al. 2009). These hadal species are more commonly observed on the abyssal plains and bathyal slopes, thus they are not endemic to the hadal zone but at the deeper limit of their habitat.

Many biodiversity studies of the seafloor focus on animals at the macrofaunal scale (too small for most photographs, but with a body length larger than 0.25 mm). Crustaceans are diverse in benthic macrofaunal samples at all depths, although they are often outnumbered by the polychaete worms, which frequently constitute up to three-quarters of the total macrofauna at all depths (Gage and Tyler 1992; but see Brandt et al. 2018). As depth increases, crustaceans tend to become proportionally more common, for instance, in the northeast Atlantic, the relative abundance of crustaceans in the macrofauna increased from 2% at 208 m to 33% at 3,673 m (Flach and Heip 1996). At hadal depths of the Puerto Rico Trench, they may even become proportionally more abundant compared to other taxa, including polychaetes (Brandt et al. 2018).

Amphipods are the most common peracarid crustaceans in coastal sediments, but as depth increases, isopods and tanaidaceans become increasingly significant (Gage and Tyler 1992, Wilson and Ahyong 2015) (Fig. 16.2). Isopoda are one of the numerically dominant crustacean taxa in deep benthic assemblages. Multiple families occur eurybathically (Brandt et al. 2009, Poore and Bruce 2012; see Chapter 7) but have their core distribution (highest diversity) at great depth (Brandt et al. 2009) and several isopod families are endemic to deep waters, particularly of the suborder Asellota (Hessler and Wilson 1983, Brandt et al. 2009). Tanaidaceans are successful at all depths, 8 families occur from shelf to abyss, 11 families are predominantly on the shelf, 3 families predominantly on the slope to abyss, and 4 families largely confined to abyssal and hadal depths (Błażewicz-Paszkowycz et al. 2012). They become proportionally more common in deep water (especially the tanaidomorphs) and are often the most abundant peracarid at abyssal and hadal depths (Fischer and Brandt 2015, De Smet el at. 2017) (Fig. 16.2C).

Meiofaunal animals (generally smaller than 0.25 mm) are abundant at all depths but become proportionally more common with increasing depth (Flach et al. 1999). A transect across the continental slope of the Goban Spur found meiofaunal densities approximately 50 times higher than macrofauna densities on the shelf, whereas in the abyss meiofauna densities were more than 1,000 times higher (Flach et al. 1999). This change in ratio was due to a significant decrease in macrofauna densities with increasing water depth, whereas the meiofauna densities stayed more or less at the same level. Harpacticoid copepods are usually the second most abundant metazoan

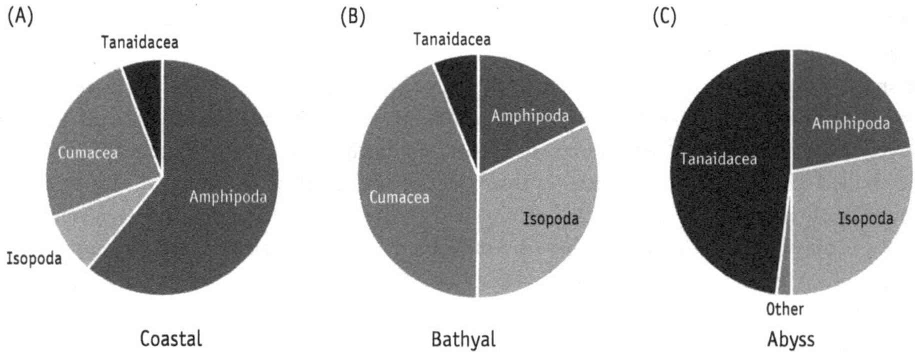

Fig. 16.2
Relative abundance of peracarid crustacean taxa in macrofaunal samples collected from: (A) Port Phillip Bay, Australia, 2–23 m, using grab and 0.5 mm sieve; (B) upper slope off East Greenland, 186–1802, using epibenthic sledge and 0.5 mesh; (C) abyss of the Kuril-Kamchatka area, NW Pacific, 4869–5413 m, using box corer and 0.3 mm sieve. A, data from Poore et al. (1975); B, from Brandt and Schnack (1999); C, from Fisher and Brandt (2015).

animals in meiofaunal samples after nematodes, which generally constitute more than 75% of individuals in a sample (Gage and Tyler 1992, Zeppilli et al. 2013). The other major meiofaunal crustacean taxon is Ostracoda, although it is relatively uncommon at all depths (Gage and Tyler 1992, Rohal et al. 2014).

PATTERNS OF BIOMASS, ABUNDANCE, AND BODY SIZE

The biomass of animals is a relatively tangible quantity to measure and predict, and the exponential decrease of standing stock (biomass and abundance per unit area) from the continental shelf to the abyss is perhaps the most well-established bathymetric trend (Rex et al. 2006, Rex and Etter 2010, Wei et al. 2010b, 2012). Biomass is directly related to food availability (Wei et al. 2010b). In the ocean, most life (except isolated chemosynthetic systems) depends on the production of energy in surface waters from phytoplankton. This carbon sinks down through the water column in the form of particulate organic carbon (POC) flux. POC flux declines with depth as organic matter is remineralized and consumed in the water column. Most phytoplankton biomass produced in surface waters is remineralized in the upper few hundred meters of the ocean, and only about 1–3% of surface production reaches the deep seafloor (De La Rocha et al. 2007).

The decline of food availability with depth results in two patterns: first, the density (number of individuals per unit area) declines, and second, the body size of organisms tends to decrease (Rex et al. 2006). The overall pattern is an exponential decline in biomass with depth and with distance from the coastline (Wei et al. 2010b). For instance, in the well-studied region of the Goban Spur in the northeast Atlantic, quantitative box-core samples along a transect from the shelf to abyssal depths showed an exponential decline in macrofaunal density from ~8,000 individuals m^{-2} at 185 m depth to ~550 individuals m^{-2} at 4,460 m depth (Flach and Heip 1996). There is now a sizable literature of studies on abundance and biomass at local and regional scales (Wei et al. 2010b). Global scale meta-analysis of this research has confirmed that a decline in both biomass and abundance of animals with depth is a general phenomenon at global scales (Rex et al. 2006, Wei et al. 2010b).

Average body size is smaller in the deep sea than at shallower depths (Wei et al. 2010b), despite the observation that for many taxa the largest representatives occur at great depth (McClain et al. 2015). Declines in body size can be seen in the groups of crustaceans that dominate at different depths. Decapods that dominate the shelf are replaced by macrofaunal peracarids at bathyal depths, and meiofaunal harpacticoids dominate at the very deepest hadal depths (Hessler and Wilson 1983, Wei et al. 2012, Wilson and Ahyong 2015). The rate of decline in biomass with depth varies between megafaunal to meiofaunal size groups (Wei et al. 2010b). In the northwest Pacific in the vicinity of the Kuril Trench, sampling between 490 and 7,090 m found harpacticoid abundance does not decrease with depth but was highest at 1,000 m (Kitahashi et al. 2013). However, even the meiofauna get smaller with depth: the meiofaunal animals of the trenches are approximately one-third smaller than their bathyal relatives (Danovaro et al. 2002). It is widely accepted that declines in body size with depth are caused by the exponential decline in nutrient availability (Rex et al. 2006). Small body size is advantageous in the food-limited deep sea as smaller organisms require less energy and can thus maintain viable populations. In contrast, where there are no constraints on food in shallow water, being large provides the benefit of greater metabolic efficiency, competitive advantage, resource exploitation, and predator avoidance (Wei et al. 2012).

A compelling contradiction to the average decline in body size with depth is the phenomenon of deep-sea gigantism. A number of well-known giant crustaceans live in the deep sea, including the lysianassoid amphipod *Eurythenes gryllus* and cirolanid isopods of the genus *Bathynomus*. The causes of gigantism remain unclear. Timofeev (2001) proposed that deep-sea gigantism in crustaceans is a consequence of larger cell sizes obtained under cold temperatures. Deep-sea gigantism in crustaceans may also reflect decreases in temperature leading to longer life spans and thus larger sizes for indeterminate growers (Timofeev 2001). Many giant crustaceans are scavengers, including the largest known isopod *Bathynomus giganteus*, which reaches 50 cm in length (McClain et al. 2015). In aquaria, *Bathynomus giganteus* can survive for eight weeks between feedings (McClain et al. 2015). Large quantities of lipid reserves and fat bodies may serve as an adaptation to opportunistic feeding in a habitat with ephemeral food resources (McClain et al. 2015). Thus, the large size of crustaceans like *Bathynomus* (and other peracarid crustaceans) may be an adaptation for scavenging in an environment with sporadic food. Larger size allows for greater fat reserves to be maintained and also confers a greater foraging area (McClain et al. 2015). Large-sized scavengers may have advantages in locating and reaching widely dispersed food items (Gage and Tyler 1992).

Population Structure within Species

Within-species depth-size trends have been widely examined for fishes with apparent body size increases with depth (Macpherson and Duarte 1991, Mindel et al. 2016). Potentially, deep-living species start life in shallower waters due to more favorable conditions (i.e., food supply, temperature, and pressure), then descend as they grow. Ontogenetic structuring with depth has been observed in many hadal amphipods, where early instars and juveniles are known to reside exclusively at the shallower end of their bathymetric distribution (Blankenship et al. 2006, Eustace et al., 2013, 2016). However, intra-specific clines in body size do not appear to be a general pattern for crustaceans. A study of decapods from Le Danios Bank (northeast Atlantic) found no general pattern of mean size and depth. Some species were larger below 600 m (*Polycheles typhlops, Munida tenuimana*), but many other species showed no change (Cartes et al. 2007). In the Gulf of Mexico, geryonid crabs of the genus *Chaceon* are smaller at the deeper end of their range. This species segregates by size and sex across depth on the upper continental slope. Females are most abundant in the shallow depths, while males are most abundant on the deeper slope. It has been suggested that the shallow distribution of females favors larval success because of the warmer and food-rich surface waters, while competition between males pushes smaller males into deeper waters (Lindberg and Lockhart 1993). Many deep-water crustaceans show changes in sex ratio with depth, but again, there is no general pattern; in some species, females become more abundant with depth (Follesa et al. 2007, Eustace et al. 2016, Papiol and Hendrickx 2016); in others, it is the males (Lindberg and Lockhart 1993).

SPECIES RICHNESS AND DIVERSITY GRADIENTS

Describing and explaining bathymetric gradients of biodiversity is a major goal of much deep-sea research, and the question of whether deep-sea sediments have more species than shallow waters has generated considerable debate (Gray 1997, 2000, Flach and de Bruin 1999; reviewed by Carney 2007 and McClain and Schlacher 2015). Much of this debate centers on the method used to estimate the number of species in a community, which can be surprisingly difficult to measure (Carney 2007, Maurer and McGill 2011). Epibenthic sleds are effective at sampling the minute and sparsely distributed animals living in the deep sea, but the area of seafloor they sample can be highly variable. Thus, rarefaction is often used to standardize species richness. The Sanders-Hulbert expected number of species $E(Sn)$ measures the expected number of species in an n-individual sample and remains the most widely used diversity index in deep-sea ecology. Some ecologists have criticized the widespread use of $E(Sn)$ in deep-sea analyses, especially when applied to quantitative samples because it is highly dependent on faunal density and evenness (May 1993, Gray 2000).

Macrofauna

In the North Atlantic, there is good evidence that species richness (the number of species per area) and species diversity ($E(Sn)$) of macrofauna are higher on the slope than the shelf but then decrease in the abyss (Rex and Etter 2010). A unimodal pattern of diversity is supported by extensive quantitative box-core sampling in the western North Atlantic (Grassle and Maciolek 1992), as reviewed by Rex and Etter (2010).

Macrofaunal crustaceans also show higher diversity and richness on the slopes of the North Atlantic compared to the shelf. Flach and de Bruin (1999) analyzed macrofaunal crustaceans at seven stations between ~200 and ~4,500 m in the Goban Spur region of the northeast Atlantic. Rarefaction curves show that crustacean species diversity $E(S40)$ (Fig. 16.3A) and species richness were highest on the upper slope at 1,425 m, although these estimates were based on only three or four replicates of variably sized box corers, and so should be taken with some caution. The low diversity of crustaceans and other macrofaunal animals on the shelf of the North Atlantic has been attributed to high food input and an unpredictable environment resulting in dominance of a few opportunistic species that may lower species richness (Flach and de Bruin 1999, Levin et al. 2001). In contrast, in the Sea of Japan in the northern Pacific, diversity and species richness of peracarid crustaceans decrease between 455 m and 3,660 m (Golovan et al. 2013) (Fig. 16.3B). Low abyssal diversity is commonly explained by insufficient food to support viable populations of many species (Rex and Etter 2010), although in this region, low diversity was attributed to a buildup of organic matter at the base of the slope (Golovan et al. 2013). The unusual history of this young and isolated deep-sea basin may also explain the low abyssal diversity of peracarids, which in Pleistocene glacial periods became stagnant and anoxic (Golovan et al. 2013).

A unimodal depth-diversity pattern does not appear to be universal (Menot et al. 2010). Many studies of species richness show a decline in total macrofauna from shelf to slope, including studies from the Norwegian Sea (reviewed by Rex and Etter 2010), the northeast Mediterranean

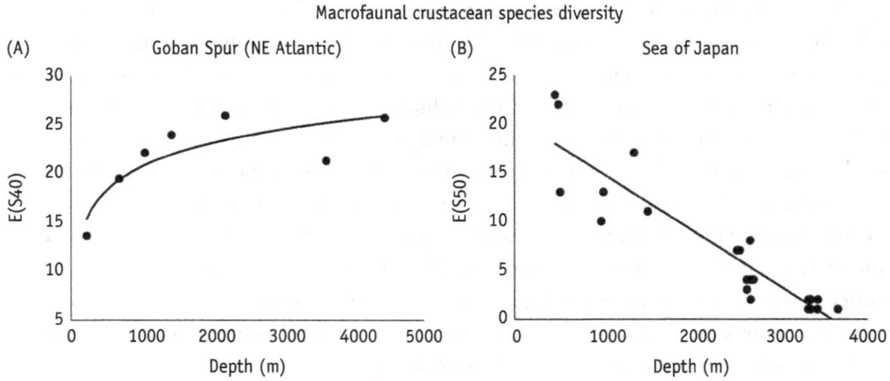

Fig. 16.3
Crustacean species diversity estimated as the expected number of species per *n* individuals along a depth gradient at the Gobun Spur (northeast Atlantic) and the Sea of Japan. From Flach and de Bruin (1999), with permission from Elsevier, with data from Golovan et al. (2013).

(Tselepides et al. 2000), the eastern Pacific (Oliver et al. 2011), and the eastern Indian Ocean (McCallum et al. 2015). Other analyses show no apparent relationship between depth and diversity (Palma et al. 2005).

Bathymetric patterns of diversity and richness vary between the different taxonomic groups of crustaceans. For instance, in the Gulf of Mexico (175–3,720 m), isopods show a unimodal pattern of species richness, but cumaceans, amphipods, and harpacticoid copepods declined in species richness (Baguley et al. 2006, Haedrich et al. 2008,). Haedrich et al. (2008) found that a variety of factors relating to food, habitat, pollution, and location were correlated with species richness, and these correlates varied considerably from taxon to taxon. These results support the view that no common environmental factor is responsible for driving patterns of species richness but rather multiple factors, which vary from group to group. Nonetheless, Haedrich et al. (2008) found the highest correlations were related to food availability. Amphipods and cumacean richness were correlated with meiofaunal biomass, and isopod richness was strongly correlated to POC, both corresponding to food availability.

Isopods have been a focus of many studies of deep-sea biodiversity and often feature in discussions of depth and diversity. Although most studies of isopod biodiversity suggest high bathyal species richness, the exact nature of the pattern tends to vary depending on the region examined: in the northwest and the southwest Atlantic, isopod diversity and species richness simply increased with depth, with maximum species richness at depths of more than 3,000 m (Wilson 1998) (Fig. 16.4). In the Arctic (Svavarsson 1997), the Gulf of Mexico (Wilson 2008), and the Southern Ocean (Ellingsen et al. 2007), isopod species richness was unimodal, with the depth of maximum species richness varying between 500 and 4,000 m. However, for southeast Australia, Poore et al. (1994) found no correlation between depth and species richness (Fig. 16.4F). That study included data from a number of transects which increased the variability, but along one of the transects, species richness appeared to decline with depth.

Amphipods comprise a large proportion of the macrofaunal abundance at all depths (Hessler and Wilson 1983) (Fig. 16.2). Depth-diversity patterns for this group vary significantly between regions. In the North Atlantic, south of Iceland, amphipod diversity (measured as $E(S_{100})$) increased with depth from the continental shelf to 1,200 m (Weisshappel and Svavarsson 1998), but in the Norwegian Sea (Weisshappel and Svavarsson 1998) and in Arctic waters off northeast Greenland to the north (Brandt 1995), there was no clear pattern over the same depth range (Weisshappel and Svavarsson 1998). Weisshappel and Svavarsson (1998) suggest historical disturbance in the Arctic caused by glacial periods may account for this difference between regions, whereby amphipods of the deep North Atlantic may have had a much longer time span for diversifying than their counterparts in the Arctic. A high diversity of amphipods at slope depths has also been observed on the New Zealand continental margins, where family-level amphipod diversity increased from the shelf to 1,200 m (Knox et al. 2012). In contrast, in the Gulf of Mexico, species richness of amphipods gradually decreased over the depth range 175–3,720 m (Haedrich et al. 2008).

Cumaceans follow a unimodal pattern of diversity in the western North Atlantic with peak diversity ($E(S_{50})$) between 2,000 and 4,000 m (Rex and Etter 2010). In the Gulf of Mexico, species richness of cumaceans gradually decreased over a depth range of 175–3,720 m (Haedrich et al. 2008). The contrasting results observed in these two studies could reflect regional differences or could result from different indices used to measure biodiversity.

For many groups of crustaceans, there are few studies to draw on to make generalizations about depth and biodiversity. Although tanaidaceans are thought to be highly diverse in the deep sea (Błażewicz-Paszkowycz et al. 2012), few analyses have examined bathymetric gradients of tanaidacean biodiversity alone. However, one analysis of 11 sled samples from the Ross Sea in the Southern Ocean showed species richness increased with depth between ~300 and ~3,500 m (Pabis et al. 2015).

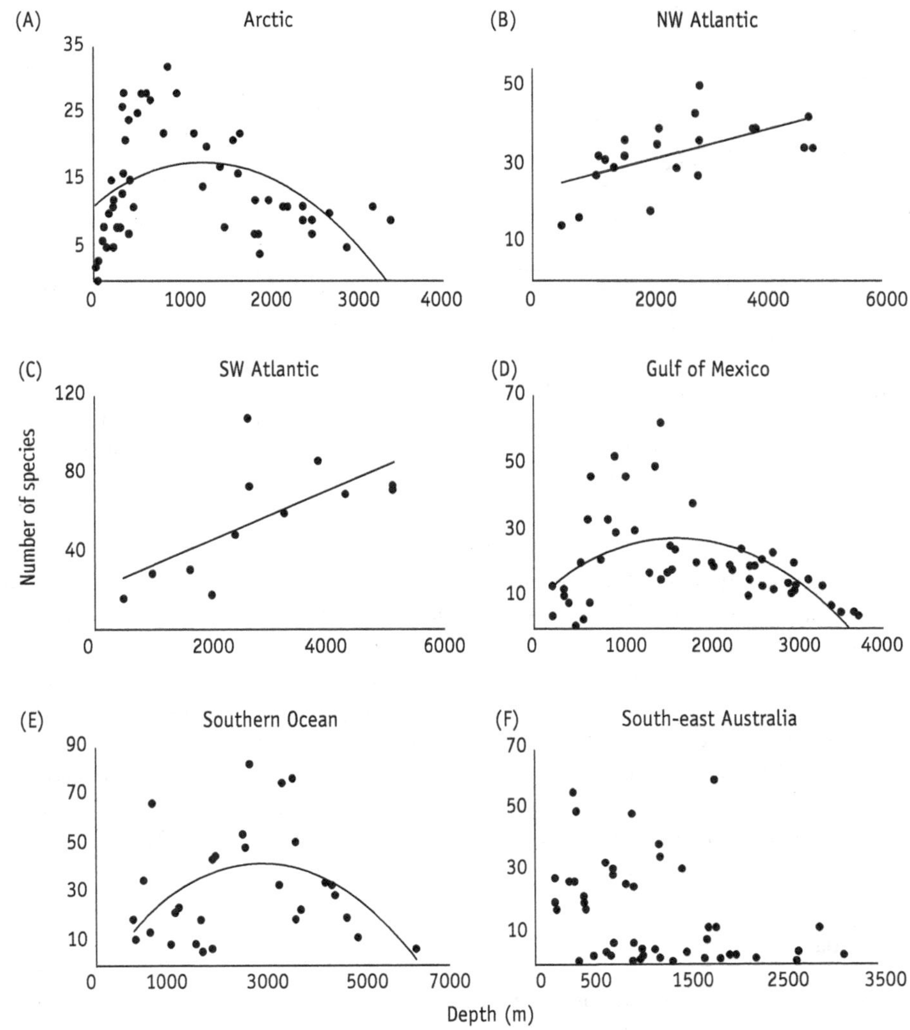

Fig. 16.4
Isopod species richness as a function of depth. Species richness is the number of species in a sample. Data are from epibenthic sleds (unless otherwise stated) collected from: (A) the Arctic using a box corer; (B, C) the northwestern Atlantic and the southwest Atlantic; (D) the Gulf of Mexico using a box corer; (E) the Southern Ocean; and (F) southeast Australia. A, from Svavarsson (1997), with permission from Springer Nature; B, C, data from Wilson (1998); D, data from Wilson (2008); E, from Ellingsen et al. (2007), with permission from Springer Nature; F, data from Poore et al. (1994).

Meiofauna

Diversity-depth patterns for the meiofauna are not well studied (Rex and Etter 2010). In the northern Gulf of Mexico, harpacticoid species richness (species per sample) declined linearly with increasing depth, while species diversity ($E(S_{30})$) was unimodal with maximum diversity at approximately 1,200 m (Baguley et al. 2006) (Fig. 16.5). A study of the Kuril Trench sampling between 490 and 7,090 m found the most harpacticoid genera in samples between 1,000 and 2,000 m (Kitahashi et al. 2013), while measures of harpacticoid diversity based on Shannon diversity and the expected

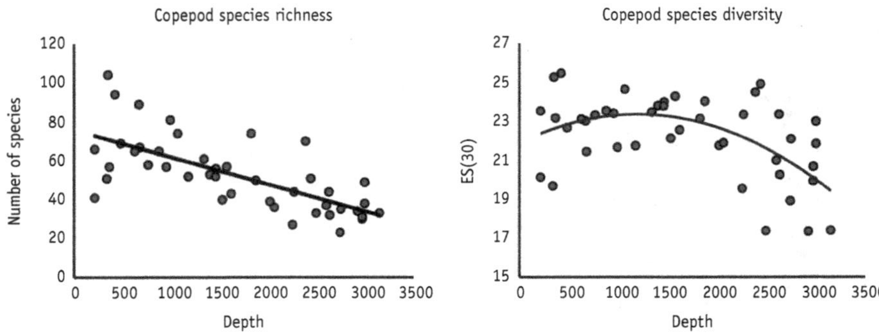

Fig. 16.5.
Harpacticoid copepods from the northern Gulf of Mexico. (A) Species richness is the number of species in pooled replicate cores adjusted to the number per 10 cm². (B) species diversity is the expected number of harpacticoid species per 30 individuals ($E(S_{30})$), for pooled replicate core samples. Modified from Baguley et al. (2006), with permission from Elsevier.

number of genera (rarefaction) indicated maximum diversity in much deeper waters (3,000–6,000 m). These contrasting results illustrate that using different biodiversity indices results in different patterns with depth.

Species diversity measures like $E(Sn)$ are an expression or index of some relation between the number of species and the number of individuals. They incorporate the evenness of a community as well as the species richness. High evenness occurs when species are equal or virtually equal in abundance, which tends to be the case in deep-sea communities. In many groups of crustaceans, including cumaceans and amphipods, species richness is often highest on the shelf (Haedrich et al. 2008), while maximum species diversity ($E(Sn)$) occurs in deeper waters (Weisshappel and Svavarsson 1998, Rex and Etter 2010).

Megafauna

Studies of megafaunal samples seldom use $E(Sn)$ and tend to report on species richness as a total of all species in a sample. On the well-studied slope off New England (northwest Atlantic), the greatest number of decapod species in demersal trawls was found at depths between 238 and 650 m. This was in contrast to fishes and echinoderms, for which there were more species at intermediate depths (1,380–1,947 m; Haedrich et al. 1980).

Analyses of decapod species richness often use the overlapping ranges of species within a sector of depth. This estimate is different from a count of species in a sample, as it produces an estimate of the regional species pool. This approach is possible with decapods but rarely with other crustaceans (but see Wilson 2008) because decapod taxonomy is better documented and sampling is more intense, often as a result of fisheries research. The decapod fauna of the Gulf of Mexico, including the abyssal plains, is well known (Wicksten and Packard 2005). Compiling known decapod records into polygons shows that richness at both species and higher levels declines with depth: 26 families and 37 genera on the upper slope, 17–18 families and 23 genera on the deep slope, and only 10 families and 13 genera on the abyssal plain (Wicksten and Packard 2005). Rosa et al. (2012) compiled data on decapod crustaceans from the intertidal to abyssal seafloor off the Madeira Archipelago in the subtropical northeast Atlantic Ocean. They found an exponential decline in decapod species richness with depth. The peak in richness was observed on the shelf in less than 50 m depth (Fig. 16.4). Some

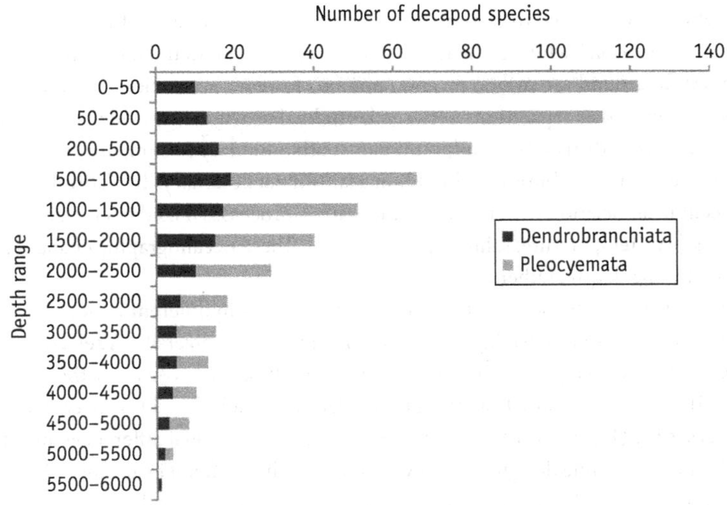

Fig. 16.6.
Decapod species richness from the intertidal to the abyss off the Madiera Archipelago in the subtropical northeast Atlantic Ocean. The suborders Dendrobranchiata and Pleocyemata show different patterns reflecting their different lifestyles. From Rosa et al. (2012), with permission from Elsevier.

groups of decapods are more species-rich on the slope. Species from the suborder Dendrobranchiata show a unimodal pattern of richness with peaks in the upper bathyal (Rosa et al. 2012) (Fig. 16.6).

Anomuran squat lobsters are another group that has diversified at bathyal depths. A compilation of squat lobster records from the taxonomic literature showed that maximum species richness occurs on the upper slope (<1,400 m) with the peak of richness between 200 and 600 m (Schnabel et al. 2011).

SPECIES DISTRIBUTIONS AND FAUNAL TURNOVER

Do Depth "Zones" Exist?

Most species are distributed across a limited range of depth. Early naturalists on pioneering voyages such as the *Challenger* expedition observed differences between two extreme biotas, that on the shelves and that in the abyss, with a zone of transition between (Beddard 1886). Thus began attempts to define depth boundaries separating distinct bathymetric zones and to determine if these zones were regional phenomena or if they show horizontal continuity at ocean-wide scales. The distribution of isopods from the order Asellota was used in early attempts to draw boundaries between faunal zones from intertidal to abyssal depths (Menzies et al. 1973). The exact depths of boundaries varied widely throughout the oceans. In the polar regions, the typically abyssal Asellota genera have distributions that extended up on the shelf. Conversely, in tropical waters, many shelf genera extend down beyond the shelf break. Based on their observations, Menzies et al. (1973) strongly objected to drawing boundaries between zones, preferring the identification of depths of maximum faunal change.

Macpherson (2003) compiled information on the latitudinal and depth distributions of more than 6,000 fish and invertebrate species, including decapods and stomatopods, from the Atlantic between 80° N and 70° S. He looked for depths and latitudes with a high number of species range endpoints, which may indicate the boundaries of biogeographic zones. In the arctic regions, there was hardly any "zonation," while in the lower-latitude regions, he found a high number of species range endpoints at around 100 m, 300 m, and 1,000 m. However, separations between depth zones were not as pronounced as separations between latitudinal provinces, as evidenced by the low proportion of species with distribution endpoints at a depth boundary as compared to the higher proportion of species with distribution endpoints at a latitudinal boundary. The latitudinal boundaries tended to occur in association with major oceanographic processes linked to changes in water mass characteristics (e.g., temperature, salinity, productivity). These oceanographic changes may not be as pronounced along depth gradients.

Efforts to survey and characterize the seafloor assemblages in different regions have increased, often with the aim of better managing and conserving marine biodiversity. A selection of zonation studies is presented here (Fig. 16.7, Table 16.1) with an emphasis on those using the distributions of crustaceans. There was substantial variation in the depths of high faunal turnover, even within the same regional setting (Fig. 16.7). This is likely to be the result of regional differences and also an artefact of inconsistent sampling design and analysis across studies (Menot et al. 2010). A meta-analysis of faunal turnover studies by Menot et al. (2010) found that except for the shelf–slope transition, there was little similarity in the depth of high turnover, while a literature review by Carney (2005) identified a consistent faunal change at 1,000 m on the upper slope. The underlying causes of the shelf–slope transition are unclear but could be due to the stratification of ocean currents which moderate larval transport (Carney 2005). High faunal turnover at 1,000 m correlates with dramatic changes in temperature and food availability (Carney 2005).

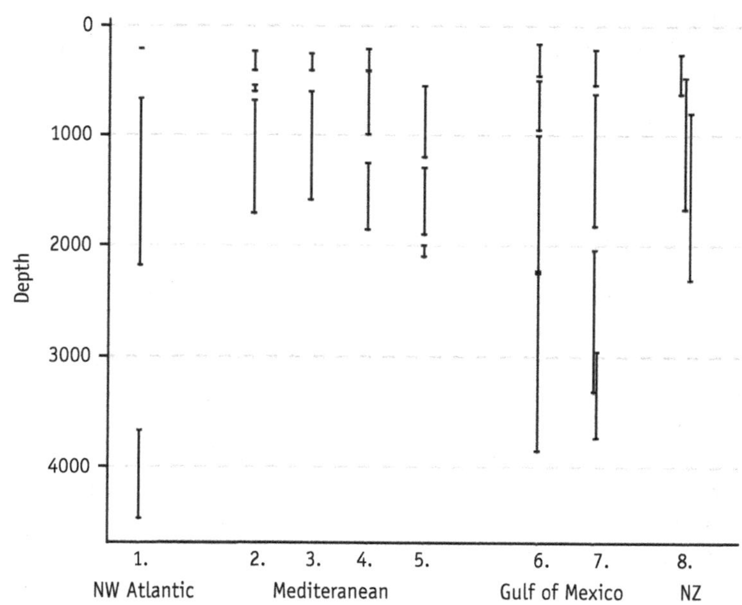

Fig. 16.7.
The results of eight studies of community similarity/turnover from five regions that are relatively well studied. Vertical lines indicate extent of samples with similar community structure; blank areas indicate gaps in sampling. See Table 16.1 for details of the studies.

Table 16.1. Description of Studies Presented in Figure 16.4, Which Analyze Faunal Similarity/Turnover[a]

	Region	Location	Depth range (m)	Taxa	Sampling device	Method to group samples	Reference
1	Northern Atlantic	Goban Spur-Porcupine Seabight	208–4,470	Macrofaunal crustaceans	Box core	Cluster	Flach and de Bruin 1999
2	Mediterranean	SW Balearic islands	230–1,714	Decapoda	Beam trawl	MDS	Maynou and Cartes 2000
3	Mediterranean	SW Balearic Islands	249–1,622	Peracarids	Suprabenthic sled	Cluster and MDS	Cartes et al. 2003
4	Mediterranean	Catalan Sea	208–1,859	Peracarids	Suprabenthic sled	Cluster and MDS	Cartes et al. 2003
5	Mediterranean	Catalan Sea	552–2,261	Decapoda	Beam trawl	Cluster	Cartes and Sardà 1993
6	Gulf of Mexico	Northern GOM	159–3,850	Megafauna	Beam trawl	Cluster	Pequegnat 1983
7	Gulf of Mexico	Northern GOM	200–3,700	Macrofauna	Box core	Cluster	Wei et al. 2010a
8	Chatham rise	New Zealand	237–2,039	Megafauna	Epibenthic sled	Cluster	McKnight and Probert, 1997

[a]Studies from well-studied regions that included crustacean data were selected.
MDS = Multidimensional scaling.

Species Distributions

Aside from the question of whether boundaries between distinct depth zones exist, it is clear that the faunal turnover occurring along depth gradients is very high. Species typically inhabit broad latitudinal ranges and narrow depth ranges. For example, the squat lobster *Agononida rubrizonata* has a distribution typical of many squat lobsters in the Indo-West Pacific. It has a wide-ranging geographic distribution from Taiwan to southern Australia and east to Vanuatu but only occurs between 250 and 750 m depth (Poore and Andreakis 2012).

Similarly, when the distributions of multiple species are examined, assemblages are often defined by depth and occur across a wide geographic distribution. On the continental margin of Australia, the decapod community on the temperate outer shelf is more similar to the tropical shelf fauna 22° further north than to the fauna on the adjacent seafloor 300 m deeper (McCallum et al. 2013). Molecular research has also shown that small bathymetric changes are more significant at promoting population differentiation than geographic distance. The giant amphipod *Eurythenes gryllus* sensu lato was once considered to occur in every ocean from bathyal to hadal depths (but see d'Udekem d'Acoz and Havermans 2015). An analysis of genetic diversity at a global scale identified at least nine putative species-level clades with a clear distinction between bathyal and abyssal clades occurring at around 3,000 m (Havermans et al. 2013). In contrast to high genetic divergence seen across depth, some clades showed genetic homogeneity across enormous distances, with one clade comprising specimens from both Arctic and Antarctic regions.

Although most crustacean species have narrow depth ranges, some species occur across both shallow and deep waters. In the Gulf of Mexico, some isopod species extend from the shelf break down to almost 3,000 m (Wilson 2008). Some tanaidaceans have similar depth ranges (Larsen 2005). Many are likely to be species complexes, as has been discovered for geographically wide-ranging deep-sea species (reviewed in Brandt et al. 2012). The limited sampling and the rarity of deep-sea fauna means that there are fewer specimens available to discriminate species morphologically.

Molecular data can help determine species boundaries (Brandão et al. 2010, Brix et al. 2014). Early work on Antarctic ostracods of the genus *Macroscapha* reported five circum-antarctic and/or eurybathic species. However, taxonomic revision and recent genetic results suggest that there were more than 20 species, each with restricted geographical and depth distributions (Brandão et al. 2010). Conversely, genetic analyses of 19 different Antarctic lysianassoid amphipod species found that most species were genetically homogeneous among specimens from remote sampling sites (Havermans et al. 2011). Two of these species, *Abyssorchomene plebs* and *Pseudorchomene coatsi*, showed very low intraspecific genetic distance between specimens from the continental shelf to abyssal depths. *A. plebs* is an abundant and highly motile bentho-pelagic species, possibly forming suprabenthic swarms able to move rapidly, which could explain the high gene flow between bathymetrically remote populations (Havermans et al. 2011). There are also examples of crustaceans which brood their larvae and have low motility but show high genetic connectivity across broad depths. For instance, the janiroidean isopod species *Macrostylis roaldi* lacks adaptations for swimming and shows almost no genetic variation between specimens collected at an Antarctic shelf site (~500 m) and 1,000 m deeper (1,500 m) (Riehl and Kaiser 2012). In another study on genetic differentiation of this family, both depth as well as geographic distance correlated with population differentiation. Although species have been found that occurred across 3,000 m depth both at abyssal as well as hadal sites, genetic differentiation was high enough to suggest limited genetic exchange between the populations inhabiting the different depth zones (Riehl et al. 2018).

Crustaceans living at greater depths tend to have larger depth ranges than those living in the shallows (Macpherson 2003). In one analysis, decapods living within the first 50 m had a median depth range of approximately 150 m, while abyssal species had ranges of thousands of meters, a pattern more pronounced in benthic than pelagic species (Rosa et al. 2012). In the hadal trenches, amphipods have depth ranges of more than 3,000 m (Blankenship et al. 2006). The relative change in environmental variables such as pressure and temperature experienced by shallow-water fauna exceeds that experienced by deep-sea species. For example, an organism descending from the surface to 10 m undergoes a doubling in pressure (1 to 2 bar), while a descent from 6,000 to 11,000 m will experience less than a doubling in pressure (600 to 1,100 bar; Jamieson et al. 2010).

EXPLAINING BATHYMETRIC PATTERNS

Depth is a spatial variable, which in itself has no physical effect on animals but is a strong proxy for abiotic factors such as temperature, pressure, and light penetration that are likely to influence the distribution and diversity of crustaceans. As these physical variables are often highly correlated, it can be difficult to isolate their independent influences.

Productivity and Food Supply

Animals on the seafloor are almost entirely dependent on sinking food particles from the euphotic zone above, and the decline in food availability from shallow to deep waters is a major driver of bathymetric faunal patterns. It is directly linked to declining biomass and has also been used to explain depth-diversity gradients (Levin et al. 2001). A unimodal productivity-diversity relationship appears to be most apparent at global scales encompassing both extremes of the productivity scale (Leduc et al. 2012).

A lack of food may be a barrier to distribution into deep water (reviewed in Carney 2005). Most abyssal crustaceans have been assumed to be deposit feeders (Wilson and Ahyong 2015). Suspension-feeding crustaceans, such as the filter-feeding isopod families Arcturidae and Antarcturidae (Wilson 1999) and barnacles (Cirripedia), are rare at abyssal depth, as are specialized carnivores (Wilson and Ahyong 2015). It may be an adaptation to these low-food conditions that some isopods feed on protists (e.g., Brökeland et al. 2010, Riehl et al. 2016). Conversely, species adapted to low food availability in deep waters may be excluded from shallow waters by competition and predation. For example, a shift from predation at shelf depths to deposit-feeding on the slope has been reported for Mediterranean decapod crustaceans (Cartes 1998). Evidence of food availability driving community structure can be seen in a number of regional settings. The Mediterranean is considered to be one of the most oligotrophic regions, and energy availability is a likely driver of community patterns (Maynou and Cartes 2000, Tecchio et al. 2011). The west is more productive than the oligotrophic east. Decapod species which are characteristic of deep waters appear to occur higher up the slope in the east than in the west, likely due to the lower food availability (Maynou and Cartes 2000). Similarly, in the Gulf of Mexico, an analysis of macrofaunal community structure found that POC flux was the best predictor of faunal similarity both with depth and along an east-to-west gradient (Wei et al. 2010a).

Temperature

With increasing depth, water temperature decreases until a minimum is reached below a thermocline at around 2°C. The deep-sea floor below this is fed with cool and well-oxygenated waters from polar regions. Temperature is one of the best correlates of species diversity in the ocean (Woolley et al. 2016). Ostracods are the only commonly fossilized metazoan group in deep-sea sediments, and the fossil record shows that species diversity was low during glacial and high during interglacial periods (Yasuhara et al. 2009). A number of mechanisms have been hypothesized to explain the relationship between species diversity and temperature: higher temperatures lead to an increase in metabolic rates, which may promote higher rates of speciation; richness varies according to the tolerance of individual species for different sets of climatic conditions (reviewed by Currie et al. 2004, Yasuhara and Danovaro 2016).

Temperature is fundamental to physiological processes, and thermoclines have long been considered barriers to dispersal in the ocean (Carney 2005). A good example of the response to temperature is found in isopods from the southeast Australian slope (Poore et al. 1994). In four transects between 200 and 2,000 m, clustering of sites resulted in a shelf break assemblage (200–520 m) and a deeper group of slope species (>400 m). The depth of overlap between the two groups dramatically shallowed at the southernmost transect off the Tasmanian coast, a depth consistent with a cold thermocline. Along the continental margin of Western Australia, temperature was the best predictor of decapod community structure (McCallum et al. 2013), as it was for benthic invertebrate community structure across two different continental margins off New Zealand (Compton et al. 2013).

In Lithodidae, lineage-specific temperature thresholds appear responsible for differences in the distribution of two subfamilies (Hall and Thatje 2009). They correlate with the effects of temperature on early life-history stages, while adults may have wider temperature tolerance than indicated by their actual distributions (Hall and Thatje 2009).

Differences in the ability to regulate magnesium levels in the hemolymph have been identified as potential causes for differential capabilities of decapod Natantia and Reptantia to colonize cold waters. That is because down-regulation of Mg^{2+} concentration promotes cardiac and thus ventilator performance countering cold-induced failure of oxygen supply to the tissues. Caridean shrimps seem to be able to regulate magnesium concentrations well, thus guaranteeing oxygen supply to their tissues in the cold. This has allowed them to colonize the Antarctic (Frederich et al. 2001) and should have important ramifications for deep-sea colonization as well. Many crabs, on the contrary, seem to lack this ability, which may be the reason for their rarity in both the high Antarctic (Frederich et al. 2001) and the abyss.

In regions that lack strong gradients in temperature, crustaceans still show faunal turnover with depth. Antarctic waters are uniformly cold (0.8–2.5°C) compared to the temperature gradients experienced on temperate and tropical margins. While many Antarctic crustaceans have wider depth ranges than observed in temperate waters (Brey et al. 1996), distinct shelf, slope, and abyssal communities have been defined using data on selected benthic taxa including isopods and ostracods (Kaiser et al. 2011). Conversely, the Mediterranean is uniformly warm across a wide depth range. The basin reaches a maximum depth of 5,000 m, but temperature does not drop below 13°C due to the strait of Gibraltar, which isolates the Mediterranean from the global circulation of cooled deep water. Despite the uniformity, trawl data from the western Mediterranean show continuous species turnover of decapods with depth (Cartes and Sardà 1993, Maynou and Cartes 2000, Fanelli et al. 2007). This is further evidence that temperature alone cannot explain the distribution of species along bathymetric gradients.

Pressure

Hydrostatic pressure is the only environmental variable in the ocean universally correlated with depth, increasing by 1 atmosphere (atm) every 10 m (Pradillon and Gaill 2007). Experiments manipulating pressure in the laboratory to examine the physiological responses of organisms showed that high pressure causes macromolecular damage and affects the functionality of membranes and the equilibrium and rate constants of reactions. In the short term, this stops cellular growth and causes the expression of stress proteins (reviewed by Pradillon and Gaill 2007). An organism can adapt to some pressure change, but these physiological adaptations elevate metabolic demand. This metabolic demand does not appear to change the oxygen requirement of cold and deep-sea organisms (Childress 1995); rather these organisms have functional adaptations, for instance, increased enzyme concentration, adoption of enzymes with greater efficacy, and inclusion of modulator compounds that facilitate enzyme reactions (Seibel and Drazen 2007).

Pressure and temperature likely interact on cell function to limit the bathymetric range of a species. Relatively moderate pressure increases are equivalent to relatively large decreases in temperature (Brown and Thatje 2014). Experiments on the bathyal lysianassoid amphipod *Stephonyx biscayensis* found that the pressure tolerance of this species is temperature-dependent: pressure equivalent to 2,000 m depth was tolerated at 1°C and 3°C, pressure equivalent to 2,500 m depth was tolerated at 5.5°C, and at 10°C pressure equivalent to 3,000 m depth was tolerated. This corresponds with the observed distribution of the species, which does not extend beyond 2,000 m and temperatures of 3°C (Brown and Thatje 2011). Experiments on the temperate shallow-water hermit crab *Pagurus cuanensis* found that this species can tolerate pressures greater than those found in its natural bathymetric range (Thatje et al. 2010). Temperature probably limits this hermit crab to shallow waters as low temperatures had a large effect on the metabolic rate.

Oxygen Minimum Zones

Although much of the ocean has oxygen values near saturation, there are extensive midwater regions, oxygen minimum zones (OMZs), where oxygen is depleted (Helly and Levin 2004). OMZs are encountered at sea bottoms of ~50–1,000 m regionally restricted near upwelling areas (Helly and Levin 2004), though they may have had much greater distribution in the past (Rogers 2000). The lack of oxygen has profound effects on benthic diversity, with only specialized taxa able to survive (Rogers 2000, Levin 2003). Since present-day OMZs may have existed for thousands of years, it has been suggested that they eventually lead to vicariance on continental slopes (Rogers 2000). Few crustaceans can tolerate the extreme conditions of OMZs (Levin 2003). When oxygen values fall to between 0.1 and 0.5 ml l^{-1}, high densities of polychaetes dominate macrofaunal communities (Levin et al. 2000), while meiofaunal communities are dominated by nematodes with much reduced densities of harpacticoid copepods and nauplii (Veit-Köhler et al. 2009). An exception are certain ampeliscid amphipods, which are often found in high numbers within OMZs (Palma et al. 2005, Oliver et al. 2011). These amphipods may possess physiological mechanisms that enhance oxygen uptake and perhaps sulfide tolerance (Levin 2003). Megafaunal crustaceans appear to be absent in the cores of most OMZs where oxygen concentrations falls below 0.15 ml l^{-1}. However, extraordinarily high densities of crustaceans and echinoderms can be found near OMZ lower boundaries (Levin 2003). Decapods known to occur in these low-oxygen, high-nutrient conditions include shrimp (*Plesionika nesisi, Benthesicymus altus*), squat lobsters (*Munidopsis* cf. *hystrix, Munidopsis scobina*), and spider crabs (*Encephaloides armstrongi*) (Wishner et al. 1995, Smallwood et al. 1999, Creasey et al. 2000). A number of squat lobsters occur at high densities within low-oxygen environments (i.e., *Munida quadrispina, Pleuroncodes monodon,* and *P. planipes*) (Lovrich and Thiel 2011). Research on squat lobsters has shown that these species can oxyregulate, maintaining rates of oxygen consumption within low-oxygen environments. This ability changes ontogenetically, with the oldest/largest animals being best able to survive the extreme conditions (Lovrich and Thiel 2011).

Substratum and Sediment Particle Size

Descending down-slope, the diversity of substrata decreases. Depths feature slopes, canyons, trenches, and terraces, providing a diversity of substrata. While at abyssal depths, rolling plains of soft sediments form the dominant habitat. The sediment is derived from terrigenous weathering or is biogenic and can reach several hundred meters in thickness (Ramirez-Llodra et al. 2010).

Sediment size decreases with depth from sandy sediments on the coast to fine muddy sediments on the abyssal plains. On continental margins, sediment is often composed of a mix of sediment particle sizes, which may increase habitat complexity. Sediment particle size diversity has been used to explain bathymetric gradients of species richness at both small and large spatial scales (Etter and Grassle 1992, Leduc et al. 2012). The size of sediment particles affects the available physical habitat and can affect feeding where deposit feeders selectively ingest particular size fractions of the sediment.

The dominance of fine soft sediment at abyssal depths may in part explain the success of interstitial and endobenthic groups of crustaceans, such as tube-dwelling harpacticoid copepods (Thistle 2001) and several groups of burrowing peracarids such as the Tanaidomorpha and isopods of the family Macrostylidae (Wilson and Ahyong 2015). Both groups can be highly abundant and diverse and are adapted to a tube-dwelling and digging lifestyle, respectively. Although mean particle size is correlated with taxonomic composition, density, and feeding mode, this property alone has not been isolated as a driving factor in bathymetric gradients of species richness (Levin et al. 2001).

Light

Daylight penetrates only the upper 1,000 m and becomes progressively dimmer and bluer with depth (Warrant and Locket 2004). Crustaceans at bathyal depths are often red, which in the absence of red light is perceived as black at those depths. At abyssal depths, many crustaceans do not have eyes, and this loss has occurred in many groups independently (Wilson and Ahyong 2015). In contrast, many benthic crustaceans living in the twilight zone (150–1,000 m) have larger eyes more receptive to blue light (Frank et al. 2012). Zoea larvae of the hydrothermal-vent crab *Bythograea thermydron* possess compound eyes characterized by a visual pigment sensitive to the blue wave-length in the mesopelagic waters, but this species develops naked-retina eyes with high sensitivity for longer wavelengths of the vent environment where it settles (Jinks et al. 2002). Some deep-sea crustaceans may be adapted for seeing bioluminescence in the deep benthic environment (Frank et al. 2012).

In clear ocean water, light is sufficient for visual predation and predator avoidance down to 1,000 m depth. Crustaceans living at these depths require locomotor capabilities at high metabolic cost (Childress 1995). Below 1,000 m, pelagic crustaceans show reduced metabolism that may be associated with the absence of predation as a selective pressure favoring high mobility (Childress 1995). Although low metabolic rates can be attributed to low temperature and food availability, Childress (1995) highlighted that different groups of animals have different rates of decline in metabolic rate. Burrowing and less active crustaceans, which may be less vulnerable to predation, show no significant decline in metabolic rate with depth beyond the effects of temperature while those groups that frequently swim show a significant decline. Childress (1995) argued that the lack of light in the deep explains the low metabolic rates in sighted groups like crustaceans.

Biotic Interactions

Competition for food resources may also influence depth distribution (McClain and Schlacher 2015). For instance, a study in the Catalan Sea showed that decapod species of similar size/morphology and feeding strategy had little bathymetric overlap (Cartes 1998) and a succession of species filling similar trophic niches at different depths.

Crustaceans often live in association with or depend on other animals. For instance, hermit crabs require a supply of gastropod shells, which are scarce in deep water. Squat lobsters of the family Chirostylidae are typically associated with deep-water corals (Baeza 2011), whose depth distributions in turn are limited by aragonite and oxygen saturation in the water (Tittensor et al. 2009). The squat lobster *Munida sarsi* is parasitized by three species of rhizocephalan barnacles (Attrill 1989). At the deeper limits of the squat lobsters range, it becomes more susceptible to infestation, but the physiological effects of depth are also experienced by the rhizocephalan parasite at the extreme of its depth range, and therefore the parasite has a less detrimental effect on the individual host in deep waters (Attrill 1989).

FUTURE DIRECTIONS

Effects of Climate Change on Bathymetric Ranges

Climate change is causing the average sea-surface temperature to rise (Collins et al. 2013). This will affect the fauna of the continental shelf and upper slope first, but eventually the whole ocean down to the abyss will warm, with projections of up to 0.1°C per decade on average for the deep ocean (Purkey and Johnson 2012). While latitudinal range shifts have been observed for shallow-water marine crustaceans in response to climate change (Neumann et al. 2013), shifts in depth ranges are also likely (Brown and Thatje 2015). Shallow-water invertebrates could be able to penetrate greater depths to escape rising temperatures but only if they are able to withstand the increasing pressure and decreasing oxygen at greater depths (Brown and Thatje 2015).

There is evidence that greater temperatures will mediate the effects of high pressure on animals, increasing hyperbaric tolerance and shifting the lower bathymetric limit downward (Brown and Thatje 2015). For example, the pressure tolerance of the shallow-water hermit crab *Pagurus cuanensis* increases from the equivalent of 200 m water depth to the equivalent of 500 m water depth with a 5°C temperature increase (Thatje et al. 2010). Based on this, the 1.0–1.5°C temperature increase predicted for 200–400 m water depth in the northern temperate zone by the year 2100 (Collins et al. 2013) will shift the lower bathymetric limit downward by 60–90 m water depth. The combined physiological effects of temperature, pressure, and oxygen concentration constrain the fundamental ecological niches of crustacean species. Conceptual models suggest that the ecological niches of tropical and temperate faunas may actually experience expansion under climate change, while polar taxa are likely to contract (Brown and Thatje 2015). Crustaceans are significantly more tolerant of increased pressure or decreased oxygen concentration than fishes but significantly less tolerant of these challenges than polychaetes, molluscs, and echinoderms (Brown and Thatje 2015). Greater consideration of depth in predictive models of species distributions is critical for understanding the impacts of climate change.

Climate change will also result in changes to primary production, increasing hypoxia and acidification. It is expected that these changes will have significant impacts on marine ecosystems at all depths (Levin and Le Bris 2015, Rogers 2015, Thresher et al. 2015). Given the importance of food supply for patterns of biomass, abundance, and diversity, changes to surface primary production is likely to have impacts on benthic communities in the deep. The functions of deep-sea food webs, particularly in relation to flows of carbon from the surface ocean to the deep sea, are poorly understood. There is a critical need to undertake observational and experimental studies to better understand the effects of changes in food supply, temperature, oxygen, and pH on the deep-sea fauna (Rogers 2015).

Biodiversity Informatics

It is the era of "Big Data," and the mode in which we address questions of biogeography is rapidly changing (Munro et al. 2015). Digitalization of natural history collections and their aggregation in databases like the Global Biodiversity Information Facility (GBIF) and the Ocean Biogeographic Information System (OBIS) hold great potential for biogeographic research (Graham et al. 2004). Now, the data from centuries of research expeditions can be re-examined at large spatial scales in light of new concepts and statistical methods (Woolley et al. 2016).

However, errors in the geographic position of locations and the identification of specimens means that data must be scrutinized and "cleaned" before use (Maldonado et al. 2015). The paucity of deep-sea data is not alleviated by aggregating data across surveys. There are significantly more data from the shelf and the coast than from the deep sea, and this spatial bias in collection effort must be carefully considered in analyses of bathymetric gradients (O'Hara et al. 2016).

CONCLUSIONS

Crustaceans often feature in biological analyses of depth gradients; in particular patterns for decapods, isopods, and harpacticoid copepods are relatively well documented. Average body size becomes smaller, despite observations of deep-sea gigantism in some groups. Fundamental questions about the relationship between seafloor depth and benthic biodiversity remain incompletely understood. Most critically, the notion of peak species richness at bathyal depths is yet to be established but may generally apply to some crustaceans such as isopods.

Molecular data has revealed that many crustacean species with apparently wide depth distributions are in fact multiple narrow-ranged species. However, eurybathic distributions have been confirmed in the Antarctic where temperature gradients are less distinct. The causes of bathymetric patterns must involve multiple factors, but the steep decline in food availability appears to be a major driver, especially at abyssal depths. Experimental observations of some crustacean species in the laboratory have shown that the interaction between pressure and temperature effects on cell function is likely to act together in limiting the bathymetric range of a species. These experiments have been successful in predicting the observed maximum depth of occurrence for these species and have been used to make future predictions on species distributions under climate change.

REFERENCES

Attrill, M.J. 1989. A rhizocephalan (Crustacea: Cirripedia) infestation of the deep-sea galatheid *Munida sarsi* (Crustacea; Decapoda), the effects on the host and the influence of depth upon the host-parasite relationship. Journal of Zoology 217:663–682.

Baeza, A.J. 2011. Squat lobsters as symbionts and in chemo-autotrophic environments. Pages 249–270 *in* G.C.B. Poore, S.T. Ahyong, and J. Taylor, editors. Biology of Squat Lobsters. CRC Press, Boca Raton, FL.

Baguley, J.G., P.A. Montagna, W. Lee, L.J. Hyde, and G.I. Rowe. 2006. Spatial and bathymetric trends in Harpacticoida (Copepoda) community structure in the Northern Gulf of Mexico deep-sea. Journal of Experimental Marine Biology and Ecology 330:327–341.

Beddard, F.E. 1886. Zoology Part XLVIII. Report on the Isopoda collected by H.M.S. Challenger during the years 1873–76. Second part. Report of the Scientific Results of the Voyage of H.M.S. Challenger During the Years 1873–76 17:1–178.

Blankenship, L.E., A.A. Yayanos, D.B. Cadien, and L.A. Levin. 2006. Vertical zonation patterns of scavenging amphipods from the hadal zone of the Tonga and Kermadec Trenches. Deep-Sea Research Part I—Oceanographic Research Papers 53:48–61.

Blazewicz-Paszkowycz, M., R.N. Bamber, and G. Anderson. 2012. Diversity of Tanaidacea (Crustacea: Peracarida) in the world's oceans; How far have we come? PloS ONE 7:e33068.

Boudreau, S.A., and B. Worm. 2012. Ecological role of large benthic decapods in marine ecosystems: a review. Marine Ecology Progress Series 469:195–213.

Brandt, A. 1995. Peracarid fauna (Crustacea, Malacostraca) of the northeast water polynya off Greenland: documenting close benthic-pelagic coupling in the Westwind Trough. Marine Ecology Progress Series 121:39–51.

Brandt, A., and K. Schnack. 1999. Macrofaunal abundance at 79 degrees N off East Greenland: opposing data from epibenthic-sledge and box-corer samples. Polar Biology 22:75–81.

Brandt, A., K. Linse, and M. Schüller. 2009. Bathymetric distribution patterns of Southern Ocean macrofaunal taxa: Bivalvia, Gastropoda, Isopoda and Polychaeta. Deep-Sea Research Part I: Oceanographic Research Papers 56: 2013–2025.

Brandt, A., M. Blazewicz-Paszkowycz, R.N. Bamber, U. Muhlenhardt-Siegel, M.V. Malyutina, S. Kaiser, C. De Broyer, and C. Havermans. 2012. Are there widespread peracarid species in the deep sea (Crustacea: Malacostraca)? Polish Polar Research 33:139–162.

Brandt, A., I. Frutos, S. Bober, S. Brix, N. Brenke, T. Guggolz, N. Heitland, M. Malyutina, U. Minzlaff, T. Riehl, E. Schwabe, A.C. Zinkann, and K. Linse. 2018. Composition of abyssal macrofauna along the Vema Fracture Zone and the hadal Puerto Rico Trench, northern tropical Atlantic. Deep Sea Research Part II: Topical Studies in Oceanography 148:35–44.

Brandão, S.N., J. Sauer, and I. Schön. 2010. Circumantarctic distribution in Southern Ocean benthos? A genetic test using the genus *Macroscapha* (Crustacea, Ostracoda) as a model. Molecular Phylogenetics and Evolution 55:1055–1069.

Brix, S., J. Svavarsson, and F. Leese. 2014. A multi-gene analysis reveals multiple highly divergent lineages of the isopod *Chelator insignis* (Hansen, 1916) south of Iceland. Polish Polar Research 35:225–242.

Brey, T., C. Dahm, M. Gorny, M. Klages, M. Stiller, and W.E. Arntz. 1996. Do Antarctic benthic invertebrates show an extended level of eurybathy? Antarctic Science 8:3–6.

Brökeland, W., G. Guðmundsson, and J. Svavarsson. 2010. Diet of four species of deep-sea isopods (Crustacea: Malacostraca: Peracarida) in the South Atlantic and the Southern Ocean. Marine Biology 157:177–178.

Brown A, and S. Thatje. 2011. Respiratory Response of the Deep-Sea Amphipod Stephonyx biscayensis Indicates Bathymetric Range Limitation by Temperature and Hydrostatic Pressure. PLOS ONE 6: e28562.

Brown, A., and S. Thatje. 2014. Explaining bathymetric diversity patterns in marine benthic invertebrates and demersal fishes: physiological contributions to adaptation of life at depth. Biological Reviews 89:406–426.

Brown, A., and S. Thatje. 2015. The effects of changing climate on faunal depth distributions determine winners and losers. Global Change Biology 21:173–180.

Carney, R.S. 2005. Zonation of deep biota on continental margins. Pages 211–278 *in* R.N. Gibson, R.J.A. Atkinson, and J.D.M. Gordon, editors. Oceanography and Marine Biology: An Annual Review. CRC Press, Boca Raton, FL.

Carney, R.S. 2007. Use of diversity estimations in the study of sedimentary benthic communities. Pages 139–172 *in* R.N. Gibson, R.J.A. Atkinson, and J.D.M. Gordon, editors. Oceanography and Marine Biology, Vol. 45. CRC Press, Boca Raton, FL.

Cartes, J.E. 1998. Feeding strategies and partition of food resources in deep-water decapod crustaceans (400–2300 m). Journal of the Marine Biological Association of the United Kingdom 78:509–524.

Cartes, J.E., and F. Sardà. 1993. Zonation of deep-sea decapod fauna in the Catalan Sea (Western Mediterranean). Marine Ecology Progress Series 94:27–34.

Cartes, J.E., D. Jaume, and T. Madurell. 2003. Local changes in the composition and community structure of suprabenthic peracarid crustaceans on the bathyal Mediterranean: influence of environmental factors. Marine Biology 143:745–758.

Cartes, J.E., A, Serrano, F. Velasco, S. Parra, and F. Sanchez. 2007. Community structure and dynamics of deep-water decapod assemblages from Le Danois Bank (Cantabrian Sea, NE Atlantic): influence of environmental variables and food availability. Progress in Oceanography 75:797–816.

Childress, J.J. 1995. Are there physiological and biochemical adaptations of metabolism in deep-sea animals? Trends in Ecology and Evolution 10:30–36.

Collins, M., R. Knutti, J. Arblaster, J.L. Dufresne, T. Fichefet, P. Friedlingstein, X. Gao, W. Gutowski, T. Johns, and G. Krinner. 2013. Long-term climate change: projections, commitments and irreversibility. Pages 1029–1136 *in* T.Q. Stocker, D. Qin, G.K. Plattner, M. Tignor, S.K. Allen, J. Boschung, A. Nauels, Y. Xia, V. Bex, and P.M. Midgley, editors. Climate Change 2013: The Physical Science Basis. Contribution of Working Group I to the Fifth Assessment Report of the Intergovernmental Panel on Climate Change. Cambridge University Press, Cambridge, UK.

Compton, T.J., D.A. Bowden, C.R. Pitcher, J.E. Hewitt, and N. Ellis. 2013. Biophysical patterns in benthic assemblage composition across contrasting continental margins off New Zealand. Journal of Biogeography 40:75–89.

Currie, D.J., G.G. Mittelbach, H.V. Cornell, R. Field, J.F. Guégan, B.A. Hawkins, D.M. Kaufman, J.T. Kerr, T. Oberdorff, E. O'Brien, and J.R.G. Turner. 2004. Predictions and tests of climate-based hypotheses of broad-scale variation in taxonomic richness. Ecology Letters 7:1121–1134.

Creasey, S., A. Rogers, P. Tyler, J. Gage, and D. Jollivet. 2000. Genetic and morphometric comparisons of squat lobster, *Munidopsis scobina* (Decapoda: Anomura: Galatheidae) populations, with notes on the phylogeny of the genus Munidopsis. Deep Sea Research Part II: Topical Studies in Oceanography 47:87–118.

Danovaro, R., C. Gambi, and N. Della Croce. 2002. Meiofauna hotspot in the Atacama Trench, eastern South Pacific Ocean. Deep-Sea Research Part I: Oceanographic Research Papers 49:843–857.

De La Rocha, C.L., and U. Passow. 2007. Factors influencing the sinking of POC and the efficiency of the biological carbon pump. Deep-Sea Research Part II: Topical Studies in Oceanography 54:639–658.

De Smet, B., E. Pape, T. Riehl, P. Bonifácio, L. Colson, and A. Vanreusel. 2017. The community structure of deep-sea macrofauna associated with polymetallic nodules in the eastern part of the Clarion-Clipperton Fracture Zone. Frontiers in Marine Science 4:103.

d'Udekem d'Acoz, C., and C. Havermans. 2015. Contribution to the systematics of the genus *Eurythenes* SI Smith in Scudder, 1882 (Crustacea: Amphipoda: Lysianassoidea: Eurytheneidae). Zootaxa 3971.1:1–80.

Ellingsen, K.E., A. Brandt, B. Ebbe, and K. Linse. 2007. Diversity and species distribution of polychaetes, isopods and bivalves in the Atlantic sector of the deep Southern Ocean. Polar Biology 30:1265–1273.

Etter, R.J., and J.F. Grassle. 1992. Patterns of species diversity in the deep sea as a function of sediment particle size diversity. Nature 360:576–578.

Eustace, R.M., N.M. Kilgallen, N.C. Lacey, and A.J. Jamieson. 2013. Population structure of the hadal amphipod *Hirondellea gigas* (Amphipoda: Lysianassoidea) from the Izu-Bonin Trench. Journal of Crustacean Biology 33:793–801.

Eustace, R.M., H. Ritchie, N.M. Kilgallen, S.B. Piertney, and A.J. Jamieson. 2016. Morphological and ontogenetic stratification of abyssal and hadal *Eurythenes gryllus* sensu lato (Amphipoda: Lysianassoidea) from the Peru–Chile Trench. Deep Sea Research Part I: Oceanographic Research Papers 109: 91–98.

Fanelli, E., F. Colloca, and G. Ardizzone. 2007. Decapod crustacean assemblages off the west coast of central Italy (western Mediterranean). Scientia Marina 71:19–28.

Fischer, V., and A. Brandt. 2015. Composition of the abyssal infauna of the Kuril–Kamchatka area (NW Pacific) collected with a box corer. Deep Sea Research Part II: Topical Studies in Oceanography 111:26–33.

Flach, E., and W. de Bruin. 1999. Diversity patterns in macrobenthos across a continental slope in the NE Atlantic. Journal of Sea Research 42:303–323.

Flach, E., and C. Heip. 1996. Vertical distribution of macrozoobenthos within the sediment on the continental slope of the Goban Spur area (NE Atlantic). Marine Ecology Progress Series 141:55–66.

Flach, E., J. Vanaverbeke, and C. Heip. 1999. The meiofauna: macrofauna ratio across the continental slope of the Goban Spur (north-east Atlantic). Journal of the Marine Biological Association of the United Kingdom 79:233–241.

Follesa, M.C., S. Cabiddu, A. Gastoni, and A. Cau. 2007. On the reproductive biology of the deep-sea lobster, *Polycheles typhlops* (Decapoda, Palinura, Polychelidae), from the central-western Mediterranean. Crustaceana 80:839–846.

Frank, T.M., S. Johnsen, and T.W. Cronin. 2012. Light and vision in the deep-sea benthos: II. Vision in deep-sea crustaceans. Journal of Experimental Biology 215:3344–3353.

Frederich, M., F. Sartoris, H.O. Poèrtner. 2001. Distribution patterns of decapod crustaceans in polar areas: a result of magnesium regulation? Polar Biology 24:719–723.

Gage, J.D., and P.A. Tyler. 1992. Deep-Sea Biology: A Natural History of Organisms at the Deep-Sea Floor. Cambridge University Press, Cambridge, UK.

Golovan, O.A., M. Blazewicz-Paszkowycz, A. Brandt, L.L. Budnikova, N.O. Elsner, V.V. Ivin, A.V. Lavrenteva, M.V. Malyutina, V.V. Petryashov, and L.A. Tzareva. 2013. Diversity and distribution of peracarid crustaceans (Malacostraca) from the continental slope and the deep-sea basin of the Sea of Japan. Deep-Sea Research Part II: Topical Studies in Oceanography 86–87:66–78.

Graham, C.H., S. Ferrier, F. Huettman, C. Moritz, and A.T. Peterson. 2004. New developments in museum-based informatics and applications in biodiversity analysis. Trends in Ecology and Evolution 19: 497–503.

Grassle, J.F., and N.J. Maciolek. 1992. Deep-sea species richness—regional and local diversity estimates from quantitative bottom samples. American Naturalist 139:313–341.

Gray, J.S. 1997. Gradients in marine biodiversity. Pages 94–121 *in* R.F.G. Ormond, J.D. Gage, and M.V. Angel, editors. Marine Biodiversity: Patterns and Processes. Cambridge Univerity Press, Cambridge, UK.

Gray, J.S. 2000. The measurement of marine species diversity, with an application to the benthic fauna of the Norwegian continental shelf. Journal of Experimental Marine Biology and Ecology 250:23–49.

Haedrich, R.L., G.T. Rowe, and P.T. Polloni. 1980. The megabenthic fauna in the deep-sea south of New-England, USA. Marine Biology 57:165–179.

Haedrich, R.L., J.A. Devine, and V.J. Kendall. 2008. Predictors of species richness in the deep-benthic fauna of the northern Gulf of Mexico. Deep Sea Research Part II: Topical Studies in Oceanography 55:2650–2656.

Hall, S., and S. Thatje, 2009. Global bottlenecks in the distribution of marine crustacea: temperature constraints in the family Lithodidae. Journal of Biogeography 36:2125–2135.

Havermans, C., Z.T. Nagy, G. Sonet, C. De Broyer, and P. Martin. 2011. DNA barcoding reveals new insights into the diversity of Antarctic species of *Orchomene* Sensu Lato (Crustacea: Amphipoda: Lysianassoidea). Deep Sea Research Part II: Topical Studies in Oceanography 58:230–241.

Havermans, C., G. Sonet, C. d'Udekem d'Acoz, Z.T. Nagy, P. Martin, S. Brix, T. Riehl, S. Agrawal, and C. Held. 2013 Genetic and morphological divergences in the cosmopolitan deep-sea amphipod *Eurythenes gryllus* reveal a diverse abyss and a bipolar species. PLoS ONE 8: e74218.

Helly, J.J., and L.A. Levin. 2004. Global distribution of naturally occurring marine hypoxia on continental margins. Deep Sea Research Part I: Oceanographic Research Papers 51:1159–1168.

Hessler, R., and G. Wilson. 1983. The origin and biogeography of malacostracan crustaceans in the deep sea 227–254 in R. W. Sims, J. H. Price, and P. E. S. Whalley, editors. Evolution, time and space: the emergence of the biosphere. Academic Press, London and New York.

Jamieson, A.J., T. Fujii, M. Solan, A.K. Matsumoto, P.M. Bagley, and I.G. Priede. 2009. First findings of decapod crustacea in the hadal zone. Deep-Sea Research Part I: Oceanographic Research Papers 56:641–647.

Jamieson, A.J., T. Fujii, D.J. Mayor, M. Solan, and I.G. Priede. 2010. Hadal trenches: the ecology of the deepest places on Earth. Trends in Ecology and Evolution 25:190–197.

Jinks, R.N., T.L. Markley, E.E. Taylor, G. Perovich, A.I. Dittel, C.E. Epifanio, T.W. Cronin. 2002. Adaptive visual metamorphosis in a deep-sea hydrothermal vent crab. Nature 420:68–70.

Kaiser, S., H.J. Griffiths, D.K.A. Barnes, S.N. Brandao, A. Brandt, and P.E. O'Brien. 2011. Is there a distinct continental slope fauna in the Antarctic? Deep-Sea Research Part II: Topical Studies in Oceanography 58:91–104.

Kitahashi, T., K. Kawamura, S. Kojima, and M. Shimanaga. 2013. Assemblages gradually change from bathyal to hadal depth: a case study on harpacticoid copepods around the Kuril Trench (north-west Pacific Ocean). Deep-Sea Research Part I: Oceanographic Research Papers 74:39–47.

Knox, M.A., I.D. Hogg, C.A. Pilditch, A.N. Lorz, and S.D. Nodder. 2012. Abundance and diversity of epibenthic amphipods (Crustacea) from contrasting bathyal habitats. Deep Sea Research Part I: Oceanographic Research Papers 62:1–9.

Larsen, K. 2005. Deep-Sea Tanaidace (Peracarida) from the Gulf of Mexico. Crustaceana Monographs 5:1–381.

Leduc, D., A.A. Rowden, D.A. Bowden, P.K. Probert, C.A. Pilditch, and S.D. Nodder. 2012. Unimodal relationship between biomass and species richness of deep-sea nematodes: implications for the link between productivity and diversity. Marine Ecology Progress Series 454:53–64.

Levin, L.A. 2003. Oxygen minimum zone benthos: adaptation and community response to hypoxia. Oceanography and Marine Biology: An Annual Review 41:1–45

Levin, L.A., and N. Le Bris. 2015. The deep ocean under climate change. Science 350:766–768.

Levin, L.A., J.D. Gage, C. Martin, and P.A. Lamont. 2000. Macrobenthic community structure within and beneath the oxygen minimum zone, NW Arabian Sea. Deep-Sea Research Part II: Topical Studies in Oceanography 47:189–226.

Levin, L.A., R.J. Etter, M.A. Rex, A.J. Gooday, C.R. Smith, J. Pineda, C.T. Stuart, R.R. Hessler, and D. Pawson. 2001. Environmental influences on regional deep-sea species diversity. Annual Review of Ecology and Systematics 32:51–93.

Lindberg, W.J., and F.D. Lockhart. 1993. Depth-stratified population structure of geryonid crabs in the eastern Gulf of Mexico. Journal of Crustacean Biology 13:713–722.

Lovrich, G.A., and M. Thiel. 2011. Ecology, physiology, feeding and trophic role of squat lobsters. Pages 183–222 *in* G.C.B. Poore, S. Poore, and J. Taylor, editors. The Biology of Squat Lobsters CSIRO Publishing, Melbourne and CRC Press, Boca Raton, FL.

Macpherson, E. 2003. Species range size distributions for some marine taxa in the Atlantic Ocean. Effect of latitude and depth. Biological Journal of the Linnean Society 80:437–455.

Macpherson, E., and C.M. Duarte. 1991. Bathymetric trends in demersal fish size—Is there a general relationship. Marine Ecology Progress Series 71:103–112.

Maldonado, C., C.I. Molina, A. Zizka, C. Persson, C.M. Taylor, J. Alban, E. Chilquillo, N. Ronsted, and A. Antonelli. 2015. Estimating species diversity and distribution in the era of Big Data: To what extent can we trust public databases? Global Ecology and Biogeography 24:973–984.

Maurer, B.A., and B.J. McGill. 2011. Measurement of Species Diversity. Biological Diversity: Frontiers in Measurement and Assessment. Oxford University Press, Oxford.

May, R.M. 1993. Marine species richness (reply). Nature 361:598.

Maynou, F., and J.E. Cartes. 2000. Community structure of bathyal decapod crustaceans off south-west Balearic Islands (western Mediterranean): seasonality and regional patterns in zonation. Journal of the Marine Biological Association of the UK 80:789–798.

McCallum, A.W., G.C.B. Poore, A. Williams, F. Althaus, and T. O'Hara. 2013. Environmental predictors of decapod species richness and turnover along an extensive Australian continental margin (13–35° S). Marine Ecology 34:298–312.

McCallum, A.W., S. Woolley, M. Błażewicz-Paszkowycz, J. Browne, S. Gerken, R. Kloser, G.C.B. Poore, D. Staples, A. Syme, J. Taylor, G. Walker-Smith, A. Williams, and R.S. Wilson. 2015. Productivity enhances benthic species richness along an oligotrophic Indian Ocean continental margin. Global Ecology and Biogeography 24:462–471.

McClain, C.R., and T.A. Schlacher. 2015. On some hypotheses of diversity of animal life at great depths on the sea floor. Marine Ecology—An Evolutionary Perspective 36:849–872.

McClain, C.R., M.A. Balk, M.C. Benfield, T.A. Branch, C. Chen, J. Cosgrove, A.D. Dove, L.C. Gaskins, R.R. Helm, and F.G. Hochberg. 2015. Sizing ocean giants: patterns of intraspecific size variation in marine megafauna. PeerJ 3:e715.

McKnight, D.G., and P.K. Probert. 1997. Epibenthic communities on the Chatham Rise, New Zealand. New Zealand Journal of Marine and Freshwater Research 31:505–513.

Menot, L., M. Sibuet, R.S. Carney, L.A. Levin, G.T. Rowe, D.S. Billett, G. Poore, H. Kitazato, A. Vanreusel, and J. Galéron. 2010. New perceptions of continental margin biodiversity. Pages 79–103 *in* A.D. McIntyre, editor. Life in the World's Oceans: Diversity, Distribution, and Abundance. Wiley-Blackwell, Oxford, UK.

Menzies, R.J., R.Y. George, and G.T. Rowe. 1973. Abyssal Environment and Ecology of the World Oceans. John Wiley, New York.

Mindel, B.L., T.J. Webb, F.C. Neat, and J.L. Blanchard. 2016. A trait-based metric sheds new light on the nature of the body size-depth relationship in the deep sea. Journal of Animal Ecology 85:427–436.

Munro, C., J.P. Morris, A. Brown, C. Hauton, and S. Thatje. 2015. The role of ontogeny in physiological tolerance: decreasing hydrostatic pressure tolerance with development in the northern stone crab *Lithodes maja*. Proceedings of the Royal Society B: Biological Sciences 282:20150577.

Neumann, H., I. de Boois, I. Kroncke, and H. Reiss. 2013. Climate change facilitated range expansion of the non-native angular crab *Goneplax rhomboides* into the North Sea. Marine Ecology Progress Series 484:143–153.

O'Hara, T.D., T.A. Schlacher, A.A. Rowden, and D.P. Tittensor. 2016. Data analysis considerations. Pages 386–403 *in* M.R. Clark, M. Consalvey and A. A. Rowden, editors. Biological Sampling in the Deep Sea. John Wiley, Chichester, UK.

Oliver, J., K. Hammerstrom, E. McPhee-Shaw, P. Slattery, J. Oakden, S. Kim, and S.I. Hartwell. 2011. High species density patterns in macrofaunal invertebrate communities in the marine benthos. Marine Ecology 32:278–288.

Pabis, K., P. Jozwiak, A.N. Lorz, K. Schnabel, and M. Blazewicz-Paszkowycz. 2015. First insights into the deep-sea tanaidacean fauna of the Ross Sea: species richness and composition across the shelf break, slope and abyss. Polar Biology 38:1429–1437.

Palma, M., E. Quiroga, V. Gallardo, W.E. Arntz, D. Gerdes, W. Schneider, and D. Hebbeln. 2005. Macrobenthic animal assemblages of the continental margin off Chile (22° to 42°S). Journal of the Marine Biological Association of the UK 85:233–245.

Papiol, V., and M.E. Hendrickx. 2015. Community structure of deep-water decapod crustaceans below the oxygen minimum zone in the south-east Gulf of California and analysis of environmental drivers. Marine and Freshwater Research 67:1862–1879.

Pequegnat, W.E. 1983. The Ecological Communities of the Continental Slope and Adjacent Regimes of the Northern Gulf of Mexico. Minerals Management Service, United States Department of the Interior Gulf of Mexico Office, Metairie, LA.

Poore, G.C.B., and N. Andreakis. 2012. The *Agononida incerta* species complex unravelled (Crustacea: Decapoda: Anomura: Munididae). Zootaxa 3492:1–29.

Poore, G.C.B., and N.L. Bruce. 2012. Global diversity of marine isopods (except Asellota and crustacean symbionts). PLoS ONE 7:e43529.

Poore, G.C.B., J. Just, and B. Cohen. 1994. Composition and diversity of Crustacea Isopoda of the southeastern Australian continental slope. Deep-Sea Research 41:677–693.

Poore, G.C.B., L. Avery, M. Błażewicz-Paszkowycz, J. Browne, N.L. Bruce, S. Gerken, C. Glasby, E. Greaves, A. W. McCallum, D. Staples, A. Syme, J. Taylor, G. Walker-Smith, M. Warne, C. Watson, A. Williams, R.S. Wilson, and S. Woolley. 2015. Invertebrate diversity of the unexplored marine western margin of Australia: taxonomy and implications for global biodiversity. Marine Biodiversity 45: 271–286.

Poore, G.C.B., S.F. Rainer, R.B. Spies, and E. Ward. 1975. The zoobenthos program in Port Phillip Bay, 1969–73. Fisheries and Wildlife Paper, Victoria 7:1–78.

Pradillon, F. and F. Gaill. 2007. Pressure and life: some biological strategies. Reviews in Environmental Science and Bio/Technology, 6:181–195.

Purkey, S.G., and G.C. Johnson. 2012. Global contraction of Antarctic bottom water between the 1980s and 2000s. Journal of Climate 25:5830–5844.

Ramirez-Llodra, E., A. Brandt, R. Danovaro, B. De Mol, E. Escobar, C.R. German, L.A. Levin, P.M. Arbizu, L. Menot, P. Buhl-Mortensen, B.E. Narayanaswamy, C.R. Smith, D.P. Tittensor, P.A. Tyler, A. Vanreusel, and M. Vecchione. 2010. Deep, diverse and definitely different: unique attributes of the world's largest ecosystem. Biogeosciences 7:2851–2899.

Rex, M.A., and R.J. Etter. 2010. Deep-Sea Biodiversity: Patterns and Scale. Harvard University Press, Cambridge, MA.

Rex, M.A., R.J. Etter, J.S. Morris, J. Crouse, C.R. McClain, N.A. Johnson, C.T. Stuart, J.W. Deming, R. Thies, and R. Avery. 2006. Global bathymetric patterns of standing stock and body size in the deep-sea benthos. Marine Ecology Progress Series 317:1–8.

Rogers, A.D. 2000. The role of the oceanic oxygen minima in generating biodiversity in the deep sea. Deep Sea Research Part II: Topical Studies in Oceanography 47: 119–148.

Rogers, A.D. 2015. Environmental change in the deep ocean. Annual Review of Environment and Resources 40:1–38.

Riehl, T., and S. Kaiser. 2012. Conquered from the deep sea? A new deep-sea isopod species from the Antarctic shelf shows pattern of recent colonization. PLoS ONE 7:e49354.

Riehl, T., S. Bober, I. Voltski, M.V. Malyutina, and A. Brandt. 2016. Caught in the act: an abyssal isopod collected while feeding on Komokiaceae. Marine Biodiversity 1–2.

Riehl, T., L. Lins, and A. Brandt. 2018. The effects of depth, distance, and the Mid-Atlantic Ridge on genetic differentiation of abyssal and hadal isopods (Macrostylidae). Deep Sea Research Part II: Topical Studies in Oceanography148:74–90.

Rohal, M., D. Thistle, and E.E. Easton. 2014. Meiofaunal abundances and faunal similarity on the continental rise off the coast of California. Deep Sea Research Part I: Oceanographic Research Papers 93:131–144.

Rosa, R., J. Boavida-Portugal, K. Trubenbach, M. Baptista, R. Araujo, and R. Calado. 2012. Descending into the abyss: bathymetric patterns of diversity in decapod crustaceans shift with taxonomic level and life strategies. Deep Sea Research Part I: Oceanographic Research Papers 64:9–21.

Schnabel, K.E., P. Cabezas, A. McCallum, E. Macpherson, S.T. Ahyong, and K. Baba. 2011. Worldwide distribution patterns of squat lobsters. Pages 149–182 *in* G.C.B. Poore, S.T. Ahyong, and J. Taylor, editors. The Biology of Squat Lobsters. CSIRO Publishing, Collingwood, Australia.

Seibel, B.A., and J.C. Drazen. 2007. The rate of metabolism in marine animals: environmental constraints, ecological demands and energetic opportunities. Philosophical Transactions of the Royal Society B: Biological Sciences 362:2061–2078.

Smallwood, B. J., G.A. Wolff, B.J. Bett, C.R. Smith, D. Hoover, J.D. Gage, and A. Patience. 1999. Megafauna can control the quality of organic matter in marine sediments. Naturwissenschaften 86: 320–324.

Svavarsson, J.R. 1997. Diversity of isopods (Crustacea): new data from the Arctic and Atlantic Oceans. Biodiversity and Conservation 6:1571–1579.

Tecchio, S., E. Ramírez-Llodra, F. Sardà, I. Palomera, A. Mechó, R. Pedrosa-Pàmies, and A. Sanchez-Vidal. 2011. Drivers of deep Mediterranean megabenthos communities along longitudinal and bathymetric gradients. Marine Ecology Progress Series 439:181–192.

Thatje, S., L. Casburn, and J.A. Calcagno. 2010. Behavioural and respiratory response of the shallow-water hermit crab *Pagurus cuanensis* to hydrostatic pressure and temperature. Journal of Experimental Marine Biology and Ecology 390:22–30.

Thistle, D. 2001. Harpacticoid copepods are successful in the soft-bottom deep sea. Hydrobiologia 453:255–259.

Thresher, R.E., J.M. Guinotte, R.J. Matear, and A.J. Hobday. 2015. Options for managing impacts of climate change on a deep-sea community. Nature Climate Change 5:635–639.

Timofeev, S.F. 2001. Bergmann's principle and deep-water gigantism in marine crustaceans. Biological Bulletin 28:646–650.

Tittensor, D.P., A.R. Baco, P.E. Brewin, M.R. Clark, M. Consalvey, J. Hall-Spencer, A.A. Rowden, T. Schlacher, K.I. Stocks, and A.D. Rogers. 2009. Predicting global habitat suitability for stony corals on seamounts. Journal of Biogeography 36:1111–1128.

Tselepides, A., K.N. Papadopoulou, D. Podaras, W. Plaiti, and D. Koutsoubas. 2000. Macrobenthic community structure over the continental margin of Crete (South Aegean Sea, NE Mediterranean). Progress in Oceanography 46:401–428.

Veit-Köhler, G., D. Gerdes, E. Quiroga, D. Hebbeln, and J. Sellanes. 2009. Metazoan meiofauna within the oxygen-minimum zone off Chile: Results of the 2001-PUCK expedition. Deep Sea Research Part II: Topical Studies in Oceanography 56:1105–1111.

Warrant, E.J., and N.A. Locket. 2004. Vision in the deep sea. Biological Reviews 79:671–712.

Watling, L., J. Guinotte, M. R. Clark, and C. R. Smith. 2013. A proposed biogeography of the deep ocean floor. Progress in Oceanography 111:91–112.

Wei, C.L., G.T. Rowe, G.F. Hubbard, A.H. Scheltema, G.D.F. Wilson, I. Petrescu, J.M. Foster, M.K. Wicksten, M. Chen, R. Davenport, Y. Soliman, and Y. Wang. 2010a. Bathymetric zonation of deep-sea macrofauna in relation to export of surface phytoplankton production. Marine Ecology Progress Series 399:1–14.

Wei, C.L., G.T. Rowe, E. Escobar-Briones, A. Boetius, T. Soltwedel, M.J. Caley, Y. Soliman, F. Huettmann, F. Qu, Z. Yu, C.R. Pitcher, R.L. Haedrich, M.K. Wicksten, M.A. Rex, J.G. Baguley, J. Sharma, R. Danovaro, I.R. MacDonald, C.C. Nunnally, J.W. Deming, P. Montagna, M. Levesque, J.M. Weslawski, M. Wlodarska-Kowalczuk, B.S. Ingole, B.J. Bett, D.S.M. Billett, A. Yool, B.A. Bluhm, K. Iken, and B.E. Narayanaswamy. 2010b. Global patterns and predictions of seafloor biomass using random forests. PLoS ONE 5:e15323.

Wei, C.L., G.T. Rowe, E. Escobar-Briones, C. Nunnally, Y. Soliman, and N. Ellis. 2012. Standing stocks and body size of deep-sea macrofauna: predicting the baseline of 2010 Deepwater Horizon oil spill in the northern Gulf of Mexico. Deep Sea Research Part I: Oceanographic Research Papers 69:82–99.

Weisshappel, J.B.F., and J. Svavarsson. 1998. Benthic amphipods (Crustacea: Malacostraca) in Icelandic waters: diversity in relation to faunal patterns from shallow to intermediate deep Arctic and North Atlantic oceans. Marine Biology 131:133–143.

Wicksten, M.K., and J.M. Packard. 2005. A qualitative zoogeographic analysis of decapod crustaceans of the continental slopes and abyssal plain of the Gulf of Mexico. Deep-Sea Research I 52:1745–1765.

Wilson, G.D.F. 1998. Historical influences on deep-sea isopod diversity in the Atlantic Ocean. Deep Sea Research Part II: Topical Studies in Oceanography 45:279–301.

Wilson, G.D.F. 1999. Some of the deep-sea fauna is ancient. Crustaceana 72:1019–1030.

Wilson, G.D.F. 2008. Local and regional species diversity of benthic Isopoda (Crustacea) in the deep Gulf of Mexico. Deep-Sea Research Part II: Topical Studies in Oceanography 55:2634–2649.

Wilson, G.D.F., and S.T. Ahyong. 2015. Lifestyles of the species-rich and fabulous: the deep-sea crustaceans. Pages 279–298 *in* M. Thiel and L. Watling, editors. The Natural History of the Crustacea, Vol. 2: Lifestyles and Feeding Biology. Oxford University Press, New York.

Wishner, K.F., C.J. Ashjian, C. Gelfman, M.M. Gowing, L. Kann, L.A. Levin, L.S. Mullineaux, and J. Saltzman. 1995. Pelagic and benthic ecology of the lower interface of the Eastern Tropical Pacific oxygen minimum zone. Deep Sea Research Part I: Oceanographic Research Papers 42:93–115.

Woolley, S.N., D.P. Tittensor, P.K. Dunstan, G. Guillera-Arroita, J.J. Lahoz-Monfort, B.A. Wintle, B. Worm, and T.D. O'Hara. 2016. Deep-sea diversity patterns are shaped by energy availability. Nature 533:393–396.

Yasuhara, M., and R. Danovaro. 2016. Temperature impacts on deep-sea biodiversity. Biological Reviews 91:275–287.

Yasuhara, M., G. Hunt, T.M. Cronin, and H. Okahashi. 2009. Temporal latitudinal-gradient dynamics and tropical instability of deep-sea species diversity. Proceedings of the National Academy of Sciences 106:21717–21720.

Zeppilli, D., L. Bongiorni, A. Cattaneo, R. Danovaro, and R.S. Santos. 2013. Meiofauna assemblages of the Condor Seamount (North-East Atlantic Ocean) and adjacent deep-sea sediments. Deep Sea Research Part II: Topical Studies in Oceanography 98 Part A:87–100.

17

POPULATION CONNECTIVITY AND PHYLOGEOGRAPHY OF CRUSTACEANS

Christine Ewers-Saucedo and John P. Wares

Abstract

Crustacea display diverse life histories and occur in all marine habitats. This makes them particularly useful when thinking about how we can predict geographical distribution from life history and ecology. As would be expected from such diversity, crustaceans exhibit various population connectivity patterns, from panmictic, well-connected populations to small and isolated populations. Here we ask first what can be learned from exploring crustacean phylogeography and connectivity around well-understood vicariance events with known ages. We find that vicariance events are generally useful in calibrating molecular rates of evolution, but that there is substantial variation between taxa. This variation can be linked, on the one hand, to habitat differences (which determine when gene flow between populations actually ceased) and, on the other hand, to population size differences (which determine how fast genetic differences accumulate). In a few instances, populations must have diverged much earlier or later than the hypothesized vicariance event, providing evidence of earlier or later dispersal or more ancient separation. Second, we ask when comparative studies of multiple taxa show consistent results, predictable from their similar life history, and when not. For example, species that disperse little, such as brooding peracarids, have smaller, more isolated populations than species with planktonic larvae, such as decapods. Less consistent are the patterns across biogeographic breaks. While gene flow is clearly limited across such breaks in some species, other species do not seem to perceive them. This to-date-unexplained variation challenges our understanding of marine phylogeography.

Evolution and Biogeography. Edited by Gary C. B. Poore and Martin Thiel.

INTRODUCTION

The distribution of biological diversity, at any taxonomic level, is used to elucidate the global mechanisms that have arranged, and rearranged, this diversity. Significant environmental gradients, dispersal vectors, and relatively sudden geologic changes all leave their signatures in these distributions. The contribution of molecular data, used to contrast diversity patterns within and among closely related species, has helped to identify the timescales on which these shifts occur. Though this work involves contributions from population genetics, landscape ecology, phylogenetic analysis, and understanding inheritance modes, we have come to refer to this field as "phylogeography" following the development of this integrative field (Avise 2000).

Since Darwin's *Origin of Species* (Darwin 1859), intraspecific variation among individuals and populations has been highlighted as a key component of biological diversification. Research in phylogeography has been used to (1) illuminate incipient or cryptic species (for example, recognizing that what appears to be a single isopod species is actually several; Hurtado et al. 2010, Varela and Haye 2012); (2) identify how physical forcing influences broad biogeographic patterns, as with work by Wares et al. (2001) or Ewers-Saucedo et al. (2016), showing that ocean currents can strongly influence intraspecific larval dispersal; or (3) establish the likely timing of particular geologic and/or climatic events that have sundered the distribution of a species (Baeza and Fuentes 2013).

Generally speaking, when regional intraspecific diversity is indistinguishable, there is thought to be high population connectivity, and trait divergence across the studied regions will be constrained. Spatial homogeneity does not, however, mean that all sites are equally likely to contribute diversity to other sites or to the next generation, and oceanic currents and other physical drivers may result in strong asymmetry, and only a small portion of a species' spatial range contribute most of the diversity (Pringle and Wares 2007, Wares and Pringle 2008). As connectivity becomes more limited, by whatever mechanism, different divergence measures are used to quantify this separation (Slatkin 1987, Wares 2016). Just as important for the field of phylogeography is establishing when populations began to diverge, by what mechanism, and whether the same mechanism influences a broad range of taxa (Hickerson et al. 2006, Ilves et al. 2010, Bell et al. 2012). In the early stages of the field, phylogeographic inferences were primarily descriptive (Hickerson et al. 2010). The accelerated use of coalescent theory has provided the necessary statistical framework to test specific phylogeographic hypotheses. Coalescent theory uses a sample of current alleles to model gene genealogies under virtually any complex demographic history. This allows estimation of phylogeographic parameters such as historical population sizes, divergence times, and migration rates given the stochastic timing of coalescent events (Wakeley 2008).

Here we focus on these patterns of connectivity and divergence in Crustacea, with an emphasis on marine taxa, as terrestrial and freshwater crustaceans have very distinct forcing mechanisms that have influenced their diversity (e.g., Buhay and Crandall 2005, Lee et al. 2014, Hamasaki et al. 2017; see Chapter 11). The Crustacea are particularly useful for thinking about how we can use life history, ecology, and distribution to predict how taxa are affected by particular geographic mechanisms. This is in part because large groups of Crustacea can be contrasted by their conserved life histories (e.g., dispersal capacity), which is expected to have a large impact on their phylogeography. Crustacea with benthic or sessile adults, such as shrimps, crabs, and lobsters, often have planktonic larvae. Their phylogeography might reveal little structure over large ranges of their distribution, owing to the large dispersal potential of the larvae. Similarly, wholly planktonic species, such as some copepods, should have large, well-connected populations (e.g., Blanco-Bercial et al. 2011). In contrast, peracarids such as isopods and amphipods, where all offspring are brooded and released locally, are expected to have highly distinct regional patterns of diversity and a propensity to have local isolated populations. By comparing the phylogeography of many species with different life histories to these expectations, we gain insight into oceanic and intertidal habitats that promote or restrict dispersal, as well as the environmental factors that restrict adult distribution (Sanford et al. 2006).

An anecdotal summary of phylogeographic studies in crustaceans has the risk of recognizing a great many idiosyncratic studies that may or may not be informative of broader mechanisms driving biodiversity. Here we ask first what can be learned from exploring crustacean phylogeography and connectivity around well-studied geographic vicariance events and comparing information from these calibrating events. Second, we ask in what instances do comparative studies of multiple taxa show consistent results, predictable from their similar life history, and in what instances are there surprises. These surprises should be further evaluated to understand how distinct ecology, or inherent plasticity, shapes phylogeographic patterns.

VICARIANCE AND PHYLOGEOGRAPHY

One component of the field of phylogeography that requires integration with other sciences is the dating of specific events that lead to vicariance, i.e., the separation of one biological distribution into two or more distributions with limited gene flow or connectivity. This can happen quite quickly, as with a river capture event that transplants organisms from one river drainage into another where the species previously did not exist. Though most attention for this mechanism has been done studying fish species' distributions (Burridge et al. 2006, Waters et al. 2007), these events likely lead to among-drainage population structure in other taxa including crayfishes (McDowall 2005, Larson and Williams 2015). More commonly, attention has been paid to rising land masses (or interglacial sea levels) that separate populations. Here we review some of the most commonly studied regions for assessing phylogeographic separation and the processes for calibrating molecular divergence, along with components that add to uncertainty in dating.

The Isthmus of Panama (Land)

The most prominent example of biogeographic vicariance for marine organisms (and one often used to calibrate the separation of populations in more recent events) involves the Isthmus of Panama, the land bridge between North and South America that isolated the Caribbean from the Tropical Eastern Pacific. Though there has been recent controversy in the geologic dating of this closure, integrative data summarized by O'Dea et al. (2016) corroborated that the Isthmus separated the two ocean basins approximately 3 million years ago (mya). It is this date that has been used in the molecular era to calibrate the rate of divergence of genetic markers, assuming neutral evolution. One of the earliest and most comprehensive studies of vicariance in this region examined 15 geminate pairs of species of snapping shrimp (*Alpheus*). Knowlton and Weigt (1998) showed substantial variation in the divergence, measured as model-corrected genetic distances between individuals within each pair of transisthmian sister species in the cytochrome oxidase I (COI) mitochondrial gene region, and concluded that the divergence of these taxa could not have been simultaneous.

From these data, Knowlton and Weigt (1998) concluded that there was a habitat dependency that influenced observed divergence patterns, e.g., shallow-water and mangrove habitat species were isolated from one another later than species from deeper water (Fig. 17.1). Under this assumption, only the species pairs with the lowest sequence divergences could be used as an estimate of the rate of molecular evolution at this gene, and the other germinate species pairs were perhaps isolated from one another prior to the final closure of the Isthmus. The implications of this study for our ability to understand rates of molecular divergence, and thus use those rates to infer more recent or subtle patterns, were of obvious importance to the field. Recent technical advances, however, have provided additional tools for evaluating divergence in light of how molecular divergence between two populations are influenced by ancestral and contemporary effective population sizes (Hickerson et al. 2006, Hurt et al. 2009). Recognizing the variance in divergence patterns contributed by different population sizes and demographies, Hurt et al. (2009) established that most of the *Alpheus* divergences were likely simultaneous with the closure of the Isthmus, but two of the eight species pairs evaluated likely diverged prior to that geologic event. This is probably not an uncommon factor in marine biogeography; similar pre-isthmian divergences are noted between species of the isopod genus *Excirolana* (Hurtado et al. 2016).

Even with advanced use of analyses based on coalescent theory (Wakeley 2008), studies comparing large numbers of transisthmian species pairs have suggested that caution is warranted for calibrating molecular divergences, as some species pairs may have separated prior to the closure of the Isthmus (Knowlton and Weigt 1998, Marko 2002), and there may have been elevated sea levels approximately 2 mya that allowed some taxa to cross the isthmus and become isolated subsequently (Cronin and Dowsett 1996, Hickerson et al. 2006). Although authors have continued to use the 3 mya date for estimating divergence rates in crustaceans (O'Dea et al. 2016, Wares et al. 2009), it has

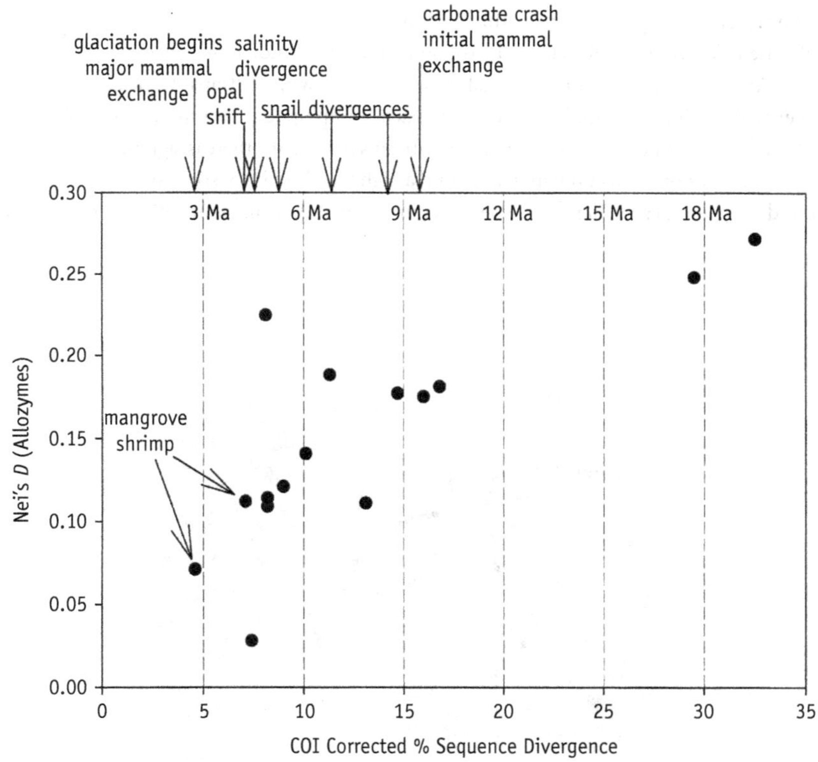

Fig. 17.1
Relation between corrected percent sequence divergence and allozyme differentiation (Nei's D) for 15 independent pairs of transisthmian sister species of the shrimp genus *Alpheus*. Maximum likelihood divergence estimates were corrected for among-site rate variation by means of a four-category gamma distribution. Just below the top axis are approximate dates of divergence based on COI data, with the assumption that the most similar pair diverged at approximately 3 Ma. Various biological (Collins 1989, Cunningham and Collins 1994) and geological (Farrell et al. 1995) events related to the rise of the Isthmus, whose timing has been independently estimated based on the geological record, are positioned above the top axis to show their relationship to the timing of shrimp divergences based on molecular rate calibrations. The two most distant pairs appear to be the survivors of lineages that diverged about the time the Tethyan connection between Atlantic and Pacific was severed. Figure reprinted from Knowlton and Weigt (1998).

become clear that it is important to test this assumption. Likely better estimates of the timing of divergence and variation among species pairs has followed statistical evaluation that multiple taxa were isolated by the same mechanism at the same time, and then this event can be used to accurately calibrate rates of divergence (Hurt et al. 2009, Hickerson et al. 2006).

The Trans-Arctic Interchange (Barrier Opening then Closing)

A biogeographic exchange and vicariance event, occurring over a similar time period as the closure of the Panamanian isthmus, involves the "trans-Arctic interchange" of the late Pliocene (Vermeij 1991a, 1991b, Ingolfsson 1992). In this scenario, more similar to the river capture events noted above, Pliocene warming of the Arctic regions allowed a largely asymmetric movement of the marine biota from the Pacific Ocean into the Atlantic Ocean (Fig. 17.2). With subsequent Pleistocene glacial cycles, the interchange is thought to have originated approximately 3.5 mya and ended around 2.8 mya (Vermeij 1991a), and this set of dates has been used to calibrate rates of molecular evolution in crustaceans (cladocerans, isopods, barnacles) and other taxa (Wares and Cunningham 2001, Durbin et al. 2008). These between-basin calibrations can then be used to date later divergence events among clades or populations within each basin.

Laakkonen (2015) and colleagues added complementary mitochondrial sequence data to complete trans-Arctic comparisons for additional crustacean taxa; with the trans-Arctic comparisons noted above, her work included multiple species of barnacles, amphipods, shrimps, and cladocerans. From these contrasts, it becomes clear that, as with the Isthmus of Panama, there is significant variation in the apparent divergence times of compared taxa, with a high likelihood that some of the taxa have crossed the Arctic during interglacial Pleistocene periods (an observation consistent with data from the urchins

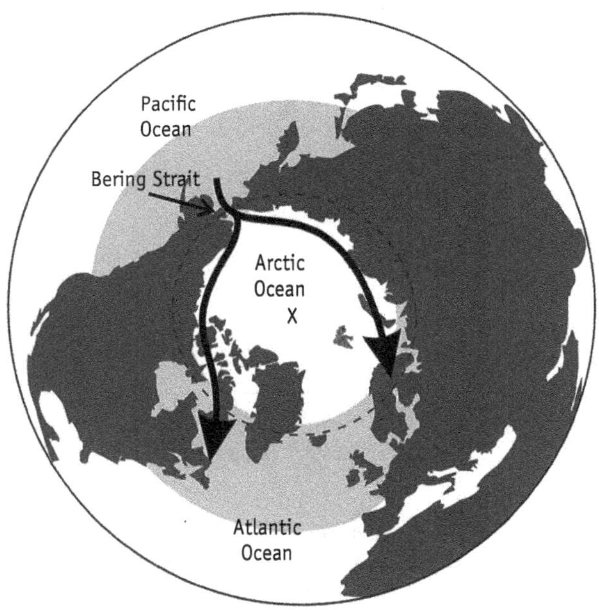

Fig. 17.2.
The amphi-arctic distribution pattern for marine taxa (grey). Arrows indicate the main direction of trans-Arctic dispersal, which has been highly asymmetric, i.e., mostly from the Pacific to the Atlantic, in the time frame from the Pliocene to the present (see Vermeij 1991). Figure reprinted from Laakkonen (2015).

Strongylocentrotus pallidus [Palumbi and Kessing 1991] and sacoglossan mollusks [Ellingson and Krug 2006)]. Indeed, as could be expected given the nature of the isolating mechanisms (a land bridge versus repeated glacial/interglacial cycles), the distribution of divergence estimates from mitochondrial data include a number of much more recent Pacific–Atlantic connections in amphiboreal marine organisms, including the barnacle *Balanus crenatus*, the shrimp *Pandalus borealis*, amphipod *Gammarus setosus*, and cladocerans *Pleopsis polyphaemoides* and *Podon leukcartii*, all of which have genetically isolated populations in the North Atlantic and North Pacific (Laakkonen 2015 and references therein).

Mediterranean—A Shallow Shelf Breached by Sea and Other Peninsular Breaches (Shallow Water)

As with the trans-Arctic interchange, it is confidently known that Atlantic marine taxa breached the land barrier at the Strait of Gibraltar about 5.3 mya, but complex shelf morphology, salinity gradients, and oceanic forcing have established a complex series of hypothesized boundaries to dispersal that affect several crustacean species (Patarnello et al. 2007). In particular in the contrast of Atlantic and Mediterranean populations, the depth and location of sampling is key to identifying pattern. Distinct mechanisms are associated with the Strait of Gibraltar itself, the "Almeria-Oran Front" (AOF) between southern-central Spain and Morocco, and a biogeographic separation between the eastern and western Mediterranean at the Siculo-Tunisian Strait (Deli et al. 2016). On either side of these barriers, water bodies circulate with different hydrological, physical, and chemical characteristics.

In large part, then, the goal of phylogeographic research in the Mediterranean is assessing these barriers for taxa that have only been in the Mediterranean for 5.3 mya or less, establishing an upper (but not lower) bound on each contrast. Patarnello et al. (2007) reviewed more than 70 studies, as well as the physical oceanography of the region, and found some shared patterns across the Strait of Gibraltar and the AOF, with evidence for limited gene flow across these proximal boundaries over the past 1.2–1.8 mya. This isolation concords with climate fluctuations that are thought to have generated boundaries to gene flow at this time. However, the Norway lobster *Nephrops norvegicus* and the European lobster *Homarus gammarus*, with similar larval and habitat biology, have contrasting structure throughout this region, with *H. gammarus* showing strong differentiation between the Atlantic and Mediterranean (Triantafyllidis et al. 2005), whereas *N. norvegicus* does not (Stamatis et al. 2004). Given a larger number of contrasts, Patarnello et al. (2007) concluded that even closely related taxa (these lobsters are not closely related) may vary in their response to climate variation, with some remaining in refugial (isolated) populations and others being extirpated and recolonized from the Atlantic, leading to very distinct population genetic patterns.

More recently, Garcia-Merchán et al. (2012) co-analyzed data from seven decapod species from several families and variation in habitat depths. They concluded that there was a strong boundary effect at the Strait of Gibraltar for two species, with a third showing strong structure in a different region of the western Mediterranean. Their work suggested distinct effects of habitat depth on population structure in these decapods, with shallow-water species showing greater structure overall. However, this work and other studies have noted significant species-by-species variation in overall phylogeographic patterning, suggesting that many more studies will be necessary to reach general conclusions about the effects of life history on phylogeographic structure. Some taxa do not show strong "breaks" in genetic distribution but instead latitudinal and/or longitudinal clines in genetic structure (as in the shrimp *Palaemon elegans*; Deli et al. 2016). The intertidal barnacle *Chthamalus montagui*, on the other hand, reinforces our understanding of Atlantic-Mediterranean isolation in the mid-Pleistocene (Pannacciulli et al. 2017), and so the generality of this major influence on crustacean phylogeography should be a primary focus for research in this region.

COMPARATIVE PHYLOGEOGRAPHY ALONG COASTS/REGIONS

Contrasting with instances of geologic or other physical separation and instances where divergence has been, in part, calibrated by those abiotic events, there is a rapidly growing literature evaluating the effects of habitat isolation because of depth, temperature, ocean currents, and other environmental factors. Often what is extraordinary in the phylogeography of crustaceans is that patterns emerge in organisms with great larval dispersal potential, and the contrast with those that have more limited dispersal potential often leads to the strongest insights.

Deep Water, Strong Currents

As noted above, the features that isolate populations are not only the physical, geologic, or glacial features, but also temporally stable oceanic forcing, including elevated water temperatures (as with the Florida peninsula reaching into tropical waters and isolating subtropical/temperate populations on each coast) and strong currents over deep water. An excellent example of such a feature is known as "Wallace's Line," named for the faunal boundary between Asian and Australian terrestrial taxa. Huxley (1868) proposed a modification of Wallace's Line, grouping much of the Philippines into the Asian zoogeographical region instead of the Australian region (see map in Fig. 17.3). This biogeographic boundary, known as "Huxley's Line," not only separates terrestrial fauna and flora, but the deep channel here between continental shelves also has major effects on marine fauna. The commercially important giant river prawn *Macrobrachium rosenbergii* has been described as having "eastern" and "western" forms concordant with these shelf boundaries, and de Bruyn et al. (2004) highlighted a reciprocally monophyletic set of populations across this line with a divergence associated with the late Miocene, c. 12 to 5 mya. The same phylogeographic pattern, including Miocene divergence estimates, is shared by the deep-sea squat lobster *Uroptychus naso* species complex (Poore and Andreakis 2011) (Fig. 17.3).

In the same region, Barber et al. (2000) explored the genetic structure of the mantis shrimp *Haptosquilla pulchella*, a reef crustacean with long planktonic larval dispersal (4–6 weeks) that should have huge dispersal potential in this region. Their work showed incredibly deep phylogeographic divergence (with an index of differentiation Fst of 0.821) between populations to the north and south of the Flores and Java Seas, along a line that is effectively perpendicular to Wallace's Line. Overall, this region of the Indo-West Pacific, often called the "coral triangle" for the tremendous diversity found on coral reefs and surrounding seas, has very complex and sometimes idiosyncratic patterns of isolation, as noted in the contrast of these two studies. This complexity carries over to freshwater and terrestrial crustaceans, which have themselves in part had their diversity driven by ancient oceanic forcing and sea level shifts (Bernardes et al. 2017).

Regions of complex land-mass structure, such as the Indo-West Pacific and the Caribbean, have been hypothesized to have greater propensity to harbor cryptic diversity (Hellberg 1998). Similar to the cases described above, there are deep channels with fast-flowing currents that also appear to drive diversity patterns in the Caribbean. Most notably, the Mona Passage (separating the islands of Hispaniola and Puerto Rico) serves as a primary connection between the Atlantic and Caribbean; though other connections are apparent, this division explains a significant portion of marine invertebrate structure in the Caribbean/Western Atlantic (Baums et al. 2006). An analysis of sponge-commensal amphipods (genus *Leucothoe*) did not, however, support this division, even though the sponge host itself did (DeBiasse et al. 2016). Instead, host sponge and amphipods shared a well-defined split across the Florida Straits. There are also contrasting results between two fiddler crab species (Laurenzano et al. 2016)—one in which there is structure associated with Mona Passage (*Minuca rapax*) and one in which there is not

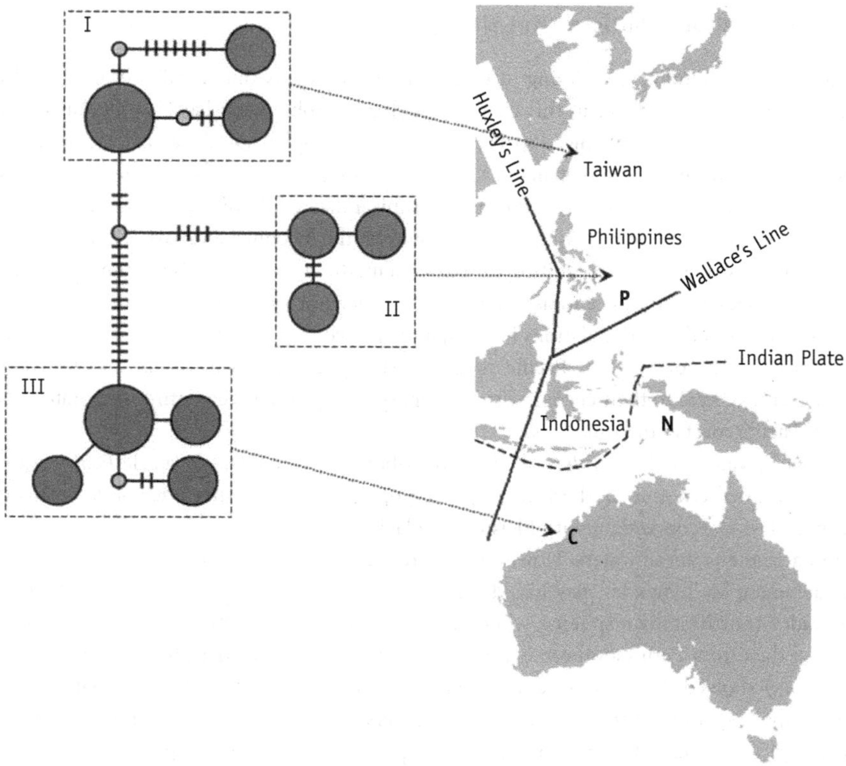

Fig. 17.3.
Median-joining network reconstruction based on 555 bp of the mitochondrial COI sequence marker. Circles represent haplotypes; circle size is proportional to the frequency of that haplotype; bars across lines connecting haplotypes indicate base changes. C, N, and P represent the type localities of the three species, *U. cyrano*, *U. naso*, and *U. pinocchio*, respectively. Modified figure reprinted in modified form from Poore and Andreakis (2011).

(*Leptuca leptodactyla*)—emphasizing that crustacean diversity does not always tell the same geographic diversity story as other taxa.

Overall the Caribbean is quite diverse, and single taxa have been split into multiple discrete lineages in recent phylogeographic work in crustaceans. Zardus and Hadfield (2005) showed that the barnacle *Chthamalus proteus* exhibits at least four distinct lineages, two that are found throughout the Caribbean and two that are endemic to localities in Panama and Brazil. Similarly, the snapping shrimp *Alpheus armillatus* contains at least three lineages in the northern Caribbean and West Atlantic Ocean, which probably denote reproductively isolated species (Matthews 2006). Only one of those lineages is present in the Gulf of Mexico, whereas outside the Gulf of Mexico, lineages do not segregate geographically. However, within each lineage there is strong evidence of population genetic differentiation between Florida, Puerto Rico, and Jamaica. North–south divisions in diversity are apparent in the porcelain crab *Petrolisthes armatus* (Hiller and Lessios 2017), and regional subdivisions are apparent in the barnacle *C. angustitergum* (Wares et al. 2009); although there is plenty of work in other marine invertebrates (Baums et al. 2006, Díaz-Ferguson et al. 2011) and fishes (Taylor and Hellberg 2003, Cowen et al. 2006), more exploration of this region for the known canonical breaks should be evaluated at higher resolution (Díaz-Ferguson et al. 2010a, 2011).

Amphi-Tropical Distributions (Hot Water)

Thanks to understanding the timing of events (following approximate calibration of molecular evolution as described earlier in this chapter) phylogeographic work has also illuminated major historical climate shifts that allowed species to disperse through tropical waters to achieve "amphi-tropical" distributions. Dawson et al. (2011) used a thorough analysis of spatial genetic diversity in the mole crab *Emerita analoga* to tie the amphi-tropical divergence of these abundant shore crabs to Pleistocene glacial periods, taking into consideration the recognized variation in estimates from transisthmian species pairs. In addition to statistical methods to evaluate the timing of divergence of major lineages, other typical tools in modern phylogeography include the ability to quantify or assess the likelihood that populations have expanded or contracted demographically. The data for *E. analoga* in the northeastern Pacific suggest a strong expansion in recent epochs (between ~250 kya and 50 kya) to their current distribution, spanning almost the entire temperate and subtropical Pacific coast of the Americas.

Similarly, Van Syoc (1994) described the distribution of the goose barnacle *Pollicipes elegans*, noting that northern and southern hemisphere populations have a net nucleotide sequence divergence of ~1.2%, concordant with Pleistocene glacial periods that may have allowed *Pollicipes* to expand across equatorial waters. However, a more recent analysis of this case suggested that distinct demographic dynamics may have been important. By including samples of *P. elegans* from El Salvador (northern hemisphere), where the genetic diversity is actually much more similar to sequence data from Peru (southern hemisphere) than Mexico (northern hemisphere), Marchant et al. (2015) showed that it is more likely that this barnacle has tropical origins with significant habitat gaps (e.g., the Central American Gap; Meyers et al. 2013) and reduced population densities in tropical regions. Their analysis, based on approximate Bayesian computation methods (e.g., Beaumont et al. 2002), helped discriminate between related demographic hypotheses to show that what may appear to be a simple vicariance event often requires careful evaluation of whether there are any intermediate populations.

Environmental Gradients

In a growing number of instances, crustacean phylogeographic analysis has identified significant patterns of cryptic diversity along continental margins with no immediately obvious boundaries to dispersal or gene flow. It is in these cases that variation in life history traits often becomes an important cue for establishing the origins and maintenance of such patterns.

Ecological gradients can play a role not only in restricting the range of species but also in limiting gene flow between differentially adapted ecotypes of the same species. The Baltic Sea has a strong west-to-east salinity gradient. It is also a very young marine habitat, colonized predominantly from the North Sea less than 8,000 years ago, with only few colonizers originating from other ocean basins. Limited migration between North and Baltic seas populations in concert with brackish-water adaptations provide the opportunity for population differentiation (Johanneson and Andre 2006). In addition, recent colonization and low migration rates or strong selection on all but a few genotypes should lead to lower genetic diversity in the Baltic Sea. Fulfilling this expectation, the ostracod *Cyprideis torosa* and the amphipod *Gammarus zaddachi* show lower genetic diversity in the Baltic Sea than in the North Sea (Bulnheim and Scholl 1981, Sywula et al. 1995; re-analyzed in Johanneson and Andre 2006). In contrast, the amphipod *Gammarus salinus*, the opossum shrimp *Mysis mixta*, and brown shrimp *Crangon crangon* show no signs of lower diversity in the Baltic Sea than in the North Sea, despite low dispersal potential of the first two species. Interestingly, populations of all of the aforementioned species except *C. crangon* are differentiated between the

North and Baltic seas (Johanneson and Andre 2006). While the strong salinity gradient could clearly be responsible for this differentiation, so could limited gene flow between the North and Baltic seas. The prevalence of genetic differentiation in the absence of lower genetic diversity hints at adaptive processes, but further functional studies are needed to assess the underlying causes.

The complex geography, environmental gradients, and glacial history of the northeastern Atlantic is itself something that is being reconstructed in part through phylogeographic analysis. Periods of glaciation, especially the Last Glacial Maximum (LGM, 21 kya), have left strong signatures on the current phylogeography of temperate, subpolar, and polar crustaceans. During periods of glaciation, habitat was limited to warmer regions beyond the ice sheets and small periglacial pockets. Distinct glacial refugia were isolated from each other, accumulating unique genetic diversity in the inhabited populations. With receding ice sheets, these populations were able to extend their range. This contraction–expansion cycle often left a distinct population genetic signature: the populations in the historical glacial refugia have high genetic diversity, whereas recently recolonized regions have lower diversity. In the green crab *Carcinus maenas*, high genetic diversity is observed in populations in three areas, separated by populations with lower diversity (Fig. 17.4). These areas likely correspond to three glacial refugia because the haplotype network shows three geographically localized clades: (1) a boreal clade distributed mostly in the eastern North Sea, (2) a Faroes/Iceland clade, and (3) a small southern clade distributed widely in southern Europe. This pattern is largely echoed in work on the brooding isopod *Idotea balthica* (Wares 2001a, Bell 2009); more recent work on *C. maenas* also showed significant components of adaptive spatial structure within the native European populations (Tepolt and Palumbi 2015).

Diversity can be equally high across refugial and recolonized regions if dispersal is high and gene flow occurs unhindered across the current range of a species. The current range of the brown shrimp *Crangon crangon* extends from the White Sea and Iceland down to Morocco, into the Baltic, the Mediterranean, and the Black Sea (Holthuis 1980). While the Mediterranean and Black Sea populations are highly structured, the Atlantic population, ranging from Portugal to Iceland and the Baltic Sea, is genetically homogeneous (Luttikhuizen et al. 2008). The authors compared two possible explanations for this genetic homogeneity: (1) gene flow is sufficient to prevent population genetic structure from building up, and (2) current gene flow is restricted but populations have not yet diverged because colonization has occurred too recently. They concluded that the second scenario is more likely because gene flow is restricted between populations in the Mediterranean Sea, suggestive of low dispersal potential for *C. crangon* at large (Luttikhuizen et al. 2008). If this were the case, diversity should be lower in the recently recolonized habitats, which it is not. Thus, we hypothesize that in addition to recent recolonization of the northern range of *C. crangon* following the LGM, gene flow is (or was during recolonization) high enough to spread all haplotypes.

The phylogeography of *C. crangon* highlights that high levels of gene flow homogenize populations efficiently, making it impossible to pinpoint the location of glacial refugia based on genetic data alone. Dambach et al. (2012) approached this problem by coupling climatic reconstructions with current species distribution models to predict the location and extent of glacial refugia for three circum-Antarctic benthic shrimp species. While the shallow-water *Chorismus antarcticus* and *Notocrangon antarcticus* were limited to a drastically reduced habitat during the LGM, the deep-water *Nematocarcinus lanceopes* found refuge in the Southern Ocean deep sea. The modeling results are in accordance with genetic diversity patterns available for C. *antarcticus* and *N. lanceopes*. While high levels of gene flow homogenize populations in both species, the deep-water species *N. lanceopes* has an overall higher genetic diversity, indicative of a less severe bottleneck during the LGM (Raupach et al. 2010). This supports the hypothesis that habitat contraction at the LGM resulted in a loss of genetic diversity especially in shallow-water benthos.

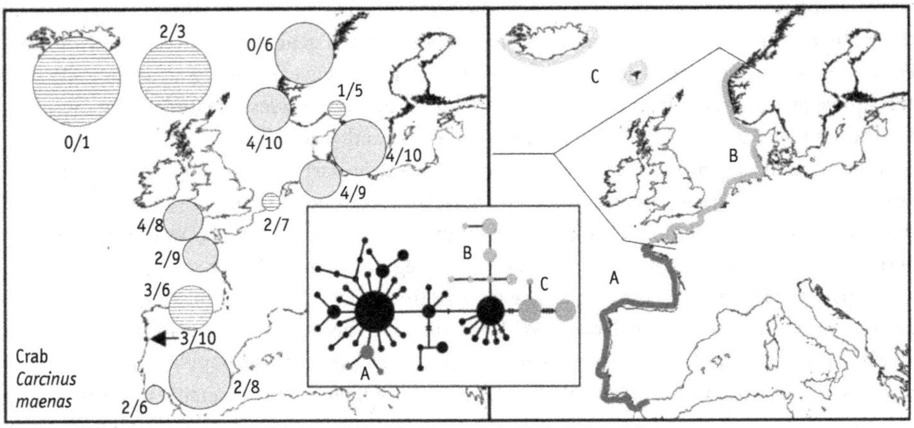

Fig. 17.4
Diversity plot and haplotype distributions for the intertidal green crab *Carcinus maenas* based on analyses of mitochondrial cytochrome oxidase I (COI) sequences in Roman and Palumbi (2004). The map on the left shows intrapopulation haplotype diversity after rarefaction resampling to account for differences in population sample sizes. Only populations with at least nine sampled individuals were analyzed. In the bubble plots, circle sizes are proportional to deviation from the mean for all populations of that species; grey fill indicates diversity above the mean and hatched fill shows diversity below the mean so that the least diverse sample is indicated by the largest hatched circle and the most diverse sample by the largest grey circle. The proportion of private haplotypes for that population (number of private haplotypes/total number of haplotypes) is shown beside each bubble or bracketed set of bubbles. Maps on right show distributions of key haplotypes or clades, coded in different grey tones to match the haplotype network (black fill is the default, and the distribution of black-coded haplotypes is not shown). Modified figure reprinted from Maggs et al. (2008).

Rafting

Especially for some crustaceans without planktonic larvae, rafting is an important potential dispersal vector. Rafting dispersal has particular characteristics, which may differ substantially from planktonic larval dispersal (Thiel and Haye 2006). Dispersal frequencies and distances can vary considerably: journeys may be frequent and short, intermittent, and of intermediate distance or episodic and extremely long. Population connectivity increases with rafting frequency. Accumulation of rafts in convergence zones increases the number of individuals traveling together, reducing the risk of small and genetically impoverished founder populations. This is particularly important for episodic long-distance rafting events.

Rafting on southern bull kelp *Durvillaea* spp. led to a circum-Antarctic distribution of two brooders, the isopod *Limnoria stephenseni* and the amphipod *Parawaldeckia kidderi* (Nikula et al. 2010). The populations around Antarctica are, however, genetically differentiated, highlighting the sporadic nature of long-distance rafting success, and the resulting low levels of population connectivity (Fig. 17.5). Nikula et al. (2010) compare the two rafters with its host, whose phylogeographic structure had been assessed previously (Fraser et al. 2009). While southern bull kelp is less diverse than the two peracarids, their phylogeographic patterns are comparable, giving credence to the conclusion that rafting on kelp led to the circum-Antarctic distribution of the two brooding crustaceans.

Haye et al. (2012) tested population genetic predictions of rafting in two Chilean isopod species, *Limnoria quadripunctata* and *L. chilensis*. Frequent and intermittent rafting should lead to low levels of population differentiation and a lack of isolation by distance (IBD). IBD was indeed absent

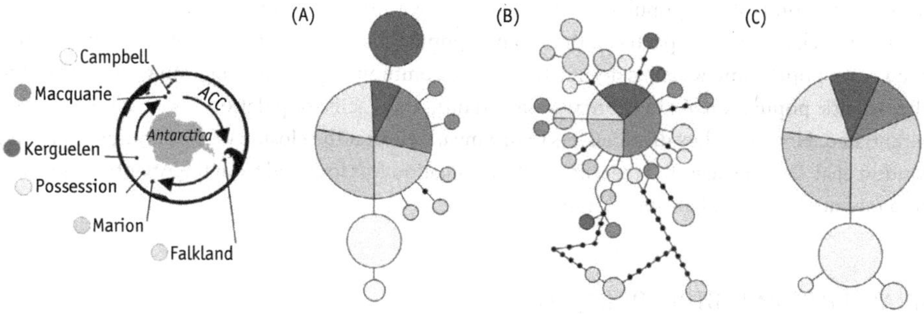

Fig. 17.5
Haplotype networks for two brooding rafters and their host kelp with circum-Antarctic distribution. The haplotype networks are unrooted statistical parsimony networks of mitochondrial cytochrome oxidase subunit I (COI) haplotypes sampled in subantarctic island populations of southern bull kelp crustacean epifauna: (a) isopod *L. stephenseni*; (b) amphipod *P. kidderi*. (c) A network for their hosts, the brown macroalga, *D. antarctica*, was redrawn after Fraser et al. (2009) to encompass the same island populations. Each circle/pie represents a distinct mitochondrial haplotype; a line connecting two haplotypes corresponds to one inferred mutational step separating them. Small black circles indicate hypothetical, unsampled haplotypes. Area and grey tones indicate sampling frequency and locations of each haplotype, respectively; area of the smallest grey circle corresponds to one copy of a haplotype. Island grey tones used in the halotype networks. ACC, Antarctic Circumpolar Current. Modified figure reprinted from Nikula et al. (2010).

from both species. A common haplotype was found in almost all populations, suggesting relatively frequent or recent rafting events. While populations separated by thousands of kilometers had low levels of population differentiation, populations of *L. quadripunctata* from fjord and open ocean locations were significantly differentiated, suggesting a potential for current flow to influence dispersal patterns.

A few species of the isopod genus *Idotea* raft occasionally on seagrasses and macroalgae, while the pelagic *I. metallica* is an obligate rafter (Ingólfsson 1992). Originating from the North American Atlantic coast, it is thought to be introduced to North West European waters by the Gulf Stream (Naylor 1957). The Gulf Stream represents a major dispersal vector, moving algae and debris from Eastern North America to Europe and from Africa to the Caribbean. Only as the North Sea has warmed in recent decades have these populations become reproductive (Franke et al. 1998, Gutow and Franke 2003, Thiel and Gutow 2005). *Idotea balthica,* on the other hand, is a nearshore benthic species, that occasionally rafts on macroalgae at high densities. It has high population differentiation along the North American East coast (Wares et al. 2007, Bell 2009), typical for a brooding species, but no differentiation along the European coast (Wares 2001b). Interestingly, one haplotype common in Europe is also found on the North American coast, indicative of a long-distance rafting event from Europe to North America (Wares 2001b).

Finally, a massive study of the coastal isopod *Excirolana braziliensis* from Atlantic and Pacific, temperate and tropical coasts of the Americas brings us full circle with respect to how long-distance dispersal and major land mass boundaries interact to generate cryptic diversity in crustaceans (Hurtado et al. 2016); as noted in the above section on the Isthmus of Panama, there is evidence for lineages divergent between the Atlantic and Pacific populations that greatly predate the closure of the Isthmus of Panama. These isopods, brooding offspring as all peracarids, also exhibit very high genetic divergences of populations even within limited spatial regions, notably with a population in the northern Gulf of California being ~9% divergent at the COI locus from two other locations that were only 180–210 km distant. The pattern differs somewhat in the cosmopolitan coral shrimp *Stenopus hispidus,* which has an extended planktonic larval phase. As expected for a crustacean with a long planktonic phase, population differentiation within the Atlantic or Pacific Ocean was low (Dudoit et al. 2018). Unexpectedly, the Atlantic population is much more genetically diverse than the Pacific population, which is dominated by one common haplotype, suggesting an older age for the Atlantic population. As in *E. braziliensis*, Atlantic and Pacific populations were significantly differentiated. However, they diverged less than 2 mya, well after the closure of the Isthmus of Panama. Dudoit et al. (2018) suggested that the Pacific population was founded from the Atlantic population by a rare dispersal event around South Africa.

GRADIENTS AND BIOGEOGRAPHY

Larval dispersal is a main source of population connectivity for marine crustaceans that have benthic adults. Dispersal is expected to be high in species with planktonic larvae, leading to low levels of population differentiation. The longer the planktonic phase, the less population structure is expected. Conversely, species without planktonic larvae, e.g., brooding peracarids, should have low dispersal potential, high population differentiation, and low genetic diversity due to drift in small isolated populations. Several studies have assessed these predictions. Kelly and Palumbi (2010) and Haye et al. (2014) tested the effect of larval dispersal ability on population connectivity across biogeographic breaks. Such biogeographic breaks are well suited to test phylogeographic predictions, as we might expect coinciding phylogeographic and biogeographic boundaries. Moreover, phylogeographic patterns can inform us about biogeographic processes. Haye et al. (2014, page 2) explained this concisely:

> Phylogeographic breaks of species with low dispersal potential reflect primarily historical processes whereas those of species with high dispersal potential reflect more recent or current gene flow restrictions. The rationale behind these patterns is that phylogeographic structure can persist many generations for species with restricted gene flow, whereas high gene flow species can bear a phylogeographic break only if a local barrier is currently restricting dispersal, whatever the history of the barrier.

They compared the phylogeography of eight marine invertebrates along the Chilean coast, three of which were crustaceans. The Pacific mole crab *Emerita analoga* and the porcellanid crab *Petrolisthes violaceus* have planktonic larvae that remain in the plankton for more than 90 and 25 days, respectively. The amphipod *Orchestoidea tuberculata*, on the other hand, is a typical peracarid brooder without planktonic larvae. The Chilean coast is characterized by a major biogeographic break at about 30–33° S. As expected, the crabs had high genetic diversity and low population differentiation, with no significant breaks in their distribution, whereas the amphipod showed the opposite pattern. Haye et al. (2014) concluded that the 30–33° S biogeographic break at the Chilean coast is a historic discontinuity, rather than a current one. More recent studies called this conclusion into question. *Notochthamalus scabrosus* for example, a widely dispersing acorn barnacle, recaptures the 30–33° S break (Zakas et al. 2009, Laughlin et al. 2012). The acorn barnacle *Austromegabalanus psittacus* also experiences moderate genetic structure associated with this break (Pappalardo et al. 2016), while the barnacle *Jehlius cirratus* does not (Zakas et al. 2009, Guo and Wares 2017). These findings suggest a complex and species-specific response to this biogeographic break. For *N. scabrosus*, however, this "break" appears to coincide with the tail of the distribution of the southern lineage of this species, with the overall pattern likely maintained by ongoing selection interacting with ocean currents and upwelling (Ewers-Saucedo et al. 2016). This study in particular emphasizes the importance of range-wide sampling to fully understand diversity patterns within species, coupled with sampling larger numbers of molecular markers to better integrate with environmental data.

The variation among crustacean taxa in response to coastal currents and upwelling noted above may be better studied with much larger numbers of species to better evaluate how life history and ecology determine the response of an organism to environmental gradients. Kelly and Palumbi (2010) included 50 coastal invertebrate species from the Northeastern Pacific rocky intertidal in their study, 21 of which were crustaceans. The majority of crustaceans had planktonic larvae. Only three isopods from the genus *Idotea* brooded their offspring, all of which had strong genetic structure across the investigated range. In contrast to the findings by Haye et al. (2014), several species with long-lived planktonic larvae also showed significant genetic differentiation: the barnacles *Balanus glandula* (also see Sotka et al. 2004) and *Pollicipes polymerus*, the crabs *Hemigrapsus nudus* and all four investigated species of the hermit crab genus *Pagurus*, as well as the copepod *Tigriopus californicus*. These species are genetically differentiated on each side of Cape Mendocino. This barrier is not impenetrable, however, as many other species with planktonic larvae, including the other crustaceans, have no discernible structure. Kelly and Palumbi (2010) further investigated possible causes for these different responses to the phylogeographic break and found a marked increase in genetic structure among species with planktonic larvae inhabiting the high- to mid-intertidal zone. However, one of the highest intertidal crustaceans, the acorn barnacle *Chthamalus dalli*, does not display phylogeographic structure across this range (Wares and Castaneda 2005). Thus, this field of research still holds many unanswered questions: as noted above, the confamilial and largely codistributed barnacles *N. scabrosus* (Ewers-Saucedo et al. 2016) and *J. cirratus* (Guo and Wares 2017) also respond very differently to the broad environmental gradient of the Chilean coast. Altogether there are many intriguing parallels in the diversity patterns and ocean dynamics of the Chilean and North American Pacific coasts, making these regions unusual replicate systems for exploring phylogeography and biogeography (Navarrete et al. 2008).

Coastal upwelling can have significant effects on gene flow. It can disrupt gene flow of coastal crustaceans by transporting dispersing stages offshore, making it impossible for them to return to suitable habitat. It can also lead to an increase of gene flow when the offshore currents aid the transport of dispersing larvae across local retention zones. This appears to be the case in the acorn barnacle *Balanus glandula*, a common well-studied intertidal barnacle along the West coast of North America. Barshis et al. (2011) compared population connectivity of populations along the Oregon coast between 2001 and 2008 and found population structure to be larger in years of weak upwelling than in years of strong upwelling (Fig. 17.6). Their data suggest that low upwelling intensity maintains more local larvae close to shore, thereby shaping the genetic structure and settlement rate of recruitment year classes. This result is in concordance with the observation of Roughgarden (1988) that larval recruitment is higher in years of weak upwelling, again suggesting that more larvae are maintained close to shore. Similarly, Petersen (2007) showed a significant shift in mitochondrial haplotype frequencies in the estuarine crab *Hemigrapsus oregonensis* associated with changing regional regimes of upwelling along this coast.

One region where phylogeographic analyses have begun to illuminate novel transition patterns is on the east coast of North America. While the major biogeographic transitions on this coast are typically associated with Cape Hatteras and Cape Cod (Pappalardo et al. 2015), population genetic analyses of many species, including many crustaceans, are beginning to show that the coastal region between Chesapeake Bay and Long Island Sound harbors a significant number of transitional patterns (Wares 2002, Díaz-Ferguson et al. 2010b, Altman et al. 2013). Key examples include the amphipod *Amphithoe longimana*, where phylogeographic analysis splits populations into two very distinct clades, with individuals from the Delaware coast being found in both clades (Sotka et al. 2003). It is notable that the goal of identifying cryptic genetic diversity within taxa is in part because these discrete populations are likely to be ecologically or functionally distinct as well; in the case of *A. longimana* there are striking differences in chemical tolerances to compounds generated by algal habitats between northern and southern populations of this amphipod. Govindarajan et al. (2015)

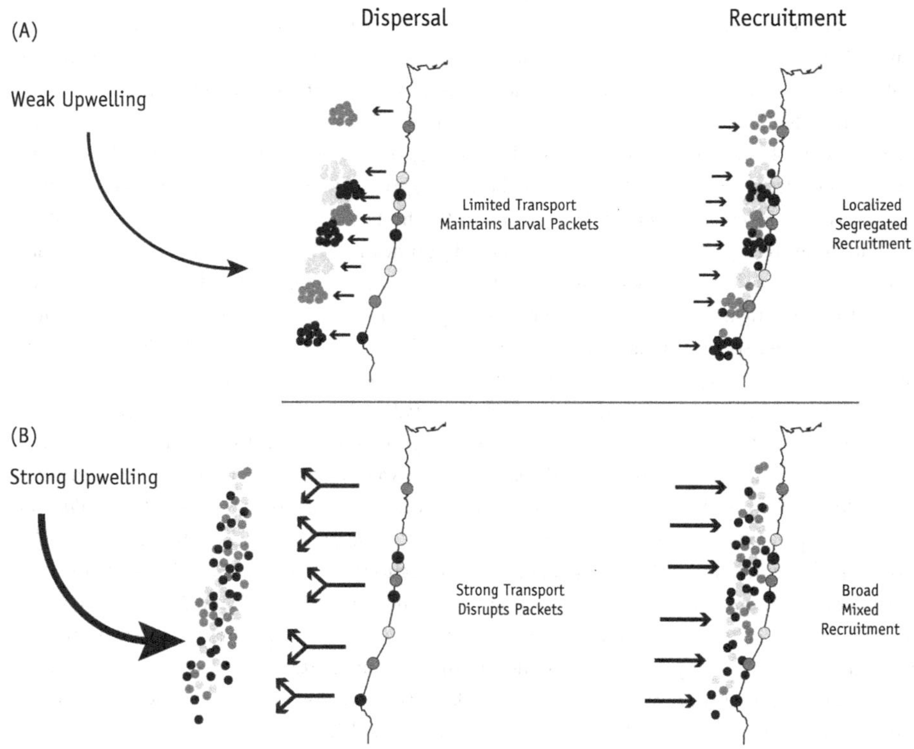

Fig. 17.6
Graphical representation of the hypothetical relationship between upwelling strength, transport, and level of mixing of larval dispersal "packets." (A) Conditions under weaker upwelling illustrating reduced transport and retention of segregation among larval "packets" throughout the dispersal and recruitment phases. (B) Conditions under strong upwelling with greater transport levels, promoting disruption and mixing of larval "packets," which results in a more homogenized pool of recruits in years with stronger upwelling conditions. Large circles in different shades of grey and black represent population locations, with smaller circles of the same shade representing hypothetical larval offspring. Modified figure reprinted from Barshis et al. (2011).

also highlighted an intriguing clinal pattern of mitochondrial diversity across this region in the barnacle *Chthamalus fragilis*. As with the strong clinal diversity in *B. glandula* noted above (and in Sotka et al. 2004, Wares and Cunningham 2005), the limited distribution of particular lineages within this species is likely indicative of physiological or ecological limits to the distribution of what should otherwise be a broadly dispersing species.

PHYLOGEOGRAPHY AND TAXONOMY

An important contribution of phylogeography in identifying discrete populations within a taxonomic entity is to help taxonomists return to the organisms to identify other traits that may guide the separation and naming of formerly cryptic species. Population genetics and phylogeography have certainly aided crustacean taxonomy in taxa with few external traits that are used for identification; genetic work on tropical eastern Pacific chthamalid barnacles identified very distinct populations (Hedgecock 1979, Wares 2001b, Wares et al. 2009, Meyers et al. 2013) that were subsequently assessed for formal phenotypic descriptions (Pitombo and Burton 2007, Chan et al. 2016). Similarly, the validity of two sympatric hermit crab species from the Western Atlantic differing only in walking leg color was initially reassured with genetic barcoding, then followed up by detailed morphological comparisons (Negri et al. 2012). In this sense, at times phylogeography erases itself: once the species are described, the pattern becomes one of biogeographic relevance instead. At other times, the subsequent taxonomic effort becomes one of formally recognizing that there is greater ecological or trait diversity within an ecosystem than had been previously recognized (Chan and Cheang 2015, Wu et al. 2015, Shahdadi et al. 2018).

Other examples of phylogeographic analysis leading to taxonomic revision include the exploration of otherwise difficult habitats; the importance of recognizing isolated and perhaps functionally distinct populations requires fully integrative analysis of genetic and morphometric data (e.g., Held 2003, Brandt et al. 2014). Watanabe et al. (2018) obtained stalked neolepadid barnacles from deep-sea vents in the Southern, Pacific, and Indian oceans, and the genetic divergences of populations of *Neolepas* identified were sufficient to be reciprocally monophyletic with high statistical support for the separation of sites separated by thousands of kilometers; in this case the barnacles from the Indian Ocean have been named *Neolepas marisindica* (Watanabe et al. 2018). Integrative phylogeographic studies have also revealed cryptic diversity in the squat lobster *Agononida incerta,* which was reported from across the Southwestern Pacific by Poore and Andreakis (2012, 2014). Careful morphological examination in combination with molecular data analysis revealed that this species complex contains at least 10 species, each with a distinct geographic distribution. The geographic sampling locality of the holotype for *Agononida incerta* helped clarify which species should keep this name and which species needed to receive new names (Poore and Andreakis 2014). These results provide key evidence about dispersal pathways and barriers to gene flow in the deep sea.

Finally, though we have focused primarily on marine ecoregions, new diversity has also been discovered and described using phylogeographic approaches in subterranean freshwater amphipods. Seven new species of *Niphargus* have been described using multiple loci and coalescent delimitation approaches with amphipods that are morphologically distinguishable only into sub-groups (Delić et al. 2017); to identify the species themselves, the authors argued that DNA-based approaches are necessary. This is a recent development in systematics that is aided by phylogeographic data and highlights a key difficulty in studying crustacean groups that are morphologically static: often an individual can only be identified to species if the source location is known or if there are genetic data as well.

FUTURE OPPORTUNITIES AND CHALLENGES

The field of phylogeography has successfully identified causes and consequences of population connectivity using single-locus mitochondrial markers. These genetic markers, especially fragments of mitochondrial genes, are easy to amplify in almost all crustaceans and have relatively high mutation rates and genetic diversity. This provides the power to discern phylogeographic scenarios. However, single mitochondrial markers might not tell the complete, or even the true, phylogeographic story. Nuclear markers should complement phylogeographic studies. They would allow the quantification of hybridization among closely related species and the investigation of adaptation at the DNA level (Bowen et al. 2014). To date, most nuclear markers used in phylogeographic studies have been microsatellites. However, these markers have not gained the same popularity as mitochondrial sequence markers because of their tedious development and implementation. Novel genomic approaches, such as RAD-Seq (restriction site–associated DNA marker sequencing) are much less time-consuming in their application. By sequencing a small, random subset of the genome, we can reconstruct a species' phylogeography with more confidence and in greater detail (Tepolt and Palumbi 2015).

A molluscan example highlights the value of nuclear markers for phylogeographic studies. The Baltic Sea, a young brackish-water basin connected to the North Sea via Skagerrak and Kattegat, is inhabited by two blue mussel species of the genus *Mytilus* (Riginos and Cunningham 2005). *Mytilus edulis* is an Atlantic and North Sea species, which can currently migrate into the Baltic Sea, while *M. trossulus* is a Pacific species, which reached the Baltic Sea from the White Sea during the Pleistocene or Holocene. Mitochondrial markers indicate that individuals in the Baltic Sea belong almost exclusively to *M. edulis*. Nuclear allozymes, however, reveal that the Eastern Baltic Sea is dominated by *M. trossulus* alleles, while North Sea and Skagerrak are dominated by *M. edulis* alleles. This indicates asymmetric introgression of different parts of the genome: the mitochondrial genome of *M. edulis* has taken over in the Baltic Sea, while allozymes experience a barrier to gene flow between the North and Baltic seas. The most likely explanation for this pattern is adaptation to low salinity at those enzymes that show significant genetic clines between the North and Baltic seas.

Another promising avenue is the integration of methods to infer phylogeography. Ewers-Saucedo et al. (2016), for example, combined population genomics with oceanographic modeling and larval ecology to show that selection plays a role in maintaining the geographic distribution of two genetically distinct clades. Larval sampling, recruitment studies, or transplant experiments could strengthen this conclusion further. The historical distribution of species, e.g., the location of glacial refugia, is often inferred from centers of genetic diversities (Maggs et al. 2008). Additionally, historical species distribution models can confirm that these geographic centers of genetic diversity were indeed suitable habitats during glaciation (Dambach et al. 2012). Lastly, Hameed et al. (2016) used a Bayesian framework to model population connectivity from fecundity and recruitment data of the porcelain crab *Petrolisthes cinctipes*. Despite its long larval dispersal phase, the Bayesian analysis suggests high levels of larval retention and significant small-scale population structure. Population genetic or genomic analyses could corroborate this finding.

Based on growing evidence, we propose a larger focus on anthropogenic effects in shaping recent phylogeography. Climate change, habitat modifications, and species invasions can alter the distribution and genetic diversity of single species and whole communities. The volcano barnacle *Tetraclita rubescens* has expanded its range northward by several hundred kilometers since the 1970s. Dawson et al. (2010) investigated which mechanisms regulate this new northern range limit. Three mechanisms—genetic impoverishment, maladaptation, or a physical barrier to dispersal—can limit a species' range in theory. Genetic impoverishment would be apparent if populations at the range edge had lower diversity than populations at the center of the distribution; under maladaptation, the opposite pattern would be expected, as gene flow would be too high to allow for locally adapted alleles; and a physical barrier to dispersal should lead to a sudden decline in barnacle abundance. To identify which of these hypotheses is most likely to limit *T. rubescens'* distribution, the authors compared geographic variation in abundance with abiotic and biotic patterns and described genetic variation in a fragment of the mitochondrial cytochrome c oxidase subunit I gene, the mitochondrial noncoding region, and 9 microsatellite loci from 27 locations across the species range. They found very high gene flow, high genetic diversity, and a physical environmental gradient coincident with the northern range limit. These results are most concordant with the maladaptation hypothesis.

Phylogeographic analyses can also be used to understand invasion histories and source populations. The caprellid *Caprella scaura*, native to the western Indian Ocean, has successfully invaded the Mediterranean and Atlantic coasts since 2005. Mitochondrial sequence data collected by Cabezas et al. (2014) suggest that Iberian populations may derive from at least two sources. The high genetic divergence and population subdivision found between non-native Iberian populations together with a high level of genetic diversity in some populations indicate multiple geographical sources and introduction points for this caprellid amphipod. In contrast, the invasion history of the cosmopolitan barnacle *Amphibalanus improvisus* could not be clearly reconstructed, despite the use of both mitochondrial sequence markers and nuclear mitosatellites (Wrange et al. 2016). These authors concluded that frequent long-distance dispersal of adult barnacles through shipping and other anthropogenic activities has eroded population structure between distant locations (Wrange et al. 2016). On a small scale (within the Baltic Sea), however, larval dispersal might play an important role in shaping its phylogeography.

CONCLUSIONS

Crustaceans are one of the most abundant taxa in the marine environment. They display a diversity of life histories and occur in all habitats. Therefore, it is not surprising that they also exhibit a diverse array of population connectivity patterns, from panmictic, well-connected populations to small and isolated populations. This allows inference of drivers of their phylogeographic patterns as well as discerning historical from ecological causes. In this regard, multi-species phylogeographic studies and studies including oceanographic modeling have provided the biggest insights to date. We propose that several topics will contribute to refocus phylogeography: (1) population-genomic studies, which allow the identification of geographically driven selection, (2) studies integrating multiple approaches, especially population genetics and coalescent theory with oceanographic and species distribution modeling, and (3) phylogeographic studies focusing on the anthropogenic effects for phylogeography, such as range expansions and species invasions.

ACKNOWLEDGMENTS

We are indebted to the editors of this volume, Martin Thiel and Gary Poore, for giving us the opportunity to write this chapter, for their patience and help revising. Thanks to Risto Väinölä for permission to use information from Hanna Laakkonen's dissertation. We would also like to thank Dan Barshis, Gary Poore, Christine Maggs, and Raisa Nikula for letting us use their instructive figures.

REFERENCES

Altman, S., J.D. Robinson, J.M. Pringle, J.E. Byers, and J.P. Wares. 2013. Edges and overlaps in Northwest Atlantic phylogeography. Diversity 5:263–275.

Avise, J.C. 2000. Phylogeography: The History and Formation of Species. Harvard University Press, Cambridge MA.

Baeza, J.A., and M.S. Fuentes. 2013. Phylogeography of the shrimp *Palaemon floridanus* (Crustacea: Caridea: Palaemonidae): a partial test of meta-population genetic structure in the wider Caribbean. Marine Ecology 34:381–393.

Barber, P.H., S.R. Palumbi, M.V. Erdmann, and M.K. Moosa. 2000. A marine Wallace's line? Nature 406:692–693.

Barshis, D.J., E.E. Sotka, R.P. Kelly, A. Sivasundar, B.A. Menge, J.A. Barth, and S.R. Palumbi. 2011. Coastal upwelling is linked to temporal genetic variability in the acorn barnacle *Balanus glandula*. Marine Ecology Progress Series 439:139–150.

Baums, I.B., C.B. Paris, and L.M. Chérubin. 2006. A bio-oceanographic filter to larval dispersal in a reef-building coral. Limnology and Oceanography 51:1969–1981.

Beaumont, M.A., W. Zhang, and D.J. Balding. 2002. Approximate Bayesian computation in population genetics. Genetics 162:2025–2035.

Bell, R.C., J.B. MacKenzie, M.J. Hickerson, K.L. Chavarria, M. Cunningham, S. Williams, and C. Moritz. 2012. Comparative multi-locus phylogeography confirms multiple vicariance events in co-distributed rainforest frogs. Proceedings of the Royal Society B: Biological Sciences 279:991–999.

Bell, T.M. 2009. The Maintenance of Genetic Diversity in the North Atlantic Isopod, *Idotea balthica*. University of Georgia, Athens.

Bernardes, S.C., A.R. Pepato, T. Rintelen, K. Rintelen, T.J. Page, H. Freitag, and M. Bruyn. 2017. The complex evolutionary history and phylogeography of *Caridina typus* (Crustacea: Decapoda): long-distance dispersal and cryptic allopatric species. Scientific Reports 7:9044.

Blanco-Bercial, L., F. Álvarez-Marqués, and A. Bucklin. 2011. Comparative phylogeography and connectivity of sibling species of the marine copepod *Clausocalanus* (Calanoida). Journal of Experimental Marine Biology and Ecology 404:108–115.

Bowen, B.W., K. Shanker, N. Yasuda, M. Celia, M.C.M.D. Malay, S. von der Heyden, G. Paulay, L.A. Rocha, K.A. Selkoe, and P.H. Barber. 2014. Phylogeography unplugged: comparative surveys in the genomic era. Bulletin of Marine Science 90:13–46.

Brandt, A., S. Brix, C. Held, and K.C. Terue. 2014. Molecular differentiation in sympatry despite morphological stasis: deep-sea *Atlantoserolis* Wägele, 1994 and *Glabroserolis* Menzies, 1962 from the South-West Atlantic (Crustacea: Isopoda: Serolidae). Zoological Journal of the Linnean Society 172:318–359.

Buhay, J.E., and K.A. Crandall. 2005. Subterranean phylogeography of freshwater crayfishes shows extensive gene flow and surprisingly large population sizes. Molecular Ecology 14:4259–4273.

Bulnheim, H.-P., and A. Scholl. 1981. Genetic variation between geographic populations of the amphipods *Gammarus zaddachi* and *G. salinus*. Marine Biology 64:105–115.

Burridge, C.P., D. Craw, and J.M. Waters. 2006. River capture, range expansion, and cladogenesis: The genetic signature of freshwater vicariance. Evolution 60:1038–1049.

Cabezas, M.P., R. Xavier, M. Branco, A.M. Santos, and J.M. Guerra-García. 2014. Invasion history of *Caprella scaura* Templeton, 1836 (Amphipoda: Caprellidae) in the Iberian Peninsula: multiple introductions revealed by mitochondrial sequence data. Biological Invasions 16:2221–2245.

Chan, B.K.K., and C.C. Cheang. 2015. A new *Chthamalus* (Crustacea: Cirripedia) from the *challengeri* subgroup on Taiwan rocky intertidal shores. Zootaxa 4000:547–558.

Chan, B.K.K., H.-N. Chen, P.R. Dando, A.J. Southward, and E.C. Southward. 2016. Biodiversity and biogeography of Chthamalid barnacles from the North-Eastern Pacific (Crustacea Cirripedia). PLOS ONE 11:e0149556.

Collins, T.M. 1989. Rates of Mitochondrial DNA Evolution in Trans-Isthmian Geminate Species. Yale University, New Haven, CT.

Cowen, R.K., C.B. Paris, and A. Srinivasan. 2006. Scaling of connectivity in marine populations. Science 5760:522–527.

Cronin, T.M., and T.M. Dowsett. 1996. Biotic and oceanographic response to the Pliocene closing of the central American isthmus. Pages 76–104 *in* J.B.C. Jackson, A.F. Budd, and A.G. Coates, editors. Evolution and Environment in Tropical America. University of Chicago Press, Chicago.

Cunningham, C. W., and T. M. Collins. 1994. Developing model systems for molecular biogeography: vicariance and interchange in marine invertebrates. Pages 405–433 *in* B. Schierwater, B. Streit, G. P. Wagner, and R. DeSalle, editors. Molecular Ecology and Evolution: Approaches and Applications. Springer, Berlin.

Dambach, J., S. Thatje, D. Rödder, Z. Basher, and M.J. Raupach. 2012. Effects of Late-Cenozoic glaciation on habitat availability in Antarctic Benthic Shrimps (Crustacea: Decapoda: Caridea). PLOS ONE 7:e46283.

Darwin, C. 1859. The Origin of Species by Means of Natural Selection, or the Preservation of Favored Races in the Struggle for Life. John Murray, London.

Dawson, M.N., P.H. Barber, L.I. Gonzalez-Guzman, R.J. Toonen, J.E. Dugan, and R.K. Grosberg. 2011. Phylogeography of *Emerita analoga* (Crustacea, Decapoda, Hippidae), an eastern Pacific Ocean sand crab with long-lived pelagic larvae. Journal of Biogeography 38:1600–1612.

Dawson, M.N., R.K. Grosberg, S. Yoel E, and S. Eric. 2010. Population genetic analysis of a recent range expansion: mechanisms regulating the poleward range limit in the volcano barnacle *Tetraclita rubescens*. Molecular Ecology 19:1585–1605.

de Bruyn, M., J.A. Wilson, and P.B. Mather. 2004. Huxley's line demarcates extensive genetic divergence between eastern and western forms of the giant freshwater prawn, *Macrobrachium rosenbergii*. Molecular Phylogenetics and Evolution 30:251–257.

DeBiasse, M.B., V.P. Richards, M.S. Shivji, and M.E. Hellberg. 2016. Shared phylogeographical breaks in a Caribbean coral reef sponge and its invertebrate commensals. Journal of Biogeography 43:2136–2146.

Deli, T., S. Fratini, L. Ragionieri, K. Said, N. Chatti, and C.D. Schubart. 2016. Phylogeography of the marbled crab *Pachygrapsus marmoratus* (Decapoda, Grapsidae) along part of the African Mediterranean coast reveals genetic homogeneity across the Siculo-Tunisian Strait versus heterogeneity across the Gibraltar Strait. Marine Biology Research 12:471–487.

Delić, T., V. Švara, C.O. Coleman, P. Trontelj, and C. Fišer. 2017. The giant cryptic amphipod species of the subterranean genus *Niphargus* (Crustacea, Amphipoda). Zoologica Scripta 46:740–752.

Díaz-Ferguson, E., R.A. Haney, J.P. Wares, and B.R. Silliman. 2010a. Population genetics of a trochid gastropod broadens picture of Caribbean Sea connectivity. PLOS ONE 5:e12675.

Díaz-Ferguson, E., R.A. Haney, J.P. Wares, and B.R. Silliman.. 2011. Genetic structure and connectivity patterns of Caribbean rocky intertidal gastropods. Journal of Molluscan Studies 76:112–118.

Díaz-Ferguson, E., J.D. Robinson, B. Silliman, and J.P. Wares. 2010b. Comparative Phylogeography of North American Atlantic Salt Marsh Communities. Estuaries and Coasts 33:828–839.

Dudoit, "Ale"alani, M. Iacchei, R.R. Coleman, M.R. Gaither, W.E. Browne, B.W. Bowen, and R.J. Toonen. 2018. The little shrimp that could: phylogeography of the circumtropical *Stenopus hispidus* (Crustacea: Decapoda), reveals divergent Atlantic and Pacific lineages. PeerJ 6:e4409.

Durbin, A., P.D.N. Hebert, and M.E.A. Cristescu. 2008. Comparative phylogeography of marine cladocerans. Marine Biology 155:1–10.

Ellingson, R.A., and P.J. Krug. 2006. Evolution of poecilogony from planktotrophy: cryptic speciation, phylogeography, and larval development in the gastropod genus *Alderia*. Evolution 60:2293–2310.

Ewers-Saucedo, C., J.M. Pringle, H.H. Sepúlveda, J.E. Byers, S.A. Navarrete, and J.P. Wares. 2016. The oceanic concordance of phylogeography and biogeography: a case study in *Notochthamalus*. Ecology and Evolution 6:4403–4420.

Franke, H.-D., L. Gutow, and M. Janke. 1998. The recent arrival of the oceanic isopod *Idotea metallica* Bosc off Helgoland (German Bight, North Sea): an indication of a warming trend in the North Sea? Helgoländer Meeresuntersuchungen 52:347.

Fraser, C.I., R. Nikula, H.G. Spencer, and J.M. Waters. 2009. Kelp genes reveal effects of subantarctic sea ice during the Last Glacial Maximum. Proceedings of the National Academy of Sciences 106:3249.

Garcia-Merchan, V.H., A. Robainas-Barcia, P. Abello, E. Macpherson, F. Palero, M. Garcia-Rodriguez, L.G. de Sola, et al. 2012. Phylogeographic patterns of decapod crustaceans at the Atlantic-Mediterranean transition. Molecular Phylogenetics and Evolution 62:664–672.

Govindarajan, A.F., F. Buksa, K. Bockrath, J.P. Wares, and J. Pineda. 2015. Phylogeographic structure and northward range expansion in the barnacle *Chthamalus fragilis*. PeerJ 3:e926.

Guo, B., and J.P. Wares. 2017. Large-scale gene flow in the barnacle *Jehlius cirratus* and contrasts with other broadly-distributed taxa along the Chilean coast. PeerJ 5:e2971.

Gutow, L., and H.-D. Franke. 2003. Metapopulation structure of the marine isopod *Idotea metallica*, a species associated with drifting habitat patches. Helgoland Marine Research 56:259.

Hamasaki, K., C. Iizuka, T. Sanda, H. Imai, and S. Kitada. 2017. Phylogeny and phylogeography of the land hermit crab *Coenobita purpureus* (Decapoda: Anomura: Coenobitidae) in the Northwestern Pacific Region. Marine Ecology 38:e12369.

Hameed, S.O., J.W. White, S.H. Miller, K.J. Nickols, and S.G. Morgan. 2016. Inverse approach to estimating larval dispersal reveals limited population connectivity along 700 km of wave-swept open coast. Proceedings of the Royal Society B: Biological Sciences 283:1833.

Haye, P.A., N.I. Segovia, N.C. Muñoz-Herrera, F.E. Gálvez, A. Martínez, A. Meynard, M.C. Pardo-Gandarillas, E. Poulin, and S. Faugeron. 2014. Phylogeographic structure in benthic marine invertebrates of the Southeast Pacific coast of Chile with differing dispersal potential. PLOS ONE 9:e88613.

Haye, P.A., Varela, and M. Thiel. 2012. Genetic signatures of rafting dispersal in algal-dwelling brooders *Limnoria spp.* (Isopoda) along the SE Pacific (Chile). Marine Ecology Progress Series 455:111–122.

Hedgecock, D. 1979. Biochemical genetic variation and evidence of speciation in *Chthamalus* barnacles of the tropical eastern Pacific Ocean. Marine Biology 54:207–214.

Held, C. 2003. Molecular evidence for cryptic speciation within the widespread Antarctic crustacean *Ceratoserolis trilobitoides* (Crustacea, Isopoda). Pages 135–139 *in* A.H.L. Huiskes, W.W.C. Gieskes, J. Rozema, R.M.L. Schorno, S.M. van der Vies, and W.J. Wolff, editors. Antarctic Biology in a Global Context. Backhuys, Leiden, The Netherlands.

Hellberg, M.E. 1998. Sympatric sea shells along the sea's shore: the geography of speciation in the marine gastropod *Tegula*. Evolution 52:1311–1324.

Hickerson, M.J., B.C. Carstens, J. Cavender-Bares, K.A. Crandall, C.H. Graham, J.B. Johnson, L. Rissler, Victoriano, and A.D. Yoder. 2010. Phylogeography's past, present, and future: 10 years after Avise, 2000. Molecular Phylogenetics and Evolution 54:291–301.

Hickerson, M.J., E. Stahl, and H. Lessios. 2006. Test for simultaneous divergence using approximate Bayesian computation. Evolution 60:2435–2453.

Hiller, A., and H.A. Lessios. 2017. Phylogeography of *Petrolisthes armatus*, an invasive species with low dispersal ability. Scientific Reports 7:3359.

Holthuis, L.B. 1980. FAO species catalogue. Volume 1: Shrimps and prawns of the world. An annotated catalogue of species of interest to fisheries. Food and Agriculture Organization of the United Nations, Rome.

Hurt, C., A. Anker, and N. Knowlton. 2009. A multilocus test of simultaneous divergence across the Isthmus of Panama using snapping shrimp in the genus *Alpheus*. Evolution 63:514–530.

Hurtado, L.A., M. Mateos, G. Mattos, S. Liu, P.A. Haye, and P.C. Paiva. 2016. Multiple transisthmian divergences, extensive cryptic diversity, occasional long-distance dispersal, and biogeographic patterns in a marine coastal isopod with an amphi-American distribution. Ecology and Evolution 6:7794–7808.

Hurtado, L.A., M. Mateos, and C.A. Santamaria. 2010. Phylogeography of supralittoral rocky intertidal *Ligia* Isopods in the Pacific region from central California to central Mexico. PLOS ONE 5:e11633.

Huxley, T.H. 1868. On the classification and distribution of the Alectoromorphae and Heteromorphae. Proceedings of the Zoological Society of London 294–319.

Ilves, K., W. Huang, J.P. Wares, and M.J. Hickerson. 2010. Colonization and/or mitochondrial selective sweeps across the North Atlantic intertidal assemblage revealed by multi-taxa approximate Bayesian computation. Molecular Ecology 19:4505–4519.

Ingolfsson, A. 1992. The origin of the rocky shore fauna of Iceland and the Canadian maritimes. Journal of Biogeography 19:705–712.

Johannesson, K., and C. André. 2006. Invited review: Life on the margin: genetic isolation and diversity loss in a peripheral marine ecosystem, the Baltic Sea. Molecular Ecology 15:2013–2029.

Kelly, R., and S. Palumbi. 2010. Genetic structure among 50 species of the Northeastern Pacific Rocky Intertidal Community. PLOS ONE 5:e8594.

Knowlton, N., and L.A. Weigt. 1998. New dates and new rates for divergence across the Isthmus of Panama. Proceedings of the Royal Society of London. Series B: Biological Sciences 265:2257–2263.

Laakkonen, H. 2015. Phylogeography of Amphi-Boreal Marine Fauna. University of Helsinki, Helsinki, Finland.

Larson, E.R., and B.W. Williams. 2015. Historical biogeography of *Pacifastacus* crayfishes and their Branchiobdellidan and Entocytherid ectosymbionts in western North America. Pages 404–447 *in* T. Kawai, Z. Faulkes, and G. Scholtz, eds. Freshwater Crayfish: A global overview. CRC Press, New York.

Laughlin, K.M., C. Ewers, and J.P. Wares. 2012. Mitochondrial lineages in *Notochthamalus scabrosus* as indicators of coastal recruitment and interactions. Ecology and Evolution 2:1584–1591.

Laurenzano, C., T.M. Costa, and C.D. Schubart. 2016. Contrasting patterns of clinal genetic diversity and potential colonization pathways in two species of Western Atlantic fiddler crabs. PLOS ONE 11:e0166518.

Lee, T.R., S.Y. Ho, G.D. Wilson, and N. Lo. 2014. Phylogeography and diversity of the terrestrial isopod *Spherillo grossus* (Oniscidea: Armadillidae) on the Australian East Coast. Zoological Journal of the Linnean Society 170:297–309.

Luttikhuizen, P.C., J. Campos, J. van Bleijswijk, K.T. Peijnenburg, and H.W. van der Veer. 2008. Phylogeography of the common shrimp, *Crangon crangon* (L.) across its distribution range. Molecular Phylogenetics and Evolution 46:1015–1030.

Maggs, C.A., R. Castilho, D. Foltz, C. Henzler, M.T. Jolly, J. Kelly, J. Olsen, K.E. Perez, W. Stam, R. Väinölä, F. Viard, and J. Wares. 2008. Evaluating signatures of glacial refugia for North Atlantic benthic marine taxa. Ecology 89:S108–S122.

Marchant, S., A.L. Moran, and P.B. Marko. 2015. Out-of-the tropics or trans-tropical dispersal? The origins of the disjunct distribution of the gooseneck barnacle *Pollicipes elegans*. Frontiers in Zoology 12:39.

Marko, P.B. 2002. Fossil calibration of molecular clocks and the divergence times of geminate species pairs separated by the Isthmus of Panama. Molecular Biology and Evolution 19:2005–2021.

Mathews, L.M. 2006. Cryptic biodiversity and phylogeographical patterns in a snapping shrimp species complex. Molecular Ecology 15:4049–4063.

McDowall, R.M. 2005. Historical biogeography of the New Zealand freshwater crayfishes (Parastacidae, *Paranephrops spp.*): restoration of a refugial survivor? New Zealand Journal of Zoology 32:55–77.

Meyers, M., M.S. Pankey, and J.P. Wares. 2013. Genealogical approaches to the temporal origins of the Central American Gap: speciation and divergence in Pacific *Chthamalus*. Revista Biologia Tropical 61:75–88.

Navarrete, S.A., B.R. Broitman, and B.A. Menge. 2008. Interhemispheric comparison of recruitment to intertidal communities: pattern persistence and scales of variation. Ecology 89:1308–1322.

Naylor, E. 1957. The occurrence of *Idotea metallica* Bosc in British waters. Journal of the Marine Biological Association of the United Kingdom 36:599–602.

Negri, M., L.G. Pileggi, and F.L. Mantelatto. 2012. Molecular barcode and morphological analyses reveal the taxonomic and biogeographical status of the striped-legged hermit crab species *Clibanarius sclopetarius* (Herbst, 1796) and *Clibanarius vittatus* (Bosc, 1802) (Decapoda: Diogenidae). Invertebrate Systematics 26:561–571.

Nikula, R., C. Fraser, H. Spencer, and J. Waters. 2010. Circumpolar dispersal by rafting in two subantarctic kelp-dwelling crustaceans. Marine Ecology Progress Series 405:221–230.

O'Dea, A., H.A. Lessios, A.G. Coates, R.I. Eytan, S.A. Restrepo-Moreno, A.L. Cione, L.S. Collins, A. de Queiroz, D.W. Farris, R.D. Norris, R.F. Stallard, M.O. Woodburne, O. Aguilera, M.-P. Aubry, W.A. Berggren, A.F. Budd, M.A. Cozzuol, S.E. Coppard, H. Duque-Caro, S. Finnegan, G.M. Gasparini, E.L. Grossman, K. G. Johnson, L.D. Keigwin, N. Knowlton, E.G. Leigh, J.S. Leonard-Pingel, P.B. Marko, N.D. Pyenson, P.G. Rachello-Dolmen, E. Soibelzon, L. Soibelzon, J.A. Todd, G.J. Vermeij, and J.B. C. Jackson 2016. Formation of the Isthmus of Panama. Science Advances 2:e1600883.

Palumbi, S.R., and B.D. Kessing. 1991. Population biology of the trans-arctic exchange: MtDNA sequence similarity between Pacific and Atlantic sea urchins. Evolution 45:1790–1805.

Pannacciulli, F.G., F. Maltagliati, C. de Guttry, and Y. Achituv. 2017. Phylogeography on the rocks: the contribution of current and historical factors in shaping the genetic structure of *Chthamalus montagui* (Crustacea, Cirripedia). PLOS ONE 12:e0178287.

Pappalardo, P., F.B. Pitombo, P.A. Haye, and J.P. Wares. 2016. A rose by any other name: systematics and diversity in the Chilean giant barnacle *Austromegabalanus psittacus* (Molina, 1782) (Cirripedia). Journal of Crustacean Biology 36:180–188.

Pappalardo, P., J.M. Pringle, J.P. Wares, and J.E. Byers. 2015. The location, strength, and mechanisms behind marine biogeographic boundaries of the east coast of North America. Ecography 38:722–731.

Patarnello, T., F.A.M.J. Volckaert, and R. Castilho. 2007. Pillars of Hercules: is the Atlantic-Mediterranean transition a phylogeographical break? Molecular Ecology 16:4426–4444.

Petersen, C.H. 2007. Historical demography and contemporary spatial genetic structure of an estuarine crab in the northeast Pacific (*Hemigrapsus oregonensis*). Marine Biology 150:1289–1300.

Pitombo, F.B., and R.S. Burton. 2007. Systematics and biogeography of Tropical Eastern Pacific *Chthamalus* with descriptions of two new species (Cirripedia, Thoracica). Zootaxa 1574:1–30.

Poore, G.C.B., and N. Andreakis. 2012. The *Agononida incerta* species complex unravelled (Crustacea: Decapoda: Anomura: Munididae). Zootaxa 3492:1–29.

Poore, G.C.B., and N. Andreakis. 2014. More species of the *Agononida incerta* complex revealed by molecules and morphology (Crustacea: Decapoda: Anomura: Munididae). Zootaxa 3860:201–225.

Poore, G.C.B., and N. Andreakis. 2011. Morphological, molecular and biogeographic evidence support two new species in the *Uroptychus naso* complex (Crustacea: Decapoda: Chirostylidae). Molecular Phylogenetics and Evolution 60:152–169.

Pringle, J.M., and J.P. Wares. 2007. Going against the flow: maintenance of alongshore variation in allele frequency in a coastal ocean. Marine Ecological Progress Series 335:69–84.

Raupach, M.J., S. Thatje, J. Dambach, P. Rehm, B. Misof, and F. Leese. 2010. Genetic homogeneity and circum-Antarctic distribution of two benthic shrimp species of the Southern Ocean, *Chorismus antarcticus* and *Nematocarcinus lanceopes*. Marine Biology 157:1783–1797.

Riginos, C., and C.W. Cunningham. 2005. Local adaptation and species segregation in two mussel (*Mytilus edulis* x *Mytilus trossulus*) hybrid zones. Molecular Ecology 14:381–400.

Roman, J.O.E., and S.R. Palumbi. 2004. A global invader at home: population structure of the green crab, *Carcinus maenas*, in Europe. Molecular Ecology 13:2891–2898.

Roughgarden, J., S. Gaines, and H. Possingham. 1988. Recruitment dynamics in complex life cycles. Science 241:1460–1466.

Sanford, E., S.B. Holzman, R.A. Haney, D.M. Rand, and M.D. Bertness. 2006. Larval tolerance, gene flow, and the northern geographic range limit of fiddler crabs. Ecology 87:2882–2894.

Shahdadi, A., P.J.F. Davie, and C.D. Schubart. 2018. Systematics and phylogeography of the Australasian mangrove crabs *Parasesarma semperi* and *P. longicristatum* (Decapoda : Brachyura : Sesarmidae) based on morphological and molecular data. Invertebrate Systematics 32:196–214.

Slatkin, M. 1987. Gene flow and the geographic structure of natural populations. Science 236:787.

Sotka, E.E., J.P. Wares, J.A. Barth, R.K. Grosberg, and S.R. Palumbi. 2004. Strong genetic clines and geographical variation in gene flow in the rocky intertidal barnacle *Balanus glandula*. Molecular Ecology 13:2143–2156.

Sotka, E.E., J.P. Wares, and M.E. Hay. 2003. Geographic and genetic variation in feeding preference for chemically defended seaweeds. Evolution 57:2262–2276.

Stamatis, C., A. Triantafyllidis, K.A. Moutou, and Z. Mamuris. 2004. Mitochondrial DNA variation in northeast atlantic and mediterranean populations of Norway lobster, *Nephrops norvegicus*. Molecular Ecology 13:1377–1390.

Sywula, T., I. Głazewska, R.C. Whatley, and A. Moguilevsky. 1995. Genetic differentiation in the brackish-water ostracod *Cyprideis torosa*. Marine Biology 121:647–653.

Taylor, M.S., and M.E. Hellberg. 2003. Genetic evidence for local retention of pelagic larvae in a Caribbean reef fish. Science 299:107–109.

Tepolt, C.K., and S.R. Palumbi. 2015. Transcriptome sequencing reveals both neutral and adaptive genome dynamics in a marine invader. Molecular Ecology 24:4145–58.

Thiel, M., and L. Gutow. 2005. The ecology of rafting in the marine environment. I. The floating substrata. Oceanography and Marine Biology: An Annual Review 42:181–264.

Thiel, M., and P.A. Haye. 2006. The ecology of rafting in the marine environment. III. Biogeographical and evolutionary consequences. Oceanography and Marine Biology: An Annual Review 48:323–429.

Triantafyllidis, A., A.P. Apostolidis, V. Katsares, E. Kelly, J. Mercer, M. Hughes, K. Jorstad, et al. 2005. Mitochondrial DNA variation in the European lobster (*Homarus gammarus*) throughout the range. Marine Biology 146:223–235.

Van Syoc, R.J. 1994. Genetic divergence between subpopulations of the eastern Pacific goose barnacle *Pollicipes elegans*: mitochondrial cytochrome c subunit 1 nucleotide sequences. Molecular Marine Biology and Biotechnology 3:338–346.

Varela, A.I., and P.A. Haye. 2012. The marine brooder *Excirolana braziliensis* (Crustacea: Isopoda) is also a complex of cryptic species on the coast of Chile. Revista Chilena de Historia Natural 85:495–502.

Vermeij, G. 1991a. Anatomy of an invasion: the trans-Arctic interchange. Paleobiology 17:281–307.

Vermeij, G. 1991b. When biotas meet: understanding biotic interchange. Science 253:1099–1104.

Wakeley, J. 2008. Coalescent theory: an introduction. Roberts & Co. Publishers, Greenwood Village, Colorado.

Wares, J.P. 2001a. Patterns of speciation inferred from mitochondrial DNA in North American *Chthamalus* (Cirripedia: Balanomorpha: Chthamaloidea). Molecular Phylogenetics and Evolution 18:104–16.

Wares, J.P. 2001b. Intraspecific variation and geographic isolation in *Idotea balthica* (Isopoda: Valvifera). Journal of Crustacean Biology 21:1007–1013.

Wares, J.P. 2002. Community genetics in the Northwestern Atlantic intertidal. Molecular Ecology 11:1131–1144.

Wares, J.P. 2016. Population structure and gene flow. Pages 327–331 *in* R. M. Kliman, editor. Encyclopedia of Evolutionary Biology, Vol. 3. Academic Press, Oxford, UK.

Wares, J.P., and A.E. Castaneda. 2005. Geographic range in *Chthamalus* along the west coast of North America. Journal of the Marine Biological Association of the United Kingdom 85:327–331.

Wares, J.P., and C.W. Cunningham. 2001. Phylogeography and historical ecology of the northern Atlantic intertidal. Evolution 55:2455–2469.

Wares, J.P., and C.W. Cunningham. 2005. Diversification predates glaciation in *Balanus glandula*. The Biological Bulletin 208:60–68.

Wares, J.P., S. Daley, R. Wetzer, and R.J. Toonen. 2007. An evaluation of cryptic lineages of *Idotea balthica* (Isopoda: Idoteidae): morphology and microsatellites. Journal of Crustacean Biology 27:643–648.

Wares, J.P., S. Gaines, and C.W. Cunningham. 2001. A comparative study of asymmetric migration events across a marine biogeographic boundary. Evolution 55:295–306.

Wares, J.P., M.S. Pankey, F. Pitombo, L.G. Daglio, and Y. Achituv. 2009. A "shallow phylogeny" of shallow barnacles (*Chthamalus*). PLOS ONE 4:e5567.

Wares, J.P., and J.M. Pringle. 2008. Drift by drift: effective population size is limited by advection. BMC Evolutionary Biology 8:235.

Watanabe, H.K., C. Chen, D.P. Marie, K. Takai, K. Fujikura, and B.K.K. Chan. 2018. Phylogeography of hydrothermal vent stalked barnacles: a new species fills a gap in the Indian Ocean "dispersal corridor" hypothesis. Royal Society Open Science 5.

Waters, J.M., D.L. Rowe, S. Apte, T.M. King, G.P. Wallis, L. Anderson, R.J. Norris, D. Craw, and C. P. Burridge. 2007. Geological dates and molecular rates: rapid divergence of rivers and their biotas. Systematic Biology 56:271–282.

Wrange, A.-L., G. Charrier, A. Thonig, M. Alm Rosenblad, A. Blomberg, J.N. Havenhand, P.R. Jonsson, and C. André. 2016. The story of a hitchhiker: population genetic patterns in the invasive barnacle *Balanus* (*Amphibalanus*) *improvisus* Darwin 1854. PLOS ONE 11:e0147082.

Wu, T.H., L.M. Tsang, B.K.K. Chan, and K.H. Chu. 2015. Cryptic diversity and phylogeography of the island-associated barnacle *Chthamalus moro* in Asia. Marine Ecology: An Evolutionary Perspective 36:368–378.

Zakas, C., J. Binford, S.A. Naverette, and J.P Wares. 2009. Restricted gene flow in Chilean barnacles reflects an oceanographic and biogeographic transition zone. Marine Ecology Progress Series 394:165–177.

Zardus, J.D., and M.G. Hadfield. 2005. Multiple origins and incursions of the Atlantic barnacle *Chthamalus proteus* in the Pacific. Molecular Ecology 14:3719–3733.

18

OUTLOOK: CRUSTACEANS IN THE ANTHROPOCENE

Karolina Bącela-Spychalska, Gary C. B. Poore, and
Michał Grabowski

Abstract

Since the mid-20th century we have been living in a new geological epoch, Anthropocene, characterized by an overwhelming impact of human activity on the Earth's ecosystems, leading to mass species extinction by habitat destruction, pollution, global climate warming, and homogenization of biota by intra- and intercontinental transfer of species. Crustaceans are among the most diverse and species-rich animal groups inhabiting predominantly aquatic ecosystems, listed as among the most threatened ecosystems. Global threats include ocean and freshwater acidification, eutrophication, pesticide, hormone and antibiotic load, coastline modification, habitat destruction, overharvesting, and the introduction of invasive species. Many crustaceans are threatened by human-induced modifications of habitats, while others are themselves threats—crustaceans are among the most common invasive species. Those non-indigenous species, when established and integrated, become important components of existing communities, strongly influencing other components directly and indirectly, including by species replacement. They are a threat mostly to species with similar ecological niches, most often to other crustaceans. It is hard to be optimistic about the future of crustacean biodiversity. We may rather expect that growing human pressure will variously further accelerate the non-natural dispersal and extinction rate.

INTRODUCTION

With industrialization, novel threats to the environment emerged, particularly pollution from toxins, pesticides, medicines, and fertilizers; overharvesting; and the by-products of globalization, such as invasive species and disease. More recently, the effects of widespread threats associated with global climate change appeared: warming temperatures, increasing rates of sea-level rise,

Evolution and Biogeography. Edited by Gary C. B. Poore and Martin Thiel.

ocean acidification, and UV exposure, for example, are now well documented. Habitat destruction resulting from human population pressure is still among the most important pressures. This chapter describes these changes and addresses how they have impacted crustaceans throughout the period now becoming popularly known as the Anthropocene (Waters et al. 2016).

PHYSICAL ENVIRONMENTAL CHANGE

Natural coastlines are highly dynamic systems, where terrestrial, freshwater, marine, and atmospheric processes interact to provide continuous changes. Coastal areas include some of the most productive and valued ecosystems of the world (Costanza et al. 1997). Direct access to coastlines and waterbodies of any kind has always been a stimulus for humanity to settle and one of the most important factors behind development of human societies and civilizations. Inland springs are a resource of drinking waters, while rivers, lakes, and sea coastal waters serve as transportation routes and provide a wealth of fresh, healthy food. River deltas and flood plains have always been fertile agricultural lands, while seashores and lagoons served as inexhaustible sources of salt, a basis of fortunes, and the reason for wars since the beginning of human societies. The demand for water and associated resources, as well as the need for safe harbors, has been growing ever since, accelerating dramatically during the Industrial Revolution. The 20th century brought another phenomenon in the form of mass seaside tourism and recreation, resulting in the development of huge resorts and increased concentrations of people, sometimes unregulated. As a result of these processes, at least 40% of the world's population now lives within 100 km of the coast, applying considerable pressure on the physical environment. As a consequence, some of these regions are among the most heavily degraded, with human pressure on coastlines increasing (Millenium Ecosystem Assessment 2005). These physical changes to aquatic habitats are just one of the threats to crustacean communities. Another is less direct—increases in water temperature as a consequence of global warming.

Habitat Destruction

The loss or conversion of habitat is the most serious and widespread threat to coastal habitats and was the first recognized. Aquatic systems have often been drained, dredged, or converted to harbors, fish or shrimp farms, upland habitat, artificial substrates, or open water. Estimates of habitat loss depend on the accuracy of historical baselines and range widely by habitat and region. Biogenic coastal ecosystems have been particularly impacted since the removal of the foundation species eliminates the entire habitat (Crain et al. 2009). Crustaceans—mainly shrimps, crabs, amphipods, isopods, and tanaidaceans, as well as taxa with smaller individuals such as copepods—are associated with threatened coastal ecosystems such as mangroves, seagrass meadows, and coral reefs (Hogarth 2007, Dubinsky and Stambler 2010). The increasing degradation of such habitats will inevitably result in the extinction of most of these species.

The modification of freshwater ecosystems results from river regulation, gravel excavation, or construction of dams, reservoirs, and hydroelectric power stations (Dudgeon et al. 2006, Dudgeon 2012). Due to the limited size of most freshwater reservoirs, they are particularly susceptible. This is a serious problem in areas with limited water resources and high endemism of freshwater biodiversity, such as the Iberian, Apennine, and Balkan peninsulas in Europe, Asia Minor, North Africa, southwestern North America, western South America, and Australia. Spring water capture, lowered water table as a result of overuse of water for agricultural and industrial needs, or building a dam may completely destroy fragile and spatially limited patches of habitats together with all associated fauna.

A good example of such a threat associated with damming is particularly visible in the rapidly developing countries of the Balkan Peninsula in Europe, where the growing economy and tourism leads to increased anthropogenic pressure on karst springs, rivers, lakes, and sea coasts. However, the effects of such pressure are as yet scarcely studied; a prominent case may be the closure of a large dam on the Gojacka Dobra River in central Croatia in 2010, which posed a threat to the population of native endemic amphipod species. In the pre-disturbance period, coexistence of two species, the endemic moss specialist *Echinogammarus cari* and the widespread microhabitat generalist *Gammarus fossarum*, was promoted by differential distribution between two dominating types of microhabitats, moss and stony substratum. Mosses were occupied predominantly by *E. cari*, while *G. fossarum* dominated on stony substratum, an excellent illustration of spatial niche partitioning between two species of river amphipods. The dam closure caused an increased load of suspended particles that accumulated in mosses and contributed to the rapid population decline of *E. cari* at downstream sites. The more tolerant *G. fossarum* was much less affected, exhibiting no drastic population decline and even an increase in population size after disturbance events at the examined sites (Zganec and Gottstein 2009, Zganec 2012). In marine coastal habitats, alienation of shallow water systems for fish and shrimp farming has a significant impact. The consequences are a loss of native biodiversity and associated eutrophication (see below).

Warming

Climate monitoring over the last century and long-term reconstruction of the climate over the past millennia indicate that the Earth has been warming recently and doing so at an unprecedented rate (Hughes et al. 2000, Sharma et al. 2016). Weather patterns and changes in the physical features of the Earth's surface, such as a decrease in the extent of glaciers and sea ice, appear to be consistent with those predicted under global warming models (Hughes 2000, Foster et al. 2017). A nearly 10% decrease of sea ice in polar regions and 10–20% reduction of the surface area of glaciers in the European Alps have been observed since the 1980s, while since the mid-1980s, the Alpine glaciers have shrunk by 30–40% (Maslanik et al. 1996, Haeberli and Beniston 1998). It is noteworthy that the warming of ocean waters, the world's most important thermal stabilizer, has been observed all around the globe, including the Atlantic, Indian, and Pacific oceans as well as near the poles. The highest warming rate has been estimated as equivalent to 1°C per year, even in deep-sea water layers (Epstein et al. 1998). Accordingly, global sea level has risen by 10–25 cm over the past century or so (Fig. 18.1a,b; IPCC 2007). This warming trend and its effects are inevitably associated with the concentration of carbon dioxide (CO_2) in the atmosphere, increasing since the beginning of the Industrial Revolution (IPCC 2007).

According to Hughes (2000), such human-induced climate change may affect biodiversity in many ways, generalized into four categories: (1) the alteration of metabolic and developmental rates, including growth, photosynthesis, respiration, etc., (2) shifts in distribution ranges of organisms, given that a 3°C change in mean annual temperature corresponds to a shift in isotherms of c. 300–400 km in the temperate climatic zone or c. 500 m in altitude, (3) changes in phenology that may influence ecological interactions between species with synchronized life cycles, and (4) enforcing microevolutionary changes in species with short generation times and rapid population growth. Cahill et al. (2017) predicted that the effects associated with anthropogenic global warming would be a major cause of species extinctions during the next 100 years and reported at least 20 cases of animal extinctions in the wild and many more of population decline, most probably due to warming. Although their study did not identify any crustacean species directly affected by climate change, it is obviously not an indication of crustacean perseverance but rather of scarcity of crustacean case studies. Among the cases raised by Cahill et al. (2017), coral bleaching appears to be particularly hazardous for crustaceans, as numerous species from many higher taxa are coral reef specialists (see

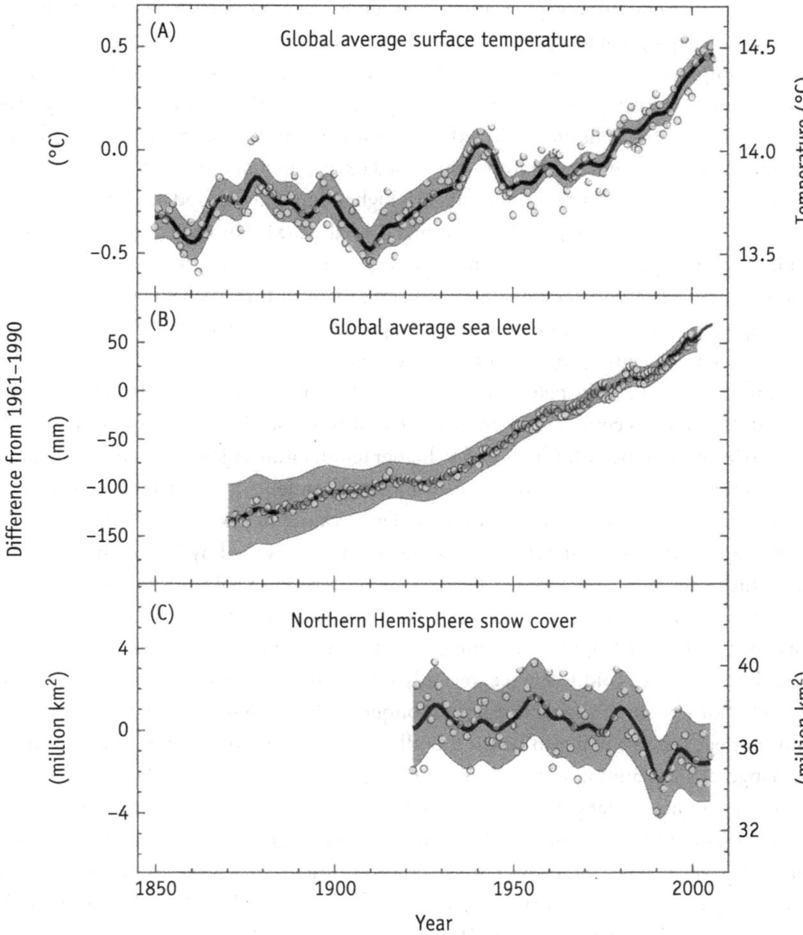

Fig. 18.1.
Observed changes in (a) global average surface temperature; (b) global average sea level from tide gauge (blue) and satellite (red) data; and (c) Northern Hemisphere snow cover for March–April. All differences are relative to corresponding averages for the period 1961–1990. Smoothed curves represent decadal averaged values while circles show yearly values. The shaded areas are the uncertainty intervals estimated from a comprehensive analysis of known uncertainties (a and b) and from the time series (c). From Figure SPM.1 in IPCC (2007), with permission from IPCC.

Chapter 8). For example, in a lagoon at Lizard Island, Australia, branching corals, which harbor the highest diversity of coral associates, comprised 13% of live coral cover at the lagoon, with 83% affected by bleaching (Great Barrier Reef Marine Park Authority 2017). A recent study at that site showed a significant decline in density of the crab *Trapezia cymodoce*, an obligate symbiont of the *Pocillopora* corals, on bleached colonies after five weeks, compared to the healthy ones. Also, egg clutches of the crabs collected from bleached hosts were 40% smaller than those from healthy hosts, indicating a significant reduction in fecundity (Stella et al. 2011).

Striking is the case of polar regions where, particularly in Antarctica, unique benthic and planktonic communities have evolved during the long-term cooling trend, which began with a sharp temperature drop ~40 Ma, in the Eocene (Aronson et al. 2009). The most speciose and abundant crustaceans in the Antarctic are peracarids (amphipods and isopods have been well studied). Although some eurytopic and opportunistic species exist in these groups, the high endemicity of species of amphipods and isopods indicate particular vulnerability to environmental change. Thus, the response to rising water temperatures is expected at the species level. Being brooders, their migration potential is limited. Altered physiology has an impact on species' feeding habits and requirements. At higher temperatures, increased oxygen demand reduces the aerobic scope of animals, while accelerated metabolism results in a higher demand for food, leaving fewer resources for growth and reproduction (for details see Ingels et al. 2012). For example, populations of the planktonic amphipod genus *Themisto*, living in water just a couple of degrees warmer, have a higher respiration rate, grow faster and attain sexual maturity earlier, and are of smaller size than those living at lower temperatures (Auel and Ekau 2009). An experimental study on the effects of temperature and salinity on vital activities, such as response to food odor, righting, swimming, and reburying of the Antarctic isopod *Serolis polita*, suggested that the species is vulnerable to environmental changes and its ability to cope with them is limited. It was also shown that the isopod was more vulnerable to lower salinities when exposed to higher temperatures (Janecki et al. 2010). Significant effects of elevated temperatures on metabolism and development were observed also in case of key marine planktonic copepods, such as *Calanus finmarchicus* in the Arctic (Mayor et al. 2015, Weydmann et al. 2017). The Antarctic shallow waters can be invaded by king crabs (i.e., *Paralomis birsteini*), which each year are recorded in shallower waters and finally (in the next few decades) may reach the unique seafloor communities of Antarctica now free from such shell-crushing predators, leading to catastrophic consequences to this unique community (Aronson et al. 2015). Larval lecithotrophy and cold tolerance, recently observed in subantarctic species and other life-history adaptations potentially enable them to conquer polar ecosystems (Thatje et al. 2005).

Freshwater ecosystems will also be profoundly affected by global climate change. This is particularly dangerous to running waters in mountainous areas, which are known to be particularly vulnerable. It is predicted that global warming will reorganize species composition and community structure along the river continuum, with the most profound effects such as significant reductions in population size of headwater species, resulting eventually in a loss of genetic diversity. In contrast, the expansion of more euryoecious species will result in the establishment of non-native macroinvertebrates in the lower reaches (Domisch et al. 2011). Empirical results concerning the effect of climate change on freshwater crustaceans are fewer than those on marine ones. Also, in freshwater invertebrates, including crustaceans, there is often a strong discrepancy between field observations and experimental-evolutionary work on trait changes under global warming (Stoks et al. 2014).

A recent study on the freshwater pseudothelphusid crab *Neostrengeria macropa* in rivers and streams of Colombia predicted that its thermal tolerance is wide enough to retain its present distribution range for at least the next 70 years (Hudson et al. 2016). On the other hand, the predicted increasing frequency of turbulent climatic events such as inundations leaving behind temporary water bodies may enhance the stability of metapopulations in crustaceans, e.g., branchiopods such as *Daphnia* (Pyke 2005, Altermatt et al. 2008). Also, the potential effect of climate change on crustacean invaders is ambiguous. For example, Maazouzi et al. (2011) used experimental data to suggest that the fast-spreading amphipod invader, the killer shrimp *Dikerogammarus villosus*, is more vulnerable to higher temperatures than the native *Gammarus pulex*, and that its further spread will not be promoted by global warming. But Hupało et al. (2018) suggested that the species' tolerance to thermal stress depends on the geographical origin of the population. A similar contradiction was predicted for other invaders, such as the signal crayfish *Pacifastacus leniusculus* in Europe (Kernan 2015). Interestingly, some authors have suggested that in some groups, such as Anomura, global warming may pose a threat to diversification in marine taxa but, at the same time, will enhance speciation in freshwater lineages (Davis et al. 2016).

CHEMICAL ENVIRONMENTAL CHANGE

Global chemical threats associated with ongoing industrialization and resulting global warming include (1) ocean and freshwater acidification, (2) eutrophication, and (3) pesticide, hormone, and antibiotic pollution.

Acidification

Ocean and freshwater acidification is a major problem worldwide. Normally, sea water is slightly alkaline (pH > 7, usually ~8.25). At these values sea water is supersaturated with calcium carbonate and similar minerals critical for building shells and skeletons of many marine organisms, including crustaceans. However, excessive CO_2 emissions produced by human activities such as combustion of fossil fuels, deforestation, soil erosion, and animal agriculture have resulted in a 40% increase in its atmospheric concentration, from ~280 ppm in 1750 (the beginning of the Industrial Revolution) to 406 ppm in early 2017 (Jacobson 2005, IPCC 2007). Atmospheric CO_2 dissolves in seawater and fresh water, where a portion reacts with water to form carbonic acid, thereby lowering the pH (Fig. 18.2). As a result, the ability of water to maintain calcium carbonate in solution decreases and its concentration becomes too low for construction of skeletons (IPCC 2007). As atmospheric CO_2 concentration has increased since the beginning of the Industrial Revolution, the pH of oceanic surface waters has fallen by 0.15 pH units, from 8.25 to 8.10 (IPCC 2007), a 35% increase in acidity. Predictions of future CO_2 levels indicate that the acidity of oceanic surface waters will increase by

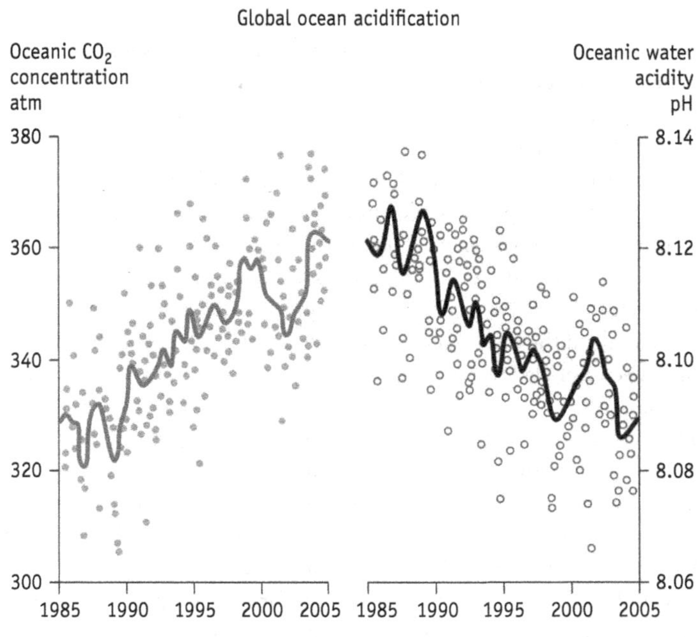

Fig. 18.2.
As carbon dioxide concentrations in the atmosphere increase, so do concentrations in the ocean, with resultant acidification as a natural chemical process. With permission from GRID-Arendal (http://www.grida.no/graphicslib/detail/global-ocean-acidification_162f).

150% by the end of the 21st century. Such conditions have not been experienced by the ocean for at least the last 20 million years (Zeebe 2012, Mora et al. 2013).

The negative effects of acidification on marine organisms, such as planktonic algae, corals, and mollusks, have been well documented (Raven et al. 2005). However, its impact on crustaceans is still ambivalent. For example, studies on planktonic copepods showed that reduced pH influenced neither feeding rate nor growth (Pedersen et al. 2013, 2014). It seems that they are already well adapted to a pH gradient as an adaptation to vertical migration, upwelling episodes, or mixing conditions in estuarine waters (Weydmann et al. 2012). Thus, while these copepods seem to have a capacity to adapt to future pH decline such as may occur with ocean acidification, other copepods that never leave surface waters showed reduced adult and nauplius survival in similar conditions (Lewis et al. 2013). Larval development of the widespread barnacle *Amphibalanus improvisus* and related species was not significantly affected by reduced pH, even at the levels projected over the next 150 years (Pansch et al. 2013, 2014).

No detrimental physiological effect of reduced pH was observed in crabs or shrimps (Appeltans et al. 2012), although some shrimp species had a prolonged zoeal stage under such conditions (Bechmann et al. 2011). Interestingly, red rock shrimp *Lysmata californica* may even build more calcified exoskeletons at lowered pH at the cost of reduced body transparency. This may adversely impact the species' biomechanical properties and its ability to avoid predators, through either camouflage or escape behavior (Taylor et al. 2015). In brine shrimp *Artemia sinica* lowered pH had a negative effect on hatching success, growth, survival rate, expression of selected genes (involved in development, immunity, cellular activity, etc.), and biological activity of several key enzymes (involved in antioxidant responses, physiological reactions, etc.) (Zheng et al. 2015). Whiteley (2011) pointed out that although the sensitivity of marine crustaceans to ocean acidification is poorly understood, acidification stress impacts predominantly slow-moving, relatively inactive crustacean species as they have low circulating protein levels and low buffering capacities. Such species, living often in low-energy environments, such as deep-sea and polar habitats, are particularly vulnerable. However, as may be seen from often contradictory results, more detailed studies are needed to check vulnerability beyond the limitations of lab experiments, especially in combination with other environmental changes.

Relatively less is known about acidification in freshwater ecosystems where pH is more variable and highly dependent on local geological and climatic conditions. Usually, freshwater lakes, ponds, and rivers have a pH of 6–8.5, but in dystrophic lakes and bog pools it may be <4. A summary by Moiseenko (2005) reported that some organisms such as freshwater zooplankton, including cladocerans, copepods, and some mysids, accumulate excessive amounts of harmful elements (e.g., Al, Hg, Cd, Zn, Pb) in an acid environment. This is due primarily to the low concentration of calcium and the resulting increased uptake of other elements. However, there are also acidophilic species of cladocerans and copepods that thrive at pH < 4.5 (summarized by Moiseenko 2005).

Amphipods such as species of *Hyalella* and *Gammarus* are generally absent from waters of pH < 6.5 and, thus, are sensitive to acid rains or other acidifying sources. Laboratory and field studies proved that water acidification is especially harmful to gametes, juveniles, and molting individuals of amphipods and crayfish (Engblom and Lingdell 1984, France and Stokes 1987, Davies 1989). North American crayfishes of the genus *Orconectes*, e.g., *O. virilis*, *O. rusticus*, and *O. propinquus*, avoid pH < 5.6, while the lowest limit for the European noble crayfish *Astacus astacus* is at pH 6.0. Recent studies on crayfish showed that low pH negatively affects survival, growth, and cuticle building and survival during infection by microsporidian parasites (Haddaway et al. 2013). It was also found that in acidified water crayfish show reduced behavioral responses to food odor (Tierney and Atema 1986). On the other hand, the common freshwater isopod *Asellus aquaticus* is a common and abundant inhabitant of acidified lakes and streams.

The effect on aquatic invertebrates can be through chronic exposure to stable low pH level or through sudden pH drops (pH shock), which, combined with the mobilization of metal ions, may destroy the aquatic fauna (Moiseenko 2005). Organisms living on the surface and in the littoral zone are more vulnerable to occasional acidification than those living in bottom sediments, burrows, and the deepest parts of a lake where pH levels are resistant to abrupt changes (Raddum 1980). The long-term effect of acidification on crustacean community structure resulting from the differential impact on individual species is still an open question.

Eutrophication

Eutrophication, or more precisely hypertrophication, is the enrichment of a water body with an excess of nutrients, such us phosphorus (predominantly in fresh waters) or nitrogen (predominantly in marine waters). This induces, e.g., the growth of phytoplankton (algal blooms) or macroalgae, resulting in reduced water clarity and light penetration, increased sedimentation and oxygen consumption, oxygen depletion in the bottom water, and sometimes the death of fish and benthic animals. Particularly problematic are blooms of blue-green algae (Cyanobacteria) that produce neuro- or hepatotoxins harmful to a wide range of organisms, including humans. In Europe, regional seas such as the Baltic and Mediterranean seas are now adversely affected by eutrophication, with climate change expected to intensify these adverse impacts (HELCOM 2009, Viaroli et al. 2015).

In the sea and estuaries, increased phytoplankton primary production results in, among other things, the dominance of microbial food webs over linear planktonic food chains, with non-siliceous phytoplankton species prevailing over diatom species and changes to zooplankton structure from crustacean-dominated to one composed mostly of gelatinous species, such as jellyfish and ctenophores (Taminen and Kuosa 1998; for a discussion see Condon et al. 2012). One effect of the progressing eutrophication may be the replacement of large planktonic crustacean species with small ones, as was observed in Japanese coastal waters where, due to eutrophication, the large copepods *Paracalanus* sp., *Calanus sinicus* and *Corycaeus* spp. were superseded by the small *Oithona davisae* (Turner 2004).

In fresh water, eutrophication causes crustacean-dominated zooplankton to suffer extensive qualitative and quantitative changes with significant responses to temporal variations of total phosphorus shown in the size and biomass of copepods and cladocerans (Pinto-Coehlo et al. 2005). However, the effect of eutrophication on size distribution of zooplankton is far less understood in freshwater reservoirs, possibly because the fish community is also affected. It has been argued that top-down forces can cause major shifts in the response patterns of zooplankton to eutrophication (Hulot et al. 2000). Benthic crustaceans cope with oxygen depletion to varying degrees (from days to a month), depending on species. Fish and crustaceans have higher oxygen requirements than less mobile animals. Generally, oxygen deficiency in sediments forces mobile benthic invertebrates to move to the surface where they are more exposed to predation. However, if as a result of algal biomass decay there is an oxygen demand and hydrogen sulfide is released, all organisms die immediately (Korpinnen and Bonsdorff 2015). For example, such lethal and sublethal effects of algal blooms on crustaceans have been shown in a wide range of key marine planktonic copepods, such as *Acartia* spp. and *Calanus* spp. (Landsberg 2002).

Pollution

The Anthropocene has seen an increasingly diverse cocktail of chemicals introduced to the aquatic environment. Pesticide pollution includes chemicals used to kill or control pests in agriculture, mainly herbicides, insecticides, fungicides, nematocides, and rodenticides. Insecticides are also used in silviculture. Pesticides arrive in aquatic ecosystems mostly with surface or groundwater flow from areas where they were applied or from point sources such as factories or improperly secured storage places (Damalas and Eleftherohorinos 2011, Schäfer et al. 2011). Rivers, lakes, estuaries, and coastal waters are heavily affected due to their relative proximity to pollution sources. Marine environments in particular suffer from antifouling chemicals that prevent the settlement of sessile fauna, such as mussels, polychaetes, and barnacles, on vessel hulls (van Wezel and van Vlaardingen 2004). Studies on the influence of herbicides and insecticides on aquatic ecosystems have concentrated on fish and other vertebrates (Damalas and Eleftherohorinos 2011). Pesticide-related fauna kills have been noted, involving not only fish but also amphibians, turtles, water birds, and mussels (Damalas and Eleftherohorinos 2011, Schäfer et al. 2011). Fewer studies have involved crustaceans. Acute toxicity tests have showed that crustaceans are 10 times more sensitive to insecticides than fish (Maltby et al. 2005). Thus, it can be assumed that some pesticides have negative effects even at low concentrations. Several insecticides and herbicides show neurotoxic activity, and they heavily affect olfactory senses in crustaceans (Olsén 2011). For example, males of marine corophiid amphipods and crabs had a reduced ability to locate females in the presence of pesticides such as antifouling agents or copper compounds (Krång 2007). Freshwater crayfish exposed to pesticides lost the ability to track food sources (Wolf and Moore 2002), while in cladocerans that normally, in reaction to predator scent, grow larger "helmets" protecting them from being eaten, pesticides diminish such changes in morphology (Sakamoto et al. 2006). Another issue is the synergistic, hard-to-predict effect of several pesticides co-occurring in an aquatic environment, as in the case of the freshwater cladocerans, whose mobility was effectively hampered by a combination of pesticides more than others (Nørgaard and Cedergreen 2010). A thorough comparative summary and discussion of the effects of pesticides and other chemicals on crustacean and fish behavior was provided by Olsén (2011).

The overuse of medicines by humans and in food production is nowadays having a serious impact not only on human health but also on other organisms (Brodin et al. 2014, Gaw et al. 2014). This threat is particularly serious in aquatic ecosystems such as rivers, lakes, and coastal waters where these substances accumulate from community waste (Fig. 18.3). For example, it has been known for more than two decades that some groups of compounds used in medicine disrupt endocrine systems as well as physiological and ontogenetical processes that are hormonally steered (deFur and Williams 2015). The well-known endocrine disrupters are xenoestrogens, such as 17α-ethinylestradiol (EE2), the synthetic sex hormone that imitates estrogen in most contraceptive drugs. Other xenoestrogens are widely used as pesticides (e.g., Atrazine, Endosulfan), compounds in food and beverage containers (e.g., Bisphenol A), compounds of plastic elements in electronics and textiles (polybrominated biphenyls), or animal growth promoters in agriculture (Zeranol, banned in EU). All are commonly detected in aquatic environments (Mnif et al. 2011) and have hormone-disruptive effects on both wildlife and humans, particularly of reproductive processes. Few studies have dealt with effects of xenoestrogens on crustaceans. Watts et al. (2001, 2002) showed that EE2 and Bisphenol A change the sex ratio in freshwater *Gammarus pulex* to a female-dominated population but that did not affect mating behavior. In another amphipod *Hyalella azteca*, exposure to different concentrations of EE2 resulted in cases of hermaphroditism, disturbed maturation of the germ cells, and disturbed spermatogenesis, even in F1 generation (Vandenbergh et al. 2003). Jaser et al. (2003) found out that exposure to EE2 caused higher mortality of the newly hatched juveniles of cladocerans and generally reduced the duration of juvenile stages. However, while a range of xenoestrogens and other vertebrate steroids (androgens, progestins, and corticosteroids) have been detected in invertebrates, including crustaceans, their impact remains unknown (Janer and Porte 2007).

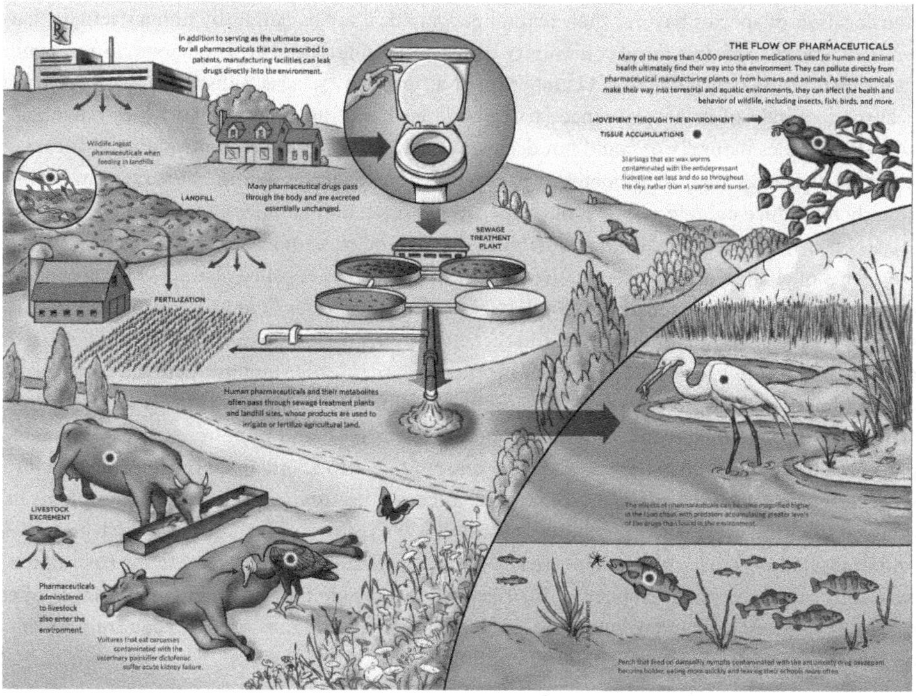

Fig. 18.3.
The flow of pharmaceuticals. Many of the more than 4,000 prescription medications used for human and animal health ultimately find their way into the environment. They can pollute directly from pharmaceutical manufacturing plants or from humans and animals. As these chemicals make their way into terrestrial and aquatic environments, they can affect the health and behavior of wildlife, including insects, fish, birds, and more. Drawing courtesy of Al Granberg. See color version of this figure in centerfold.
This figure was published online: https://www.the-scientist.com/?articles.view/articleNo/43615/title/Drugging-the-Environment/

Antibiotics are commonly used in shrimp (penaeid) aquaculture even though they are known to accumulate in shrimp tissue and to have an adverse effect on shrimp metabolism (Tu et al. 2010). Their use is also associated with environmental and human health problems, including bacterial resistance, persistence of disease in the aquatic environment, and pollution of the aquatic environment. The most frequently used include oxytetracycline, florfenicol, sarafloxacin, and enrofloxacin (Roque et al. 2001, Soto-Rodríguez et al. 2006). Also, antibiotics used in human medicine pollute the environment and alter physiological processes of crustaceans. For example, aminoglycoside antibiotics such as neomycin and streptomycin block calcium action potentials in crustacean muscles and calcium channel currents in pituitary glands (Suarez-Kurtz 1989). In another study, florfenicol, cephalexin, and dihydrostreptomycin were found to induce egg abortion, to reduce the number of mobile juveniles produced, and to significantly reduce population growth rate in cladocerans (Martins et al. 2013).

BIOLOGICAL TRANSLOCATION

The Anthropocene witnesses the alteration of natural inland aquatic and marine biogeographic units through the accelerated transfer of biota across vast distances. Biological invasions, understood as translocation of species beyond their natural geographical range caused by human activity, have led to drastic changes in natural community structures posing, together with habitat loss, a significant threat to biological diversity (Lambertini et al. 2011). A "species, subspecies, or lower taxon occurring outside of its natural range (past or present) and dispersal potential (i.e., outside the range it occupies naturally or could not occupy without direct or indirect introduction or care by humans) as well as any part, gametes or propagule of such species that might survive and subsequently reproduce are referred to as alien species (=exotic, non-indigenous, non-native, foreign)" (IUCN p. 4–5). Those that "become established in natural or semi-natural ecosystems or habitat, are agents of change, and threaten native biological diversity" are called alien invasive species (or non-indigenous invasive species, NIS) (IUCN p. 4). In a broader definition of an alien invasive species, the negative economic impact and threat to human health caused by the introduced taxon is taken into account (e.g., McNeely et al. 2001). Crustaceans, after mollusks, are the most prominent and pervasive invertebrate taxa introduced intentionally and unintentionally to new regions (Gherardi 2007). Crustaceans, some long noteworthy as food and more recently as ornamental pets, have been deliberately transported from one continent to another, but most crustacean invasive species are accidental and of no commercial interest. The number of crustacean transplants has doubled since the 1990s, and the trend is upward (Seebens et al. 2017). Due to their high ecological importance, many crustacean species are key elements of the ecosystem forming the basis of food webs and/or as ecosystem engineers. Their settlement outside natural geographical ranges often cause negative environmental consequences, while in other cases they provide economic benefits, both having considerable biogeographic implications.

Freshwater Ecosystems

Human activity has promoted the introduction and spread of freshwater crustaceans for many years. Both intentional and unintentional introductions are of great importance. Inland waters are particularly vulnerable to colonization by alien species. Globally, ~190 alien freshwater crustacean species are listed (Aleksandrov et al. 2007, Holdich and Pöckl 2007, Gherardi et al. 2009, Xu et al. 2012, USGG 2016, CABI 2017). Among them, decapods with 42 species of crayfish, copepods (46 species), and amphipods (37 species) dominate. Interestingly, these records come mostly from Europe (117 species) and North America (83 species), compared with 10 species recorded in Asia (with only two records from China), 5 in South America, 4 each in Africa and Hawaii, 3 in Australia, and one in Madagascar (Jones et al. 2009). Elton (1958) suggested that the number of invasive taxa reflects the directions of increasing trade and travel. Indeed, the two regions with the highest gross domestic product (GDP) are the European Union (EU) and the United States (EuroStat 2014), and the biggest trade has been observed within the EU (20% of the global trade) and between EU, the United States, and China. Recently, the direction of trade has changed due to the increasing commercial exchange between Asia and other regions such as Africa and Central and South America (Selfin and Hope 2011).

Obviously, such changes will influence the direction and intensity of future biological invasions. On the other hand, the number of navigable canals connecting previously isolated water basins (Fig. 18.4) may explain the disproportionate spread of a number of freshwater invasive crustaceans. Indeed, the network of navigable canals is best developed in Europe, connecting all sea basins (Bij de Vaate et al. 2002, Panov et al. 2009) and serving as migratory corridors ("highways") for freshwater fauna (Bij de Vaate et al. 2002, Galil et al. 2007, Leuven et al. 2009, Panov et al. 2009). The colonization process is enhanced by intensive shipping; the dispersal rate for some species can be as high as 460 km/year (Leuven et al. 2009). This enables extensive exchange of biota within the continent. Out of the 117 alien crustacean species recorded in fresh waters of Europe, 87 are of European origin, mainly from the Ponto-Caspian region (Black, Azov, and Caspian seas) (Aleksandrov et al.

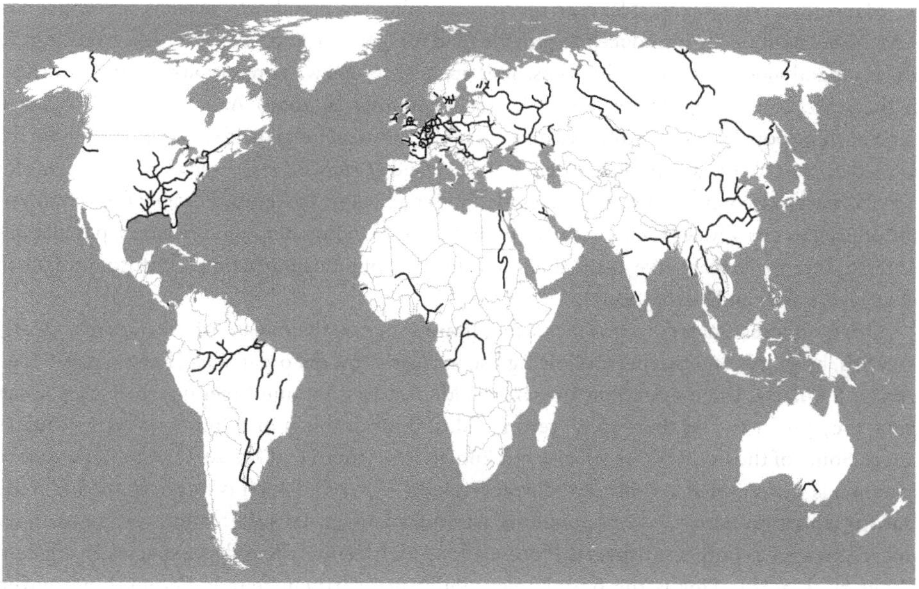

Fig. 18.4.
Navigable inland waterways that serve as migratory corridors for invasive freshwater species. Figure courtesy of David Edwards-May (http://inlandwaterwaysinternational.org/).

2007, Holdich and Pöckl 2007, Gherardi et al. 2009). Many invertebrates, including crustaceans, may be transported via weeds or other structures entangled on boats' motors and trailers, including recreational ones (e.g., Rothlisberger et al. 2010).

Notably, the first crustacean introduction took place in the 12th century, when monastic orders introduced the European crayfish *Austropotamobius pallipes* into Irish fresh waters from France (Gouin et al. 2003). In the 16th century, the second intentional introduction of crayfish took place in Sweden and Finland, where the Swedish King John III imported the noble crayfish *Astacus astacus* from Russia (Westman 1973). The next wave of introductions happened in the 19th century when the narrow-clawed crayfish *Pontastacus leptodactylus* was imported from the Ponto-Caspian region to most European countries to replace the noble crayfish decimated by the crayfish plague, a lethal disease caused by the fungus *Aphanomyces astaci* (Oomycetes). It was optimistically hoped that the Ponto-Caspian species would not be vulnerable to the pathogen (Holdich et al. 2006). The same idea was behind the introduction of the first North American species, the spiny-cheek crayfish, *Orconectes limosus*, to European inland waters in 1890. This time, the species was indeed resistant to the pathogenic fungus, but unfortunately it served as its vector and, with the spread of this American crayfish, the crayfish plague extended (and still does) its range, causing another wave of drastic decline in European crayfish populations (Pârvulescu et al. 2012). Several more crayfish species (e.g., *Cherax destructor* from Australia, *Procambarus clarkii, Orconectes rusticus,* and *Pacifastacus leniusculus* from North America) were introduced intentionally for the live food trade and restaurant needs all over the world on both the inter- and intra-continental scale (e.g., Ackefors 1999). In the United States alone, ~30 species have been translocated, as live bait between river basins, mostly by fisherman and anglers (USGG 2016). What is more, many crayfish and crab species may move over land for long distances and spontaneously and rapidly expand their range to new regions (e.g., Ramalho and Anastacio 2015).

Many other crustaceans, at least 40 species of amphipods, mysids, cladocerans, and copepods, have been intentionally introduced to provide food for fish in newly built water reservoirs, e.g., in the 1950s and 1960s, but also in the 1990s, after opening several water reservoirs along major rivers in the Soviet Union, e.g., Dnieper and Volga (Aleksandrov et al. 2007). Another important reason for launching populations of exotic crustaceans is biocontrol of pests such as weeds or snails. In the United States, at least two species, *Orconectes causeyi* and *O. rusticus*, were used as control agents for aquatic weeds in commercially used lakes (Hobbs et al. 1989). In Kenya, *Procambarus clarkii* was introduced to many waterbodies in the 1950s and 1960s to reduce, through predation, populations of freshwater snails that serve as the intermediate hosts of human-infecting schistosomes (Mkoji et al. 1999, Foster and Harper 2007).

Often underestimated, but of increasing importance, is the release of crustaceans, mostly crayfish, by pupils and teachers following use in class as model organisms (Larson and Olden 2008). Similarly, when aquarium owners decide not to continue breeding their crustacean pets, they prefer to "let them free," getting rid of their problem in a "humane" way (Faulkes 2015). Some of those "pets" are species not known to science (Faulkes 2015). An extraordinary example is the marbled crayfish (= Marmorkrebs) (Fig. 18.5), which is raised in the pet trade but for which no native wild populations are so far known. Its systematic position is under debate, treated as both a subspecies *Procambarus fallax* f. *virginalis* and as a species *P. virginalis* (Vogt et al. 2015). This is the only known obligatory parthenogenetic decapod, moreover, with a very high reproductive potential. Every eight or nine weeks, the female releases 120 juveniles on average, three times more than many endangered Astacidae (Alwes and Scholtz 2006). The high risk of invasion is due to its popularity in the aquarium pet trade around the world (Martin et al. 2010, Faulkes 2015). Its temperature tolerance precludes massive invasion in moderate climates (Martin et al. 2010), but the species already poses a threat to native crayfish in Madagascar, where growing populations have been recorded in the wild (Jones et al. 2009, Kawai et al. 2009).

Fig. 18.5.
The aquatic pet: obligatory parthenogenetic marbled crayfish = Marmorkrebs (*Procambarus virginalis*). Photo courtesy of Piotr Gadawski. See color version of this figure in centerfold.

Marine Ecosystems

Whereas the earliest known translocations in freshwater systems were deliberate and many still are, the movement of marine crustaceans is largely unintentional, with one significant exception. The American shrimp *Penaeus vannamei* is now cultivated worldwide; its production in 2014 reached ~3.7 million tons, which is ~50% of the total crustacean trade and most of the world production of cultured penaeid shrimps (FAO 2017). Although commercial culture began in South and Central America, nowadays the biggest producers are Asian countries, with China as a leader, where the species was introduced commercially in 1998. Soon after, it was successfully introduced to other Asian countries. Each year, an upward trend is observed (in 2014 the mass of farmed *P. vannamei* had increased 13 times since 2001; FAO 2017), while production of other species (i.e., *Penaeus chinensis, P. monodon,* and *P. japonicas*) decreases (Liao and Chien 2011 and references therein). The success of *P. vannamei* cultivation can be explained by extreme euryhalinity of adults ranging from 1–2 psu to 40 psu (Menz and Blake 1980), allowing successful inland aquaculture (FAO 2017). Selective breeding programs focusing on growth and disease resistance have been established and genomic selection programs evaluated, promoting reduced losses in cultivated populations (Castillo-Juárez et al. 2015 and references therein). Interestingly, despite numerous escapes from aquaculture into non-native waters, wild reproducing populations have not yet been confirmed outside the native range of the species. Thus, the major threat caused by the cultivation of *P. vannamei* is its role as a vector of a range of infectious diseases (CABI 2017) and alienation of coastal habitats from other uses.

Massive marine translocations started with the first transoceanic voyages, becoming more serious at the beginning of the 17th century (Carlton 2011). Ship-fouling species were part of those voyages. Establishment of fouling species occurs when adults have been transported and larvae are released in new locations. Fouling on ships' hulls continues to contribute to the movement of species from one place to another, but the use of ballast water in container ships has more recently added to this traffic. A third pathway has been the transport of aquaculture species, notably oysters, between continents and coasts, with their associated epifauna.

A wide variety of crustaceans have been moved by ships. Barnacles were among the first to be recognized. So far, the coasts of the Americas have been invaded by 12 thoracican barnacle species. Three have colonized the Pacific coasts from the Atlantic— *Amphibalanus improvisus, A. eburneus,* and *A. subalbidus*—and two more came from the Indo-West Pacific—*A. amphitrite* and *A. reticulatus*. The Atlantic coasts of the Americas have been colonized by six species of Pacific origin—*A. amphitrite, A. reticulatus, Balanus trigonus, B. glandula, Striatobalanus amaryllis,* and *Megabalanus coccopoma*. One species, *A. subalbidus*, spread from the Western North Atlantic to the Western South Atlantic (Carlton 2011). There are five more cryptogenic species of uncertain status, meaning they could be native or alien. This transoceanic human-mediated dispersal of barnacles was first remarked on in the 19th century, but many species seemed to have been overlooked, and museum collections may give us a more information on the numbers of introduced barnacle species (e.g., Biccard and Griffiths 2016).

Some barnacles have spread via ship transport so effectively that now they have worldwide distribution. One of the most prominent examples is *Amphibalanus improvisus* present in the eastern and western North Atlantic, coasts of European seas (including Black and Caspian seas), west coast of Africa, Red Sea, coasts of Indonesia, Malaysia, Australia, New Zealand, China, Japan, and Pacific coast of North and South America (CABI 2017). It started to spread with vessels and was one of the earliest newcomers in many areas, possibly including South America, which is treated as its native range. *Amphibalanus improvisus* was one of the earliest crustacean introductions to European coastline waters, first recorded in 1844 (Leppäkoski et al. 2002) in the Vistula Lagoon of the Baltic Sea. However, archeological records from Belgium (the North Sea) suggest an estimated time of arrival in the 17th century (Kerckhof 2002). Moreover, fossil records from Spain suggest its wide distribution in the Pliocene on both Atlantic coasts, and, as possibly shown for the bivalve *Mya arenaria* (Lasota et al. 2004), it disappeared from European waters during the last glaciation (Kerckhof 2002). This may be a case of recolonization. This species has been found attached to the carapaces of other invasive species such as crabs (e.g., *Eriocheir sinensis*), which enhances its dispersal potential (Naser et al. 2015).

Lately a new and surprising pathway for barnacles as invasive species has been discovered. Tøttrup et al. (2010) recorded adult barnacles of two species, *Fistulobalanus pallidus* and *F. albicostatus*, attached to plastic leg rings on the Lesser Black-backed Gull (*Larus fuscus*) in Northern Europe. The gull is a long-distance palaearctic migrant breeding in the temperate regions and migrating to the subtropics for the non-breeding period. The two species of barnacles inhabit mangroves and estuaries where the bird could have spent time feeding, allowing the barnacle cyprids to settle.

Ballast water is now a prominent vector for crustacean intercontinental transport, especially those with larval stages. But adults of many species, besides barnacles, can be transported on the outside of ships. The most celebrated example is the European shore crab *Carcinus maenas*, one of the world's most successful aquatic invaders, having established on every continent with temperate shores and being listed among the 100 World's Worst Invasive Alien Species (GISD 2016). The record of the species outside its native range came from the early 19th century from North America and the Red Sea, before the opening of the Suez Canal, suggesting its introduction with a solid ballast or/and hull fouling (Carlton and Cohen 2003). Further widening of its geographic range in late 19th century to Australia, Cape Town, Japan, Nova Scotia, and Argentina was likely due to translocation of larvae in ballast water (Darling et al. 2008). The planktonic larval stage may survive up to 80 days depending on temperature, and this duration is sufficient for successful transport in ballast tanks (Cohen et al. 1995) between ports thousands of kilometers apart.

The fact that crustaceans can be transported as adults is evidenced by the occurrence of exotic peracarids which without larval stage. The list of marine amphipod species introduced worldwide is long. In the Mexican part of the Gulf of Mexico, 11 species of invasive peracarids are listed (Winfield et al. 2011). Some spread successfully all around the world—e.g., *Caprella mutica* and *C. scaura*, which in a few decades colonized the coasts of the Atlantic and Pacific, including Oceania and the south coast of Africa. They are also recorded in the Mediterranean, where numerous populations occur (Cook et al. 2007, Guerra-García et al. 2011). Mostly, their dispersal was associated with the fouling on commercial ship hulls, and molecular analysis revealed that their global distribution is most likely a result of several introductions from multiple sources (Ashton et al. 2008). In Australia, an extensive study revealed numerous alien species introduced to Port Phillip Bay since first visited by Europeans in 1802 (Hewitt et al. 2004). Three species of caprellids (*Caprella equilibra, C. penantis,* and *C. scaura*), three species of corophiid amphipods (*Corophium acherusicum, Co. insidiosum,* and *Co. sextonae*), and two isopods (*Paracerceis sculpta* and *Cirolana harfordi*) were recorded as introduced. Novel introductions continue to be discovered since this study, e.g., the Japanese shrimp *Crangon uritai* (Taylor and Komai 2011).

Lately, the role of tourist marine traffic was also underlined as a very effective vector of such fouling amphipod species (Marchini et al. 2016). Especially vulnerable are places isolated for millions of years but now, being of high tourist interest, coveted for their secluded spots. Such an example is Antarctica, which was free of invasive crustaceans until 2003 when the first male and female majid spider crab *Hyas araneus* were collected at the Antarctic Peninsula (Tavares and De Melo 2004).

The opening of the Suez Canal in 1869 provided a new, almost unidirectional, route for invasion of a wide variety and increasing number of species of marine biota. Today, the Mediterranean Sea is home to 821 alien species (Galil et al. 2016, Zenetos et al. 2017), with 751 regarded as threatening local communities (Galil et al. 2017). Two-thirds of these intruders entered Mediterranean waters through the Suez Canal, including at least 64 crustacean species (Galil 2011). This number is greater each year as projects to widen the canal are still in progress, with the last one finished in 2015 when an additional 72.4-km-long lane called the "New Suez Canal" was opened. Such procedures may easily result in higher propagule pressure of non-native species in the Mediterranean Sea (Galil et al. 2017). In contrast, the freshwater nature of the Panama Canal, the biggest interoceanic "shortcut," does not allow marine species to cross freely from one side to another. Here, ship-mediated transfer, specifically with ballast waters, plays a major role in the exchange of biota, not the Canal itself (Galil et al. 2007).

Consequences

The translocation of crustaceans results in unpredictable geographic range expansion. The "leaders" in this are present on five continents: the estuarine-marine European shore crab *Carcinus maenas* (Carlton and Cohen 2003, Tepolt and Somero 2014) and the red swamp crayfish *Procambarus clarkii* (Gherardi 2006). Depending on the way and dynamics (single vs. multiple introductions, one or several source points of introduction) of the invasion process and the abiotic and biotic factors met in the newly colonizing regions, populations of alien species may face very different situations. First, they may be completely isolated from the source populations, and, by the forces of natural selection, a rapid speciation can be expected (Lee 2002). Although the phenomenon is of high importance, it is still scarcely studied. The copepod *Eurytemora affinis*, colonizing the Pacific coast of the United States, has been demonstrated to be reproductively isolated from the native Atlantic populations (Lee 2000). Most studies concentrate on the effects that can be observed at a genetic level, such as bottleneck, founder effect, or genetic drift. Diminishing genetic diversity was observed in several crustacean species, including amphipods, cladocerans, and mysids, but unexpectedly no negative effect on viability was found in invasive populations (Rewicz et al. 2017). This can be explained by the reduction (purging) of recessive deleterious alleles in a population that inbreeds, which may result in high fitness and promote the invasion success of populations with low genetic diversity. On the other hand, an example of significant increase in genetic richness by admixture and hybridization of different invasive lineages of the same or closely related species was observed in *C. maenas* after multiple introduction events from different sources (Darling et al. 2008).

Evidence is mounting for adaptations in invasive crustacean species in response to selective pressures of new environments, e.g., to a different temperature or salinity regime. The North American freshwater crayfish *P. clarkii*, inhabiting waters of temperatures oscillating from 21°C to 30°C in its native range, adapted to much colder conditions and was discovered to reproduce in springs with an average temperature of ~13°C in a colonized area (Peruzza et al. 2015). The brackish water copepod *E. affinis* has invaded independently numerous fresh waters in recent years. Its physiological tolerance and plasticity, increased body fluid regulation, and evolutionary shifts in ion transporter activity and expression has evolved during the invasion process, both in nature and in experimental conditions (Lee 2016).

Another important outcome of the invasion process may be the loss of typical parasites and/or acquisition of new ones (Parasite Release Hypothesis) (Lagrue 2017). Parasite release has been found in several crab species, thereby promoting their invasion success. The European shore crab is the best example, but in general most alien crustaceans are accompanied by a number of their parasites with different pathogenicity and variety of life cycles (Lagrue 2017). On the one hand, these parasites may regulate the population of their invasive host, serving as biocontrol agents, but there are also examples of the positive effects of the presence of the "hitchhikers" on the host. Sex-ratio-distorting parasites, such as the microsporidian *Fibrillanosema crangonycis* infecting the amphipod *Crangonyx pseudogarcilis*, invasive in European inland waters, may promote its invasion by the overproduction of female offspring (Slothouber Galbreath et al. 2010).

Parasites infecting crustaceans range from viruses to metazoans, such as trematodes or acanthocephalans (Lagrue 2017). They also act as intermediate hosts for a range of helminths, including *Paragonimus* spp., that may also be pathogenic to people (Holdich et al. 2006, Loureiro et al. 2015). In the cases of host shift to native species, the parasites may pose a threat to local fauna. One of the most spectacular examples is the crayfish plague that spread together with the North American crayfish, which is resistant to this pathogen but serves as its vector. This oomycete appeared to be lethal to European crayfishes and led to the extinction of many populations of vulnerable species (Holdich et al 2009). There are many more parasites and pathogens introduced with their hosts that can act in a similar way, e.g., white spot syndrome virus (WSSV), a virus causing vibrosis, and the microsporidium *Thelohania contejeani*. The spread and impact of many of these pathogens in new hosts is hard to predict and is often ignored as they have low or even no impact on the typical host (Roy et al. 2016).

Some invasive parasites are themselves crustaceans, copepods especially (e.g., Raibaut et al. 1998). Several species were recorded as introduced into the Mediterranean and switched to native fish hosts (El-Rashidy and Boxshall, 2009, 2014). Other common alien crustacean parasites are rhizocephalans, parasitic barnacles that were suggested to regulate the invasive crab population through increased mortality, castration of the host, and inducing morphological and behavioral changes inhibiting the host's reproduction (Kuris 1997). However, several studies did not confirm a decrease of host populations of the invasive crab *Charybdis longicollis*, introduced to the Mediterranean around 40 years earlier than its parasitic barnacle *Heterosaccus dollfusi* (Innocenti et al. 2017). Thus, the role of this parasite in regulating host populations is still under question.

Invasive crustaceans have considerable impact on native crustacean species in newly colonized ecosystems. The successful non-indigenous crustacean species, when established and integrated, become important components of an existing community and may strongly influence each other directly and indirectly (Gherardi 2007). Their presence also affects socioeconomic and human health. Surprisingly, in contrast to the estimate that 70% of native North American freshwater species extinctions during the 20th century are a result of biological invasion, most alien crustaceans do not exhibit noticeable and significant impacts in colonized waters (Gherardi et al. 2009, Bugnot et al. 2015). However, those that do lead to multilevel catastrophic changes in the colonized ecosystems, including species displacement and the reconstruction of food webs (e.g., Rodriguez et al. 2005; Fig. 18.6). According to Gherardi et al. (2009) such changes are the effects of:(1) competition and predation resulting in the outnumbering of native fauna (e.g., *Dikerogammarus villosus*; Rewicz et al. 2014), (2) hybridization or introgression with native species leading to the consequent reduction of genetic diversity (e.g., *Orconectes rusticus* with *O. propinquus*; Olden et al. 2006), (3) transmission of parasites and diseases (e.g., North American crayfishes being a vector of fungi *Aphanomyces astaci*), (4) habitat modification and changes to ecosystem functioning, and (5) damage to socioeconomic, recreation, and human health and well-being. In Germany alone, the costs of the damage of waterways due to the burrowing activity of the Chinese mitten crab *Eriocheir sinensis* are estimated to add up to ~€80 million since 1912. Another top invasive species, *C. maenas* causes $22.6 million of damage per year through its predation of shellfish on the east coast of the United States (GISD 2016).

The ecological impacts are especially dangerous when the aliens meet endangered species or colonize areas with a high level of endemism or ecological importance and quality. According to the IUCN Red List of Threatened Species, only one species, *Pacifastacus nigrescens*, has gone extinct due to the introduction of two alien invasive crayfish species (Schuster et al. 2010). However, 160 more species are threatened as a result of biological invasions, including negative interactions with alien

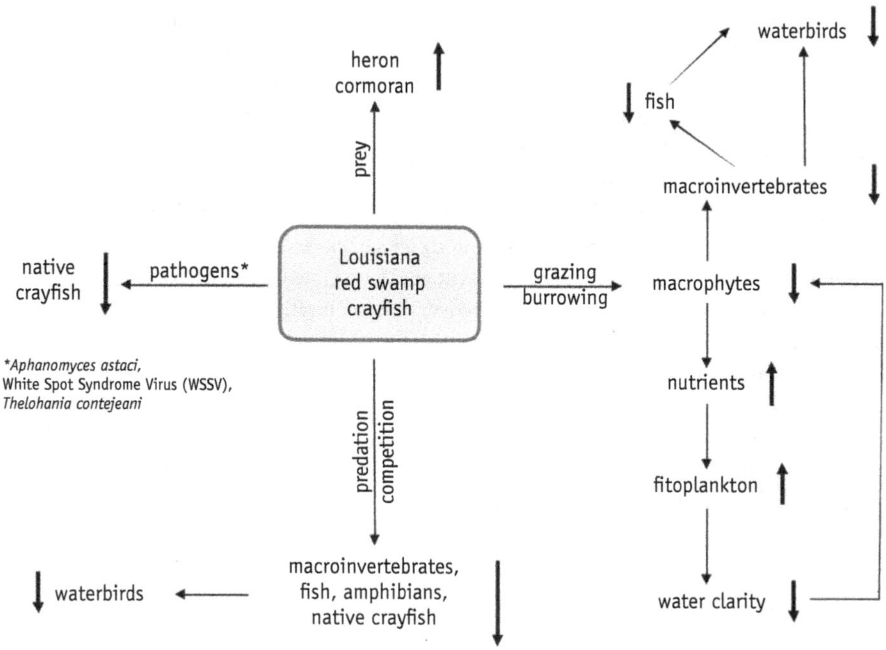

Fig. 18.6.
Multilevel impact of one of the most prominent invasive crustaceans, red swamp crayfish, *Procambarus clarkii*. Data from Rodriguez et al. (2005).

crustaceans. In Australia, a continent with a high degree of endemism, the invasive crayfish *Cherax caini* may lead to the extinction of its close relative, the critically endangered *C. tenuimanus*, in the last river system where a wild population of the latter still exists (Guildea et al. 2015). Reproductive isolation between the two species is imperfect, so hybridization and introgression will lead to the deterioration of the genetic pool of the native species and its replacement by hybrid individuals. In addition, this invasive crayfish is very competitive and displaces the native one from preferred habitats (Guildea et al. 2015).

At a time of increasing interest in biological invasions, we lack studies focused on their impact (Galil et al. 2016). Most studies consider the influence of those species that reach extraordinary densities in colonized waters. The isopod *Cirolana harfordi* occurs in great densities in Sydney Harbour, having the second highest biomass of species inhabiting oyster beds (Bugnot et al. 2015). However, in spite of its abundance and being a scavenger and detritus feeder that may have decreased organic matter, no impact on the local community was detected. On the other hand, we have examples of invasive crustaceans that lead to drastic changes in the native communities, including native species displacement (e.g., *Caprella mutica*, mantis shrimp *Erugosquilla massavensis*) (Shucksmith et al. 2009, Galil 2011). While studies are few, species interactions can be strongly modulated by abiotic factors. Ros et al. (2015) showed in their comprehensive field-based study that the success of invasive *Caprella scaura* and its ability to outcompete native *C. equilibra* was strongly influenced by salinity and temperature. Regardless of the fact that both species are euryhaline, invasive species dominated in the Mediterranean coasts where salinities and temperatures are higher than in European coasts of the Atlantic. This study stresses that generalizations based on single studies may be misleading. Nevertheless, there are some "flag" invasive species known from their multilevel negative impact such as the crabs *Carcinus maenas* or *Eriocheir sinensis*, both listed among 100 of the

World's Worst Invasive Alien Species (GISD 2016). In both cases, the introduction of the species has resulted in a decline of the number and biomass of native species through direct and indirect interactions such as predation, competition, and herbivory (Klassen and Locke 2007, Bentley 2011). Moreover, they significantly increased bank erosion by burrowing activity impacting the nutrient cycle in the ecosystem. However, both species are possibly more notorious because of their negative economic impact on fisheries and aquaculture.

Nowadays there are no marine ecosystems free from invasive species. The Mediterranean Sea is the richest in the number of alien species among all marine ecosystems worldwide (Galil 2011), followed by Port Phillip Bay in Australia (Hewitt et al. 2004), which may have overtaken the San Francisco Bay (Cohen and Carlton 1995). Moreover, the colonization process of those water bodies is an old one, dating from at least 300 years of human transcontinental movement. Such ecosystems are already highly degraded and disturbed by biological pollutants (i.e., invasive species). These ecosystems no longer function in a "natural, virgin" way, and in accordance with the "invasional meltdown hypothesis," they are more vulnerable to future introductions that can also lead to an acceleration of the negative impact of the newcomers (Simberloff and von Holle 1999). To stop the invasion seems unrealistic and impossible, but efforts should be made to lower the dynamic of the process as long as possible.

FUTURE PROSPECTS

It is hard to provide any optimistic conclusion about the future of crustacean biodiversity. We may rather expect that growing human pressure will, through the various means discussed above, further accelerate the non-natural dispersal and extinction rate. This will only worsen the situation we face now, with the possibility of future extinctions. Many species disappear before we even get to know them. The protection of particular species may help only in the cases of some spectacular, big, and commercially important species such as decapods: some lobsters, crayfish, and crabs. In most cases, the current and future strategies for slowing the decline of the threatened species, particularly in fresh waters, should include the sustainable management of their habitats and the collection of more baseline data on their diversity, population and distribution patterns, and conservation status. These measures, including multiproxy and holistic approaches to biodiversity protection, should be introduced in the case of marine ecosystems. However, due to the large spatial scale and global political issues, this may be much harder to implement.

ACKNOWLEDGMENTS

We thank Martin Thiel for his valuable editorial comments.

REFERENCES

Ackefors, H. 1999. The positive effects of established crayfish introductions in Europe. Pages 49–61 *in* F. Gherardi and D.M. Holdich, editors. Crayfish in Europe as Alien Species. How to Make the Best of a Bad Situation? Balkema, Rotterdam, The Netherlands.

Aleksandrov, B., A. Boltachev, T. Kharchenko, A. Liashenko, M. Son, P. Tsarenko, and V. Zhukinsky. 2007. Trends of aquatic alien species invasions in Ukraine. Aquatic Invasions 27:215–242.

Altermatt, F., V.I. Pajunen, and D. Ebert. 2008. Climate change affects colonization dynamics in a metacommunity of three *Daphnia* species. Global Change Biology 14:1209–1220.

Alwes, F., and G. Scholtz. 2006. Stages and other aspects of the embryology of the parthenogenetic Marmorkrebs (Decapoda, Reptantia, Astacida). Development Genes and Evolution 216:169–184.

Appelhans, Y.S., J. Thomsen, C. Pansch, F. Melzner, and M. Wahl. 2012. Sour times: seawater acidification effects on growth, feeding behaviour and acid–base status of *Asterias rubens* and *Carcinus maenas*. Marine Ecology Progress Series 459:85–98.

Aronson, R.B., R.M. Moody, L.C. Ivany, D.B. Blake, J.E. Werner, and A. Glass. 2009. Climate change and trophic response of the Antarctic bottom fauna. PLOS ONE 4:e4385.

Aronson, R.P., K.E. Smith, S.C. Vos, J.B. McClintock, M.O. Amsler, P.O. Moksnes, D.S. Ellis, J. Kaeli, H. Singh, J.W. Bailey, J.C. Schiferl, R. van Woesik, M.A. Martin, B.V. Steffel, M.E. Deal, S.M. Lazarus, J.N. Havenhand, R. Swalethorp, S. Kjellerup, and S. Thatje. 2015. No barrier to emergence of bathyal king crabs on the Antarctic shelf. Proceedings of the National Academy of Science USA 112:12997–13002.

Ashton, G.V., M.I. Stevens, M.C. Hart, D.H. Green, E.J. Cook, and K.J. Willis. 2008. Mitochondrial DNA reveals multiple Northern Hemisphere introductions of *Caprella mutica* (Crustacea, Amphipoda). Molecular Ecology 7:1293–1303.

Auel, H., and W. Ekau. 2009. Distribution and respiration of the high-latitude pelagic amphipod *Themisto gaudichaudi* in the Benguela Current in relation to upwelling. Progress in Oceanography 83:237–241.

Bechmann, R.K., I.C. Taban, S. Westerlund, B.F. Godal, M. Arnberg, S. Vingen, A. Ingvarsdottir, and T. Baussant. 2011. Effects of ocean acidification on early life stages of shrimp (*Pandalus borealis*) and mussel (*Mytilus edulis*). Journal of Toxicology and Environmental Health Part A 74:424–438.

Bentley, M.G. 2011. The global spread of the Chinese Mitten Crab *Eriocheir sinensis*. Pages 107–127 *in* B.S. Galil, P.F. Clark, and J.T. Carlton, editors. In the Wrong Place—Alien Marine Crustaceans: Distribution, Biology and Impacts. Springer, Dordrecht, The Netherlands.

Biccard, A., and C.L. Griffiths. 2016. Additions to the barnacle (Crustacea: Cirripedia) fauna of South Africa. African Zoology 51:99–116.

Bij de Vaate, A., K. Jażdżewski, H.A.M. Ketelaars, S. Gollasch, and G. Van der Velde. 2002. Geographical patterns in range extension of Ponto-Caspian macroinvertebrate species in Europe. Canadian Journal of Fisheries and Aquatic Sciences 59:1159–1174.

Brodin, T., S. Piovano, J. Fick, J. Klaminder, M. Heynen, and M. Jonsson. 2014. Ecological effects of pharmaceuticals in aquatic systems—impacts through behavioural alterations. Philosophical Transactions of the Royal Society B 369:20130580.

Bugnot, A.B., R.A. Coleman, W.F. Figueira, and E.M. Marzinelli. 2015. Community-level impacts of the invasive isopod *Cirolana harfordi*. Biological Invasions 17:1149–1161.

Cahill, A.E., M.E. Aiello-Lammens, M.C. Fisher-Reid, X. Hua, C.J Karanewsky, H. Yeong Ryu, G.C. Sbeglia, F. Spagnolo, J.B. Waldron, O. Warsi, and J.J. Wiens. 2017. How does climate change cause extinction? Proceedings of the Royal Society B 280:20121890.

Carlton, J.T. 2011.The global dispersal of marine and estuarine crustaceans. Pages 3–23 *in* B.S. Galil, P.F. Clark, and J.T. Carlton, editors. In the Wrong Place—Alien Marine Crustaceans: Distribution, Biology and Impacts. Springer, Dordrecht, The Netherlands.

Carlton, J.T., and A.N. Cohen. 2003. Episodic global dispersal in shallow water marine organisms: the case history of the European shore crabs *Carcinus maenas* and *C. aestuarii*. Journal of Biogeography 30:1809–1820.

Castillo-Juárez, H., G.R. Campos-Montes, A. Caballero-Zamora, and H.H. Montaldo. 2015. Genetic improvement of Pacific white shrimp [*Penaeus* (*Litopenaeus*) *vannamei*]: perspectives for genomic selection. Frontiers in Genetics 6:93.

Centre for Agriculture and Biosciences International (CABI). 2017. Invasive Species Compendium. CAB International, Wallingford, England. www.cabi.org/isc.

Cohen, A.N., and J.T. Carlton. 1995. Nonindigenous aquatic species in a United States estuary: a case study of the biological invasions of the San Francisco Bay and delta. A report of the United States fish and wildlife service. The National Sea Grant College Program, Connecticut Sea Grant, Washington. http://www.nfrcg.gov/nas/sfinvade.htm.

Cohen, A.N., J.T. Carlton, and M.C. Fountain. 1995. Introduction, dispersal and potential impacts of the green crab *Carcinus maenas* in San Francisco Bay, California. Marine Biology 122:225–238.

Condon, R.H., W.M. Graham, C.M. Duarte, C.A. Pitt, C.H. Lucas, S.H.D. Haddock, C.R. Sutherland, K.L. Robinson, M.N. Dawson, M.B. Decker, C.E. Mills, C.E. Purcell, A. Malej, H. Mianzan, S. Uye, S. Gelcich, and L.P. Madin. 2012. Questioning the rise of gelatinous zooplankton in the world's oceans. BioScience 62:160–169.

Cook, E.J., M. Jahnke, F. Kerckhof, D. Minchín, M. Faasse, K. Boos, and G. Ashton. 2007. European expansion of the introduced amphipod *Caprella mutica* Schurin 1935. Aquatic Invasions 2:411–421.

Costanza, R., R. dArge, R. deGroot, S. Farberparallel, M. Grasso, B. Hannon, K. Limburgstar, S. Naeem, RV. O'Neill, J. Paruelo, R.G. Raskin, P. Sutton, and M. van den Belt. 1997. The value of the world's ecosystem services and natural capital. Nature 387:253–260.

Crain, C.M., B.S. Halpern, M.W. Beck, and C.V. Kappel. 2009. Understanding and managing human threats to the coastal marine environment. Annals of the New York Academy of Sciences 1162:39–62.

Damalas, C.A., and I.G. Eleftherohorinos. 2011. Pesticide exposure, safety issues, and risk assessment indicators. International Journal of Environmental Research and Public Health 8:1402–1419.

Darling, J.A., M.J. Bagley, J. Roman, C.K. Tepolt, and J.B. Geller. 2008. Genetic patterns across multiple introductions of the globally invasive crab genus *Carcinus*. Molecular Ecology 17:4992–5007.

Davies, I.J. 1989. Population collapse of the crayfish *Orconectes virilis* in response to experimental whole-lake acidification. Canadian Journal of Fisheries and Aquatic Sciences 46:910–922.

Davis, K., J. Hill, T. Astrop, and M. Wills. 2016. Global cooling as a driver of diversification in a major marine clade. Nature Communications 7:13003.

deFur, P.L., and L.E. Williams. 2015. Endocrine-disrupting chemicals. Pages 461–476 *in* E. Chang and M. Thiel, editors. The Natural History of the Crustacea, Vol. 4: Physiology. Oxford University Press, New York.

Domisch, S., S.C. Jahnig, and P. Haase. 2011. Climate-change winners and losers: stream macroinvertebrates of a submontane region in Central Europe. Freshwater Biology 56:2009–2020.

Dubinsky, Z., and N. Stambler. 2010. Coral Reefs: An Ecosystem in Transition. Springer, Dordrecht, The Netherlands.

Dudgeon, D. 2012. Threats to freshwater biodiversity globally and in the Indo-Burma Biodiversity Hotspot. Pages 1–12 *in* D.J., Allen, K.G. Smith, and W.R.T. Darwall, editors. The Status and Distribution of Freshwater Biodiversity in Indo-Burma. IUCN, Cambridge, UK.

Dudgeon, D., A.H. Arthington, M.O. Gessner, Z.I. Kawabata, D.J. Knowler, C. Lévêque, R.J. Naiman, A.H. Prieur-Richard, D. Soto, M.L.J. Stiassny, and C.A. Sullivan. 2006. Freshwater biodiversity: importance, threats, status and conservation challenges. Biological Reviews 81:163–182.

El-Rashidy, H.H., and G.A. Boxshall. 2009. Parasites gained: alien parasites switching to native hosts. Journal of Parasitology 95:1326–1329.

El-Rashidy, H.H., and G.A. Boxshall. 2014. A new parasitic copepod (Cyclopoida: Bomolochidae) from a ponyfish (Leiognathidae) caught in Egyptian Mediterranean waters, with a review of hosts and key to species of *Nothobomolochus*. Systematic Parasitology 87:111–126.

Elton, C. 1958. The Ecology of Invasions by Animals and Plants. Methuen, London.

Engblom, E., and P. Lingdell. 1984. The mapping of short-term acidification with the help of biological pH indicators. Institute of Freshwater Research Drottningholm 61:60–68.

Epstein, P., H.F. Diaz, S. Elias, G. Gabherr, and N.E. Graham. 1998. Biological and physical signs of climate change: focus on mosquito-borne diseases. Bulletin of the American Meteorological Society 78:410–417.

EuroStat. 2014. EU Transport in Figures. Statistical Pocketbook. Publications Office of the European Union, Luxembourg.

FAO. 2017. Cultured Aquatic Species Information Programme—*Penaeus vannamei* (Boone, 1931). http://www.fao.org/fishery/culturedspecies/Penaeus_vannamei/en.

Faulkes, Z. 2015. The global trade in crayfish as pets. Crustacean Research 44:75–92.

Foster, J., and D. Harper. 2007. Status and ecosystem interactions of the invasive Louisianan red swamp crayfish *Procambarus clarkii* in East Africa. Pages 91–101 *in* F. Gherardi, editor. Biological Invaders in Inland Waters: Profiles, Distribution, and Threats. Springer, Dordrecht, The Netherlands.

Foster, G.L., D.L. Royer, and D.J. Lunt. 2017. Future climate forcing potentially without precedent in the last 420 million years. Nature Communications 8:14845.

France, R.L., and P.M. Stokes. 1987. Life stage and population resistance and tolerance of *Hyalella azteca* (Amphipoda) to low pH. Canadian Journal of Fisheries and Aquatic Sciences 44:1102–1111.

Galil, B.S. 2011. The alien crustaceans in the Mediterranean Sea: an historical review. Pages 377–401 *in* B.S. Galil, P.F. Clark, and J.T. Carlton, editors. In the Wrong Place—Alien Marine Crustaceans: Distribution, Biology and Impacts. Springer, Dordrecht, The Netherlands.

Galil, B.S., S. Nehring, and V.E. Panov. 2007. Waterways as invasion highways—impact of climate change and globalization. Pages 59–74 *in* W. Nentwig, editor. Biological Invasions. Ecological Studies 193. Springer, Berlin.

Galil, B.S., A. Marchini, and A. Occhipinti-Ambrogi. 2016. East is east and West is west? Management of marine bioinvasions in the Mediterranean Sea. Estuarine, Coastal and Shelf Science 201:7–16.

Galil, B.S., A. Marchini, A. Occhipinti-Ambrogi, and H. Ojaveer. 2017. The enlargement of the Suez Canal—Erythraean introductions and management challenges. Management of Biological Invasions 8:141–152.

Gaw, S., K.V. Thomas, and T.H. Hutchinson. 2014. Sources, impacts and trends of pharmaceuticals in the marine and coastal environment. Philosophical Transactions of the Royal Society B 369:20130572.

Gherardi, F. 2006. Crayfish invading Europe: the case study of *Procambarus clarkii*. Marine and Freshwater Behaviour and Physiology 39:175–191.

Gherardi, F. 2007. Biological invasions in inland waters: an overview. Pages 3–25 *in* F. Gherardi, editor. Biological Invaders in Inland Waters: Profiles, Distribution, and Threats. Springer, Dordrecht, The Netherlands.

Gherardi, F., S. Gollasch, D. Minchin, S. Olenin, and V.E. Panov. 2009. Alien invertebrates and fish in European inland waters. Pages 81–92 *in* DAISIE, editors. Handbook of Alien Species in Europe. Springer, Dordrecht, The Netherlands.

Global Invasive Species Database (GISD). 2016. http://www.iucngisd.org.

Gouin, N., F. Grandjean, S. Pain, C. Souty-Grosset, and J. Reynolds. 2003. Origin and colonization history of the white-clawed crayfish, *Austropotamobius pallipes*, in Ireland. Heredity 91:70–77.

Great Barrier Reef Marine Park Authority. 2017. Final Report: 2016 Coral Bleaching Event on the Great Barrier Reef. GBRMPA, Townsville, Australia.

Guerra-García, J.M., M. Ros, A. Dugo-Cota, V. Burgos, A.M. Flores-León, E. Baeza-Rojano, M.P. Cabezas, and J. Núñez. 2011. Geographical expansion of the *invader Caprella scaura* (Crustacea: Amphipoda: Caprellidae) to the East Atlantic coast. Marine Biology 158:2617–2622.

Guildea, C., Y. Hitchen, R. Duffy, P.J. Dias, J.M. Ledger, M. Snow, and W.J. Kennington. 2015. Introgression threatens the survival of the critically endangered freshwater crayfish *Cherax tenuimanus* (Decapoda: Parastacidae) in the wild. PLoS ONE 10:e0121075.

Haeberli, W., and M. Beniston. 1998. Climate change and its impacts on glaciers and permafrost in the Alps. Ambio 27:258–265.

Haddaway, N., R. Mortimer, M. Christmas, and A. Dunn. 2013. Effect of pH on growth and survival in the freshwater crayfish *Austropotamobius pallipes*. Freshwater Crayfish 19:53–62.

HELCOM. 2009. Eutrophication in the Baltic Sea—an integrated thematic assessment of the effects of nutrient enrichment and eutrophication in the Baltic Sea region. Baltic Sea Environment Proceedings 115B.

Hewitt, C.L., M.L. Campbell, R.E. Thresher, R.B. Martin, S. Boyd, B.F. Cohen, D.R. Currie, M.F. Gomon, M.J. Keough, J.A. Lewis, M.M. Lockett, N. Mays, M.A. McArthur, T.D. O'Hara, G.C.B. Poore, D.J. Ross, M.J. Storey, J.E. Watson, and R.S. Wilson. 2004. Introduced and cryptogenic species in Port Phillip Bay, Victoria, Australia. Marine Biology 144:183–202.

Hobbs, H.H. III, J.P. Jass, and J.V. Huner. 1989. A review of global crayfish introductions with particular emphasis on two North American species (Decapoda, Cambaridae). Crustaceana 56:299–316.

Hogarth, P.J. 2007. The Biology of Mangroves and Seagrasses. Oxford University Press, New York.

Holdich, D.M., and M. Pöckl. 2007. Invasive crustaceans in European inland waters Holdich. Pages 29–75 *in* F. Gherardi, editor. Biological Invaders in Inland Waters: Profiles, Distribution and Threats. Springer, Dordrecht, The Netherlands.

Holdich, D.M., P. Haffner, P. Noël, J. Carral, L. Füreder, F. Gherardi, Y. Machino, J. Madec, M. Pöckl, P. Śmietana, T. Taugbøl, and E. Vigneux. 2006. Species files. Pages 49–129 *in* C. Souty-Grosset, D.M. Holdich, P.Y. Noël, J.D. Reynolds, and P. Haffner, editors. Atlas of Crayfish in Europe. Patrimoines Naturels 63, Museum National d'Histoire Naturelle, Paris.

Holdich, D.M., J.D. Reynolds, C. Souty-Grosset, and P.J. Sibley. 2009. A review of the ever increasing threat to European crayfish from non-indigenous crayfish species. Knowledge and Management of Aquatic Ecosystems 11:1–46.

Hudson, D.M., V. Brittain, and G. Phillips. 2016. Behavioral response to temperature change by the freshwater crab *Neostrengeria macropa* (H. Milne Edwards, 1853) (Brachyura: Pseudothelphusidae) in Colombia. Journal of Crustacean Biology 36:287–294.

Hughes, L. 2000. Biological consequences of global warming: Is the signal already apparent. Trends in Ecology and Evolution 15:56–61.

Hulot, F.D., G. Lacroix, F. Lescher-Moutoue, and M. Loreau. 2000. Functional diversity governs ecosystem response to nutrient enrichment. Nature 405:340–344.

Hupało, K., W. Riss, M. Grabowski, J. Thiel, K. Bącela-Spychalska, and E.I. Meyer. 2018. Climate change as a possible driver of invasion and differential in HSP70 expression in two genetically distinct populations of the invasive killer shrimp, *Dikerogammarus villosus*. Biological Invasions:1–13.

Ingels, J., A. Vanreusel, A. Brandt, A.I. Catarino, B. David, C. De Ridder, P. Dubois, A.J. Gooday, P. Martin, F. Pasotti, and H. Robert. 2012. Possible effects of global environmental changes on Antarctic benthos: a synthesis across five major taxa. Ecology and Evolution 2:453–485.

Innocenti, G., G. Stasolla, M. Goren, N. Stern, Y. Levitt-Barmats, A. Diamant, and B. Galil. 2017. Going down together: invasive host, *Charybdis longicollis* (Leene, 1938) (Decapoda: Brachyura: Portunidae) and invasive parasite, *Heterosaccus dollfusi* Boschma, 1960 (Cirripedia: Rhizocephala: Sacculinidae) on the upper slope off the Mediterranean coast of Israel. Marine Biology Research 13:229–236.

Intergovernmental Panel on Climate Change (IPCC). 2007. Climate Change 2007: Synthesis Report. Contribution of Working Groups I, II and III to the Fourth Assessment Report of the Intergovernmental Panel on Climate Change. Core Writing Team, R.K. Pachaurim, and A. Reisinger, editors. IPCC, Geneva, Switzerland.

International Union for Conservation of Nature (IUCN). 2000. Guidelines for the prevention of biodiversity loss caused by alien invasive species. IUCN, Gland, Switzerland.

Jacobson, M.Z. 2005. Studying ocean acidification with conservative, stable numerical schemes for nonequilibrium air-ocean exchange and ocean equilibrium chemistry. Journal of Geophysical Research 110:D07302.

Janecki, T., A. Kidawa, and M. Potocka. 2010. The effects of temperature and salinity on vital biological functions of the Antarctic crustacean *Serolis polita*. Polar Biology 33:1013–1020.

Janer, G., and C. Porte. 2007. Sex steroids and potential mechanisms of non-genomic endocrine disruption in invertebrates. Ecotoxicology 16:145–160.

Jaser, W., G.F. Severin, U. Jutting, I. Juttner, K.W. Schramm, and A. Kettrup. 2003. Effects of 17 alpha-ethinylestradiol on the reproduction of the cladoceran species *Ceriodaphnia reticulata* and *Sida crystallina*. Environment International 28:633–638.

Jones, J.P.G., J.R. Rasamy, A. Harvey, A. Toon, B. Oidtmann, M.H. Randrianarison, N. Raminosoa, and O.R. Ravoahangimalala. 2009. The perfect invader: a parthenogenic crayfish poses a new threat to Madagascar's freshwater biodiversity. Biological Invasions 11:1475–1482.

Kawai, T., G. Scholtz, S. Morioka, F. Ramanamandimby, C. Lukhaup, and Y. Hanamura. 2009. Parthenogenetic alien crayfish (Decapoda: Cambaridae) spreading in Madagascar. Journal of Crustacean Biology 29:562–567.

Kerckhof, F. 2002. Barnacles (Cirripedia, Balanomorpha) in Belgian waters, an overview of the species and recent evolutions, with emphasis on exotic species. Bulletin de l'Institut Royal des Sciences Naturelles de Belgique. Biologie 72 (Suppl.):93–104.

Kernan, M. 2015. Climate change and the impact of invasive species on aquatic ecosystems. Aquatic Ecosystem Health and Management 18:321–333.

Klassen, G., and A. Locke. 2007. A Biological Synopsis of the European Green Crab, *Carcinus maenas*. Canadian Manuscript Report of Fisheries and Aquatic Science, Fisheries and Oceans, New Brunswick, Canada.

Korpinnen, S., and E. Bonsdorff. 2015. Eutrophication and hypoxia: impacts of nutrient and organic enrichment. Pages 202–243 *in* T.P. Crowe and C.L.J. Frid, editors. Marine Ecosystems: Human Impacts on Biodiversity, Functioning and Services. Cambridge University Press, Cambridge, UK.

Krång, A.S. 2007. Naphthalene disrupts pheromone induced mate search in the amphipod *Corophium volutator* (Pallas). Aquatic Toxicology 85:9–18.

Kuris, A.M. 1997. Conceptual framework for biocontrol of introduced marine pests. Proceedings of the First International Workshop on the Demography, Impacts and Management of Introduced Populations of the European Crab, *Carcinus maenas*. Center for Research on Introduced Marine Pests. Technical Reports 11:66–68.

Lagrue, C. 2017. Impacts of crustacean invasions on parasite dynamics in aquatic ecosystems: a plea for parasite-focused studies. International Journal for Parasitology: Parasites and Wildlife 6:364–374.

Lambertini, M., J. Leape, J. Marton-Lefevre, R.A. Mitter-Meier, M. Rose, J.G. Robinson, S.N. Stuart, B. Waldman, and P. Genovesi. 2011. Invasives: a major conservation threat. Science 333:404–405.

Landsberg, J.H. 2002. The effects of harmful algal blooms on aquatic organisms. Reviews in Fisheries Science 10:113–390.

Larson, E.R., and J.D. Olden. 2008. Do schools and golf courses represent emerging pathways for crayfish invasions? Aquatic Invasions 3:465–468.

Lasota, R., H. Hummel, and M. Wolowicz. 2004. Genetic diversity of European populations of the invasive soft-shell clam *Mya arenaria* (Bivalvia). Journal of the Marine Biological Association of the United Kingdom 84:1051–1056.

Lee, C.E. 2000. Global phylogeography of a cryptic copepod species complex and reproductive isolation between genetically proximate "populations." Evolution 54:2014–2027.

Lee, C.E. 2002. Evolutionary genetics of invasive species. Trends in Ecology and Evolution 17: 286–391.

Lee, C.E. 2016. Evolutionary mechanisms of habitat invasions, using the copepod *Eurytemora affinis* as a model system. Evolutionary Applications 9:248–270.

Leppäkoski, E., S. Olenin, and S. Gollasch. 2002. The Baltic Sea—a field laboratory for invasion biology. Pages 253–259 *in* E. Leppäkoski, S. Gollasch, and S. Olenin, editors. Invasive Aquatic Species of Europe. Distribution, Impacts and Management. Springer, Dordrecht, The Netherlands.

Leuven, R.S.E.W., G. van der Velde, I. Baijens, J. Snijders, C. van der Zwart, H.J.R. Lenders, and A. Bij de Vaate. 2009. The river Rhine: a global highway for dispersal of aquatic invasive species. Biological Invasions 11:1989–2008.

Lewis, C.N., K.A. Brown, L.A. Edwards, G. Cooper, and H.S. Findlay. 2013. Sensitivity to ocean acidification parallels natural pCO_2 gradients experienced by Arctic copepods under winter sea ice. Proceedings of the National Academy of Sciences of the United States of America 110:E4960–E4967.

Liao, I.C., and Y.H. Chien. 2011. The Pacific white shrimp, *Litopenaeus vannamei*, in Asia: the world's most widely cultured alien crustacean. Pages 489–519 *in* B.S. Galil, P.F. Clark, and J.T. Carlton, editors. In the Wrong Place—Alien Marine Crustaceans: Distribution, Biology and Impacts. Springer, Dordrecht, The Netherlands.

Loureiro, T.G., P.M.S.G. Anastácio, P.B. Araujo, C. Souty-Grosset, and A.M. Pereira. 2015. Red swamp crayfish: biology, ecology and invasion—an overview. Nauplius 23:1–19.

Maazouzi, C., C. Piscart, F. Legier, and F. Hervant. 2011. Ecophysiological responses to temperature of the "killer shrimp" *Dikerogammarus villosus*: Is the invader really stronger than the native *Gammarus pulex*? Comparative Biochemistry and Physiology Part A: Molecular and Integrative Physiology 159:268–274.

Maltby, L., N. Blake, T.C.M. Brock, and P.J. Van den Brink. 2005. Insecticide species sensitivity distributions: importance of test species selection and relevance to aquatic ecosystems. Environmental Toxicology and Chemistry 24:379–388.

Marchini, A., J. Ferrario, and E. Nasi. 2016. Arrival of the invasive amphipod *Grandidierella japonica* to the Mediterranean Sea. Marine Biodiversity Records 9:38.

Martin, P., H. Shen, G. Füllner, and G. Scholtz. 2010. The first record of the parthenogenetic Marmorkrebs (Decapoda, Astacida, Cambaridae) in the wild in Saxony (Germany) raises the question of its actual threat to European freshwater ecosystems. Aquatic Invasions 5:397–403.

Martins, A., J. Vieira, F. Marques, and L. Guilhermino. 2013. Effects of Three Antibiotics on the Reproduction of the Crustacean Cladoceran *Daphnia magna*. 9º Congresso Ibérico e 6º Iberoamericano de Contaminação e Toxicologia Ambiental, Valencia, Spain.

Maslanik, J.A., M.C. Serreze, and R.G. Barry. 1996. Recent decreases in Arctic summer ice cover and linkages to atmospheric circulation anomalies. Geophysical Research Letters 23:1677–1680.

Mayor, D.J., U. Sommer, K.B. Cook, and M.R. Viant. 2015. The metabolic response of marine copepods to environmental warming and ocean acidification in the absence of food. Scientific Reports 5:13690.

McNeely, J.A., H.A. Mooney, L.E. Neville, P. Schei, and J.K. Waage. 2001. A Global Strategy on Invasive Alien Species. IUCN, Gland and Cambridge, UK.

Menz, A., and B.F. Blake. 1980. Experiments on the growth of *Penaeus vannamei* Boone. Journal of Experimental Marine Biology and Ecology 48: 99–111.

Millenium Ecosystem Assessment. 2005. Ecosystems and Human Well-being: Synthesis Report. Island Press, Washington, DC.

Mkoji, G.M., B.V. Hofkin, A.M. Kuris, A. Stewart-Oaten, B.N. Mungai, J.H. Kihara, F. Mungai, J. Yundu, J. Mbui, J.R. Rashid, C.H. Kariuki, J.H. Ouma, D.K. Koech, and E.S. Loker. 1999. Impact of the crayfish *Procambarus clarkii* on *Schistosoma haematobium* transmission in Kenya. American Journal of Tropical Medicine and Hygiene 61:751–759.

Mnif, W., A.B.H. Hassine, A. Bouaziz, A. Bartegi, O. Thomas, and B. Roig. 2011. Effect of endocrine disruptor pesticides: a review. International Journal of Environmental Research and Public Health 8:2265–2303.

Moiseenko, T.I. 2005. Effects of acidification on aquatic ecosystems. Russian Journal of Ecology 36:93–102.

Mora, C., C.L. Wei, A. Rollo, T. Amaro, A.R. Baco, D. Billett, L. Bopp, Q. Chen, M. Collier, R. Danovaro, A.J. Gooday, B.M. Grupe, P.R. Halloran, J. Ingels, D.O.B. Jones, L.A. Levin, H. Nakano, K. Norling, E. Ramirez-Llodra, M. Rex, H.A. Ruhl, C.R. Smith, A.K. Sweetman, A.R. Thurber, J.F. Tjiputra, P. Usseglio, L. Watling, T. Wu, and M. Yasuharae. 2013. Biotic and human vulnerability to projected changes in ocean biogeochemistry over the 21st century. PLOS Biology 11:e1001682.

Naser, M.D., P.S. Rainbow, P.F. Clark, A.G. Yasser, and D.S. Jones. 2015. The barnacle *Amphibalanus improvises* (Darwin, 1854), and the mitten crab *Eriocheir*: one invasive species getting off on another! BioInvasions Records 4:205–209.

Nørgaard, K.B., and N. Cedergreen. 2010. Pesticide cocktails can interact synergistically on aquatic crustaceans. Environmental Science and Pollution Research 17:957–967.

Olden, J.D., J.M. McCarthy, J.T. Maxted, W.W. Fetzer, and M.J. Vander Zanden. 2006. The rapid spread of rusty crayfish (*Orconectes rusticus*) with observations on native crayfish declines in Wisconsin (U.S.A.) over the past 130 years. Biological Invasions 8:1621–1628.

Olsén, K.H. 2011. Effects of pollutants on olfactory mediated behaviors in fish and crustaceans. Pages 507–529 *in* T. Breithaupt and M. Thiel, editors. Chemical Communication in Crustaceans. Springer, New York.

Panov, V.E., B.G. Alexandrov, K. Arbaciauskas, R. Binimelis, G.H. Copp, M. Grabowski, F. Lucy, R.S. Leuven, S. Nehring, M. Paunović, V. Semenchenko, and M.O. Son. 2009. Assessing the risks of aquatic species invasions via European inland waterways: from concepts to environmental indicators. Integrated Environmental Assessment and Management 5:110–126.

Pansch, C., A. Nasrolahi, Y.S. Appelhans, and M. Wahl. 2013. Tolerance of juvenile barnacles (*Amphibalanus improvises*) to warming and elevated pCO_2. Marine Biology 160: 2023–2035.

Pansch, C., I. Schaub, J. Havenhand, and M. Wahl. 2014. Habitat traits and food availability determine the response of marine invertebrates to ocean acidification. Global Change Biology 20:765–777.

Pârvulescu, L., A. Schrimpf, E. Kozubíková, S. Cabanillas Resino, T. Vrålstad, A. Petrusek, and R. Schultz. 2012. Invasive crayfish and crayfish plague on the move: first detection of the plague agent *Aphanomyces astaci* in the Romanian Danube. Diseases of Aquatic Organisms 98:85–94.

Pedersen, S.A., B.H. Hansen, D. Altin, and A.J. Olsen. 2013. Medium-term exposure of the North Atlantic copepod *Calanus finmarchicus* (Gunnerus, 1770) to CO_2-acidified seawater: effects on survival and development. Biogeosciences 10:7481–7491.

Pedersen, S.A., O.J. Hakedal, I. Salaberria, A. Tagliati, L.M. Gustavson, B.M. Jenssen, A.J. Olsen, and D. Altin. 2014. Multigenerational exposure to ocean acidification during food limitation reveals consequences for copepod scope for growth and vital rates. Environmental Science and Technology 48:275–284.

Peruzza, L., F. Piazza, C. Manfrin, C.L. Bonzi, S. Battistella, and P.G. Giulianini. 2015. Reproductive plasticity of a *Procambarus clarkii* population living 10°C below its thermal optimum. Aquatic Invasions 10:199–208.

Pinto-Coelho, R.M., J.F. Bezerra-Neto, and C.A. Morais-Jr. 2005. Effects of eutrophication on size and biomass of crustacean zooplankton in a tropical reservoir. Brazilian Journal of Biology 65:325–338.

Pyke, C.R. 2005. Assessing climate change impacts on vernal pool ecosystems and endemic branchiopods. Ecosystems 8:95–105.

Raddum, G.G. 1980. Comparison of benthic invertebrates in lakes of different acidity. International Conference on the Ecological Impacts of Acid Precipitation, Oslo, SNSF Project: 330–331.

Raibaut, A., C. Combes, and F. Benoit. 1998. Analysis of the parasitic copepod species richness among Mediterranean fish. Journal of Marine Systems 15:185–206.

Ramalho, R.O., and P.M. Anastacio. 2015. Factors inducing overland movement of invasive crayfish (*Procambarus clarkii*) in a ricefield habitat. Hydrobiologia 746:135–146.

Raven, J., K. Caldeira, H. Elderfield, O. Hoegh-Guldberg, P.S. Liss, U. Riebesell, J. Sheperd, C. Turley, and A. Watson. 2005. Ocean acidification due to increasing atmospheric carbon dioxide. Royal Society Policy Document 12/05.

Rewicz, T., M. Grabowski, C. MacNeil, and K. Bącela-Spychalska. 2014. The profile of a "perfect" invader—the case of killer shrimp, *Dikerogammarus villosus*. Aquatic Invasions 9:267–288.

Rewicz, T., R. Wattier, T. Rigaud, M. Grabowski, T. Mamos, and K. Bącela-Spychalska. 2017. The killer shrimp, *Dikerogammarus villosus*, invading European Alpine lakes: a single main source but independent founder events with an overall loss of genetic diversity. Freshwater Biology 62:1036–1051.

Rodríguez, C.F., E. Bécares, M. Fernández-Aláez, and C. Fernández-Aláez. 2005. Loss of diversity and degradation of wetlands as a result of introducing exotic crayfish. Biological Invasions 7:75–85.

Roque, A., A.A. Molina, M.C. Bolán, and G.B. Gómez. 2001. In vitro susceptibility to 15 antibiotics of vibrios isolated from penaeid shrimps in Northwestern Mexico. International Journal of Antimicrobial Agents 17:383–387.

Ros, M., M. Vázquez-Luis, and J.M. Guerra-García. 2015. Environmental factors modulating the extent of impact in coastal invasions: The case of a widespread invasive caprellid (Crustacea: Amphipoda) in the Iberian Peninsula. Marine Pollution Bulletin 98:247–258.

Rothlisberger, J., W.L. Chadderton, J. McNulty, and D. Lodge. 2010. Aquatic invasive species transport via trailered boats: what is being moved, who is moving it, and what can be done? Fisheries 35:121–132.

Roy, H.E., H. Hesketh, B.V. Purse, J. Eilenberg, A. Santini, R. Scalera, G.D. Stentiford, T. Adriaens, K. Bącela-Spychalska, D. Bass, K.M. Beckmann, P. Bessell, J. Bojko, O. Booy, A.C. Cardoso, F. Essl, Q. Groom, C. Harrower, R. Kleespies, A.F. Martinou, M.M. van Oers, E.J. Peeler, J. Pergl, W. Rabitsch, A. Roques, F. Schaffner, S. Schindler, B.R. Schmidt, K. Schönrogge, J. Smith, W. Solarz, A. Stewart, A. Stroo, E. Tricarico, K.M.A. Turvey, A. Vannini, M. Vilà, S. Woodward, A.A. Wynns, and A.M. Dunn. 2016. Alien pathogens on the horizon: opportunities for predicting their threat to wildlife. Conservation Letters 10:379–498.

Sakamoto, M., K.H. Chang, and T. Hanazato. 2006. Inhibition of development of anti-predator morphology in the small cladoceran *Bosmina* by an insecticide: impact of an anthropogenic chemical on prey-predator interactions. Freshwater Biology 51:1974–1983.

Schäfer, R.B., P.J. van den Brink, and M. Liess. 2011. Impacts of pesticides on freshwater ecosystems. Pages: 111–137 *in* F. Sánchez-Bayo, P.J. van den Brink, and R.M. Mann, editors. Ecological Impacts of Toxic Chemicals. Francisco Sanchez-Bayo, Bentham, Australia.

Schuster, G.A., C.A. Taylor, and J. Cordeiro. 2010. *Pacifastacus leniusculus*. The IUCN Red List of Threatened Species 2010: e.T153648A4526314.

Seebens, H., T.M. Blackburn, and F. Essl. 2017. No saturation in the accumulation of alien species worldwide. Nature Communications 8:14435.

Selfin, Y., and D. Hope. 2011. Future of World Trade. Top 25 Sea and Air Freight Routes in 2030. Economic Views. PriceWaterhouseCoopers, London.

Sharma, S., J.J. Magnuson, R.D. Batt, L.A. Winslow, J. Korhonen, and Y. Aono. 2016. Direct observations of ice seasonality reveal changes in climate over the past 320–570 years. Scientific Reports 6:25061.

Shucksmith, R., E.J. Cook, D.J. Hughes, and M.T. Burrows. 2009. Competition between the non-native amphipod *Caprella mutica* and two native species of caprellids *Pseudoprotella phasma* and *Caprella linearis*. Journal of the Marine Biological Association of the United Kingdom 89:1125–1132.

Simberloff, D., and B. von Holle. 1999. Positive interactions of nonindigenous species: invasional meltdown? Biological Invasions 1:21–32.

Slothouber Galbreath, J.G.M., J.E. Smith, J.J. Becnel, R.K. Butlin, and A.M. Dunn. 2010. Reduction in post-invasion genetic diversity in *Crangonyx pseudogracilis* (Amphipoda: Crustacea): a genetic bottleneck or the work of hitchhiking vertically transmitted microparasites? Biological Invasions 12:191–209.

Soto-Rodríguez, S., M. Armenta, and B. Gomez-Gil. 2006. Effects of enrofloxacin and florfenicol on survival and bacterial population in an experimental infection with luminescent *Vibrio campbellii* in shrimp larvae of *Litopenaeus vannamei*. Aquaculture 255:48–54.

Stella, J.S., P.L. Munday, and G.P. Jones. 2011. Effects of coral bleaching on the obligate coral-dwelling crab *Trapezia cymodoce*. Coral Reefs 30:719–727.

Stoks, R., A.N. Geerts, and L. De Meester. 2014. Evolutionary and plastic responses of freshwater invertebrates to climate change: realized patterns and future potential. Evolutionary Applications 7:42–55.

Suarez-Kurtz, G. 1989. Effects of aminoglycoside antibiotics on calcium action potentials and calcium channel currents. Acta Physiologica et Pharmacologica Latinoamericana 39:407–418.

Tamminen T., and H. Kuosa. 1998. Eutrophication in planktonic ecosystems: food web dynamics and elemental cycling. Developments in Hydrobiology 127, Springer Science+Business Media, Dordrecht, Netherlands.

Tavares, M., and G.A.S. De Melo. 2004. Discovery of the first known benthic invasive species in the Southern Ocean: the North Atlantic spider crab *Hyas araneus* found in the Antarctic Peninsula. Antarctic Science 16:129–131.

Taylor, J., and T. Komai. 2011. First record of the introduced sand shrimp species *Crangon uritai* (Decapoda: Caridea: Crangonidae) from Newport, Port Phillip Bay, Victoria, Australia. Marine Biodiversity Records 4:e22.

Taylor, J.R.A., J.M. Gilleard, M.C. Allen, and D.D. Deheyn. 2015. Effects of CO_2-induced pH reduction on the exoskeleton structure and biophotonic properties of the shrimp *Lysmata californica*. Scientific Reports 5:10608.

Tepolt, C.K., and G.N. Somero. 2014. Master of all trades: thermal acclimation and adaptation of cardiac function in a broadly distributed marine invasive species, the European green crab, *Carcinus maenas*. Journal of Experimental Biology 217:1129–1138.

Thatje, S., K. Anger, J.A. Calcagno, G.A. Lovrich, H.O. Pörtner, and W.E. Arntz. 2005. Challenging the cold: crabs reconquer the Antarctic. Ecology 86:619–625.

Tierney, A.J., and J. Atema. 1986. Effects of acidification on the behavioral response of crayfishes (*Orconectes virilis* and *Procambarus acutus*) to chemical stimuli. Aquatic Toxicology 9:1–11.

Tøttrup, A.P., B.K.K. Chan, H. Koskinen, and J.T. Høeg. 2010. "Flying barnacles": implications for the spread of non-indigenous species. Biofouling 26:577–582.

Tu, H.T., F. Silvestre, N.T. Phuong, and P. Kestemont. 2010. Effects of pesticides and antibiotics on penaeid shrimp with special emphases on behavioral and biomarker responses. Environmental Toxicology and Chemistry 29:929–938.

Turner, J.T. 2004. The importance of small planktonic copepods and their roles in pelagic marine food webs. Zoological Studies 43:255–266.

U.S. Geological Survey (USGG). 2016. Nonindigenous Aquatic Species Database. Gainesville, Florida. http://nas.er.usgs.gov.

Vandenbergh, G.F., D. Adriaens, T. Verslycke, and C.R. Janssen. 2003. Effects of 17a-ethinylestradiol on sexual development of the amphipod *Hyalella azteca*. Ecotoxicology and Environmental Safety 54:216–222.

van Wezel, A.P., and P. van Vlaardingen. 2004. Environmental risk limits for antifouling substances. Aquatic Toxicology 66:427–444.

Viaroli, P., D. Nizzoli, M. Pinardi, E. Soana, and M. Bartoli. 2015. Eutrophication of the Mediterranean Sea: a watershed-cascading aquatic filter approach. Rendiconti Lincei 26:13–23.

Vogt, G., C. Falckenhayn, A. Schrimpf, K. Schmid, K. Hanna, J. Panteleit, M. Helm, R. Schulz, and F. Lyko. 2015. The marbled crayfish as a paradigm for saltational speciation by autopolyploidy and parthenogenesis in animals. Biology Open 4:1583–1594.

Waters, C.N., J. Zalasiewicz, C. Summerhayes, A.D. Barnosky, C. Poirier, A. Gałuszka, A. Cearreta, M. Edgeworth, E.C. Ellis, M. Ellis, C. Jeandel, R. Leinfelder, J.R. McNeill, D.D. Richter, W. Steffen, J. Syvitski, D. Vidas, M. Wagreich, M. Williams, A. Zhisheng, J. Grinevald, E. Odada, N. Oreskes, and A.P. Wolfe. 2016. The Anthropocene is functionally and stratigraphically distinct from the Holocene. Science 351: aad2622.

Watts, M.M., D. Pascoe, and K. Carroll. 2001. Survival and precopulatory behaviour of *Gammarus pulex* (L.) exposed to two xenoestrogens. Water Resources 35:2347–2352.

Watts, M.M., D. Pascoe, and K. Carroll. 2002. Population responses of the freshwater amphipod *Gammarus pulex* (L.) to an environmental estrogen, 17α-ethinylestradiol. Environmental Toxicology and Chemistry 21:445–450.

Westman, K. 1973. The population of the crayfish *Astacus astacus* in Finland and the introduction of the American crayfish *Pacifastacus leniusculus* Dana. Freshwater Crayfish 1:41–55.

Weydmann, A., J.E. Søreide, S. Kwasniewski, and S. Widdicombe. 2012. Influence of CO_2-induced acidification on the reproduction of a key Arctic copepod *Calanus glacialis*. Journal of Experimental Marine Biology and Ecology 428:39–42.

Weydmann, A., W. Walczowski, J. Carstensen, and S. Kwaśniewski. 2017. Warming of the subarctic waters accelerates development of a key marine zooplankton *Calanus finmarchicus*. Global Change Biology 24:172–183

Whiteley, N.M. 2011. Physiological and ecological responses of crustaceans to ocean acidification. Marine Ecology Progress Series 430:257–271.

Winfield, I., S. Cházaro-Olvera, M. Ortiz, and U. Palomo-Aguayo. 2011. Updated checklist of marine invasive species of amphipods (Peracarida: Gammaridea and Corophiidea) from Mexico. Revista de Biología Marina y Oceanografía 46:349–361.

Wolf, M.C., and P.A. Moore. 2002. The effects of the herbicide metolachlor on the perception of chemical stimuli by *Orconectes rusticus*. Journal of the North American Benthological Society 21:457–467.

Xu, H.G., S. Qiang, P. Genovesi, H. Ding, J. Wu, L. Meng, Z.M. Han, J.L. Miao, B.S. Hu, J.Y. Guo, H.Y. Sun, C. Huang, J.C. Lei, Z.F. Le, X.P. Zhang, S.P. He, Y. Wu, Z. Zheng, L. Chen, V. Jarosik, and P. Pyšek. 2012. An inventory of invasive alien species in China. Neobiota 15:1–26.

Zeebe, R.E. 2012. History of seawater carbonate chemistry, atmospheric CO_2, and ocean acidification. Annual Review of Earth and Planetary Sciences 40:141–165.

Zenetos, A., M.E. Çinar, F. Crocetta, D. Golani, A. Rosso, G. Servello, N. Shenkar, X. Turon, and M. Verlaque. 2017. Uncertainties and validation of alien species catalogues: the Mediterranean as an example. Estuarine, Coastal and Shelf Science 191:171–187.

Zganec, K. 2012. The effects of water diversion and climate change on hydrological alteration and temperature regime of karst rivers in central Croatia. Environmental Monitoring and Assessment 184:5705–5723.

Zganec, K., and S. Gottstein. 2009. The river before damming: distribution and ecological notes on the endemic species *Echinogammarus cari* (Amphipoda: Gammaridae) in the Dobra River and its tributaries, Croatia. Aquatic Ecology 43:105–115.

Zheng, C.Q., J. Jeswin, K.L. Shen, M. Lablche, K.J. Wang, and H.P. Liu. 2015. Detrimental effect of CO_2-driven seawater acidification on a crustacean brine shrimp, *Artemia sinica*. Fish and Shellfish Immunology 43:181–190.

INDEX